Synthesis of Naturally Occurring Nitrogen Heterocycles from Carbohydrates

El Sayed H. El Ashry, PhD, DSc
Chemistry Department
Faculty of Science, Alexandria University
Alexandria, Egypt

Ahmed El Nemr, PhD
Environmental Division
National Institute of Oceanography and Fisheries
Alexandria, Egypt

Blackwell
Publishing

Editorial Offices:
Blackwell Publishing Ltd, 9600 Garsington Road, Oxford OX4 2DQ, UK
Tel: +44 (0)1865 776868
Blackwell Publishing Professional, 2121 State Avenue, Ames, Iowa 50014-8300, USA
Tel: +1 515 292 0140
Blackwell Publishing Asia, 550 Swanston Street, Carlton, Victoria 3053, Australia
Tel: +61 (0)3 8359 1011

First published 2005 by Blackwell Publishing Ltd

Library of Congress Cataloging-in-Publication Data

El Ashry, El Sayed H.
Synthesis of naturally occurring nitrogen heterocycles from carbohydrates / El Sayed H. El Ashry, Ahmed El Nemr.– 1st ed.
p. cm.
Includes bibliographical references and index.
ISBN 1-4051-2934-4 (hardback : alk. paper)
1. Heterocyclic compounds–Synthesis. 2. Carbohydrates. 3. Natural products–synthesis.
I. El-Nemr, Ahmed. II. Title.

QD400.5.S95E52 2005
547′.590459–dc22

2004018684

ISBN 1-4051-2934-4

A catalogue record for this title is available from the British Library

Set in 10/12 pt Times
by TechBooks
Printed and bound in India
by Gopsons Papers Ltd, Noida

The publisher's policy is to use permanent paper from mills that operate a sustainable forestry policy, and which has been manufactured from pulp processed using acid-free and elementary chlorine-free practices. Furthermore, the publisher ensures that the text paper and cover board used have met acceptable environmental accreditation standards.

For further information on Blackwell Publishing, visit our website:
www.blackwellpublishing.com

Contents

Preface

Carbohydrates, such as starch, are extensively used as feedstocks by the chemical industry; similarly, derivatized carbohydrates are increasingly used by organic chemists as starting materials in the synthesis of chiral heterocyclic compounds. The aim of this book is to review the recent literature dealing with the use of carbohydrates as raw materials in the synthesis of naturally occurring nitrogen heterocycles. Although carbohydrates have been used for the synthesis of other types of heterocycles, we have limited our review to their use in the synthesis of naturally occurring nitrogen heterocycles. This limitation was dictated by the extremely large number of publications that has appeared on the subject during the last two decades and our desire to give the reader as much information as possible in the confines of our book. We have not merely cited references for a given synthesis but instead have given as much detail as was possible on the experimental conditions used. The text contains six main chapters arranged according to the size and complexity of their heterocyclic rings, ranging from five- to seven-membered rings and from single to multiple fused rings. The book gives enough information on the synthesis of the compounds to enable a chemist to design a multistep synthesis. It cites the different approaches to the synthesis of naturally occurring nitrogen heterocycles in a format that enables the reader to make comparisons with other methods and make decisions on whether to use a certain procedure, modify it or devise a new synthetic methodology. In summary, the book is not a mere list of the conversion methods cited in the literature, but rather a rational discussion of these methods. Of course, the large volume of literature cited has dictated that some references be discussed in less detail than some readers would have liked, but we hope that they will understand our difficulty and forgive us. We feel that the added information in our reference book will be of greatest value to chemist in both industry and academia, and to researchers and graduate students in the fields of organic chemistry, medicinal chemistry, heterocyclic chemistry, natural product chemistry and glycochemistry.

We thank Professor N. Rashed (Egypt), Professor H.S. El Khadem (United States), Professor T. Tsuchiya (Japan) and Professor S. Abdo (Egypt) for comments and useful suggestions. In addition, thanks are due to Professor R.R. Schmidt (Germany) for his valuable discussions and for providing library facilities. The AvH and DFG are highly appreciated for their partial supports. Our appreciation goes also to the members of El Ashry group for their help and efforts.

E.S.H. El Ashry
A. El Nemr

Author details

El Sayed H. El Ashry was born in 1942 in Elmahal Alkobra, Egypt. He studied chemistry at Alexandria University (BSc 1963, MSc 1966, PhD 1969 and DSc 1997). He has been a visiting professor at Tokyo Institute of Technology, Ohio State University, Michigan Technological University, New York State University, Darmstadt Institute of Organic Chemistry, UmAlqura University and Konstant University. He has given lectures at various universities, institutes, companies and conferences around the world. He is currently a Professor of Organic Chemistry at Alexandria University after being the head of the department for the last four years. He has supervised more than 70 MSc and PhD students and published about 300 publications and review articles in highly renowned journals in the field of carbohydrates and nucleosides, a major area of research in the series 'Heterocycles from Carbohydrate Precursors'. He also edits various international journals. He has received many awards of recognition and distinction: in particular 'Excellence' and '1st class Ribbon of Science and Arts' awards from the President of Egypt.

Ahmed El Nemr was born in 1962 in El Behera, Egypt. He received his bachelor's degree in chemistry in 1984 and his master's degree in organic chemistry from Alexandria University under the supervision of Professor E.S.H. El Ashry, after which he was awarded his PhD in Engineering in Applied Chemistry by Keio University, Yokohama, Japan. He worked for six years as a researcher at the Institute of Bioorganic Chemistry, Kawasaki, Japan, with Professor Tsutomu Tsuchiya. He is now an Associate Professor at the National Institute of Oceanography and Fisheries, Alexandria, Egypt. He is the head of Egyptian National Oceanography Data Centre (ENODC). His research interests involve carbohydrate chemistry, natural products, and organic as well as inorganic pollutants in marine environment.

List of abbreviations and acronyms used in this book

Ac	acetyl
AIBN	azobis(isobutronitrile)
All	allyl
An	anisyl
Ar	aryl
9-BBN	9-borabicyclo[3.3.l]nonane
BMS	borane–dimethyl sulfide complex
Bn	benzyl
Boc	*tert*-butoxycarbonyl
Boc-L-Val-OH	*tert*-butoxycarbonyl protected L-valine
BOM	benzyloxymethyl
BOP-Cl	*N,N*-bis(2-oxo-3-oxazolidinyl)phosphinic chloride
Bz	benzoyl
CAN	ceric ammonium nitrate
Cbz	benzyloxy carbonyl
Cp	cyclopentadienyl
CSA	camphorsulfonic acid
DABCO	1,4-diazabicyclooctane
DBU	1,8-diazabicyclo[5.4.0]undec-7-ene
DCC	dicyclohexylcarbodiimide
DCE	dichloroethane
DCH	dicyclohexano
DDQ	2,3-dichloro-5,6-dicyano-1,4-benzoquinone
DEAD	diethyl azodicarboxylate
DEPC	diethylphosphorocyanide
DET	diethyl tartrate
DHAP	dihydroxyacetone phosphate
DHP	dihydropyran
DHQ-CLB	Sharpless asymmetric dihydroxylation reagent
DIAD	diisopropyl azodicarboxylate
DIBAL-H	diisobutylaluminum hydride
DIPEA	*N,N*-diisopropylethylamine

DIPT	diisopropyl tartrate
DMAP	4-dimethylaminopyridine
DMF	dimethylformamide
DMP	2,2-dimethoxypropane
DMPU	*N*,*N'*-dimethylpropylene urea
DMS	dimethyl sulfide
DMSO	dimethyl sulfoxide
DNAP	dinitroaminopyridine
DPPA	diphenylphosphoryl azide
EDAC	1-(3-dimethylaminopropyl)-3-ethylcarbodiimide hydrochloride
FDP	fructose-1,6-diphosphate
FVP	pyrolysis
HMDS	hexamethyldisilazane
HMPA	hexamethylphosphoramide
HOBT	1-hydroxybenzotriazole hydrate
IDCP	iodonium dicollidine perchlorate
Im	imidazole
IRA	Amberlite IRA
KHMDS	potassium hexamethyldisilazane
LDA	lithium diisopropylamide
Lev	levulinoyl
LHMDS	lithium hexamethyldisilazide
LPTS	2,6-luidinium *p*-toluenesulfonate
L-Selectride	lithium tri-*sec*-butylborohydride
***m*-CPBA**	*m*-chloroperoxybenzoic acid
MEM	methoxyethoxymethyl
***mm*TrCl**	monomethoxytrityl chloride
MOM	methoxymethyl
MPM	*p*-methoxybenzyl
Ms	mesyl
MS	molecular sieves
Naph	naphthyl
NBS	*N*-bromosuccinimide
NCS	*N*-chlorosuccinimide
NIS	*N*-iodosuccinimide
NMDA	*N*-methyl-D-aspartate
NMO	*N*-methylmorpholine *N*-oxide
NSA	2-naphthalenesulfonic acid
PASE	phosphate esters of acid phosphatase

PCC	pyridinium chlorochromate
PDC	pyridinium dichromate
Pf	9-phenyl fluoren-9-yl
Ph	phenyl
Phth	phthalyl
Piv	trimethylacetyl
PM	phenylmenthyl
PMB	*p*-methoxybenzyl
PPL	porcine pancreatic lipase
PPTs	pyridinium *p*-toluenesulfate
Pr	propyl
***p*-TsOH**	*p*-toluenesulfonic acid
Py	pyridine
RAMA	rabbit muscle aldolase, D-fructose-1,6-bisphosphate aldolase
rt	room temperature
TBAF	tetrabutyl ammonium fluoride
TBDPS	*t*-butyl diphenylsilyl
TBHP	*t*-butyl hydroperoxide
TBS	*t*-butyl dimethylsilyl
TEA	triethylamine
TEMPO	2,2,6,6-tetramethyl-piperidinooxy, free radical
TEOC	*O*-(2-(trimethylsilyl)ethyl)carbamate
TES	triethylsilyl
Tf	trifluoromethylsulfonyl
TFA	trifluoroacetic acid
TFAA	trifluoroacetic anhydride
Th	thiazole
THF	tetrahydrofuran
TIPS	triisopropylsilyl
TMEDA	*N*,*N*,*N*′,*N*′-tetramethylethylenediamine
TMS	trimethylsilyl
TMSOTf	trimethylsilyltriflate
Tol	tolyl
TPAP	tetra-*n*-propyl ammonium perruthenate
TPS	*t*-butyl diphenylsilyl
Tr	trityl
Ts	tosyl
TSNSO	*N*-sulfinyl-*p*-toluenesulfonamide
Z	benzyloxycarbonyl

Introduction

Carbohydrates are widely distributed in nature and constitute the largest renewable biomasses available. As such, they are considered by many as one of the most promising feedstock for the industrial preparation of many organic chemical compounds. Carbohydrates, resembling the largest class of resources, have attracted the attention to be used as substitutes for the crude oil, natural gas or coal[1–8] in developing materials for trendsetting technologies.[6–8] Moreover, many pharmaceutical agents incorporate nitrogen heterocyclic rings, which has attracted attention toward their synthesis. The use of carbohydrates as starting materials for the synthesis of heterocyclic compounds has long been a subject of interest in our laboratory, where significant efforts were devoted to the exploration of novel routes for the synthesis of nitrogen heterocyclic compounds from nitrogen derivatives of carbohydrates. The skeleton and functionalities in the hydrazones and bishydrazones derived from carbohydrates have been found to be of high synthetic potential, particularly as precursors for acyclic nucleosides[9–11] and heterocyclic compounds.[12–15] Thus, a great deal of work has been done on the transformations of hydrazones and bishydrazones into heterocyclic compounds. Since the subject of the book is naturally occurring nitrogen heterocycles, only a quick information on the role of hydrazones and osazones as precursors for heterocyclic compounds will be given herein, which could be of potential value in using such approaches in the synthesis of naturally occurring nitrogen heterocycles. This can be exemplified by the synthesis of the important starting material, acetonide of L-glyceraldehyde,[16] which has utilized the readily available dehydro-L-ascorbic acid monophenylhydrazone.[17] Much attention has been drawn to the synthesis of pyrazoles[18–74] and pyrazolines [75–97] via various routes.[18] Isoxazolines,[98,99] 1,2,3-triazoles,[100–149] 1,2,4-triazoles,[150–153] oxa- and thia-diazoles as well as dioxalanes,[154–172] tetrazoles,[173,174] pyridazines,[47,175–177] 1,2,4-triazines,[50,53,92,178–187] pyrrolotriazines,[188] triazolotriazinoindoles,[189–193] pyrazolopyrazoles,[194] fused pyridazines,[195] pyrimidines,[196] quinoxalines,[197–221] and pyridopyrazines,[222] as well as condensed triazolo ring systems,[223–233] condensed diazines,[234–244] and condensed quinoxalines[245–247] have been synthesized. Moreover, resulting periodate oxidation of the polyol residues linked to heterocycles followed by using the aldehyde or carboxylic acid functionalities for building heterocycles led to the synthesis of various types of biheterocycles.[248–257]

Rationale and arrangement of the topics

The naturally occurring nitrogen heterocycles possess a wide range of biological and medicinal properties. Consequently, the design of synthetic schemes for naturally occurring heterocycles is highly desirable. Among the useful features of carbohydrates is the ability of their nitrogen derivatives to incorporate part or all of the nitrogen atoms into the heterocyclic rings they form. Another useful feature is the availability of their multiple chiral centers in nearly all the possible configurations and their ability to retain most of their configuration during conversion to heterocycles. The resulting chirons (optically active synthons) can readily be used as versatile intermediates in the synthesis of naturally occurring heterocycles. Consequently, some monographs and reviews related to the topic became available.[258–262] However, as a result of the tremendous achievement in the field we have found that the topic needs a comprehensive review to help the reader in getting ready information on the topic. The synthetic approaches described in this book are discussed in six chapters arranged according to the size of the heterocyclic rings and the number of heteroatoms in the ring. Accordingly, the chapters are arranged into (1) five-membered nitrogen heterocycles, (2) five-membered heterocycles with two heteroatoms, (3) six-membered nitrogen heterocycles, (4) seven-membered nitrogen heterocycles, (5) fused nitrogen heterocycles and (6) multifused heterocycles. Three- and four-membered nitrogen heterocycles are not treated in separate chapters but are included in the appropriate fused heterocycles.

References

1. Lichtenthaler, F.W.; Mondel, S. *Pure Appl. Chem.* **69** (1997) 1853.
2. Lichtenthaler, F.W. *Carbohydrates as Organic Raw Materials;* VCH: Weinheim, 1991.
3. Cottier, L.; Descotes, G. *Trends Heterocycl. Chem.* **2** (1991) 233.
4. Oikawa, N.; Müller, C.; Kunz, M.; Lichtenthaler, F.W. *Carbohydr. Res.* **309** (1998) 269.
5. Baumann, H.; Bühler, M.; Fochem, H.; Hirsinger, F.; Zoebelein, H.; Falbe, J. *Angew. Chem.* **100** (1988) 41; *Angew. Chem., Int. Ed. Engl.* **27** (1988) 41.
6. Schiweck, H.; Rapp, K.; Vogel, M. *Chem. Ind. (London)* (1988) 228.
7. Clark, M.G. *Chem. Ind. (London)* (1985) 258.
8. Jeffrey, G.A. *Acc. Chem. Res.* **19** (1986) 168.
9. El Ashry, E.S.H.; El Kilany, Y. In *Advances in Heterocyclic Chemistry*, Vol. 67; Katritzky, A.R., Ed.; Academic Press: New York, 1996; p. 391.
10. El Ashry, E.S.H.; El Kilany, Y. In *Advances in Heterocyclic Chemistry*, Vol. 68; Katritzky, A.R., Ed.; Academic Press: New York, 1997; p. 1.
11. El Ashry, E.S.H.; El Kilany, Y. In *Advances in Heterocyclic Chemistry*, Vol. 69; Katritzky, A.R., Ed.; Academic Press: New York, 1998; p. 129.
12. El Ashry, E.S.H. *Ascorbic Acid, Chemistry Metabolism and Uses*, 1 ed. (*Advances in Chemistry Series*, Vol. 200); Seib, P.A., Tolbert, M., Eds.; American Chemical Society: Washington, DC, 1982; p. 179.
13. El Ashry, E.S.H.; Mousaad, A.; Rashed, N. In *Advances in Heterocyclic Chemistry*, Vol. 53; Katritzky, A.R., Ed.; Academic Press: New York, 1992; p. 233.
14. El Ashry, E.S.H.; Rashed, N.; Taha, M.; Ramadan, E. In *Advances in Heterocyclic Chemistry*, Vol. 59; Katritzky, A.R., Ed.; Academic Press: New York, 1994; p. 39.
15. El Ashry, E.S.H.; Rashed, N.; Ramadan, E. In *Advances in Heterocyclic Chemistry*, Vol. 61; Katritzky, A.R., Ed.; Academic Press: New York, 1994; p. 207.
16. El Ashry, E.S.H.; Rashed, N.; Mousaad, A. *J. Chin. Chem.* **41** (1994) 591.
17. El Ashry, E.S.H.; El Kilany, Y.; Abdel Hamid, H. *Carbohydr. Res.* **172** (1988) 308.
18. Schmidt, R.R.; Heermann. D. *Chem. Ber.* **114** (1981) 2825.
19. Schmidt, R.R.; Kary, J.; Guilliard, W. *Chem. Ber.* **110** (1977) 2433.
20. Schmidt, R.R.; Guilliard, W.; Heermann, D. *Liebigs Ann. Chem.* (1981) 2309.
21. Kett, W.C.; Batley, M.; Redmond, J.W. *Carbohydr. Res.* **299** (1997) 129.
22. Logue, M.W.; Sarangan, S. Proc. N.D. *Acad. Sci.* **34** (1980) 6; *Chem. Abstr.* **93** (1980) 95562s.
23. Logue, M.W.; Sarangan, S. *Nucleosides Nucleotides* **1** (1982) 89.
24. Kozikowski, A.P.; Goldstein, S. *J. Org. Chem.* **48** (1983) 1139.
25. Klein, U.; Steglich, W. *Liebigs Ann. Chem.* (1989) 247.

26. Buchanan, J.G.; Ouijano, M.L.; Wightman, R.H. *J. Chem. Soc., Perkin Trans. 1* (1992) 1573.
27. Buchanan, J.G.; Dunn, A.D.; Edgar, A.R.; Hutchison, R.J.; Power, M.J.; Williams, G.C. *J. Chem. Soc., Perkin Trans. 1* (1977) 1786.
28. Buchanan, J.G.; Chacon Fuertes, M.E.; Fuertes, A.; Stobie, A.; Wightman, R.H. *J. Chem. Soc., Perkin Trans. 1* (1980) 2561.
29. Buchanan, J.G.; Chacon Fuertes, M.E.; Fuertes, A.; Stobie, A., Wightman, R.H. *J. Chem. Soc., Perkin Trans. 1* (1980) 2567.
30. Repke, D.B.; Albrecht, H.P.; Moffatt, J.G. *J. Org. Chem.* **40** (1975) 2481.
31. Cooney, D.A.; Jayaram, H.N.; Gebeyehu, G.; Belts, C.R.; Kelley, J.A.; Marquez, V.-E.; Johns, D.G. *Biochem. Pharmacol.* **31** (1982) 2133; *Chem. Abstr.* **97** (1982) 156045g.
32. Gomez Guillen, M.; Vazquez de Migue, L.M.; Valezquez Jinienez, J. *An. Quim., Ser. C* **79** (1983) 152; *Chem. Abstr.* **102** (1985) 62522e.
33. Gomez Guillen, M.; Conde Jimenez, J.L.; Podio Lora, V. *An. Quim., Ser. C* **82** (1986) 204; *Chem. Abstr.* **108** (1988) 2215j.
34. Gomez Guillen, M.; Conde Jimenez, J.L. *Carbohydr. Res.* **180** (1988) 1.
35. Gomez Guillen, M.; Hans, F.; Lasaletta Simon, J.M.; Martin Zamora, M.E. *Carbohydr. Res.* **189** (1989) 349.
36. Vazquez de Miguel, L.M.; Velazquez Jimenez, J.; Gomez Guillen, M. *An Quim., Ser. C* **87** (1991) 126; *Chem. Abstr.* **115** (1991) 159579k.
37. Gomez Guillen, M.; Lassaletta Simon, J.M. *Carbohydr. Res.* **211** (1991) 287.
38. Tronchet, J.M.J.; Jotterand, A.; Le Hong, N. *Helv. Chim. Acta* **52** (1969) 2569.
39. Tronchet, J.M.J.; Jotterand, A.; Le Hong, N.; Perret, M.F.; Thorndahl Jaccard, S.; Tronchet, J.; Chalet, J.M.; Favire, M.L.; Hausser, C.; Sebastian, C. *Helv. Chim. Acta* **53** (1970) 1484.
40. Tronchet, J.M.J.; Baehler, B., Le Hong, N.; Livio, P.F. *Helv. Chim. Acta* **54** (1971) 921.
41. Tronchet, J.M.J.; Perret, F. *Helv. Chim. Acta* **53** (1970) 648.
42. Tronchet, J.M.J.; Perret, F. *Helv. Chim. Acta* **54** (1971) 683.
43. Tronchet, J.M.J.; Jotterand, A. *Helv. Chim. Acta* **54** (1971) 1131.
44. Tronchet, J.M.J.; Perret, F. *Helv. Chim. Acta* **55** (1972) 2121.
45. Tronchet, J.M.J.; Perret, F.; Barbalat Rey, F.; Nguyen Xuan, T. *Carbohydr. Res.* **46** (1976) 19.
46. Vismara, E.; Torn, G.; Pastori, N.; Marchiandi, M. *Tetrahedron Lett.* **33** (1992) 7575.
47. Paulsen, H.; Steinert, K.; Steinert, G. *Chem. Ber.* **103** (1970) 1846.
48. Farkas, J.; Flegelova, Z.; Sorm, F. *Tetrahedron Lett.* **13** (1972) 2279.
49. Just, G.; Kim, S. *Tetrahedron Lett.* **17** (1976) 1063.
50. Just, G.; Kim, S. *Can. J. Chem.* **55** (1977) 427.
51. Sato, T.; Noyori, R. *Heterocycles* **13** (1979) 141.
52. Sato, T.; Noyori, R. *Bull. Chem. Soc. Jpn.* **56** (1983) 2700.
53. Just, G.; Ramjeesingh, M.; Liak, T.G. *Can. J. Chem.* **54** (1976) 2940.
54. De Las Heras, F.G.; Chu, C.K.; Tam, S.Y.K.; Klein, R.S.; Watanabe, K.A.; Fox, J.J. *J. Heterocycl. Chem.* **13** (1976) 175.
55. Tam, S.Y.K.; Klein, R.S.; Wempen, I.; Fox, J.J. *J. Org. Chem.* **44** (1979) 4547.
56. Chu, C.K.; Watanabe, K.A.; Fox., J.J. *J. Heterocycl. Chem.* **17** (1980) 1435.
57. Chu, C.K. *Heterocycles* **22** (1984) 345.
58. Chu, C.K.J. *Heterocycl. Chem.* **21** (1984) 389.
59. Ullas, G.V.; Chu, C.K.; Ahn, M.K.; Kosugi, Y. *J. Org. Chem.* **53** (1988) 2413.
60. El Khadem, H. *Adv. Carbohydr. Chem.* **25** (1970) 351.
61. El Khadem, H.; Mohammed Ali, M.M. *J. Chem. Soc.* (1963) 4929; El Khadem, H.; El Shafei, Z.M.; Mohamed Ali, M.M. *J. Org. Chem.* **29** (1964) 1565.
62. El Khadem, H. *J. Org. Chem.* **29** (1964) 3072.
63. El Khadem, H.; El Shafei, Z.M.; Abdel Rahman, M.M. *Carbohydr. Res.* **1** (1965) 31.
64. El Khadem, H.; Abdel Rahman, M.M.A.; Sallam, M.A.E. *J. Chem. Soc. (C)* (1968) 2411.
65. El Khadem, H.; Abdel Rahman, M.M.A. *Carbohydr. Res.* **6** (1968) 470.
66. Diehi, V.; Cuny, E.; Lichtenthaler, F.W. *Heterocycles* **48** (1998) 1193.
67. El Khadem, H.; El Shafei, Z.; El Ashry, E.S.H.; El Sadek, M. *Carbohydr. Res.* **49** (1976) 185.
68. Kraus, A.; Simon, H. *Chem. Ber.* **105** (1972) 954.
69. Somogyi, L. *Carbohydr. Res.* **144** (1985) 71.
70. Somogyi, L. *Carbohydr. Res.* **152** (1986) 316.
71. Diels, O.; Meyer, R.; Onnen, O. *Ann.* **525** (1936) 94.
72. El Khadem. H.; Abdel Rahman, M.M.A. *Carbohydr. Res.* **3** (1966) 25.
73. Simon, H.; Moldenhauer, W. *Chem. Ber.* **100** (1967) 3121.
74. Avalos, M.; Babiano, R.; Cintas, P.; Jimenez, J.L.; Palacios, J.C.; Sanchez, J.B. *Tetrahedron: Asymmetry* **6** (1995) 945.
75. Ohle H. *Chem. Ber.* **67** (1934) 1750.

76. El Khadem, H.; El Ashry, E.S.H. *J. Chem. Soc. (C)* (1968) 2248.
77. El Ashry, E.S.H.; El Kilany, Y. *Chem. Ind.* (*London*) (1976) 372.
78. El Ashry, E.S.H. *Carbohydr. Res.* **52** (1976) 69.
79. El Ashry, E.S.H.; El Kilany, Y.; Singab, F. *Carbohydr. Res.* **56** (1977) 93.
80. El Sekily, M.; Mancy, S.; El Kholy, I.; El Ashry, E.S.H.; El Khadem, H.; Swartz, D.L. *Carbohydr. Res.* **59** (1977) 141.
81. El Ashry, E.S.H.; El Kholy, I.; El Kilany, Y. *Carbohydr. Res.* **59** (1977) 417.
82. El Ashry, E.S.H.; El Kilany, Y.; Singab, F. *Carbohydr. Res.* **67** (1978) 415.
83. El Sekily, M.A.; Mansy, S. *Carbohydr. Res.* **68** (1979) 87.
84. El Ashry, E.S.H.; El Kilany, Y.; Singab, F. *Carbohydr. Res.* **79** (1980) 151.
85. El Ashry, E.S.H.; El Kilany, Y.; Singab, F. *Carbohydr. Res.* **82** (1980) 25.
86. El Kilany, Y.; Rashed, N.; Mansour, M.; El Ashry, E.S.H. *J. Carbohydr. Chem.* **7** (1988) 187.
87. El Ashry, E.S.H. *J. Chem. Soc., Chem. Commun.* (1986) 1024.
88. El Ashry, E.H.S.; Nassr, M.; Singab, F. *Carbohydr. Res.* **56** (1977) 200.
89. Soliman, R.; El Ashry, E.S.H.; El Kholy, I. E.; El Kilany, Y. *Carbohydr. Res.* **67** (1978) 179.
90. El Sekily, M.; Mancy, S. *Carbohydr. Res.* **98** (1981) 148.
91. Prévost, C.; Fleury, M.C. *R. Hebd. Seances Acad. Sci.* **258** (1964) 587.
92. Fleury, M. *Bull. Soc. Chem. Fr.* (1966) 522.
93. Pollet, P.L. *Arch. J. Sci. Eng.* **6** (1981) 3.
94. El Sekily, M.A.; Mancy, S. *Carbohydr. Res.* **112** (1983) 151.
95. El Khadem, H.S.; El Ashry, E.S.H. *J. Heterocycl. Chem.* **10** (1973) 1051.
96. El Khadem, H.S.; El Ashry, E.S.H. *Carbohydr. Res.* **7** (1968) 507.
97. El Khadem, H.; Meshreki, M.H.; El Ashry, E.S.H.; El Sekily, M. *Carbohydr. Res.* **21** (1972) 430.
98. El Sekily, M.A.; Mancy, S.; Gross, B. *Carbohydr. Res.* **110** (1982) 229.
99. El Sekily, M.A.; Mancy, S. *Carbohydr. Res.* **124** (1983) 97.
100. Hann, R.M.; Hudson, C.S. *J. Am. Chem. Soc.* **66** (1944) 735.
101. Hanisch, G.; Henseke. G. *Chem. Ber.* **101** (1968) 2074.
102. Haskins, W.T.; Hann, R.M.; Hudson, C.S. *J. Am. Chem. Soc.* **68** (1946) 1766.
103. Haskins, W.T.; Hann, R.M.; Hudson, C.S. *J. Am. Chem. Soc.* **67** (1945) 939.
104. Hardegger, E.; El Khadem, H. *Helv. Chim. Acta* **30** (1947) 900.
105. Harclegger, E.; El Khadem, H. *Helv. Chim. Acta* **30** (1947) 1478.
106. El Khadem, H. In *Advances in Carbohydrates Chemistry and Biochemistry*, Vol. 25; Tipson, R.S., Horton, D., Eds.; Academic Press: New York, 1970; p. 351.
107. Regna, P.P. *J. Am. Chem. Soc.* **69** (1947) 246.
108. Haskins, W.T.; Hann, R.M.; Hudson, C.S. *J. Am. Chem. Soc.* **69** (1947) 1050.
109. Haskins, W.T.; Hann, R.M.; Hudson, C.S. *J. Am. Chem. Soc.* **69** (1947) 1461.
110. Hardegger, E.; El Khadem, H.; Schreier, E. *Helv. Chim. Acta* **34** (1951) 253.
111. Stewart, L.C.; Richtmyer, N.K.; Hudson, C.S. *J. Am. Chem. Soc.* **74** (1952) 2206.
112. Patt, J.W.; Richtmyer, N.K.; Hudson, C.S. *J. Am. Chem. Soc.* **74** (1952) 2210.
113. Karabinos, J.V.; Hann, R.M.; Hudson, C.S. *J. Am. Chem. Soc.* **75** (1953) 4320.
114. El Khadem, H. *J. Chem. Soc.* (1953) 3452.
115. Henseke, G.; Binte, H.J. *Chem. Ber.* **88** (1955)1167.
116. Richtmyer, N.K.; Bodenheimer, T.S. *J. Org. Chem.* **27** (1962) 1892.
117. El Khadem, H.; Kolkaila, A.M.; Meshereki, M.H. *J. Chem. Soc.* (1963) 3531.
118. Bishay, B.B.; El Khadem, H.; El Shafei, Z.M. *J. Chem. Soc.* (1963) 4980.
119. Richtmyer, N.K. *Carbohydr. Res.* **17** (1971) 401.
120. Hann, R.M.; Hudson, C.S. *J. Org. Chem.* **9** (1944) 470.
121. Hassid, W.C.; Doudoroff, M.; Baker, H.A. *Arch. Biochem.* **14** (1947) 29; *Chem. Abstr.* **42** (1948) 524h.
122. Hassid, W.Z.; Doudoroff, M.; Potter, A.L.; Baker, H.A. *J. Am. Chem. Soc.* **70** (1948) 306.
123. Haskins, W.T.; Hann, R.M.; Hudson, C.S. *J. Am. Chem. Soc.* **70** (1948) 2288.
124. Stodola, F.H.; Koepsell, H.J.; Sharp, E.S. *J. Am. Chem. Soc.* **74** (1952) 3202.
125. Thompson, A.; Wolform, M.L. *J. Am. Chem. Soc.* **76** (1954) 5173.
126. Stodola, F.H.; Sharpe, E.S.; Koespell, H.J. *J. Am. Chem. Soc.* **78** (1956) 2514.
127. Rutherford, D.; Richtmyer, N.K. *Carbohydr. Res.* **11** (1969) 341.
128. Ettel, V.; Liebster, *J. Collect. Czech. Chem. Commun.* **14** (1949) 80; *Chem. Abstr.* **43** (1949) 7548c.
129. Teuber, H.J.; Jellinek, G. *Chem. Bem.* **85** (1952) 95.
130. Henseke, G. *Acta Chim. Acad. Sci. Hung.* **12** (1957) 173.
131. El Khadem, H.; El Shafei, Z.M. *J. Chem. Soc.* (1958) 3117.
132. El Khadem, H.; El Shafei, Z.M.; *J. Chem. Soc.* (1959) 1655.
133. El Khadem, H.; El Shafei, Z.M.; Mohammed, Y.S. *J. Chem. Soc.* (1960) 3993.
134. Bishay, B.B.; El Khadem, H.; El Shafei, Z.M.; Meshreki, M.H. *J. Chem. Soc.* (1962) 3154.

135. El Khadem, H.; El Shafei, Z.M.; Meshreki, M.H. *J. Chem. Soc.* (1965) 1524.
136. El Khadem, H.; El Shafei, Z.M.; Meshreki, M.H. *J. Chem. Soc.* (1961) 2957.
137. Mester, L.; Weygand, F. *Bull. Soc. Chim. Fr.* (1960) 350.
138. Mester, L. *Bull. Soc. Chim. Fr.* (1962) 381.
139. Weygand, F.; Grisebach, H.; Kirchner, K.D.; Haselhorst, M. *Chem. Ber.* **88** (1955) 487.
140. Henseke, G.; Winter, M. *Cherm. Ber.* **93** (1960) 45.
141. El Khadem, H.S. *Carbohydr. Res.* **313** (1998) 225.
142. Hardegger, E.; Schreier, E. *Helv. Chim. Acta* **35** (1952) 232.
143. Hardegger, E.; Schreier, E. *Helv. Chim. Acta* **35** (1952) 623.
144. El Khadem, H.; Schreier, E.; Stohr, G.; Hardegger, E. *Helv. Chim. Acta* **35** (1952) 993.
145. Sallam, M.A.E.; Hegazy, E.I.A. *Carbohydr. Res.* **95** (1981) 177.
146. Fatiadi, A.J. *Carbohydr. Res.* **20** (1971) 179.
147. El Khadem, H.; Shaban, M.A.E. *Carbohydr. R*es. **2** (1966) 178.
148. El Khadem, H.; Shaban, M.A.E. *J. Chem. Soc. (C)* (1967) 519.
149. El Khadem, H.; Shaban, M.A.E.; Nassr, M.A.M. *J. Chem. Soc. (C)* (1968) 1465.
150. Jung, K.H.; Schmidt, R.R.; Heermann, D. *Chem. Ber.* **114** (1981) 2834.
151. El Ashry, E.S.H.; Awad, L.F.; Winkler, M. *J. Chem. Soc., Perkin. Trans. 1* (2000) 829.
152. Awad, L.F.; El Ashry, E.S.H. *Carbohydr. Res.* **312** (1998) 9.
153. Ram, V.J. *Arch. Pharm.* **311** (1978) 968.
154. Somogyi, L. *Carbohydr. Res.* **64** (1978) 289.
155. El Ashry, E.S.H.; Soliman, R.; Mackawy, K. *Carbohydr. Res.* **72** (1979) 305.
156. Abdel Rahman, M.M.; El Ashry, E.S.H.; Abdallah, A. A.; Rashed, N. *Carbohydr. Res.* **73** (1979) 103.
157. Somogyi, L. *Carbohydr. Res.* **75** (1979) 325.
158. El Khadern, H.; Shaban, M.A.E.; Nassr M.A.M. *Carbohydr. Res.* **13** (1970) 470.
159. El Khadem, H.; Shaban, M.; Nassr, M.A.M. *Carbohydr. Res.* **23** (1972) 103.
160. Nassr, M.A.M. *Org. Prep. Proceed. Int.* **15** (1983) 329.
161. El Ashry, E.S.H.; El Kilany, Y.; Abdallah, A.A.; Mackawy, K. *Carbohydr. Res.* **113** (1983) 273.
162. El Kilany, Y.; Awad, L.; Makawy, K.; El Ashry, E.S.H. *Alex. J. Pharm. Sci.* **2** (1988) 139.
163. Shaban, M.A.E.; El Ashry, E.S.H.; Nassr, M.A.M.; Reinhold, V.N. *Carbohydr. Res.* **42** (1975) C1.
164. Tronchet, J.M.J.; Moskalyk, R.E. *Helv. Chim. Acta* **55** (1972) 2816.
165. Just, G.; Ranjeesingh, M. *Tetrahedron Lett.* **16** (1975) 985.
166. Just, G.; Chalard Faure, B. *Can. J. Chem.* **54** (1976) 861.
167. El Ashry, E.S.H.; Abdel Rahman, M.A.; El Kilany, Y.; Rashed. N. *Carbohydr. Res.* **163** (1987) 123.
168. Holmberg. B. *Ark. Kemi* **7** (1954) 513; *Chem. Abstr.* **50** (1956) 239b.
169. El Ashry, E.S.H.; Nassr, M.A.M.; El Kilany, Y.; Mousaad, A. *J. Prakt. Chem.* **328** (1986) 1.
170. El Ashry, E.S.H.; Nassr, M.M.A.; El Kilany, Y.; Mousaad, A. *Bull. Chem. Soc. Jpn.* **60** (1987) 3405.
171. Shaban, M.A.E.; Iskander, M.F.; El-Badry, S.M. *Pharmazie* **52** (1997) 350.
172. Zemplen, G.; Mester, L.A. *Chem. Ber.* **86** (1953) 697.
173. Zemplen, G.; Mester, L.; Eckhart, E. *Chem. Ber.* **86** (1953) 472.
174. Zemplen, G.; Mester, L. *Magy. Tud. Akad. Kern. Tudo, Oszt. Kozl.* **3** (1953) 7; *Chem. Abstr.* **49** (1955) 6242e.
175. Maeba, I.; Suzuki, M.; Hara, O.; Takeuch, T.; Iijimar, T.; Furukawa, H. *J. Org. Chem.* **52** (1987) 4521.
176. Lopez Aparicio, F.J.; Plaza Lopez-Espinosa, M.T.; Robles Diaz, R. *An. Quim., Ser. C* **80** (1984) 156.
177. El Ashry, E.S.H.; Awad, L.F.; Abdel Hamid, H.; El Kilany, Y. *Z. Naturforsch.* **54B** (1999) 1061.
178. Sadikun, A.B.; Davies, D.I, Kenyon, R.F. *J. Chem. Soc., Pekin Trans. 1* (1981) 2299.
179. Just, G.; Ouellet, R. *Can. J. Chem.* **54** (1976) 2925.
180. Sato, T.; Ito, R.; Hayakawa, Y.; Noyori, R. *Tetrahedron Lett.* **19** (1978) 1829.
181. Sato, T.; Hayakawa, Y.; Noyori, R. *Bull. Chem. Soc. Jpn.* **57** (1984) 2515.
182. Bobek, M.; Farkas, J.; Sorm, F. *Collect. Czech. Chem. Commun.* **31** (1966) 1414; *Chem. Abstr.* **64** (1966) 19744a.
183. Bobek, M.; Farkas, J.; Sorm, F. *Tetrahedron Lett.* **7** (1966) 3115.
184. Bobek, M.; Farkas, J.; Sorm, F. *Collect. Czech. Chem. Commun.* **32** (1967) 3572; *Chem. Abstr.* **68** (1968) 49952e.
185. Bobek, M.; Farkas, J.; Sorm, F. *Collect. Czech. Chem. Commun.* **34** (1969) 1637; *Chem. Abstr.* **71** (1969) 22282m.
186. Bobek, M.; Farkas, J.; Sorm, F. *Tetrahedron Lett.* **9** (1968) 1543.
187. Swartz, D.L.; El Khadem, H.S. *Carbohydr. Res.* **112** (1983) C1.
188. Hayashi, M.; Araki, A.; Maeba, I. *Heterocycles* **34** (1992) 569.
189. Mousaad, A.; Abdel Hamid, H.; El Nemr, A.; El Ashry, E.S.H. *Bull. Chem. Soc. Jpn.* **65** (1992) 546.
190. Rashed, N.; El Nemr, A.; El Ashry, E.S.H. *Spectrosc. Lett.* **26** (1993) 1817.
191. Mousaad, A.; Rashed, N.; Ramadan, E.; El Ashry, E.S.H. *Spectrosc. Lett.* **27** (1994) 677.

192. El Ashry, E.S.H.; Rashed, N.; Abdel Hamid, H.; Ramadan, E. *Z. Naturforsch* **52B** (1997) 873.
193. Rashed, N.; Abdel Hamid, H.; Ramadan, E.; El Ashry, E.S.H. *Nucleosides Nucleotides* **17** (1998) 1373.
194. Hanisch, G.; Henseke. G. *Chem. Ber.* **101** (1968) 4170.
195. El Khadem, H.; El Shafei, Z.M.; El Sekily, M. *J. Org. Chem.* **22** (1972) 3523.
196. Schmidt, R.R.; Guilliard, W.; Karg, J. *Chem. Ber.* **110** (1977) 2445.
197. Ohle, H.; Iltgen. A. *Chem. Ber.* **76** (1943) 1.
198. Buu-Hoi, N.P.; Vallat, J.N.; Saint-Ruf, G.; Lambelin, G. *Chim. Ther.* **6** (1971) 245; *Chem. Abstr.* **76** (1972) 3794s.
199. Ohle, H.; Hielscher. M. *Chem. Ber.* **74** (1941) 13.
200. Ohle, H.; Melkonian. G.A. *Chem. Ber.* **74** (1941) 279.
201. Ohle, H.; Melkonian. G.A. *Chem. Ber.* **74** (1941) 398.
202. Ohle, H.; Liebig, R. *Chem. Ber.* **75** (1942) 1536.
203. Ohle, H.; Kruyff, J.J. *Chem. Ber.* **77** (1944) 507.
204. Henseke, G.; Lemke, W. *Chem. Ber.* **91** (1958) 1605.
205. Courtois, J.E.; Ariyoshi, U. *Ann. Pharm. Fr.* **16** (1958) 385; *Chem. Abstr.* **53** (1959) 8003a.
206. Weygand, F.; Fehr. K.; Klebe, J. *Z. Naturforsch.* **14B** (1959) 217.
207. Von Saltza, M.: Dutcher, J.D.; Reid, J.; Wintersteiner, O. *J. Org. Chem.* **28** (1963) 999.
208. Henseke, G. *Z. Chem.* **6** (1966) 329.
209. Teichmann, B.; Himmelspach, K.; Westplial, O. *Z. Chem.* **11** (1971) 380.
210. Teichmann, B.; Himmelspach, K.; Westphal, O. *J. Prakt. Chem.* **314** (1972) 877.
211. Henseke, G.; Dose, W.; Dittrich, K. *Angew. Chem.* **69** (1957) 479.
212. Henseke, G.; Dittrich, K. *Chem. Ber.* **92** (1959) 1550.
213. El Ashry, E.S.H.; El Kholy, I.E.; El Kilany, Y. *Carbohydr. Res.* **60** (1978) 303.
214. El Ashry, E.S.H.; Abdel Rahman, M.M.; Nassr, M.A.; Amer, A. *Carbohydr. Res.* **67** (1978) 403.
215. El Ashry, E.S.H.; Abdel Rahman, M.M.; Rashed, N.; Amer, A. *Carbohydr. Res.* **67** (1978) 423.
216. El Ashry, E.S.H.; Rahman, M.A.; Labib, G.H.; El Massry, A.M.; Mofti, A. *Carbohydr. Res.* **152** (1986) 339.
217. El Ashry, E.S.H.; El Kilany, Y.; Mousaad, A. *Carbohydr. Res.* **163** (1987) 262.
218. Awad, L.; Mousaad, A.; El Ashry, E.S.H. *J. Carbohydr. Chem.* **8** (1989) 756.
219. Mousaad, A.; Awad, L.; El Shimy, N.; El Ashry, E.S.H. *J. Carbohydr. Chem.* **8** (1989) 733.
220. Somogyi, L. *Carbohydr. Res.* **229** (1992) 89.
221. Rashed, N.; Abdel Hamid, H.; Shoukry, M.M. *Heterocycles* **36** (1993) 961.
222. Henseke, G.; Lehmann, D. *Angew. Chem., Int. Ed. Engl.* **3** (1964) 802.
223. El Ashry, E.S.H.; Abdel Rahman, M.M.; Mancy, S.; El Shafei, Z.M. *Acta Chim. Acad. Sci. Hung.* **95** (1977) 409.
224. Pollet, P.; Gelin, S. *Synthesis* (1979) 977.
225. Rashed, N.; Ibrahim, E.I.; El Ashry, E.S.H. *Carbohydr. Res.* **254** (1994) 295.
226. Turk, C.; Svete, J.; Golobic, A.; Golic, L.; Stanovnik, B. *J. Heterocycl. Chem.* **35** (1998) 513.
227. Svete, J.; Golic, L.; Stanovnik, B. *J. Heterocycl. Chem.* **34** (1997) 1115.
228. Shaban, M.A.E.; Taha, M.A.M. *J. Carbohydr. Chem.* **10** (1991) 757.
229. Rashed, N.; Mousaad, A.; Saleh, A. *Alex. J. Pharm. Sci.* **7** (1993) 171.
230. El Ashry, E.S.H.; El Kilany, Y.; Rashed, N.; Assafir, H. In *Advances in Heterocyclic Chemistry*, Vol. 75; Katritzky, A.R., Ed.; Academic Press: New York, 1999; p. 79.
231. Shaban, M.A.E.; Taha, M.A.M.; Nassr, A.Z.; Morgaan, A.E.A. *Pharmazie* **50** (1995) 784.
232. Shaban, M.A.E.; Taha, M.A.M.; Nassr, A.Z. *Pharmazie* **51** (1996) 707.
233. Rashed, N.; Shoukry, M.; El Ashry, E.S.H. *Bull. Chem. Soc. Jpn.* **67** (1994) 149.
234. El Ashry, E.S.H.; El Kilany, Y. *Carbohydr. Res.* **80** (1980) C23.
235. El Khadem, H.; El Shafei, Z.M.; Sallam, M.E. *Carbohydr. Res.* **18** (1971) 147.
236. Percival, E.G.V. *J. Chem. Soc.* (1938) 1384.
237. Henseke, G.; Müller, V.; Baiseke, G. *Chem. Ber.* **91** (1958) 2270.
238. El Khadem, H.; Abdel Rahman, M.M. *J. Org. Chem.* **31** (1966) 1178.
239. El Kourashy, A.; Mousaad, A.; El Ashry, E.S.H. *Alex. J. Pharm. Sci.* **4** (1990) 123.
240. Stam, C.; El Ashry, E.S.H.; El Kilany, Y.; Vander Plas, H.C. *J. Heterocycl. Chem.* **17** (1980) 617.
241. Ogura, H.; Sakaguchi, M.; Nakata, K.; Hida, N.; Takeuchi H. *Chem. Pharm. Bull.* **29** (1981) 629.
242. Alder, C.; Curtius, H.C.; Datta, S.; Viscontini, M. *Helv. Chim. Acta* **73** (1990) 1058.
243. Viscontini, M.; Bosshard, R. *Helv. Chim. Acta* **73** (1990) 337.
244. Soyka, R.; Pfeiderer, W.; Prewo, R. *Helv. Chim. Acta* **73** (1990) 808.
245. Rashed, N.; Abdel Hamid, H.; El Ashry, E.S.H. *Carbohydr. Res.* **243** (1993) 399.
246. El Ashry, E.S.H.; Abdel Hamid, H.; El Kilany, Y. *Heterocycl. Commun.* **2** (1996) 325.
247. Henseke G.; Lemke, W. *Chem. Ber.* **91** (1958) 101.
248. El Ashry, E.S.H.; Abdel Rahman, M.M.; Hazah, A.; Singab, F. *Sci. Pharm.* **48** (1980) 13.

249. Awad, L.F.; Abdel Rahman, M.M.; Zakaria, M.; El Ashry, E.S.H. *Alex. J. Pharm. Sci.*, **3** (1989) 119.
250. El Ashry, E.S.H.; El Kilany, Y.; Abdel Hamid, H. *Gazz. Chim. Ital.* **116** (1986) 721.
251. El Kilany, Y.; Rashed, N.; Mansour, M.; Abdel Rahman, M.A.; El Ashry, E.S.H. *J. Carbohydr. Chem.* **7** (1988) 199.
252. Awad, L.; El Kilany, Y. *Bull. Chem. Soc. Ethiop.* **3** (1989) 97.
253. Awad, L.F. *Carbohydr. Res.* **326** (2000) 34.
254. El Ashry, E.S.H.; Nassr, M.; Shoukry, M. *Carbohydr. Res.* **83** (1980) 79.
255. El Ashry, E.S.H.; Abdel Rahman, M.M.A.; El Kilany, Y.; Amer, A. *Carbohydr. Res.* **87** (1981) C5.
256. El Ashry, E.S.H.; El Kilany, Y.; Amer, A. *Heterocycles* **26** (1987) 2101.
257. Sallam, M.A.E.; Mostafa, M.A.; Hussein, N.A.R.; Townsend, L.B. *Alex. J. Pharm. Sci.* **4** (1990) 18.
258. Hanessian, S. *Total Synthesis of Natural Products: The Chiron Approach*; Pergamon Press: New York, 1983.
259. Hale, K.J., *Monosaccharides: Use in the Asymmetric Synthesis of Natural Products in Rodd's Chemistry of Carbon Compounds*, Vol. IE/F/G (Second Supplements to the second edition); Sainsbury, M., Ed.; Elsevier: Amsterdam, 1993.
260. Boons, G.-J.; Hale, K.J. *Organic Synthesis with Carbohydrates*; Blackwell Science: Malden, MA, 2000.
261. Zechmeister, L. In *Progress in the Chemistry of Organic Natural Products*; Herz, W., Grisebach, H., Kirby, G.W., Eds.; Springer-Verlag: New York, 1980.
262. Witczak, Z.J.; Nieforth, K.A. Eds. *Carbohydrates in Drug Design*; Marcel Dekker: New York, 1997.

1 Five-membered nitrogen heterocycles

As the title denotes, this chapter deals with the conversion of carbohydrate derivatives into five-membered heterocyclic compounds containing nitrogen. The types of target compounds are (1) hydroxymethylpyrrolidines, (2) carboxypyrrolidines, (3) aralkyl pyrrolidines and (4) aryl pyrrolidines, as well as other heterocycles that are grouped under the title miscellaneous (5). Although many of these naturally occurring compounds and their stereoisomers and analogues have been synthesized from noncarbohydrates, the synthesis of the naturally occurring five-membered nitrogen heterocycles from only carbohydrate will be discussed in this chapter.

1.1 Hydroxymethylpyrrolidines

Because of their 'sugar-like' structure it is not surprising that most syntheses of the naturally occurring hydroxymethylpyrrolidines utilize carbohydrates as starting materials. Pentoses, hexoses and their derivatives are often used; three chiral centers are usually required. Nitrogen is introduced in the synthetic sequence as azide, followed by reduction to the respective amine that can be intramolecularly cyclized. This part contains three groups of compounds: 2-hydroxymethylpyrrolidines, dihydro-2-hydroxymethyl pyrrole (nectrisine) and 2,5-dihydroxymethylpyrrolidines. Each of the first and third groups contains five natural compounds isolated from different sources and characterized by having glycosidase inhibition properties.

1.1.1 *2-Hydroxymethylpyrrolidines*

1,4-Dideoxy-1,4-imino-D-arabinitol [(2*R*,3*R*,4*R*)-2-hydroxymethyl pyrrolidine-3,4-diol, DAB1, **1**] has been found in both *Arachniodes standishii*[1,2] and *Angylocalyx boutiqueanus*[3] and it is a potent inhibitor of yeast α-glucosidase (50% inhibition at 1.8×10^{-7} M)[4–7] and mouse gut disaccharidases to different degrees.[8] Compound **1** inhibits the hydrolysis of sinigrin and progoitrin from mustard and cabbage aphid *Brevicoryne brassicae*.[9] It also inhibits phloem unloading and/or utilization of sucrose, resulting in insufficient sucrose transport from cotyledons to roots and hypocotyls.[10] The mechanism of insect antifeedant activity of **1** has been studied[11] and it was found that it may be carcinogenic to rodents.[12] The enantiomer 1,4-dideoxy-1,4-imino-L-arabinitol [(2*S*,3*S*,4*S*)-2-hydroxymethyl pyrrolidine-3,4-diol, LAB1, **2**] occurs as a component of bacterial lipopolysaccharides[13,14] but it shows a weaker inhibition of α-glucosidase (50% inhibition at 1.0×10^{-5} M)[15,16] and exhibits several biological activities.[17–20] 1,4-Dideoxy-1,4-imino-D-ribitol [(2*R*,3*R*, 4*S*-2-hydroxymethyl pyrrolidine-3,4-diol, **3**] has been isolated from *Morus* spp.[21,22] 1,4-Dideoxy-1,4-imino-L-xylitol [(2*S*,3*R*,4*R*)-2-hydroxymethyl pyrrolidine-3,4-diol, **4**] was isolated from diatom cell walls[23] and *Amanita vitosa* mushrooms.[24] 2-Hydroxymethyl-3-hydroxypyrrolidine [(2*R*,3*S*)-2-hydroxymethyl pyrrolidin-3-ol, CYB3, **5**] was isolated from legume *Castanospermum australe* and it has no significant biological activity.[25]

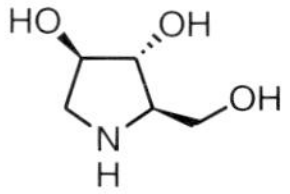

(2*R*,3*R*,4*R*)-2-Hydroxymethyl pyrrolidine-3,4-diol (DAB1)
1

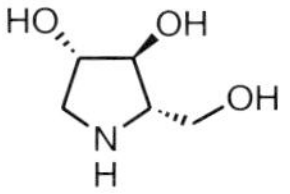

(2*S*,3*S*,4*S*)-2-Hydroxymethyl pyrrolidine-3,4-diol (LAB1)
2

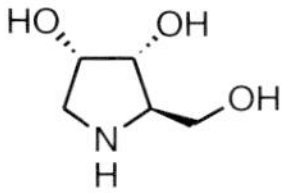

(2*R*,3*R*,4*S*)-2-Hydroxymethyl pyrrolidine-3,4-diol
3

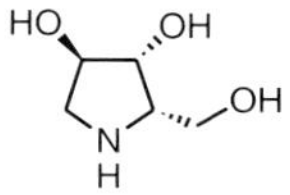

(2*S*,3*R*,4*R*)-2-Hydroxymethyl pyrrolidine-3,4-diol
4

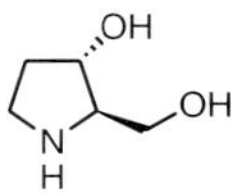

(2*R*,3*S*)-2-Hydroxymethyl pyrrolidin-3-ol (CYB3)
5

Syntheses of natural polyhydroxypyrrolidines from noncarbohydrate and their unnatural analogues from carbohydrate and noncarbohydrate have been reported.[26–86] Herein, the synthesis of the natural analogues from carbohydrate building blocks will be reviewed.

1.1.1.1 *Synthesis from* D*-glucose* A stereoselective synthesis of DAB1 (**1**) from D-glucose has been reported (Scheme 1).[87] Diacetone D-glucose (**6**) was benzylated to give the fully protected furanose, which underwent acid hydrolysis of the terminal isopropylidene group followed by periodate oxidation, sodium borohydride reduction, mesylation and then

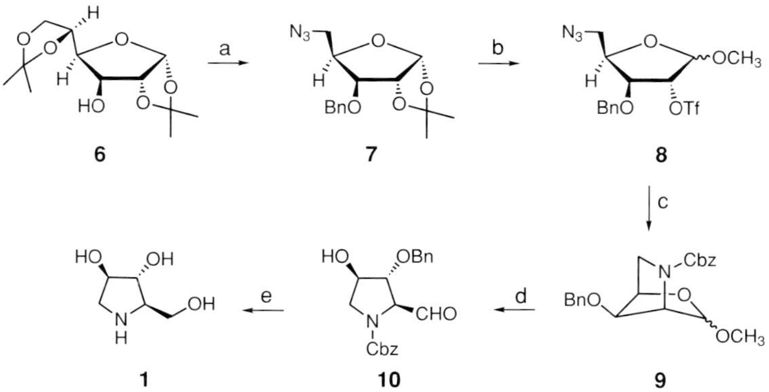

Scheme 1 (*a*) 1. THF, NaH, Bu_4NI, 0°C, BnBr, rt to 50°C, 2 h, 97%; 2. CH_3OH–AcOH–H_2O (1:1:1), 50°C, 16 h, 87%; 3. $NaIO_4$, 10% aqueous EtOH, 3 h, CH_2Cl_2; then 20% aqueous EtOH, $NaBH_4$, 8 h at rt, 88%; 4. Py, 0°C, MsCl, rt, 2 h, 94%; 5. NaN_3, DMF, 70°C for 12 h, 97%. (*b*) 1. AcCl, CH_3OH, 0°C, 36 h, α (38%), β (44%); 2. Py, Tf_2O, –50 to −30°C, 1 h, α (92%), β (78%). (*c*) 1. EtOAc, rt, 5%, Pd *on* C, H_2, α (95%), β (92.5%); 2. 3:2 mixture of ether and aqueous $NaHCO_3$, CbzCl, rt, 12 h, α (76%), β (90%). (*d*) 1:1 mixture of TFA and H_2O, 92%. (*e*) 1. EtOH, $NaBH_4$, 15 min, 98%; 2. AcOH, H_2, Pd black, 18 h, 97.6%.

replacement of the mesyloxy group with azide ion to afford the azide **7**. Compound **7** was treated with methanolic hydrogen chloride, followed by triflation of C-2 hydroxyl group to give the corresponding triflate **8**. Hydrogenation of **8** followed by protection of the resulting bicyclic compound with benzyl chloroformate afforded the carbamate **9**. Subsequent hydrolysis with TFA gave the key intermediate **10**. Reduction of **10** with sodium borohydride followed by removal of the carbamate and *O*-benzyl protecting groups by hydrogenolysis in acetic acid gave DAB1 (**1**) in 33% from **6**.

Synthesis of 1,4-dideoxy-1,4-imino-L-xylitol (**4**) has been achieved from D-glucose (Scheme 2).[88] Borane-reductive ring opening of the benzylidene ring in compound **11**, obtained from D-glucose,[89] afforded **12** (90%). Reduction of **12** with sodium borohydride produced the corresponding triol **13** (73%), which was subjected to periodate oxidation to give the cyclic hemiacetal **14** (92%). Hydrogenation of **14** over palladium led to the formation of 3,4-dihydroxypyrrolidine **4**·HCl.

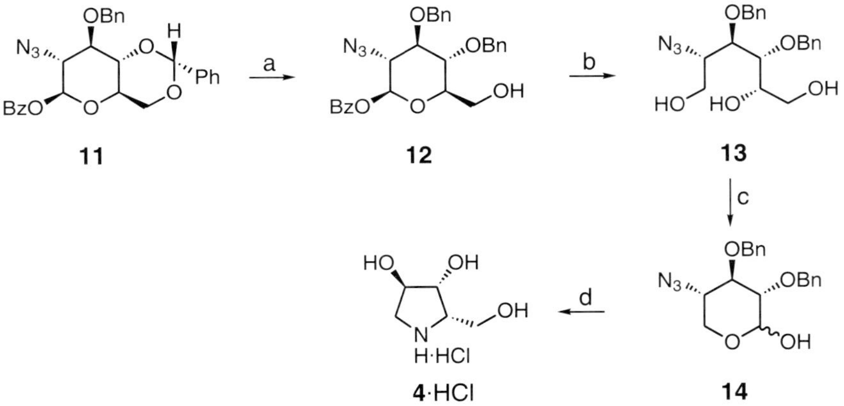

Scheme 2 (*a*) BH_3, THF, 15 mol% $V(O)(OTf)_2$, CH_2Cl_2, rt, 3 h, 90%. (*b*) $NaBH_4$, CH_3OH, 73%. (*c*) $NaIO_4$, CH_3OH, 92%. (*d*) 10% Pd *on* C, H_2, EtOH, 1 N HCl, 94%.

1.1.1.2 *Synthesis from* D-*mannose* A stereoselective synthesis of 1,4-dideoxy-1,4-imino-L-xylitol (**4**) from D-mannose has been reported (Scheme 3).[90] A pyridine solution of D-mannose containing iron(III) triflate or iron(III) chloride was irradiated with a high-pressure mercury lamp in a Pyrex vessel, while oxygen gas was bubbled through to afford after acetylation the aldopentose derivative **15**, which was treated with aluminum chloride in aqueous methanol to afford 1,2,3-tri-*O*-acetyl-D-arabinopyranose **16** in 24% overall yield from D-mannose. Triflation of **16** followed by treatment with sodium azide gave 1,2,3-tri-*O*-acetyl-4-azido-4-deoxy-L-xylopyranose (**17**) in 55% yield. Deacetylation of **17** with potassium carbonate in methanol followed by catalytic hydrogenation gave **4** in 77% yield.

1.1.1.3 *Synthesis from* L-*arabinose* Synthesis of 1,4-dideoxy-1,4-imino-L-arabinitol (**2**) from methyl β-L-arabinopyranoside has been reported (Scheme 4).[91] The double inversion involving the introduction of the azide function at C-4 has been effected in

Scheme 3 (*a*) Py, $FeTf_3$, $h\nu$, 8 h, O_2; then Ac_2O, rt, 14 h. (*b*) $AlCl_3$, CH_3OH, H_2O, rt, 36 h, 24% from D-mannose. (*c*) 1. Tf_2O, CH_2Cl_2, Py, 0°C, 1 h; 2. NaN_3, DMF, 15-crown-5, rt, 24 h, 55%. (*d*) 1. 10% aqueous CH_3OH, K_2CO_3, 0°C, 15 min, AcOH; 2. 10% Pd *on* C, H_2, 3 days; then Dowex 50X8-100 (H^+) resin, 77%.

two steps by reacting methyl 2,3-di-*O*-benzoyl-β-L-arabinoside (**18**) with triphenylphosphine and 2,4,5-tribromoimidazole to form methyl 2,3-di-*O*-benzoyl-4-bromo-4-deoxy-α-D-xylopyranoside (**19**). This bromide was reacted with sodium azide to give 4-azido-4-deoxy-L-arabinoside (**20**). Debenzoylation with methanolic sodium methoxide gave methyl 4-azido-4-deoxy-β-L-arabinopyranoside (**21**), whose acid hydrolysis and catalytic hydrogenation gave **2**.

Scheme 4 (*a*) Ph_3P, 2,4,5-tribromoimidazole. (*b*) NaN_3, DMF. (*c*) $NaOCH_3$, CH_3OH. (*d*) 1. H_3O^+; 2. H_2, Pd, Amberlite CG-400 (OH^-) resin.

1.1.1.4 *Synthesis from D-xylose* The synthesis of 1,4-dideoxy-1,4-imino-L-arabinitol (**2**) can also be achieved from D-xylose (Scheme 5).[92] Thus, methyl β-D-xylopyranoside (**22**) has been treated with 2-methoxypropene followed by triflation with trifluoromethane sulfonic anhydride to afford the triflate **23**. The latter underwent S_N2 displacement with sodium azide followed by acid hydrolysis to produce the azide **24**. Reduction of **24** afforded **2**, via the intermediates **25–27**, in 21% overall yield from **22**.

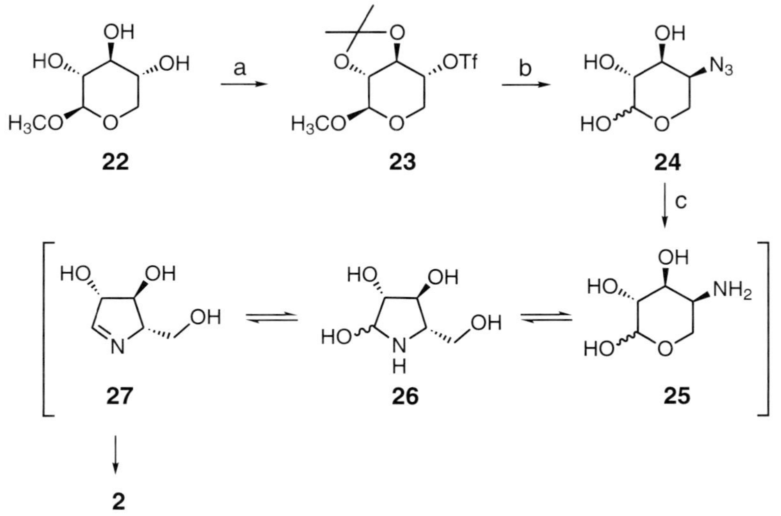

Scheme 5 (*a*) 1. DMF, 4 M HCl in CH_3OH, 2-methoxypropene, 60°C, 2 h; then rt, overnight, 72%; 2. CH_2Cl_2, Py, –50°C, Tf_2O, 45 min at –25°C. (*b*) 1. DMF, NaN_3, rt, 2 h, 45% for two steps; 2. AcOH, 2 M H_2SO_4, 95°C, 3 h; then $NaHCO_3$, pH 4, 65%. (*c*) 0.1 M aqueous HCl, 10% Pd *on* C, H_2, rt, 6 h, 100%.

DAB1 (**1**) and LAB1 (**2**) can be alternatively synthesized from D-xylose (Scheme 6).[15] The acetonide **28**,[93] obtained from D-xylose, was triflated, followed by S_N2 displacement with azide ion and subsequent removal of the isopropylidene group to give **29**. Selective

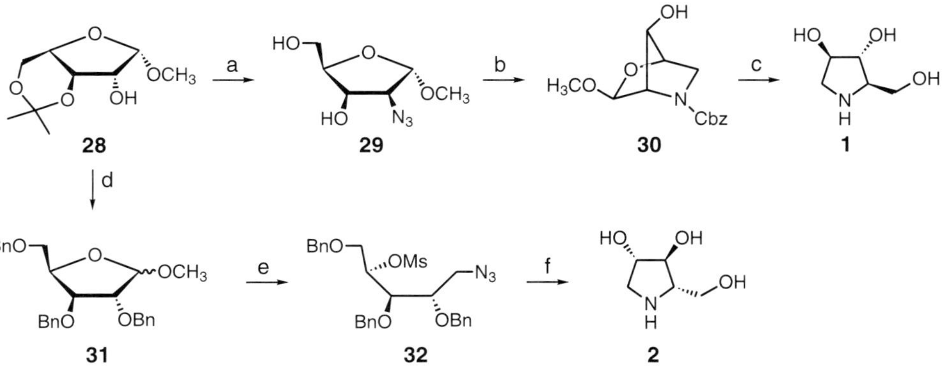

Scheme 6 (*a*) 1. Tf_2O, Py, CH_2Cl_2; 2. NaN_3, DMF, 100°C, 12 h, 76% for two steps; 3. Dowex 50W8X resin, CH_3OH, rt, 4 h, 83%. (*b*) 1. *p*-TsCl, Py, 0°C; 2. H_2, Pd black, EtOH, NaOAc, 50°C; CbzCl, ether, H_2O containing $NaHCO_3$, 36% for three steps. (*c*) 1. TFA–H_2O (4:1); 2. $NaBH_4$, EtOH; 3. H_2, Pd black, AcOH, 65%. (*d*) 1. BnBr, NaH; 2. Dowex 50W8X (H^+) resin; 3. BnBr, NaH, *t*-Bu_4NI, THF, 45% for three steps. (*e*) 1. TFA, H_2O; 2. $NaBH_4$, EtOH, rt, 1 h, 87%; 3. MsCl, Py, 90%; 4. NaN_3, DMF, 66%. (*f*) 1. H_2, Pd black, EtOH; 2. ion-exchange chromatography, 48%.

tosylation of the primary hydroxyl group in **29** followed by azide reduction and subsequent cyclization with sodium acetate and protection with benzyl chloroformate afforded the carbamate **30**. Hydrolysis of **30** by aqueous TFA followed by reduction of the resulting aldehyde, removal of the carbamate protecting group and purification by ion-exchange chromatography gave **1** in 15% overall yield from **28**.

On the other hand, the xylofuranoside **28** was benzylated, followed by removal of the isopropylidene group and subsequent benzylation to give the tribenzylated derivative **31**. Acid hydrolysis of **31** followed by sodium borohydride reduction of the resulting lactol and subsequent mesylation and then selective nucleophilic displacement of the primary mesylate by sodium azide in DMF afforded the azido mesylate **32**. Reduction of the azide was accompanied by cyclization and deprotection to afford **2** in 11% overall yield from **28**.

1.1.1.5 *Synthesis from* D*-threose* Synthesis of DAB1 (**1**) has been carried out by conversion of the D-threose derivative **33**,[94] readily available from D-(−)-diethyl tartrate, to the aminonitrile **34** as an inseparable diastereomeric mixture (Scheme 7).[95] Subsequent deprotection with TBAF gave the alcohol **35** (quantitative). Esterification of **35** with *p*-toluenesulfonyl chloride afforded **36** (84%), which was treated with TFA–H_2O–THF

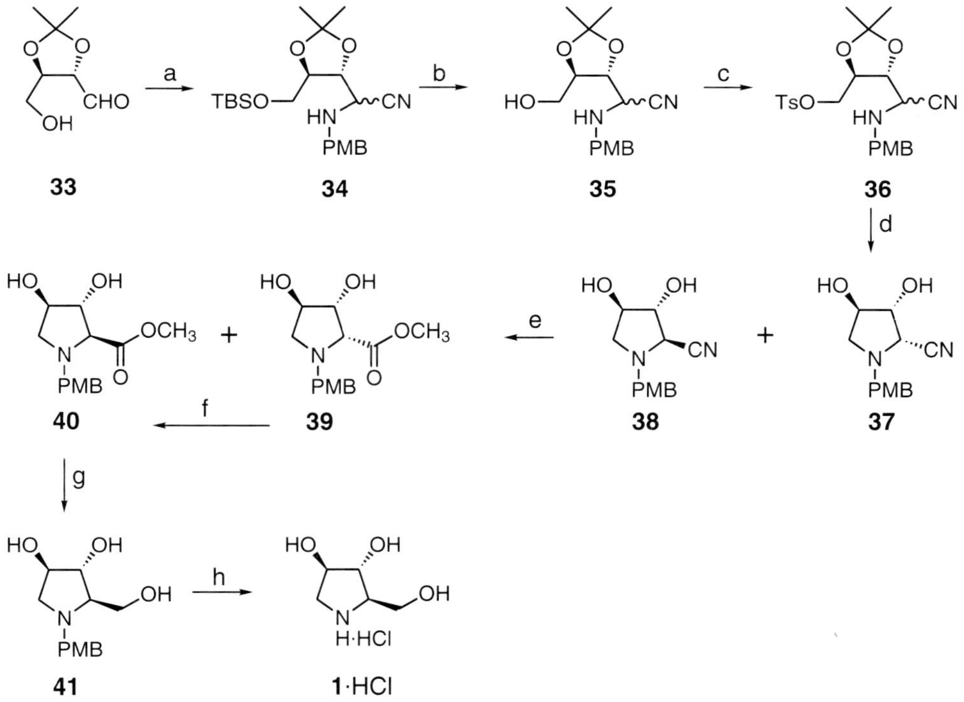

Scheme 7 (*a*) 1. TBSCl, imidazole, CH_2Cl_2; 2. *p*-$(CH_3O)C_6H_4CH_2NH_2$, $(EtO)_2P(O)CN$, THF, 86.7%. (*b*) TBAF, THF, quantitative. (*c*) *p*-TsCl, Py, 84%. (*d*) TFA–H_2O–THF (5:1:1), 70–75°C. (*e*) $NaOCH_3$, CH_3OH; then 2 N HCl, 87%. (*f*) Same as (*e*), 65–70°C, 2 h; then 2 N HCl, 78%. (*g*) $NaBH_4$, EtOH, 89%. (*h*) 1. H_2, 20%, $Pd(OH)_2$ *on* C, HCO_2H, EtOH; 2. conc. HCl, 94%.

to afford the cyclized isomeric mixture **37** and **38** in a ratio of 4:1 (74%). Subsequent treatment with sodium methoxide in methanol gave a chromatographically separable mixture of the methyl esters **39** (21%) and **40** (28%) in addition to recovery of the starting material (48.7%), which could be recycled. Treatment of **39** with sodium methoxide in methanol afforded a 1:1 mixture of **39** and **40**. Reduction of **40** with sodium borohydride gave the alcohol **41** (89%). Removal of the PMB group from **41** by catalytic hydrogenolysis provided **1**, which was conveniently isolated as its crystalline hydrochloride by treatment with conc. HCl (94%). Its enantiomer LAB1 (**2**) was synthesized from L-(+)-diethyl tartrate following the same set of reactions previously described for **1**.

1.1.1.6 *Synthesis from* D-*lyxonolactone* A synthesis of 1,4-dideoxy-1,4-imino-L-arabinitol (**2**) from D-lyxonolactone (**42**) (Scheme 8)[17] was achieved by benzylidenation, followed by mesylation to afford **43** in 80% yield from **42**. Lithium borohydride reduction of **43** followed by treatment with potassium carbonate afforded the epoxide **44** (72%), which underwent triflation of the free hydroxyl group followed by S_N2 displacement with azide ion to furnish the azidoepoxide **45** (92%). Hydrogenation of **45** followed by ring closure of the resulting amine using tetrabutylammonium iodide, via the intermediate iodoalcohol, afforded the pyrrolidine **46**. Finally, removal of the benzylidene group with H_2SO_4 afforded **2**, which was isolated as the hydrochloride salt in 21% overall yield from **42**.

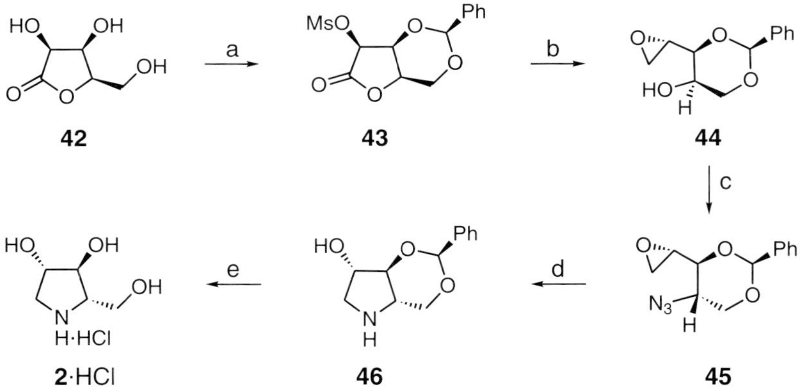

Scheme 8 (*a*) 1. PhCHO, conc. HCl, 0°C, 3 h, 96%; 2. MsCl, Py, rt, 3 h, 83%. (*b*) 1. $LiBH_4$, THF, –68°C to rt, 16 h, 90%; 2. K_2CO_3, CH_3OH, rt, 24 h, 80%. (*c*) 1. Tf_2O, Py, CH_2Cl_2, –30°C, 2 h, 100%; 2. NaN_3, DMF, 0°C, 2 h, 92%. (*d*) 1. H_2, 5% Pd *on* C, CH_3OH, 30 min, 62%; 2. Bu_4NI, THF, reflux, 72 h, 58%. (*e*) 1. 0.1 N H_2SO_4, 100°C, 3 h, 90%; 2. CH_3OH, 12 M HCl, 10 min, 84%.

1.1.1.7 *Synthesis from* D-*gulonolactone* Synthesis of 1,4-dideoxy-1,4-imino-D-ribitol (**3**) from D-gulonolactone has been reported (Scheme 9).[96] D-Gulonolactone was treated with DMP to produce diacetone D-gulonolactone (**47**), which underwent $LiAlH_4$ reduction followed by mesylation to furnish **48** in 74% yield from D-gulonolactone. Heating of **48** with benzylamine afforded the protected pyrrolidine **49**. Treatment of **49** with aqueous acetic acid followed by periodate oxidation of the terminal diol and subsequent borohydride reduction afforded the pyrrolidine **50**. Compound **50** was debenzylated and then deacetonated to produce **3** in 29% overall yield from D-gulonolactone.

D-Gulonolactone → (a) → **47** → (b) → **48** → (c) → **49** → (d) → **50** → (e) → **3**

Scheme 9 (*a*) Acetone, DMP, *p*-TsOH, rt, 2 days; then anhydrous Na_2CO_3, 85%. (*b*) 1. $LiAlH_4$, THF, rt, 30 min, 87%; 2. MsCl, DMAP, Py, rt, 2 h, 100%. (*c*) $BnNH_2$, 60–70°C, 60 h, 77%. (*d*) 1. 80% aqueous AcOH, 50°C, 48 h, 93%; 2. $NaIO_4$, EtOH–H_2O (5:1), rt, 20 min; then $NaBH_4$, 0°C, 30 min, 71%. (*e*) EtOH, H_2, 10% Pd *on* C, rt, 2 h; then 50% aqueous TFA, rt, 24 h, 78%.

References

1. Furukawa, J.; Okuda, S.; Saito, K.; Hatanaka, S.-I. *Phytochemistry* **24** (1985) 593.
2. Jones, D.W.C.; Nash, R.J.; Bell, E.A.; Williams, J.M. *Tetrahedron Lett.* **26** (1985) 3125.
3. Nash, R.J.; Bell, E.A.; Williams, J.M. *Phytochemistry* **24** (1985) 1620.
4. Fleet, G.W.J.; Karpas, A.; Dwek, R.A.; Fellows, L.E.; Tyms, A.S.; Petursson, S.; Namgoong, S.K.; Ramsden, N.G.; Smith, P.W.; Son, J.C.; Wilson, F.; Witty, D.R.; Jacob, G.S.; Rademacher, T.W. *FEBS Lett.* **237** (1988) 128.
5. Meng, Q.; Hesse, M. *Helv. Chim. Acta* **74** (1991) 445.
6. Asano, N.; Nash, R.J.; Molyneux, R.J.; Fleet, G.W.J. *Tetrahedron: Asymmetry* **11** (2000) 1645.
7. El Ashry, E.S.H.; Rashed, N.; Shobier, A.H.S. *Pharmazie* **55** (2000) 331.
8. Scofield, A.M.; Fellows, L.E.; Fleet, G.W.J.; Nash, R.J. *Life Sci.* **39** (1986) 645.
9. Scofield, A.M.; Rossiter, J.T.; Witham, P.; Kite, G.C.; Nash, R.J.; Fellows, L.E. *Phytochemistry* **29** (1990) 107.
10. Aoki, T.; Hatanaka, S. *Phytochemistry* **30** (1991) 3197.
11. Simmonds, M.S.J.; Blaney, W.M.; Fellows, L.E. *J. Chem. Ecol.* **16** (1990) 3167.
12. Rosenkranz, H.S.; Klopman, G. *Carcinogenesis* (*London*) **11** (1990) 349.
13. Volk, W.A.; Galanos, C.; Luderitz, O. *Eur. J. Biochem.* **17** (1970) 223.
14. Galanos, C.; Lüdertiz, O.; Rietschel, E.T.; Westphal, O. *Int. Rev. Biochem.* **14** (1977) 239.
15. Fleet, G.W.J.; Nicholas, S.J.; Smith, P.W.; Evans, S.V.; Fellows, L.E.; Nash, R.J. *Tetrahedron Lett.* **26** (1985) 3127.
16. Fleet, G.W.J.; Smith, P.W. *Tetrahedron* **42** (1986) 5685.
17. Behling, J.R.; Campbell, A.L.; Babiak, K.A.; Ng, J.S.; Medich, J.; Farid, P.; Fleet, G.W.J. *Tetrahedron* **49** (1993) 3359.
18. Axamawaty, M.T.H.; Fleet, G.W.J.; Hannah, K.A.; Namgoong, S.K.; Sinnott, M.L. *Biochem. J.* **266** (1990) 245.
19. Robinson, K.M.; Rhinehard, B.L.; Ducep, J.-B.; Danzin, C. *Drugs Future* **17** (1992) 705.
20. Karpas, A.; Fleet, G.W.J.; Dwek, R.A.; Petursson, S.; Namgoon, S.K.; Jacob, G.S.; Rademacher, T.W. *Proc. Nat. Acad. Sci. USA*, **85** (1988) 9229.
21. Asano, N.; Tomioka, E.; Kizu, H.; Matsui, K. *Carbohydr. Res.* **253** (1994) 235.

22. Asano, N.; Oseki, K.; Tomioka, E.; Kizu, H.; Matsui, K. *Carbohydr. Res.* **259** (1994) 243.
23. Nakajima, T.; Volcani, B.E. *Science* **164** (1969) 1400.
24. Buku, A.; Faulstich, H.; Wieland, T.; Dabrowski, J. *Proc. Natl. Acad. Sci. USA* **77** (1980) 2370.
25. Nash, R.J.; Bell, E.A.; Fleet, G.W.J.; Jones, R.H.; Williams, J.M. *J. Chem. Soc., Chem. Commun.* (1985) 738.
26. Fleet, G.W.J.; Gough, M.J.; Smith, P.W. *Tetrahedron Lett.* **25** (1984) 1853.
27. Austin, G.N.; Baird, P.D.; Fleet, G.W.J.; Peach, J.M.; Smith, P.W.; Watkin, D.J. *Tetrahedron* **43** (19870) 3095.
28. Setoi, H.; Kayakiri, H.; Takeno, H.; Hashimoto, M. *Chem. Pharm. Bull.* **35** (1987) 3995.
29. Ikota, N.; Hanaki, A. *Chem. Pharm. Bull.* **35** (1987) 2140.
30. Buchanan, J.G.; Edgarand, A.R.; Hewitt, B.D. *J. Chem. Soc., Perkin Trans. 1* (1987) 2371.
31. Buchanan, J.G.; Jigajinni, V.B.; Singh, G.; Wightman, R.H. *J. Chem. Soc., Perkin Trans. 1* (1987) 2377.
32. Bashyal, B.P.; Fleet, G.W.J.; Gough, M.J.; Smith, P.W. *Tetrahedron* **43** (1987) 3083.
33. Fleet, G.W.J.; Son, J.C.; Green, D.S.C.; Di-Bello, I.C.; Winchester, B. *Tetrahedron* **44** (1988) 2649.
34. Ziegler, T.; Straub, A.; Effenberger, F. *Angew. Chem.* **100** (1988) 737.
35. Hughes, P.; Clardy, J. *J. Org. Chem.* **54** (1989) 3260.
36. Thaning, M.; Wistrand, L.-G. *Acta Chem. Scand.* **43** (1989) 290.
37. Hirai, Y.; Chintani, M.; Yamazaki, T.; Momose, T. *Chem. Lett.* (1989) 1449.
38. Ikota, N.; Hanaki, A. *Chem. Pharm. Bull.* **37** (1989) 1087.
39. Ikota, N. *Chem. Pharm. Bull.* **37** (1989) 3399.
40. Pederson, R.L.; Wong, C.H. *Heterocycles* **28** (1989) 477.
41. Von der Osten, C.H.; Sinskey, A.J.; Barbas, C.F.; Pederson, R.L.; Wang, Y.F.; Wong, C.H. *J. Am. Chem. Soc.* **111** (1989) 3924.
42. Buchanan, J.G.; Lumbard, K.W.; Sturgeon, R.J.; Thompson, D.K.; Wightman, R.H. *J. Chem. Soc., Perkin Trans. 1* (1990) 699.
43. Duréault, A.; Greck, C.; Depezay, J.-C. *J. Carbohydr. Chem.* **9** (1990) 121.
44. Takahata, H.; Banba, Y.; Momose, T. *Tetrahedron: Asymmetry* **1** (1990) 763.
45. Hanessian, S.; Ratovelomanana, V. *Synlett* (1990) 501.
46. Moss, W.O.; Bradbury, R.H.; Hales, N.J.; Gallagher, T. *J. Chem. Soc., Chem. Commun.* (1990) 51.
47. Wehner, V.J.; Jaeger, V. *Angew. Chem.* **102** (1990) 1180.
48. Hung, R.R.; Straub, J.A.; Whitesides, G.M. *J. Org. Chem.* **56** (1991) 3849.
49. Kajimoto, T.; Chen, L.; Liu, K.K.C.; Wong, C.H. *J. Am. Chem. Soc.* **113** (1991) 9009.
50. Witte, J.F.; McClard, R.W. *Tetrahedron Lett.* **32** (1991) 3927.
51. Hamada, Y.; Hara, O.; Kawai, H.; Kohno, Y.; Shioiri, T. *Tetrahedron* **47** (1991) 8635.
52. Moss, W.O.; Bradbury, R.H.; Hales, N.J.; Gallagher, *J. Chem. Soc., Perkin Trans 1* (1992) 1901.
53. Thaning, M.; Wistrand, L.-G. *Acta Chem. Scand.* **46** (1992) 194.
54. Maggini, M.; Prato, M.; Ranelli, M.; Scorrano, G. *Tetrahedron Lett.* **33** (1992) 6537.
55. Hassan, M.E. *Grazz. Chim. Ital.* **122** (1992) 7.
56. Takano, S.; Moriya, M.; Ogasawara, K. *Tetrahedron: Asymmetry* **3** (1992) 681.
57. Jurczak, J.; Prokopowicz, P; Golebiowski, A. *Tetrahedron Lett.* **34** (1993) 7107.
58. McCaig, A.E.; Wightman, R.H. *Tetrahedron Lett.* **34** (1993) 3939.
59. Blanco, M.J.; Sardina, F.J. *Tetrahedron Lett.* **35** (1994) 8493.
60. Goli, D.M.; Cheesman, B.V.; Hassan, M.E.; Lodaya, R.; Slama, J.T. *Carbohydr. Res.* **259** (1994) 219.
61. Griffat-Brunet, D.; Langlois, N. *Tetrahedron Lett.* **35** (1994) 2889.
62. Huwe, C.M.; Blechert, S. *Tetrahedron Lett.* **36** (1995) 1621.
63. Huwe, C.M.; Blechert, S. *Synthesis* (1997) 61.
64. Huang, Y.; Dalton, D.R. *J. Org. Chem.* **62** (1997) 372.
65. Dötz, K.H.; Klumpe, M.; Nieger, M. *Chem. Eur. J.* **5** (1999) 691.
66. Evans, D.A.; Weber, A.E. *J. Am. Chem. Soc.* **109** (1987) 7151.
67. Kurokawa, N.; Ohfune, Y. *Tetrahedron* **49** (1993) 6195.
68. Remuzon, P. *Tetrahedron* **52** (1996) 13803.
69. Nájera, C.; Yus, M. *Tetrahedron: Asymmetry* **10** (1999) 2245.
70. Razavi, H.; Polt, R. *J. Org. Chem.* **65** (2000) 5693.
71. Sifferlen, T.; Defoin, A.; Streith, J.; Le Nouen, D.; Tarnus, C.; Dosbaa, I.; Foglietti, M.-J. *Tetrahedron* **56** (2000) 971.
72. Mechelke, M. F.; Meyers, A.I. *Tetrahedron Lett.* **41** (2000) 9377.
73. Hulme, A.N.; Montgomery, C.H.; Henderson, D.K. *J. Chem. Soc., Perkin Trans. 1* (2000) 1837.
74. Humphrey, A.J.; Parsons, S.F.; Smith, M.E.B.; Turner, N.J. *Tetrahedron Lett.* **41** (2000) 4481.
75. Langlois, N.; Rakotondradany, F. *Tetrahedron* **56** (2000) 2437.
76. Kumareswaran, R.; Hassner, A. *Tetrahedron: Asymmetry* **12** (2001) 3409.
77. Schieweck, F.; Altenbach, H.-J. *J. Chem. Soc., Perkin Trans. 1* (2001) 3409.

78. Francisco, C.G.; Freire, R.; Gonzárez, C.C.; León, E.I.; Riesco-Fagundo, C.; Suárez, E. *J. Org. Chem.* **66** (2001) 1861.
79. Filichev, V.V.; Brandt, M.; Pedersen, E.B. *Carbohydr. Res.* **333** (2001) 115.
80. Chen, X.; Du, D.-M.; Hua, W.-T. *Tetrahedron: Asymmetry* **13** (2002) 43.
81. Pham-Huu, D.-P.; Gizaw, Y.; BeMiller, J. N.; Petrus, L. *Tetrahedron Lett.* **43** (2002) 383.
82. Taylor, C.M.; Barker, W.D.; Weir, C.A.; Park, J.H. *J. Org. Chem.* **67** (2002) 4466.
83. Ewing, W.R.; Joullie, M.M. *Heterocycles* **27** (1988) 2843.
84. Dell'Uomo, N.; Cristina Di Giovanni, M.; Misiti, D.; Zappia, G.; Monache, G. D. *Tetrahedron: Asymmetry* **7** (1996) 181.
85. Hulme, A.N.; Curley, K.S. *J. Chem. Soc., Perkin Trans. 1* (2002) 1083.
86. El Ashry, E.S.H.; El Nemr, A. *Carbohydr. Res.* **338** (2003) 2265.
87. Fleet, G.W.J.; Witty, D.R. *Tetrahedron: Asymmetry 1* (1990) 119.
88. Wang, C.-C.; Luo, S.-Y.; Shie, C.-R.; Hung, S.-C. *Org. Lett.* **4** (2002) 847.
89. Hung, S.-C.; Thopate, S.R.; Chi, F.-C.; Chang, S.-W.; Lee, J.-C.; Wang, C.-C.; Wen, Y.-S. *J. Am. Chem. Soc.* **123** (2001) 3153.
90. Hosaka, A.; Ichikawa, S.; Shindo, H.; Sato, T. *Bull. Chem. Soc. Jpn.* **62** (1989) 797.
91. Jones, D.W.C.; Nash, R.J.; Arthur Bell, E.; Williams, J.M. *Tetrahedron Lett.* **26** (1985) 3125.
92. Naleway, J.J.; Raetz, C.R.H.; Anderson, L. *Carohydr. Res.* **179** (1988) 199.
93. Baker, B.R.; Schaub, R.E.; Williams, J.H. *J. Am. Chem. Soc.* **77** (1955) 7.
94. Iida, H.; Yamazaki, N.; Kibayashi, C. *J. Org. Chem.* **52** (1987) 3337.
95. Kim, Y.J.; Kido, M.; Bando, M.; Kitahara, T. *Tetrahedron* **53** (1997) 7501.
96. Fleet, G.W.J.; Son, J.C. *Tetrahedron* **44** (1988) 2637.

1.1.2 *Dihydro-2-hydroxymethylpyrrole (nectrisine)*

Nectrisine (FR 900483, [2*R*,3*R*,4*R*]-3,4-dihydro-2-hydroxymethyl-2*H*-pyrrole-3,4-diol, **1**) is a fungal metabolite isolated from *Nectria lucida* F-4490[1,2] and obtained from the fermentation broth of actinomycete *Kitasatosporia kifunense*. Nectrisine enhances the activity of the mouse immune system *in vitro* and exhibits a competitive action against immunosuppressive factor produced in the serum of tumor-bearing mice. It has the capacity to restore the depression of lymphocytes[3] to a normal level,[1] where concanavalin A-stimulated lymphocyte proliferation has been suppressed by the addition of immunosuppressive factor. It exhibits a potent inhibition against Baker's yeast α-glucosidase (IC_{50} 8.0×10^{-8} M) and α-mannosidase enzymes.[1] The synthesis of **1** and its ribo analogue **2** has been reported.[4]

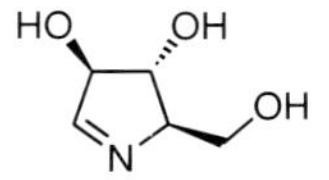

Nectrisine (FR 900483)
[2*R*,3*R*,4*R*]-3,4-Dihydro-2-hydroxymethyl-2*H*-pyrrole-3,4-diol
1

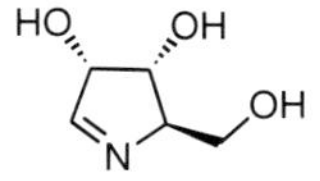

[2*R*,3*R*,4*S*]-3,4-Dihydro-2-hydroxymethyl-2*H*-pyrrole-3,4-diol
2

1.1.2.1 *Synthesis from* D-*glucose* A stereospecific synthesis of nectrisine (**1**) has been achieved from diacetone D-glucose by conversion to **3**[5] (Scheme 1).[6,7] Catalytic reduction of **3** followed by acylation with trifluoroacetic anhydride afforded the trifluoroacetyl amides **4** (82%) and **5** (13%). Removal of the isopropylidene group from **4** afforded **6**, which was subjected to oxidation with $NaIO_4$ followed by removal of the benzyl group and subsequent deacylation of the formyl and trifluoroacetyl groups to produce **1**.

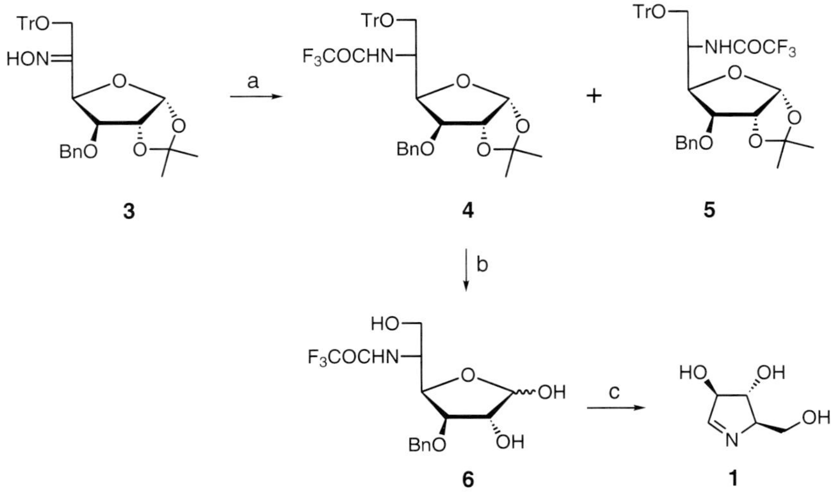

Scheme 1 (*a*) 1. H_2, Raney nickel, NH_4OH, CH_3OH; 2. TFAA, NEt_3, CH_2Cl_2, **4** (82%), **5** (13%). (*b*) 75% aqueous TFA, 82%. (*c*) 1. $NaIO_4$, THF, H_2O; 2. H_2, Pd black, HCO_2H, CH_3OH; 3. 1 N NaOH.

D-Glucal has been used as a starting material for the synthesis of nectrisine (**1**) (Scheme 2).[8] The glucal derivative **7** was obtained from D-glucal in two steps in 40% overall yield by monosilylation of the primary hydroxyl group with TBSCl followed by benzylation of the two secondary hydroxyl groups. Treatment of **7** with *m*-CPBA followed by desilylation with fluoride ion and subsequent selective bromination and silylation afforded **8**, which was treated with zinc to give the aldehyde **9**. Reaction of **9** with sodium borohydride led to the reduction of the aldehyde and subsequent silyl transfer to give **10**. Triflation followed by reaction with sodium azide led to inversion at C-2 to afford **11**. Ozonolysis of **11** followed by desilylation and acetylation afforded **12**. Hydrogenation of **12** in the presence of palladium and Al_2O_3 followed by acylation with TFAA and subsequent debenzylation using Pearlman's catalyst[9] and 1 N NaOH treatment afforded nectrisine (**1**).

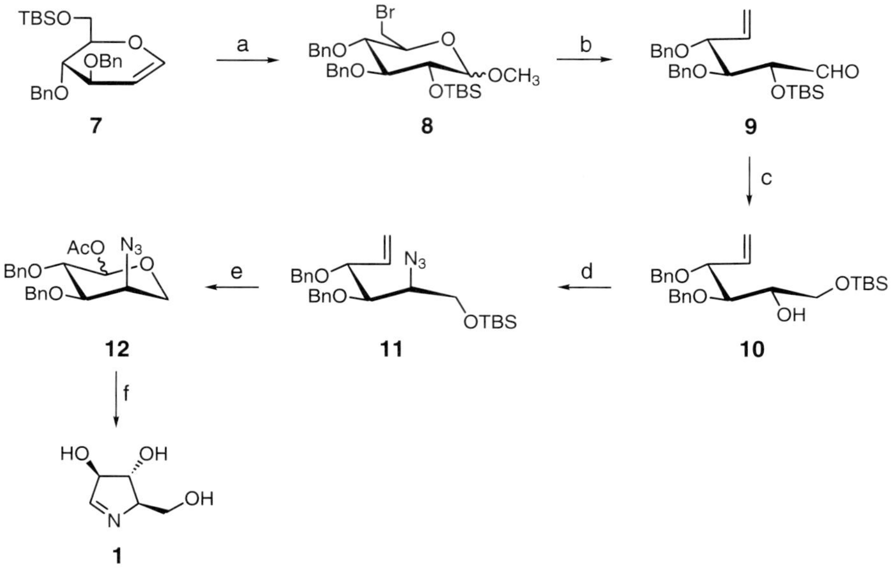

Scheme 2 (*a*) 1. *m*-CPBA, CH_3OH; 2. TBAF, THF; 3. Ph_3P, CBr_4, Py; 4. TBSCl, imidazole, 50–60% for four steps. (*b*) Zn, aqueous EtOH, 75%. (*c*) $NaBH_4$, 76%. (*d*) 1. Tf_2O, Py, CH_2Cl_2; 2. NaN_3, DMF, rt, 40–45% for two steps. (*e*) 1. O_3, CH_2Cl_2, –78°C; 2. TBAF, THF; 3. Ac_2O, Py, 80% for three steps. (*f*) 1. H_2, Pd, Al_2O_3; 2. TFAA, 100% for two steps; 3. Pearlman's catalyst; 4. 1 N NaOH; Dowex acidic resin.

1.1.2.2 *Synthesis from D-arabinose* Synthesis of nectrisine (**1**) has been reported from commercially available 2,3,5-tri-*O*-benzyl-D-arabinose (Scheme 3).[10] It was first converted to the D-lyxose derivative **13**.[11] Hydrazinolysis of the phthalimide group in **13** and treatment of the resulting amine with trifluoroacetic anhydride provided the trifluoroacetamide **14**. Dihydroxylation of the olefinic bond with OsO_4 and oxidative cleavage with $NaIO_4$ of the resulting diol led to an aldehyde, which was cyclized to give **15** upon standing. Deprotection of **15** to **16** could be effected in high yield by BCl_3 treatment at low temperature. Hydrolysis of the trifluoroacetamide with dil. NaOH and concomitant dehydration followed by ion-exchange chromatography completed the synthesis of **1** in 18% overall yield from 2,3,5-tri-*O*-benzyl-D-arabinose.

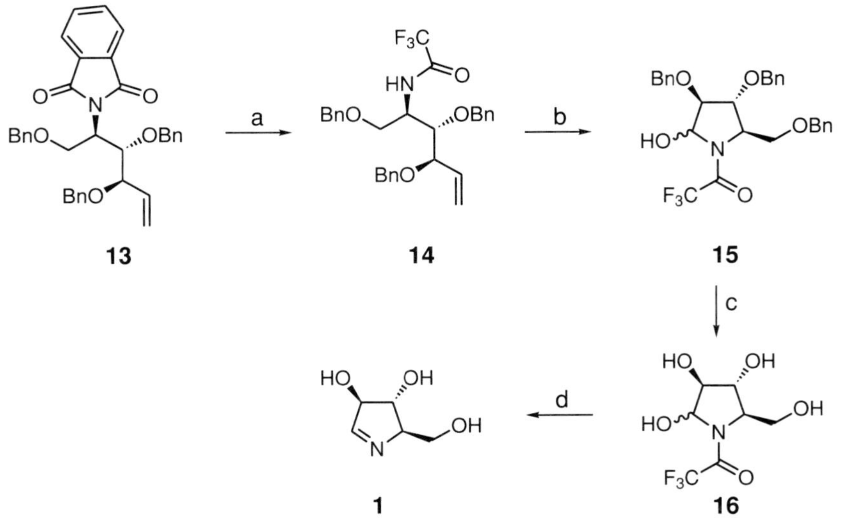

Scheme 3 (*a*) 1. $N_2H_4{\cdot}H_2O$, EtOH, reflux, 1.5 h; 2. TFAA, NEt_3, 0°C, rt, 2 h, 72% for two steps. (*b*) 1. OsO_4, NMO, THF, acetone, H_2O, rt, 48 h; 2. $NaIO_4$, THF, acetone, H_2O, rt, 1.5 h, 98% for two steps. (*c*) BCl_3, CH_2Cl_2, –78°C, 2 h; then –40°C, 16 h, 96%. (*d*) 0.5 M NaOH, rt, 30 min; then AcOH to pH 4, 96%.

1.1.2.3 *Synthesis from* D*-glyceraldehyde* D-Glyceraldehyde acetonide (**17**) has been used for the synthesis of a protected nectrisine **23** (Scheme 4).[12] Compound **17** was converted into 3,3-diethoxy-2-hydroxypropanal (**18**), which underwent transketolase-mediated condensation with hydroxypyruvate **19** to afford the triol **20** in 56% yield. Silylation of **20** using TBSOTf and NEt_3 (74%) followed by treatment with hydroxylamine gave the oxime **21** in 82% yield. Reduction of **21** using Raney nickel afforded the diastereomeric mixture of

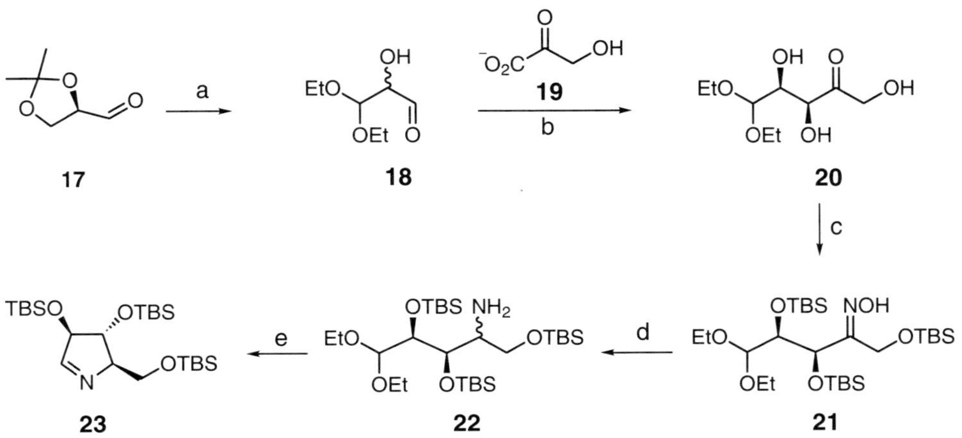

Scheme 4 (*a*) 1. TEMPO, EtOAc, toluene, 0°C, $Bn(CH_3)_3NCl$, NaBr, aqueous $NaHCO_3$, 2. 1.1 M NaOCl, 15 h, 17%. (*b*) Transketolase, thiamine pyrophosphate Mg^{2+}, pH 7.0 (pH stat). (*c*) 1. TBSOTf, NEt_3; 2. $NH_2OH{\cdot}HCl$, $KHCO_3$, CH_3OH. (*d*) H_2, Raney nickel. (*e*) 1. TMSI; 2. SiO_2 chromatography.

amines **22** (65%), which upon cyclization by treatment with iodotrimethylsilane gave a 3:2 mixture of cyclic imines from which the major diastereomer **23**, bearing the stereochemistry in **1**, was isolated. However, treatment of **23** under a range of desilylation conditions (e.g. TBAF; AcOH–H_2O–THF; fluoride resin; HF–acetonitrile) failed to yield a pure sample of nectrisine (**1**).

1.1.2.4 *Synthesis from L-threitol* The L-threitol derivative **24**,[13] obtained from D-(–)-diethyl tartarate in three steps and 90% overall yield, was used as a starting material for the synthesis of nectrisine (**1**) (Scheme 5).[14,15] Swern oxidation of **24** produced the L-threose derivative **25**, which was transformed[16] into the aminonitrile **26** in 96% overall yield from **24**, as an inseparable diastereomeric mixture. Removal of the silyl protecting group from **26** followed by oxidation of the resulting primary hydroxyl group with TPAP[17] afforded the lactam **27**, which was treated with sodium methoxide to produce the methyl ester **28** in 62% yield from **26**. Lithium borohydride reduction of **28** afforded a chromatographically separable mixture of the lactams **29** and **30** in a ratio of 56:44 and 87% total yield. Silylation

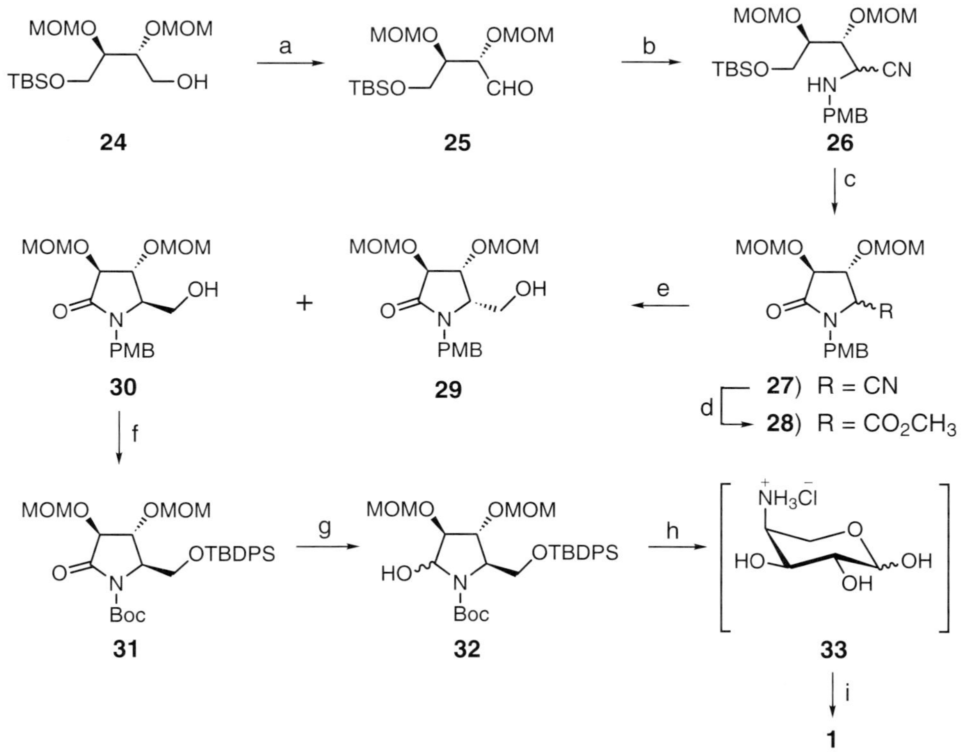

Scheme 5 (*a*) $(COCl)_2$, DMSO, CH_2Cl_2, –78°C; then NEt_3. (*b*) *p*-$(CH_3O)C_6H_4CH_2NH_2$, $(EtO)_2P(O)CN$, THF, 96% for two steps. (*c*) 1. TBAF, THF, 87%; 2. TPAP, NMO, MS 4 Å, CH_2Cl_2. (*d*) $NaOCH_3$, 0°C to rt; then 1 N HCl, 71% for two steps. (*e*) $LiBH_4$, THF, 0°C to rt, 87%. (*f*) 1. TBDPSCl, imidazole, DMF, 96%; 2. $(NH_4)_2Ce(NO_3)_6$, CH_3CN, H_2O, 0°C, 84%; 3. NEt_3, $(Boc)_2O$, DMAP, CH_2Cl_2, quantitative. (*g*) $LiEt_3BH$, THF, –78°C, 93%. (*h*) 6 N HCl, THF, 50°C, 2 h, 80%. (*i*) Dowex 1X2 (OH^-) resin, 90%.

of the primary alcohol of **30** with TBDPSCl followed by replacement of the N-protecting group with the more electron-withdrawing and easily removable Boc group afforded **31** in 81% yield from **30**. Reduction of the imide **31** afforded **32** (93%), which underwent removal of the protecting groups with 6 N HCl to give the amino sugar **33**,[18] which was followed by ion-exchange chromatography to furnish **1** in 54% overall yield from **30**.

References

1. Shibata, T.; Nakayama, O.; Tsurumi, Y.; Okuhara, M.; Terano, H.; Kohsaka, M. *J. Antibiot.* **41** (1988) 296.
2. Asano, N.; Nash, R.J.; Molyneux, R.J.; Fleet, G.W.J. *Tetrahedron: Asymmetry* **11** (2000) 1645.
3. Iwami, M.; Nakayama, O.; Terano, M.; Kohsaka, M.; Aoki, H.; Imanaka, H. *J. Antibiot.* **40** (1987) 612.
4. Witte, J.F.; McClard, R.W. *Tetrahedron Lett.* **32** (1991) 3927.
5. Inone, S.; Tsuruoka, T.; Ito, T.; Niida, T. *Tetrahedron* **23** (1968) 2125.
6. Kayakiri, H.; Takase, S.; Seloi, H.; Uchida, I.; Terano, H.; Hashimoto, M. *Tetrahedron Lett.* **29** (1988) 1725.
7. Kayakiri, H.; Nakamura, K.; Takase, S.; Seloi, H.; Uchida, I.; Terano, H.; Hashimoto, M.; Tada, T.; Koda, S. *Chem. Pharm. Bull.* **39** (1991) 2807.
8. Chen, S.-H.; Danishefsky, S.J. *Tetrahedron Lett.* **31** (1990) 2229.
9. Pearlman, W.M. *Tetrahedron Lett.* **8** (1967) 1663.
10. Bosco, M.; Bisseret, P.; Bouix-Peter, C.; Eustache, J. *Tetrahedron Lett.* **42** (2001) 7949.
11. Bouix, C.; Bisseret, P.; Eustache, J. *Tetrahedron Lett.* **39** (1998) 825.
12. Humphrey, A.J.; Parsons, S.F.; Smith, M.E.B.; Turner, N.J. *Tetrahedron Lett.* **41** (2000) 4481.
13. Kuwahara, S.; Moriguchi, M.; Miyagawa, K.; Konno, M.; Kodama, O. *Tetrahedron* **51** (1995) 8809.
14. Kim, Y.J.; Kitahara, T. *Tetrahedron Lett.* **38** (1997) 3423.
15. Kim, Y.J.; Takatsuki, A.; Kogoshi, N.; Kitahara, T. *Tetrahedron* **55** (1999) 8353.
16. Harusawa, S.; Hamada, Y.; Shioiri, T. *Tetrahedron Lett.* **20** (1979) 4663.
17. Griffith, W.P.; Ley, S.V.; Whitcombe, G.P.; White, A.D. *J. Chem. Soc., Chem. Commun.* (1987) 1625.
18. Naleway, J.J.; Raetz, C.R.H.; Anderson, L. *Carbohydr. Res.* **179** (1988) 199.

1.1.3 *2,5-Dihydroxymethylpyrrolidines*

2,5-Imino-D-mannitol (2*R*,5*R*-dihydroxymethyl-3*R*,4*R*-dihydroxypyrrolidine, DMDP, **1**) was isolated from *Derris ellipica*[1,2] and showed a potent inhibition of viral glycoprotein processing glycosidase.[3,4] It has antiviral, antifeedant and nematicidal activities.[5] 2*S*,5*R*-Dihydroxymethyl-3*R*,4*R*-dihydroxypyrrolidine (**2**), the *C*-5-epimer of **1**, is a potent inhibitor of a number of glucosidases.[6–8] 2*R*,5*S*-Dihydroxymethyl-3*R*,4*R*-dihydroxypyrrolidine (**3**) is a xylose isomerase inhibitor.[9] The structurally related natural products (+)-2,5-imino-2,5,6-trideoxy-*manno*-heptitol (**4**) and (+)-2,5-imino-2,5,6-trideoxy-*gulo*-heptitol (**5**) have been isolated from *Hyacinthus orientalis* and were found to display interesting specific glycosidase inhibitory properties.[10,11] The structure of **1** has been characterized by X-ray diffraction.[12] A number of unnatural analogues of **1** have been synthesized from carbohydrates.[13–15] Various carbohydrate derivatives were used for the synthesis of such group of compounds.

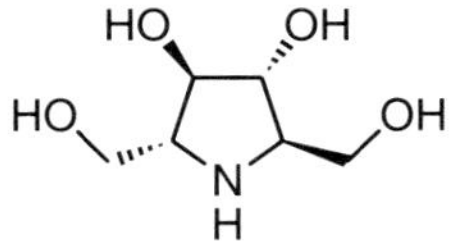

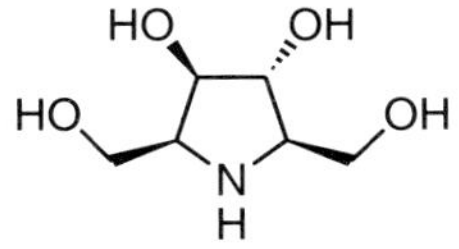

1.1.3.1 *Synthesis from D-glucose*

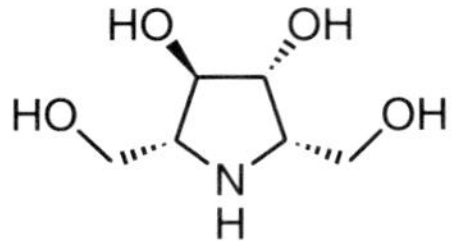

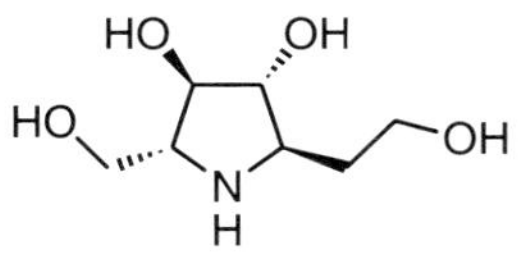

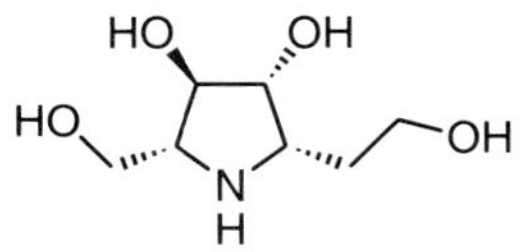

The intermediate 2-azido-3-*O*-benzyl-2-deoxy-α-D-mannofuranoside (**11**), prepared from diacetone D-glucose (**6**), has been used for the synthesis of 2*R*,5*R*-dihydroxymethyl-3*R*,4*R*-dihydroxypyrrolidine (**1**) (Scheme 1).[16,17] Compound **6** was benzylated, followed by deprotection of the side chain acetonide and subsequent reaction with dimethyl carbonate to give the carbonate **7** in 80% overall yield. The carbonate protecting group is stable to acid, and thus treatment of **7** with methanol in the presence of acidic ion-exchange resin caused a cleavage of the isopropylidene group to give a mixture of β- **8** and α-furanosides **9** (92%) in a ratio of approximately 2:1. Triflation of **9** followed by treatment with sodium azide in DMF at 50°C gave the manno-azide derivative **10** in 88% yield. Removal of the carbonate group from **10** was accomplished by a catalytic amount of sodium methoxide in methanol at room temperature to give the key intermediate **11**. Selective benzoylation of the primary hydroxyl group in **11** followed by treatment with methanesulfonyl chloride and subsequent cyclization gave the epoxide **12**, with inversion of configuration at C-5. Hydrogenation of the azido epoxide **12** followed by benzyloxycarbonylation led to the formation of two products, the major one (43% yield)

was the cyclized carbamate **13** and the minor one was **14**. Hydrogenation of **13** gave, after neutralization and purification by ion-exchange chromatography, **1** in 70% yield. Selective *p*-toluenesulfonylation of the diol **11** in pyridine followed by hydrogenation of the azide group to the corresponding amine and subsequent treatment with sodium acetate in ethanol gave a bicyclic imine, which could be isolated cleanly as the benzyl carbamate **15**; it was used in the synthesis of polyhydroxylated piperidines.

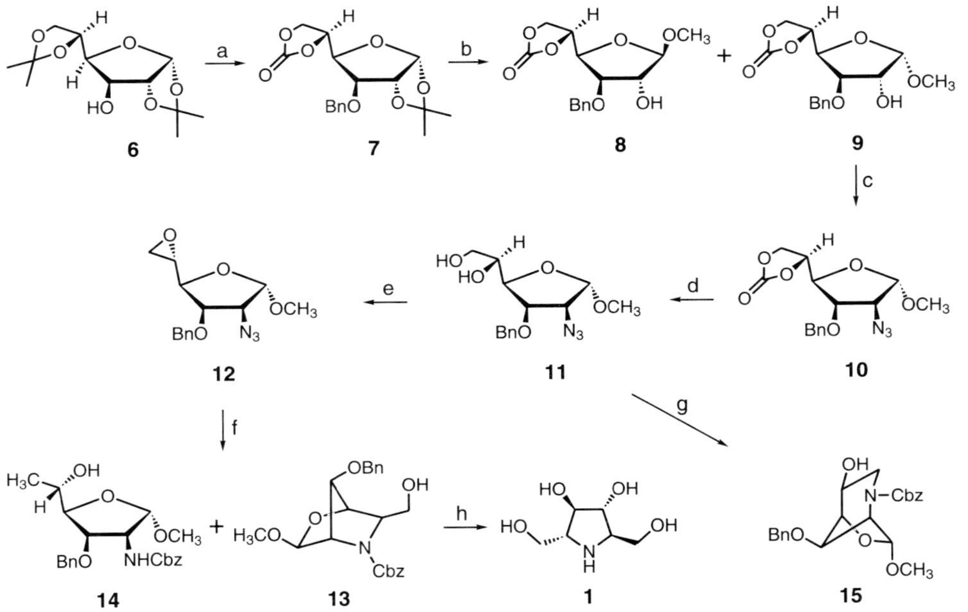

Scheme 1 (*a*) 1. BnBr, NaH, THF, Bu_4NI, reflux, 45 min; 2. HCl–H_2O–CH_3OH (1.1:20:200), rt, 20 h; 3. dimethyl carbonate, $NaOCH_3$, reflux, 3 h, 80%. (*b*) CH_3OH, Dowex 50WXH resin, 12 h, 92%, 1:2 of α/β. (*c*) 1. Tf_2O, –30°C, CH_2Cl_2, Py; 2. DMF, NaN_3, 50°C, 24 h, 88%. (*d*) $NaOCH_3$, CH_3OH, rt, 6 h, 80%. (*e*) 1. BzCl, Py, rt, 4 h, 81%; 2. MsCl, Py, rt, 4 h, 94%; $NaOCH_3$, DMF, 50°C, 3 h, 82%. (*f*) H_2, Pd black, EtOH, 1 h; 2. $NaHCO_3$, ether; 3. CbzCl, rt, 2 h, **13** (43%), **14** (24%). (*g*) 1. *p*-TsCl, Py, 0°C, 12 h, 95%; 2. Pd black, EtOH, H_2, rt, 12 h; then NaOAc, heated, 50°C, 12 h; then CbzCl, $NaHCO_3$, rt, 2 h, 65%. (*h*) AcOH, H_2, Pd black, 13 h, Amberlite CG-120 (H^+) resin, 70%.

A stereoselective synthesis of 2*R*-hydroxymethyl-5*R*-methoxymethyl-3*R*,4*R*-dihydroxypyrrolidine (**20**) from D-glucose has been reported (Scheme 2).[18] 5-Azido-5-deoxy-1,2-*O*-isopropylidene-α-D-glucofuranose (**16**)[19] was treated with chlorodimethyl(1,1,2-trimethylpropyl)silane followed by methoxymethylation at O-3 to produce **17**. Removal of the O-6 silyl group with TBAF followed by treatment of the resulting alcohol with iodomethane furnished the methyl ether **18**. Acid hydrolysis of the 1,2-*O*-isopropylidene group in **18** followed by isomerization using glucose isomerase gave the D-fructose analogue **19**. Catalytic hydrogenation of **19** followed by purification with Amberlite CG-50 (H^+) resin afforded **20** in 23% overall yield from **16**.

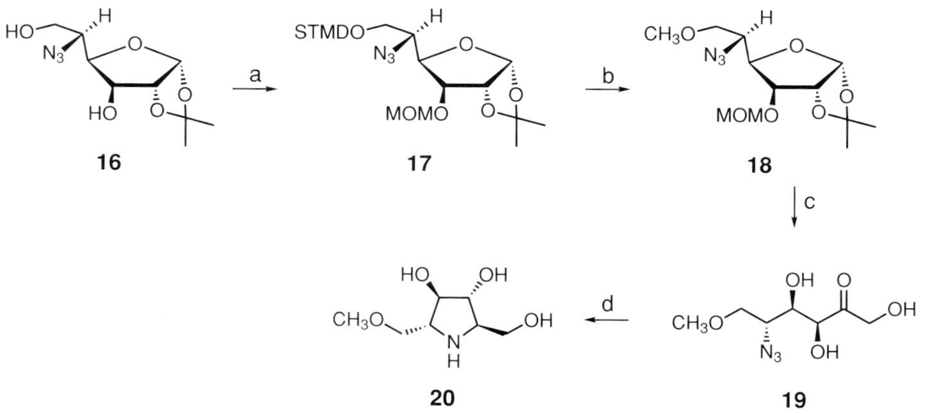

Scheme 2 (*a*) 1. DMF, imidazole, chlorodimethyl(1,1,2-trimethylpropyl)silane, rt; 2. CH_2Cl_2, NEt_3, MOMCl, 95% for two steps. (*b*) 1. THF, TBAF, 50°C, 90%; 2. THF, DMF, NaH, CH_3I, 89%. (*c*) 1. CH_3CN–H_2O, Amberlite IR-120 (H^+) resin, 45°C, 77%; 2. $MgSO_4$, immobilized glucose isomerase (Sweetzyme TEC 5.3.1.5), 65°C, 5 h, 61%. (*d*) 1. CH_3OH, 10% Pd *on* C, H_2, rt, 14 h; 2. Amberlite CG-50 (H^+) resin, 0.05 M aqueous NH_3, 65%.

1.1.3.2 *Synthesis from* D*-glucosamine* 2S,5*S*-Dihydroxymethyl-3*R*,4*R*-dihydroxypyrrolidine (**23**) was synthesized from D-glucosamine via compound **21** (Scheme 3).[20] Treatment of **21** with DMP followed by tosylation produced the tosylate **22**, which was cyclized with NaOEt to afford the imino sugar **23**.

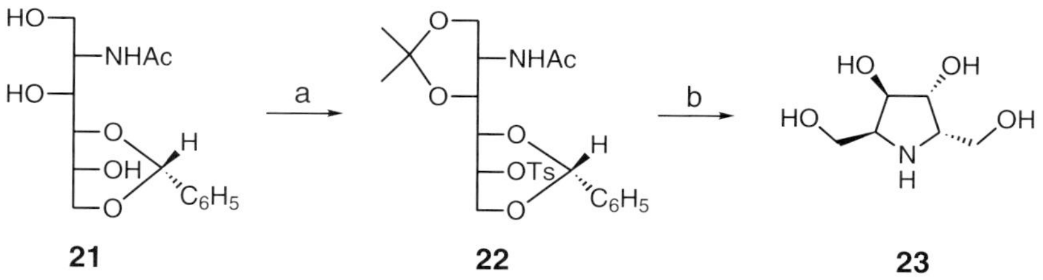

Scheme 3 (*a*) 1. DMP, H^+, 75%; 2. *p*-TsCl, Py, 95%. (*b*) NaOEt, EtOH, 89%; acid hydrolysis.

1.1.3.3 *Synthesis from* D*-fructose* D-Fructose has been used for the synthesis of 2*R*,5*S*-dihydroxymethyl-3*R*,4*R*-dihydroxypyrrolidine (**3**) and its analogues (Scheme 4).[21] Thus, microbial oxidation of D-fructose led to 5-keto-D-fructose (**24**),[22] which was condensed with Ph_2CHNH_2 to afford a mixture of **25**, **26** and **27** in a ratio of 86:8:6. Removal of the benzhydryl group from **25** by hydrogenation in the presence of $Pd(OH)_2$ afforded **3** in 91% yield.

1.1.3.4 *Synthesis from* L*-sorbose* The first total synthesis of DMDP (**1**) has been reported from L-sorbose (Scheme 5).[23] L-Sorbose was converted into 3,4-di-*O*-acetyl-1,2-*O*-isopropylidene-5-*O*-tosyl-α-L-sorbose (**28**) in three steps.[24] Nucleophilic displacement of the tosyloxy group with azide ion afforded the azido derivative **29**. Removal of the protecting groups from **29** with sodium methoxide followed by acidic ion-exchange resin afforded **30**, which upon catalytic hydrogenation produced the pyrrolidine **1** in 43% overall yield from **28**.

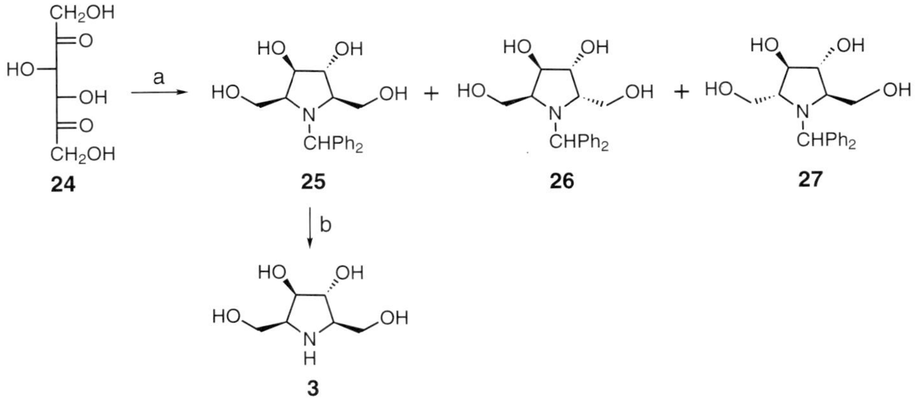

Scheme 4 (*a*) Ph_2CHNH_2, $NaCNBH_3$, CH_3OH, 68%. (*b*) 20% $Pd(OH)_2$, H_2, 91%.

Alternatively, a mixture of 5-azido-5-deoxy-D-fructose (**30**) and 5-azido-5-deoxy-L-sorbose (**31**) (Scheme 5),[25] obtained chemoenzymatically as it will be shown later, can be separated by chromatography, after acetonation, as 5-azido-5-deoxy-D-fructose 1,2-acetonide (**33**) in 22% yield and 5-azido-5-deoxy-L-sorbose 1,2-acetonide (**32**) in 56% yield. Treatment of **33** with Dowex (H^+) resin in ethanol afforded **30**, which upon hydrogenation furnished **1** in 92% yield.

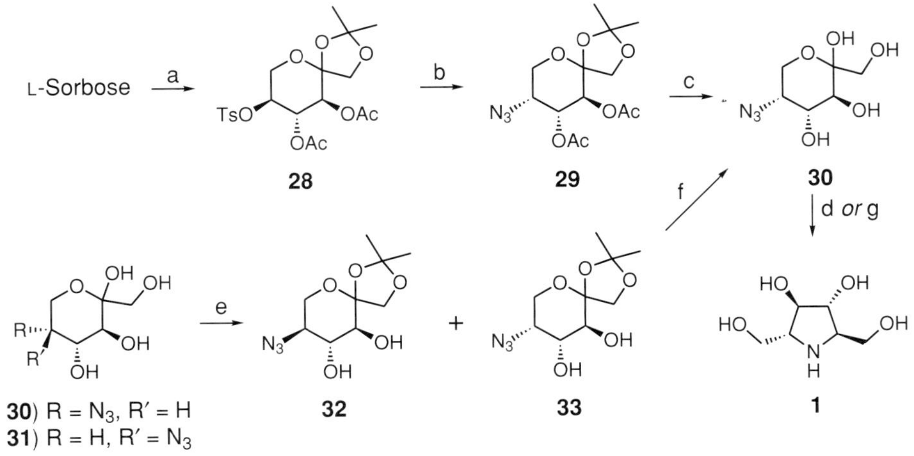

Scheme 5 (*a*) 1. 2-Methoxypropene, *p*-TsOH, acetone, rt; 2. *p*-TsCl, Py; 3. Ac_2O, Py. (*b*) LiN_3, DMF, 74%. (*c*) 1. $NaOCH_3$, CH_3OH, rt, 1 h, 63%; 2. acidic ion-exchange resin, 93%. (*d*) H_2, 10% Pd *on* C, EtOH, 100%. (*e*) DMP, *p*-TsOH, acetone, **32** (56%), **33** (22%), chromatography. (*f*) Dowex (H^+) resin, EtOH, H_2O, 92%. (*g*) H_2, $Pd(OH)_2$ *on* C, EtOH, H_2O, 92%.

Various pyrrolidines were prepared by using the chemoenzymatic approach. Thus, 2*R*,5*S*-dihydroxymethyl-3*R*,4*R*-dihydroxypyrrolidine (**3**) was prepared from the L-sorbose derivative **36** (Scheme 6).[26] Thus, periodate oxidation of 2-azido-2-deoxythreitol (**34**) led to 2-azido-3-hydroxypropanal (**35**), which was treated with DHAP and FDP aldolase to give **36**. Enzymatic hydrolysis of the phosphate group in **36** with acid phosphatase afforded 5-azido-5-deoxy-L-sorbose (**31**). Finally, compound **31** was reductively cyclized to furnish **3**.

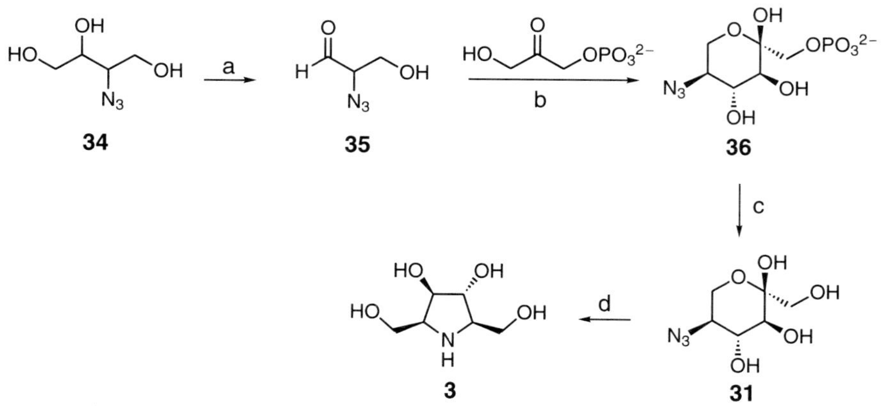

Scheme 6 (*a*) $NaIO_4$, 0°C, 5 min. (*b*) DHAP, FDP aldolase, 2 days. (*c*) Acid phosphatase, pH 4.7, 37°C, 36 h, 78%. (*d*) H_2, Pd 50 psi, 1 day, 97%.

The pyrrolidines were also prepared by using chemoenzymatic strategy for constructing the required azido-sugars (Scheme 7).[27] Enzymatic aldol condensation of 2-azido-3-hydroxy propanal (**35**) and DHAP gave a diastereoisomeric mixture of **30** and **31**, which

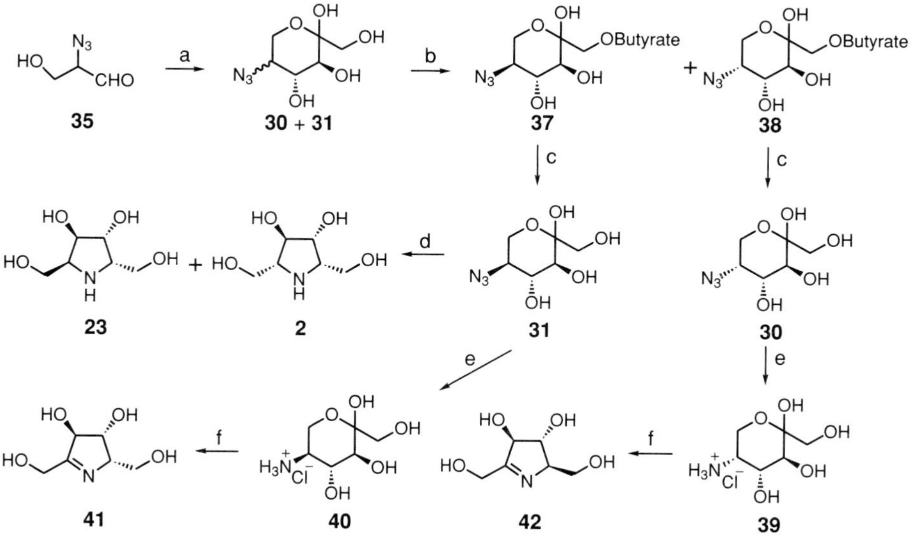

Scheme 7 (*a*) 1. DHAP, rabbit muscle aldolase, pH 6.7, 25°C; 2. acid phosphatase, pH 5.0, 37°C, 78% for two steps. (*b*) Vinyl butyrate, PPL, THF, 85%. (*c*) IRA-400 (OH^-), CH_3OH, 99%. (*d*) 10% Pd *on* C, H_2 50 psi, **2**:**27** (90:10); *or* 5% Rh-alumina, H_2 15 psi, **2**:**27** (98:2). (*e*) H_2 (1 atm), Pd *on* C, aqueous HCl, quantitative. (*f*) Excess NaOH.

underwent enzymatic butyrylation of the primary hydroxyl group to afford a mixture of L-sorbose derivative **37** and D-fructose derivative **38** whose chromatographic separation followed by removal of the butyrate group furnished the azides **31** and **30**, respectively. Catalytic hydrogenation of **30** and **31** in aqueous HCl afforded the salt **39** and **40**, respectively, which upon treatment with excess NaOH gave the corresponding dehydropyrrolidines **42** and **41**. On the other hand, hydrogenation of azide **31** over 5% Rh-alumina gave **2** and **23** in a ratio 98:2.

1.1.3.5 *Synthesis from D-arabinose* A synthesis of 2*R*,5*R*-dihydroxymethyl-3*R*,4*R*-dihydroxypyrrolidine (**1**) from D-arabinose has been achieved (Scheme 8).[28] Benzylation of methyl D-arabinofuranoside (**43**) with benzyl bromide afforded the tribenzyl derivative **44**, which was treated with acetic acid to give compound **45**. Condensation of **45** with $Ph_3P{=}CH_2$ afforded the olefin **46**, which underwent oxidation of the secondary hydroxyl group to give **47**. Subsequent condensation with hydroxylamine hydrochloride gave **48**, which was subjected to $LiAlH_4$ reduction followed by protection of the resulting amine to produce **49**. Intramolecular cyclization of **49** afforded lactam **51** via intermediate **50**. Basic hydrolysis of **51** afforded **52**, which underwent removal of the benzyl groups to afford **1**.

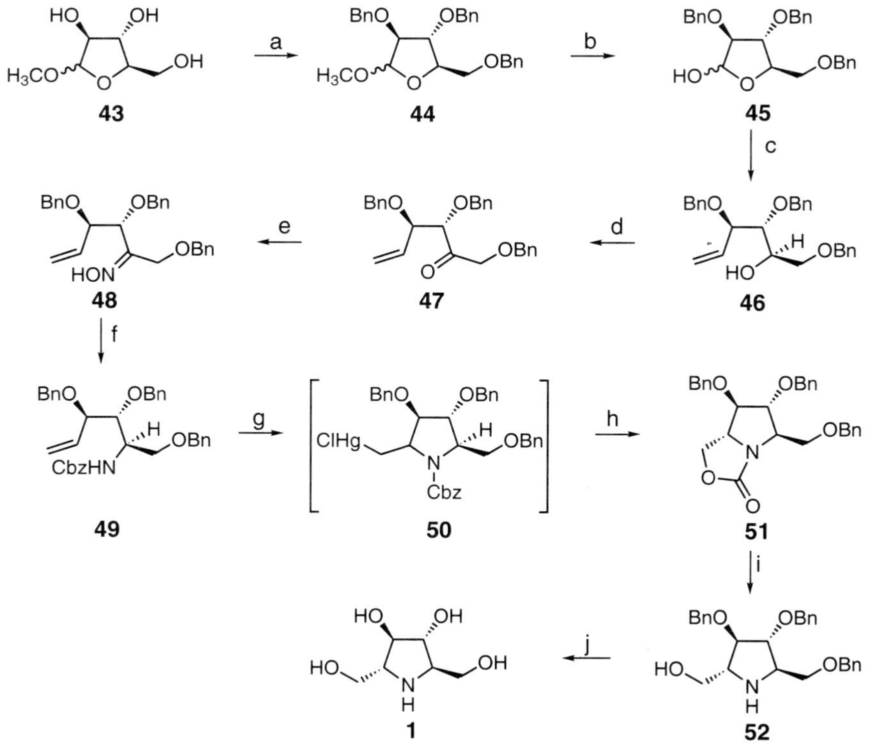

Scheme 8 (*a*) 1. KH, DMF–THF (4:1); 2. BnBr. (*b*) 80% aqueous AcOH, 50°C, 1 h. (*c*) $Ph_3P^+CH_3Br^-$, *n*-BuLi, toluene, 40°C, 48 h. (*d*) 1. DCC, DMSO; 2. pyridinium trifluoroacetate. (*e*) $H_2NOH{\cdot}HCl$, $KHCO_3$, CH_3OH. (*f*) $LiAlH_4$, ether; CbzCl, K_2CO_3. (*g*) 1. $Hg(OAc)_2$, THF, 50°C, 24 h; 2. KCl, H_2O. (*h*) I_2, AcOH, rt, dark, 18 h. (*i*) KOH (50%), EtOH, reflux, 18 h. (*j*) H_2, Pd *on* C, EtOH.

1.1.3.6 *Synthesis from* L*-xylose* Syntheses of (+)-2,5-imino-2,5,6-trideoxy-*manno*-heptitol (**4**) and (+)-2,5-imino-2,5,6-trideoxy-*gulo*-heptitol (**5**) from L-xylose have been reported (Scheme 9).[29] Glycosidation of L-xylose with methanol–HCl gave a mixture of methyl xylofuranosides, which was benzylated, followed by hydrolysis and then reaction with benzylamine to give the corresponding glycosylamine **53**. Subsequent treatment of **53** with allylmagnesium chloride gave a mixture of diastereoisomers **54** and **55** (40:60), which were separated and each of them was subjected to intramolecular cyclization via their mesyl derivatives to give pyrrolidines **57** and **56**, respectively. Ozonolysis of **57** and **56** carried out on the respective sulfate salts, to avoid the amine oxidation, afforded the corresponding aldehydes, which were reduced with sodium borohydride to give **58** and **59**, respectively. Removal of the protecting groups gave **4** and its *C*-5-epimer **5**.

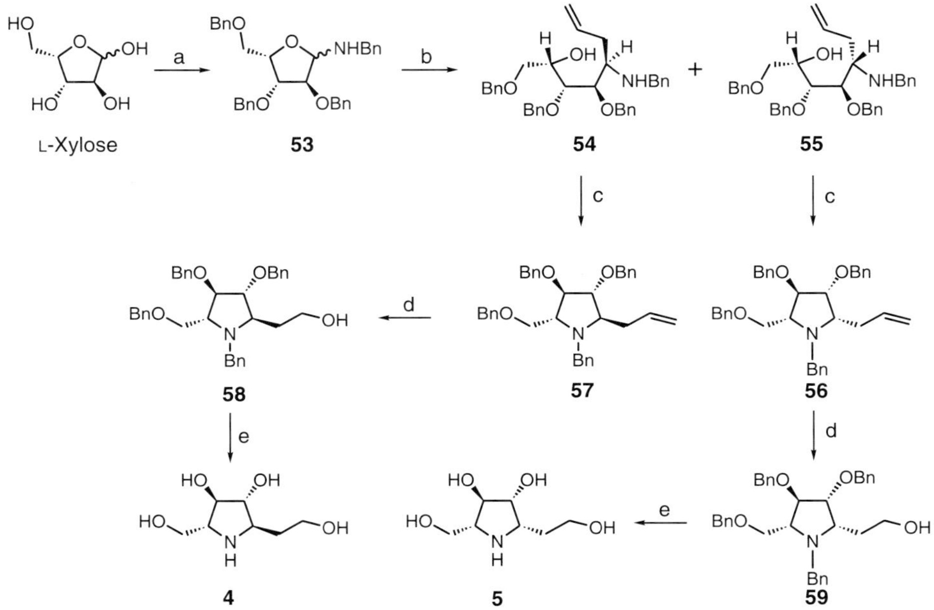

Scheme 9 (*a*) 1. HCl, CH_3OH, 100%; 2. BnBr, $Ba(OH)_2$, DMF, 40%; 3. aqueous HCl, dioxane, reflux, 50%; 4. $BnNH_2$, CH_2Cl_2, MS 4 Å, 98%. (*b*) AllMgCl, THF, 0°C, 91%. (*c*) MsCl, Py, **57** (79%), **56** (97%). (*d*) H_2SO_4, O_3; then $NaBH_4$, CH_3OH, ~74%. (*e*) 10% Pd *on* C, HCO_2NH_4, CH_3OH, 60°C, ~40%.

1.1.3.7 *Synthesis from* D*-iditol* A short enantioselective synthesis of **64**, the enantiomer of **1**, has been reported from the D-iditol derivative **60** (Scheme 10).[30] Removal of the isopropylidene group of **60** and protection of the resulting tetraol with MOMCl gave the 2,5-di-*O*-benzyl-D-iditol derivative **61** in 65% overall yield. Hydrogenation of **61** over Pearlman's catalyst followed by mesylation of the resulting diol afforded the dimesylate **62** in 80% overall yield. Heating of **62** with benzylamine effected a stereoselective cyclization with complete inversion of configuration at both C-2 and C-5 to furnish the azasugar **63** in 76% yield. Removal of the protecting groups from **63** using TFA and hydrogenation followed by subsequent purification with Dowex 1X8-200 resin afforded **64** in 63% overall yield from **63**.

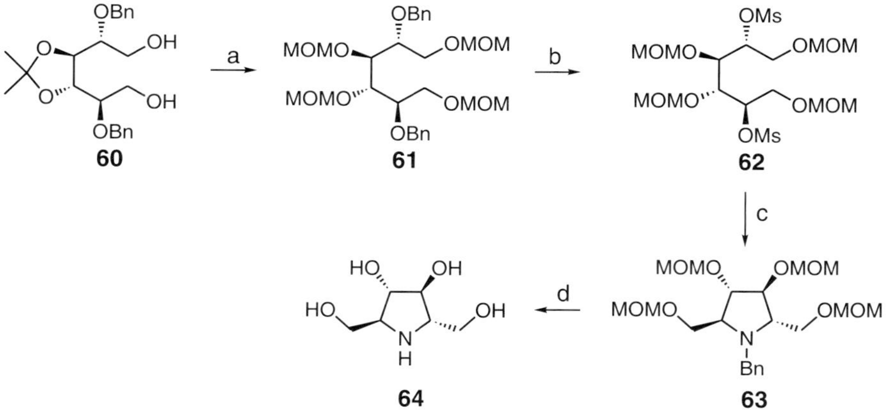

Scheme 10 (*a*) 1. Dowex 50WX8 resin, CH_3OH; 2. MOMCl, $(i\text{-}Pr)_2NEt$, CH_2Cl_2, 81%. (*b*) 1. H_2, $Pd(OH)_2$, CH_3OH; 2. MsCl, NEt_3, CH_2Cl_2, 80%. (*c*) $BnNH_2$, 120°C, 18 h, 70%. (*d*) 1. TFA, H_2O; 2. H_2, $Pd(OH)_2$, AcOH; 3. Dowex 1X8-200 resin, 63%.

1.1.3.8 *Synthesis from* D*-mannitol* Syntheses of 2*R*,5*R*-dihydroxymethyl-3*R*,4*R*-dihydroxypyrrolidine (**1**) and 2*S*,5*S*-dihydroxymethyl-3*R*,4*R*-dihydroxypyrrolidine (**23**) starting from D-mannitol have been done (Schemes 11 and 12).[31] Regioselective opening of the bisepoxide **65** by sodium benzoxide afforded the 1,6-di-*O*-benzyl-L-iditol derivative **66**, which was tosylated and the isopropylidene group was hydrolyzed to afford the diol **67**. Subjection of **67** to acid-catalyzed benzylation by benzyl trichloroacetimidate produced **68**. Cyclization of **68** with benzylamine afforded the pyrrolidine **69**, which upon deprotection of the benzyl groups followed by ion-exchange chromatography furnished **1** in 20% overall yield from **65** (Scheme 11).

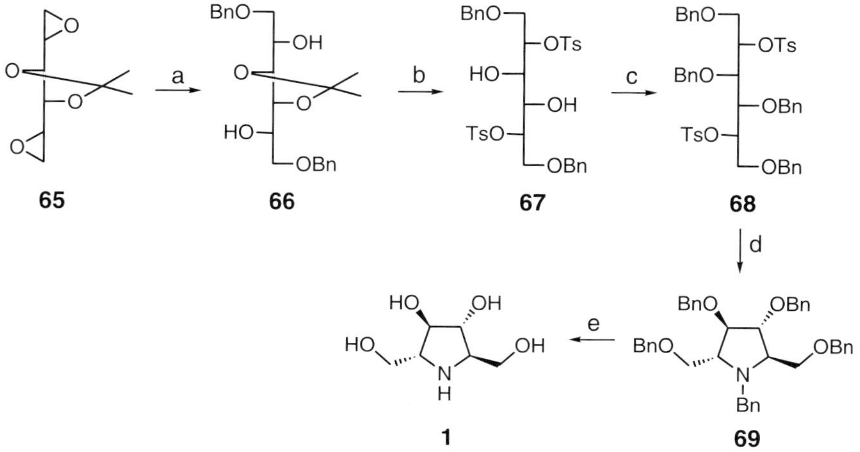

Scheme 11 (*a*) NaH, BnOH, DMF, 20°C, 24 h, 57%. (*b*) 1. *p*-TsCl, NEt_3, DMAP, CH_2Cl_2, 84%; 2. TFA–H_2O (9:1), 0°C, 2 h, 86%. (*c*) $Cl_3CC(NH)OBn$, CH_2Cl_2–C_6H_{12} (1:2), CF_3SO_3H, 25°C, 5 h, 75%. (*d*) $BnNH_2$, 120°C, 12 h, 79%. (*e*) 1. H_2, Pd black, AcOH; 2. Dowex 50WX8 resin, 80%.

Selective benzylation of the two primary hydroxyl groups of **70** afforded the tetrabenzyl derivative **71** (Scheme 12), which was subjected to mesylation of the C-2 and C-5 hydroxy groups to give the dimesylate **72**. Cyclization of **72** with benzylamine afforded 2,5-dideoxy-2,5-*N*-benzylimino-1,3,4,6-tetra-*O*-benzyl-L-iditol (**73**). Hydrogenation of **73** removed the benzyl groups to give, after neutralization and purification by ion-exchange chromatography, **23** in 37% overall yield from **70**.

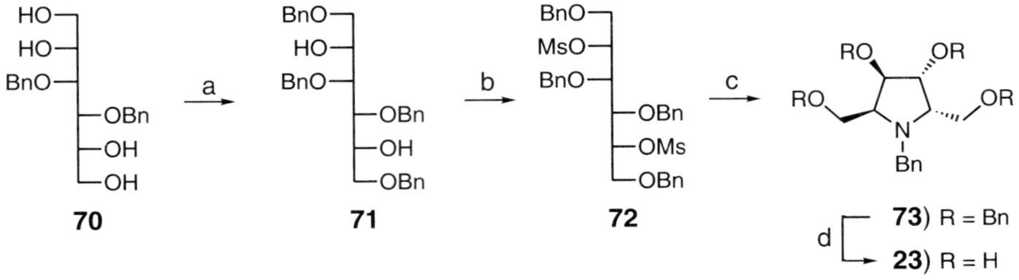

Scheme 12 (*a*) 1. Bu_2SnO, toluene, reflux, 10 h; 2. BnBr, Bu_4NI, 70°C, 12 h, 74%. (*b*) MsCl, NEt_3, DMAP, CH_2Cl_2, 80%. (*c*) $BnNH_2$, 120°C, 18 h, 78%. (*d*) 1. H_2, Pd black, AcOH; 2. Dowex 50WX8 resin, 80%.

Alternatively, the synthesis of **23** from D-mannitol has been started by acetalation with benzaldehyde[32] to give 1,3:4,6-di-*O*-benzylidene-D-mannitol (**74**) (58%), which was triflated to form the ditriflate **75** (95%) (Scheme 13).[33,34] Nucleophilic substitution of the ditriflate groups in **75** with anhydrous hydrazine afforded compound **77**, which was hydrogenated over Raney nickel to give the protected pyrrolidine **78** (100%), which upon acid hydrolysis gave **23**.

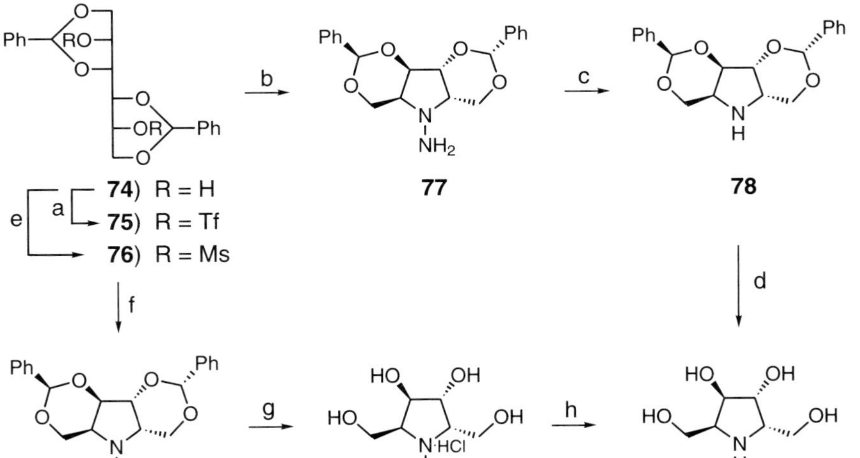

Scheme 13 (*a*) Tf_2O, THF, Py, −5°C, 1.5 h at 0°C, 95%. (*b*) NH_2NH_2, THF, rt, 20 h, 93%. (*c*) H_2, Raney nickel, THF, EtOH, rt, 24 h, 100%. (*d*) TFA (60%), rt, 48 h, 88%. (*e*) MsCl, NEt_3, CH_2Cl_2, 0°C, 2 h, quantitative. (*f*) $BnNH_2$, 135°C, 8 h, 94%. (*g*) HCl, CH_3OH, rt, 2 days, 98%. (*h*) 5% Pd *on* C, H_2, HCl, EtOH, rt, 10 h, 86%.

A similar methodology using mesylate groups instead of the triflate groups to produce **23** from D-mannitol has also been reported[35] (Scheme 13). Mesylation of **74** gave the 2,5-di-*O*-mesylate derivative **76**, which upon heating in benzylamine afforded the tetra-substituted pyrrolidine **79**. Successive treatment of **79** with conc. HCl in methanol provided the *N*-benzyl-tetraol hydrochloride salt **80**, which on hydrogenation produced **23**.

A synthesis of **3** starting from a conformationally flexible D-mannitol *N*-Boc bis-aziridine derivative **81** has been reported (Scheme 14).[36] The cyclic carbamate-protected pyrrolidine **82** was obtained from **81**[37] via the regioselective bis-aziridine ring opening with Li_2NiBr_4, followed by Ag^+-promoted intramolecular substitution of the bromide by the *N*-Boc group in 75% overall yield. Nitrous acid deamination of **82** with isoamyl nitrite led, in 50% yield, to a 1:1 mixture of cyclic carbamate protected pyrrolidines **84** and **85** via the intermediate **83**. Complete deprotection of the mixture of **84** and **85** gave **3**.

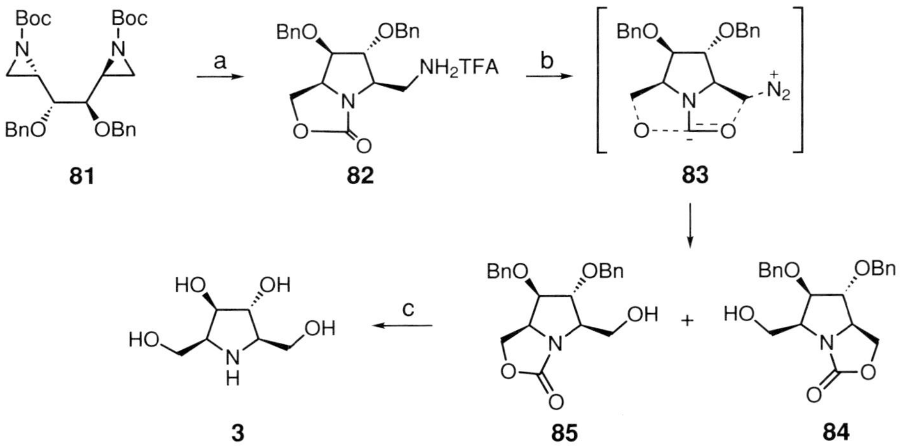

Scheme 14 (*a*) Li_2NiBr_4, Ag^+, Ref. 37, 75%. (*b*) Isoamyl nitrite, NEt_3, THF, 60°C, 1 h, **84** (30%), **85** (50%). (*c*) 1. H_2, Pd black, AcOH; 2. K_2CO_3, CH_3OH, reflux, 95%.

1.1.3.9 *Synthesis from* D-*glucosamic acid* A chirospecific synthesis of 2*R*,5*S*-dihydroxymethyl-3*R*,4*R*-dihydroxypyrrolidine (**2**) from D-glucosamic acid has been achieved (Scheme 15).[38] The D-glucosamic acid was converted to **86**,[39] which was protected with benzyl chloroformate followed by removal of the terminal isopropylidene group with Dowex 50WX8 resin to afford the diol **87**. Selective silylation with TBSCl followed by mesylation afforded **88**. Reduction of **88** with $LiAlH_4$ gave the corresponding alcohol, which was then subjected to acid hydrolysis and subsequently hydrogenated to afford the dihydroxypyrrolidine **2** in 46% overall yield from **86**.

1.1.3.10 *Synthesis from* D-*glyconolactone* D-Glucono-1,5-lactone was used for the synthesis of 2*S*,5*S*-dihydroxymethyl-3*R*,4*R*-dihydroxypyrrolidine (**3**) (Scheme 16).[40] D-Glucono-1,5-lactone was converted into azide **89** (95%),[41,42] which was subjected to sequential reactions involving azide reduction, protection of the resulting amine with $(Boc)_2O$, reduction of the ester group and acetylation of the resulting hydroxyl group to afford **90**. Selective removal of the terminal isopropylidene group followed by selective

D-Glucosamic acid → a → 86 → b → 87 → c → 88 → d → 2

Scheme 15 (*a*) Ref. 39. (*b*) 1. CbzCl, Na_2CO_3, CH_2Cl_2, rt, 30 min, 97%; 2. Dowex 50WX8 resin, 90% CH_3OH, 93%. (*c*) 1. TBSCl, imidazole, DMF, rt, 10 h, 94%; 2. MsCl, NEt_3, 0°C, 1 h, 97%. (*d*) 1. $LiAlH_4$, THF, 0°C to rt, 3 h, 90%; 2. Dowex 50WX8 resin, CH_3OH; then 10% Pd *on* C, H_2, 3 h; then NEt_3, reflux, 2 h; Dowex 50W8X resin, 62%.

mesylation of the primary hydroxyl group and subsequent treatment with sodium hydroxide and silylation of the primary hydroxyl group afforded the epoxide **91**. Removal of the Boc and isopropylidene groups from **91** followed by removal of the TBS group afforded **3** in 29% overall yield from D-glucono-1,5-lactone.

D-Gluconolactone → a → 89 → b → 90 → c → 91 → d → 3

Scheme 16 (*a*) Refs. 41 and 42, 95%. (*b*) 1. 10% Pd *on* C, H_2, EtOAc, rt, 1 h; 2. $(Boc)_2O$, CH_3OH, NEt_3, rt, 20 min, 93% for two steps; 3. $LiAlH_4$, THF, 0°C to rt, 13 h, 95%; 4. Ac_2O, Py, rt, 15 h, 93%. (*c*) 1. Dowex 50WX8 resin, 90% CH_3OH, rt, 18 h, 98%; 2. MsCl, NEt_3, CH_2Cl_2, −10°C, 97%, 5 min; 3. NaOH, CH_3OH, rt, 98%, 5 min; 4. TBSCl, imidazole, DMF, rt, 12 h, 92%. (*d*) 1. $AlCl_3$, $LiAlH_4$, ether, rt to reflux; 2. Dowex 50WX8 resin, 90% CH_3OH, reflux.

Azide displacement of the triflate group in the manno 2-*O*-triflate **93** under thermodynamic controlled conditions afforded the manno-azide **94**, with overall retention of configuration at C-2 (Scheme 17).[43] Alternatively, **94** may be prepared from the open chain azidoester **89**,[41,42,44] readily derived from D-glucono-1,5-lactone, by hydrolysis in aqueous TFA to give the azide lactone **92** and subsequent protection of the side chain diol as its acetonide. Protection of the C-3 hydroxyl group of the manno-azide **94** as its silyl ether **95** followed by removal of the side chain acetonide gave the corresponding diol; the C-6 primary hydroxyl group was then selectively protected. Subsequent triflation of the C-5 hydroxyl group followed by reduction of the azide group gave a nonisolable C-2 amine, which, on treatment with sodium acetate in acetonitrile, underwent spontaneous intramolecular S_N2 displacement of the C-5 triflate to afford the [2.2.1] bicycle **96**. Ring opening of the bicyclic lactone **96** with methylamine or *n*-butylamine in THF followed by deprotection afforded the pyrrolidine amides **97** (80%) and **98** (86%), respectively. On the other hand, reduction of the lactone **96** followed by acid hydrolysis of the silyl protecting groups afforded **3** (Scheme 17).

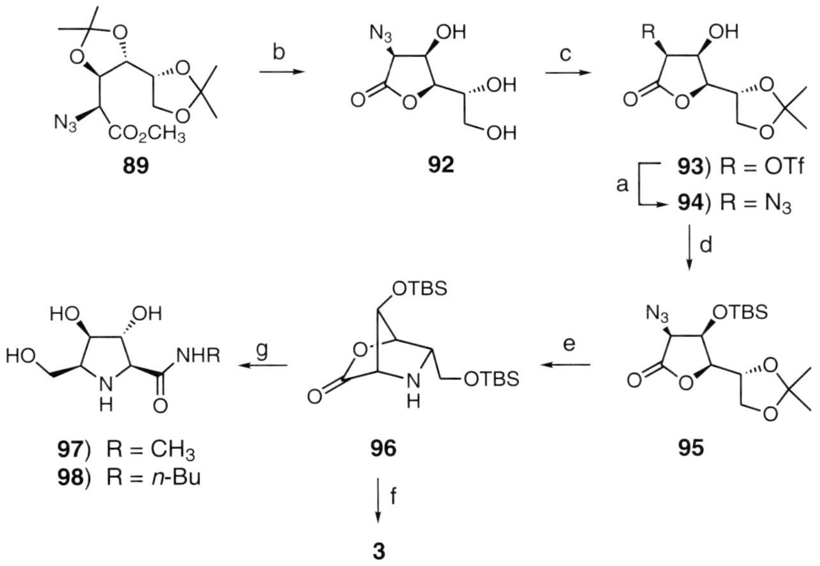

Scheme 17 (*a*) NaN_3, DMF, 40 h, 82%, thermodynamic conditions. (*b*) TFA–H_2O (3:2), 96%. (*c*) Acetone, CSA, 91%. (*d*) TBSOTf, Py, CH_2Cl_2, 75%. (*e*) 1. AcOH–H_2O (4:1), 84%; 2. TBSCl, imidazole, DMF, 71%; 3. Tf_2O, pyridine, CH_2Cl_2, 95%; 4. H_2, Pd black, EtOAc; 5. NaOAc, CH_3CN, 86%. (*f*) 1. $LiHBEt_3$, THF; 2. 1% HCl in CH_3OH, 78%. (*g*) 1. CH_3NH_2, THF; *or* *n*-$BuNH_2$, THF; 2. 1% HCl in CH_3OH, **97** (80%), **98** (86%).

A similar sequence of reactions was done on L-gulono-1,4-lactone to give **99**. Azide displacement of the 2-*O*-triflate in **99** under kinetic controlled conditions gave the ido-azide **100**, which was treated with *tert*-butyldimethylsilyl triflate to give the fully protected azido lactone **101** (Scheme 18).[43] Sequential treatment of **101** with aqueous acetic acid, *tert*-butyldimethylsilyl chloride and triflic anhydride in pyridine afforded the C-5 triflate **102** in 71% overall yield. Reduction of the azide group yielded the monocyclic amino triflate **103** (99%), which was treated with sodium acetate in methanol to give, under spontaneous cyclization, the proline ester **104** in 74% yield. Reduction of the ester **104** with lithium triethylborohydride in THF and subsequent removal of the silyl group gave **3**.

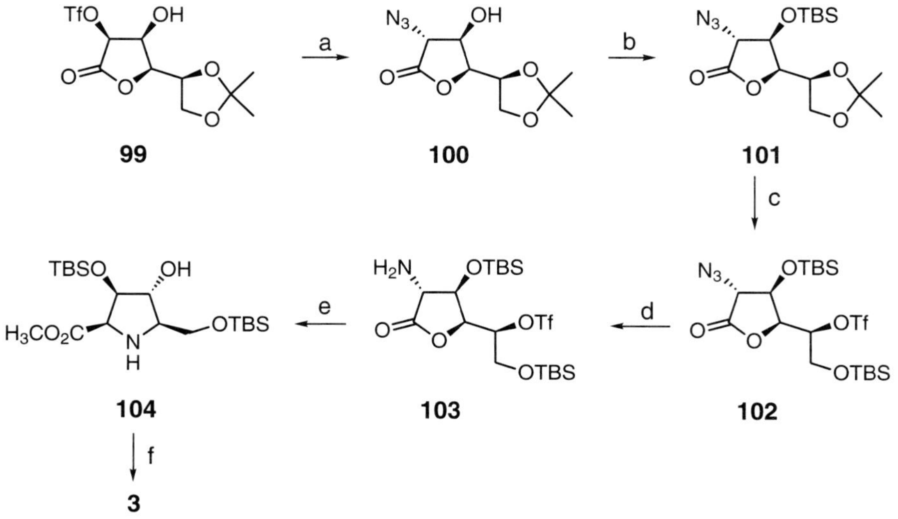

Scheme 18 (*a*) NaN_3, DMF, 2.5 h, 82%, kinetic conditions. (*b*) TBSOTf, Py, CH_2Cl_2, 85%. (*c*) 1. AcOH–H_2O (4:1), 99%; 2. TBSCl, imidazole, DMF, 76%; 3. Tf_2O, Py, CH_2Cl_2, 95%. (*d*) H_2, Pd black, EtOAc, 99%. (*e*) NaOAc, CH_3OH, 74%. (*f*) 1. $LiHBEt_3$, THF; 2. 1% HCl in CH_3OH, 90%.

References

1. Walter, A.; Jodot, J.; Dardenne, G.; Marlier, M.; Casimir, J. *Phytochemistry* **15** (1976) 747.
2. Evans, S.V.; Fellows, L.E.; Shing, T.K.M.; Fleet, G.W.J. *Phytochemistry* **24** (1985) 1953.
3. Elbein, A.D.; Mitchell, M.; Sanford, B.A.; Fellow, L.E.; Evans, S.V. *J. Biol. Chem.* **259** (1984) 12409.
4. Fellows, L.E. *Pestic. Sci.* **17** (1986) 602.
5. Asano, N. *J. Enzyme Inhib.* **15** (2000) 215.
6. Kevin, K.C.; Tesuya, K.; Lihren, C.; Ziyang, Z.; Yoshitaka, I.; Chi-Huey, W. *J. Org. Chem.* **56** (1991) 6280.
7. El Ashry, E.S.H.; Rashed, N.; Shobier, A.H.S. *Pharmazie* **55** (2000) 331.
8. Compain, P.; Martin, O.R. *Bioorg. Med. Chem.* **9** (2001) 3077.
9. Collyer, C.A.; Goldberg, J.D.; Viehmann, H., Blow, D.M.; Ramsden, N.G.; Fleet, G.W.J.; Montgomery, F.J.; Grice, P. *Biochemistry* **31** (1992) 12211.
10. Asano, N.; Kato, A.; Miyauchi, M.; Kizy, H.; Kameda, Y.; Watson, A.A.; Nash, R.J.; Fleet, G.W.J. *J. Nat. Prod.* **61** (1998) 625.
11. Lamotte-Brasseur, P.J.; Dupont, L.; Dideberg, O. *Acta. Crystallogr., Sect. B* **33** (1977) 409.
12. Asano, N.; Yasuda, K.; Kizu, H.; Kato, A.; Fan, J.-Q.; Nash, R.J.; Fleet, G.W.J.; Molyneux, R.J. *Eur. J. Biochem.* **268** (2001) 35.
13. Gautier-Lefebvre, I.; Behr, J.-B.; Guillerm, G.; Ryder, N.S. *Bioorg. Med. Chem. Lett.* (2000) 1.
14. Bosco, M.; Bisseret, P.; Bouix-Peter, C.; Eustache, J. *Tetrahedron Lett.* **42** (2001) 7949.
15. Verma, S.K.; Atanes, N.; Busto, J.H.; Thai, D.L.; Rapoport, H. *J. Org. Chem.* **67** (2002) 1314.
16. Fleet, G.W.J.; Smith, P.W. *Tetrahedron Lett.* **26** (1985) 1469.
17. Fleet, G.W.J.; Smith, P.W. *Tetrahedron* **43** (1987) 971.
18. Andersen, S.M.; Ebner, M.; Ekhart, C.W.; Gradnig, G.; Legler, G.; Lundt, I.; Stütz, A.E.; Withers, S.G.; Wrodnigg, T. *Carbohydr. Res.* **301** (1997) 155.
19. Dax, K.; Gaigg, B.; Grassberger, V.; Kölblinger, B.; Stütz, A.E. *J. Carbohydr. Chem.* **9** (1990) 479.
20. Morin, C. *Tetrahedron Lett.* **25** (1984) 3205.
21. Reitz, A.B.; Baxter, E.W. *Tetrahedron Lett.* **31** (1990) 6777.
22. Avigad, G.; Englard, S. *Meth. Enzymol.* **41** (1975) 84.
23. Card, P.J.; Hitz, W.D. *J. Org. Chem.* **50** (1985) 891.
24. Chmielewski, M.; Whistler, R.L. *J. Org. Chem.* **40** (1975) 639.
25. Hung, R.R.; Straub, J.A.; Whitesides, G.M., *J. Org. Chem.* **56** (1991) 3849.

26. Liu, K.K.-C.; Kajimoto, T.; Chen, L.; Zhong, Z.; Ichikawa, Y.; Wong, C.-H. *J. Org. Chem.* **56** (1991) 6280.
27. Takayama, S.; Martin, R.; Wu, J.; Laslo, K.; Siuzdak, G.; Wong, C.H. *J. Am. Chem. Soc.* **119** (1997) 8146.
28. Chorghade, M.S.; Cseke, C. *Pure Appl. Chem.* **66** (1994) 2211.
29. Behr, J.-B.; Guillerm, G. *Tetrahedron: Asymmetry* **13** (2002) 111.
30. Colobert, F.; Tito, A.; Khiar, N.; Denni, D.; Medina, M.A.; Martin-Lomas, M.; Ruano, J.L.G.; Solladie, G. *J. Org. Chem.* **63** (1998) 8918.
31. Dureault, A.; Portal, M.; Depezay, J.C. *Synlett* (1991) 225.
32. Baggett, N.; Stribblehill, P. *J. Chem. Soc., Perkin Trans. 1* (1977) 1123.
33. Shing, T.K.M. *J. Chem. Soc., Chem. Commun.* (1987) 262.
34. Shing, T.K.M. *Tetrahedron* **44** (1988) 7261.
35. Masaki, Y.; Oda, H.; Kazuta, K.; Usui, A.; Itoh, A.; Xu, F. *Tetrahedron Lett.* **33** (1992) 5089.
36. McCort, I.; Dureault, A.; Depezay, J.C. *Tetrahedron Lett.* **37** (1996) 7717.
37. Campanini, L.; Dureault, A.; Depezay, J.C. *Tetrahedron Lett.* **36** (1995) 8015.
38. Park, K.H. *Heterocycles* **41** (1995) 8.
39. Brook, M.A.; Chan, T.H. *Synthesis* (1983) 201.
40. Lee, S.G.; Yoon, Y.-J.; Shin, S.C.; Lee, B.Y.; Cho, S.-D.; Kim, S.K.; Lee, J.-H. *Heterocycles* **45** (1997) 701.
41. Csuk, R.; Hugener, M.; Vasella, A. *Helv. Chim. Acta* **71** (1988) 609.
42. Regeling, H.; Rouville, E.D.; Chittenden, G.J.F. *Recl. Trav. Chim. Pays-Bas* **106** (1987) 461.
43. Long, D.D.; Frederiksen, S.M.; Marquess, D.G.; Lane, A.L.; Watkin, D.J.; Winkler, D.A.; Fleet, G.W.J. *Tetrahedron Lett.* **39** (1998) 6091.
44. Hubschwerlen, C. *Synthesis* (1986) 962.

1.2 2-Carboxypyrrolidines

Carboxypyrrolidines can be grouped as hydroxyprolines and bulgecins. The syntheses of 11 naturally occurring hydroxyprolines from carbohydrates are described in this section, followed by syntheses of potent β-lactam synergistic agents (also known as bulgecins), which are glycosylated carboxyhydroxymethylpyrrolidines.

1.2.1 *Hydroxyprolines*

The (2*S*,3*S*,4*S*)- (**1**) and (2*S*,3*R*,4*R*)-3,4-dihydroxyprolines (**2**) have been isolated from diatom cell walls[1] and *Amanita vitosa* mushrooms.[2,3] It is believed that dihydroxyprolines act in plants as defense agents against predators and parasites.[4] (2*S*,3*R*,4*S*)-3,

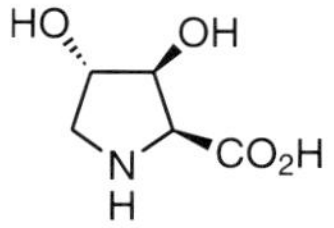

(2*S*,3*S*,4*S*)-3,4-Dihydroxyproline **1**

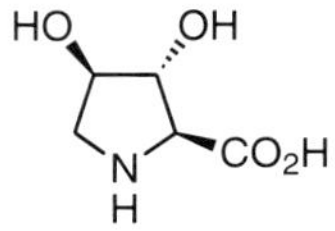

(2*S*,3*R*,4*R*)-3,4-Dihydroxyproline **2**

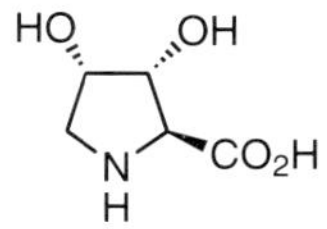

(2*S*,3*R*,4*S*)-3,4-Dihydroxyproline **3**

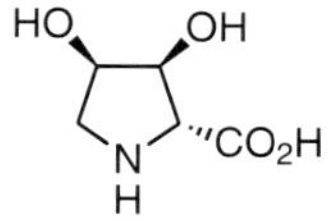

(2*R*,3*S*,4*R*)-3,4-Dihydroxyproline **4**

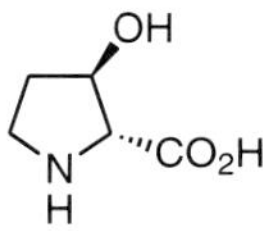

(2*R*,3*R*)-3-Hydroxyproline **5**

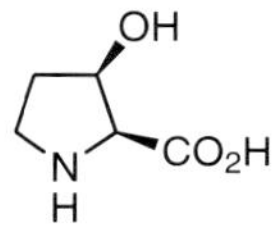

(2*S*,3*R*)-3-Hydroxyproline **6**

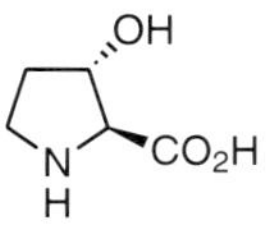

(2*S*,3*S*)-3-Hydroxyproline **7**

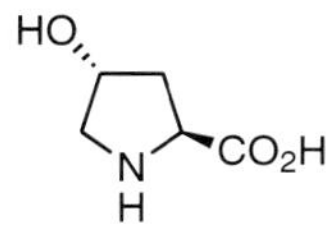

(2*S*,4*R*)-4-Hydroxyproline **8**

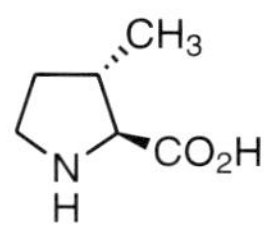

(2*S*,3*S*)-3-Methylproline **9**

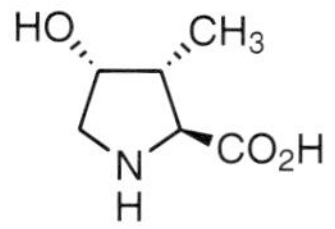

(2*S*,3*R*,4*R*)-4-Hydroxy-3-methylproline **10**

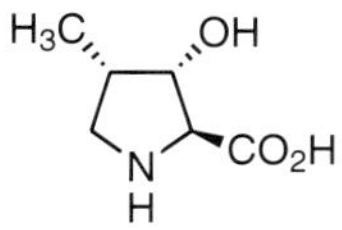

(2*S*,3*S*,4*S*)-3-Hydroxy-4-methylproline **11**

4-Dihydroxyproline (**3**) was isolated from animal adhesive protein (Mefp 1) found in the mussel *Mytilus edulis*, [5–7] and its (2*R*,3*S*,4*R*) analogue **4** was also isolated from natural sources.[8,9] (2*R*,3*R*)-3-Hydroxyproline (**5**) was isolated from dried Mediterranean sponge and telomycin,[10–12] while its (2*S*,3*R*)-isomer **6** was obtained only from telomycin.[13,14] (2*S*,3*S*)-3-Hydroxyproline (**7**) was found in naturally occurring peptides, namely mucrorin-D,[15] telomycin[16] and bovine Achilles tendon collagen.[17]

Nonproteinohenic proline derivatives **9** and **10** have been detected in the cyclic peptide scytonemin A, a metabolite of the cultured cyanophyte *Scytonema* sp. which possesses potent calcium antagonistic properties.[18] (2*S*,4*R*)-4-Hydroxyproline (**8**) and (2*S*,3*S*,4*S*)-3-hydroxy-4-methylproline (**11**) were found in echinocandin B, C and D, which were isolated from a strain of *Aspergillus ruglosus* and *Aspergillus nidulans* and characterized by their high antifungal and anti-yeast activities. [19–21]

Syntheses of natural hydroxyprolines from noncarbohydrate and their unnatural analogues from carbohydrate and noncarbohydrate have been reported.[22–76] Herein, the synthesis of the natural analogues from carbohydrate building blocks will be reviewed.

1.2.1.1 *Synthesis from* D*-glucose* Synthesis of (2*S*,3*R*,4*R*)-3,4-dihydroxyproline (**2**) from D-glucose has been reported (Scheme 1).[77] Diacetone D-glucose (**12**) was benzylated to give the fully protected derivative, which underwent acid hydrolysis of the terminal isopropylidene group followed by periodate oxidation, sodium borohydride reduction, mesylation and then replacement of the mesyloxy group with azide ion to afford the azide **13**. Treatment of **13** with methanolic hydrogen chloride followed by triflation of C-2 hydroxyl group gave the corresponding triflate **14**. Hydrogenation of **14** followed by protection of the resulting bicyclic compound with benzyl chloroformate afforded the carbamate

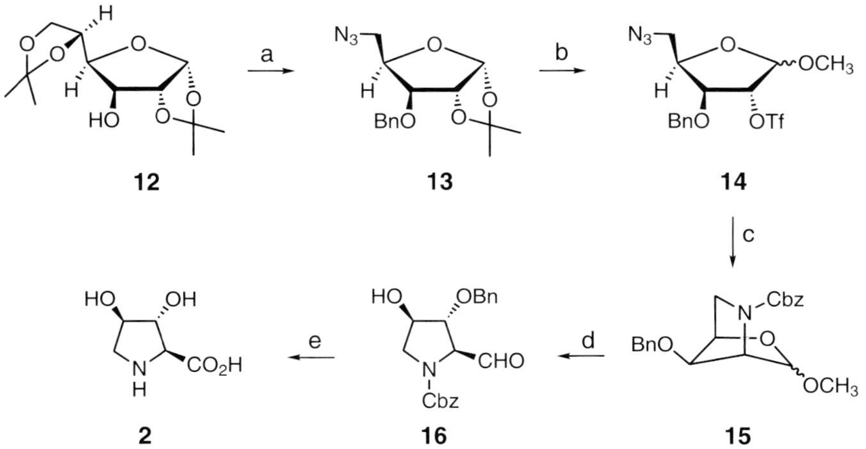

Scheme 1 (*a*) 1. THF, NaH, Bu_4NI, 0°C, BnBr, rt to 50°C, 2 h, 97%; 2. CH_3OH–AcOH–H_2O (1:1:1), 50°C, 16 h, 87%; 3. $NaIO_4$, 10% aqueous EtOH, 3 h, CH_2Cl_2; then 20% aqueous EtOH, $NaBH_4$, 8 h at rt, 88%; 4. Py, 0°C, MsCl, rt, 2 h, 94%; 5. NaN_3, DMF, 70°C for 12 h, 97%. (*b*) 1. AcCl, CH_3OH, 0°C, 36 h, α (38%), β (44%); 2. Py, Tf_2O, –50 to –30°C, 1 h, α (92%), β (78%). (*c*) 1. EtOAc, rt, 5%, Pd *on* C, H_2, α (95%), β (92.5%); 2. 3:2 mixture of ether and aqueous $NaHCO_3$, CbzCl, rt, 12 h, α (76%), β (90%). (*d*) 1:1 mixture of TFA and H_2O, 92%. (*e*) 1. H_2O–1,4-dioxane (1:1), $BaCO_3$, 0°C, Br_2, 24 h, 75%; 2. AcOH, H_2, Pd black, 48 h, 93.5%.

15. Subsequent hydrolysis with TFA gave the key intermediate **16**. Oxidation of **16** with bromine in aqueous dioxane containing barium carbonate followed by removal of the protecting groups afforded **2** in 18% yield from **12**.

1.2.1.2 *Synthesis from* D*-mannitol* The alditols have been also utilized for the synthesis of such series of compounds. Thus, D-mannitol has been used for the synthesis of 3,4-dihydroxy-L- and -D-prolines **3** and **4**, via its diisopropylidene which can be readily converted to 2,3-*O*-isopropylidene-D-glyceraldehyde (**17**)[78] (Scheme 2).[79] Lewis acid-catalyzed condensation[80,81] of **17** with *N*-(*tert*-butoxycarbonyl)-2-(*tert*-butyldimethylsilyloxy)pyrrole (**18**) led to the diastereoselective formation of lactam **19**. Treatment of **19** with TESOTf and 2,6-lutidine gave a quantitative yield of the protected lactam **20**, which was subjected to dihydroxylation using $KMnO_4$ to give the corresponding pyrrolidinone, which was directly transformed into **21** (60%) by treatment with DMP in the presence of catalytic amount of *p*-toluenesulfonic acid. Selective deprotection of the terminal acetonide in **21** furnished **22**. Subsequent oxidative cleavage with $NaIO_4$ afforded the aldehyde **23**, whose sodium borohydride reduction and subsequent protection with TBSCl afforded **25** (35% from **21**). Reduction of **25** with $LiEt_3BH$ gave the corresponding lactol,

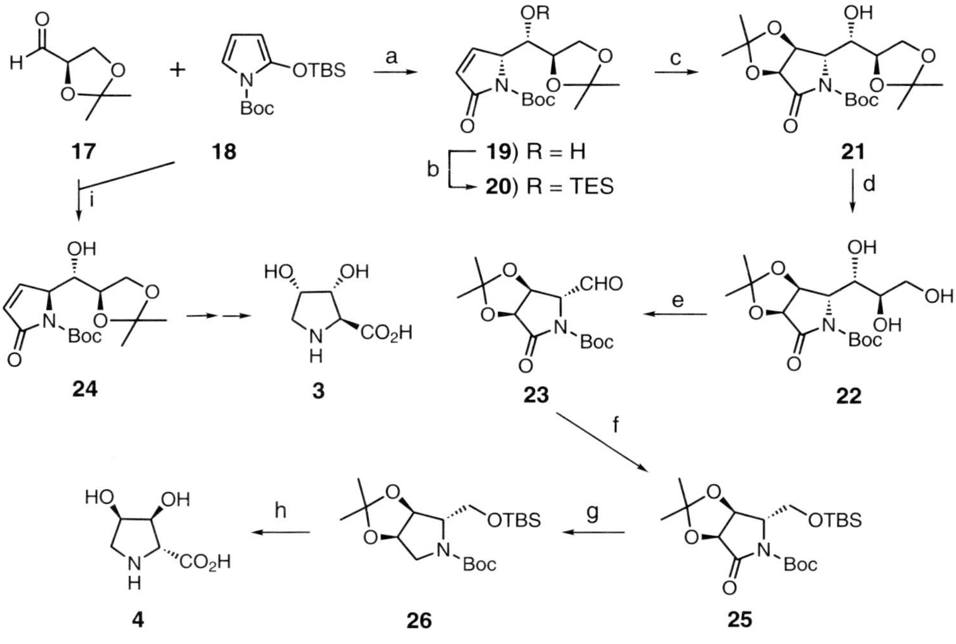

Scheme 2 (*a*) $SnCl_4$ (1.5 equiv.), Et_2O, –85°C, 80%. (*b*) TESOTf, 2,6-lutidine, CH_2Cl_2, rt, 4 h, 98%. (*c*) 1. $KMnO_4$, DCH–18-crown-6 ether, CH_2Cl_2, rt; 2. DMP, *p*-TsOH, rt, 60% for two steps. (*d*) Citric acid, CH_3OH, 40°C for 3 h; then 65°C, 2 h, 66%. (*e*) 0.65 M aqueous $NaIO_4$, SiO_2, CH_2Cl_2, rt, 68%. (*f*) 1. $NaBH_4$, THF–H_2O (3:1), –30°C; 2. TBSCl, imidazole, CH_2Cl_2, rt, 78%. (*g*) 1. $LiEt_3BH$, THF, –80°C, 98%; 2. Et_3SiH, $BF_3{\cdot}OEt_2$, CH_2Cl_2, –80°C, 65%. (*h*) 1. TBAF, THF, rt; then, $NaIO_4$, hydrated RuO_2, CH_3CN–CCl_4–H_2O–acetone (1:1:1.4:0.3), quantitative; 2. 3 N aqueous HCl, THF, rt; then Dowex (OH^-) resin, 95%. (*i*) $BF_3{\cdot}OEt_2$ (1.0 equiv.), Et_2O, –85°C, 70%.

followed by BF_3 etherate and triethylsilane as a hydride source to furnish the pyrrolidine **26** (65%). Removal of the TBS group from **26** with TBAF followed by oxidation of the resulting free hydroxymethyl group to CO_2H with $NaIO_4$–RuO_2, and finally removal of the isopropylidene group with 3 N aqueous HCl afforded **4** in 10% overall yield from **17**.

On the other hand, when BF_3 etherate was used to catalyze the condensation of D-glyceraldehyde **17** with **18**, the isomeric lactam **24** was obtained, which has been used for the synthesis of **3** in a similar sequence to that used above for the conversion of **19** to **4**.

A synthesis of (2*S*,3*S*,4*S*)-3-hydroxy-4-methylproline (**11**) from the tetraol derivatives **27**,[82,83] readily available from D-mannitol, has been achieved (Scheme 3).[84] Conversion of **27** to **28** took place in nearly quantitative yield. The 1,2-diol moiety of **28** was transformed into the epoxide with either inversion or retention of configuration to give **30** and **32**,

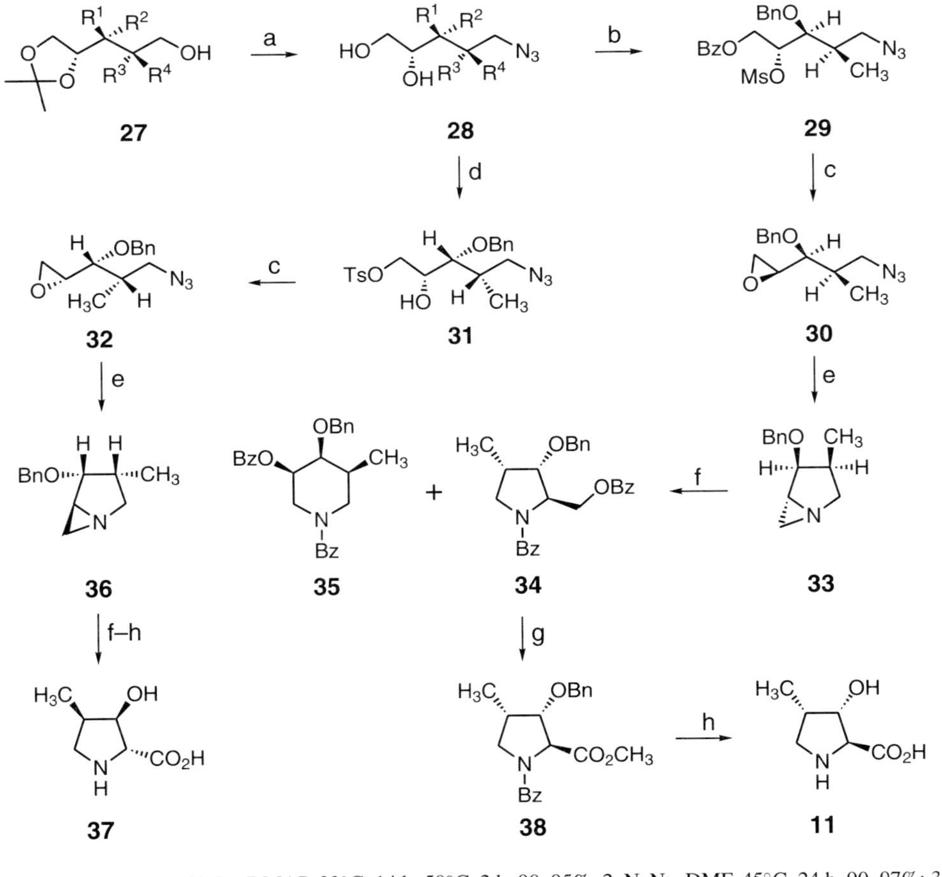

Scheme 3 (*a*) 1. MsCl, Py, DMAP, 22°C, 14 h, 50°C, 2 h, 90–95%; 2. NaN_3, DMF, 45°C, 24 h, 90–97%; 3. *p*-TsOH, CH_3OH, 45°C, 12 h, 95–99%. (*b*) 1. BzCl, Py, DMAP, 0°C, 1 h, 87–92%; 2. MsCl, Py, DMAP, 22°C, 14 h, 50°C, 2 h, 90–95%. (*c*) CH_3ONa, CH_3OH, 0°C, 30 min, 90–95%. (*d*) *p*-TsCl, Py, DMAP, 0°C, 24 h, 95–100%. (*e*) PPh_3, THF; *or* hexane, 22°C, 3 h, 65–75%. (*f*) Bz_2O, −10°C, 10 min. (*g*) 1. 2 N NaOH, 22°C, 5 h, quantitative; 2. $(COCl)_2$, DMSO, CH_2Cl_2, *i*-$PrNEt_2$, −55°C; 3. $KMnO_4$, H_2O, *t*-BuOH, pH 6, 22°C, CH_2N_2, ether, 0°C, 68%. (*h*) 1. H_2, Pd *on* C, CH_3OH, HCl, 3 bar, 22°C, 3 h, 94%; 2. 40% $NaOCH_3$, CH_3OH–H_2O (1:1), reflux 36 h; then Dowex 50X4 resin, 64%.

respectively, via the intermediates **29** and **31**. Compounds **30** and **32** were treated with triphenylphosphine under aprotic conditions to give compounds **33** and **36**, respectively. Compound **33** was treated with benzoic anhydride to give the dibenzoyl derivative **34**, in addition to **35**, which underwent debenzoylation followed by Swern oxidation of the primary hydroxyl group to afford the respective aldehyde, and then conversion[85] to the corresponding ester **38**. Subsequent debenzylation of **38** followed by ester hydrolysis and debenzoylation afforded **11**. An analogous sequence starting from **36** led to **37**.

1.2.1.3 *Synthesis from* L*-arabinono- and* L*-lyxono-lactones* Syntheses of **1** and **3** have been performed from L-arabinono- **39** and L-lyxono-lactones **40**, respectively, utilizing similar methodology (Scheme 4).[86] Starting by protection of the primary hydroxyl group with TrCl produced the corresponding derivatives **41** and **42**. Subsequent silylation gave **43** and **44**, followed by lactone reduction to produce the protected L-arabinitol **45** and L-lyxitol **46**, respectively. Then, mesylation followed by cyclization with benzylamine afforded the pyrrolidines **49** and **50** via the mesylated compounds **47** and **48**, respectively. Hydrogenation followed by N-protection of **49** (**50**) afforded the protected pyrrolidine **51** (**52**). Subsequent removal of the Tr-protecting group followed by oxidation of the primary hydroxyl group

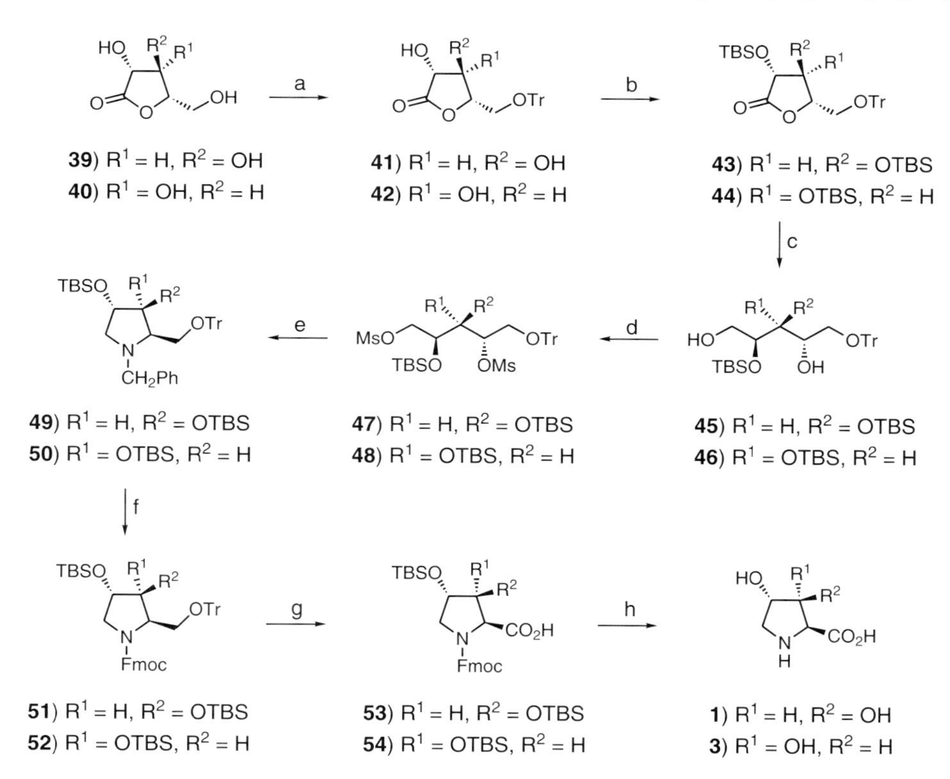

Scheme 4 (*a*) TrCl, Py, DMAP, 80°C. (*b*) TBSCl, DMF, imidazole, **43** (88%), **44** (84%). (*c*) $LiBH_4$, THF *or* $NaBH_4$, $CeCl_3$, CH_3OH, **45** (98%), **46** (73%). (*d*) MsCl, Py, DMAP, **47** (74%), **48** (99%). (*e*) $BnNH_2$, reflux, 60 h, **49** (79%), **50** (67%). (*f*) 1. H_2, Pd *on* C; 2. fluorenylmethylchloroformate (Fmoc-Cl), NEt_3, toluene, **51** (65%), **52** (90%). (*g*) 1. HCO_2H, CH_3CN; 2. $(COCl)_2$, DMSO, CH_2Cl_2; then NEt_3; 3. $NaClO_2$, C_6H_{10}, KH_2PO_4, **53** (63%), **54** (46%). (*h*) 1. TFA, CH_2Cl_2; 2. TBAF, THF; 3. Tesser's base, **1** (59%), **3** (93%).

produced the pyrrolidine **53** (**54**). Removal of the protecting groups from **53** and **54** furnished **1** and **3**, respectively.

1.2.1.4 *Synthesis from D-ribonolactone* D-Ribonolactone has been used for a stereoselective synthesis of 3,4-dihydroxyproline (**4**) starting by conversion to the benzylidene derivative **55**[87–90] in 89% yield (Scheme 5).[91–93] Replacement of the unprotected OH group with the azide ion via its *O*-triflyl derivative afforded the azide **56** with retention of configuration. Acid hydrolysis of the benzylidene group followed by selective mesylation of the primary hydroxyl group afforded **57** in 43% yield from **55**. Hydrogenation of **57** followed by treatment of the resulting aminolactone with aqueous NaOH gave **4** in 20% overall yield from ribonolactone.

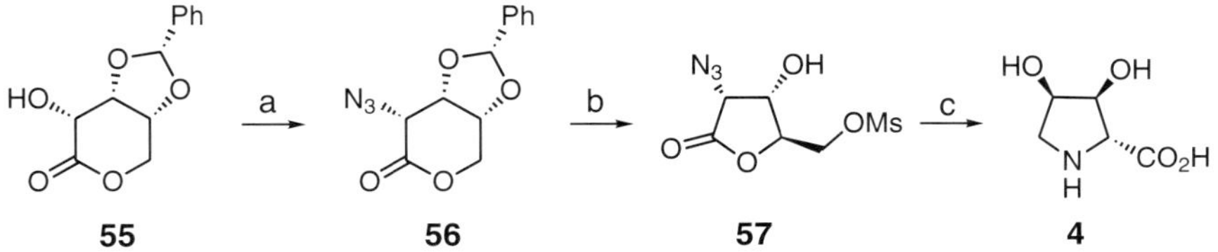

Scheme 5 (*a*) Tf_2O, Py, −10°C, NaN_3, DMF, rt, 64% for two steps. (*b*) 1. Aqueous TFA, 50°C, 94%; 2. MsCl, Py, −20°C, 71%. (*c*) 1. H_2, Pd black, EtOAc; 2. NaOH, H_2O, ion-exchange chromatography, 51%.

1.2.1.5 *Synthesis from D-gulonolactones* Total synthesis of 3,4-dihydroxyproline (**3**) from D-gulonolactone has been reported (Scheme 6).[94] D-Gulonolactone was treated with DMP to produce diacetone D-gulonolactone (**58**), which underwent $LiAlH_4$ reduction

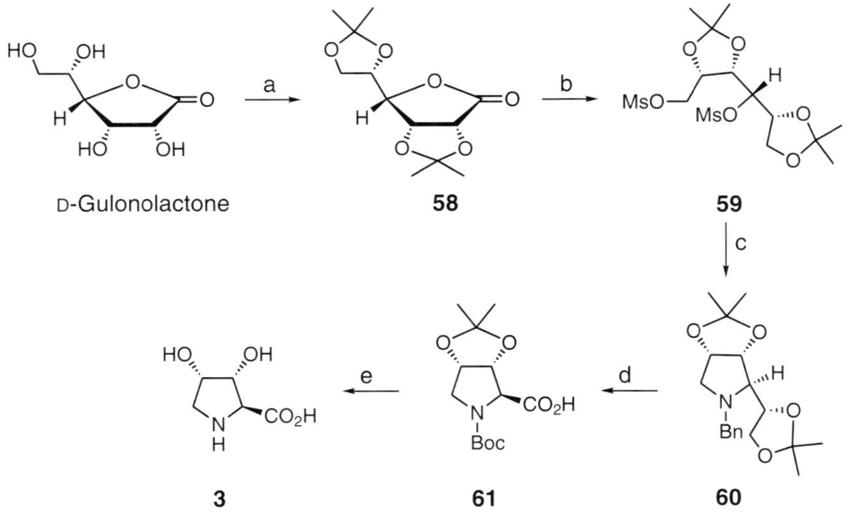

Scheme 6 (*a*) Acetone, DMP, *p*-TsOH, rt, 2 days; then anhydrous Na_2CO_3, 85%. (*b*) 1. $LiAlH_4$, THF, rt, 30 min, 87%; 2. MsCl, DMAP, Py, rt, 2 h, 100%. (*c*) $BnNH_2$, 60–70°C, 60 h, 77%. (*d*) 1. 80% aqueous AcOH, 50°C, 48 h, 93%; 2. EtOH, H_2, 10% Pd *on* C, rt, 2 h, 93%; 3. Py, di-*tert*-butyl dicarbonate, rt, 2.5 h, 75%; 4. EtOH–H_2O (5:2), $NaIO_4$, rt, 10 min; 5. *t*-BuOH, C_6H_{10}, $NaClO_2$, KH_2PO_4, H_2O, rt, overnight, 75%. (*e*) 80% aqueous TFA, rt, 23 h; then chromatography using Dowex 50X8-100 (H^+) resin, elution with 0.5 M NH_4OH, 81%.

followed by mesylation to furnish **59** in 74% yield from D-gulonolactone. Heating of **59** with benzylamine afforded the protected pyrrolidine **60**, which was subjected to five steps including removal of the terminal isopropylidene and benzyl protecting groups, oxidation and N-protection to produce **61**, which underwent complete deprotection to produce **3** in 22% overall yield from D-gulonolactone.

Alternatively, **3** was obtained from the Fmoc derivative **62**, resulting from the hydrogenation of the pyrrolidine **60**[90,94,95] followed by treatment with Fmoc-Cl (Scheme 7).[96] Acid hydrolysis of **62** gave **63** and **64**. Oxidation of **63** with sodium periodate followed by sodium chlorite furnished **65** in 86% yield, whose deprotection with TFA followed by Tesser's base afforded **3**.

Scheme 7 (*a*) 1. H_2, 10% Pd *on* C, rt, overnight, 88%; 2. Fmoc-Cl, toluene, NEt_3, rt, 15 h, 84%. (*b*) 70%, EtOH, conc. HCl, 60°C, 2.5 h, 47%. (*c*) 1. $NaIO_4$, EtOH, H_2O, rt, 15 min; 2. $NaClO_2$, C_6H_{10}, KH_2PO_4, *t*-BuOH, rt, 12 h, 86% for two steps. (*d*) 1. TFA, CH_2Cl_2, overnight; 2. Tesser's base, rt, 1 h; then Dowex (H^+) resin, elution with 0.5 M NH_4OH.

1.2.1.6 *Synthesis from D-gluconolactone* The enantiomerically pure 3-hydroxy-L-proline **7** has been prepared from D-glucono-δ-lactone (Scheme 8).[97] The diol **66**,[98–100] obtained from D-glucono-δ-lactone in five steps and 70% overall yield, underwent periodate oxidation followed by reduction and subsequent mesylation to produce the mesylate **67**, which was treated with LiI to produce iodide **68**. Dealkoxyhalogenation of **68** and subsequent silylation afforded (2*S*,3*S*)-2-amino-3-silyloxy-4-pentenoate **69** (85%), which underwent complete hydroboration with BMS and the resulting organoborane as oxidized with alkaline hydrogen peroxide to give 5-hydroxypentanoate **70** (70%). Cyclization of **70** was done through mesylation followed by intramolecular amination to produce the proline ester **71**. Removal of the protecting groups from **71** led to **7**.

The intermediate **67** in the last scheme was also used for the synthesis of 3,4-dihydroxyproline (**2**) (Scheme 9).[101] Thus, the mesylate **67** was refluxed with iodine to give **2** presumably via the intermediates **72** and **73**.

1.2.1.7 *Synthesis from D-glucoronolactone* A synthesis of the dihydroxyproline **76** from the lactone **74** has been reported (Scheme 10).[102,103] The lactone **74** was converted into the azide **75** in four steps. Hydrogenation of **75** in water in the presence of palladium black afforded **76** in 12% overall yield from the lactone **74**.

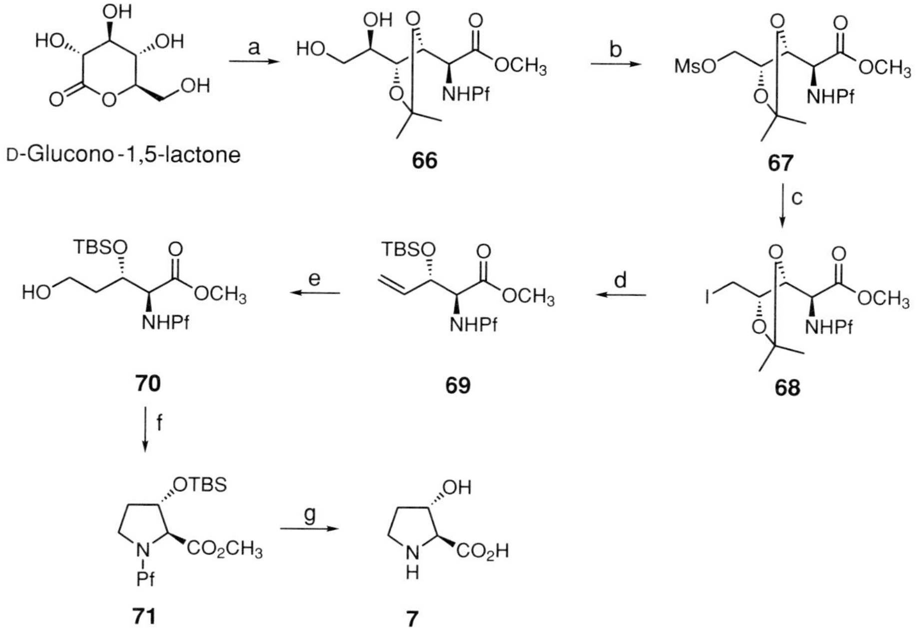

D-Glucono-1,5-lactone → (a) 66 → (b) 67 → (c) 68 → (d) 69 → (e) 70 → (f) 71 → (g) 7

Scheme 8 (*a*) Refs. 98–100. (*b*) 1. $NaIO_4$, $NaBH_4$, EtOH, rt, 98%; 2. MsCl, NEt_3, THF, 0°C, 98%. (*c*) LiI, DMF, 80°C, 95%. (*d*) 1. *n*-BuLi, THF, –40°C, 85%; 2. TBSCl, imidazole, DMF, rt, 98%. (*e*) BMS, THF, 0°C, 70%. (*f*) MsCl, NEt_3, THF, 0°C, 87%. (*g*) H_2, Pd *on* C, CH_3OH, 60°C, Dowex 50WX8 resin, THF, H_2O, reflux, 76%.

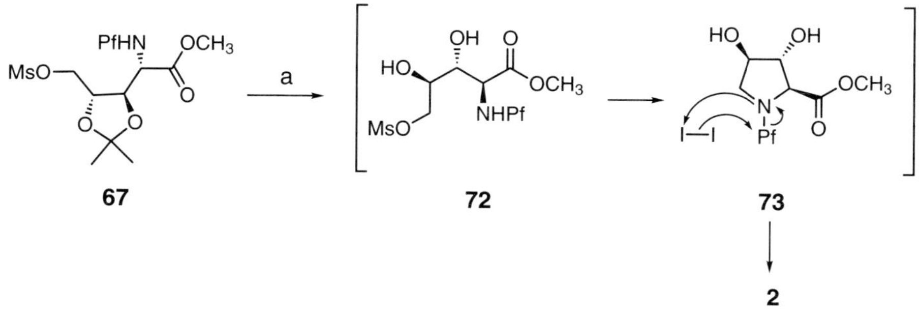

67 → (a) [72 → 73] → 2

Scheme 9 (*a*) 60% (w/w) I_2, CH_3OH, reflux, 4 h; Dowex 50W8X resin, CH_3OH, reflux, 3 h.

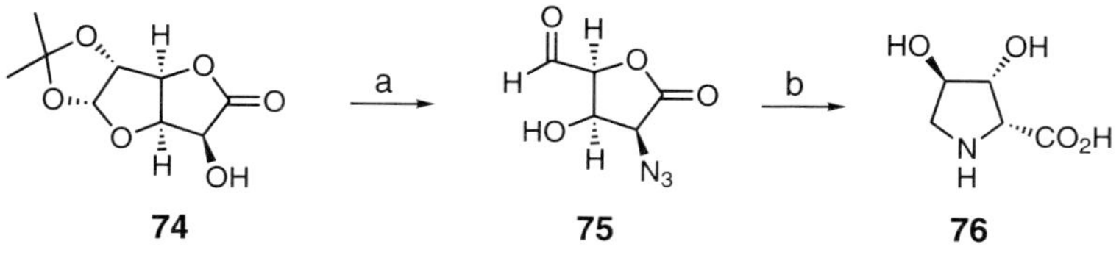

74 → (a) 75 → (b) 76

Scheme 10 (*a*) 1. Tf_2O, CH_2Cl_2, Py, –40°C, 5 h, 99%; 2. NaN_3, DMF, –20°C, 2.5 h, 83%; 3. TFA, H_2O, rt, 3.5 h; then $NaIO_4$, EtOH, H_2O, rt, 25 min. (*b*) H_2O, H_2, Pd black, rt, 4 days; then Dowex 50 (H^+) resin, 12% from **74**.

1.2.1.8 *Synthesis from* D*-xylonolactone* A synthesis of (2*S*,3*R*,4*R*)-3,4-dihydroxyproline (**2**) from 2,5-dibromo-2,5-dideoxy-D-xylono-1,4-lactone (**77**) has been reported (Scheme 11).[104] Compound **77** was reacted with sodium azide to give a 5:1 mixture of azides **78** and **79** in 95% yield. Hydrogenation of **78** afforded 2-amino-5-bromo-2,5-dideoxy-D-lyxono-1,4-lactone (**80**), which upon treatment with aqueous $Ba(OH)_2$ led to spontaneous cyclization of the amino acid **81** to afford **2** in 60% yield.

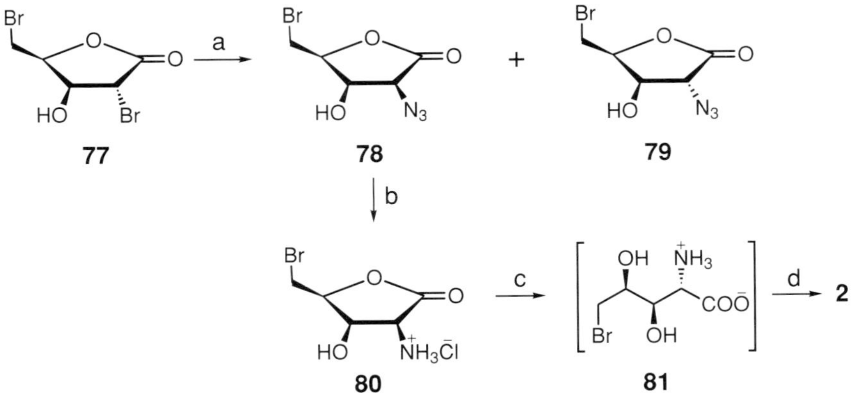

Scheme 11 (*a*) NaN_3, DMF, 25°C, 24 h, 95%. (*b*) Aqueous HCl, 5% Pd *on* C, 50% aqueous dioxane, 3 h, H_2, 34%. (*c*) Aqueous $Ba(OH)_2$, pH 9, 3 h. (*d*) Amberlite IR-120 (H^+) resin, eluted NH_4OH (5%), 60% for two steps.

References

1. Nakajima, T.; Volcani, B.E. *Science* **164** (1969) 1400.
2. Buku, A.; Faulstich, H.; Wieland, T.; Dabrowski, J. *Proc. Natl. Acad. Sci. USA* **77** (1980) 2370.
3. Karl, J.-U.; Wieland, T. *Liebigs Ann. Chem.* (1981) 1445.
4. Fellows, L.E. *Chem. Br.* (1987) 842.
5. Taylor, S.W.; Waite, J.H.; Ross, M.M.; Shabanowitz, J.; Hunt, D.F. *J. Am. Chem. Soc.* **116** (1994) 10803.
6. Waite, J.H.; Tanzer, M.L. *Science* **212** (1981) 1038.
7. Waite, J.H.; Housley, T.J.; Tanzer, M.L. *Biochemistry* **24** (1985) 5010.
8. Ohfune, Y.; Kurokawa, N. *Tetrahedron Lett.* **26** (1985) 5307.
9. Mauger, A.B. *J. Nat. Prod.* **59** (1996) 1205.
10. Irreverre, F.; Morita, K.; Robertson, A.V.; Witkop, B. *J. Am. Chem. Soc.* **85** (1963) 2824.
11. Sheehan, J.C.; Whitner, J.G. *J. Am. Chem. Soc.* **84** (1962) 3980.
12. Ogle, J.D.; Arlinghaus, R.B.; Logan, M.A. *J. Biol. Chem.* **237** (1962) 3667.
13. Irreverre, F.; Morita, K.; Robertson, A.V.; Witkop, B. *Biochem. Biophys. Res. Commun.* **8** (1962) 453.
14. Irreverre, F.; Morita, K.; Ishii, S.; Witkop, B. *Biochem. Biophys. Res. Commun.* **9** (1963) 69.
15. Tschesche, R.; Samuel, T.D.; Uhlendorf, J.; Fehlhaber, H.-W. *Chem. Ber.* **105** (1972) 316.
16. Sheehan, J.C.; Mania, D.; Nakamura, S.; Stock, J.A.; Maeda, K. *J. Am. Chem. Soc.* **90** (1968) 462.
17. Wolff, J.S.; Ogle, J.D.; Logan, M.A. *J. Biol. Chem.* **241** (1966) 1300.
18. Moore, R.E.; Helms, G.L.; Niemczura, W.P.; Patterson, G.M.L.; Tomer, K.B.; Gross, M.L. *J. Org. Chem.* **53** (1988) 1298.
19. Benz, F.; Knüsel, F.; Nüesch, J.; Treichler, H.; Voser, W.; Nyfeler, R.; Keller-Schierlein, W. *Helv. Chim. Acta* **57** (1974) 2459.
20. Keller-Juslén, C.; Kuhn, M.; Loosli, H.-R.; Petcher, T.J.; Weber, H.P.; von Wartburg, A. *Tetrahedron Lett.* **17** (1976) 4147.
21. Wehner, V.; Jäger, V. *Angew. Chem., Int. Ed. Engl.* **29** (1990) 1169.

22. Morita, K.; Irreverre, F.; Sakiyama, F.; Witkop, B. *J. Am. Chem. Soc.* **85** (1963) 2832.
23. Fleet, G.W.J.; Gough, M.J.; Smith, P.W. *Tetrahedron Lett.* **25** (1984) 1853.
24. Saito, S.; Matsumoto, S.; Sato, S.; Inaba, M.; Moriwake, T. *Heterocycles* **24** (1986) 2785.
25. Kurokawa, N.; Ohfune, Y. *J. Am. Chem. Soc.* **108** (1986) 6041.
26. Austin, G.N.; Baird, P.D.; Fleet, G.W.J.; Peach, J.M.; Smith, P.W.; Watkin, D.J. *Tetrahedron* **43** (1987) 3095.
27. Setoi, H.; Kayakiri, H.; Takeno, H.; Hashimoto, M. *Chem. Pharm. Bull.* **35** (1987) 3995.
28. Ikota, N.; Hanaki, A. *Chem. Pharm. Bull.* **35** (1987) 2140.
29. Buchanan, J.G.; Edgarand, A.R.; Hewitt, B.D. *J. Chem. Soc., Perkin Trans. 1* (1987) 2371.
30. Buchanan, J.G.; Jigajinni, V.B.; Singh, G.; Wightman, R.H. *J. Chem. Soc., Perkin Trans. 1* (1987) 2377.
31. Bashyal, B.P.; Fleet, G.W.J.; Gough, M.J.; Smith, P.W. *Tetrahedron* **43** (1987) 3083.
32. Ziegler, T.; Straub, A.; Effenberger, F. *Angew. Chem.* **100** (1988) 737.
33. Ikota, N.; Hanaki, A. *Heterocycles* **27** (1988) 2535.
34. Hughes, P.; Clardy, J. *J. Org. Chem.* **54** (1989) 3260.
35. Thaning, M.; Wistrand, L.-G. *Acta Chem. Scand.* **43** (1989) 290.
36. Hirai, Y.; Chintani, M.; Yamazaki, T.; Momose, T. *Chem. Lett.* (1989) 1449.
37. Pederson, R.L.; Wong, C.H. *Heterocycles* **28** (1989) 477.
38. Von der Osten, C.H.; Sinskey, A.J.; Barbas, C.F.; Pederson, R.L.; Wang, Y.F.; Wong, C.H. *J. Am. Chem. Soc.* **111** (1989) 3924.
39. Hanessian, S.; Ratovelomanana, V. *Synlett* (1990) 501.
40. Moss, W.O.; Bradbury, R.H.; Hales, N.J.; Gallagher, T. *J. Chem. Soc., Chem. Commun.* (1990) 51.
41. Wehner, V.J.; Jaeger, V. *Angew. Chem.* **102** (1990) 1180.
42. Hung, R.R.; Straub, J.A.; Whitesides, G.M. *J. Org. Chem.* **56** (1991) 3849.
43. Kajimoto, T.; Chen, L.; Liu, K.K.C.; Wong, C.H. *J. Am. Chem. Soc.* **113** (1991) 9009.
44. Witte, J.F.; McClard, R.W. *Tetrahedron Lett.* **32** (1991) 3927.
45. Arakawa, Y.; Yoshifuji, S. *Chem. Pharm. Bull.* **39** (1991) 2219.
46. Hamada, Y.; Hara, O.; Kawai, H.; Kohno, Y.; Shioiri, T. *Tetrahedron* **47** (1991) 8635.
47. Moss, W.O.; Bradbury, R.H.; Hales, N.J.; Gallagher, *J. Chem. Soc., Perkin Trans 1* (1992) 1901.
48. Thaning, M.; Wistrand, L.-G. *Acta Chem. Scand.* **46** (1992) 194.
49. Maggini, M.; Prato, M.; Ranelli, M.; Scorrano, G. *Tetrahedron Lett.* **33** (1992) 6537.
50. Jurczak, J.; Prokopowicz, P.; Golebiowski, A. *Tetrahedron Lett.* **34** (1993) 7107.
51. Knight, D.W.; Cooper, J.; Gallager, P.T. *J. Chem. Soc., Perken Trans. 1* (1993) 1313.
52. Blanco, M.J.; Sardina, F.J. *Tetrahedron Lett.* **35** (1994) 8493.
53. Goli, D.M.; Cheesman, B.V.; Hassan, M.E.; Lodaya, R.; Slama, J.T. *Carbohydr. Res.* **259** (1994) 219.
54. Herdeis, C.; Hubmann, H.P. *Tetrahedron: Asymmetry* **5** (1994) 119.
55. Huwe, C.M.; Blechert, S. *Tetrahedron Lett.* **36** (1995) 1621.
56. Evans, D.A.; Weber, A.E. *J. Am. Chem. Soc.* **109** (1987) 7151.
57. Kurokawa, N.; Ohfune, Y. *Tetrahedron* **49** (1993) 6195.
58. Remuzon, P. *Tetrahedron* **52** (1996) 13803.
59. Nájera, C.; Yus, M. *Tetrahedron: Asymmetry* **10** (1999) 2245.
60. Sifferlen, T.; Defoin, A.; Streith, J.; Le Nouen, D.; Tarnus, C.; Dosbaa, I.; Foglietti, M.-J. *Tetrahedron* **56** (2000) 971.
61. Humphrey, A.J.; Parsons, S.F.; Smith, M.E.B.; Turner, N.J. *Tetrahedron Lett.* **41** (2000) 4481.
62. Langlois, N.; Rakotondradany, F. *Tetrahedron* **56** (2000) 2437.
63. Schieweck, F.; Altenbach, H.-J. *J. Chem. Soc., Perkin Trans. 1* (2001) 3409.
64. Francisco, C.G.; Freire, R.; Gonzárez, C.C.; León, E.I.; Riesco-Fagundo, C.; Suárez, E. *J. Org. Chem.* **66** (2001) 1861.
65. Filichev, V.V.; Brandt, M.; Pedersen, E.B. *Carbohydr. Res.* **333** (2001) 115.
66. Chen, X.; Du, D.-M.; Hua, W.-T. *Tetrahedron: Asymmetry* **13** (2002) 43.
67. Pham-Huu, D.-P.; Gizaw, Y.; BeMiller, J.N.; Petrus, L. *Tetrahedron Lett.* **43** (2002) 383.
68. Durand, J.-O.; Larcheveque, M.; Petit, Y. *Tetrahedron Lett.* **39** (1998) 5746.
69. Graziani, L.; Porzi, G.; Sandri, S. *Tetrahedron: Asymmetry* **7** (1996) 1341.
70. Demange, L.; Menez, A.; Dugave, C. *Tetrahedron Lett.* **39** (1998) 1169.
71. Croce, P.D.; La Rosa, C. *Tetrahedron: Asymmetry* **13** (2002) 197.
72. El Ashry, E.S.H.; El Nemr, A. *Carbohydr. Res.* **338** (2003) 2265.
73. Poupardin, O.; Greck, C.; Genêt, J.-P. *Synlett* (1998) 1279.
74. Ban, F.; Gauld, J.W.; Boyd, R.J. *J. Phys. Chem. A* **104** (2000) 8583.
75. Jenkins, C.L.; Bretscher, L.E.; Guzei, I.A.; Raines, R.T. *J. Am. Chem. Soc.* **125** (2003) 6422.
76. Martin, R.; Alcon, M.; Pericas, M.A.; Riera, A. *J. Org. Chem.* **67** (2002) 6896.
77. Fleet, G.W.J.; Witty, D.R. *Tetrahedron: Asymmetry 1* (1990) 119.

78. Schmid, C.R.; Bryant, J.D. *Org. Synth.* **72** (1995) 6.
79. Zanardi, F.; Battistini, L.; Nespi, M.; Rassu, G.; Spanu, P.; Cornia, M.; Casiraghi, G. *Tetrahedron: Asymmetry* **7** (1996) 1167.
80. Casiraghi, G.; Rassu, G.; Spanu, P.; Pinna, L. *J. Org. Chem.* **57** (1992) 3760.
81. Rassu, G.; Casiraghi, G.; Spanu, P.; Pinna, L.; Gasparri Fava, G.; Belicchi Ferrari, M.; Pelosi, G. *Tetrahedron: Asymmetry* **3** (1992) 1035.
82. Mulzer, J.; De Lasalle, P.; Freissler, A. *Liebigs Ann. Chem.* (1986) 1152.
83. Mulzer, J.; Angermann, A.; Münch, W. *Liebigs Ann. Chem.* (1986) 825.
84. Mulzer, J.; Becker, R.; Brunner, E. *J. Am. Chem. Soc.* **111** (1989) 7500.
85. Abiko, A.; Roberts, J.C.; Takemasa, T.; Masamune, S. *Tetrahedron Lett.* **27** (1986) 4536.
86. Taylor, C.M.; Barker, W.D.; Weir, C.A.; Park, J.H. *J. Org. Chem.* **67** (2002) 4466.
87. Zinner, H.; Voight, H.; Voight, J. *Carbohydr. Res.* **7** (1968) 38.
88. Baggett, N.; Buchanan, J.G.; Fatah, M.Y.; Cullough, K.J.; Webber, J.M. *J. Chem. Soc., Chem. Commun.* (1985) 1826.
89. Chen, S.Y.; Joullie, M.M. *Tetrahedron Lett.* **24** (1983) 5027.
90. Chen, S.Y.; Joullie, M.M. *J. Org. Chem.* **49** (1984) 2168.
91. Dho, J.C.; Fleet, G.W.J.; Peach, J.M.; Prout, K.; Smith, P.W. *Tetrahedron Lett.* **27** (1986) 3203.
92. Baird, P.D.; Dho, J.C.; Fleet, G.W.J.; Peach, J.M.; Prout, K.; Smith, P.W. *J. Chem. Soc., Perkin Trans. 1* (1987) 1785.
93. Dho, J.C.; Fleet, G.W.J.; Peach, J.M.; Prout, K.; Smith, P.W. *Tetrahedron Lett.* **27** (1986) 5203.
94. Fleet, G.W.J.; Son, J.C. *Tetrahedron* **44** (1988) 2637.
95. Herdeis, C.; Aschenbrenner, A.; Kirfel, A.; Schwabenlander, F. *Tetrahedron: Asymmetry* **8** (1997) 2421.
96. Weir, C.A.; Taylor, C.M. *J. Org. Chem.* **64** (1999) 1554.
97. Lee, W.H.; Kang, J.E.; Yang, M.S.; Kang, K.Y.; Park, K.H. *Tetrahedron* **57** (2001) 10071.
98. Kim, J.H.; Lee, W.S.; Yang, M.S.; Park, K.H. *J. Chem. Soc., Perkin Trans. 1* (1998) 2877.
99. Csuk, R.; Hugener, M.; Vasella, A. *Helv. Chim. Acta* **71** (1988) 609.
100. Gerspacher, M.; Rapoport, H. *J. Org. Chem.* **56** (1991) 3700.
101. Kim, J.H.; Lee, W.S.; Yang, M.S.; Lee, S.G.; Park, K.H. *Synlett* (1999) 614.
102. Bashyal, B.P.; Chow, H.-F.; Fleet, G.W.J. *Tetrahedron Lett.* **27** (1986) 3205.
103. Bashyal, B.P.; Chow, H.-F.; Fellows, L.E.; Fleet, G.W.J. *Tetrahedron* **43** (1987) 415.
104. Bols, M.; Lundt, I. *Acta Chem. Scand.* **46** (1992) 298.

1.2.2 *Bulgecins*

The bulgecins A (**1**), B (**2**), C (**3**), SQ-28504 (**4**) and SQ-28546 (**5**) are potent β-lactam synergists found in the culture broth of *Pseudomonas acidophila*, *Pseudomonas mesoacidophila*[1–3] and chromobacterium violaceum.[4] They are substituted glucosides of (–)-bulgecinine (**6**), which is a proline derivative. These compounds introduced characteristic morphological change called *bulge formation* in Gram-negative bacteria in cooperation with the β-lactam antibiotics such as sulfazecin (**7**) or isosulfazecin (**8**). These antibiotics are also produced from *P. acidophila* strain G-6302 and *P. mesoacidophila* strain SB-72310.[5–7] The activity of these antibiotics was effectively enhanced as a result of the bulge formation. However, the sole use of bulgecins did not show antibacterial activity at all.[1]

Bulgecins, bulgecinine and their analogues have been synthesized from noncarbohydrates as starting materials.[8–20] Their methods of synthesis from carbohydrates will be presented below.

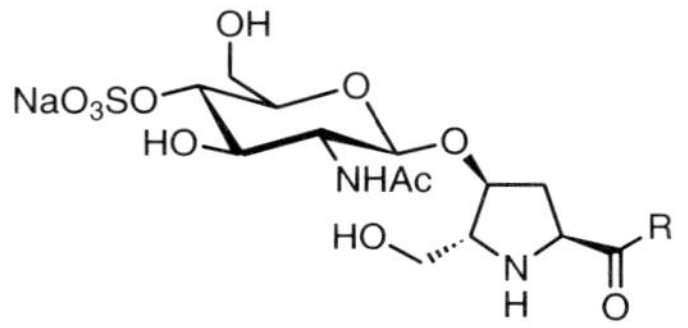

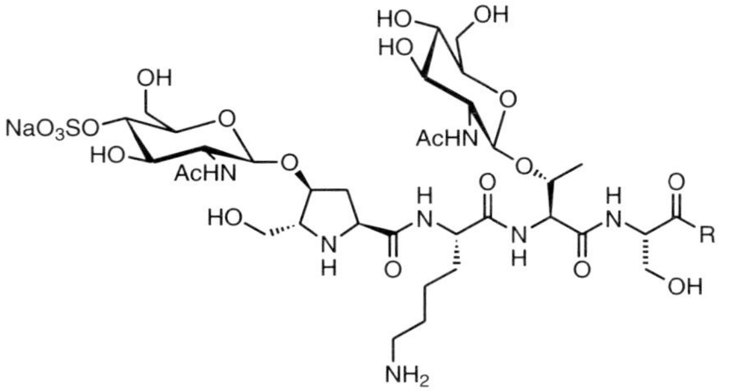

1) Bulgecin A, R = $NHCH_2CH_2SO_3H$
2) Bulgecin B, R = $NHCH_2CH_2CO_2H$
3) Bulgecin C, R = OH

4) SQ-28504, R = OH
5) SQ-28546, R =

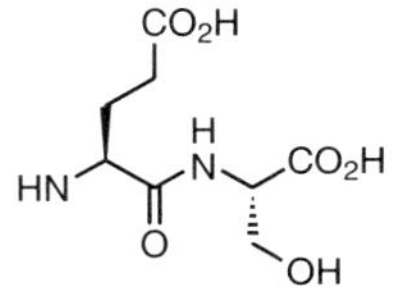

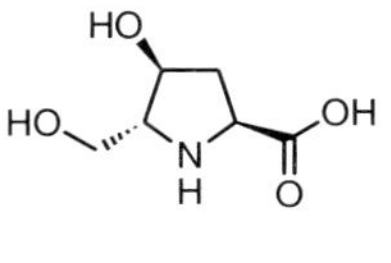

Bulgecinine
[(2*S*,4*S*,5*R*)-4-hydroxy-5-hydroxymethylproline]
6

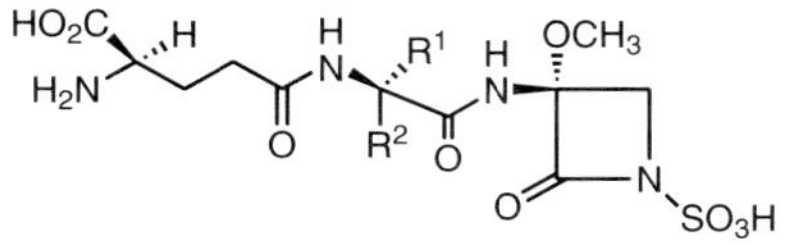

7) Sulfazecin, R^1 = H, R^2 = CH_3
8) Isosulfazecin, R^1 = CH_3, R^2 = H

1.2.2.1 *Synthesis from* D-*glucose* Bulgecinine (**6**) was synthesized stereospecifically by using D-glucose as a chiral precursor (Scheme 1).[21] The 3-deoxy-D-glucose derivative **9** was obtained by $LiAlH_4$ reduction of methyl 4,6-*O*-benzylidene-2,3-di-*O*-tosyl-α-D-glucopyranoside.[22] Tosylation of the free hydroxyl group at C-2 of **9** followed by substitution with azide ion and subsequent hydrogenolysis of the azide group followed by protection of the resulting amino group furnished **10** in 59% overall yield from **9**. Protection of the free hydroxyl groups in **10** followed by acid hydrolysis of the glycosidic linkage and then oxidation produced the lactone **11** in 24% overall yield. Methanolysis of **11** afforded **12** (100%), which was chlorinated[23] to afford the chloro derivative **13** (43%). Hydrogenolysis of **13** under acidic conditions followed by treatment with saturated $Ba(OH)_2$ solution gave **6** in 4.5% overall yield from **9**.

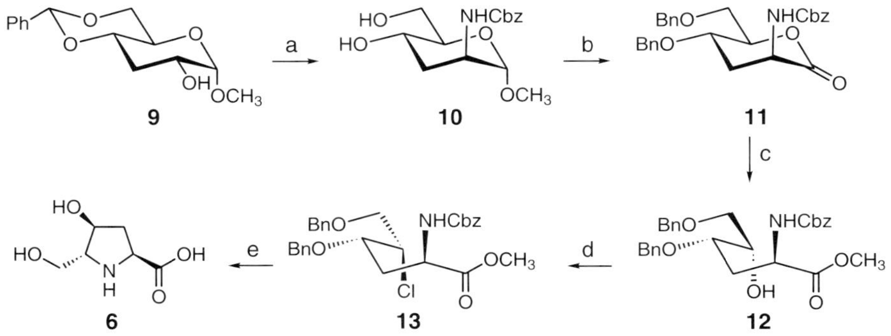

Scheme 1 (*a*) 1. *p*-TsCl, Py, 88%; 2. NaN_3, DMF, 73%; 3. H_2, Pd black, CH_3OH, HCl, quantitative; 4. *N*-benzyloxycarbonyloxysuccinimide, NEt_3, CH_3OH, 92%. (*b*) 1. BnBr, NaOH, DMF, 61%; 2. conc. HCl, AcOH, 66%; 3. PDC, CH_2Cl_2, 59%. (*c*) CH_3OH, reflux, quantitative. (*d*) PPh_3, CCl_4, 43%. (*e*) 1. H_2, Pd black, CH_3OH, conc. HCl, quantitative; 2. sat. $Ba(OH)_2$, pH 9.0, 75%.

Synthesis of bulgecinine has also been achieved from the 4-hydroxyproline **14** (Scheme 2),[24] which upon esterification, N-protection and subsequent inversion of configuration of C-4, by esterification using the Mitsunobu reaction, gave **15**. Anodic oxidation[25] gave the

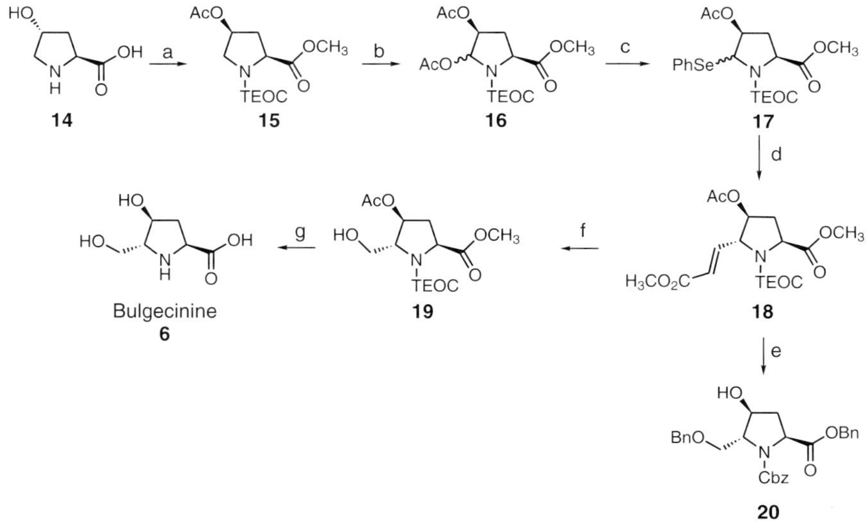

Scheme 2 (*a*) 1. CH_3OH, $SOCl_2$, 100%; 2. TEOC–N_3, NEt_3, CH_3CN, 90%; 3. PPh_3, DEAD, AcOH, THF, 65%. (*b*) 1. Et_4NOTs, CH_3OH, graphite electrodes, 5.5 F/mol; then Ac_2O, NEt_3, CH_2Cl_2, 64%; 2. Ac_2O, AcOH, H_2SO_4, 77%. (*c*) PhSeH, TsOH, 86%. (*d*) (*E*)- *or* (*Z*)-$CH_3O_2CCH{=}CHSnBu_3$, $(Bu_3Sn)_2$, 250 W sunlamp, Pyrex filter, 67%. (*e*) 1. TFA; then CbzCl, aqueous $NaHCO_3$, CH_2Cl_2, 82%; 2. O_3, CH_3OH, CH_2Cl_2; $NaBH_4$, 97%; 3. BnBr, Ag_2O, CH_2Cl_2, reflux, 91%; 4. BnOH, Ti(O*i*-Pr)$_4$, 110°C, 72%. (*f*) O_3, CH_3OH, CH_2Cl_2, $NaBH_4$, 83%. (*g*) NaOH, CH_3OH; then TBAF, 50%.

5-methoxy compound as a mixture of diastereoisomers followed by acetolysis to afford **16**, which was converted, by reaction with benzeneselenol under acidic conditions, into the 5-phenylseleno compound **17**. Irradiation of the selenide **17** in the presence of methyl (*Z*)- or (*E*)-2-(tributylstannyl)acrylate gave the radical substitution product **18**. Ozonolysis of the α,β-unsaturated ester **18** followed by reduction with sodium borohydride afforded the (5*R*)-hydroxymethyl compound **19** (83%). Subsequent removal of the protecting groups gave bulgecinine (**6**). On the other hand, the bulgecinine derivative **18** was also converted into the *N*-Cbz amino acid derivative **20** as an acceptor for the total synthesis of **3** shown in Scheme 3.

The required donor for the total synthesis of bulgecin C (**3**) was prepared from 3,4,6-tri-*O*-acetyl-D-glucal by conversion into 2-azido glucoside **21** (Scheme 3).[26] Benzylation with benzyl bromide followed by acid hydrolysis and subsequent selective benzylation of the primary hydroxyl group using bis(tributyltin)oxide and benzyl bromide produced **22**. Benzoylation followed by O-1 deprotection with fluoride ion and then treatment with trichloroacetonitrile in the presence of DBU[27–31] afforded the trichloroacetimidate **23**. Coupling of **23** with **20** in the presence of boron trifluoride etherate produced both the desired β-glycoside **24** (42%) and the α-anomer **25** (13%). Reduction of the azido group in **24** with AcSH to the corresponding acetamide derivative followed by debenzoylation afforded **26**, which underwent sulfation with pyridine–sulfur trioxide complex followed by debenzylation to produce **3**.

Scheme 3 (*a*) 1. BnBr, NaH, Bu_4NI, DMF, 0°C, 84%; 2. TFA, H_2O, 0°C, 74%; 3. $(Bu_3Sn)_2O$, toluene, reflux, 80°C, 4 h, 4. BnBr, Bu_4NBr, DMF, 16 h, 82% for two steps. (*b*) 1. BzCl, NEt_3, DMAP, CH_2Cl_2, 0°C, 4 h, 89%; 2. TBAF, AcOH, THF, 1 h, 96%; 3. Cl_3CCN, DBU, –40°C, CH_2Cl_2, MS 4 Å, 3 h. (*c*) $BF_3{\cdot}OEt_2$, CH_2Cl_2, MS 4 Å, 4 h, –40°C. (*d*) 1. AcSH, 24 h, 80%; 2. KOH, H_2O, CH_3OH, 74%. (*e*) 1. $SO_3{\cdot}Py$, DMF, 65%, 1 h; then NEt_3, acetone, 0°C, AG 50WX4 ion-exchange resin, 95%; 2. HCO_2H, Pd black, CH_3OH, 30 min, 70%.

1.2.2.2 *Synthesis from D-glucuronolactone* Synthesis of bulgecinine from D-glucuronolactone derivative **27** has been reported (Scheme 4).[32,33] Triflation of the lactone **27**[34,35] followed by nucleophilic displacement of the triflate group with azide ion and then hydrogenation and protection with benzyl chloroformate produced **28**. Conversion of **28** to the unstable aldehyde **29** followed by reduction *in situ* by sodium borohydride afforded the crystalline diol **30**, with no epimerization at C-5. Selective mesylation of **30** gave the mesylate

31. Hydrogenation of **31** followed by treatment with ethanolic potassium hydroxide led to an intramolecular cyclization to give pipecolic acid **34** in 73% yield and bulgecinine (**6**) in 7% yield. On the other hand, protection of the primary hydroxyl group in **30** with *tert*-butyldimethylsilylchloride followed by hydrogenation and subsequent reprotection of the amine afforded the carbamate **32**. Mesylation of **32** gave **33** in 80% overall yield from **30**. Hydrogenolysis of the carbamate group followed by intramolecular cyclization and subsequent removal of the silyl group afforded **6** in 50% overall yield from **30**.

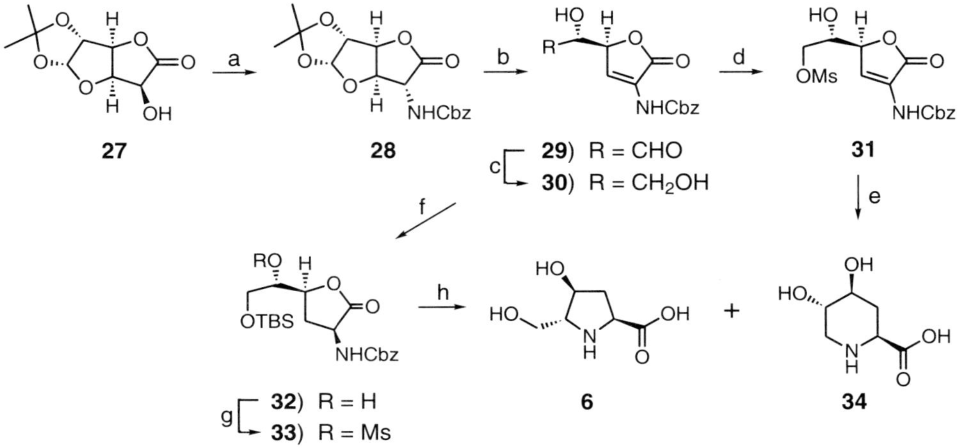

Scheme 4 (*a*) 1. Tf_2O, CH_2Cl_2, Py, –40°C, 5 h, 99%; 2. NaN_3, DMF, –20°C, 2.5 h, 83%; 3. 10% Pd *on* C, EtOAc, rt, 6 h; then CbzCl, EtOAc, 0°C, aqueous $NaHCO_3$, 30 min, 75%. (*b*) $NaOCH_3$, CH_3OH, 2 min. (*c*) $NaBH_4$, 0°C, 10 min, 91%. (*d*) MsCl, Py, –15°C, 30 min, 80%. (*e*) 1. EtOAc, Py, H_2, Pd black, 20°C, 4 h; 2. KOH, EtOH, H_2O, 20°C, 1 h; then Dowex 50 (H^+) resin, 80%. (*f*) 1. DMF, CH_2Cl_2, DNAP, TBSCl, 20°C, 12 h, 91%; 2. EtOAc, Py, Pd black, rt, 24 h; then aqueous $NaHCO_3$, CbzCl, 90%. (*g*) DMAP, MsCl, Py, rt, 14 h, 98%. (*h*) 1. EtOAc, EtOH, Pd black, rt, 48 h; 2. EtOH, $NaHCO_3$, rt, 24 h; 3. 5% HCl, THF, 4 h; Dowex 50 (H^+) resin, 56% for three steps.

References

1. Imada, A.; Kintaka, K.; Nakao, M.; Shinagawa, S. *J. Antibiot.* **35** (1982) 1400.
2. Shinagawa, S.; Maki, M.; Kintaka, K.; Imada, A.; Asai, M. *J. Antibiot.* **38** (1985) 17.
3. Shinagawa, S.; Kasahara, F.; Wada, Y.; Harada, S.; Asai, M. *Tetrahedron* **40** (1984) 3465.
4. Cooper, R.; Unger, S. *J. Org. Chem.* **51** (1986) 3942.
5. Imada, A.; Kitano, K.; Kintaka, K.; Muroi, M.; Asai, M. *Nature* **289** (1981) 590.
6. Asai, M.; Haibara, K.; Muroi, M.; Kintaka, K.; Kinshi, T. *J. Antibiot.* **34** (1981) 621.
7. Kintaka, K.; Haibara, K.; Asai, M.; Imada, A. *J. Antibiot.* **34** (1981) 1081.
8. Maeda, M.; Okazaki, F.; Murayama, M.; Tachibana, Y.; Aoyagi, Y.; Ohta, A. *Chem. Pharm. Bull.* **45** (1997) 962.
9. Wakamiya, T.; Yamanoi, K.; Kanou, K.; Shiba, T. *Tetrahedron Lett.* **28** (1987) 5887.
10. Jackson, R.F.W.; Rettie, A.B. *Tetrahedron Lett.* **34** (1993) 2985.
11. Hirai, Y.; Terada, T.; Amemiya, Y.; Momose, T. *Tetrahedron Lett.* 33 (1992) 7893.
12. Panday, S.K.; Langlois, N. *Synth. Commun.* **27** (1997) 1373.
13. Ohta, T.; Hosoi, A.; Nozoe, S. *Tetrahedron Lett.* **29** (1988) 329.
14. Ohfune, Y.; Hori, K.; Sakaitani, M. *Tetrahedron Lett.* **27** (1986) 6079.
15. Madau, A. Porzi, G.; Sandri, S. *Tetrahedron: Asymmetry* **7** (1996) 825.

16. Fehn, S.; Burger, K. *Tetrahedron: Asymmetry* **8** (1997) 2001.
17. Holt, K.E.; Swift, J.P.; Smith, M.E.B.; Taylor, S.J.C.; McCague, R. *Tetrahedron Lett.* **43** (2002) 1545.
18. Jurczak, J.; Krasinski, A. *Tetrahedron Lett.* **42** (2001) 2019.
19. Oppolzer, W.; Moretti, R.; Zhou, C. *Helv. Chim. Acta* **77** (1994) 2363.
20. Graziani, L.; Porzi, G.; Sandri, S. *Tetrahedron: Asymmetry* **7** (1996) 1341.
21. Wakamiya, T.; Yamanoi, K.; Nishikawa, M.; Shiba, T. *Tetrahedron Lett.* **26** (1985) 4759.
22. Vis, E.; Karrer, P. *Helv. Chim. Acta* **37** (1954) 378.
23. Calzada, J.G.; Hooz, J. *Org. Synth.* **54** (1974) 63.
24. Barrett, A.G.M.; Pilipauskas, D. *J. Org. Chem.* **55** (1990) 5194.
25. Shono, T.; Matsumura, Y.; Tsubata, K. *Org. Synth.* **63** (1985) 206.
26. Barrett, A.G.M.; Pilipauskas, D. *J. Org. Chem.* **56** (1991) 2787.
27. Schmidt, R.R. *Pure Appl. Chem.* **61** (1989) 1257.
28. Schmidt, R.R.; Michel, J. *Angew. Chem., Int. Ed. Engl.* **19** (1980) 731.
29. Grundler, G.; Schmidt, R.R. *Liebigs Ann. Chem.* (1984) 1826.
30. Tavecchia, P.; Trumtel, M.; Veyrieres, A.; Sinay, P. *Tetrahedron Lett.* **30** (1989) 2533.
31. Kinzy, W.; Schmidt, R.R. *Liebigs Ann. Chem.* (1985) 1537.
32. Bashyal, B.P.; Chow, H.-F.; Fleet, G.W.J. *Tetrahedron Lett.* **27** (1986) 3205.
33. Bashyal, B.P.; Chow, H.-F.; Fleet, G.W.J. *Tetrahedron* **43** (1987) 423.
34. Kitihara, T.; Ogawa, T.; Naganuma, T.; Matsui, M. *Agric. Biol. Chem.* **38** (1974) 2189.
35. Cauk, R.; Honig, H.; Nimpt, J.; Weidmann, H. *Tetrahedron Lett.* **21** (1980) 2135.

1.3 2-Aralkyl pyrrolidines

Two naturally occurring 2-aralkyl pyrrolidines are included in this part: The first one is the antibiotic anisomycin, which has various important therapeutic values. The second one is preussin, which exhibits antifungal activity.

1.3.1 *(−)-Anisomycin*

The antibiotic (–)-anisomycin (**1**) was first isolated from two S*treptomyces* species, *Str. grieolus* and *Str. roseochromogenes*.[1] It was also found in two related strains, *Streptomyces* sp. No. 638[2] and *Streptomyces* SA 3097.[3] Anisomycin has a broad activity against certain pathogenic protozoa and strains of fungi,[1–11] and is effective in the treatment of amoebic dysentery, tricomonas vaginitis[12,13] as well as plant fungicide.[14] This alkaloid blocks the aminoacyl-sRNA transfer reaction in protein biosynthesis and exhibits a remarkably selective inhibition of peptide chain elongation on 60S eukaryotic ribosomes.[15–17] It has high antitumor activity *in vitro*, with IC_{50} values in the nanomolar range.[3] Anisomycin may be used in a synergistic fashion with a cyclin-dependent protein kinase inhibitor to kill carcinoma cells.[18,19] It has a widespread use as a tool in molecular biology, where it inhibits protein synthesis[20,21] and activates JNK and p38 kinases.[22] The X-ray crystallographic analysis[23] of anisomycin has been studied and the absolute configuration was established by chemical correlation studies.[11–19] Several syntheses of anisomycin and its analogues from noncarbohydrate derivatives have been reported.[24–43]

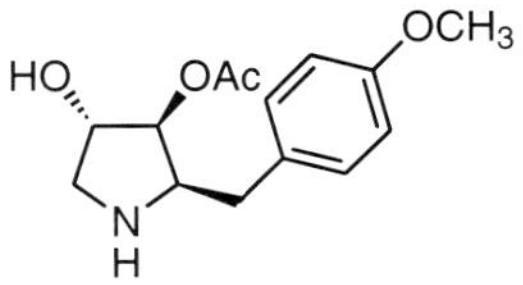

HO OAc OCH3 N H

(–)-Anisomycin

1

1.3.1.1 *Synthesis from* D*-galactose* (−)-Anisomycin (**1**) has been synthesized[44] from D-galactose by conversion firstly to ethyl 2,3-di-*O*-benzyl-β-D-galactofuranoside (**2**)[45–47] in 70% overall yield (Scheme 1). Periodate oxidation of **2** followed by treatment with (4-methoxyphenyl)magnesium bromide afforded a mixture of epimeric alcohols **3**. The latter underwent ionic deoxygenation with triethylsilane in the presence of TFA and subsequent acid hydrolysis to produce the anomeric mixture **4**. Compound **4** was condensed with hydroxylamine hydrochloride to afford the *E*- and *Z*-oximes **5**. Treatment of **5** with methanesulfonyl chloride in pyridine caused dehydration and simultaneous O-mesylation to afford the nitrile **6**. Nitrile reduction with BH_3 proceeded with cyclization to give **7**. This underwent catalytic hydrogenation in the presence of formic acid to afford (–)-deacetylanisomycin (**8**). Compound **8** was subjected to a five-step sequence[2,24,48–50] to produce **1**.

1.3.1.2 *Synthesis from* L*-arabinose* (–)-Anisomycin (**1**) was prepared from 2,3,5-tri-*O*-benzyl-β-L-arabinofuranose (**9**) (Scheme 2).[51] Compound **9** was treated with benzylamine

D-Galactose →(a) **2** (OH, HO, OEt, BnO, OBn) →(b) **3** (H_3CO, OEt, HO BnO, OBn) ↓(c) **4** (H_3CO, OH, BnO, OBn) ←(d) **5** (H_3CO, OH, =NOH, BnO, OBn) ↓(e) **6** (H_3CO, OMs, CN, BnO, OBn) →(f) RO, OR, N, H, OCH_3 →(h) **1**

g: **7**) R = Bn → **8**) R = H

Scheme 1 (*a*) Refs. 45–47. (*b*) 1. EtOH–H_2O (1:1), $NaIO_4$, rt, 3 h, 95%; 2. 4-$CH_3OC_6H_4MgBr$, ether, boiling, 2 h, 87%. (*c*) 1. Et_3SiH, CH_2Cl_2, TFA, rt, 48 h, 76%; 2. 80% acetic acid, reflux, overnight, 97%. (*d*) $HONH_2{\cdot}HCl$, Py, EtOH, reflux, 5 h, 86%. (*e*) Py, MsCl, rt, 1 h; then 60–70°C, 2 h, 71%. (*f*) 1 M BH_3 in THF, reflux, 3 h; then 2 M HCl, reflux, 20 min, 65%. (*g*) 10% Pd *on* C, H_2, EtOH, HCO_2H, sonication, 90 min, 77%. (*h*) 1. CH_2Cl_2, Na_2CO_3, CbzCl, 2.5 h, 72%; 2. DMF, imidazole, TBSCl, rt, 1 h, 80%; 3. Ac_2O, Py, rt, 3 days, 96%; 4. 0°C, THF, 1 M TBAF, 30 min, 85%; 5. EtOH, 10% Pd *on* C, H_2, 15 min, 95%.

9 (OH, BnO, BnO, OBn) →(a) **10** (NHBn, BnO, BnO, OBn) →(b) **11** (OH, BnHN, H, OCH_3, BnO, BnO, OBn) ↓(c) Bn, N, O, RO, OR, OCH_3 ←(e) R, N, HO, OH, OCH_3 ←(g) **1**

d: **12**) R = Bn → **13**) R = H

f: **14**) R = H → **15**) R = Cbz

Scheme 2 (*a*) $BnNH_2$, CH_2Cl_2, MS 4 Å, quantitative. (*b*) 4-$CH_3OC_6H_4CH_2MgCl$, –78 to 0°C, THF, 78%. (*c*) PCC, MS 4 Å, CH_2Cl_2, 63%. (*d*) Pd black, HCO_2H, CH_3OH, 99%. (*e*) 1. $LiAlH_4$, THF, 91%; 2. Pd black, HCO_2H, CH_3OH. (*f*) CbzCl, $NaHCO_3$, CH_3OH, 77% for two steps. (*g*) 1. DMF, imidazole, TBSCl, rt, 1 h, 80%; 2. Ac_2O, Py, rt, 3 days, 96%; 3. 0°C, THF, 1 M TBAF, 30 min, 85%; 4. EtOH, 10% Pd *on* C, H_2, 15 min, 95%.

to afford the furanosylamine **10**, which was smoothly reacted with *p*-methoxybenzylmagnesium chloride at low temperature to provide the adduct **11**.[52] Oxidative degradation of **11** with PCC afforded the functionalized lactam **12**. Removal of the *O*-benzyl groups followed by reduction of the resulting lactam **13** with $LiAlH_4$ and subsequent hydrogenation afforded the corresponding amino alcohol intermediate **14**, which was treated with CbzCl in the presence of $NaHCO_3$ to give the carbamate **15**. Conversion of **15** into **1** has been achieved as reported.[50–52]

1.3.1.3 *Synthesis from* D*-ribose* D-Ribose was also used for the synthesis of (–)-anisomycin (**1**) and its phenyl analogue (Scheme 3).[53,54] Thus, 2,3-*O*-isopropylidene-D-ribose (**16**)[55] was treated with the required Grignard reagents to give the triols **17** (77%) and **18** (70%); only the D-allo stereoisomer could be isolated in each case. Periodate oxidation of **17** and **18** gave the corresponding hemiacetals **19** and **20**, respectively. Reaction of **19** and **20** with hydroxylamine hydrochloride in pyridine furnished the corresponding oximes **21** and **22**, which upon treatment with methanesulfonyl chloride in pyridine gave the corresponding nitriles **23** and **24**. $LiAlH_4$ reduction of **23** and **24** led to the corresponding pyrrolidines **25** and **26** in 48 and 42% overall yield, respectively, from the hemiacetals **19** and **20**. Acid hydrolysis of the isopropylidene group in **25** and **26** followed by treatment with HBr in glacial acetic acid gave the bromoacetates (**27**, **29**) and (**28**, **30**), which were converted into the epoxides **31** and **32** by treatment with aqueous methanolic potassium hydroxide. The epoxides **31** and **32** were reacted with allyl alcohol in the presence of perchloric acid to give the allyl ethers (**33** and **34**), which were N-benzylated, O-acetylated and finally hydrogenated to produce **35** and **1** in an overall yield 10 and 7%, respectively, from **16**.

1.3.1.4 *Synthesis from* L*-threose* (–)-Anisomycin (**1**) has been prepared from the L-threo-furanose **36** (Scheme 4).[50] 2,3-*O*-Bis(methoxymethyl)-L-threo-furanose (**36**), obtained from diethyl L-tartarate, was treated with (4-methoxybenzyl)magnesium chloride to furnish a mixture of the two diastereomers, xylo **37** and lyxo **38**, in a ratio of 79:21. This mixture was selectively benzylated on the primary hydroxyl group followed by Swern oxidation of the secondary hydroxyl group to produce the ketone **39**, which upon reduction with $Zn(BH_4)_2$ afforded the desired lyxo isomer **40** in 40% yield from **36**. Removal of the benzyl group from **40** furnished **38** that underwent mesylation followed by S_N2 displacement of the primary mesylate group with sodium azide to produce the azide **41** whose hydrogenation gave **42**. Removal of the MOM groups from **42** with conc. HCl followed by protection of the secondary amine with benzyl chloroformate afforded the carbamate **15**. Selective silylation of the 4-hydroxy group of **15** followed by acetylation of C-3 hydroxyl group afforded the carbamate **43**, which upon removal of the Cbz and TBS groups furnished **1**.

1.3.1.5 *Synthesis from* L*-threitol* The L-threitol was used for the syntheses of deacetylanisomycin and its derivative via a highly selective addition of organolithium or Grignard reagents to the L-threose imine **47** (Scheme 5).[55] Thus, the aldehyde **46** was prepared[56] in two steps from the commercially available 2-*O*-benzyl-L-threitol **44** via conversion to **45**. Condensation of **46** with benzylamine gave quantitatively the *N*-benzyl imine **47**, which was treated with 3 equiv. of organolithium compound to afford the aminotriols **48** and **49** in a ratio of 95:5. The diastereomer **48** was transformed into 2-substituted *trans*-dihydroxypyrrolidine **50** in 77% yield, either by an intramolecular Mitsunobu reaction or by cyclization using the

Scheme 3 (*a*) $ArCH_2MgCl$, THF, rt, 1 h, **17** (77%), **18** (70%). (*b*) 1. $NaIO_4$, EtOH, H_2O, 30 min, **19** (93%), **20** (76%). (*c*) Py, $NH_2OH{\cdot}HCl$, 3 h, **21** (99%). **22** (94%). (*d*) Py, MsCl, 1 h, 60°C, **23** (91%), **24** (88%). (*e*) Ether, $LiAlH_4$, 2 h; then EtOAc, 1 h, **25** (54%), **26** (51%). (*f*) 1. CH_3OH, 1 M HCl, reflux, 3 h; 2. glacial AcOH, HBr, 50°C, 1 h. (*g*) CH_3OH, H_2O, KOH, 10 min, **31** (68%), **32** (70%). (*h*) Allyl alcohol, $CHCl_3$, perchloric acid (70%), 60°C, 36–48 h, **33** (63%), **34** (67%). (*i*) 1. $CHCl_3$, BnBr, NEt_3, 2 h; then Ac_2O, 60°C, 5 h; 2. CH_3OH, 2 M HCl, 10% Pd *on* C, reflux, 48 h; then 3 h under H_2, **35** (66%), **1** (68%).

PPh_3 and CCl_4 in NEt_3.[57,58] Hydrogenation of the anisylmethyl derivative **50** afforded the deacetylanisomycin hydrobromide **14** in 34% overall yield, which could be converted[51] to anisomycin (**1**) in 45% yield.

(–)-Anisomycin was also prepared from L-threitol derivative **51** in 17–20% overall yield (Scheme 6).[59] Swern oxidation of **51**[60] followed by Wittig methylenation, acidic

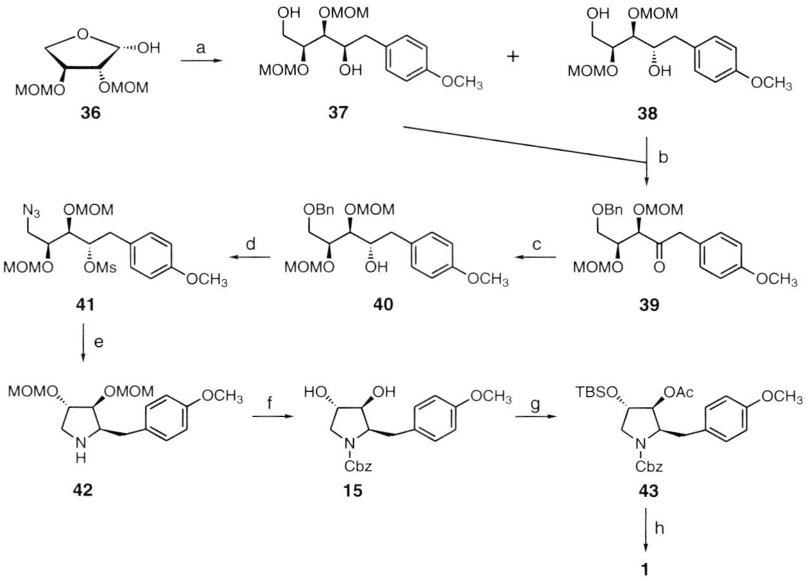

Scheme 4 (*a*) 4-$CH_3OC_6H_4CH_2MgCl$, THF, rt, 14 h, 69%, **37**:**38**, 79:21. (*b*) 1. *n*-Bu_4NBr, 6 N NaOH, CH_2Cl_2, BnCl, CH_2Cl_2, rt to 60°C, 24 h, 69%; 2. $(COCl)_2$, DMSO, CH_2Cl_2, –78°C; then NEt_3, 93%. (*c*) $Zn(BH_4)_2$, ether, 0°C, 10 min to rt, 50 min, 91%. (*d*) 1. CH_3OH, 10% Pd *on* C, H_2, 1 h, 100%; 2. NEt_3, CH_2Cl_2, MsCl, 10 min, 87%; 3. NaN_3, DMF, 80°C, 0.5 h, 45%. (*e*) CH_3OH, 10% Pd *on* C, H_2, 95%. (*f*) 1. CH_3OH–HCl–H_2O (2:1:1), reflux, 20 h, 81%; 2. CbzCl, CH_2Cl_2, Na_2CO_3, 2.5 h, 72%. (*g*) 1. DMF, imidazole, TBSCl, rt, 1 h, 80%; 2. Ac_2O, Py, rt, 3 days, 96%. (*h*) 1. 0°C, THF, 1 M TBAF, 30 min, 85%; 2. EtOH, 10% Pd *on* C, H_2, 15 min, 95%.

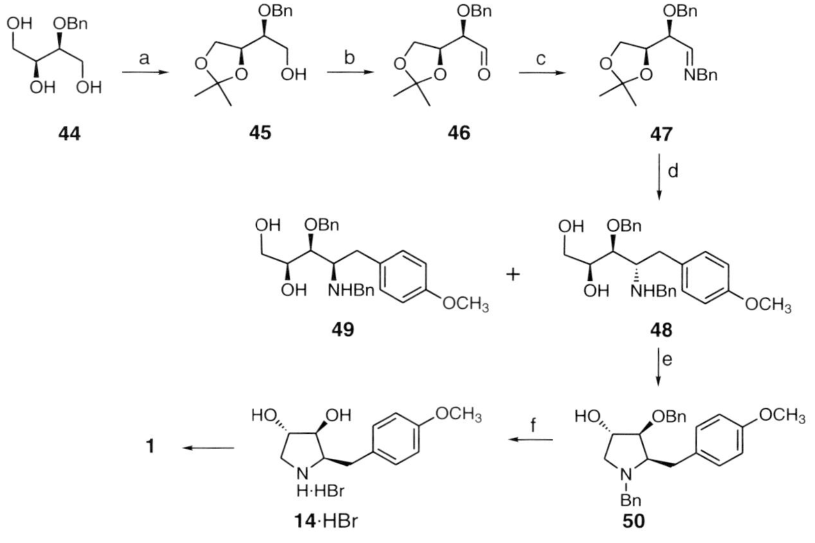

Scheme 5 (*a*) Ref. 56. (*b*) CrO_3, Py, 74%. (*c*) Al_2O_3 (63–200 μm), $BnNH_2$, rt, 1 h, quantitative. (*d*) 1. RLi, ether, –78°C, 15 min; *or* RMgX, ether, 0°C to rt, 2–5 h; aqueous NH_4Cl; 2. HCl, H_2O–dioxane (1:1), 62%. (*e*) PPh_3, DEAD, Py, 0°C, 1.5 h; H_2O, LiOH, dioxane, 80°C; *or* PPh_3, CCl_4, NEt_3, DMF, rt, CH_3OH. (*f*) H_2 (4 bar), Pd *on* C, CH_3OH, HCl, rt, 3 days; *or* H_2 (4 bar), $Pd(OH)_2$ *on* C, CH_3OH, rt, 2 days; HBr, rt, 1 day, 99%.

hydrolysis and protection with CCl_3CN afforded the olefin **52**, whose reaction with iodine in the presence of sodium hydrogencarbonate gave 4.5:1 mixture of dihydro-1,3-oxazine **53** and oxazoline **54**. Hydrolysis of the mixture followed by N-protection with $(Boc)_2O$ afforded a 12:1 mixture of the carbamates **55** and **56**. Alternatively, when **52** was reacted with iodine monobromide in the presence of potassium carbonate, only the six-membered heterocycle **53** was obtained that could be transformed into a 37:1 mixture of **55** and **56**. Subsequent separation and isopropylidenation of **55** followed by LDA furnished a 3:1 mixture of aziridines **57** and **58**. Reaction of the mixture with 4-methoxyphenylmagnesium bromide in the presence of copper(I) bromide–dimethyl sulfide complex in toluene followed by acidic hydrolysis and N-protection produced the carbamate **59**. The pyrrolidine ring was formed from **59** by using DEAD and triphenylphosphine in the presence of PPTs to produce **60**. Silylation of **60** and then acetylation afforded **61**, which underwent acidic hydrolysis to furnish **1**.

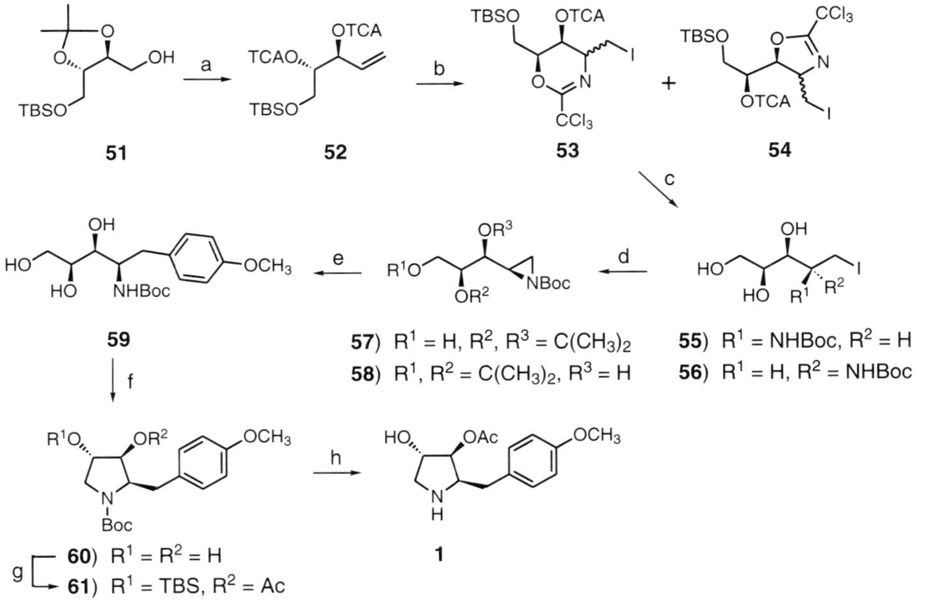

Scheme 6 (*a*) 1. $(COCl)_2$, DMSO, CH_2Cl_2, –78°C; then NEt_3; 2. $Ph_3P^+CH_3I^-$, *n*-BuLi, THF; 3. 2 M HCl, THF, 20°C, 78% for three steps; 4. CCl_3CN, DBU, CH_3CN, –30°C. (*b*) I_2, $NaHCO_3$, CH_3CN, 0°C, 4.5:1 mixture of **53** and **54**; *or* IBr, K_2CO_3, CH_3CH_2CN, –60°C. (*c*) 1. 6 M HCl, CH_3OH, 20°C; 2. $(Boc)_2O$, $NaHCO_3$, CH_3OH, 0°C, 75% for four steps. (*d*) 1. Acetone, TFA, 20°C; 2. LDA, THF, –20°C, 77% for two steps. (*e*) 1. 4-$CH_3OC_6H_4MgBr$, CuBr, DMS, toluene, –30°C; 2. TFA, 20°C; 3. $(Boc)_2O$, $NaHCO_3$, CH_3OH, 0°C; 81% for three steps. (*f*) DEAD, Ph_3P, PPTs, THF, 0°C. (*g*) 1. TBSCl, imidazole, DMF, 20°C; 2. Ac_2O, DMAP, NEt_3, CH_2Cl_2, 20°C, 56% from **59**. (*h*) 6 M HCl, CH_3OH, 20°C, 88%.

(–)-Deacetylanisomycin (**14**) has been synthesized from **62** by transformation into the primary amide **63** by reaction with ammonia in ethanol, followed by dehydration with trifluoroacetic anhydride to give the nitrile **64** (Scheme 7).[61] Treatment of **64** with 4-$CH_3OC_6H_4CH_2MgCl$ followed by $BnNH_2$ and subsequent reduction with sodium borohydride afforded the diastereomeric *N*-benzylamines **65** and **66** in 81:19 ratio and 80%

overall yield from **64**. Compound **65** was deprotected to give **67**, mesylated, deisopropylidenated and then intramolecularly cyclized to furnish *N*-benzyl deacetylanisomycin **68**. Hydrogenation of **68** gave **14**.

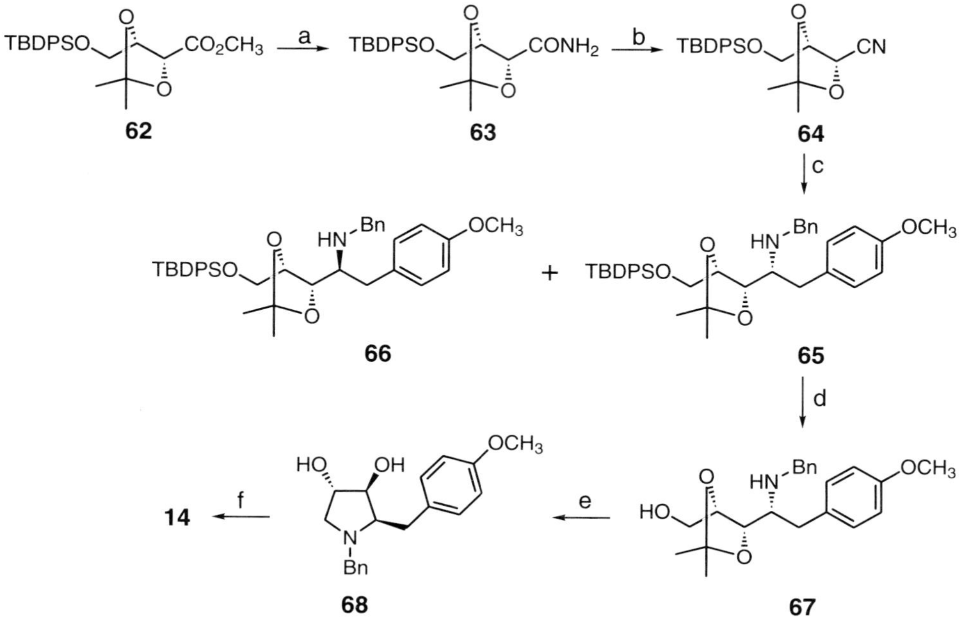

Scheme 7 (*a*) NH_3, EtOH, 90%. (*b*) TFAA, Py, CH_2Cl_2, 87%. (*c*) 1. 4-$CH_3OC_6H_4CH_2MgCl$, ether; 2. $BnNH_2$, CH_3OH; 3. $NaBH_4$, 80% for three steps. (*d*) TBAF, THF, 80%. (*e*) 1. MsCl, NEt_3, cat. DMAP, CH_2Cl_2; 2. 10% HCl, THF; then $NaHCO_3$, 80% for two steps. (*f*) H_2, 10% Pd *on* C, AcOEt, 90%.

1.3.1.6 *Synthesis from* D*-mannitol* An enantioconvergent synthesis[62] of (–)-anisomycin (**1**) employed (*R*)- (**69**) and (*S*)- (**77**) enantiomers of epichlorohydrin[63] that could be obtained from D-mannitol (Schemes 8 and 9). (*R*)-Epichlorohydrin (**69**) was first transformed to (*R*)-*O*-benzylglycidol (**71**) in 60% overall yield, by treatment with benzyl alcohol in the presence of boron trifluoride etherate followed by cyclization of the resulted chlorohydrin **70**. Then, **71** was treated with 4-methoxyphenyllithium to give **72** (98%), which on catalytic debenzylation followed by benzylidenation gave the benzylidene acetal **73**. The latter was treated with *N*-bromosuccinimide to give the bromobenzoate **74**, whose methanolysis in the presence of potassium carbonate afforded the epoxide **75** in 67% overall yield. Alternatively, treatment of (*S*)-epichlorohydrin (**77**) with 4-methoxyphenyllithium in the presence of copper(I) cyanide afforded the chlorohydrin **76**, which was immediately exposed to methanolic potassium carbonate to give (*S*)-(4-methoxybenzyl)oxirane (**75**) in 74% overall yield.

Treatment of **75** with lithium acetylide ethylenediamine complex afforded the acetylene derivative **78** (85%), which was transformed into the vinyl alcohol **79** by partial hydrogenation using Lindlar catalyst. Employing the Mitsunobu reaction, compound **79** was transformed into the phthalimide **80**, which was converted into the benzamide **82** (64%) via the primary amine **81** by sequential deacylation and benzoylation. When the

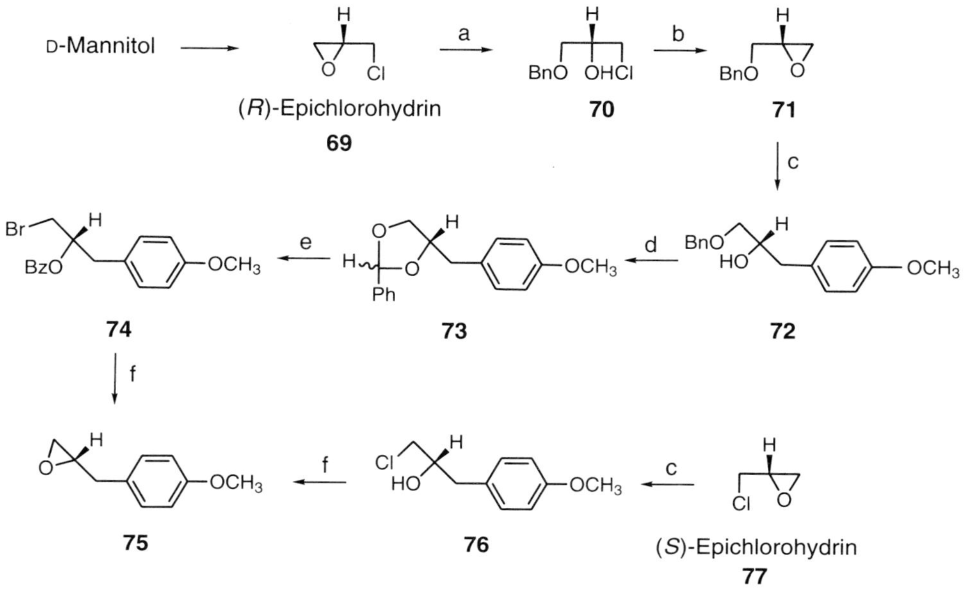

Scheme 8 (*a*) $BF_3{\cdot}OEt_2$, BnOH, 50°C. (*b*) NaOH, H_2O, Et_2O. (*c*) 4-Bromoanisole, *n*-BuLi, CuCN, THF, –78°C. (*d*) 1. H_2, $Pd(OH)_2$, CH_3OH; 2. PhCHO, *p*-TsOH, benzene, reflux. (*e*) NBS, CCl_4. (*f*) K_2CO_3, CH_3OH.

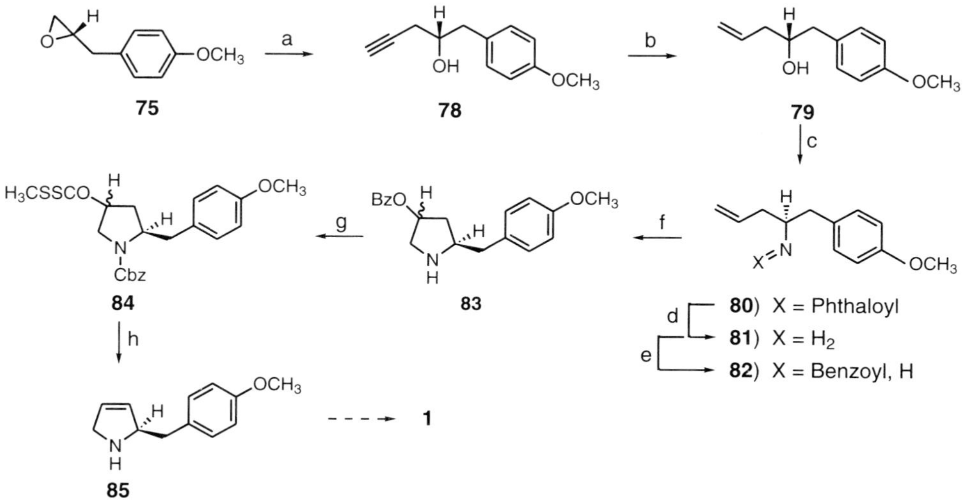

Scheme 9 (*a*) Lithium acetylide ethylenediamine complex, DMSO, rt. (*b*) H_2, Pd, $CaCO_3$, AcOEt. (*c*) Phthalimide, diisopropyl azodicarboxylate, PPh_3, THF, –20°C. (*d*) H_2NNH_2, EtOH, reflux. (*e*) BzCl, NEt_3, CH_2Cl_2. (*f*) I_2, H_2O, CH_3CN. (*g*) 1. CbzCl, NEt_3, CH_2Cl_2; 2. K_2CO_3, CH_3OH; 3. CS_2, NaOH, *n*-Bu_4NHSO_4; then CH_3I, benzene, 87%. (*h*) 1. ODB, reflux, 70%; 2. NaOH, $(CH_2OH)_2$, 120°C, 89%.

amide **82** was exposed to 3 equiv. of iodine in aqueous acetonitrile, 2-(4-methoxybenzyl)-4-benzoyloxypyrrolidine (**83**) was obtained in 90% yield in a single step as a 2:1 mixture of epimers at C-4 center. The mixture of **83** was N-protected, debenzoylated and the resulting hydroxyl group was converted into the xanthate **84** in 87% overall yield. Thermolysis of **84** in *o*-dichlorobenzene followed by alkaline hydrolysis furnished the secondary amine **85** in 89% yield, a precursor for the natural (–)-anisomycin (**1**).[24,64]

References

1. Sobin, B.A.; Tanner, F.W., Jr., *J. Am. Chem. Soc.* **76** (1954) 4053.
2. Buchanan, J.G.; Maclean, K.A.; Wightman, R.H.; Paulsen, H. *J. Chem. Soc., Perkin Trans. 1* (1985) 1463.
3. Hosoya, Y.; Kameyama, T.; Naganawa, H.; Okami, Y.; Takeuchi, T. *J. Antibiot.* **46** (1993) 1300.
4. Lynch, J.E.; English, A.R.; Bauck, H.; Deligianis, H. *Antibiot. Chemother.* **4** (1954) 844.
5. Frye, W.W.; Mule, J.G.; Swartzwelder, C. *Antibiot. Ann.* (1955) 820.
6. Armstrong, T.; Santa Maria, O. *Antibiot. Ann.* (1955) 824.
7. Grollman, A.P. *J. Biol. Chem.* **242** (1967) 3226.
8. Jiimenez, A.; Vazquez, D. In *Antibiotics*, Vol. 5; Hahn, F.E., Ed.; Springer Verlag: Berlin, 1979; pp. 1–19.
9. Beereboom, J.J.; Butler, K.; Pennington, F.C.; Solomons, I.A. *J. Org. Chem.* **30** (1965) 2334.
10. Butler, K. *J. Org. Chem.* **33** (1968) 2136.
11. Wong, C.M. *Can. J. Chem.* **46** (1968) 1101.
12. Korzybski, T.; Kowszyk-Gindifer, Z.; Kurytowicz, W. In *Antibiotics*, Vol. 1; American Society of Microbiology: Washington, DC, 1978; p. 343.
13. Santander, V.M.; Cue, A.B.; Diaz, J.G.H.; Balmis, F.J.; Miranda, G.G.; Urbina, E.; Portilla, J.; Plata, A.A.; Zapata, H.B.; Munoz, V.A.; Abreu, L.M. *Rev. Invest. Bol. Univ. Guadalajara 1* (1961) 94.
14. Windholz, M., Ed. *The Merck Index*, 9th ed.; Merck: Rahway, NJ, 1976; p. 91.
15. Grollman, A.P. *Proc. Natl. Acad. Sci. USA* **56** (1966) 1867.
16. Vasquez, D. *FEBS Lett.* **40** (1974) 63.
17. Barbacid, M.; Vasquez, D. *J. Mol. Biol.* **84** (1974) 603.
18. Van der Bosch, J.; Rueller, S.; Schlaak, M. German Patent DE 19,744,676, 1999; *Chem. Abstr.* **130** (1999) 291581.
19. Rueller, S.; Stahl, C.; Kohler, G.; Eickhoff, B.; Breder, J.; Schlaak, M.; Van der Bosch, J. *Clin. Cancer Res.* **5** (1999) 2714.
20. Dudai, Y. *Nature* **406** (2000) 686.
21. Nader, K.; Schafe, G.E.; Le Doux, J.E. *Nature* **406** (2000) 722.
22. Törocsik, B.; Szeberényi, J. *Eur. J. Neurosci.* **12** (2000) 527.
23. Schaefer, J.P.; Wheatley, P.J. *J. Org. Chem.* **33** (1968) 166.
24. Schumacher, D.P.; Hall, S.S. *J. Am. Chem. Soc.* **104** (1982) 6076.
25. Hall, S.S.; Loebenberg, D.; Schumacher, D.P. *J. Med. Chem.* **26** (1983) 469.
26. Jegham, S.; Das, B.C. *Tetrahedron Lett.* **29** (1988) 4419.
27. Ballini, R.; Marcantoni, E.; Petrini, M. *J. Org. Chem.* **57** (1992) 1316.
28. Tokuda, M.; Fujita, H.; Miyamoto, T.; Suginome, H. *Tetrahedron* **49** (1993) 2413.
29. Shi, Z.C.; Lin, G.Q. *Tetrahedron: Asymmetry* **6** (1995) 2907.
30. Kim, G.; Hong, H.W.; Lee, S.H. *Bull. Korean Chem. Soc.* **19** (1998) 37.
31. Kim, G.; Hong, H.W.; Lee, S.H. *Bull. Korean Chem. Soc.* **20** (1999) 321.
32. Delair, P.; Brot, E.; Kanazawa, A.; Greene, A.E. *J. Org. Chem.* **64** (1999) 1383.
33. Oida, S.; Ohki, E. *Chem. Pharm. Bull.* **17** (1969) 1405.
34. Verheyden, J.P.H.; Richardson, A.C.; Bhatt, R.S.; Grant, B.D.; Flitch, W.L.; Moffat, J.G. *Pure Appl. Chem.* **50** (1978) 1363.
35. Shono, T.; Kise, N. *Chem. Lett.* (1987) 697.
36. Takahata, H.; Banba, Y.; Tajima, M.; Momose, T. *J. Org. Chem.* **56** (1991) 240.
37. Ikota, N. *Heterocycles* **41** (1995) 983.
38. Han, G.; LaPorte, M.G.; Mclntosh, M.C.; Weinreb, S.M.; Parvez, M. *J. Org. Chem.* **61** (1996) 9483.
39. Huang, P.Q.; Wang, S.L.; Ruan, Y.P.; Gao, J.X. *Nat. Prod. Lett.* **11** (1998) 101.
40. Schwartdt, O.; Veith, U.; Gaspard, C.; Jager, V. *Synthesis* (1999) 1473.
41. Wang, Y.; Ma, D. *Tetrahedron: Asymmetry* **12** (2001) 725.
42. Hulme, A.N.; Rosser, E.M. *Org. Lett.* **4** (2002) 265.

43. Chang, M.-Y.; Chen, S.-T.; Chang, N.-C. *Heterocycles* **60** (2003) 1203.
44. Baer, H.H.; Zamkanei, M. *J. Org. Chem.* **53** (1988) 4786.
45. Thiem, J.; Wessel, H.-P. *Justus Liebigs Ann. Chem.* (1983) 2173.
46. Green, J.W.; Pacsu, E. *J. Am. Chem. Soc.* **59** (1937) 1205.
47. Pascu, E. *Methods Carbohydr. Chem.* **2** (1963) 354.
48. Felner, I.; Schenker, K. *Helv. Chim. Acta* **53** (1970) 754.
49. Wong, C.M.; Ruccini, J.; Chang, I.; Te Raa, J.; Schwenk, R. *Can. J. Chem.* **47** (1969) 2421.
50. Iida, H.; Yamazaki, N.; Kibayashi, C. *J. Org. Chem.* **51** (1986) 1069.
51. Yoda, H.; Nakajima, T.; Yamazaki, H.; Takabe, K. *Heterocycles* **41** (1995) 2423.
52. Nagai, I.; Gaudino, J.J.; Wilcox, C.S. *Synthesis* (1992) 163.
53. Buchanan, J.G.; Maclean, K.A.; Paulsen, H.; Wightman, R.H. *J. Chem. Soc., Chem. Commun.* (1983) 486.
54. Hughes, N.A.; Speakman, P.R.H. *Carbohydr. Res. 1* (1965) 171.
55. Veith, U.; Schwardt, O.; Jäger, V. *Synlett* (1996) 1181.
56. Valverde, S.; Herradon, B.; Martin-Lomas, M. *Tetrahedron Lett.* **26** (1985) 3731.
57. Appel, R.; Kleinslück, R.; *Chem. Ber.* **107** (1974) 5.
58. Rassu, G.; Casiraghi, G.; Pinna, L.; Spanu, P.; Ulgheri, F. *Tetrahedron* **49** (1993) 6627.
59. Kang, S.H.; Choi, H.-W. *Chem. Commun.* (1996) 1521.
60. Savage, I.; Thomas, E.J. *J. Chem. Soc., Chem. Commun.* (1989) 717.
61. Hutin, P.; Haddad, M.; Larchevêque, M. *Tetrahedron: Asymmetry* **11** (2000) 2547.
62. Takano, S.; Iwabuchi, Y.; Ogasawara, K. *Heterocycles* **29** (1989) 1861.
63. Baldwin, J.J.; Raab, A.W.; Mensler, K.; Arison, B.H.; McClure, D.E. *J. Org. Chem.* **43** (1978) 4876.
64. Meyers, A.I.; Dupre, B. *Heterocycles* **25** (1987) 113.

1.3.2 *(+)-Preussin*

(+)-Preussin (L-657,398, **1**) was first isolated in 1988 from the fermentation broths of *Aspergillus ochraceus* ATCC 22947[1,2] and then from *Preussia* sp.[3] It possesses a broad spectrum of potent antifungal activity against both filamentous fungi and yeasts, significantly broader than the structurally related pyrrolidine anisomycin.[1] Many syntheses of (+)-preussin and its analogues from noncarbohydrates as starting materials have been reported;[4–12] those from carbohydrates will be presented in this review.

(+)-Preussin
1

1.3.2.1 *Synthesis from D-glucose*

The first total synthesis of (+)-preussin (**1**) was reported from D-glucose (Scheme 1).[13] Epoxyfuranose **2**, obtained from D-glucose,[14,15] underwent Grignard reaction to give the alcohol **3**. Tosylation of **3** followed by nucleophilic

Scheme 1 (*a*) Refs. 14 and 15. (*b*) PhMgCl, CuI, THF. (*c*) 1. *p*-TsCl, Py, 45°C, 95%; 2. NaN_3, DMSO, 80°C, **4** (8%), **5** (90%). (*d*) 1. Anhydrous HCl, CH_3OH, β (84%), α (16%); 2. Tf_2O, Py, CH_2Cl_2, –30°C, β (95%), α (98%). (*e*) 1. H_2, Pd black, EtOAc, rt, β (65%), α (70%); 2. $ClCO_2CH_3$, Py, CH_2Cl_3, 0°C, (86%). (*f*) 1. 0.5 M HCO_2H, THF, H_2O, reflux; 2. *n*-$C_8H_{17}P^+Ph_3I^-$; *n*-BuLi, THF, HMPA. (*g*) 1. H_2, Pd *on* C, EtOH, 100%; 2. $LiAlH_4$, THF, reflux, 100%.

substitution of the resulting tosylate with azide ion afforded a chromatographically separable mixture of **4** (8%) and **5** (90%). Methanolysis of **5** gave a mixture of anomeric furanosides, which was triflated to afford **6**. Reduction of the azido group of **6** to its corresponding primary amine led to an intramolecular nucleophilic displacement of the triflate, causing cyclization to methoxy bicyclic amine, whose protection with methyl chloroformate gave **7**. Hydrolysis of **7** with formic acid followed by Wittig reaction afforded a mixture of Z-(81%) and E- (9%) isomers **8**. Hydrogenation of both isomers followed by reduction with $LiAlH_4$ afforded **1**.

1.3.2.2 *Synthesis from* D*-mannose* Synthesis of an intermediate precursor to (+)-preussin from D-mannose has been reported (Scheme 2).[16] The epoxide **9**, obtained from D-mannose in 60% overall yield, was treated with PhMgCl to give the secondary alcohol **10**. The trifluoromethane sulfonate ester of **10** was treated with sodium azide to give **11**. Reduction of **11** followed by benzyloxycarbonylation of the resulting amine gave the carbamate **12**. Removal of the anomeric protecting group from **12** gave **13** whose ionic cyclization with $PhIO–I_2$ afforded the cyclic derivative **14**. Treatment of **14** with allyltrimethylsilane in the presence of $BF_3{\cdot}OEt_2$ gave **15** and **17** (65%) in a ratio >95:5, whereas a combination of $BF_3{\cdot}OEt_2$ and TMSOTf led to a mixture of **16** and **18** in 92% yield but the diastereoselectivity decreased to 70:30. The carbon chain in **16** was extended by oxidative cleavage of the alkene

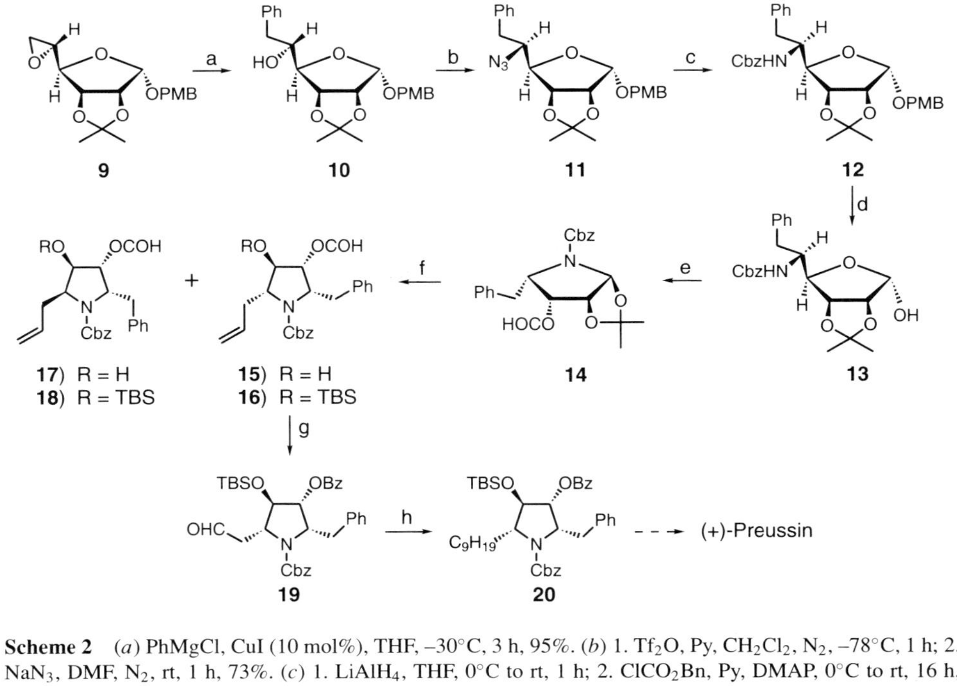

Scheme 2 (*a*) PhMgCl, CuI (10 mol%), THF, –30°C, 3 h, 95%. (*b*) 1. Tf_2O, Py, CH_2Cl_2, N_2, –78°C, 1 h; 2. NaN_3, DMF, N_2, rt, 1 h, 73%. (*c*) 1. $LiAlH_4$, THF, 0°C to rt, 1 h; 2. $ClCO_2Bn$, Py, DMAP, 0°C to rt, 16 h, 75%. (*d*) DDQ, $CH_2Cl_2–H_2O$ (19:1), N_2, rt, 2 h, 85%. (*e*) PhIO (3 mmol), I_2 (1 mmol), CH_2Cl_2, rt, 2 h, 78%. (*f*) $CH_2CHCH_2Si(CH_3)_3$; Lewis acid (4 mmol); *or* $BF_3{\cdot}OEt_2$, TBSOTf, 92%. (*g*) 1. $NaCO_3$, CH_3OH, rt, 45 min; 2. BzCl, Py, rt, 16 h; 3. OsO_4, MNO, H_2O, acetone, *t*-BuOH; 4. H_2O, $NaIO_4$. (*h*) 1. $CH_3(CH_2)_6{}^+PPh_3I^-$, BuLi, THF, –78°C, 10 min; 2. 10% Pd *on* C, H_2 (1 atom), EtOH, rt, 41% from **16**.

to the aldehyde **19** followed by Wittig reaction and hydrogenation to give **20** in 41% overall yield from **16**.

1.3.2.3 *Synthesis from D-arabinose* (+)-Preussin (**1**) was also synthesized from 2,3,5-tri-*O*-benzyl-β-D-arabinofuranose (**21**)[18–20] by conversion to the *N*-*p*-methoxybenzyl lactam **22** (Scheme 3).[17] Treatment of **22** with CAN followed by $(Boc)_2O$ gave lactam **23**. Removal of the benzyl groups from **23** followed by regioselective acylation with PhOCSCl and subsequent radical deoxygenation afforded the lactam **24**. Compound **24** was then silylated followed by Grignard reaction with nonylmagnesium bromide and subsequent reductive deoxygenation with Et_3SiH to afford the lactam **25**, which underwent desilylation followed by reduction of the carbamate with $LiAlH_4$ to give **1** in 18% overall yield from **21**.

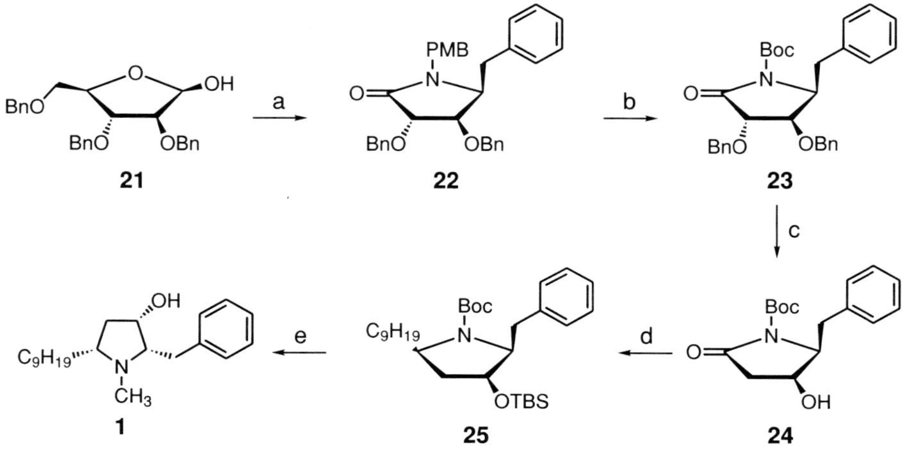

Scheme 3 (*a*) 1. $PMBNH_2$, PhH, MS 4 Å, reflux, 100%; 2. BnMgCl, –78°C, THF; 3. PCC, MS 4 Å, CH_2Cl_2, 59% for two steps. (*b*) 1. $Ce(NH_4)_2(NO_2)_6$, CH_3CN, H_2O, 78%; 2. $(Boc)_2O$, NEt_3, DMAP, CH_2Cl_2, 100%. (*c*) 1. Pd black, HCO_2H, CH_3OH, 100%; 2. PhOCSCl, Py, DMAP, CH_3CN; 3. Bu_3SnH, AIBN, toluene, 90°C, 72% for two steps. (*d*) 1. TBSCl, imidazole, DMF, 91%; 2. $C_9H_{19}MgBr$, –78°C, THF; 3. Et_3SiH, $BF_3{\cdot}OEt_2$, –40 to –30°C, CH_2Cl_2, 67% for two steps. (*e*) TBAF, THF, 97%; $LiAlH_4$, THF, –78°C, 92%.

References

1. Schwartz, R.E.; Liesch, J.; Hensens, O.; Zitano, L.; Honeycutt, S.; Garrity, G.; Fromtling, R.A.; Onishi, J.; Monaghan, R.L. *J. Antibiot.* **41** (1988) 1774.
2. Schwartz, R.E.; Onishi, J.C.; Monaghan, R.L.; Liesch, J.M.; Hensens, O.D. U.S. Patent 4,847,284, 1989.
3. Johnson, J.H.; Philippson, D.W.; Kahle, A.D. *J. Antibiot.* **42** (1989) 1184.
4. Kanazawa, A.; Gillet, S.; Delair, P.; Greene, A.E. *J. Org. Chem.* **63** (1998) 4663.
5. Deng, W.; Overman, L.E. *J. Am. Chem. Soc.* **116** (1994) 11241.
6. Shimazaki, M.; Okazaki, F.; Nakajima, F.; Ishikawa, T.; Ohta, A. *Heterocycles* **36** (1993) 1823.
7. Overhand, M.; Hecht, S.M. *J. Org. Chem.* **59** (1994) 4721.
8. McGrane, P.L.; Livinghouse, T. *J. Am. Chem. Soc.* **115** (1993) 11485.
9. Verma, R.; Ghosh, S.K. *J. Chem. Soc., Perkin Trans. 1* (1999) 265.
10. Verma, R.; Ghosh, S.K. *Chem. Commun.* (1997) 1601.
11. Okue, M.; Watanabe, H.; Kitahara, T. *Tetrahedron* **57** (2001) 4107.
12. Krasiński, A.; Gruza, H.; Jurczak, J. *Heterocycles* **54** (2001) 581.

13. Pak, C.S.; Lee, G.H. *J. Org. Chem.* **56** (1991) 1128.
14. Barton, D.H.R.; McCombie, S.W. *J. Chem. Soc., Perkin Trans. 1* (1975) 1574.
15. Szabo, P.; Szabo, L. *J. Chem. Soc.* (1964) 5139.
16. De Armas, P.; Garcia-Tellado, F.; Marrero-Tellado, J.J.; Robles, J. *Tetrahedron Lett.* **39** (1998) 131.
17. Yoda, H.; Yamazaki, H.; Takabe, K. *Tetrahedron: Asymmetry* **7** (1996) 373.
18. Yoda, H.; Nakajima, T.; Yamazaki, H.; Takabe, K. *Heterocycles* **41** (1995) 2423.
19. Lay, L.; Nicotra, F.; Paganini, A.; Pangrazio, C.; Panza, L. *Tetrahedron Lett.* **34** (1993) 4555.
20. Hashimoto, M.; Terashima, S. *Chem. Lett.* (1994) 1001.

1.4 2-Aryl pyrrolidines

Codonopsinine and codonopsine are aryl pyrrolidines isolated from natural sources and exhibit hypotensive pharmacological activity. L-Threitol is the only carbohydrate used in the synthesis of these aryl pyrrolidines.

1.4.1 *Codonopsinine and codonopsine*

Codonopsinine and codonopsine were isolated in 1969 from *Codonopsis clematidea* by a Russian group[1,2] and exhibit hypotensive pharmacological activity with no effect on the central nervous system in animal tests.[3] Their structural characterization[4,5] revealed that they are simple 1,2,3,4,5-penta-substituted pyrrolidine alkaloids,[6] whose absolute configurations were firstly determined to be as in **1** and **2**, respectively, based on analyses of their ^{1}H NMR coupling constants using the Karplus equation. Later study[7–11] unambiguously determined the stereochemistry of the natural (−)-codonopsinine antibiotic to possess the 2*R*,3*R*,4*R*,5*R* configuration as depicted in **3** instead of **1**, thus establishing structure **4** to be (+)-codonopsinine. The same study also led to stereochemical revision of codonopsine from **2** to **6** and allows the absolute structure of the levorotatory natural product to be assigned as **5**, since the stereostructure of codonopsine has been claimed to be identical with that of codonopsinine. In addition, the structure of (–)-codonopsine (**5**) was confirmed by using X-ray crystallographic analysis.[10] Syntheses of codonopsinine and codonopsine analogues from noncarbohydrates as starting materials have been reported.[7,10,11]

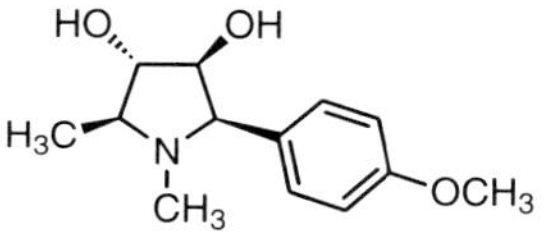

1

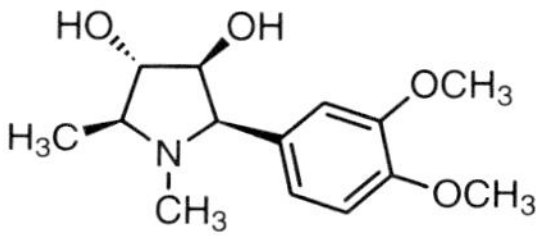

2

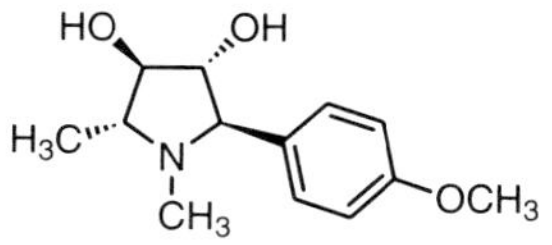

(–)-Codonopsinine
(Natural)
3

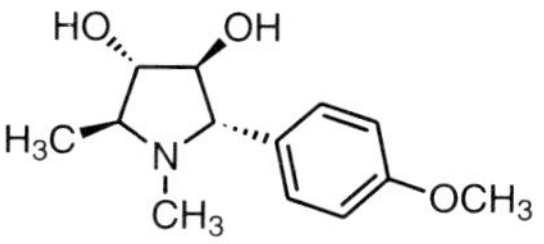

(+)-Codonopsinine
4

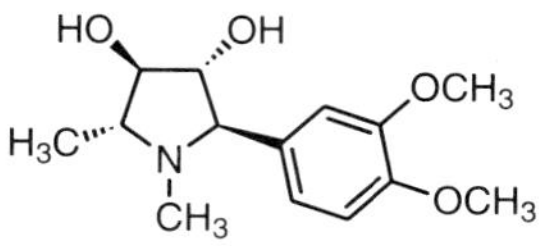

(–)-Codonopsine
(Natural)
5

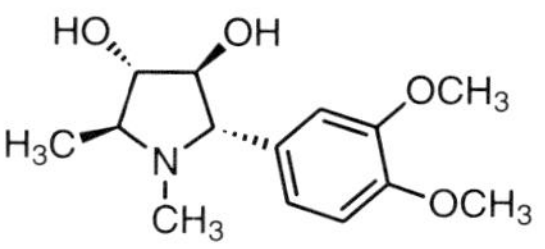

(+)-Codonopsine
6

The first total synthesis of the enantiomerically pure (+)-form **1** starting from L-threitol led to the assignment of the absolute configuration for the natural (–)-codonopsinine as **3** (Schemes 1–4).[8,9] The L-threitol derivative **7**, obtained from L-tartaric acid in 55% overall yield, underwent Swern oxidation to give the aldehyde **8** in 82% yield, which was treated with *p*-methoxyphenylmagnesium bromide to produce a 3.3:1 mixture of the two diastereomeric alcohols **9** and **10**. This mixture was treated, without separation, with phthalimide under Mitsunobu conditions to furnish a separable 1:1 mixture of the isomers **11** and **12**. Removal of the benzyl group from **12** by catalytic hydrogenation followed by Swern oxidation of the resulting hydroxyl group afforded the aldehyde **13**, which was then treated with methylmagnesium bromide to produce the threo alcohol **14**. Removal of the phthaloyl group with hydrazine hydrate followed by protection of the resulting amine with benzyl chloroformate and subsequent mesylation of the secondary hydroxyl group afforded the mesylate **15**. Catalytic hydrogenolysis of **15** led to intramolecular cyclization, which was followed by N-methylation to give **16**. Removal of the protecting groups from **16** with aqueous HCl gave **1**.

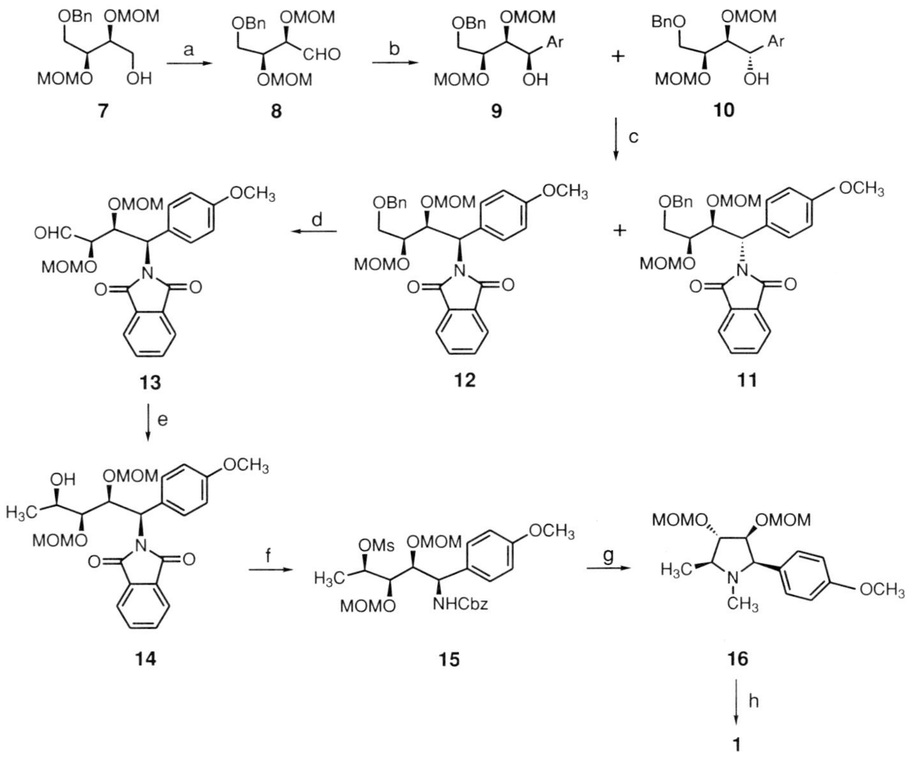

Scheme 1 (*a*) $(COCl)_2$, DMSO, NEt_3, CH_2Cl_2, –78°C, 82%. (*b*) *p*-$CH_3C_6H_4MgBr$, THF, –10°C to rt, 14 h, 83%. (*c*) HNPhth (2.5 equiv.), DEAD, Ph_3P, THF, rt, 14 h, 64%. (*d*) 1. Pd *on* C, H_2, CH_3OH, 70%; 2. $(COCl)_2$, DMSO, NEt_3, CH_2Cl_2, –78°C, 83%. (*e*) CH_3MgBr, Et_2O, –78°C to rt, 14 h, 62%. (*f*) 1. $NH_2NH_2{\cdot}H_2O$, EtOH, reflux; then CbzCl, aqueous Na_2CO_3, CH_2Cl_2, 0°C, 100%; 2. MsCl, NEt_3, CH_2Cl_2, 0°C, 10 min. (*g*) 1. Pd *on* C, H_2, CH_3OH; 2. aqueous HCHO, Pd *on* C, H_2, CH_3OH. (*h*) Aqueous HCl, CH_3OH, 50°C, 2.5 h.

Removal of the benzyl group from **11** followed by Swern oxidation afforded the aldehyde **17**, which was treated with CH_3MgBr to give the separable aldehydes **18** and **19**. The major product **19** was treated with hydrazine hydrate followed by protection of the resulting amine with Cbz group to give **20**. Subsequent mesylation led to **21**, which underwent intramolecular cyclization to give **22**. N-Methylation of **22** gave **23**, which was subjected to acid to remove the MOM protecting groups to give codonopsinine isomer **4** (Scheme 2).

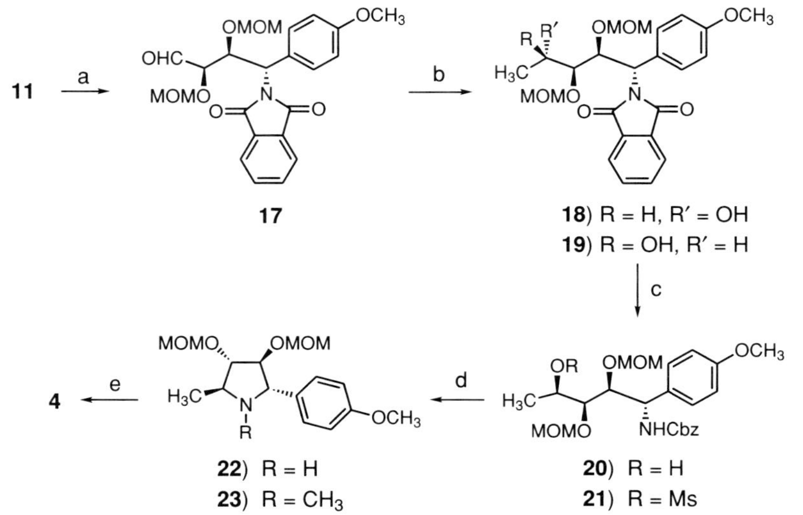
11
a
17
b
18) R = H, R′ = OH
19) R = OH, R′ = H
c
20) R = H
21) R = Ms
d
22) R = H
23) R = CH3
e
4

Scheme 2 (*a*) 1. H_2, CH_3OH, 10% Pd on C, 1 h, 71%; 2. $(COCl)_2$, DMSO, NEt_3, CH_2Cl_2, –78°C, 89%. (*b*) CH_3MgBr, THF, –10°C to rt, 14 h, **18** (17%), **19** (62%). (*c*) 1. $NH_2NH_2{\cdot}H_2O$, EtOH, reflux; then CbzCl, aqueous Na_2CO_3, CH_2Cl_2, 0°C, 93%; 2. MsCl, NEt_3, CH_2Cl_2, 0°C, 10 min, 72%. (*d*) 1. H_2, Pd *on* C, CH_3OH, 1 h, 75%; 2. H_2, aqueous HCHO, 10% Pd *on* C, CH_3OH, 30 min, 60%. (*e*) Aqueous HCl, CH_3OH, 50°C, 2.5 h, 97%.

Similarly, compound **18** was converted into codonopsinine isomer **26** via intermediates **24** and **25** (Scheme 3).

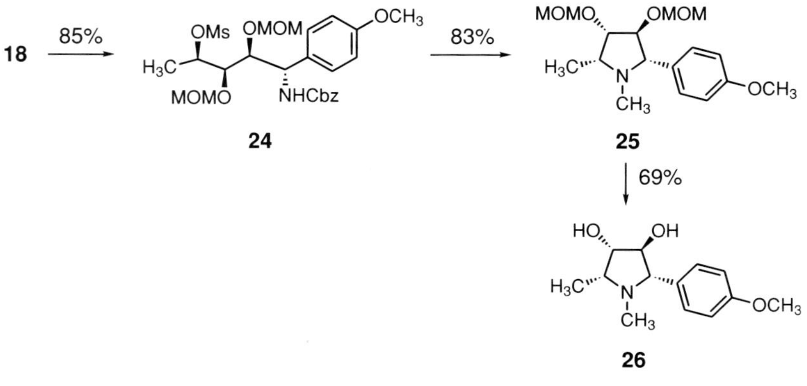
18
85%
24
83%
25
69%
26

Scheme 3

Inversion at C-4 in compound **14** gave compound **28**, which was similarly converted into codonopsinine isomer **31** via intermediates **29** and **30** (Scheme 4).

Scheme 4

References

1. Matkhalikova, S.F.; Malikov, V.M.; Yunusov, S.Y. *Khim. Prir. Soedin* **5** (1969) 607; *Chem. Abstr.* **73** (1970) 25712d.
2. Matkhalikova, S.F.; Malikov, V.M.; Yunusov, S.Y. *Khim. Prir. Soedin* **5** (1969) 30; *Chem. Abstr.* **71** (1969) 13245z.
3. Khanov, M.T.; Sultanov, M.B.; Egorova, M.R. *Farmakol. Alkaloidov Serdech. Glikoyidov.* (1971) 210; *Chem. Abstr.* **77** (1972) 135091r.
4. Matkhalikova, S.F.; Malikov, V.M.; Yunusov, S.Y. *Khim. Prir. Soedin* **5** (1969) 606.
5. Matkhalikova, S.F.; Malikov, V.M.; Yagudaev, M.R.; Yunusov, S.Y. *Khim. Prir. Soedin* **7** (1971) 210.
6. Yagudaev, M.R.; Matkhalikova, S.F.; Malikov, V.M.; Yunusov, S.Y. *Khim. Prir. Soedin* **8** (1972) **495**; *Chem. Abstr.* **77** (1972) 164902m.
7. Iida, H.; Yamazaki, N.; Kibayashi, C.; Nagase, H. *Tetrahedron Lett.* **27** (1986) 5393.
8. Iida, H.; Yamazaki, N.; Kibayashi, C. *Tetrahedron Lett.* **26** (1985) 3255.
9. Iida, H.; Yamazaki, N.; Kibayashi, C. *J. Org. Chem.* **52** (1987) 1956.
10. Wang, C.-L.J.; Calabrese, J.C. *J. Org. Chem.* **56** (1991) 4341.
11. Yoda, H.; Nakajima, T.; Takabe, K. *Tetrahedron Lett.* **37** (1996) 5531.

1.5 Miscellaneous

Three groups are discussed in this part: detoxins, gualamycin and lactacystin. All of them incorporate the hydroxymethyl pyrrolidine nucleus in a modified or rather complex manner.

1.5.1 *Detoxins*

The detoxin complex[1–8] is a collection of 12 depsipeptides **1–12**, and they are metabolites produced by *Streptomyces caespitosus* var. *detoxicus* 7072 GC_1. Detoxins are selective antagonists of the antibiotic blasticidin S (**15**) against *Bacillus cereus*.[9] Thus, the blasticidin S inhibits the virulent fungus *Piriculuria oryzae*, which caused rice blast disease in Japan,[10] but its curative effect required dosages that caused phytotoxicity. This phytotoxicity was greatly reduced when detoxin complex was administered with blasticidin S, without diminishing

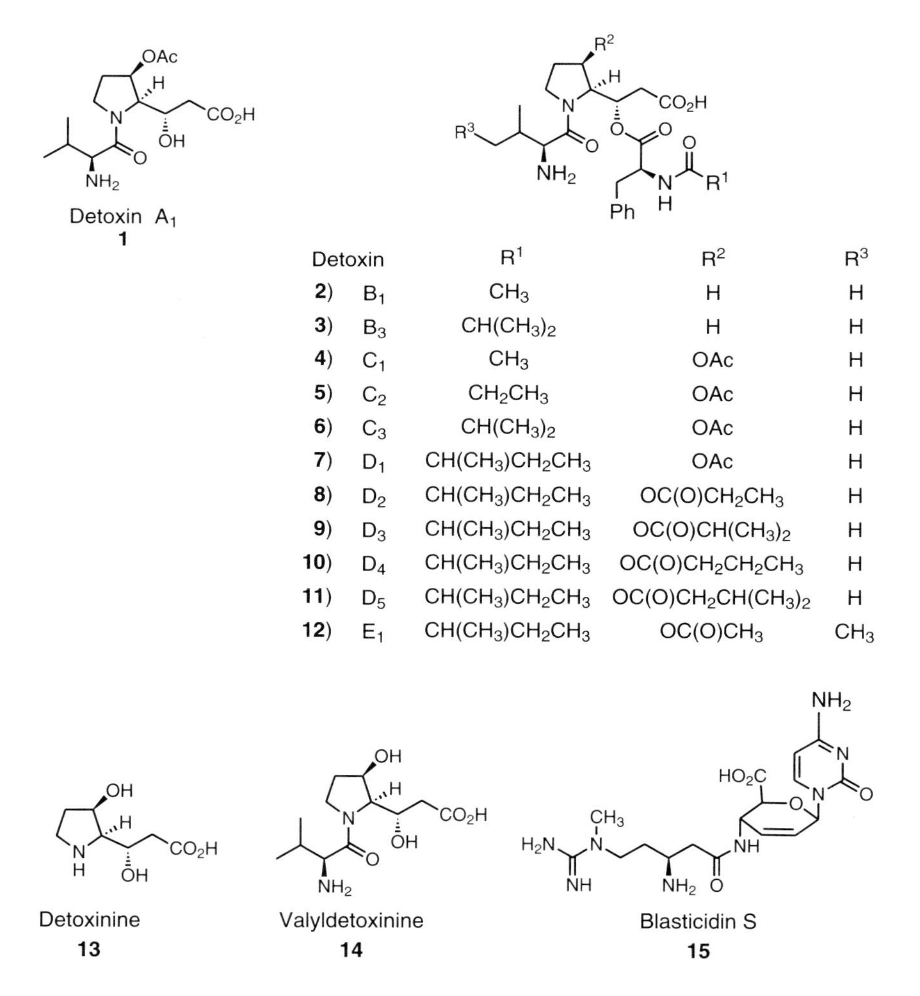

Detoxin A_1
1

Detoxin		R^1	R^2	R^3
2)	B_1	CH_3	H	H
3)	B_3	$CH(CH_3)_2$	H	H
4)	C_1	CH_3	OAc	H
5)	C_2	CH_2CH_3	OAc	H
6)	C_3	$CH(CH_3)_2$	OAc	H
7)	D_1	$CH(CH_3)CH_2CH_3$	OAc	H
8)	D_2	$CH(CH_3)CH_2CH_3$	$OC(O)CH_2CH_3$	H
9)	D_3	$CH(CH_3)CH_2CH_3$	$OC(O)CH(CH_3)_2$	H
10)	D_4	$CH(CH_3)CH_2CH_3$	$OC(O)CH_2CH_2CH_3$	H
11)	D_5	$CH(CH_3)CH_2CH_3$	$OC(O)CH_2CH(CH_3)_2$	H
12)	E_1	$CH(CH_3)CH_2CH_3$	$OC(O)CH_3$	CH_3

Detoxinine
13

Valyldetoxinine
14

Blasticidin S
15

the effectiveness of the drug against *P. oryzae*. Moreover, the *in vivo* studies showed that its administration decreased eye irritation caused by the antibiotic, together with a remarkable decrease of conjunctivitis in rats.[3] Detoxins have detoxification effect against the antibiotic in both animal and plant cells. The most active component among detoxins has been characterized as detoxin D_1 (**7**), which includes the unusual β-hydroxy-γ-imino acid, (−)-detoxinine [(2*S*,3*R*,1′*S*)-2-(2′-carboxy-1′-hydroxyethyl)-3-hydroxypyrrolidine, **13**], as the crucial subunit.[1,8]

Syntheses of detoxin and detoxinine from noncarbohydrates have been reported by different groups,[11–17] and those from carbohydrates are shown in this part.

1.5.1.1 *Synthesis from* D*-glucose* Total syntheses of (+)-valyldetoxinine (**14**) and (−)-detoxin D_1 (**7**) were achieved from diacetone D-glucose (Schemes 1 and 2).[18–20] The diacetone D-glucose was oxidized with pyridinium chlorochromate and the resulting ketone

Scheme 1 (*a*) 1. PCC, MS 3 Å, CH_2Cl_2; 2. $NaBH_4$, EtOH, 82%; 3. MsCl, Py, 95%; 4. Dowex 50X4-400 resin, dioxane, CH_3OH, H_2O, 0°C, 59%; *or* aqueous H_2SO_4, 74%. (*b*) 1. Ph_3P, CBr_4, THF; 2. NaN_3, DMF, 90% for two steps; *or* Ph_3P, CBr_4, LiN_3, DMF, 96%. (*c*) 1. Pd *on* C, H_2; *or* Raney nickel, H_2; 2. NaOAc, EtOH, reflux; 3. CbzCl, NEt_3, THF, 50% from **17**; *or* CbzCl, H_2O, acetone, Na_2CO_3, 78% from **17**. (*d*) ImI_3, imidazole, Ph_3P, toluene, 99%. (*e*) *n*-Bu_3SnH, benzene, AIBN, 97%. (*f*) 1. Dowex 50X4-400 resin, dioxane, H_2O, 40°C; 2. $NaIO_4$, dioxane, H_2O, 0°C; 3. $NaBH_4$, CH_3OH, 0°C to rt, 95%. (*g*) 1. TBSCl, imidazole, DMF, 0°C to rt, 98%; 2. AcOH, H_2O, THF, 0°C to rt, 83%; 3. SO_3·Py, CH_2Cl_2, NEt_3, DMSO, 0°C to rt, 80%. (*h*) $LiCH_2CO_2t$-Bu, THF, −78°C, 87%. (*i*) 1. H_2, Pd *on* C, CH_3OH, rt; 2. Boc-L-Val-OH, DCC, HOBT, CH_2Cl_2, 0°C to rt, 75%. (*j*) TBAF, THF, 0°C, 96%. (*k*) 1. Dry HCl, EtOAc, rt, 89%; 2. ion-exchange chromatography, 92%.

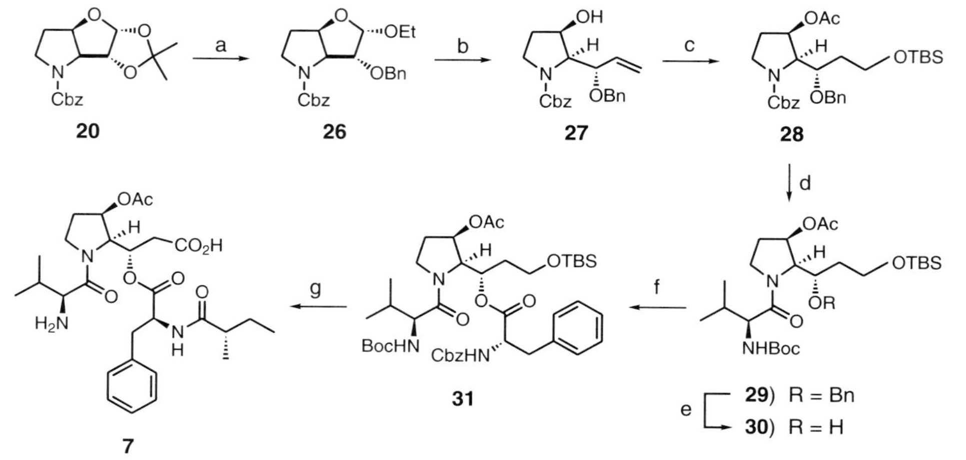

Scheme 2 (*a*) 1. EtOH, 15% HCl, Et_2O, 94%; 2. BnBr, KH, DMF, 89%. (*b*) 1. TFA, H_2O, 90%; 2. $Ph_3P{=}CH_2$, THF, 71%. (*c*) 1. $[(CH_3)_2CHC(CH_3)_2]_2BH$; then H_2O_2, NaOH, 83%; TBSCl, NEt_3, DMAP, CH_2Cl_2; then Ac_2O, NEt_3, 91%. (*d*) 1. Raney nickel, H_2, EtOAc, CH_3OH; 2. Boc-Cl, NEt_3, CH_2Cl_2, 90%. (*e*) H_2, Pd black, EtOH. (*f*) Cbz-phenylalanine, DCC, DMAP, CSA, CH_2Cl_2, 88%. (*g*) 1. Pd *on* C, H_2, EtOAc, CH_3OH; 2. (*S*)-2-methylbutyric acid, BOP, DIPEA, CH_2Cl_2, 70%; 3. HOAc, THF, H_2O, 99%; 4. TFAA, DMSO, NEt_3, CH_2Cl_2; then 1 M $KMnO_4$, 5% $NaHPO_4$; 5. TFA, CH_2Cl_2; then ion exchange.

was then reduced with sodium borohydride to give the α-D-allofuranose, which was mesylated followed by selective hydrolysis of the terminal isopropylidene group to produce **16**. Conversion of the primary alcohol in **16** into the corresponding bromide followed by displacement of the bromide with azide gave the intermediate **17**. Reduction of **17** into a primary amine was followed by intramolecular cyclization and subsequent protection of the resulting secondary amine with benzyl chloroformate to afford the pyrrolidinol **18**. Compound **18** was deoxygenated at C-5, via reductive radical cleavage of a halide, to give compound **20** via iodide **19** in 42% overall yield from diacetone D-glucose.

The isopropylidene group in **20** was hydrolyzed and the product was subjected to the reaction with sodium metaperiodate. Subsequent reduction of the resulting aldehyde afforded the diol **21** in 95% yield from **20**. Protection of both hydroxyl groups in **21** as silyl ethers followed by selective removal of that one on the primary position and subsequent oxidation gave the aldehyde **22**. Reaction of **22** with $LiCH_2CO_2t$-Bu afforded only one diastereomer **23**. Catalytic hydrogenation of **23** and subsequent coupling with *tert*-butoxycarbonyl-protected L-valine afforded **24**. Treatment of **24** with fluoride ion afforded **25**, which was treated with dry HCl to afford the corresponding valyldetoxinine hydrochloride, which was purified by ion-exchange chromatography to afford (+)-valyldetoxinine (**14**) in 59% yield from **23**.

On the other hand, treatment of **20** with dry HCl followed by ethanol and then benzylation of the generated secondary hydroxyl group gave **26**. Hydrolysis of the ethyl glycoside **26** with aqueous TFA and subsequent treatment with methylenetriphenylphosphorane afforded the olefin **27**. Compound **27** was treated with $[(CH_3)_2CHC(CH_3)_2]_2BH$, followed by oxidation and subsequent protection of the resulting primary hydroxyl group as silyl ether and then acetylation of the secondary hydroxyl group, to give the fully protected compound **28**. Hydrogenolysis of **28** followed by coupling with Boc-valine gave **29**, whose debenzylation gave compound **30** that coupled with Cbz-L-phenylalanine to afford the depsipeptide **31**. Removal of the Cbz group by catalytic hydrogenolysis followed by coupling with (*S*)-2-methylbutyric acid and then removal of the TBS with acid and subsequent oxidation of the generated hydroxyl group gave a carboxylic acid group, which upon removal of the Boc group gave detoxin D_1 (**7**) in 16% overall yield from **20**.

Benzyl 3-*O*-benzyl-2-deoxy-4,6-*O*-isopropylidene-α,β-D-arabino-hexopyranoside (**32**) and methyl 3-*O*-benzyl-2-deoxy-4,6-*O*-isopropylidene-α-D-ribohexopyranoside were prepared[21,22] from D-glucose and used as precursors for the syntheses of **13** and its analogue **39**, respectively (Scheme 3).[23] The crucial step was the formation of the pyrrolidine ring in **35** from **34**, which was effected by sodium borohydride reduction followed by alkaline treatment. The pyrrolidine **35** was then coupled with benzyloxycarbonyl-L-valine via an active ester or by the DCC method to give **36**, which was converted to the lactone **37** by acid hydrolysis followed by oxidation with pyridinium chlorochromate. Finally, the lactone **37** was hydrogenated to give **38**, a precursor for **13**.

Scheme 3 (*a*) 1. Aqueous AcOH; 2. MsCl *or* *p*-TsCl, Py. (*b*) NaCN, DMSO. (*c*) 1. $NaBH_4$, $CoCl_2$, CH_3OH; 2. KOH, CH_3OH. (*d*) Cbz-L-Val-OH. (*e*) 1. HCl; 2. PCC, CH_2Cl_2. (*f*) H_2, Pd *on* C.

1.5.1.2 *Synthesis from L-ascorbic acid* (–)-Detoxinine (**13**) was also prepared from L-ascorbic acid (Scheme 4).[24] The α-hydroxy ester **40**,[25] obtained from L-ascorbic acid, underwent inversion of configuration at C-2 followed by reduction of the resulting ester group to produce the diol **41** (76%). Selective protection of the primary hydroxyl group followed by nucleophilic displacement of the secondary hydroxyl group via its mesylate with sodium azide afforded the azide **42** (76%). Hydrogenolysis of **42** followed by treatment with *N*-(benzyloxycarbonyloxy)succinimide gave the *N*-Cbz derivative **43** (97%), which was treated with fluoride ion to produce **44** (99%). Swern oxidation of **44** followed by olefination of the resulting aldehyde with phosphonate according to Horner–Wadsworth–Emmons reaction[26] afforded the α,β-unsaturated ester **45** (*Z*/*E* 16:1) in 93% yield.

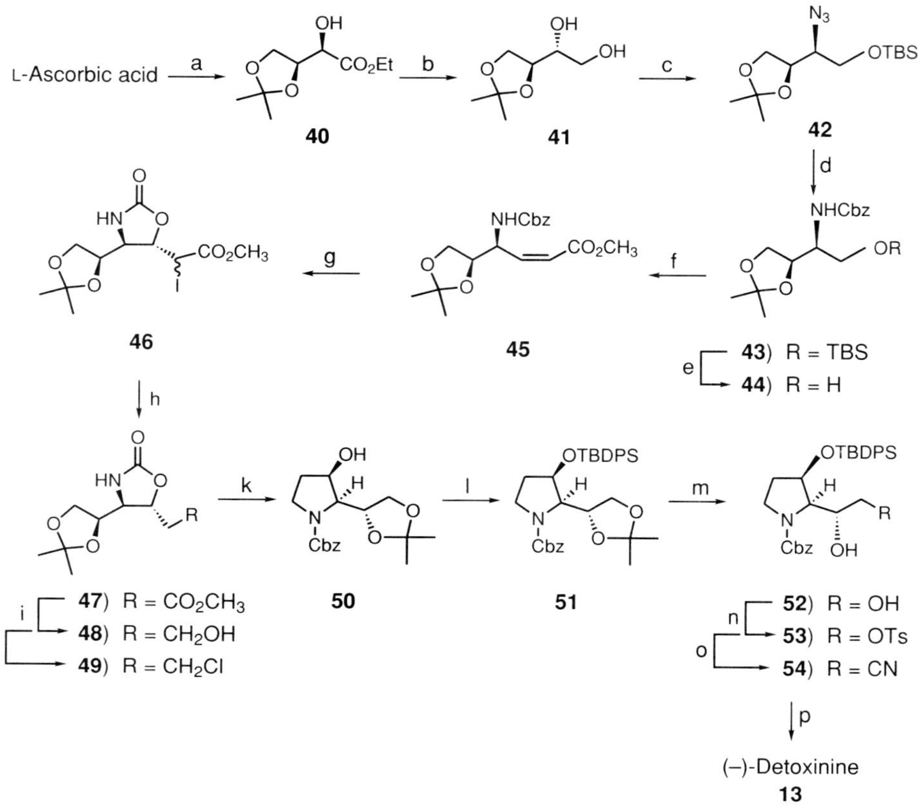

Scheme 4 (*a*) Ref. 25. (*b*) 1. Ph_3P, benzoic acid, THF, DIAD, –5°C, 30 min, rt, 24 h, 83%; 2. $LiAlH_4$, THF, –15°C to rt, 1 h; then reflux, 3 h; then EtOAc, 0°C, 30 min, 92%. (*c*) 1. THF, TBSCl, imidazole, –5°C, 3 h; 2. CH_2Cl_2, NEt_3, MsCl, –5°C to rt, 1 h; 3. DMF, NaN_3, 95°C, 10 h, 76% for three steps. (*d*) 1. EtOH, H_2, 10% Pd *on* C, 6 h; 2. THF, NEt_3, *N*-(benzyloxycarbonyloxy)succinimide, overnight, rt, 97% for two steps. (*e*) TBAF, THF, 0°C to rt, 2 h, 99%. (*f*) 1. $(COCl)_2$, CH_2Cl_2, DMSO, –63°C; then *N*,*N*-diisopropylethyl amine, 15 min; 2. 18-crown-6, $(CF_3CH_2O)_2POCH_2CO_2CH_3$, THF, –65°C, $KN(TMS)_2$, 20 min; then added the aldehyde at –78°C, 45 min, 93% for two steps. (*g*) AgOTf, $NaHCO_3$, CH_3CN, I_2, 2 h, rt, 81%. (*h*) 1. Bu_3SnH, AIBN, toluene, reflux, 3 h, 93%. (*i*) THF, $LiAlH_4$, –15°C, 3 h; then EtOAc, –15°C, 10 min, 95%. (*j*) Py, Ph_3P, CCl_4, rt to 45°C, 15 h, 98%. (*k*) 1. CH_3OH–H_2O (3:1), NaOH, 75%, overnight; 2. NEt_3, dibenzyl dicarbonate, rt, overnight, 68%. (*l*) Imidazole, THF, TBDPSCl, rt, overnight, 98%. (*m*) 1. $FeCl_3$, SiO_2, $CHCl_3$, rt, 16 h, 93%. (*n*) CH_2Cl_2, Py, *p*-TsCl, –5°C, 10 h, 86%. (*o*) NaCN, DMF, 45°C, 4 h, 83%. (*p*) EtOAc, H_2, 5% Pd *on* C, 5 h; then 4 N HCl, 50–75°C, 2.5 h; then 12 h; then Dowex 50X8-200 (H^+) resin, eluting 1 N NH_4Cl, 68%.

Iodocyclocarbamation of **45** afforded the epimeric iodo oxazolidin-2-ones **46**. This mixture was subjected to reductive removal of the iodo group under radical-induced conditions to give the ester **47** (93%), which was reduced to furnish the alcohol **48** followed by chlorination[27] to produce **49** (98%). Subsequent treatment of **49** with NaOH led to cleavage of the cyclic urethane, followed by displacement of the chlorine by nitrogen to give the corresponding pyrrolidine, isolated as the *N*-Cbz derivative **50** (68%). The free hydroxyl group in **50** was protected as the TBS to give **51 (98%)**, which underwent selective removal of the acetonide protecting group[28] to produce the diol **52** (93%). Selective tosylation of the primary hydroxyl group in **52** afforded **53** (86%), which was converted into the nitrile **54** (83%) via the nucleophilic displacement with NaCN. Catalytic hydrogenolysis of **54** followed by acid hydrolysis furnished **13** in 24.7% overall yield from **40**.

References

1. Kakinurna, K.; Otakc, N.; Yoncharli, H. *Tetrahedron Lett.* **13** (1972) 2509.
2. Otake, N.; Kakinuma, K.; Yoneharn, H. *Agric. Biol. Chem.* **37** (1973) 2777.
3. Yonehara, H.; Seto, H.; Shimazu, A.; Aizawa, S.; Hidaka, T.; Kakinuma, K.; Otake, N. *Agric. Biol. Chem.* **37** (1973) 2771.
4. Kakinurna, K.; Otake, N.; Yonehara, H. *Agric. Biol. Chem.* **38** (1974) 2529.
5. Otake, N.; Furihata, K.; Kakinuma, K.; Yonehara, H. *J. Antibiot.* **27** (1974) 484.
6. Ogita, T.; Seto, H.; Otake, N.; Yonehara, H. *Agric. Biol. Chem.* **45** (1981) 2605.
7. Otake, N.; Ogita, T.; Seto, H.; Yonehara, H. *Experientia* **37** (1981) 926.
8. Kakinuma, K.; Otake, N.; Yonchara, H. *Tetrahedron Lett.* **21** (1980) 167.
9. Yonehara, H.; Seto, H.; Aizawa, S.; Hidaka, T.; Shimazu, A.; Otake, N. *J. Antibiot.* **21** (1968) 369.
10. Takeuchi, S.; Hirayara, K.; Ueda, K.; Sakai, H.; Yonehara, H. *J. Antibiot.* **11** (1985) 1.
11. Ewing, W.R.; Harris, B.D.; Bhat, K.L.; Joullié, M.M. *Tetrahedron* **42** (1986) 2421.
12. Kogen, H.; Kadokawa, H.; Kurabayashi, M. *J. Chem. Soc., Chem. Commun.* (1990) 1240.
13. Ohfune, Y.; Nishio, H. *Tetrahedron Lett.* **25** (1984) 4133.
14. Takahata, H.; Banba, Y.; Tajima, M.; Momose, T. *J. Org. Chem.* **56** (1991) 240.
15. Harris, B.D.; Bhat, K.L.; Joullié, M.M. *Heterocycles* **24** (1986) 1045.
16. Ewing, W.R.; Joullié, M.M. *Heterocycles* **27** (1988) 2843.
17. Li, W.-R.; Han, S.-Y.; Joullié, M.M. *Heterocycles* **36** (1993) 359.
18. Li, W.-R.; Han, S.-Y.; Joullié, M.M. *Tetrahedron Lett.* **33** (1992) 3595.
19. Li, W.-R.; Han, S.-Y.; Joullié, M.M. *Tetrahedron* **49** (1993) 785.
20. Han, S.-Y.; Liddel, P.A.; Joullié, M.M. *Synth. Commun.* **18** (1988) 275.
21. Inglis, G.R.; Schwarz, J.C.P.; McLaren, L. *J. Chem. Soc.* (1962) 1014.
22. Hough, L.; Richardson, A.C. In *Rodd's Chemistry for Carbon Compounds*, Vol. 1F, 2nd ed.; Coffey, S., Ed.; Elsevier: Amsterdam, 1967; p. 367.
23. Kakinuma, K. *Tetrahedron Lett.* **21** (1980) 167.
24. Delle Monache, G.; Misiti, D.; Zappia, G. *Tetrahedron: Asymmetry* **10** (1999) 2961.
25. Abushanab, E.; Vemishetti, P.; Leiby, R.W.; Singh, H.K.; Mikkileneni, A.B.; Wu, D.C.-J.; Saibaba, R.; Panzica, R.P. *J. Org. Chem.* **53** (1988) 2598.
26. Maryanoff, B.E.; Reitz, A.B. *Chem. Rev.* **89** (1989) 863.
27. Anisuzzaman, A.K.M.; Whistler, R.L. *Carbohydr. Res.* **61** (1978) 511.
28. Kim, K.S.; Song, Y.H.; Lee, B.H.; Hahn, C.S. *J. Org. Chem.* **51** (1986) 404.

1.5.2 *Gualamycin*

Gualamycin (**1**) is a water-soluble acaricide isolated from the culture broth of *Streptomyces* sp. NK11687.[1] The absolute configuration was mainly confirmed by enantiospecific synthesis of the corresponding disaccharide and pyrrolidine-aglycone portions.[2,3]

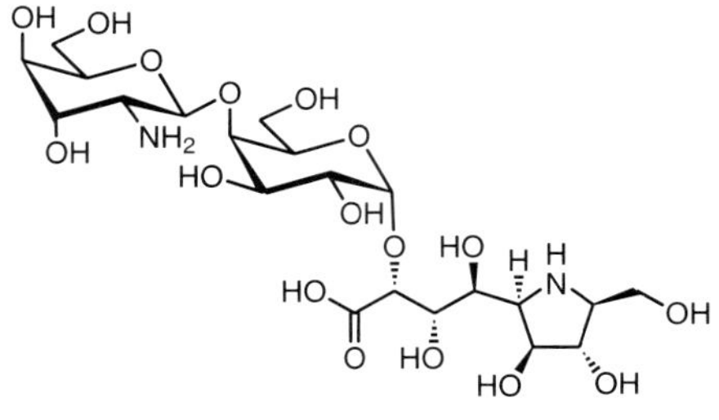

Gualamycin

1

The first synthesis of gualamycin (**1**) was achieved from **2** by conversion to the glycosyl acceptor **7** whose glycosidation gave **1** (Schemes 1 and 2).[4] Thus, the di-*O*-benzylidene derivative **7** was synthesized from compound **4**,[5,6] which was derived from the azido compound **2** through the key intermediate **3**. Treatment of **4** with Na_2CO_3 in methanol gave the δ-lactam **5**, whose benzylidenation provided **6** in 54% overall yield. Silylation of **6** with TMSCl followed by acetylation and desilylation gave **7**.

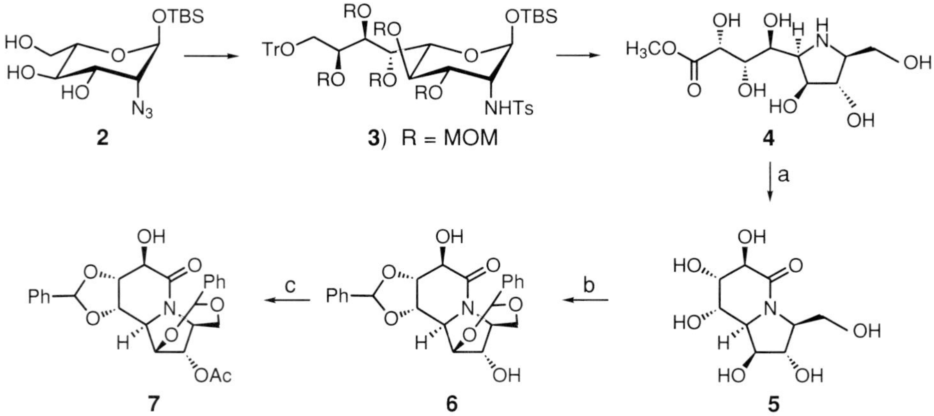

Scheme 1 (*a*) Na_2CO_3, CH_3OH, 5 h. (*b*) PhCHO, $ZnCl_2$, 2.5 h, 54%. (*c*) 1. TMSCl, DIPEA, CH_2Cl_2, 1 h; 2. Ac_2O, Py, 2 days; 3. TBAF, THF, 30 min, 70%.

The glycosyl donor **11** was prepared from phenyl-1-thio-galactoside **8**[7] (Scheme 2). Thus, treatment of **8** with PhCHO and HCO_2H followed by O-benzylation afforded the respective protected 4,6-*O*-benzylidene derivative. De-O-benzylidenation with 2 M HCl in dioxane followed by selective O-benzoylation with BzCl and pyridine gave **9** (81%). The alcohol **9** was glycosylated with the protected bromide **10** to give the glycosyl donor **11** (42%). Coupling of **11** with the acceptor **7** provided exclusively the desired α-glycoside **12**.[8] Hydrogenolysis of **12** followed by treatment with 40% CH_3NH_2 in methanol and subsequent hydrolysis gave the dihydrochloride of gualamycin (**1**) in 86% yield.

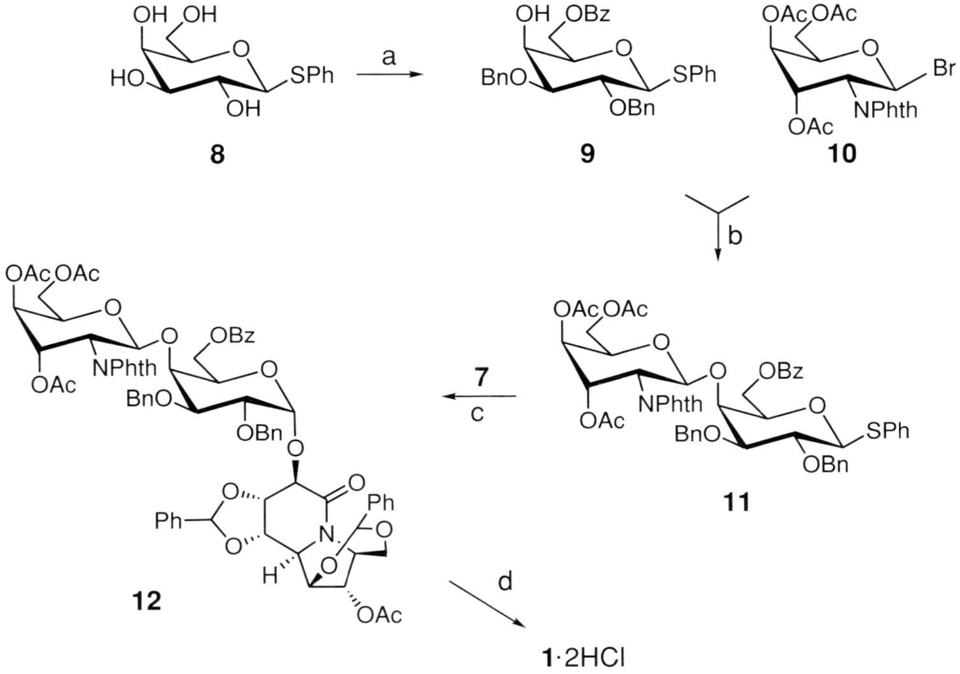

Scheme 2 (*a*) 1. PhCHO, HCO_2H, 30 min; 2. BnBr, NaH, DMF, 3 h; 3. 2 M HCl, dioxane, 50°C, 1.5 h; 4. BzCl, Py, 2 days, 81% for three steps. (*b*) AgOTf, S-collidine, CH_2Cl_2, –40°C to rt, 15 h, 42%. (*c*) NIS, TfOH, MS 4 Å, CH_2Cl_2, –40°C, 1 h, 82%. (*d*) 1. H_2, 10% Pd *on* C, CH_3OH, 4 days, 90%; 2. 40% CH_3NH_2, CH_3OH, 76%; 3. 2 M HCl, rt, 6 days, 86%.

References

1. Tsuchiya, K.; Kobayashi, S.; Harada, T.; Kurokawa, T.; Nakagawa, T.; Shimada, N.; Kobayashi, K. *J. Antibiot.* **48** (1995) 626.
2. Tsuchiya, K.; Kobayashi, S.; Kurokawa, T.; Nakagawa, T.; Shimoda, N.; Nakamura, H.; Iitaka, Y.; Kitagawa, M.; Tatsuda, K. *J. Antibiot.* **48** (1995) 630.
3. Tatsuta, K.; Kitagawa, M.; Horiuchi, T.; Tsuchiya, K.; Shimada, N. *J. Antibiot.* **48** (1995) 741.
4. Tatsuta, K.; Kitagawa, M. *Tetrahedron Lett.* **36** (1995) 6717.
5. Tsuchiya, K.; Kobayashi, S.; Kurokawa, T.; Nakagawa, T.; Shimada, N.; Nakamura, H.; Iitaka, Y.; Kitagawa, M.; Tatsuda, K. *J. Antibiot.* **48** (1995) 630.
6. Tatsuta, K.; Kitagawa, M.; Horiuchi, T.; Tsuchiya, K.; Shimada, N. *J. Antibiot.* **48** (1995) 741.
7. Kohata, K.; Konno, T.; Meguro, H. *Tetrahedron Lett.* **21** (1980) 3771.
8. Konradsson, P.; Udodong, U.E.; Fraser-Reid, B. *Tetrahedron Lett.* **31** (1990) 4313.

1.5.3 *Lactacystin*

(+)-Lactacystin (**1**) was isolated from *Streptomyces* sp. OM-6519.[1,2] It is a nonprotein neurotrophic agent;[3–7] neurotrophic factors such as nerve growth factor are required for the survival and function of nerve cells.[8–10] Decreased availability of neurotrophic factors is thought to cause various nerve disorders including Alzheimer's disease. It inhibits cell proliferation and induces neuritogenesis and increases the intracellular cAMP level transiently in the Neuro 2A neuroblastoma cell line[11] and it is also active against sarcoma 180. Its structure has been studied by ^{1}H NMR, ^{13}C NMR and a single-crystal X-ray analysis. It has (*R*)-*N*-acetylcysteine residue and a unique pyroglutamic acid via a thioester linkage.[12] Several total syntheses of lactacystin and its analogues from noncarbohydrates have been reported.[13–24]

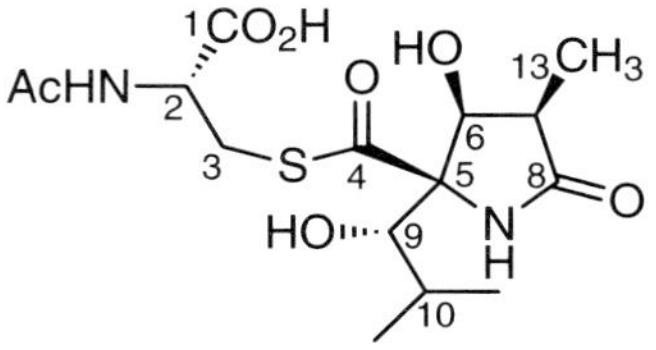

(+)-Lactacystin
1

The total synthesis of (+)-lactacystin (**1**) has been achieved via Overman rearrangement of allylic trichloroacetimidate derived from D-glucose as an effective method for chiral synthesis of α,α-disubstituted amino acid derivatives (Schemes 1 and 2).[25,26] Selective benzylation of the primary hydroxyl group in 3-deoxy-1,2-*O*-isopropylidene-3-*C*-methyl-α-D-allofuranose (**3**), obtained from diacetone D-glucose (**2**),[27] in the presence of dibutyltin oxide[28] followed by oxidation of the secondary hydroxyl group with Jones reagent and subsequent treatment of the resulting ketone with (carbethoxymethylene)triphenylphosphorane afforded the olefin **4** as a mixture of *E*- and *Z*-isomers (1:1) in 52% yield from **3**. Reduction of the ester function in **4** with DIBAL-H afforded a separable 1:1 mixture of **5***E* and **5***Z*, which are the substrates for Overman rearrangement.[29] Treatment of this mixture with trichloroacetonitrile afforded the trichloroacetimidate **6**, which, without isolation, was heated in toluene to provide **7**. Removal of the isopropylidene group from the mixture of **7** afforded **8** (19%) and **9** (72%). The major isomer **9** was treated with sodium periodate to afford **10**, which was subjected to oxidation with Jones reagent followed by removal of the *N*-trichloroacetyl and *O*-formyl groups with sodium borohydride to furnish the respective γ-lactam in 75% yield from **9**. Silylation of the free hydroxyl group followed by debenzylation with sodium in dry ammonia afforded the lactam **11**.

Moffatt oxidation of **11** gave **12**, which, without isolation, was treated with isopropylmagnesium bromide to give **11** (21%), **13** (35%) and **14** (30%). The undesired product **13** could be converted into the desired product **14** via oxidation–reduction process in 70% overall yield. Removal of the silyl protecting group from **14** with TFA afforded the lactam **15**, which underwent ozonolysis followed by selective oxidation of the resulting aldehyde to afford the carboxylic acid **16**. The latter, without isolation, was coupled with

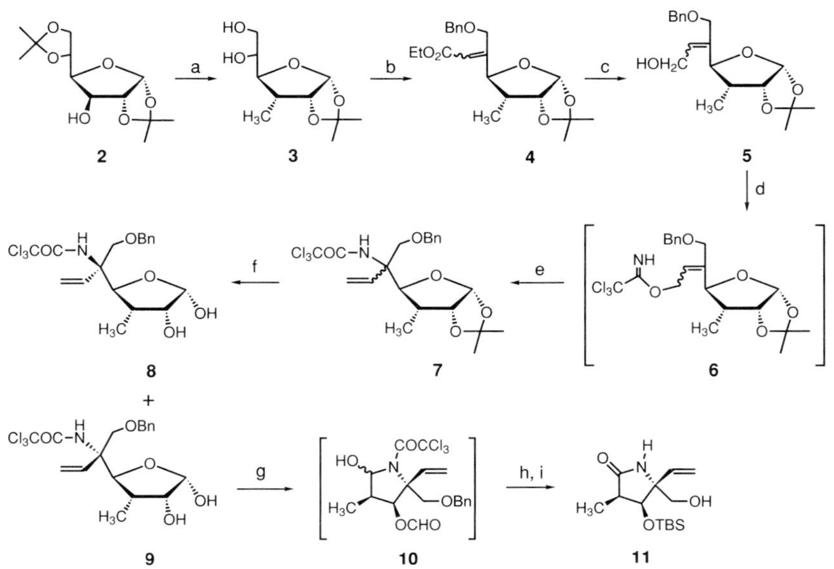

Scheme 1 (*a*) Ref. 27. (*b*) 1. *n*-Bu_2SnO, toluene, reflux, 3 h; then CeF, BnBr, DMF, rt, 18 h; then 10% aqueous KF, rt, 1 h, 66%; 2. Jones reagent (CrO_3, aqueous H_2SO_4), acetone, 0°C, 1 h; 3. (carbethoxymethylene)triphenylphosphorane, rt, 1 h; then 60°C, 15 h, 1:1 mixture of E/Z in 78% for two steps. (*c*) DIBAL-H, CH_2Cl_2, CH_3OH, –15°C, 30 min, 90% (1:1 separable **5***E* and **5***Z*). (*d*) NaH, Et_2O, –15°C, 15 min; then Cl_3CCN, –15°C, 20 min; then rt, 2 h. (*e*) Toluene, sealed tube 140°C, 89 h, inseparable mixture (4:1), 60% from 6. (*f*) TFA–H_2O (3:2), 0°C, 5 h, **8** (19%), **9** (72%). (*g*) $NaIO_4$, CH_3OH, 0°C, 1 h to rt, 5 h; then $NaIO_4$, rt, 22 h. (*h*) 1. Jones reagent, acetone, 0°C, 5 h to 5°C, 17 h; 2. CH_3OH, $NaBH_4$, 0°C to rt, 3 h; then Amberlite IR-120B (H^+) resin, 75%. (*i*) 1. 2,6-Lutidine, CH_2Cl_2, TBSOTf, rt, 43 h, 92%; 2. NH_3, –78°C, Na, THF, 30 min, 75%.

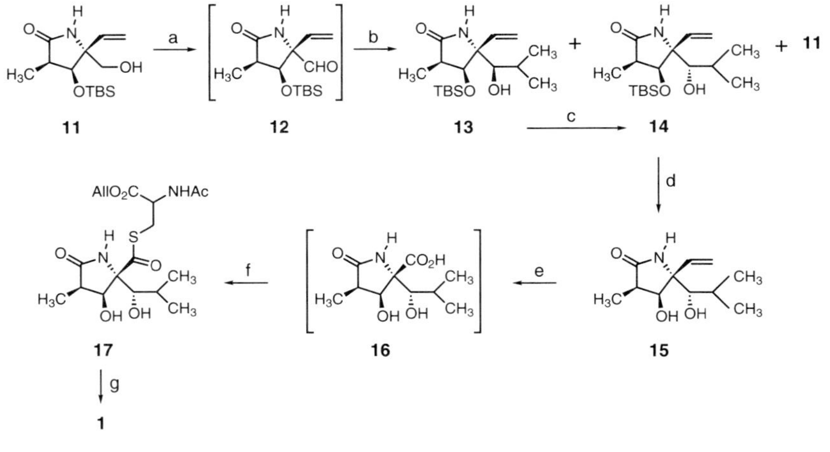

Scheme 2 (*a*) DMSO–benzene (1:1), rt, Py, TFA, DCC, rt, 5 h. (*b*) THF, *i*-PrMgBr, –15°C, 45 min, rt, 12 h, **13** (35%), **11** (21%), **14** (30%). (*c*) 1. DMSO–benzene (1:1), rt, Py, TFA, DCC, rt, 7 h, 78%; 2. triisobutyl aluminum, CH_2Cl_2, 0°C, 2 h; then rt, 3 h, **13** (7%), **14** (70%). (*d*) 1. TFA–H_2O (4:1), rt, 15 h; then 50°C, 2 h; 2. $NaOCH_3$, CH_3OH, 0°C, 3 h; then Amberlite IR-120B (H^+) resin, 76%. (*e*) O_3, CH_3OH, –78°C, DMSO, rt, 5 h; then *t*-BuOH–H_2O (1:1), $HOSO_2NH_2$, $NaH_2PO_4 \cdot 2H_2O$, $NaClO_2$, 15 min, 81%. (*f*) CH_2Cl_2, 0°C, NEt_3, bis(2-oxo-3-oxazolidinyl)phosphinic chloride, *N*-acetyl-L-cysteine allyl ester, rt, 19 h, 60% from **15**. (*g*) $(PPh_3)_4Pd$, NEt_3, HCO_2H, THF, rt, 5 h, 70%.

N-acetyl-L-cysteine allyl ester to provide the lactacystin allyl ester **17** in 60% yield from **15**. Removal of the allyl group from **17** furnished (+)-lactacystin (**1**) in 70% yield.

References

1. Omura, S.; Fujimoto, T.; Otogurro, K.; Matsuzaki, K.; Moriguchi, R.; Tanaka, H.; Sasaki, Y. *J. Antibiot.* **44** (1991) 113.
2. Omura, S. *Gene* **115** (1992) 141.
3. Levi-Montalcini, R. *Science* **237** (1987) 1154.
4. McDonald, N.Q.; Lapatto, R.; Murray-Rust, J.; Gunning, J.; Wlodawer, A.; Blundell, T.L. *Nature* **354** (1991) 411.
5. Bothwell, M. *Cell* **65** (1991) 915.
6. Rosenberg, S. *Annu. Rep. Med. Chem.* **27** (1992) 41.
7. Fenteany, G.; Standaert, R.F.; Lane, W.S.; Choi, S.; Corey, E.J.; Schreiber, S.L. *Science* **268** (1995) 726.
8. Perez-Polo. J.R. In *Cell Culture in the Neurosciences*; Bottenstein, J.E., Sato, G., Eds.; Plenum Press: New York, 1985; p. 95.
9. Hefti, F.; Weiner, W.J. *Ann. Neurol.* **20** (1986) 275.
10. Barde, Y.-A. *Neuron* **2** (1989) 1525.
11. Tanaka, H.; Katagiri, M.; Arinie, S.; Matsuzaki, K.; Inokoshi, J.; Omura, S. *Biochem. Biophys. Res. Commun.* **216** (1995) 291.
12. Omura, S.; Matsuzaki, K.; Fujimoto, T.; Kosuge, K.; Furuya, T.; Fujita, T.; Nakagawa, A. *J. Antibiot.* **44** (1991) 117.
13. Sunazuka, T.; Nagamitsu, T.; Matsuzaki, K.; Tanaka, H.; Omura, S.; Smith, A.B., III *J. Am. Chem. Soc.* **115** (1993) 5302.
14. Nagamitsu, T.; Sunazuka, T.; Tanaka, H.; Omura, S.; Sprengeleri, P.A.; Smith, A.B., III, *J. Am. Chem. Soc.* **118** (1996) 3584.
15. Uno, H.; Baldwin, J.E.; Russell, A.T. *J. Am. Chem. Soc.* **116** (1994) 2139.
16. Corey, E.J.; Choi, S. *Tetrahedron Lett.* **34** (1993) 6969.
17. Corey, E.J.; Reichared, G.A.; Kania, R. *Tetrahedron Lett.* **34** (1993) 6977.
18. Corey, E.J.; Reichared, G.A. *Tetrahedron Lett.* **34** (1993) 6973.
19. Corey, E.J.; Li, W.-D.Z. *Tetrahedron Lett.* **39** (1998) 7475.
20. Nakagawa, A.; Takahashi, S.; Uchida, K.; Matsuzaki, K.; Omura, S.; Nakamura, A.; Kurihara, N.; Nakamatsu, T.; Miyake, Y.; Take, K.; Kainosho, M. *Tetrahedron Lett.* **35** (1994) 5009.
21. Corey, E.J.; Reichared, G.A. *J. Am. Chem. Soc.* **114** (1992) 10677.
22. Corey, E.J.; Li, W.-D.Z.; Reichared, G.A. *J. Am. Chem. Soc.* **120** (1998) 2330.
23. Kang, S.H.; Jun, H.-S. *Chem. Commun.* (1998) 1929.
24. Green, M.P.; Prodger, J.C.; Hayes, C.J. *Tetrahedron Lett.* **43** (2002) 6609.
25. Chida, N.; Takeoka, J.; Tsutsmi, N.; Ogawa, S. *J. Chem. Soc., Chem. Commun.* (1995) 793.
26. Chida, N., Takeoka, J.; Ando, K.; Tsutsumi, N. Ogawa, S. *Tetrahedron* **53** (1997) 16287.
27. Rosenthal, A.; Sprinzl, M. *Can. J. Chem.* **47** (1969) 3941.
28. David, S.; Hanessian, S. *Tetrahedron* **41** (1985) 643.
29. Overman, L.E. *J. Am. Chem. Soc.* **98** (1978) 2901.

2 Five-membered heterocycles with two heteroatoms

This chapter discusses the synthesis of naturally occurring five-membered heterocycles having two heteroatoms from sugars. The source, activity and synthesis of these compounds from their carbohydrate precursors are presented for each of the following biologically active compounds: the herbicidal hydantocidin; an antitumor agent, bleomycin; a bioactive metabolite, calyculin; another antitumor antibiotic, acivicin; and an active antifungal agent, bengazole.

2.1 (+)-Hydantocidin

(+)-Hydantocidin (**1**), isolated from the fermentation broth of *Streptomyces hygroscopicus* SANK 63584,[1–3] Tu-2474[4] and A1491,[5] is the first example of natural products carrying a

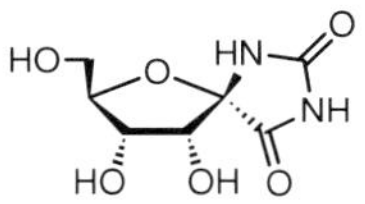

(+)-Hydantocidin
1

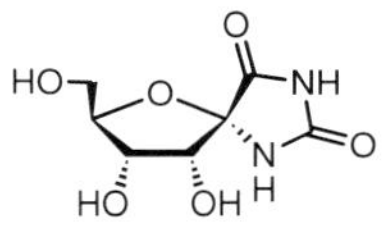

(+)-5-*epi*-Hydantocidin
2

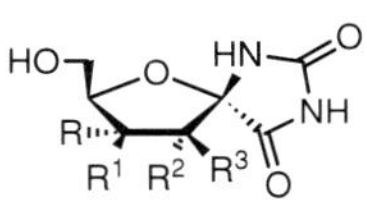

3) R = R^2 = H, R^1 = R^3 = OH
4) R = R^3 = OH, R^1 = R^2 = H
5) R = R^3 = H, R^1 = R^2 = OH

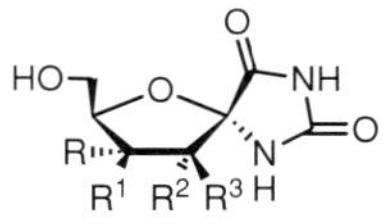

6) R = R^2 = H, R^1 = R^3 = OH
7) R = R^3 = OH, R^1 = R^2 = H
8) R = R^3 = H, R^1 = R^2 = OH

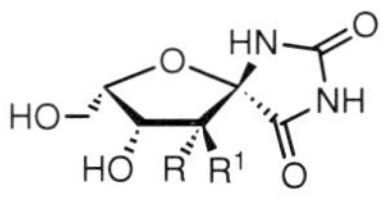

9) R = H, R^1 = OH
10) R = OH, R^1 = H

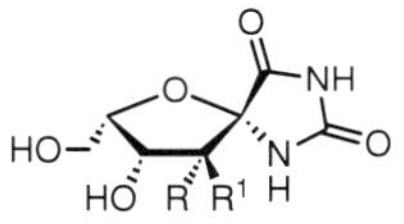

11) R = OH, R^1 = H
12) R = H, R^1 = OH

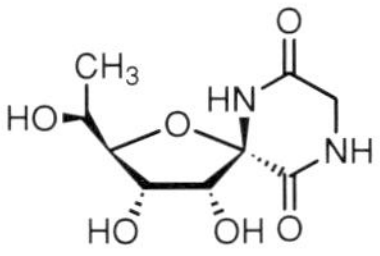

13 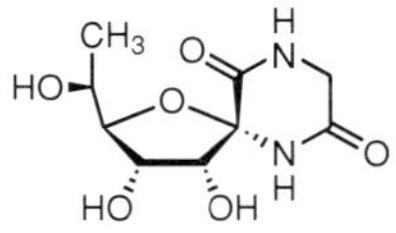 **14** 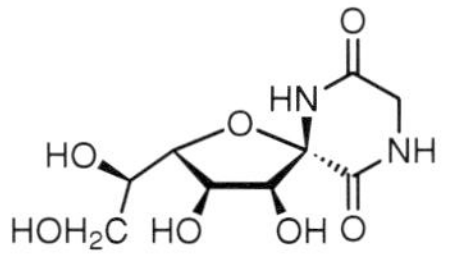 **15**

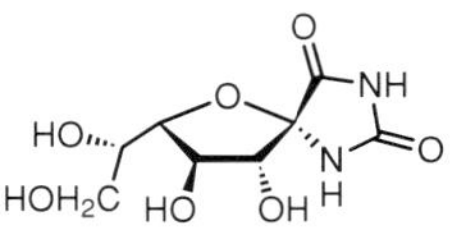

16 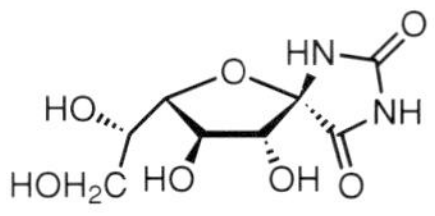 **17** 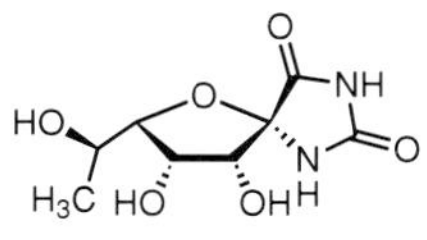 **18**

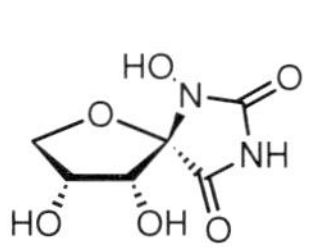

19 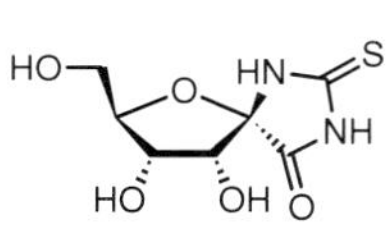 **20** 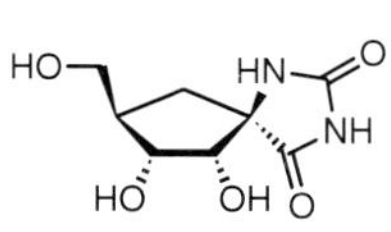 **21**

spirohydantoin nucleus at the anomeric position of D-ribofuranose. This unique structural characteristic has never been found in the family of nucleoside antibiotics.[6,7] Compound **1** exhibits a potent herbicidal activity against perennial plants which is almost equal to which of glyphosate,[8] and it has essentially no toxicity against microorganisms, fishes and animals.[9–11] The *C*-5-epimer **2** exhibits herbicidal activity to be almost 60% of **1**.[12,13] Interestingly, the herbicidal activity of hydantocidin is associated with the D-*ribo* configuration of **1**, since the other possible diastereoisomers **3-12** with four contiguous stereogenic centers were found to be devoid of activity.[14,15] The structure of **1** was determined[2] by the combination of mass and ^{1}H NMR spectra, which established the relative configuration of its asymmetric carbon atoms. Since the achievement of the first total synthesis of **1** in 1991,[16,17] which has confirmed its absolute configuration, efforts toward the synthesis of **1** and its stereoisomers[12–20] **2–12** and its analogues **13–20**,[21–37] retaining a furanosyl ring, as well as the carbocyclic analogue **21** and the pyranose analogues[38–42] **22–31** have been reported.

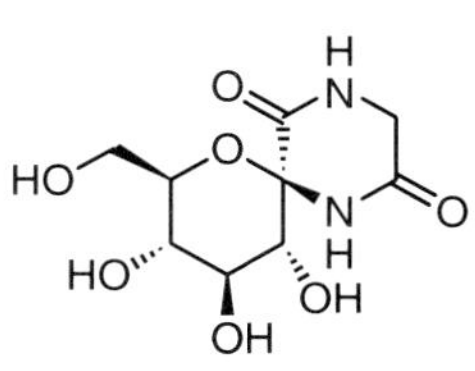

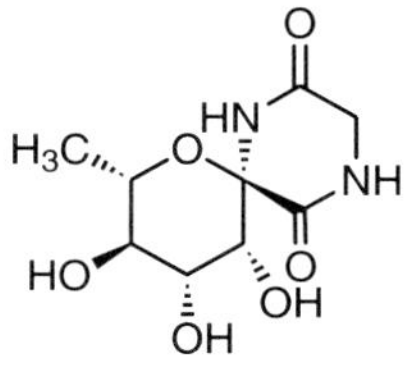

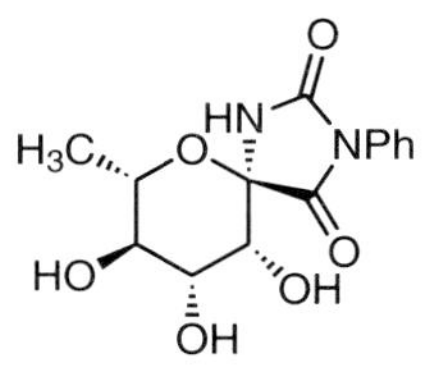

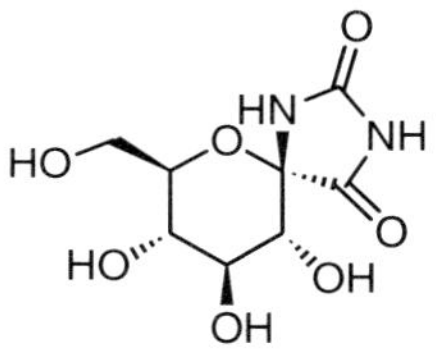

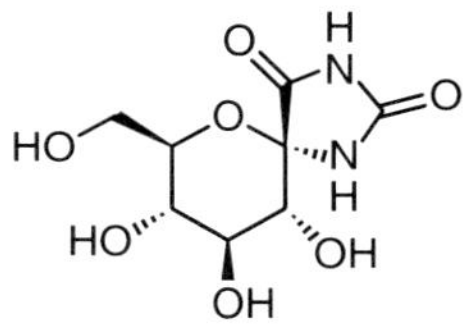

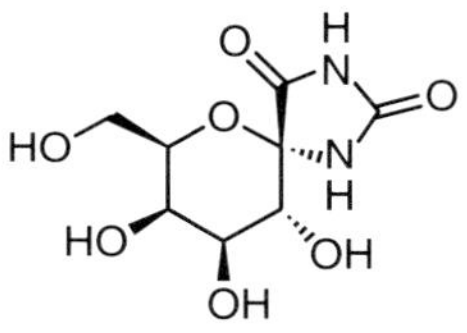

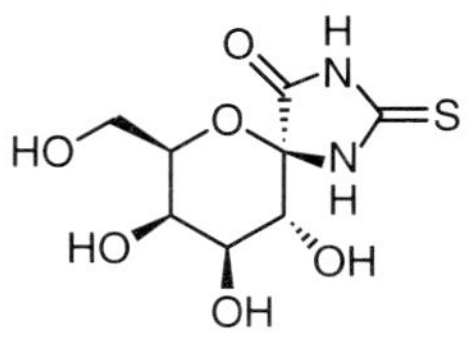

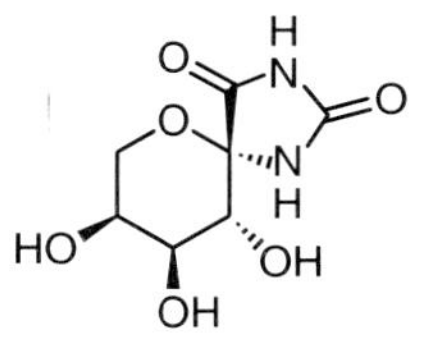

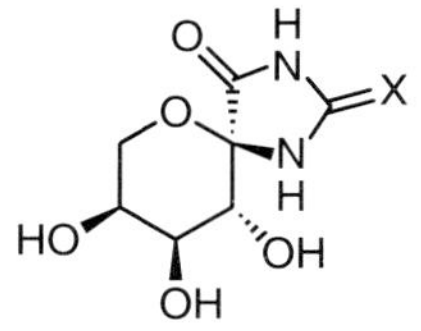

2.1.1 *Synthesis from D-fructose*

A large-scale synthesis of (+)-hydantocidin (**1**) from D-fructose has been reported (Scheme 1).[17] Treatment of D-fructose with DMP followed by oxidation of the unprotected

D-Fructose

32

33) R = H

34) R = Bn

37

36

35

38

(+)-Hydantocidin

1

Scheme 1 (*a*) 1. $HClO_4$, DMP, 58%; 2. $RuCl_3 \cdot xH_2O$, $NaIO_4$, $BnN^+Et_3Cl^-$, K_2CO_3, $CHCl_3$, H_2O, 99%; 3. $NaBH_4$, EtOH, 96%. (*b*) $HClO_4$, DMP, 72%. (*c*) BnCl, NaOH, $BnN^+Et_3Cl^-$, 99%. (*d*) $TMSN_3$, CH_3CN, TMSOTf, 97%, **36**:**35** (18:1). (*e*) 1. $(COCl)_2$, DMSO, NEt_3, CH_2Cl_2, 96%; 2. $NaClO_2$, $NaH_2PO_4 \cdot 2H_2O$, 2-methylbutene, *t*-BuOH, H_2O; 3. $ClCO_2Et$, NEt_3, THF, 0°C; then NH_3 gas, 72% for two steps. (*f*) PBu_3, CO_2 gas, CH_3CN, rt, 5 h; then Ac_2O, Py, DMAP, 90%. (*g*) 1. Dowex 50W (H^+) resin, CH_3OH, H_2O, 92%; 2. $NH_2NH_2 \cdot H_2O$, CH_3OH, 96%; 3. H_2, Pd *on* C (10%), CH_3OH, 55°C, 93%.

hydroxyl group and subsequent reduction by sodium borohydride afforded the respective isomer **32**. Isomerization of the pyranose ring in **32** gave the corresponding furanose **33**, which upon subsequent protection of the free hydroxyl group gave 6-*O*-benzyl-1,2:3,4-di-*O*-isopropylidene-D-psicofuranose **(34)** in 72% overall yield from D-fructose. Treatment of **34** with trimethylsilyl azide ($TMSN_3$), in the presence of TMSOTf, afforded predominantly the azido compounds **35** and **36** in a ratio of 1:18. The major one was transformed by a combination of Swern oxidation[43] and $NaOCl_2$ oxidation[44–46] into the corresponding carboxylic acid that was converted to the amide **37**. Treatment of **37** with tri-*n*-butylphosphine (PBu_3) in THF afforded a polar intermediate, iminophosphorane, which was cyclized in the presence of CO_2 gas to the spirohydantoin, isolated as its *N*-acetyl derivative **38**. Removal of the isopropylidene group in **38** was achieved by treatment with Dowex 50W (H^+) resin, followed by deacetylation with hydrazine monohydrate and subsequent debenzylation to furnish the final compound **1** in 27% overall yield from D-fructose.

Alternatively, (+)-hydantocidin (**1**) has been synthesized from 1,2:3,4-di-*O*-isopropylidene-D-psicofuranose (**33**)[17,47] by conversion to the hydroxylamine **39** (Scheme 2).[48] Compound **39** was transformed to *p*-methoxybenzylurea (**40**) (92%) to prevent the intramolecular acetone migration. Treatment of **40** with TMSOTf gave **41** (97%), which underwent Jones oxidation followed by spontaneous cyclization to afford the tricyclic compound **42**. Deprotection with CAN gave **43**, whose structure was established by single-crystal X-ray analysis. Reduction of **43** with the organometallic complex $Mo(CO)_6$ gave 2′,3′-isopropylidene-hydantocidin (**44**) (70%), which was subsequently deprotected with TFA in aqueous solution to afford **1** in 36% overall yield from **33**.

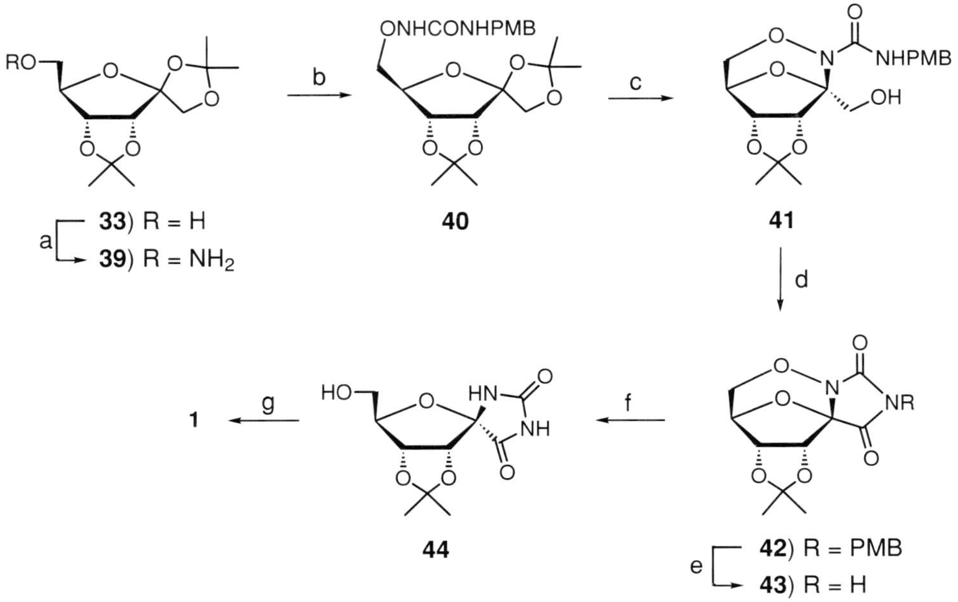

Scheme 2 (*a*) 1. *N*-Hydroxyphthalimide, Ph_3P, DEAD, THF; 2. $NH_2NH_2 \cdot H_2O$, EtOH, reflux, 82% for two steps. (*b*) PMB–N=C=O, CH_3CN, rt, 92%. (*c*) TMSOTf (0.1 equiv.), CH_3CN, 0°C to rt, 97%. (*d*) Na_2CrO_7, H_2SO_4, acetone, 70%. (*e*) CAN, CH_3CN, H_2O, 100%. (*f*) $Mo(CO)_6$, CH_3CN, H_2O, 70%. (*g*) TFA–H_2O (1:3), 0°C, quantitative.

Syntheses of (+)-hydantocidin (**1**) and its *C*-5-epimer **2** have also been achieved starting from D-fructose and proceeding through **34** via **42** (Schemes 3–5).[12,13] The critical formation of benzyl glycoside **45** was achieved by treatment of **34** with benzyl alcohol in the presence of trifluoromethanesulfonic acid. Swern oxidation of **45** followed by sodium chlorite oxidation of the resulting aldehyde afforded the carboxylic acid **46**. Formation of the *N*-acylurea derivative **47** from **46** was achieved by following stepwise reaction sequence, rather than the direct formation, which failed because of both low nucleophilicity of urea and steric hindrance of the carbonyl group in **46**.

The respective D-piscopyranose derivative **52** was prepared from **32** by acid hydrolysis of the acetonide moiety to give **48**, followed by complete benzylation to provide **49**. Then

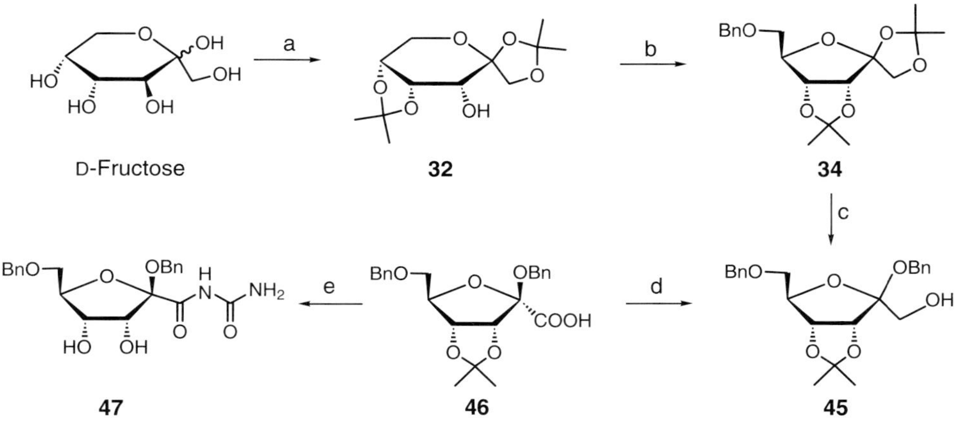

Scheme 3 (*a*) 1. H_2SO_4, acetone, rt, 73%; 2. DMSO, Ac_2O, rt, 77%; 3. $NaBH_4$, EtOH, rt, 95%. (*b*) 1. H_2SO_4, acetone, rt, 73%; 2. BnCl, $BnEt_3NCl$, aqueous NaOH, 100°C, 92%. (*c*) TfOH, BnOH, rt, 74%. (*d*) 1. $(COCl)_2$, DMSO, CH_2Cl_2, −78°C, NEt_3, 100%; 2. $NaClO_2$, $NaH_2PO_4-H_2O$, 2-methyl-2-butene, *t*-BuOH, H_2O, rt, 100%. (*e*) 1. $ClCO_2$*i*-Pr, NEt_3, THF, 0°C; NH_3 (gas), rt, 92%; 2. $(COCl)_2$, $Cl(CH_2)_2Cl$, 80°C; NH_3 (gas), rt, 89%; 3. HCl, *i*-PrOH, 90°C, 99%.

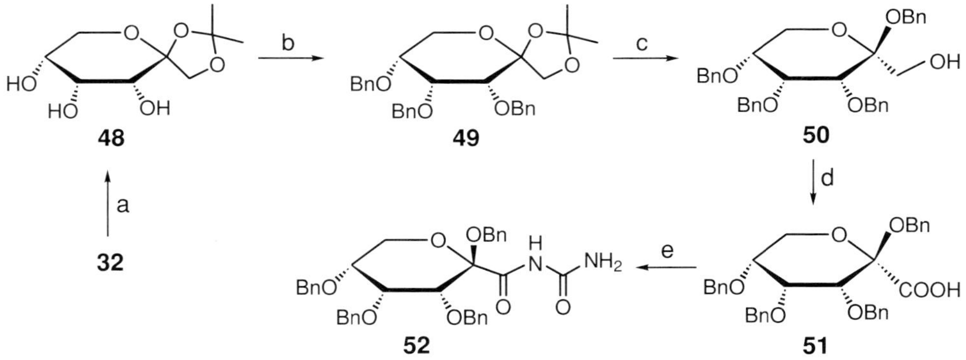

Scheme 4 (*a*) *p*-TsOH, CH_3OH, rt, 86%. (*b*) BnCl, KOH, 130°C, 100%. (*c*) TfOH, BnOH, rt, 71%. (*d*) 1. $(COCl)_2$, DMSO, CH_2Cl_2, –78°C; then NEt_3, 100%; 2. $NaClO_2$, $NaH_2PO_4 \cdot H_2O$, 2-methyl-2-butene, *t*-BuOH-H_2O, rt, 85%. (*e*) 1. $ClCO_2$*i*-Pr, NEt_3, THF, 0°C; NH_3 (gas), rt, 92%; 2. $(COCl)_2$, $Cl(CH_2)_2Cl$, 80°C; NH_3 (gas), rt, 70%.

a reaction sequence similar to that described above for the preparation of **47** was used to convert **49** into **52** via **50** and **51**.

Removal of the protecting groups from the key intermediates **47** and **52** furnished the same equilibrium mixture of the furanose and pyranose derivatives **53** and **54**. These observations can be explained to be due to the tautomerism that has taken place through the open chain intermediate **55**. Thermal treatment of this equilibrium mixture afforded **56**, which was converted to **1** and **2** in a 1:1.3 ratio and 90% yield.

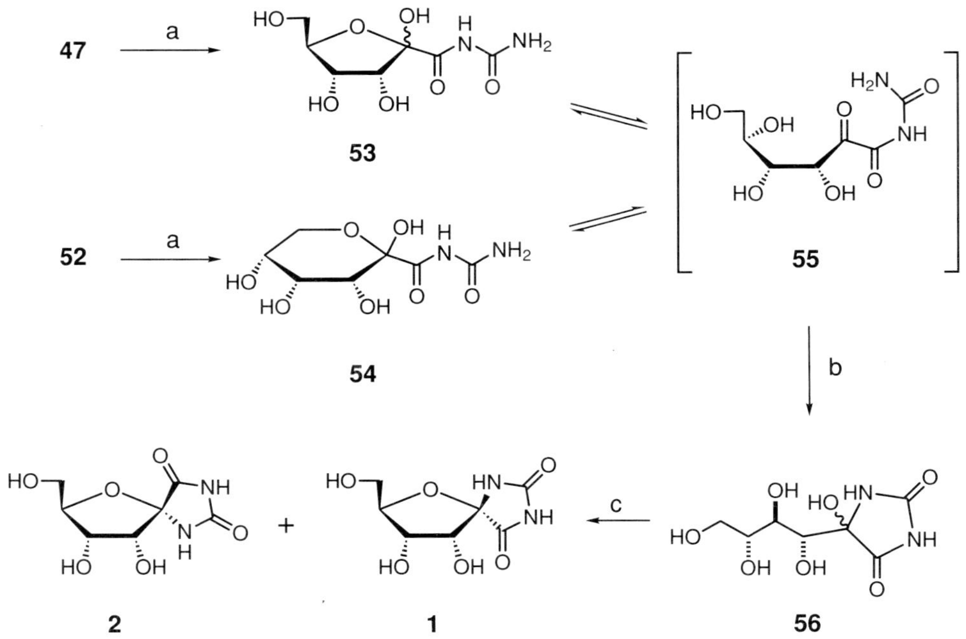

Scheme 5 (*a*) H_2 (4 atm), 10% Pd *on* C, EtOH, rt, 96% from **47**, 87% from **49**. (*b*) H_2O, 80°C, 100%. (*c*) Dowex 50X (H^+) resin, *n*-PrOH–H_2O, 45°C, 90% (**1**:**2**, 43:57).

2.1.2 *Synthesis from* D*-ribose*

Synthesis of (+)-hydantocidin (**1**) by starting with D-ribose via a stereoselective bromination of β-D-ribofuranosylamide **58** has been reported (Scheme 6).[49] The readily available 2,3,5-tri-*O*-benzoyl-β-D-ribofuranosyl cyanide[50] (**57**) was converted with manganese(IV) oxide in dichloromethane[51] to the amide **58** in 95% yield, which upon free radical bromination with NBS and benzoyl peroxide produced the α-bromo amide **59** stereoselectively. Spirocyclization of **59** was effected with freshly prepared silver cyanate at 80°C in anhydrous nitromethane to furnish a 1:2 mixture of **60** and **61** in 46% yield. The minor isomer **60** could be converted to the major one **61** with camphorsulfonic acid. Deprotection of **61** with LiOOH led to **1** in 90% yield.

2,3-Cyclohexylidene D-ribofuranose (**62**)[52] has been used for the synthesis of (+)-hydantocidin (**1**) and 5-*epi*-hydantocidin (**2**) (Schemes 7 and 8).[53,54] Treatment of **62** with sodium cyanide afforded the crystalline altrono-δ-lactone **63**.[55] The primary hydroxyl group in **63** was protected as the silyl ether and the secondary hydroxyl group was esterified with triflic anhydride to produce **64** in 92% overall yield from **63**. Reaction of the triflate **64** with sodium azide gave a mixture of the altronoazidolactone **65** (50%) and the allonoazidolactone **66** (25%). The silyl group in **65** was selectively removed with aqueous acetic acid at 60°C to give **67** (76%) together with a trace of **68**. On the other hand, treatment of **66** with aqueous acetic acid afforded **68** (47%) accompanied by 28% of **67**. However, reaction of

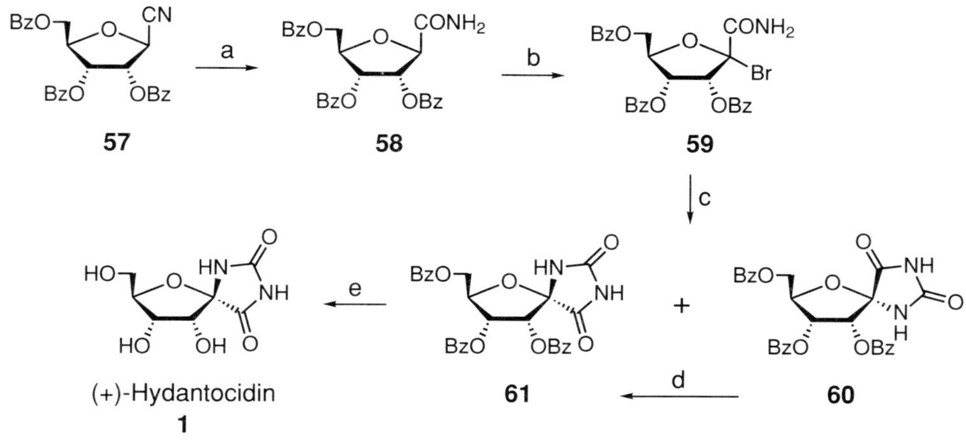

Scheme 6 (*a*) $Mn^{IV}O_2$, CH_2Cl_2, 95%. (*b*) NBS, $(PhCOO)_2$, CCl_4, 51%. (*c*) AgOCN, CH_3NO_2, 80°C, **61**:**60** in 2:1 ratio, 46%. (*d*) CSA. (*e*) LiOOH, THF, H_2O, 0°C, 90%.

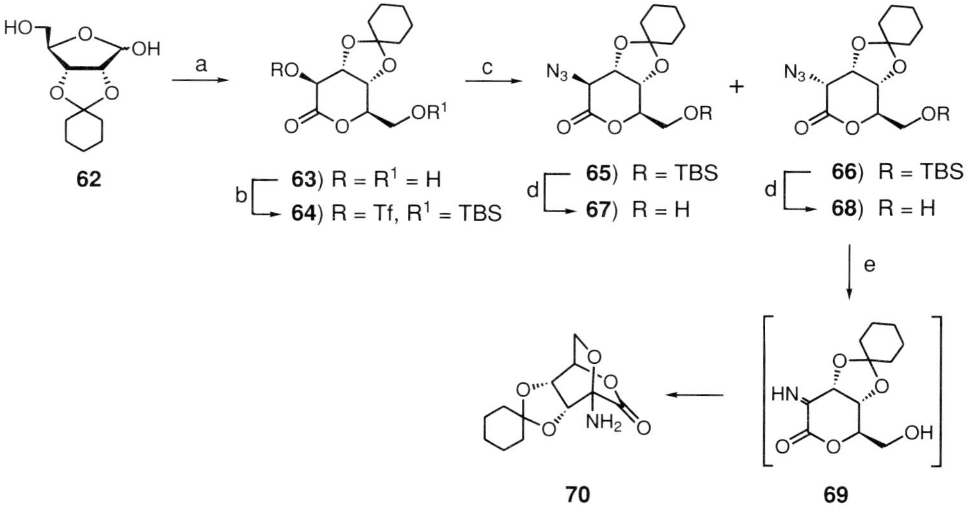

Scheme 7 (*a*) NaCN, 20% from D-ribose. (*b*) 1. TBSCl, DMF, imidazole, 92%; 2. Tf_2O, Py, CH_2Cl_2, quantitative. (*c*) NaN_3, DMF, 10 min, **65** (50%), **66** (25%). (*d*) Aqueous AcOH, 60°C, **65** gave 76% of **67** and trace of **68**, **66** gave 28% of **67** and 47% of **68**. (*e*) TPAP, MNO, CH_3CN, rt, 1 h, 62%.

67 and **68** with TPAP in the presence of morpholine-*N*-oxide gave a single product **70** in 63% yield via the intermediate imine **69**.

The amine **70** was treated with potassium cyanate in acetic acid to yield **71** (76%). Cyclization of **71** by treatment with potassium *tert*-butoxide in DMF afforded the cyclohexylidene *epi*-hydantocidin **73** via **72** in 61% yield from **70**. Acetylation of **73** gave **74**,

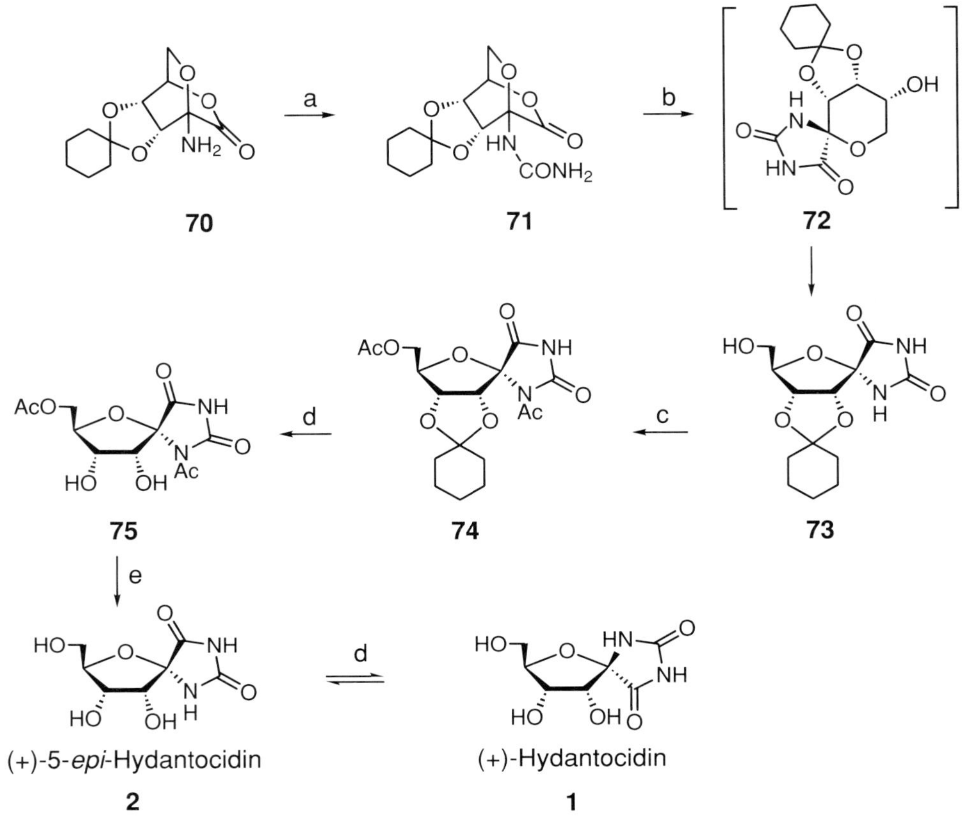

Scheme 8 (*a*) KOCN, AcOH, 60°C, 76%. (*b*) *t*-BuOK, DMF, rt, 61% for two steps. (*c*) Ac_2O, Py, DMAP, rt, 72%. (*d*) TFA–H_2O (2:3), rt. (*e*) $N_2H_4{\cdot}H_2O$, CH_3OH, rt, 70%.

whose cyclohexylidene group was then removed either by treatment with aqueous TFA or with acidic ion-exchange resin in methanol to give **75**, which upon deacetylation produced 5-*epi*-hydantocidin **2**. Significant epimerization of the spirocenter of hydantocidin occurred during the employed acidic conditions for the removal of the protecting groups.

2.1.3 *Synthesis from* D*-threose*

The first total synthesis of (+)-hydantocidin (**1**) in an optically active form has established its absolute configuration (Scheme 9).[16] The synthesis was started by aldol condensation of 4-*O*-benzyl-2,3-*O*-isopropylidene-D-threose (**78**)[56–58] and 1-*N*-acetyl-3-*N*-(4-methoxybenzyl)-hydantoin (**76**)[59,60] in the presence of potassium *tert*-butoxide to afford a mixture of *Z*- and *E*-isomers **79** (71%) and **80** (14%), respectively. Treatment of the mixture of **79** and **80** with *p*-toluenesulfonic acid afforded the cyclized product **82** and **81**. Alternatively, treatment of (2*R*,3*R*)-4-benzyloxy-2,3-epoxybutanal (**84**)[61–63] with lithium derivative of **77** afforded **85**, which upon treatment with lithium bis(trimethylsilyl)amide

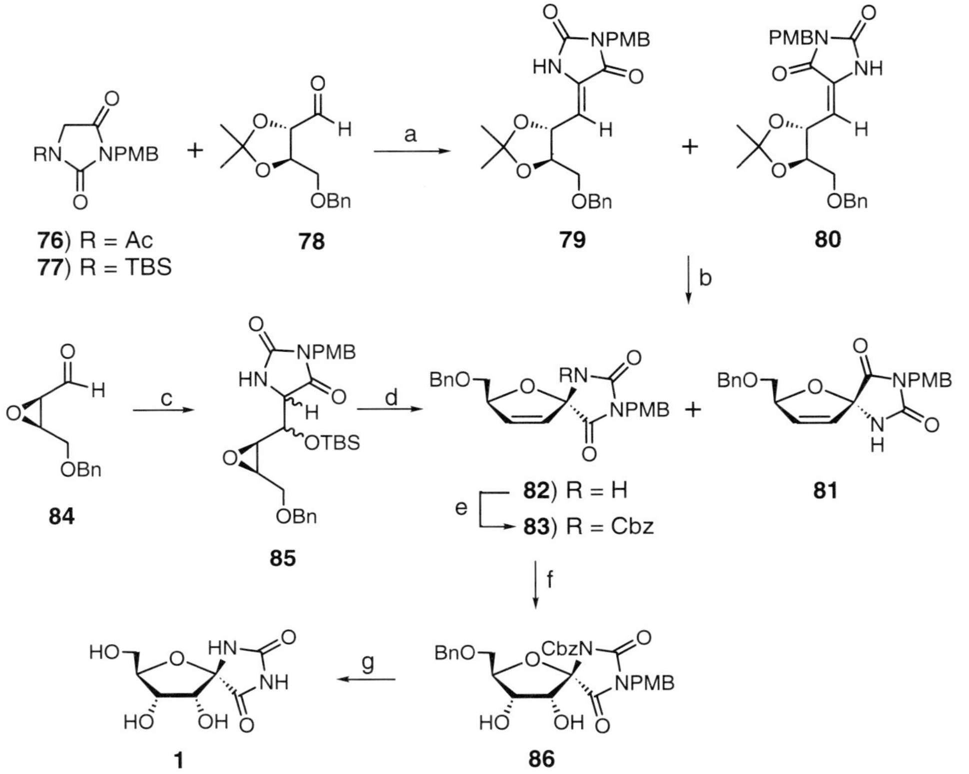

Scheme 9 (*a*) *t*-BuOK, dioxane, rt, 5 h, **79** (71%), **80** (14%). (*b*) *p*-TsOH, H_2O, MS 4 Å; reflux, 2 h, CH_2Cl_2, **79–81** (30%) and **82** (52%); **80**, **81** (23%) and **82** (38%). (*c*) **77**, THF, –78°C, 20 min, $LiN(TMS)_2$, 95%. (*d*) $LiN(TMS)_2$, THF, rt, **81**:**82** (2:1), 54%. (*e*) CbzCl, *t*-BuOK, THF, 97%. (*f*) OsO_4, acetone, MNO, *t*-BuOH, H_2O, rt, 5 days. (*g*) 1. CAN, CH_3CN, H_2O, rt, 20 min, 94%; 2. H_2 (3.5 kg/cm^2), 5% Pd *on* C, CH_3OH, 89%.

produced a mixture of **82** and **81** in a ratio of 1:2. Protection of the amide NH group in **82** with benzyl chloroformate afforded the carbamate **83**, which upon hydroxylation with osmium tetroxide produced only one stereoisomer **86** (48%) along with the recovery of **83** (50%). Treatment of **86** with CAN and subsequent hydrogenation furnished **1** in 89% yield.

2.1.4 *Synthesis from* D*-ribonolactone*

The azido aldehyde **91** as intermediate for the synthesis of (+)-hydantocidin was synthesized from ribonolactone (Scheme 10).[22] The α- or β-anomers **87** and **88** were readily obtained (76–80%) through the addition of 2-lithiothiazole to the ribonolactone and subsequent acetylation.[64,65] Their reaction with $TMSN_3$ afforded the α- and β-azides[66] **89** and **90** in a 1:3 ratio and 84% overall yield. The cleavage of the thiazole ring in the major isomer **90** by using either mercury(II) or copper(II) ion assisted hydrolysis[67] in the final step afforded the aldehyde **91** (57%).

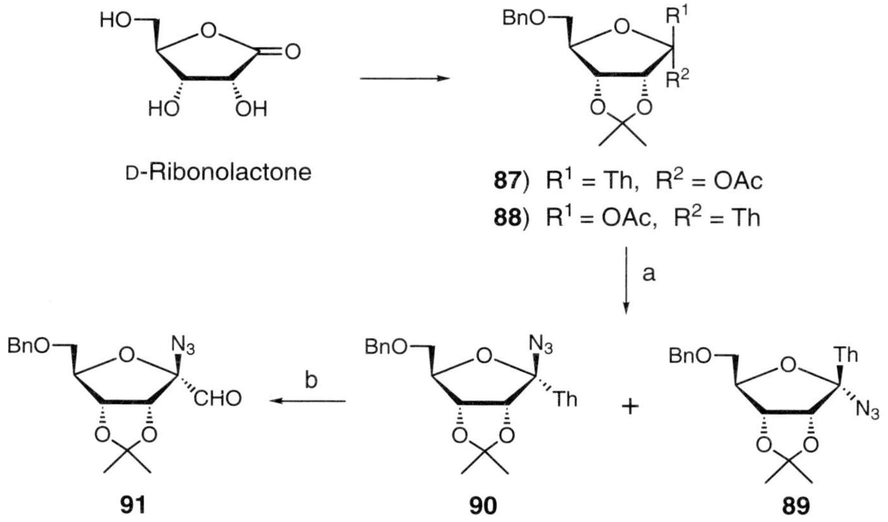

Scheme 10 (*a*) $TMSN_3$, TMSOTf, 84%. (*b*) 1. TfOMe; 2. $NaBH_4$; 3. $HgCl_2$, H_2O, 57% for three steps.

References

1. Nakajima, M.; Itoi, K.; Takamatsu, Y.; Kinoshita, T.; Okazaki, T.; Kawakubo, K.; Shindou, M.; Honma, T.; Tohjigamori, M.; Haneishi, T. *J. Antibiot.* **44** (1991) 293.
2. Haruyama, H.; Takayama, T.; Kinoshita, T.; Kondo, M.; Nakajima, M.; Haneishi, T. *J. Chem. Soc., Perkin Trans. 1* (1991) 1637.
3. Sankyo, European Patent Application 0232572A2, 1987. (See references 14 and 15).
4. Ciba Geigy AG, DE Patent 4129,616A (1990) (See references 14 and 15).
5. Mitsubishi KIasei Corp., Japanese Patent 04222589-A, 1990. (See references 14 and 15).
6. Sammes, R., In *Topics in Antibiotic Chemistry*, Vol. 6; Hardwood, E., Ed.; Wiley: New York, 1982.
7. Townsend, L.B. *Chemistry of Nucleoside and Nucleotide*, Vol. 1; Plenum Press: New York, 1988.
8. Takahashi, S.; Nakajima, M.; Kinoshita, T.; Haruyama, H.; Sugai, S.; Honma, T.; Sato, S.; Haneishi, T. *ACS Symp. Ser.* **551** (1994) 74.
9. Haneishi, T.; Nakajima, M.; Torikata, A.; Tohjigamori, M.; Kawakubo, K. Japanese Patent Application 85/2060305, Feb. 1985; *Chem. Abstr.* **115** (1991) 66804.
10. Mirza, S. CH Application 90/2929, Sept. 10, 1990; *Chem. Abstr.* **117** (1992) 8356.
11. Honma, T.; Shindo, M.; Mizukai, M.; Mio, S. Japanese Kokai Tokkyo Koho JP 04,235,102; Japanese Patent Application. 90/160,667, June 19, 1990; *Chem. Abstr.* **118** (1993) 75374n.
12. Matsumoto, M.; Kirihara, M.; Yoshino, T.; Katoh, T., Terashima, S. *Tetrahedron Lett.* **34** (1993) 6289.
13. Nakajima, N.; Matsumoto, M.; Kirihara, M.; Hashimoto, M.; Katoh, T.; Terashima, S. *Tetrahedron* **52** (1996) 1177.
14. Mio, S.; Shiraishi, M.; Sugai, S.; Haruyama, H.; Sato, S. *Tetrahedron* **47** (1991) 2121.
15. Mio, S.; Ueda, M.; Hamura, M.; Kitagawa, J.; Sugai, S. *Tetrahedron* **47** (1991) 2145.
16. Mio, S.; Ichinose, R.; Goto, K.; Sugai, S; Sato, S. *Tetrahedron* **47** (1991) 2111.
17. Mio, S.; Kitagawa, J.; Sugai, S. *Tetrahedron* **47** (1991) 2133.
18. Fairbanks, A.J.; Ford, P.S.; Watkin, D.J.; Fleet, G.W.J. *Tetrahedron Lett.* **34** (1993) 3327.
19. Fairbanks, A.J.; Fleet, G.W.J. *Tetrahedron* **51** (1995) 3881.
20. Burton, J.W.; Son, J.C.; Fairbank, A.J.; Choi, S.S.; Taylor, H.; Watkin, D.J.; Winchester, B.G.; Fleet, G.W.J. *Tetrahedron Lett.* **34** (1993) 6119.
21. Sano, H.; Mio, S.; Kitagawa, J., *Tetrahedron: Asymmetry* **5** (1994) 2233.
22. Dondoni, A.; Scherrmann, M.C.; Marra, A.; Delépine, J.-L. *J. Org. Chem.* **59** (1994) 7517.
23. Bichard, C.J.F.; Mitchell, E.P.; Wormald, M.R., Watson, K.A.; Johnson, L.N.; Zographos, S.P.; Koutra, D.D.; Oikonomakos, N.G.; Fleet, G.W.J., *Tetrahedron Lett.* **36** (1995) 2145.

24. Brondstetter, T.W.; Kim, Y.H.; Son, J.C.; Taylor, H.M.; Lilley, P.M.Q.; Watkin, D.J.; Johnson, L.N.; Oikonomakos, N.G.; Fleet, G.W.J. *Tetrahedron Lett.* **36** (1995) 2149.
25. Sano, H.; Mio, S.; Tsukaguchi, N.; Sugai, S. *Tetrahedron* **51** (1995) 1387.
26. Sano, H.; Sugai, S. *Tetrahedron* **51** (1995) 4635.
27. Hannessian, S.; Sanceau, J.Y.; Chemla, P. *Tetrahedron* **51** (1995) 6669.
28. Brondstetter, T.W.; Fuente, C.D.L.; Kim, Y.; Cooper, R.I.; Watkin, D.J.; Oikonomakos, N.G.; Johnson, L.N.; Fleet, G.W.J. *Tetrahedron* **52** (1996) 10711.
29. Estevez, J.C.; Smith, M.D.; Lane, A.L.; Crook, S.; Watkin, D.J.; Besra, G.S.; Brennan, P.J.; Nash, R.J.; Fleet, G.W.J. *Tetrahedron: Asymmetry* **7** (1996) 387.
30. Brondstetter, T.W.; Fuente, C.D.L.; Kim, Y.; Johnson, L.N.; Crook, S.; Lilley, P.M.D.Q.; Watkin, D.J.; Tsitsanou, K.E.; Zographos, S.E.; Chrysina, E.D.; Oikonomakos, N.G.; Fleet, G.W.J. *Tetrahedron* **52** (1996) 10721.
31. Estevez, J.C.; Long, D.D.; Wormald, M.R.; Dwek, R.A.; Fleet, G.J. *Tetrahedron Lett.* **36** (1995) 8287.
32. Estevez, J.C.; Smith, M.D.; Wormald, M.R.; Besra, G.S.; Brennam, P.J.; Nash, R.G.; Fleet, G.J. *Tetrahedron: Asymmetry* **7** (1996) 391.
33. Sano, H.; Mio, S.; Kitagawa, J.; Shinodu, M.; Honma, T.; Sugai, S. *Tetrahedron* **51** (1995) 12563.
34. Shiozaki, M. *Carbohydr. Res.* **335** (2001) 147.
35. Gasch, C.; Salameh, B.A.B.; Pradera, M.A.; Fuentes, J. *Tetrahedron Lett.* **42** (2001) 8615.
36. Koóš, M.; Steiner, B.; Langer, V.; Gyepesová, D.; Ďurik, M. *Carbohydr. Res.* **328** (2000) 115.
37. Postel, D.; Van Nhien, A.N.; Villa, P.; Ronco, G. *Tetrahedron Lett.* **42** (2001) 1499.
38. Sano, H.; Sugai, S., *Tetrahedron: Asymmetry* **6** (1995) 1143.
39. Krulle, T.M., Watson, K.A.; Gregoriou, M.; Johnson, L.N.; Crook, S.; Watkin, D.J.; Griffiths, R.C.; Nash, R.J.; Tsitsanou, K.E.; Zographos, S.E.; Oikonomakos, N.G.; Fleet, G.W.J. *Tetrahedron Lett.* **36** (1995) 8291.
40. Fuente, C.D.L.; Krulle, T.M.; Watson, K.A.; Gregoriou, M.; Johnson, L.N.; Tsitsanou, K.E.; Zographos, S.E.; Oikonomakos, N.G.; Fleet, G.W.J. *Synlett* (1997) 485.
41. Krulle, T.M.; Fuente, C.D.L.; Watson, K.A.; Gregoriou, M.; Johnson, L.N.; Tsitsanou, K.E.; Zographos, S.E.; Oikonomakos, N.G.; Fleet, G.W.J. *Synlett* (1997) 211.
42. Osz, E.; Sos, E.; Somsak, L.; Szilagyi, L.; Dinya, Z. *Tetrahedron* **53** (1997) 5813.
43. Mancuso, A.J.; Huang, S.L.; Swern, D. *J. Org. Chem.* **43** (1978) 2480.
44. Isobe, M.; Ichikawa, Y.; Goto, T. *Tetrahedron Lett.* **27** (1986) 963.
45. Kraus, A.G., Taschner, M.J. *J. Org. Chem.* **45** (1980) 1175.
46. Bal, B.S.; Childers, W.E., Jr., Pinnick, H.W. *Tetrahedron* **37** (1981) 2091.
47. Prisbe, E.J.; Smejakal, J.; Verheyden, J.P.H.; Moffat, J.G. *J. Org. Chem.* **41** (1976) 1836.
48. Chemla, P. *Tetrahedron Lett.* **34** (1993) 7391.
49. Harrington, P.M.; Jung, M.E. *Tetrahedron Lett.* **35** (1994) 5145.
50. Hashimoto, K.; Wakabashi, Y.; Horiie, T.; Inoue, M.; Shishiyama, Y.; Obayashi, M.; Nozaki, H. *Tetrahedron* **39** (1983) 967.
51. Cook, M.; Forbes, E.; Khan, G. *J. Chem. Soc., Chem. Commun.* (1966) 121.
52. Mori, K.; Kikuchi, H. *Liebigs Ann. Chem.* (1989) 1267.
53. Fairbanks, A.J.; Ford, P.S.; Watkin, D.J.; Fleet, G.W.J. *Tetrahedron Lett.* **34** (1993) 3327.
54. Fairbanks, A.J.; Fleet, G.W.J. *Tetrahedron* **51** (1995) 3881.
55. Bichard, C.J.F.; Fairbanks, A.J.; Fleet, G.W.J.; Ramsden, N.G.; Voget, K.; Doherty, O.; Pearce, L.; Watkin, D.J. *Tetrahedron: Asymmetry* **2** (1991) 901.
56. Hungebuhler, E.; Seebach, D. *Helv. Chim. Acta* **64** (1981) 687.
57. Mukaiyama, T.; Suzuki, K.; Yamada, T. *Chem. Lett.* (1982) 929.
58. Lee, A.W.M.; Martin, V.S.; Masamune, S.; Sharpless, K.B.; Walkar, F. J. *J. Am. Chem. Soc.* **104** (1982) 3515.
59. Finkbeiner, H. *J. Org. Chem.* **30** (1965) 3414.
60. Lopez, C.G.; Trigo, G.C. *Adv. Heterocycl. Chem.* **38** (1985) 177.
61. Howe, A.W. M.; Procter, G. *Tetrahedron Lett.* **28** (1987) 2629.
62. Takano, S., Morimoto, M.; Ogasawara, K. *Synthesis* (1984) 834.
63. Takano, S.; Kasahara, C.; Ogasawara, K. *Chem. Lett.* (1983) 175.
64. Dondoni, A.; Scherrmann, M.C. *Tetrahedron Lett.* **34** (1993) 731.
65. Dondoni, A.; Scherrmann, M.C. *J. Org. Chem.* **59** (1994) 6404.
66. Gyorgydeak, Z.; Szilagyi, L.; Paulsen, H. *J. Carbohydr. Chem.* **12** (1993) 139.
67. Dondoni, A.; Marra, A.; Perrone, D. *J. Org. Chem.* **58** (1993) 275.

2.2 Bleomycin

Bleomycin is the generic name for a family of glycopeptide-derived antitumor antibiotics elaborated by *Streptomyces verticillus*.[1] Bleomycins **1–4** are of current interest because of their activity against sequamous cell carcinomas and malignant lymphomas including Hodgkin's disease.[2–9] The first proposed[10] structure of bleomycin A_2 was revised[11] to that shown in **1** as a result of X-ray crystallographic analysis of P-3A, a presumed biosynthetic intermediate in the elaboration of bleomycin. Syntheses of different moieties of bleomycins from noncarbohydrates have been reported.[12–15] For a total synthesis of bleomycin antibiotics, moieties such as heterocyclic and acyclic amino acids were prepared from carbohydrates.

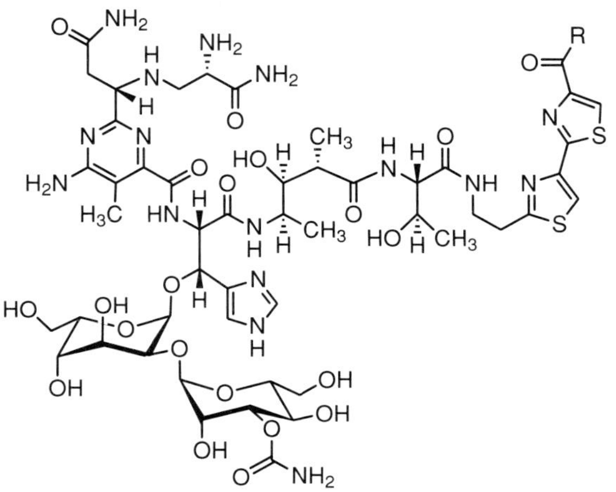

1) Bleomycin A_2, R = NH—$(CH_2)_3S^+(CH_3)_2X^-$
2) Pepleomycin, R = NH—$(CH_2)_3$NH—$CH(CH_3)C_6H_5$
3) Bleomycinic acid, R = OH
4) Bleomycin B_2, R = NH—$(CH_2)_4$NH—C(=NH)NH_2

2.2.1 *Synthesis from* D*-glucosamine*

A suitable method for the preparation of L-*erythro*-β-hydroxyhistidine has been reported from D-glucosamine as a starting material (Scheme 1).[16] Thus, 2-amino-2-deoxy-D-mannono-1,4-lactone (**5**) was obtained from D-glucosamine in 53% overall yield.[17,18] Its 2-acetyl derivative could be obtained by acetylation or directly in a single step from *N*-acetyl-D-mannosamine by oxidation with Br_2.[19] Selective oxidative cleavage of C-5—C-6 bond in the acetyl derivative of **5** with aqueous $NaIO_4$ afforded the aldehyde **6** in quantitative yield, which may exist in equilibrium with the dimer **7**. Dissolution of **6** in NH_4OAc and heating in the presence of $Cu(OAc)_2$ and excess formaldehyde afforded Cu(II) complex of *N*-acetyl-DL-*erythro*-β-hydroxyhistidine (**8**) in 25% yield, whose deacetylation gave **9**.

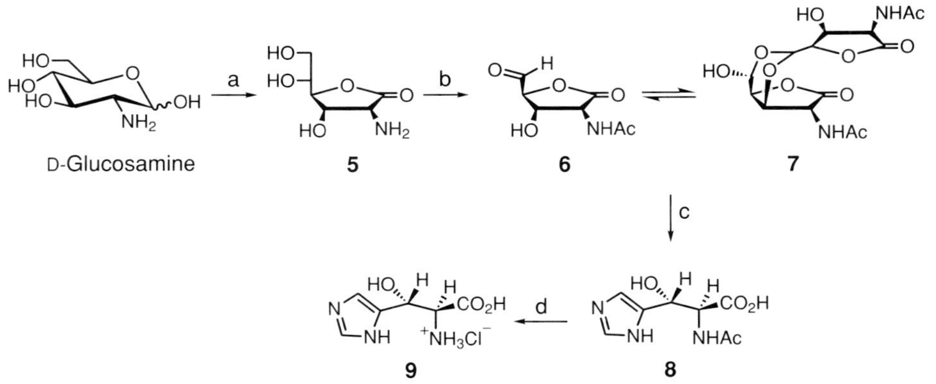

Scheme 1 (*a*) Refs. 17 and 18. (*b*) 1. Ac_2O, Dowex 1X4 (HCO_3^-) resin, 81%; 2. $NaIO_4$, H_2O, 4°C, 50 min, 100%. (*c*) 6.8 M NH_4OAc, HCHO, $Cu^{(II)}(OAc)_2$, 110°C, 3 h, 25% yield. (*d*) H_2S; then Dowex 50X8 (H^+) resin.

2.2.2 *Synthesis from* L*-rhamnose*

Synthesis of the optically pure (2*S*,3*S*,4*R*)-4-amino-3-hydroxy-2-methylvaleric acid (**18**) has been achieved from L-rhamnose by conversion to 5-deoxy-L-arabino-γ-lactone (**10**)[20,21] (Scheme 2).[22] Treatment of **10** with excess benzylamine in methanol afforded the

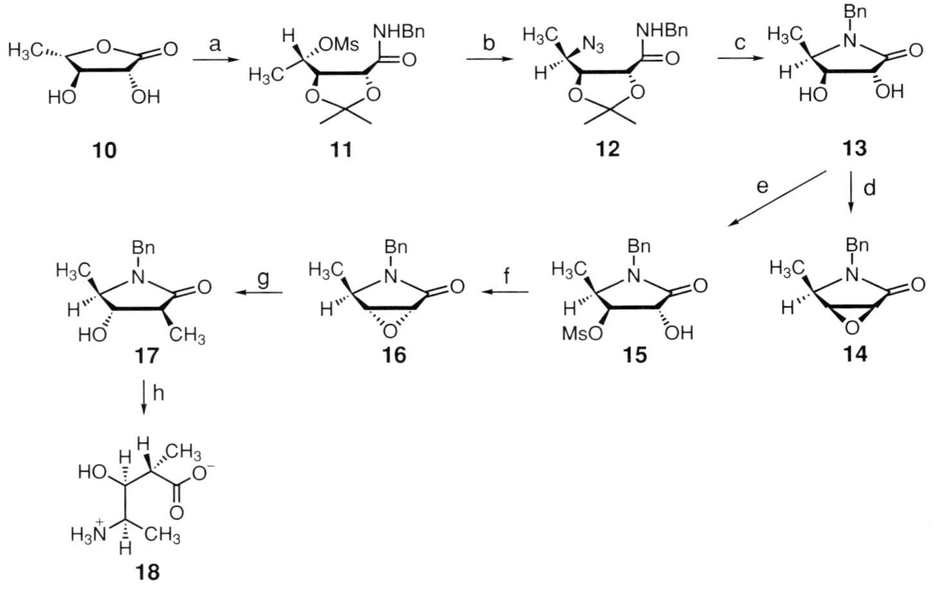

Scheme 2 (*a*) 1. $BnNH_2$, CH_3OH, reflux, 12 h, Hünig's base, 82%; 2. acetone, *p*-TsOH, 96%; 3. MsCl, Py, –20 to 0°C, 5 h, 100%. (*b*) NaN_3, DMF, 100°C, 5 h. (*c*) Hydrogenolysis, PhCHO, 61% from **10**. (*d*) THF, NaH, MsCl (2 equiv.). (*e*) *n*-BuLi, Et_2O, THF, MsCl; then isopropylmagnesium bromide (4.4 equiv.), THF, 0°C, MsCl, 18 h, 0–25°C, 70%. (*f*) *t*-BuOK, THF, 25°C, 40 min, 78%. (*g*) $(CH_3)_2CuLi$ (5 equiv.), ether, −78 to 5°C, 4 h, 57%. (*h*) Na, liquid NH_3, 84%; then 2 N HCl, reflux, 4 h, 95%.

corresponding N-benzylamide, whose acetonation and subsequent mesylation furnished **11** in 79% overall yield from **10**. S_N2 displacement of the mesyl group in **11** with azide ion produced the corresponding 4-azido derivative **12** in 92% yield, whose hydrogenolysis furnished the key intermediate **13** in 61% overall yield from **10**. Treatment of **13** with MsCl in the presence of NaH gave exclusively the undesired epoxide **14**. On the other hand, treatment of **13** with n-BuLi and MsCl followed by isopropylmagnesium bromide afforded **15** in 70% yield. The epoxide **16** was obtained in 78% yield from mesylate **15** by treatment with t-BuOK. Treatment of **16** with 5 equiv. of $(CH_3)_2CuLi$ furnished compound **17** in 57% yield, which was debenzylated with sodium in liquid ammonia followed by treatment with 2 N HCl to produce **18** in 15% overall yield from **10**.

2.2.3 *Total synthesis of bleomycin* A_2

For the total synthesis of bleomycin A_2 (**1**), the glycosyl donor **21** was prepared (Schemes 3 and 4).[23] Disaccharide **19**[24] was treated with a 3:1 mixture of acetic anhydride and acetic

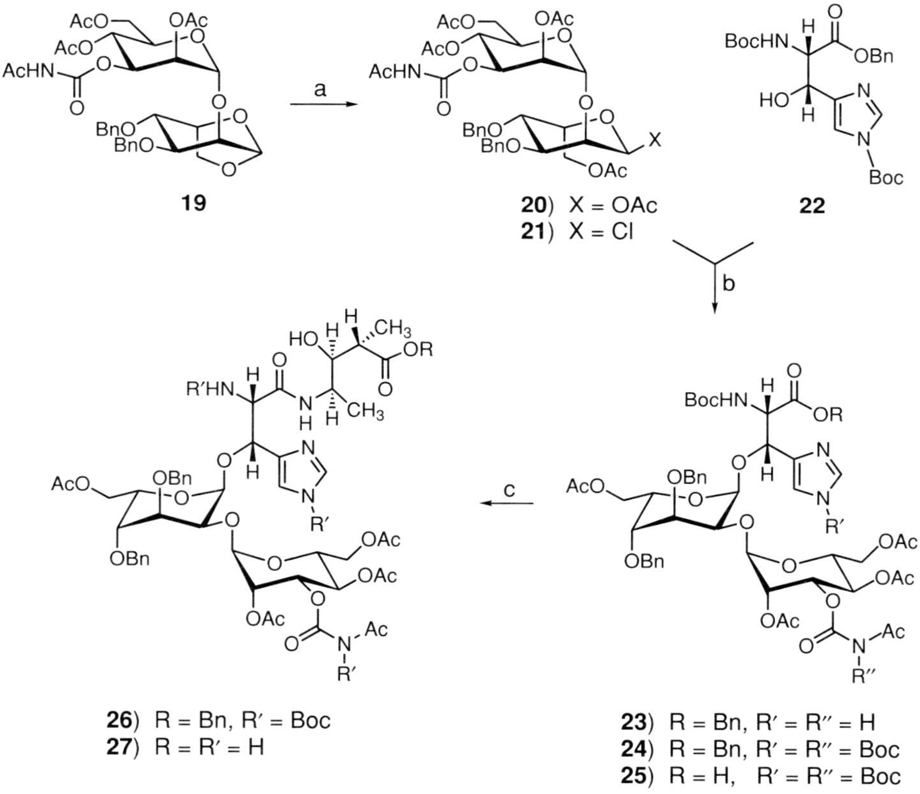

19

20) X = OAc
21) X = Cl

22

26) R = Bn, R′ = Boc
27) R = R′ = H

23) R = Bn, R′ = R″ = H
24) R = Bn, R′ = R″ = Boc
25) R = H, R′ = R″ = Boc

Scheme 3 (*a*) 1. Ac_2O–AcOH (3:1), 1% H_2SO_4, 1 h, 0°C, 100%; 2. HCl, CH_2Cl_2, 12 h, quantitative. (*b*) 1. **22**, CF_3SO_3Ag, $(CH_3)_2NCON(CH_3)_2$, $ClCH_2CH_2Cl$, 45°C, 12 h, 20–25%; 2. di-*tert*-butylpyrocarbonate, Py, 1 h, 25°C, 77%; 3. EtOAc, 5% Pd *on* C, H_2, 2 h, 45°C, 75–80%. (*c*) 1. Benzyl (2S,3S,4R)-4-amino-3-hydroxy-2-methylvalerate, CH_2Cl_2, DCC, 1-hydroxybenzotriazole, 25°C, 3 h, 77%; 2. EtOH, H_2, Pd black, 55°C, 24 h, quantitative.

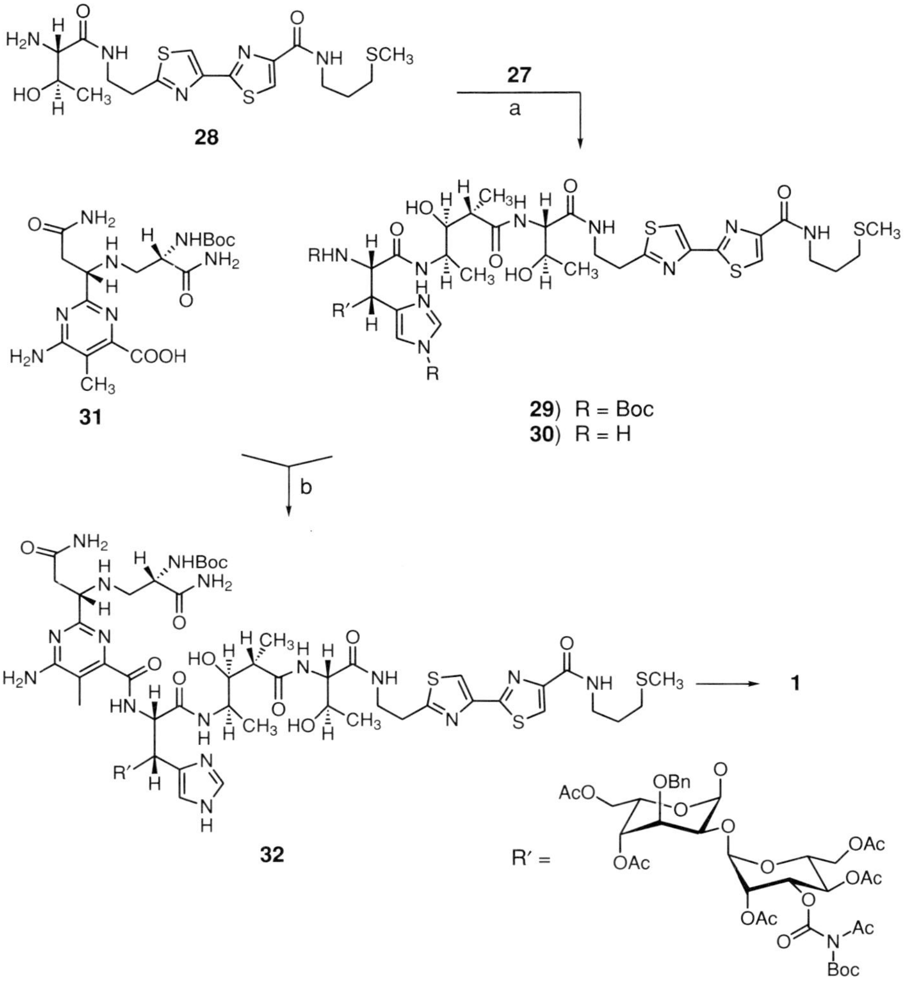

Scheme 4 (*a*) 1. DCC, 1-hydroxy benzotriazole, DMF, 25°C, 24 h, 61%; 2. DMS–TFA (3:5), 0°C, 1 h, 59%. (*b*) DMF, diphenylphosphoryl azide, 25°C, 48 h; then 0.1 M NaOH, 0°C, 22 h; then DMS–TFA (1:2), 0°C, 1 h.

acid containing 1% H_2SO_4 to give **20**, which was dissolved in dichloromethane containing HCl to produce the chloro derivative **21** in quantitative yield. Coupling of **21** with the L-*erythro*-di-*tert*-butoxycarbonyl-β-hydroxyhistidine benzyl ester[25] **22** provided **23** in 20–25% yield, which was treated with di-*tert*-butylpyrocarbonate to furnish the protected Boc derivative **24** in 77% yield. Subsequent hydrogenation of **24** effected selective removal of the benzyl ester group to produce the free carboxylate **25** in 75–80% yield. Condensation of **25** with benzyl (2*S*,3*S*,4*R*)-4-amino-3-hydroxy-2-methylvalerate[26] afforded the dipeptide analogue **26** in 77% yield, which was hydrogenated over palladium black to effect removal of the benzyl ester group and solvolysis of the *N*-Boc to produce the ester **27** in quantitative yield.

Condensation of **27** with the tripeptide derivative **28**[27] afforded the respective peptide **29** in 61% yield, which underwent removal of the Boc groups using DMS–TFA to give **30** in 59% yield, which upon coupling with Boc-pyrimidoblamic acid (**31**)[13,28] afforded **32**. Deblocking with 0.1 M NaOH followed by treatment with DMS–TFA produced bleomycin A_2 (**1**).[29]

Another total synthesis of bleomycin A_2 (**1**) has been achieved starting with the bromide **35** (Scheme 5).[30] The latter was prepared from 2-*O*-(α-D-mannopyranosyl)-α-L-gulopyranose (**33**)[31] by treatment with TBSCl followed by *N*,*N*′-carbonyldiimidazole and subsequent removal of the silyl groups and acetylation to afford **34**. Dissolving **34** in liquid ammonia followed by acetylation and treatment with HBr afforded **35**. Coupling of the pentapeptide **36**[32] with **35** in anhydrous sulfolane produced a mixture including **37**, which

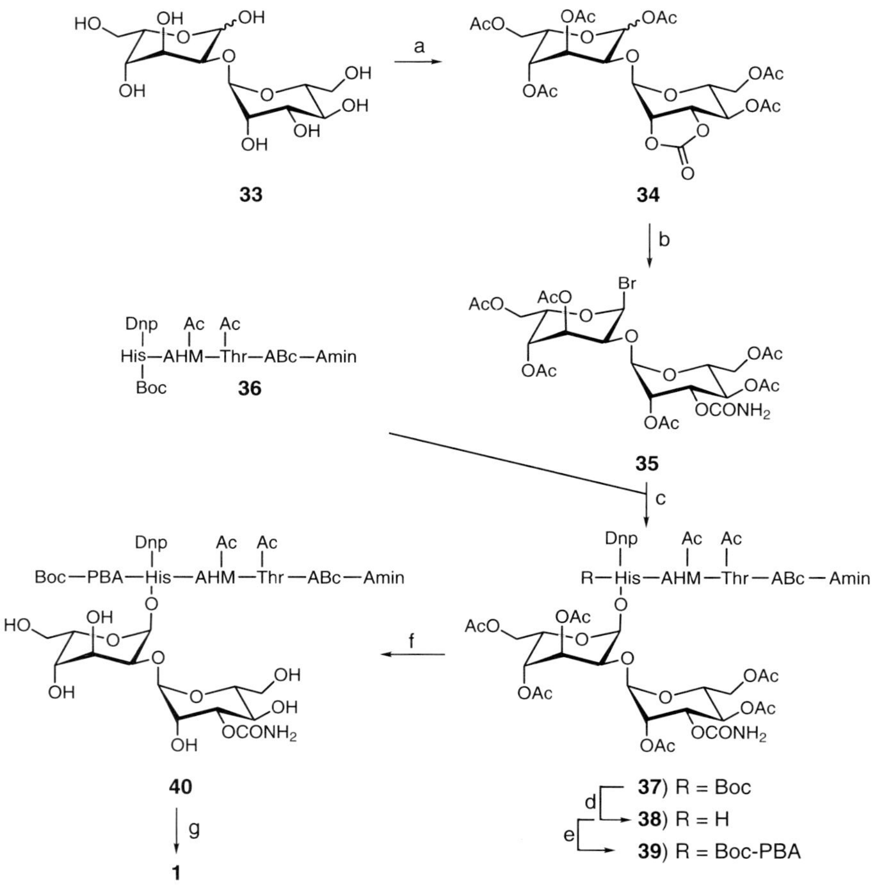

Scheme 5 (*a*) 1. TBSCl, imidazole, DMF, rt, 2 days, 71%; 2. *N*,*N*′-carbonyldiimidazole, THF, rt, overnight, 72%; 3. TBAF, THF, 65%. (*b*) 1. Liquid ammonia, –75°C to rt; 2. acetylation; 3. HBr, CH_2Cl_2, 0°C, overnight. (c) $Hg(CN)_2$, sulfoxane, MS 4 Å, 40°C to overnight, Sephadex LH-20. (*d*) TFA. (*e*) DCC–HOBt, DMF, rt, overnight. (*f*) 0.1 M NaOH–CH_3OH, rt, overnight. (*g*) TFA, 0°C, 30 min.

was used without further separation. After treatment with TFA, in order to remove the Boc protecting group, the resulting mixture containing **38** was allowed to react with **31**[14] to give **39**. The resulting product containing **39** was deprotected to afford **40**. The mixture containing **40** was treated with TFA at 0°C to produce **1**.

References

1. Umezawa, H. *Lioydia* **40** (1977) 67.
2. Umezawa, H. *Prog. Biochem. Pharmacol.* **11** (1976) 18.
3. Ichikawa, T. *Prog. Biochem. Pharamcol.* **11** (1976) 143.
4. Carter, S.K.; Blum, R.H. *Prog. Biochem. Pharmacol.* **11** (1976) 158.
5. Bonadonna, G.; Tancini, G.; Bajetta, E. *Prog. Biochem. Pharmacol.* **11** (1976) 172.
6. Bipierre, A. *Prog. Biochem. Pharmacol.* **11** (1976) 205.
7. Rathert, P.; Lutzeyer, W. *Prog. Biochem. Pharmacol.* **11** (1976) 223.
8. Koyama, G.; Nakamura, H.; Muraoka, Y.; Takita, T.; Maeda, K.; Umezawa, H.; Litaka, Y. *J. Antibiot.* **26** (1973) 109.
9. Crooke, S.T. In *Bleomycin: Current Status and New Developments*; Carter, S.K., Crooke, S.T., Umezawa, H., Eds.; Academic Press: New York, 1978; p. 1.
10. Takita, T.; Muraoka, Y.; Yoshioka, T.; Fujii, A.; Maeda, K.; Umezawa, H. *J. Antibiot.* **25** (1972) 755.
11. Takita, T.; Muraoka, Y.; Nakatani, T.; Fujii, A.; Umezawa, Y.; Naganawa, H.; Umezawa, H. *J. Antibiot.* **31** (1978) 801.
12. Boger, D.L.; Menezes, R.F. *J. Org. Chem.* **57** (1992) 4331.
13. Arai, H.; Hagmann, W.K.; Suguna, H.; Hecht, S.M. *J. Am. Chem. Soc.* **102** (1980) 6633.
14. Umezawa, Y.; Morishima, H.; Saito, S.; Takita, T.; Umezawa, H.; Kobayashi, S.; Otsuka, M.; Narita, M.; Ohno, M. *J. Am. Chem. Soc.* **102** (1980) 6630.
15. Tanaka, W.; Takita, T. *Heterocycles* **13** (1979) 469.
16. Hecht, S.M.; Rupprecht, K.M.; Jacobs, P.M. *J. Am. Chem. Soc.* **101** (1979) 3982.
17. Wolform, M.L.; Cron, M.J. *J. Am. Chem. Soc.* **74** (1952) 1715.
18. Levene, P.A. *J. Biol. Chem.* **36** (1918) 73.
19. Pravdic, N.; Fletcha, H.G., Jr., *Carbohydr. Res.* **19** (1971) 339.
20. Taylor, E.C.; Jacobi, P.A. *J. Am. Chem. Soc.* **98** (1976) 2301.
21. Andrews, P.; Hough, L.; Jones, J.K.N. *J. Am. Chem. Soc.* **77** (1955) 125.
22. Ohgi, T.; Hecht, S.M. *J. Org. Chem.* **46** (1981) 1232.
23. Aoyagi, Y.; Katano, K.; Suguna, H.; Primeau, J.; Chang, L.-H.; Hecht, S.M. *J. Am. Chem. Soc.* **104** (1982) 5537.
24. Pozsgay, V.; Ohgi, T.; Hecht, S.M.J. *J. Org. Chem.* **46** (1981) 3761.
25. Hecht, S.M.; Rupprecht, K.M.; Jacobs, P.M. *J. Am. Chem. Soc.* **101** (1979) 3982.
26. Narita, M.; Otsuka, M.; Kobayashi, S.; Ohno, M.; Umezawa, Y.; Morishima, H.; Saito, S.; Takita, T.; Umezawa, H. *Tetrahedron Lett.* **23** (1982) 525.
27. Levin, M.D.; Subrahamania, K.; Katz, H.; Smith, M.B.; Burlett, D.J.; Hecht, S.M. *J. Am. Chem. Soc.* **101** (1979) 1452.
28. Umezawa, Y.; Morishima, H.; Saito, S.; Takita, T.; Umezawa, H.; Kobayashi, S.; Otsuka, M.; Narita, M.; Ohno, M. *J. Am. Chem. Soc.* **102** (1980) 6631.
29. Roy. S.N.; Orr, G.A.; Brewer, C.F.; Horwitz, S.B. *Cancer Res.* **41** (1981) 4471.
30. Takita, T.; Umezawa, Y.; Saito, S.; Morishima, H.; Naganawa, H.; Umezawa, H.; Tsuchiya, T.; Miyake, T.; Kageyama, S.; Umezawa, S.; Muraoka, Y.; Suzuki, M.; Otsuka, M.; Narita, M.; Kobayashi, S.; Ohno, M. *Tetrahedron Lett.* **23** (1982) 521.
31. Tsuchiya, T.; Miyake, T.; Kageyama, S.; Umezawa, S.; Umezawa, H.; Takita, T. *Tetrahedron Lett.* **22** (1981) 1413.
32. Saito, S.; Umezawa, Y.; Morishima, H.; Takita, T.; Umezawa, H.; Narita, M.; Otsuka, M.; Kobayashi, S.; Ohno, M. *Tetrahedron Lett.* **23** (1982) 529.

2.3 Calyculins

Calyculins **A–H** are bioactive metabolites isolated from the sponge *Discodermia calyx*.[1,2] Calyculins are potent serine–threonine protein phosphatase (PP1 and PP2A) inhibitors[3–7] and endowed with remarkable cell membrane permeability.[8] The absolute stereochemistry of calyculins and its $C_{33}-C_{37}$ portion has been investigated.[9,10] The relative configuration of the stereocenters was determined by X-ray analysis.[1,2]

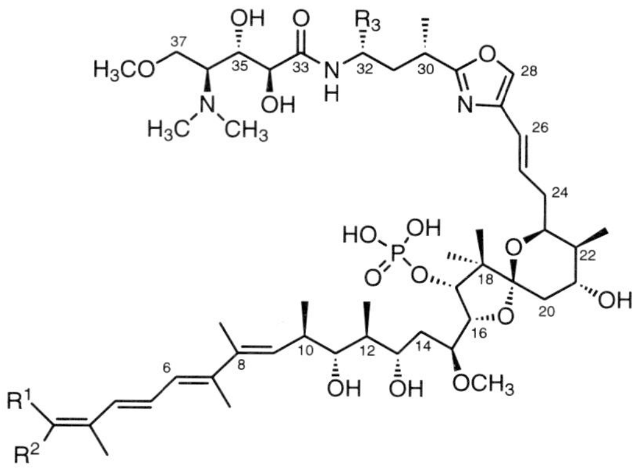

Calyculins

	R^1	R^2	R^3	
A)	CN	H	H	**E**) 6*Z*-isomer of **A**
B)	H	CN	H	**F**) 6*Z*-isomer of **B**
C)	CN	H	CH_3	**G**) 6*Z*-isomer of **C**
D)	H	CN	CH_3	**H**) 6*Z*-isomer of **D**

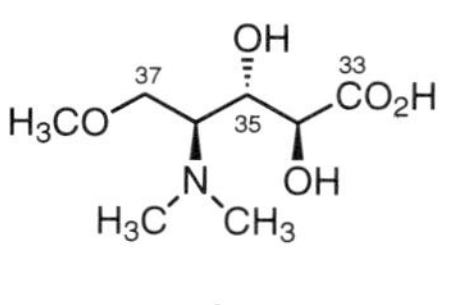

1 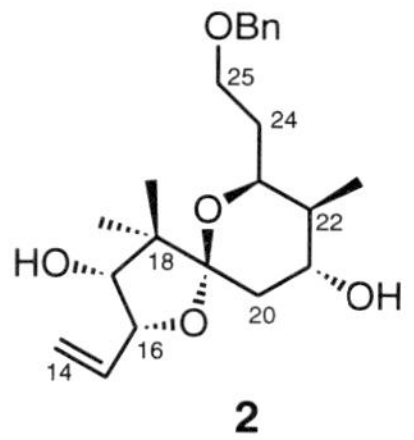 **2** 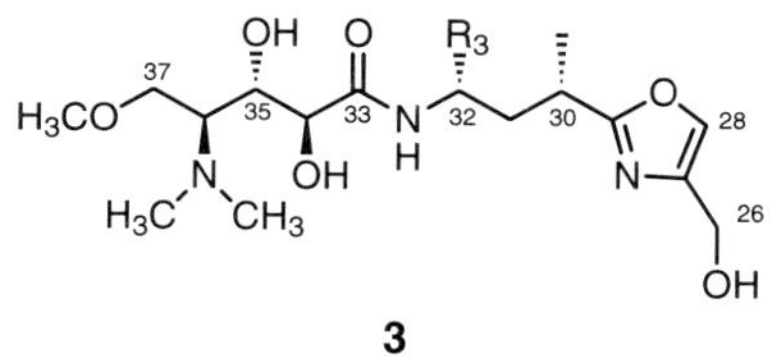 **3**

Syntheses of calyculins and their fragments from noncarbohydrates have received much attention since their isolation.[11–23] The main building blocks in the structures of calyculins, which can be deduced from a retrosynthetic analysis and can fit with the scope of this book, are the acyclic chain amino acid **1**, the spiral acetal **2**, the oxazole derivative **3** and a polyene derivative.

2.3.1 *Synthesis from* D-*lyxose*

The $C_{26}-C_{37}$ fragment **15** of calyculin **C** has been synthesized from methyl 2,3-di-*O*-benzyl-α-D-lyxofuranoside (**4**),[24] prepared from D-lyxose (Scheme 1).[25] Methylation of

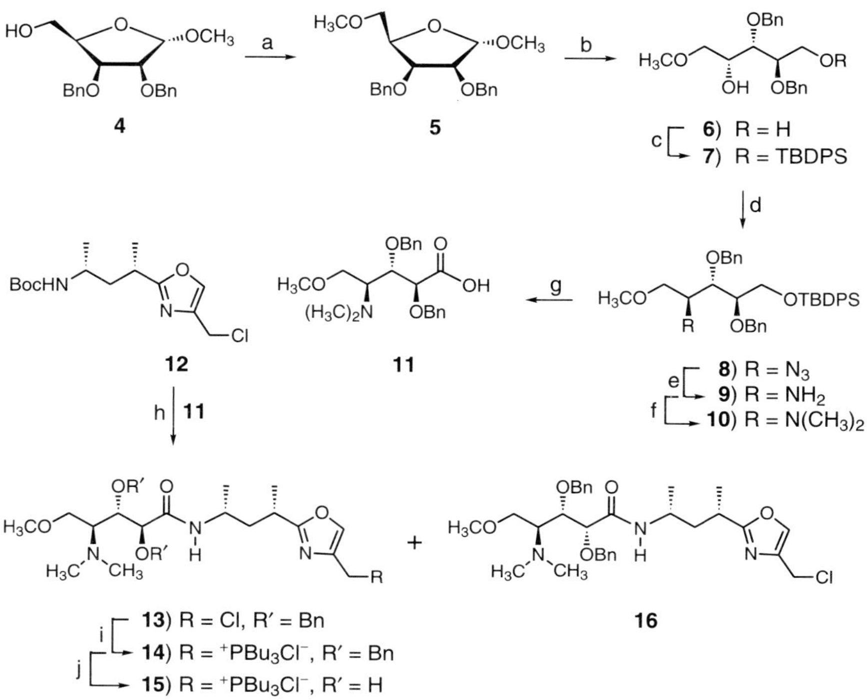

Scheme 1 (*a*) NaH, CH_3I, 81%. (*b*) 1. HCl, AcOH, 70°C; 2. $NaBH_4$, 79%. (*c*) TBDPSCl, imidazole, 100%. (*d*) 1. MsCl, NEt_3; 2. NaN_3, DMF, 100°C, 81%. (*e*) $LiAlH_4$, 58%. (*f*) HCHO, $NaBH_3CN$, AcOH, 83%. (*g*) 1. TBAF, 96%; 2. Jones oxidation, 38%; 3. $(TMS)CHN_2$, HCl, 97%. (*h*) 1. HCl, EtOAc; 2. $Al(CH_3)_3$; then **11**, 35°C, 62%. (*i*) PBu_3, DMF, 70%. (*j*) H_2, Pd *on* C, HCl, CH_3OH, 68%.

4 afforded **5** (81%), which underwent acidic hydrolysis followed by reduction of the resulting hemiacetal to produce **6**. Silylation of **6** afforded **7** in a quantitative yield that upon mesylation and subsequent azide displacement afforded **8**, whose reduction with $LiAlH_4$ gave the amine **9**. Subsequent methylation under exhaustive reductive amination conditions gave the dimethylamine **10** (83%). Desilylation of **10** with TBAF followed by Jones oxidation[26,27] and esterification afforded **11**, which was reacted with the oxazole **12** after deprotection to afford a 2.7:1 separable mixture of diastereomers **13** and **16**. Treatment of **13** with PBu_3 afforded **14** that was hydrogenated in the presence of palladium to produce the $C_{26}-C_{37}$ fragment **15**.

2.3.2 *Synthesis from* D*-gulonolactone*

Synthesis of C_{26}–C_{37} fragment **22**[28] of calyculins **A** and **B** has been done from lactone **17**,[29,30] obtained from D-gulonolactone (Scheme 2). Addition of Weinreb reagent[31] to **17** followed by mesylation and intramolecular cyclization using potassium *tert*-butoxide provided lactam **18**, with inversion of stereochemistry at the γ-position. Cleavage of the acetonide group of **18** and subsequent benzylation afforded **19**, which was converted into the dimethylamine ester **20** in 79% yield. Coupling of **20** with compound **21** provided the C_{26}–C_{37} fragment **22**.

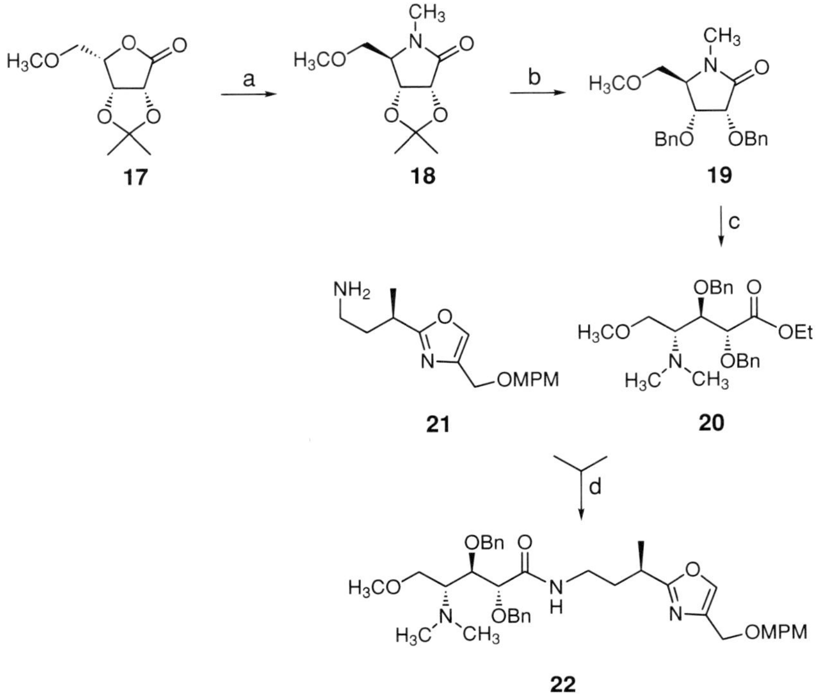

Scheme 2 (*a*) 1. $CH_3AlClNHCH_3$, benzene, rt, 99%; 2. MsCl, NEt_3, 0°C; 3. *t*-BuOK, THF, –40°C, 84% for two steps. (*b*) 1. HCl, THF; 2. NaH, BnCl, THF, 83% for two steps. (*c*) 1. Et_3OBF_4, CH_2Cl_2, 2,6-di-*t*-butylpyridine; 2. aqueous CH_2O, CH_3OH, 3. $ZnCl_2$, $NaBH_3CN$, CH_3OH, rt, 79% for three steps. (*d*) 1. LiOH, THF, H_2O; 2. DCC, DMAP, 64% for two steps.

2.3.3 *Synthesis from L-idonolactone*

L-Idonolactone has been used for the synthesis of the spiroketal part (Schemes 3–5).[32] The methyl ester **23**[33] was transformed into the methyl ketone **24**, which underwent a diastereoselective reaction with aldehyde **25**, prepared from **26**,[34] to provide a separable mixture of **27** and its epimer in 55% yield in 18:1 ratio. The aldol adduct **27** was transformed into the spiroketal **28** in 63% yield, which underwent bis-silylation followed by selective removal of the *C*-14-TBS group to provide the primary alcohol **29** in 85% yield.

Oxidation of **29** with Pr_4NRuO_4 gave the respective aldehyde, which was coupled with the lactone **30**[34] in the presence of LDA to give a diastereoisomeric mixture of the coupled product **31** in 84% yield. Barton's deoxygenation[35–37] of the secondary hydroxyl group of **31** furnished a mixture of **32** and its *C*-13-epimer in 62% yield in a 4:1 ratio. The undesired *C*-13-epimer was epimerized with CH_3Li in THF at –78°C to give the desired one **32** (63%). The acetonide **33** was obtained[38,39] in 56% overall yield from **32**. After oxidative removal of the *C*-17 MPM group[40] from **33** in 94% yield, the liberated *C*-17 alcohol **34** was converted to its bis(2-trimethylsilylethyl)phosphate triester, followed by removal of the *C*-9-TBS group to give the alcohol **35** (71%). Ozonolysis of the terminal alkene of **35** afforded the aldehyde **36** (97%), whose *C*-9-hydroxyl group was protected as the TMS derivative to yield the C_9–C_{25} spiroketal fragment **37**, which was used for the Wittig-based C_{25}–C_{26} alkenation (Scheme 4).

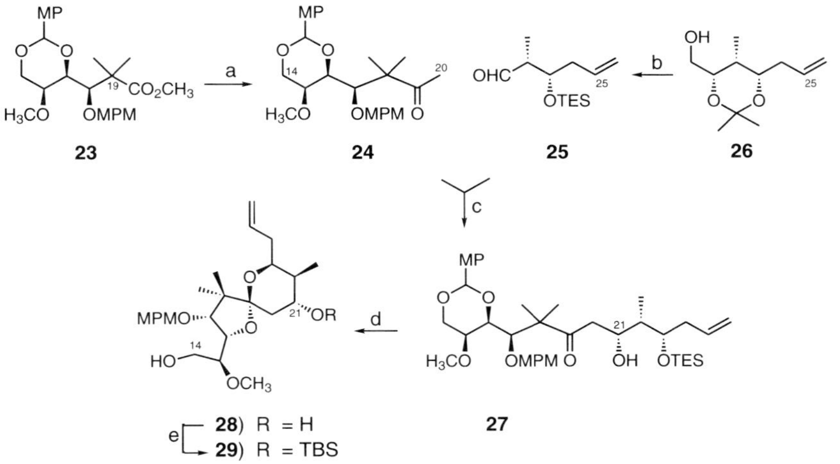

Scheme 3 (*a*) 1. DIBAL-H, CH_2Cl_2, –78°C, 88%; 2. SO_3·Py, NEt_3, DMSO, CH_2Cl_2; 3. CH_3MgBr, THF, 81% for two steps; 4. PDC, DMF, 9% *or* $TMSCH_2Li$, THF; then CH_3OH, 87%. (*b*) 1. *p*-TsOH, CH_3OH, 60%; 2. $NaIO_4$, aqueous THF; 3. TESCl, NEt_3, DMAP, CH_2Cl_2, 51%. (*c*) *t*-BuOK, THF, –78°C; then **25**. (*d*) 48% aqueous HF–CH_3CN, CH_2Cl_2, –10 to 0°C, 2 h, 63%. (*e*) 1. TBSOTf, NEt_3, CH_2Cl_2; 2. HF·Py, Py, THF, 85% for two steps.

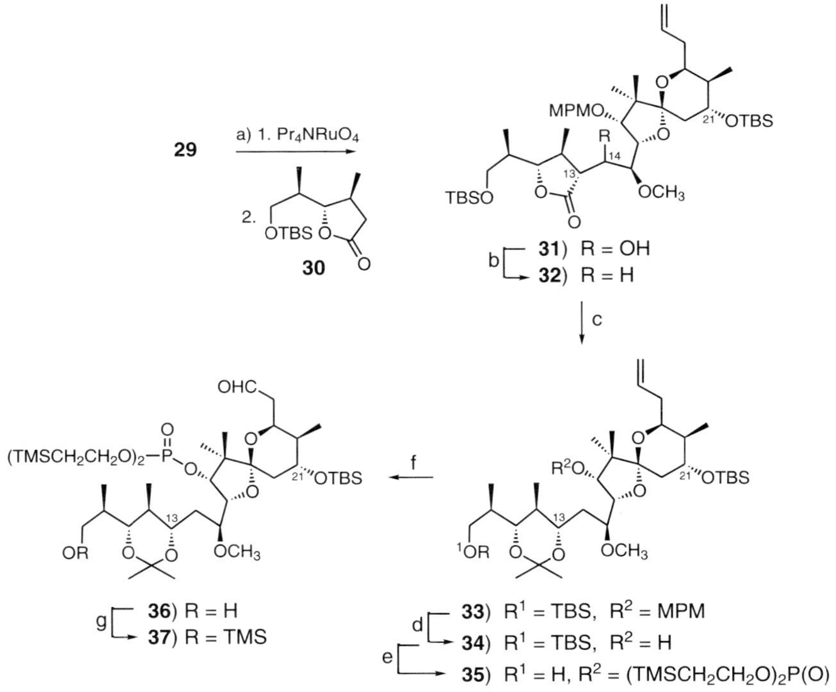

Scheme 4 (*a*) 1. Pr_4NRuO_4, NMO, MS 4 Å; 2. **30**, CH_2Cl_2, LDA, THF, –78°C, 84% for two steps. (*b*) 1. BuLi, PhOC(S)Cl, THF, 82%; 2. Bu_3SnH, AIBN, 100°C, 75%. (*c*) 1. CH_3Li, THF, –78°C; 2. 30% H_2O_2, AcOH, THF; 3. nosyl chloride (NsCl), NEt_3, THF; 4. DIBAL-H, CH_2Cl_2, –78°C; 5. DMP, PPTs, CH_2Cl_2, 51% for five steps. (*d*) DDQ, CH_2Cl_2, H_2O, 94%. (*e*) PCl_3, Py, $(CH_3)_3SiCH_2CH_2OH$; then 30% H_2O_2. (*f*) 1. HF·Py, Py, THF, 71% for two steps; 2. O_3, CH_2Cl_2, 78°C; then Ph_3P, 97%. (*g*) TMSCl, NEt_3, CH_2CL_2, 0°C.

Construction of the fragment **42** was initiated by coupling of the γ-amino acid fragment **38**[41] with the oxazole fragment **39**[42] via the DEPC method, to give the amide **40** in 90% yield (Scheme 5). Replacement of the acetonide group in **40** with Et_3Si (TES) group gave **41** whose transformation into the tributylphosphonium salt **42** was accomplished by sequential reductive methylation, reduction with $LiAlH_4$, bromination and then phosphonium salt formation. Finally, addition of the aldehyde **37** to the phosphonium salt **42** followed by the addition of LDA and then deprotection of the *C*-9-TMS group gave the fragment **43**, which could be converted to calyculin **A** in four steps.

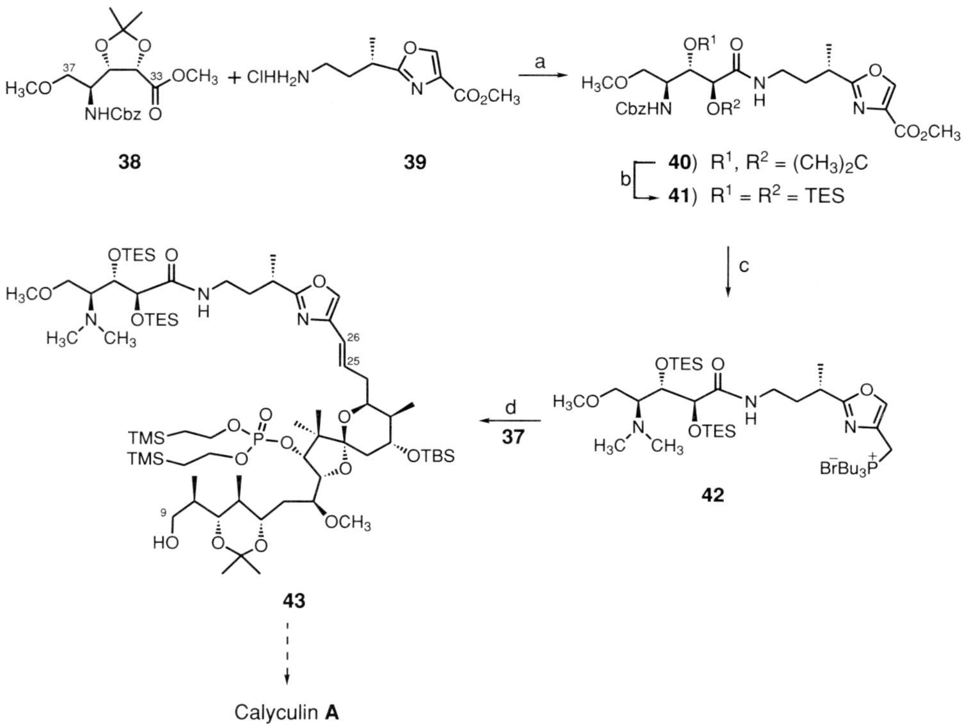

Scheme 5 (*a*) 1. Aqueous LiOH, THF, 0°C; 2. **39**, DEPC, NEt_3, DMF, 90% for two steps. (*b*) 1. CSA, CH_3OH; 2. TESOTf, 2,6-lutidine, CH_2Cl_2, 0°C, 83% for two steps. (*c*) 1. H_2, 5% Pd *on* C, aqueous HCHO, AcOH, CH_3OH, 91%; 2. $LiAlH_4$, Et_2O, –78°C, 67%; CBr_4, Ph_3P, 2,6-lutidine, CH_3CN, 75%; PBu_3, DMF, rt, 30 min. (*d*) 1. **37**, DMF, 0°C; then LDA, THF, 0°C; 2. K_2CO_3, CH_3OH, 0°C, 52% from **36**.

2.3.4 *Synthesis from* D*-ribonolactone*

Total syntheses of (+)-calyculin **A** and (−)-calyculin **B** from the ribonolactone derivative have been reported (Schemes 6 and 7).[43] Compound **44** was coupled with amine **45** to afford **46** (50% for four steps), which was converted into **47** followed by reduction with $LiAlH_4$, mesylation and then treatment with PBu_3 to afford compound **48**. This was coupled with **49** to afford **50**, which was converted to the aldehyde **51** (Scheme 6).

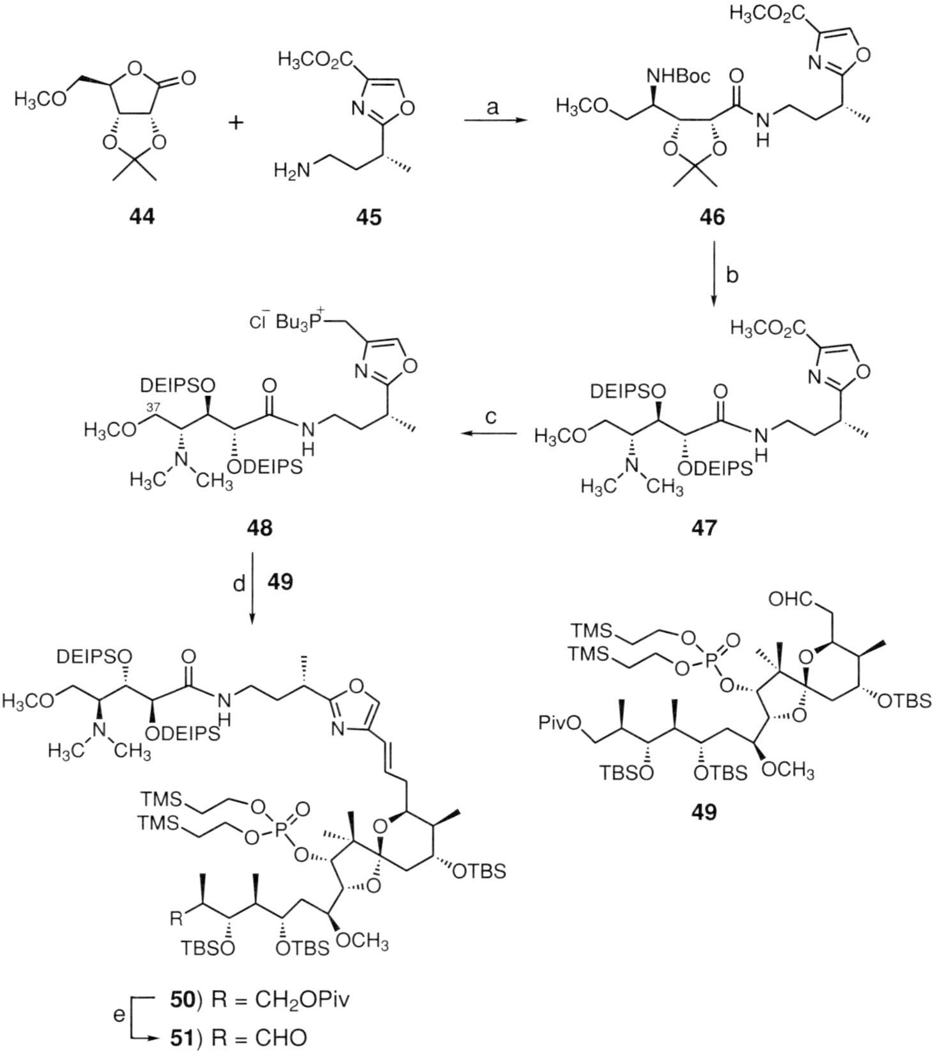

Scheme 6 (*a*) 1. LiOH, H_2O, THF; 2. DEPC, NEt_3, 75% for two steps. (*b*) 1. TMSOTf, 2,6-lutidine; 2. HCHO, $NaBH_3CN$; 3. HCl, CH_3OH; 4. DEIPSOTf, 2,6-lutidine, 50% for four steps. (*c*) 1. $LiAlH_4$, 89%; 2. MsCl, NEt_3, $BnEt_3NCl$, 79%; 3. PBu_3, 23°C, CH_3CN, THF, 95%. (*d*) LiHMDS, DMF, 0°C, 83%, 9:1 *E*/*Z*. (*e*) 1. DIBAL-H, –78°C, CH_2Cl_2, 87%; 2. TPAP, NMO, CH_2Cl_2, 84%.

Horner–Emmons olefination of **51** with phosphonate **52** followed by brief exposure to acid afforded **53** (67%), which was converted to (+)-calyculin **A** and (–)-calyculin **B** (Scheme 7).[43]

Coupling of the 4-amino-4-deoxy-ribonic acid derivative **54**, obtained from L-serine aldehyde,[44] and amine **55** furnished the amide **56**, which underwent hydrolysis of the Boc protecting group followed by N-methylation to provide the oxazole **57** (Scheme 8).[45]

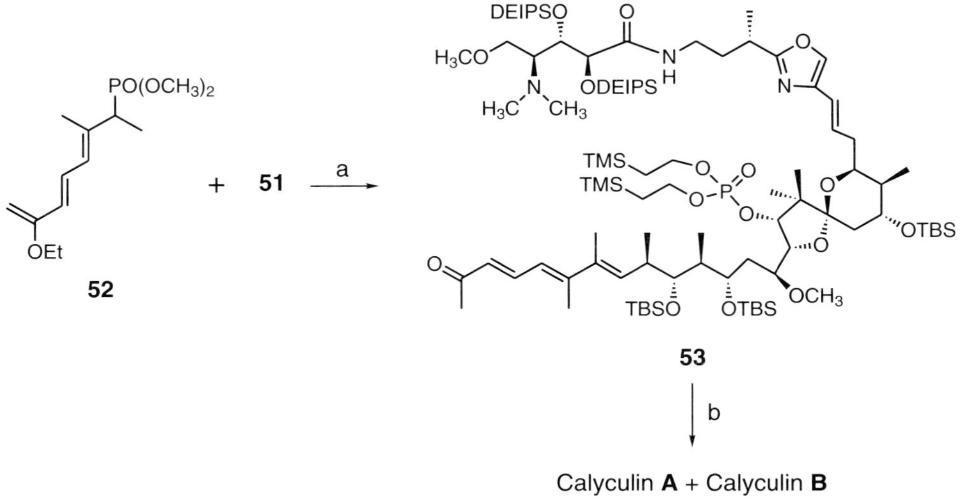

Scheme 7 (*a*) **52**, *n*-BuLi, THF, –78°C, 0.5 N HCl, 92%, 15:1 *E*/*Z*. (*b*) 1. TMSCH$_2$CN, *n*-BuLi, –78°C, 1.7:1 *E*/*Z*, 94%; 2. separation; 3. HF, CH$_3$CN, H$_2$O.

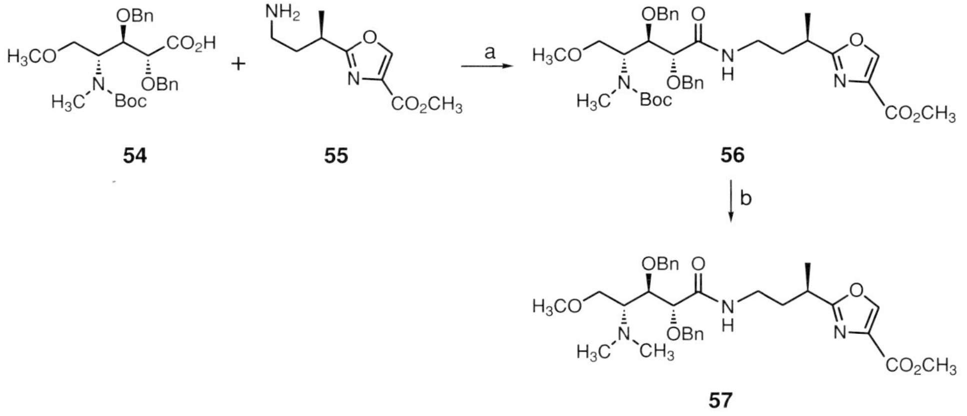

Scheme 8 (a) *N*-Hydroxybenzotriazole, DCC, DMF, 0–25°C, 78%. (*b*) 1. TFA, CH$_2$Cl$_2$, 25°C; 2. CH$_3$I, (*i*-Pr)$_2$NEt, CH$_2$Cl$_2$.

2.3.5 *Synthesis from* D*-erythronolactone*

Synthesis of γ-amino acid–oxazole fragment **68** of calyculins **A** and **B** from D-erythronolactone **58** has been reported by conversion to **59**,[46] which was subjected to oxidation reaction to afford the hemiaminal **60** (Scheme 9).[47] Acetylation of **60** furnished **61**, which was converted to ketone **62** in 88% yield. Conversion of **62** to a silyl enol ether, ozonolysis with reductive workup and O-methylation of the resultant alcohol **63** furnished γ-lactam **64**. Treatment of **64** with CAN led to **65** (60%), which was reacted with (CH$_3$)$_2$Al derivative of **66** to provide **67** (62%), which upon removal of the silyl group provided **68**.

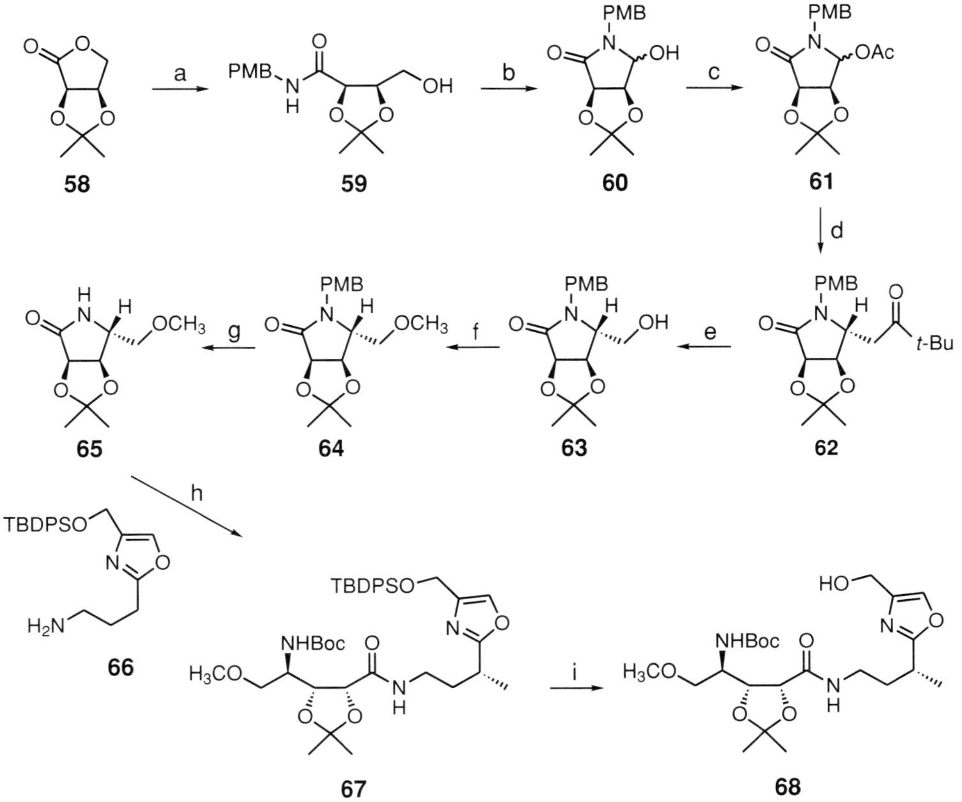

Scheme 9 (*a*) 1. $PMBNH_2$; 2. $Al(CH_3)_3$, CH_2Cl_2, rt, 89%. (*b*) $SO_3 \cdot Py$, DMSO, 78%. (*c*) Ac_2O, Py, 90%. (*d*) $TMSOC(CH_2)t$-Bu, $BF_3 \cdot Et_2O$, CH_2Cl_2, 0°C to rt, 88%. (*e*) 1. LiHMDS, TMSCl, NEt_3, THF, –20°C to rt, 74%; 2. O_3, $NaBH_4$, CH_2Cl_2, CH_3OH, –78 to 0°C, 65%. (*f*) NaH, CH_3I, 40°C, THF–DMF (2.5:1), 82%. (*g*) (0.25 M) CAN, CH_3CN, H_2O, 60%. (*h*) 1. $(Boc)_2O$, DMAP, THF, 88%; 2. $Al(CH_3)_3$, CH_2Cl_2, reflux, 62%. (*i*) TBAF, THF, 25°C, 77%.

References

1. Kato, Y.; Fusetani, N.; Matsunaga, S.; Hashimoto, K.; Fujita, S.; Furuya, T. *J. Am. Chem. Soc.* **108** (1986) 2780.
2. Kato, Y.; Fusetani, N.; Matsunaga, S.; Hashimoto, K.; Koseki, K. *J. Org. Chem.* **53** (1988) 3930.
3. Ishihara, H.; Martin, B.L.; Brautigan, D.L.; Karaki, H.; Ozaki, H.; Kato, Y.; Fusetani, N.; Watabe, S.; Hashimoto, K.; Uemura, D.; Hartshorne, D.J. *Biochem. Biophys. Res. Commun.* **159** (1989) 871.
4. Kato, Y.; Fusetani, N.; Matsunaga, S.; Hashimoto, K. *Drugs Exp. Clin. Res.* **14** (1988) 723.
5. Ishihara, H.; Ozaki, H.; Sato, K.; Hori, M.; Karaki, H.; Watabe, S.; Kato, Y.; Fusetani, N.; Hashimoto, K.; Uemura, D.; Hartshorne, D.J. *J. Pharmacol. Exp. Ther.* **250** (1989) 388.
6. Yabu, H.; Yoshino, M.; Usuki, T.; Someya, T.; Obara, K.; Ozaki, H.; Karaki, H. *Prog. Clin. Biol. Res.* **327** (1990) 623.
7. Suganuma, M.; Fujiki, H.; Furuya-Suguri, H.; Yoshizawa, S.; Yasumoto, S.; Kato, Y.; Fusetani, N.; Sugimura, T. *Cancer Res.* **50** (1990) 3521.
8. Favre, B.; Turowski, P.; Hemmings, B.A. *J. Biol. Chem.* **272** (1997) 13856.
9. Hamada, Y.; Tanada, Y.; Yokokawa, F.; Shiori, T. *Tetrahedron Lett.* **32** (1991) 5983.

10. Matsunaga, S.; Fusetani, N. *Tetrahedron Lett.* **32** (1991) 5605.
11. Duplantier, A.J.; Nantz, M.H.; Roberts, J.C.; Short, R.P.; Somfai, P.; Masamune, S. *Tetrahedron Lett.* **30** (1989) 7357.
12. Evans, D.A.; Gage, J.R. *Tetrahedron Lett.* **31** (1990) 6129.
13. Barrett, A.G.M.; Malecha, J.W. *J. Org. Chem.* **56** (1991) 5243.
14. Smith, A.B.; Duan, J.-W.; Hull, K.G.; Salvatore, B.A. *Tetrahedron Lett.* **32** (1991) 4855.
15. Zhao, Z.; Scarloto, G.R.; Armstrong, R.W. *Tetrahedron Lett.* **32** (1991) 1609.
16. Armstrong, R.W.; DeMattei, J.A. *Tetrahedron Lett.* **32** (1991) 5749.
17. Barrett, A.G.M.; Edmunds, J.J.; Hendrix, J.A.; Horita, K.; Parkinson, C. J. *J. Chem. Soc., Chem. Commun.* (1992) 1238.
18. Barrett, A.G.M.; Edmunds, J.J.; Horita, K.; Parkinson, C.J. *J. Chem. Soc., Chem. Commun.* (1992) 1236.
19. Evans, D.A.; Gage, J.R.; Leighton, J.L. *J. Am. Chem. Soc.* **114** (1992) 9434.
20. Evans, D.A.; Gage, J.R. *J. Org. Chem.* **57** (1992) 1958.
21. Evans, D.A.; Gage, J.R.; Leighton, J.L.; Kim, A.S. *J. Org. Chem.* **57** (1992) 1961.
22. Evans, D.A.; Gage, J.R.; Leighton, J.L. *J. Org. Chem.* **57** (1992) 1964.
23. Barrett, A.G.M.; Malecha, J.W. *J. Chem. Soc., Perkin Trans. 1* (1994) 1901.
24. Veeneman, G.H.; Gomes, L.J.F.; Van Boom, J.H. *Tetrahedron* **45** (1989) 7433.
25. Ogawa, A.K.; DeMattei, J.A.; Scarlato, G.R.; Fellew, J.E.; Chong, L.S.; Armstrong, R.W. *J. Org. Chem.* **61** (1996) 6153.
26. Bowden, K.; Heilbron, I.M.; Jones, E.R.H.; Weedon, B.C.L. *J. Chem. Soc.* (1946) 39.
27. Bowers, A.; Halsall, T.G.; Jones, E.R.H.; Lernin, A.J. *J. Chem. Soc.* (1953) 2548.
28. Vaccaro, H.A.; Levy, D.E.; Sawabe, A.; Jaetsch, T.; Masamune, S. *Tetrahedron Lett.* **33** (1992) 1937.
29. Hough, L.; Jones, J.K.N.; Mitchell, D.L. *Can. J. Chem.* **36** (1958) 1720.
30. Hulyalkar, R.K.; Jones, J.K.N. *Can. J. Chem.* **41** (1963) 1898.
31. Levin, J.I.; Turos, E.; Weinreb, S.M. *Synth. Commun.* **12** (1982) 989.
32. Yokkawa, F.; Hamada, Y.; Shioiri, T. *Chem. Commun.* (1996) 871.
33. Takebuchi, K.; Hamada, Y.; Shioiri, T. *Tetrahedron Lett.* **35** (1994) 5239.
34. Brown, H.C.; Bhat, K.S.; Randad, R.S. *J. Org. Chem.* **54** (1989) 1570.
35. Barton, D.H.R.; McCombie, S.W. *J. Chem. Soc., Perkin Trans. 1* (1975) 1574.
36. Barton, D.H.R.; Subramanian, R. *J. Chem. Soc., Perkin Trans. 1* (1977) 1718.
37. Robins, M.J.; Wilson, J.S. *J. Am. Chem. Soc.* **103** (1981) 932.
38. Ziegler, F.E.; Kneisley, A.; Thottathil, J.K.; Wester, R.T. *J. Am. Chem. Soc.* **110** (1988) 5434.
39. Menges, M.; Bruckner, R. *Synlett* (1993) 901.
40. Oikawa, Y.; Yoshioka, T.; Yonemitsu, O. *Tetrahedron Lett.* **23** (1982) 885.
41. Yokokawa, F.; Hamada, Y.; Shioiri, T. *Synlett* (1992) 703.
42. Yokokawa, F.; Hamada, Y.; Shioiri, T. *Synlett* (1992) 149.
43. Smith, A.B., III; Friestad, G.K.; Duan, J.J.-W.; Barbosa, J.; Hull, K.G.; Iwashima, M.; Qiu, Y.; Spoors, P.G.; Bertounesque, E.; Salvatore, B.A. *J. Org. Chem.* **63** (1998) 7596.
44. Garner, P.; Park, J.M. *J. Org. Chem.* **52** (1987) 2361.
45. Barrett, A.G.M.; Edmunds, J.J.; Hendrix, J.A.; Malecha, J.W.; Parkinson, C.J. *J. Chem. Soc., Chem. Commun.* (1992) 1240.
46. Cohen, N.; Banner, B.L.; Laurenzano, A.J.; Carozza, L. *Org. Syn.* **63** (1985) 127.
47. Smith, A.B., III; Salvatore, B.A.; Hull, K.G.; Duan, J.J.-W. *Tetrahedron Lett.* **32** (1991) 4859.

2.4 Acivicin

Acivicin (AT-125, **1**) is an antitumor antibiotic isolated in 1973 from the fermentation broths of the soil bacterium *Streptomyces sviceus*.[1] Subsequent determination of its structure and absolute configuration showed that acivicin is (α*S*,5*S*)-α-amino-3-chloro-4,5-dihydroisoxazole-5-acetic acid.[2] Acivicin significantly increases the life span of tumor (L1210 or P388) bearing mice[3] and of immune deficient mice implanted with a solid human mammary tumor.[3–6] Compound **1** has been shown to be an inhibitor of several L-glutamine amidotransferases involved in the *de novo* biosynthesis of purine and pyrimidine nucleotides; antitumor activity is believed to be a consequence of this inhibition.[1,6] Also it may have a role in the treatment of nonsmall cell lung cancer and that it may be active against colon cancer.[4,5] Acivicin is available in a large-scale production through fermentation, but unfortunately the production has been impeded by the occurrence of noneasy separable contaminants.[6] Several syntheses of acivicin from noncarbohydrates have been reported.[7–14]

The nitrone **2** was used as a precursor for the synthesis of acivicin (Scheme 1).[15,16] Cycloadditions[17] of nitrone **2** to the vinylglycine derivative **3** produced the cycloadducts **4** and **5**, which could be converted by hydrolysis with formic acid to isoxazolines **6** and **7**. Oxidation of **6** with *N*-chlorosuccinimide followed by deprotection with boron tris(trifluoroacetate) afforded acivicin (**1**).

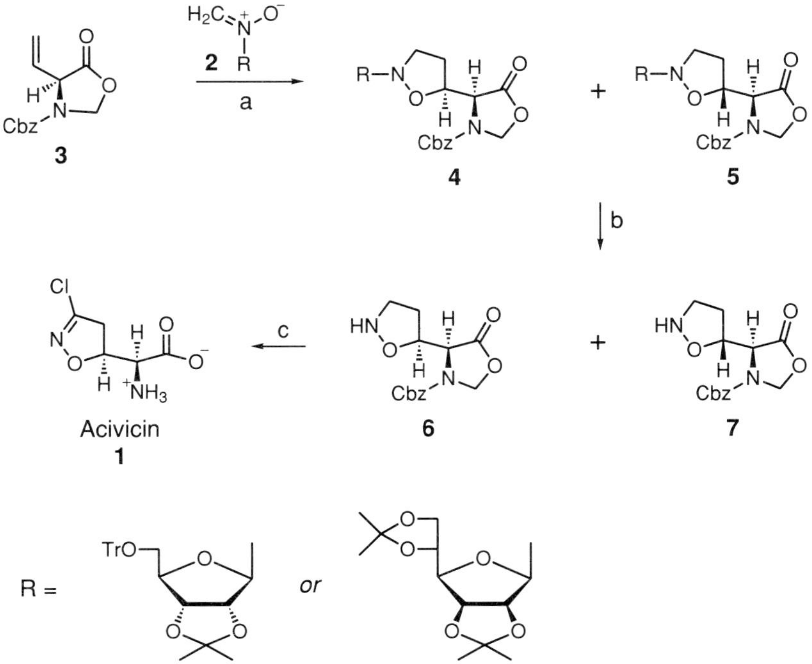

Scheme 1 (*a*) 1. HCHO; 2. $CHCl_3$, reflux, 1.5 days, 93%. (*b*) 98% HCO_2H. (*c*) 1. NCS, CH_2Cl_2; 2. boron tris(trifluoroacetate), TFA, 89%.

References

1. Hanka, L.J.; Dietz, A. *Antimicrob. Agents Chemother*. **3** (1973) 425.
2. Martin, D.G.; Duchamp, D.G.; Ghidester, C.G. *Tetrahedron Lett*. **14** (1973) 2549.
3. Earhart, R.E.; Neil, G.L. *Adv. Enzyme Regul*. **24** (1986) 179.
4. Hauchers, D.; Ovejara, A.; Johnson, R.; Bogden, A.; Neil, G. *Proc. Am. Assoc. Cancer Res*. **19** (1978) 40.
5. Hanka, L.J.; Martin, D.G.; Neil, G.L. *Cancer Chemother. Res*. **57** (1973) 141.
6. Martin, D.G.; Biles, C.; Mizsak, S.A. *J. Antibiot*. **34** (1981) 459.
7. Kelly, R.C.; Schletter, I.; Stein, S.J.; Wierenga, W. *J. Am. Chem. Soc*. **101** (1979) 1054.
8. Silverman, R.B.; Holladay, M.W. *J. Am. Chem. Soc*. **103** (1981) 757.
9. Baldwin, J.E.; Cha, J.K.; Kruse, L.I. *Tetrahedron* **41** (1985) 5241.
10. Wade, P.A.; Pillay, M.K.; Singh, S.M. *Tetrahedron Lett*. **23** (1982) 4563.
11. Vyas, D.M.; Chang, Y.; Doyle, T.W. *Tetrahedron Lett*. **25** (1984) 487.
12. Baldwin, J.E.; Kruse, L.I.; Cha, J.K. *J. Am. Chem. Soc*. **103** (1981) 942.
13. Hagedorn, A.E., III; Miller, B.J.; Nagy, J.O. *Tetrahedron Lett*. **21** (1980) 229.
14. Baldwin, J.E.; Hoskins, C.; Kruse, L.I. *J. Chem. Soc., Chem. Commun*. (1976) 795.
15. Mzengeza, S.; Yang, C.M.; Whitney, R.A. *J. Am. Chem. Soc*. **109** (1987) 276.
16. Mzengeza, S.; Whitney, R.A. *J. Org. Chem*. **53** (1988) 4074.
17. Vasella, A. *Helv. Chim. Acta* **60** (1977) 1273.

2.5 Bengazole

Bengazole A and some related compounds were isolated from marine sponges of the genus *Jaspis*, a representative member of the family of bisoxazole natural products.[1–4] Bengazole A exhibits potent in vitro antifungal activity against *Candida albicans*[2,5] and fluconazole-resistant *Candida* strains,[6] which is comparable to that of the clinical agent amphotericin B. The NMR and chiroptical studies[2] were used to establish the configuration of bengazole A (**9**).

The first total syntheses of bengazole A (**9**) and 10-*epi*-bengazole A (**10**) were achieved from D-galactose (Scheme 1).[6] Conversion of D-galactose to the aldehyde **1** (26%)[7,8] and then reaction of **1** with the lithiooxazole derivative gave the coupled products **2** and **3** (57%, in a ratio 1:7), although **2** was the required isomer for the synthesis of bengazole A (**9**). However, the inversion of configuration in **3** was done by oxidation of the secondary hydroxyl group in the mixture of **2** and **3** before separation to give the ketone **4**, which was reduced with sodium borohydride to provide **2** and **3** with a higher ratio of the

D-Galactose **1** **2** **3** **4** **5** **7** **8**

9 (10*S*) Bengazole A
10 (10*R*) *epi*-Bengazole A

Scheme 1 (*a*) Refs. 7 and 8. (*b*) *n*-BuLi, THF, hexanes, –78°C, oxazole, 57%. (*c*) 1. DEAD, Ph_3P, benzene, $p\text{-}NO_2C_6H_4CO_2H$, 70%; 2. K_2CO_3, CH_3OH, 96%. (*d*) $(COCl)_2$, DMSO, NEt_3, CH_2Cl_2, –78 to 0°C, 80%. (*e*) $NaBH_4$, CF_3CH_2OH, –20°C, 94%. (*f*) TBSOTf, 2,6-lutidine, CH_2Cl_2, 25°C, 99%. (*g*) BH_3–THF, 25°C, 30 min, THF, –78°C, *n*-BuLi, then added **6**, 40%. (*h*) 1. $n\text{-}C_{13}H_{27}COCl$, DMAP, NEt_3, CH_2Cl_2, 84%. (*i*) Aqueous HF, CH_3CN, 94%.

former (3.3:1). More conveniently, the Mitsunobu inversion of **3** followed by methanolysis gave **2** (>86% ds). Protection of **2** as the silyl ether **5** followed by reaction with 5-oxazolecarboxal-dehyde (**6**) gave a 1:1 mixture of diastereomers **7**. The synthesis of bengazole A (**9**) was completed by esterification of the epimeric alcohols **7** to give **8** that were deprotected using HF to deliver bengazole A (**9**) and *epi*-bengazole A (**10**).

References

1. Adamczeski, M.; Quiñoa, E.; Crews, P. *J. Am. Chem. Soc.* **110** (1988) 1598.
2. Searle, P.A.; Richter, R.K.; Molinski, T.F. *J. Org. Chem.* **61** (1996) 4073.
3. Rodriguez, J.; Nieto, R.M.; Crews, P. *J. Nat. Prod.* **56** (1993) 2034.
4. Rudi, A.; Kashman, Y.; Benayahu, Y.; Schleyer, M. *J. Nat. Prod.* **57** (1994) 829.
5. Antonio, J.; Molinski, T.F. *J. Nat. Prod.* **56** (1993) 54.
6. Mulder, R.J.; Shafer, C.M.; Molinski, T.F. *J. Org. Chem.* **64** (1999) 4995.
7. Shafer, C.M.; Molinski, T.F. *Carbohydr. Res.* **310** (1998) 223.
8. Shafer, C.M.; Molinski, T.F. *Tetrahedron Lett.* **39** (1998) 2903.

3 Six-membered nitrogen heterocycles

This chapter discusses naturally occurring polyhydroxylated piperidines, which can also be regarded as aza- or imino-sugars. Of particular importance are nojirimycin, mannonojirimycin, galactonojirimycin (galactostatin), fagomine, homonojirimycin and siastatin B, as well as their deoxy analogues. These compounds can be obtained from a variety of sources including carbohydrates and have been shown to be potent inhibitors of glycosidases. They inhibit many other hydrolytic enzymes and display unique isomeric specificity. Imino-sugars have a potential use in the prevention and treatment of a variety of diseases, including cancer, diabetes, viral infections such as AIDS, and influenza as well as hereditary lysosomal storage diseases. Serious efforts have been made to develop appropriate synthetic methods for these compounds as well as their unnatural analogues, starting from carbohydrates or noncarbohydrates.

3.1 Hydroxymethylpiperidines

Chapter 3 is divided into two parts: the first part discusses five types of naturally occurring hydroxymethylpiperidines that inhibit glycosidases. They are nojirimycin, mannojirimycin, galactonojirimycin, fagomine and homojirimycins. Because of their sugar-like nature, glucose, galactose and mannose are usually used as their precursors. The syntheses of these naturally occurring six-membered nitrogen heterocycles are achieved via substitution of one of the hydroxyl groups, usually the hydroxyl group at C-2 or C-6, with an azide group followed by hydrogenation and intramolecular cyclization.

3.1.1 *Nojirimycin*

(+)-Nojirimycin (**1**) was isolated from several strains of *Streptomyces*,[1–3] such as *Str. roseochromogenes* R-468, *Str. lavendulae* SF-425 and *Str. nojiriensis* sp. SF-426, as well as *Bacillus*.[4] Also, **1** was isolated from leaves of *Jacobinia subereta*.[5] It exhibits potent biological activity against drug-resistant strains of *Sarcina lutea*, *Shigella flexneri* and *Xanthomonas oryzae*.[1] Furthermore, (+)-nojirimycin (**1**) shows significant inhibitory activity against various glycosidases and glucoamylase.[6–11] The IC_{50} of the inhibition of α-glucosidase in various animals is around 10^{-3} M, but it is a poor inhibitor of exo- and endo-glucanases and related enzymes.[7] The taxonomy, fermentation,[12] structure[13] and X-ray analysis[14] of **1** produced by *Str. roseochromogenes* R-468 have been investigated.

1-Deoxynojirimycin (**2**) was produced in the culture medium when *Str. subrutilus* ATCC 27467 was grown on glucose-containing soybean medium.[15] When 1- or 2-[^{2}H]-D-glucose is used, the deuterium label appears at C-6 in **2** and the labeling pattern suggested that the first step in the biosynthesis of **2** is the isomerization of glucose to fructose. Studies with 5-[^{2}H]- and 6,6-[$^2H_2$]-D-glucose indicated that oxidation of 6-position of the glucose and fructose occurs during the biosynthesis.

1-Deoxynojirimycin[16] (**2**) has been isolated from mulberry root bark plants of the genus *Morus Mori cortex*[17–19] and named moranoline *Morus bombycis Koidz*[20] as well

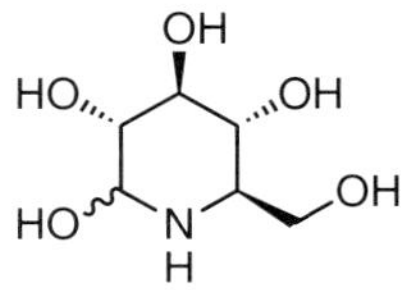

(+)-Nojirimycin
1

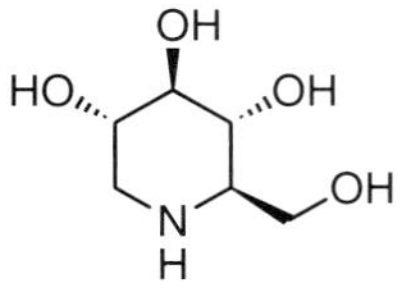

1-Deoxynojirimycin
2

as from strains of *Bacillus*.[4,21–23] 1-Deoxynojirimycin is an inhibitor of a number of glucosidases.[21–24] The first moranoline-producing *Streptomyces*[25] was identified as *Str. lavendulae* SEN-158 and deposited in the American-type culture collection, Rockville, MD, as ATCC 31434.

The mechanism of glycosidase enzyme inhibitors was studied[26]; however, there is no clear knowledge of the particular glycosidase mechanism(s), although there are two generally accepted pathways involving acid-catalyzed cleavage of (i) the exocyclic (anomeric) carbon–oxygen bond, giving cyclic oxonium ion,[27–35] and (ii) the endocyclic (ring) carbon–oxygen bond, resulting in an acyclic oxonium ion[36–38] (Scheme 1). For mannosidase inhibitors, it has been suggested that a correlation with mannofuranose is important,[39,40] but calculations indicated that the structures similar to the mannopyranosyl cation, not mannose itself,

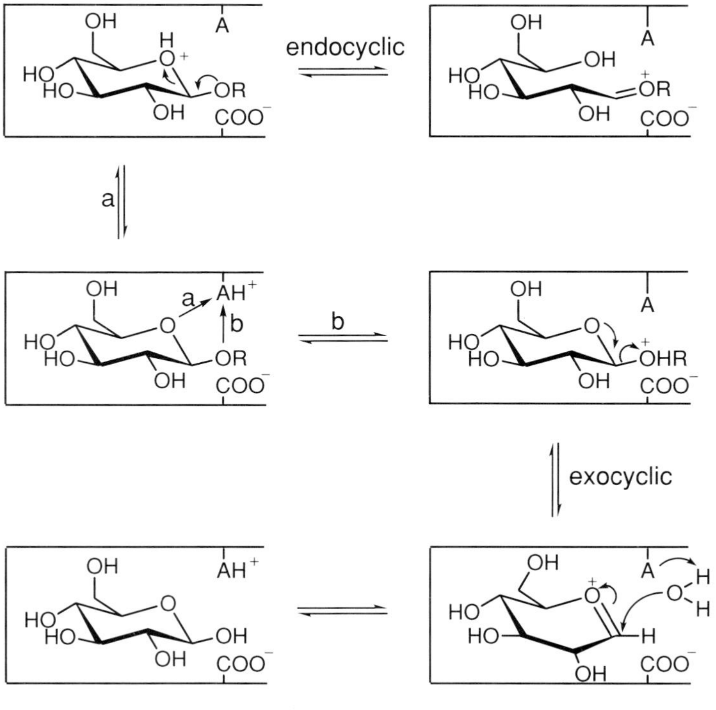

Scheme 1

exhibit the more potent activity.[26] Many efforts have been devoted to develop appropriate synthetic methods for their synthesis from noncarbohydrate.[41–56] Also many syntheses of unnatural analogues of nojirimycin have been reported.[57–179] The synthetic approaches from carbohydrates will be discussed in this chapter.

3.1.1.1 *Synthesis from* D-*glucose* Synthesis of 1-deoxynojirimycin (**2**) from D-glucose using combined chemical and microbiological methodologies has been reported[180] (Scheme 2). 1-Amino-1-deoxy-D-sorbitol (**3**), obtained from D-glucose,[181] was treated with benzyloxycarbonyl chloride to afford **4**, which was oxidized with *Gluconobacter oxydans* to give **5**. Hydrogenation of **5** in methanol and water in the presence of palladium on carbon led to the removal of the protecting group and stereoselective ring closure to afford **2** in 69% overall yield from **4**; the low yield was due to the instability of the amino-ketone.

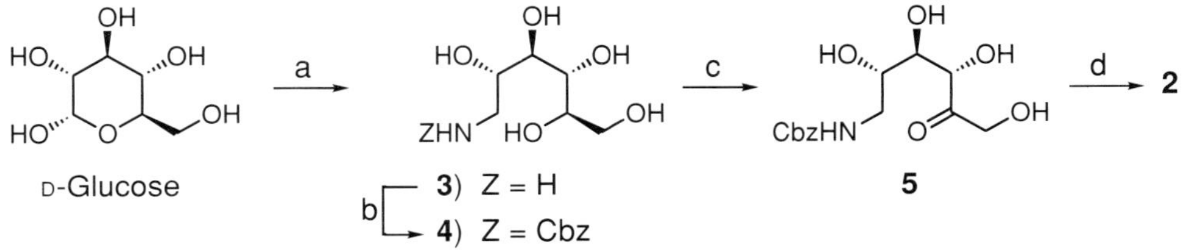

Scheme 2 (*a*) Ref. 181. (*b*) CbzCl, pH 8–10, $NaHCO_3$. (*c*) 2% Ohly yeast, 5% sorbitol K_2HPO_4, pH 6.5, KOH, 45 min at 121°C; then *Gluconobacter oxydans*, O_2, 250 mL, 24 h at 30°C, 10 L air/min, 500 rpm, 92%. (*d*) H_2, Pd *on* C, CH_3OH, 1 h, 40–50°C, 2 h, 60°C, 75%.

Methyl β-D-glucopyranoside (**6**) has been used for the synthesis of 1-deoxynojirimycin (**2**) (Scheme 3).[90,182] The peracetate of **6** was oxidized with chromium trioxide to give the keto-ester **7** in a quantitative yield, which was condensed with hydroxylamine to afford the oxime **8** as a mixture of *syn*- and *anti*-isomers in 95% yield. Deacetylation of

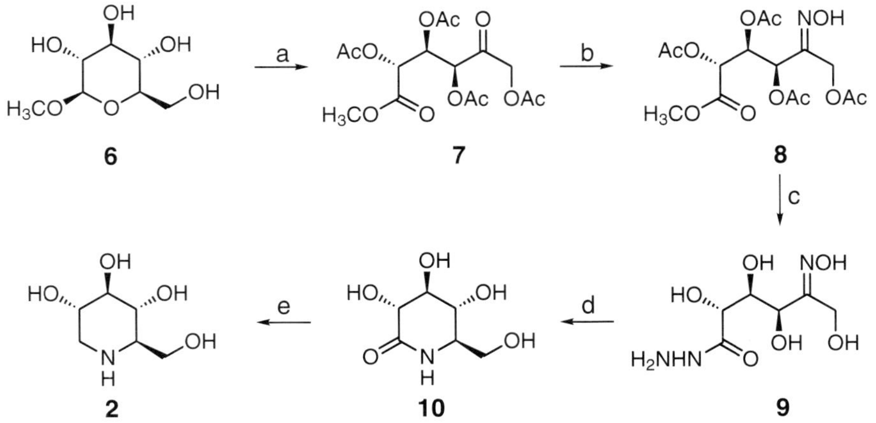

Scheme 3 (*a*) 1. Ac_2O, Py, 4 h, rt, 100%; 2. CrO_3, Ac_2O, 50°C, 2 h, quantitative. (*b*) H_2NOH, Py, 0°C, 15 min, 95%. (*c*) NH_2NH_2, 100%. (*d*) H_2, Pd *on* C, AcOH, 95%. (*e*) 1 M BH_3, THF, rt, 1.5 h; then reflux, 1.5 h.

8 with hydrazine led to the hydrazide **9** in quantitative yield, which underwent catalytic hydrogenation to provide the lactam **10**. Subsequent reduction with BH_3 gave **2**.

The readily available 2,3,4,6-*tetra-O*-benzyl-D-glucopyranose (**11**) can be used for the synthesis of both nojirimycin (**1**) and 1-deoxynojirimycin (**2**) (Scheme 4).[183] It was treated[184] with EtSH to furnish **12**, which was oxidized[185] to the corresponding ketone **13** using TPAP, while the Swern oxidation[186] method failed to produce **13**. Treatment of **13** with mercury(II) salts in the presence of methanol followed by treatment with hydroxylamine hydrochloride in the presence of pyridine afforded the oxime **14** in 73% yield. Treatment of **14** with $LiAlH_4$ in diethyl ether followed by N-protection of the resulting diastereomeric mixture of amines with di-*tert*-butyl dicarbonate furnished **15** and **16** in 65 and 15% yield, respectively. Pearlman's catalytic hydrogenation of **15** over palladium hydroxide in ethanol followed by treatment of the resulting tetrol with SO_2 in water furnished the sulfonic acid **17** in 80% yield. Conversion[187] of **17** into **1** was accomplished by treatment with Dowex 1X2 (OH^-) resin.

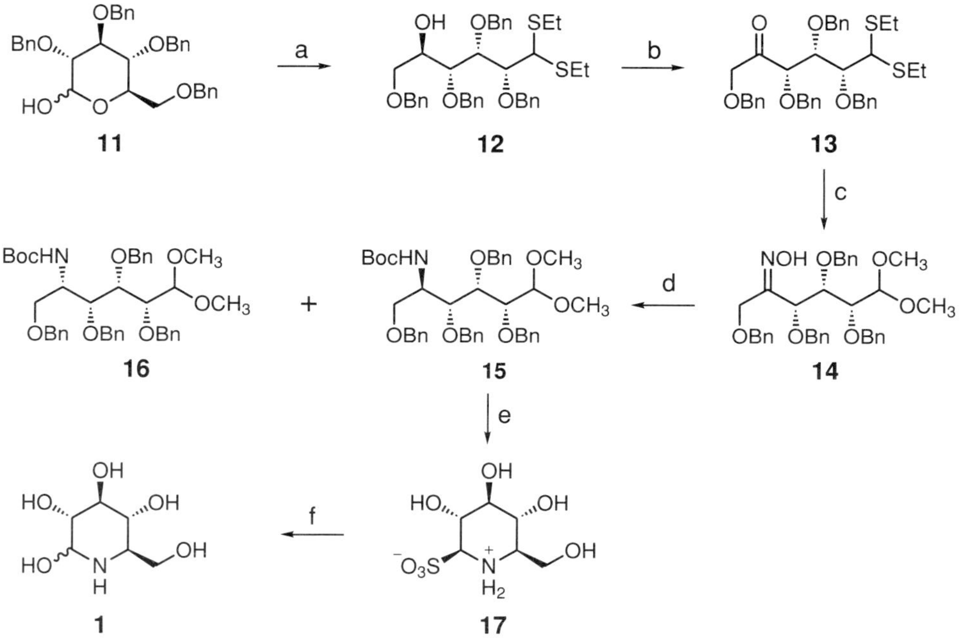

Scheme 4 (*a*) EtSH, conc. HCl, 1,4-dioxane, 56%. (*b*) TPAP, NMO, CH_2Cl_2, 81%. (*c*) 1. HgO, $HgCl_2$, CH_3OH, 81%; 2. $NH_2OH{\cdot}HCl$, Py, EtOH, 90%. (*d*) 1. $LiAlH_4$, Et_2O, rt, overnight; 2. $(Boc)_2O$, NEt_3, CH_3CN, rt, 10 min, **15** (65%), **16** (15%). (*e*) 1. $Pd(OH)_2$, H_2, EtOH, rt, 4 h; 2. SO_2, H_2O, 40°C, 2 days, 80% for two steps. (*f*) Dowex 1X2 (OH^-) resin, rt, 40 min, H_2O, 100%, quantitative.

Various approaches toward the synthesis of 1-deoxynojirimycin (**2**) have also utilized 2,3,4,6-tetra-*O*-benzyl-α-D-glucopyranose (**11**) as a starting material (Scheme 5).[188] Thus, reduction of **11** with $LiAlH_4$ in THF gave the 1,5-diol **18** in quantitative yield. Oxidation of **18** followed by a stereocontrolled reductive amination of the resulting 1,5-dicarbonyl sugar derivative **19**, using ammonium formate or *N*-butyl ammonium formate in the presence of $NaBH_3CN$ as a source of hydrogen, produced the cyclized compounds **20** and **21** in 73

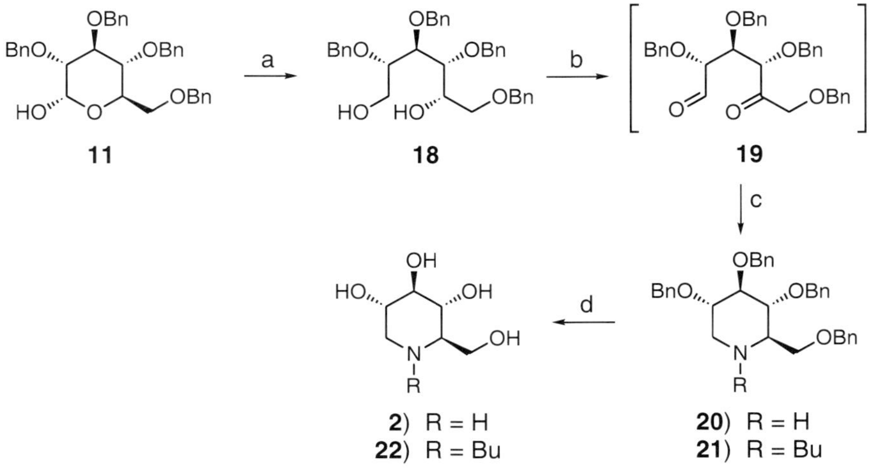

Scheme 5 (*a*) $LiAlH_4$, THF, 100%, rt, overnight. (*b*) DMSO, TFAA, CH_2Cl_2, NEt_3, –78°C. (*c*) HCO_2NH_4 or HCO_2NH_3Bu, $NaBH_3CN$, CH_3OH, MS 3 Å. (*d*) Li, NH_3, THF, –78°C, 2.5 h; then Dowex 50WX8 resin, 4 h, eluent, 1 M NH_4OH.

and 77% yield, respectively. The benzyl protecting groups in **20** and **21** were removed with lithium in ammonia and THF followed by purification with Dowex 50WX8 ion-exchange resin to give 1-deoxynojirimycin (**2**) and *N*-butyl-1-deoxynojirimycin (**22**) in 89 and 91% yield, respectively.

Treatment of **11** with methoxyamine hydrochloride in pyridine afforded the oxime **23**, which was oxidized with chromium trioxide in pyridine to afford the ketone **24** (Scheme 6).[189] Subsequent radical cyclization with Bu_3SnH in the presence of AIBN afforded a 1.4:1 mixture of the two amino alcohols **25** and **26**. Reduction of the 1,5-*trans*-methoxyamine **26**

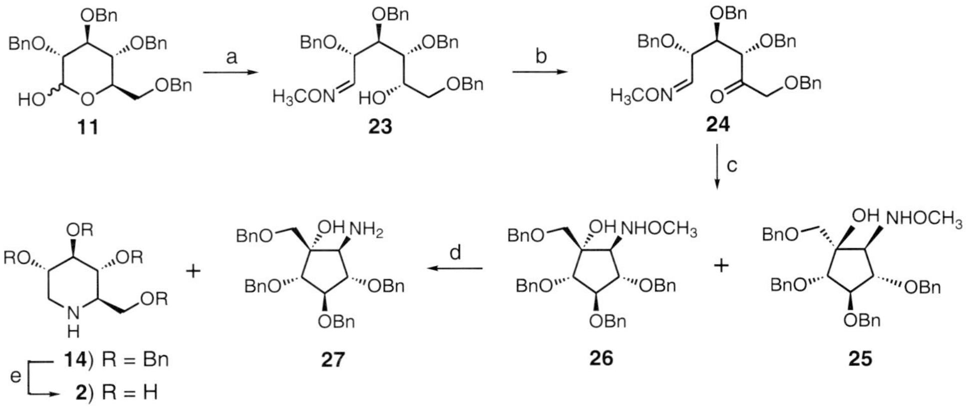

Scheme 6 (*a*) $CH_3ONH_2{\cdot}HCl$, Py, 92%. (*b*) CrO_3–Py, 79%. (*c*) Bu_3SnH, AIBN, 68%. (*d*) $LiAlH_4$. (*e*) H_2, Pd *on* C, AcOH.

with $LiAlH_4$ afforded a 5:2.3 mixture of **27** and tetra-*O*-benzyl-1-deoxynojirimycin (**14**) as a result of ring expansion. Hydrogenation[190] of **14** furnished 1-deoxynojirimycin (**2**).

Conversion of **11** to the oxime **28**[191] took place quantitatively, whose treatment with PPh_3 and CBr_4 yielded the nitrile **29** (Scheme 7).[192] Two routes have utilized **29**, in which **29** was converted into the L-*ido*-bromide **30** or iodide **31** by treatment with an excess of PPh_3, Br_2 or I_2 and imidazole in boiling toluene,[193] but in low yield because of the partial neighboring group participation of the C-2-OBn group, leading to the 2,5-anhydro-L-idononitrile **32**. Treatment of **30** or **31** with sodium azide in dimethylsulfoxide led to the tetrazole **36**. Alternatively, the second approach was done by Swern oxidation[194,195] of **29** to yield 92% of the ketone **33** whose reduction with sodium borohydride in methanol in the presence of $CeCl_3{\cdot}6H_2O$ gave the L-*ido*-hydroxynitrile **34** as the main product. Subsequent tosylation of **34** gave **35**, which was reacted with NaN_3 to give the tetrazole **36** in addition to nitrile **37** in 70 and 10% yield, respectively. Treatment of **36** with $LiAlH_4$ followed by hydrogenolytic debenzylation and purification led to 1-deoxynojirimycin (**2**).

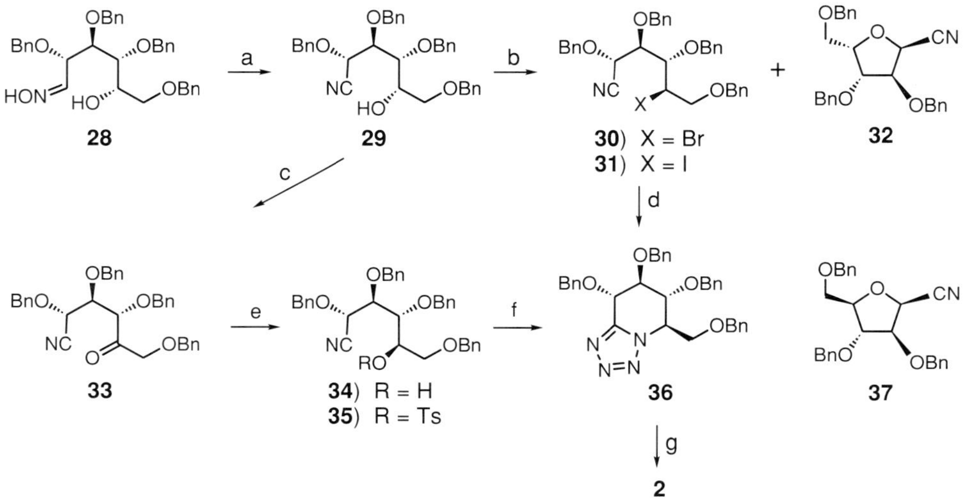

Scheme 7 (*a*) PPh_3, CBr_4, CH_3CN, rt, 20 min, 80%. (*b*) PPh_3, imidazole, Br_2 (or I_2), toluene, 110°C, 2 h, 42% of **30** or 33% of **31**. (*c*) DMSO, $(COCl)_2$, CH_2Cl_2, NEt_3, 92%. (*d*) NaN_3, DMSO, 110–125°C, 4 h, 43%. (*e*) 1. $NaBH_4$, CH_3OH, $CeCl_3{\cdot}6H_2O$, –60 to –40°C, 55 min, 86%; 2. *p*-TsCl, Py, 40–50°C, 20 h, 97%. (*f*) Same as (*d*), 195 min, 70% of **36**, 10% of **37**. (*g*) 1. $LiAlH_4$, Et_2O, reflux, 5 h, 83%; 2. H_2, 10% Pd *on* C, AcOH, rt, 15 h, 86%; 3. CH_3OH, aqueous HCl; 4. Dowex 1X8 (OH^-) resin.

Methyl α-D-glucopyranoside has also been used for the synthesis of (+)-nojirimycin (**1**) and 1-deoxynojirimycin (**2**) (Scheme 8).[196] The 2,3-di-*O*-benzyl-α-D-glucopyranoside **38**[197] was treated with *p*-toluenesulfonic acid to give the 4-O-unprotected 1,6-anhydropyranose **39** and 1,6-anhydrofuranose **40**. The latter was oxidized with pyridinium chlorochromate followed by reduction with sodium borohydride to produce **41**, which upon triflation and subsequent S_N2 displacement with sodium azide afforded **42**. Opening of the anhydro ring in **42** with acetic anhydride and TFA gave the diacetate **43**. Deacetylation of **43** followed by hydrogenolysis afforded **2** in 54% yield.

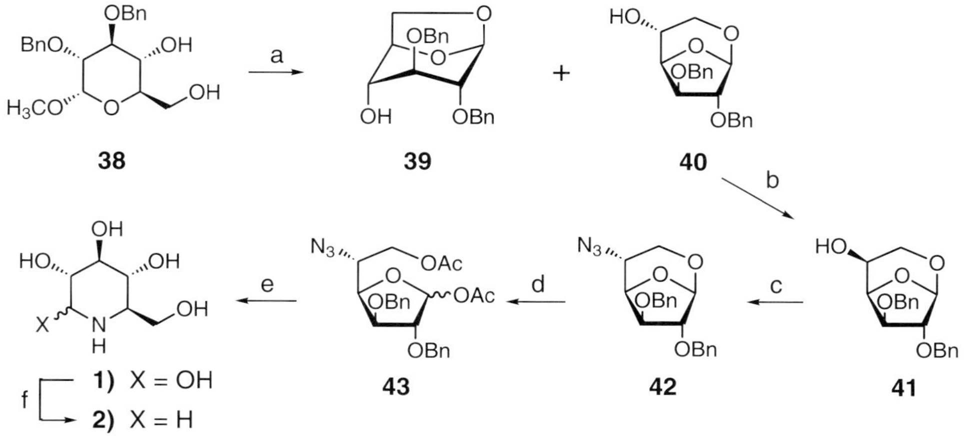

Scheme 8 (*a*) CF_3CH_2OH, *p*-TsOH *or* CCl_3CH_2OH, *p*-TsOH, benzene, 48 h. (*b*) 1. PCC, CH_2Cl_2, 4 h, 90%; 2. EtOH–dioxane (10:15), $NaBH_4$, 1.5 h, 96%. (*c*) 1. CH_2Cl_2, Py, –10°C, Tf_2O, 20 min, 91%; 2. NaN_3, DMF, 60°C, 2.5 h, 51%. (*d*) Ac_2O, TFA, 40°C, 2 h. (*e*) 1. CH_3OH, $NaOCH_3$, 2 h, 76% for two steps; 2. dioxane–H_2O, Pd *on* C, 0.1 N HCl, H_2, 1 day. (*f*) Dowex 1X2 (OH^-) resin, 54%.

The 6-bromodeoxy-tri-*O*-benzyl-pyranoside **44**[198] was heated in a mixture of acetic acid and zinc dust to give **45**, which directly underwent reductive amination using benzylamine and $NaBH_3CN$ to afford **46** (Scheme 9).[199] Intramolecular aminomercuration of **46** with mercuric trifluoroacetate afforded bromomercurials **48** (61%) and **47** (39%). Reductive

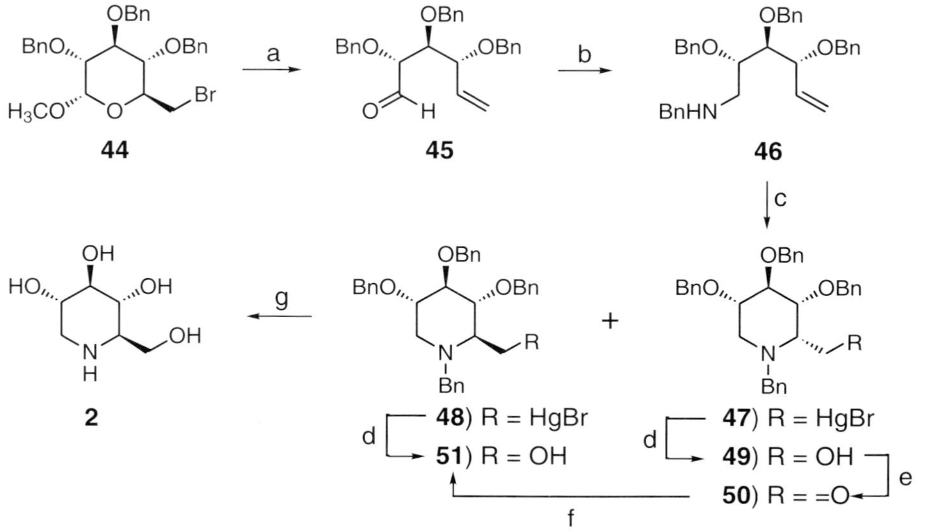

Scheme 9 (*a*) AcOH, Zn dust, reflux, 2 h. (*b*) $BnNH_2$ (16 equiv.), aqueous $NaBH_3CN$, *n*-propylalcohol–H_2O (19:1), 91% for two steps. (*c*) $Hg(CF_3CO_2)_2$, THF. (*d*) $NaBH_4$, DMF, O_2, 70%. (*e*) $(COCl)_2$, DMSO, CH_2Cl_2, –78°C; then NEt_3. (*f*) DBU, CH_2Cl_2, $NaBH_4$. (*g*) H_2, Pd *on* C, AcOH.

oxygenation of **48** using $NaBH_4$–DMF–O_2 afforded **51** (70%); similarily, **47** was converted to **49**. Hydrogenation of **51** produced 1-deoxynojirimycin hydrochloride (**2**·HCl) in 28% overall yield. On the other hand, Swern oxidation of **49** gave the aldehyde **50**, which upon epimerization followed by reduction afforded **51**, whose conversion to **2** was achieved in 35% overall yield from methyl α-D-glucopyranoside.

The synthesis of (+)-nojirimycin (**1**) from tetra-*O*-benzyl-D-glucopyranose (**11**) as depicted in Scheme 10[200] was found to be unreliable.[201] Compound **11** underwent anomeric oxidation with DMSO–Ac_2O to afford tetra-*O*-benzyl-D-glucono-1,5-lactone (**52**) in 84% yield, which was treated with liquid ammonia solution in the presence of trace amounts of Amberlite IR-120 (H^+) resin in dioxane to give the gluconolactam **53** (50%). Treatment of **53** with sodium borohydride followed by removal of the benzyl protecting groups afforded **1**. However, this procedure was reported to be unreliable.[201] Hydrogenation of the lactam **53** led to partial or complete removal of the benzyl groups, with no evidence indicating the formation of **1** or **2** in the reaction mixture. When the lactone **52** was reacted with benzylamine under a variety of conditions, the expected lactam was not detected, but the only isolated product was the amide **54**.

OBn BnO OBn HO O OBn **11** → a → OBn BnO OBn O O OBn **52** → b → OBn BnO OBn O N H OBn **53**
↓ c
CONHBn OH HO OH OH OH **54** ← d ↛ OH HO OH HO N H OH **1**

Scheme 10 (*a*) DMSO, Ac_2O, rt, 84%. (*b*) Liquid NH_3, dioxane, rt, Amberlite IR-120 (H^+) resin, 6 h, 50%. (*c*) 1. $NaBH_4$, EtOH, H^+, 80%; 2. 5% Pd *on* C, AcOH, H_2, 72%. (*d*) $PhCH_2NH_2$, $PhCH_3$, reflux, 3 h, 80%.

An efficient synthesis of (+)-nojirimycin (**1**) and 1-deoxynojirimycin (**2**) from the readily available 1,2-*O*-isopropylidene-α-D-glucofuranose (**55**) was reported (Scheme 11).[2,202–207] Selective oxidation[203,208–211] of **55** with dibutyltin oxide and bromine afforded the corresponding 5-oxo derivative **56** in 92% yield. Treatment of **56** with *O*-methylhydroxylamine hydrochloride furnished the two geometrical isomers of the oxime **57** in a ratio of 1:2.5. Reduction of **57** with $LiAlH_4$ in THF gave the gluco- **(60)** and the ido- **(61)** isomers in a 4:1 ratio. Hydrolysis of **60** in the presence of SO_2 afforded nojirimycin bisulfite (**17**). Alternatively, removal of the isopropylidene group from **57** gave the intermediate **59** whose subsequent reduction gave **17**, which was converted into **1**.

On the other hand, deisopropylidenation of **56** gave 5-keto-D-glucose (**58**) whose reductive amination with benzhydrylamine gave a mixture of 1-deoxynojirimycin derivative **62** and the L-iditol diastereomer in a ratio of 96:4 (74%). Deprotection of **62** afforded **2** (Scheme 11).[207]

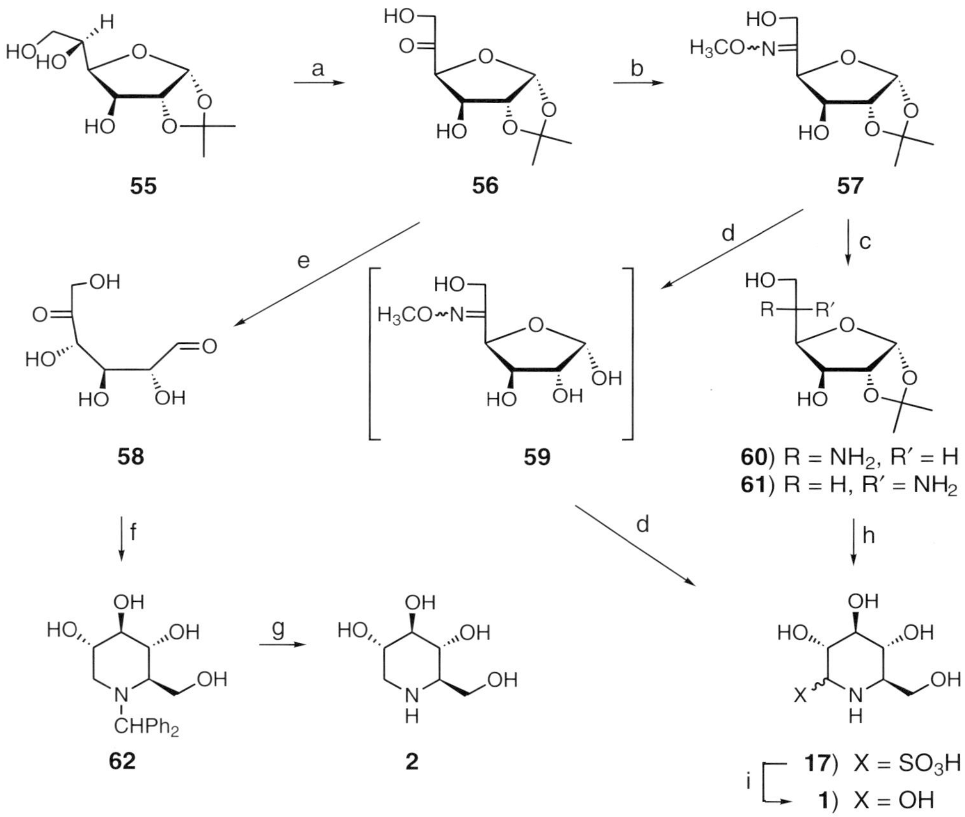

Scheme 11 (*a*) n-Bu_2SnO, CH_3OH, 0°C, Br_2, 5 min, 92%. (*b*) CH_3ONH_2HCl, $NaHCO_3$, CH_3OH, 1 h. (*c*) $LiAlH_4$, THF, reflux, 5 h, 84%, 4:1 gluco–ido. (*d*) p-TsOH, CH_3OH, 2 h; then $LiAlH_4$, H_2O, SO_2, rt, 3 days. (*e*) Dowex 50WX8 resin, 75% for three steps. (*f*) Ph_2CHNH_2, $NaBH_3CN$, CH_3OH, 74%. (*g*) 20% $Pd(OH)_2$ *on* C, H_2, 90%. (*h*) H_2O, SO_2, rt, 3 days, CH_3OH, refrigerator, overnight, 75%. (*i*) Dowex 50WX8 resin, 67% for three steps.

The hydrates **63** and **64**,[212,213] derived from the acylation of **58**, underwent reductive amination with benzylamine, under carefully controlled conditions to give a 1:1 mixture of **65** and **67**, or a 1:2 mixture of **66** and **68** in low combined yield 27 and 30%, respectively (Scheme 12).[207]

(1-^{13}C)-1-Deoxynojirimycin (**2**) has been synthesized from 5-azido-5-deoxy-1,2-*O*-isopropylidene-α-D-glucofuranose (**69**) utilizing ^{13}C-enriched potassium cyanide (Scheme 13).[214] Compound **69**[215] was treated with benzyl bromide and sodium hydride followed by acid hydrolysis of the isopropylidene group to give an anomeric mixture of partially

63) R = Ac
64) R = Bz

65) R = Ac
66) R = Bz

67) R = Ac
68) R = Bz

Scheme 12 (*a*) $NaBH_3CN$, CH_3OH, $BnNH_2$, –78°C, 1 h.

protected aldofuranose **70** in 72% overall yield from **69**. Oxidative cleavage of **70** with sodium metaperiodate in THF and water afforded the unstable aldehyde **71**, which underwent one-carbon extension with ^{13}C-enriched (99%) potassium cyanide to give a mixture of the (1-^{13}C)-substituted D-glucono **72** and D-mannono-nitriles **73** in practically quantitative yield. Saponification of the mixture of **72** and **73** afforded **75** and **74** in 48 and 13% overall yield from **70**, respectively. Reduction of the azido lactone **75** with sodium borohydride in methanol followed by hydrogenation afforded **2** in 25% overall yield from azide **69**.

69 70 71

72) R^1 = OH, R^2 = H
73) R^1 = H, R^2 = OH

(1-^{13}C)-**2** 75 74

Scheme 13 (*a*) 1. BnBr, NaH, DMF–THF (3:1), 80%; 2. TFA, CH_3CN, H_2O, 90%. (*b*) $NaIO_4$, THF, H_2O. (*c*) $K^{13}CN$, THF, H_2O, 100% for two steps; then $NaHCO_3$, 3.7:1 ratio of **72**:**73**. (*d*) H_2SO_4, H_2O, **75** (48% from **70**) and **74** (13% from **70**). (*e*) $NaBH_4$, CH_3OH, H_2, Pd *on* C, 25% from **69**.

With the use of epoxide **76** as a divergent intermediate, the syntheses of (+)-nojirimycin (**2**) by insertion of nitrogen between C-1 and C-5 with inversion of configuration at C-5 (Scheme 14) and by insertion of the nitrogen between C-2 and C-6 with inversion of configuration at C-2 have been achieved (Scheme 15).[216] Treatment of **76**, obtained from diacetone glucose,[217] with benzyl alcohol in the presence of sodium hydride followed by triflation of the secondary hydroxyl group at C-5 and subsequent S_N2 displacement of the triflate group with azide ion afforded **77** in 37% overall yield from diacetone

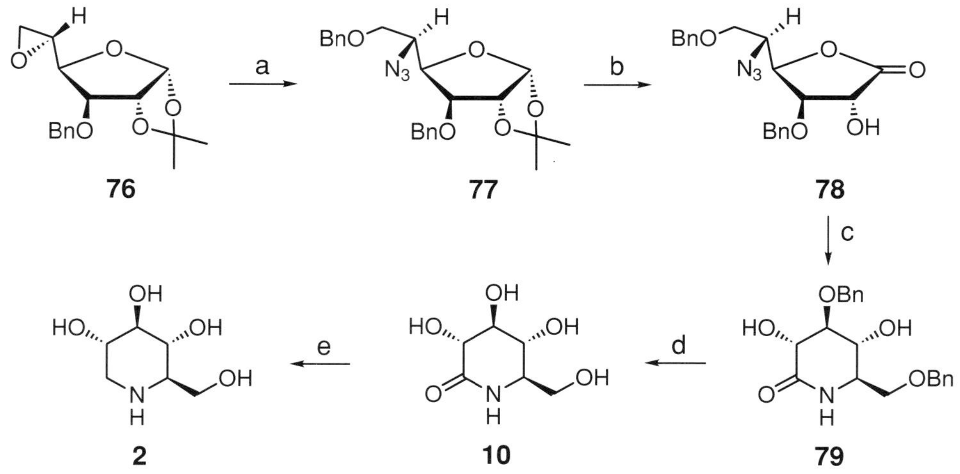

Scheme 14 (*a*) 1. NaH, BnOH, DMF, 49% from diacetone D-glucose; 2. Tf_2O, CH_2Cl_2, Py; NaN_3, DMF, 75%. (*b*) 1. Aqueous TFA; 2. Br_2, aqueous dioxane, barium benzoate, 74%. (*c*) 1. $SnCl_2$, CH_3OH; 2. K_2CO_3, CH_3OH, 56%. (*d*) H_2, EtOH, Pd black. (*e*) BH_3, THF, rt, 1.5 h; then reflux, 1.5 h.

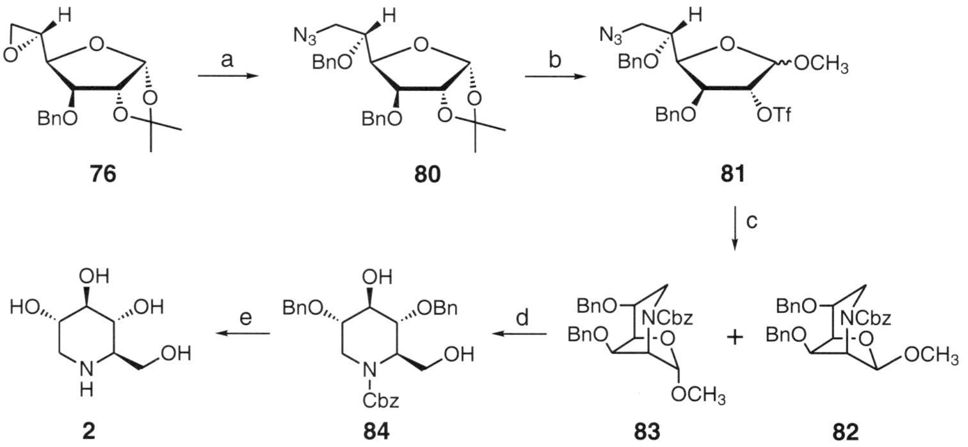

Scheme 15 (*a*) 1. NaN_3, DMF, 87%; 2. NaH, BnBr, Bu_4NI, THF, 91%. (*b*) CH_3OH, HCl, 67%; 2. Tf_2O, CH_2Cl_2, Py. (*c*) 1. $SnCl_2$, CH_3OH; 2. NaOAc, CH_3OH; 3. CbzCl, 67% for three steps. (*d*) 1. TFA, aqueous dioxane; 2. $NaBH_4$, EtOH, 49% for two steps. (*e*) H_2, AcOH, Pd black, ion-exchange chromatography.

glucose. Removal of the isopropylidene group from **77** with aqueous TFA followed by oxidation of the resulting lactol with bromine in aqueous 1,4-dioxane in the presence of barium benzoate afforded **78** in 74% yield. Tin(II) chloride reduction of azidolactone **78** followed by intramolecular cyclization by treatment with potassium carbonate furnished the dibenzyl lactam **79** (56%). Removal of the benzyl groups from **79** afforded nojirimycin δ-lactam (**10**), which could be converted into **2**.

On the other hand, treatment of the epoxide **76** with sodium azide in *N*,*N*-dimethylformamide followed by protection of the resulting secondary hydroxyl group with benzyl bromide afforded the azide **80** in 79% yield. Treatment of **80** with methanolic hydrogen chloride followed by triflation of the resulting hydroxyl group at C-2 afforded the triflate **81**. Tin(II) chloride reduction of the azidotriflates **81** followed by intramolecular cyclization with sodium acetate in methanol and subsequent protection of the resulting secondary amine with benzyl chloroformate afforded the α- and β-furanoside carbamates **83** and **82**. Hydrolysis of **82** and **83** by TFA in aqueous dioxane followed by sodium borohydride reduction of the resulting lactol furnished the protected 1-deoxynojirimcin **84** (49%). Subsequent hydrogenation of **84** afforded **2**.

The synthesis of 1-deoxynojirimycin (**2**) has been achieved from D-glucose by conversion to 2-azido-3-*O*-benzyl-2-deoxy-α-D-mannoside **85** and then **86** (Scheme 16).[218–221] Inversion of the hydroxyl group at C-5 in **86** is necessary for the synthesis of **2**; pyridinium chlorochromate oxidation of **86** followed by sodium borohydride reduction of the resulting ketone afforded **87**. Acid hydrolysis of **87** followed by borohydride reduction and subsequent removal of the protecting groups afforded **2** in 51% overall yield from **86**.

Scheme 16 (*a*) Ref. 218. (*b*) CH_2Cl_2, rt, PCC, MS 3Å, 2 h; then EtOH, $NaBH_4$, 0°C, 1 h, 78%. (*c*) 1. 50% aqueous TFA, rt, 30 min; then EtOH, $NaBH_4$, rt, 15 min, 65%; 2. AcOH, H_2, Pd black, 48 h, ion-exchange chromatography CG-400 (OH^-); then CG-120 (H^+), 100%.

3.1.1.2 *Synthesis from L-sorbose* Paulsen[222] reported the first total synthesis of 1-deoxynojirimycin (**2**) in 1967, 10 years before its isolation from natural sources, using L-sorbose as a starting material (Scheme 17). Starting by removal of the terminal isopropylidene group from 1-*O*-acetyl-2,3:4,6-di-*O*-isopropylidene-α-L-sorbofuranose (**88**) followed by tosylation of the resulting primary hydroxyl group afforded **89**, which underwent tosylate displacement with azide ion to produce **90**. Hydrogenation of azide **90** over Raney nickel led to the amine **93**, which could also be obtained from the ditosylate **91** via **92**. Removal of the isopropylidene group from **93** gave a mixture of the intermediates **94**, **95** and **96**, which were hydrogenated over platinum in water to furnish **2** and a trace amount of its isomer **97**.

Synthesis of 1-deoxynojirimycin (**2**) from 2,3-*O*-isopropylidene-α-L-sorbofuranose (**98**), obtained from L-sorbose in 71% yield,[223] was reported (Scheme 18).[224] It was converted into azide **90** in 80% yield by treatment with Ph_3P, CBr_4 and lithium azide in *N*,*N*-dimethylformamide. Removal of the acetonide moiety by acid hydrolysis afforded the 6-azido-6-deoxy-L-sorbofuranose **99** in 95% yield, which upon hydrogenation gave **2**.

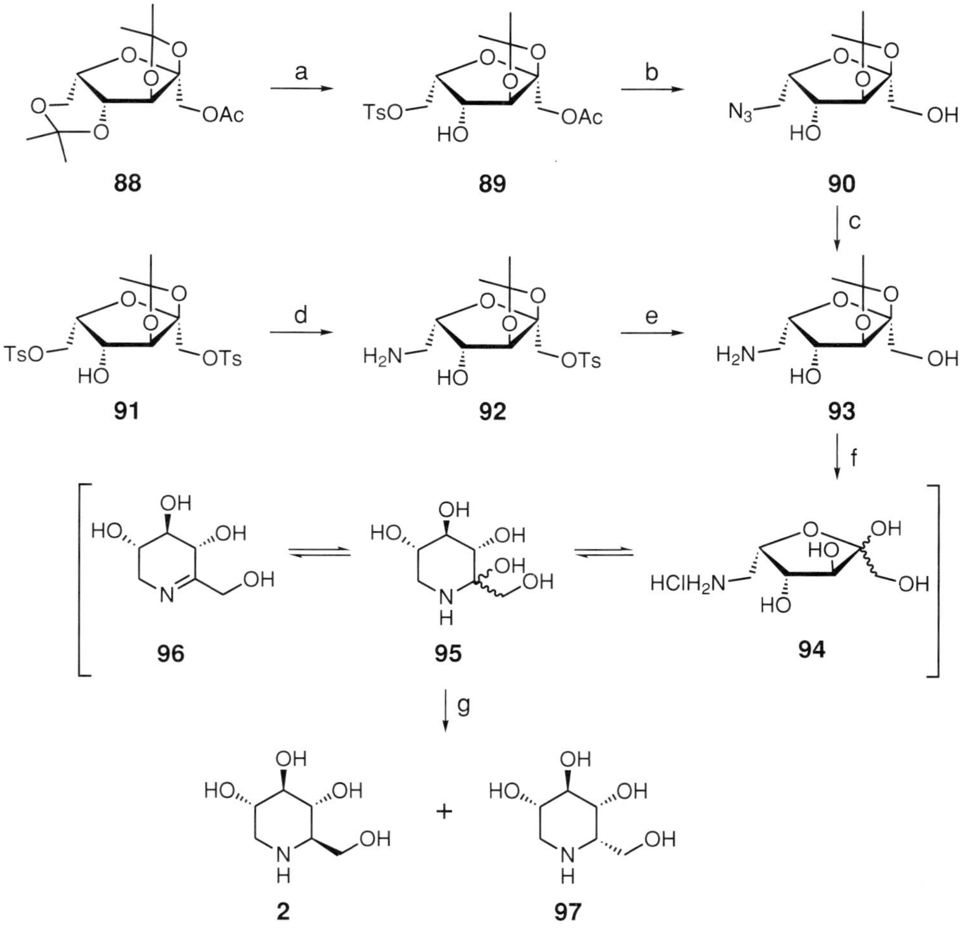

Scheme 17 (*a*) 1. AcOH, 80°C, 40 min; 2. *p*-TsCl, Py, 4 h, 80% for two steps. (*b*) 1. Na, CH_3OH, 40°C, 1 h, 100%; 2. NaN_3, DMF, 100°C, 15 h, 90%. (*c*) H_2O–CH_3OH (2:1), Raney nickel, H_2, 5 h. (*d*) 1. NaN_3, DMF, 100°C, 10 h, 90%; 2. Pt, H_2, CH_3OH, 5 h, 78%. (*e*) Na–Hg (3:1), 16 h, CH_3OH, 42%. (*f*) 1 N HCl, rt, 20 h, 70%. (*g*) 1. Amberlite IR-45 (OH^-) resin; 2. Pt, water, 5 h, H_2.

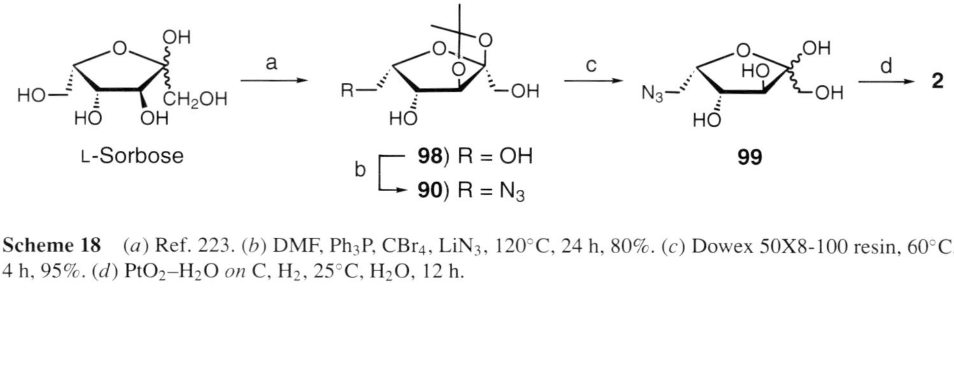

Scheme 18 (*a*) Ref. 223. (*b*) DMF, Ph_3P, CBr_4, LiN_3, 120°C, 24 h, 80%. (*c*) Dowex 50X8-100 resin, 60°C, 4 h, 95%. (*d*) PtO_2–H_2O *on* C, H_2, 25°C, H_2O, 12 h.

The 1,2-*O*-isopropylidene derivative **100**, derived from L-sorbose, was also used for the synthesis of 1-deoxynojirimycin (**2**) (Scheme 19).[225] It was prepared[226] from L-sorbose by reaction with 2,2-dimethoxypropane in the presence of stannous chloride, followed by acid hydrolysis. Selective sulfonylation of the primary hydroxyl group with 2,4,6-triisopropylbenzenesulfonyl chloride (TIBSCl) in a 1:1 mixture of triethylamine and pyridine followed by nucleophilic displacement with azide ion in DMF afforded the 6-azido-1,2-*O*-isopropylidene-L-sorbofuranose **101**. Reduction[227] of the azide function in **101** was carried out using Ph_3P in THF and the resulting amine **102** was subjected to acid hydrolysis using Dowex (H^+) resin to afford the amine intermediate **103**, which was hydrogenated in the presence of 20% palladium on carbon to afford **2** in 61% overall yield from **101** and 17.6% from L-sorbose.

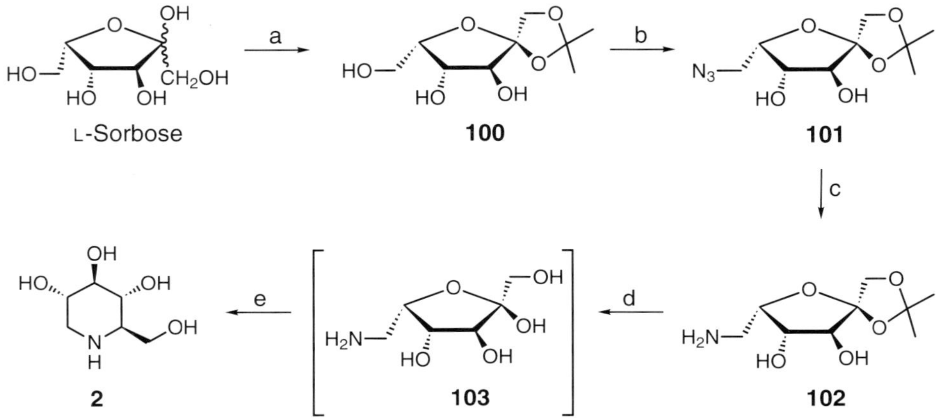

Scheme 19 (*a*) 1. DMP, $SnCl_2$, THF, reflux; 2. H_2O, H_2SO_4, 7 h, 40% for two steps. (*b*) 1. 2,4,6-TIBSCl, NEt_3–Py (1:1); 2. NaN_3, DMF, 100°C, 20 h, 72%. (*c*) 1. Ph_3P, THF; 2. H_2O. (*d*) Dowex 50WX8-200 (H^+) resin, H_2O. (*e*) H_2, 20% Pd *on* C, 72 h, H_2O; then 25% NH_3, CH_3OH, 30 min, 61% from **101**.

A chemoenzymatic synthesis of 1-deoxynojirimycin (**2**) was started by the condensation of DHAP and 3-azido-2-hydroxypropanal (**104**) catalyzed by the enzyme FDP aldolase to give **105** and **106** (Scheme 20).[228–234] Removal of the phosphate was catalyzed by phosphatase, followed by hydrogenation of the mixture over palladium on carbon to give a 4:1 mixture of **2** and its manno analogue.

The enzyme catalyzed formation of the diastereoisomeric mixture of 6-azido-6-deoxy-1-phosphates of D-fructose and L-sorbose, which were precipitated as their barium salts **107** and **108** in 70% yield. Subsequent hydrolysis of the phosphate esters with acid phosphatase under mild conditions afforded **109** and **110**. Hydrogenation of **109** and **110** afforded **2** and its manno analogue.

1-Deoxynojirimycin (**2**) and its analogues **114** and **117** have also been synthesized chemoenzymatically (Scheme 21).[161,235,236] Reaction of DHAP with (*R*)-3-azido-2-hydroxypropanal catalyzed by fuc-1-phosphate aldolase and rham-1-phosphate aldolase gave **112** and **115**, respectively, which upon removal of the phosphate group gave **117** and **118** whose reductive amination generated **114** and **117**. On the other hand, reaction of DHAP with (*S*)-**104** afforded **111**, followed by conversion to **2**.

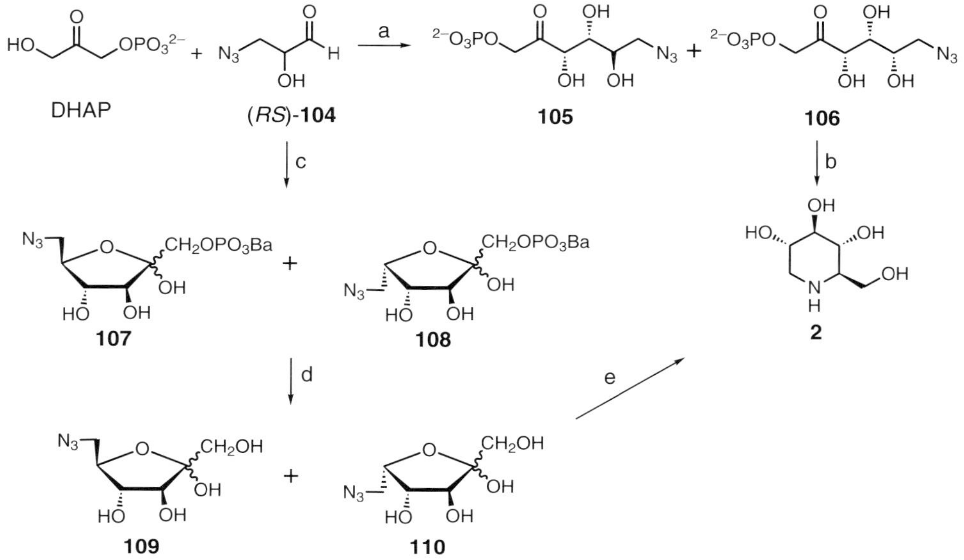

Scheme 20 (*a*) FDP aldolase, pH 2.5–6.5 (2 N NaOH), 12 h. (*b*) 1. PASE; 2. H_2, 10% Pd *on* C, 10 h; then Dowex 1 (OH^-) resin, 59%. (*c*) RAMA (EC 4.1.2.13), pH 6.5, 2 N NaOH, 25°C, 12 h; then $BaCl_2 \cdot 2H_2O$, H_2O, 1 h at 0°C, 70%. (*d*) 1. Dowex 50WX8 (H^+) resin; 2. PASE (EC 3.1.3.2), pH 4.5, 2 N NaOH, 38°C, 48 h; then $Ba(OH)_2$; 3. Dowex 1X8 (HCO_2^-) resin, 77%. (*e*) Pt *on* C, H_2, K_2CO_3, H_2O, 25°C, 12 h, **2** (65%).

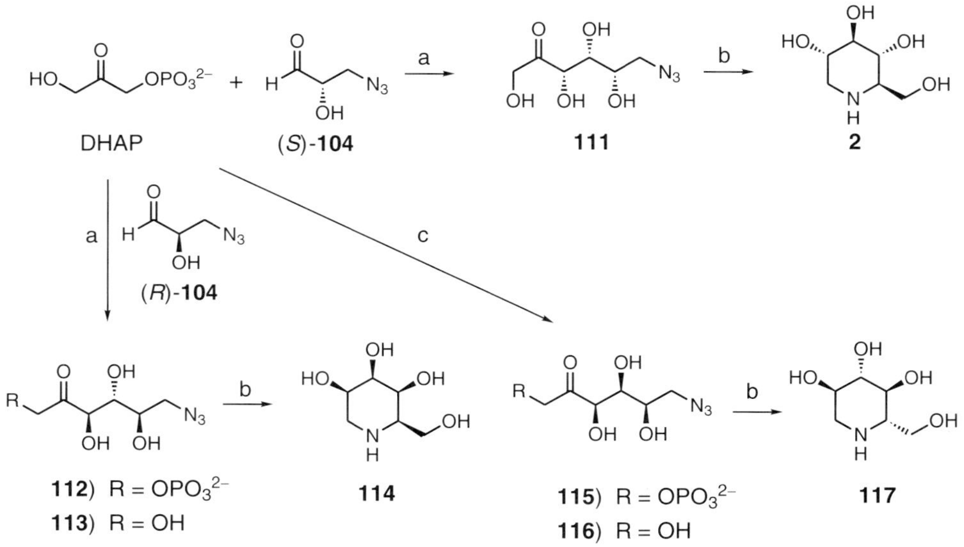

Scheme 21 (*a*) Fuc-1-phosphate aldolase. (*b*) 1. Phosphatase; 2. H_2, Pd *on* C. (*c*) Rham-1-phosphate aldolase.

3.1.1.3 *Synthesis from L-threose* Nojirimycin (**1**) and 1-deoxynojirimycin (**2**) have been synthesized from 4-*O*-(*tert*-butyldimethylsilyl)-2,3-*O*-isopropylidene-L-threose (**118**) (Scheme 22).[237] Compound **118** was treated with trimethyl phosphonoacetate to provide the *E* ester **119**, as a single isomer in 95% yield. Reduction of the ester group in **119** followed by Sharpless asymmetric epoxidation[238] of the resulting allylic alcohol afforded the *syn*-epoxide **120**. Treatment of **120** with NaN_3 and NH_4Cl in a mixture of 1,2-dimethoxyethane, 2-methoxyethanol and water followed by protection of the resulting azidodiol with MOMCl furnished the azide **121**, which underwent removal of the silyl protecting group with fluoride ion followed by mesylation of the resulting primary hydroxyl group to produce the mesylate

Scheme 22 (*a*) NaH, 0°C, benzene, trimethyl phosphonoacetate, rt, 1 h; then **118**, 1 h, 95%. (*b*) 1. CH_2Cl_2, DIBAL-H, rt, 14 h; 2. MS 4 Å, –20°C, CH_2Cl_2, titanium(IV) isopropoxide, diethyl L-tartrate, 10 min; then CH_2Cl_2, –20°C; then *tert*-butyl hydroperoxide, –20°C, 14 h, 78%. (*c*) 1. NaN_3, NH_4Cl, 1,2-dimethoxyethane–2-methoxy ethanol–H_2O (1:2:1), reflux, 6 h, 75% based on recovered **120**; 2. DIPEA, $ClCH_2OCH_3$, $CHCl_3$, reflux, 3 h, 91%. (*d*) 1. THF, TBAF, rt, 30 min, 98%; 2. NEt_3, CH_2Cl_2, MsCl, 0°C, 10 min, 94%. (*e*) 10% Pd *on* C, CH_3OH, H_2, 2 h; then NEt_3, reflux, 2 h, 92%. (*f*) HCl–CH_3OH (1:2), reflux, 1 h, 98%. (*g*) 1. CH_3OH, 10% Pd *on* C, H_2, 4 h, 86%; 2. NEt_3, dioxane, *p*-methoxybenzyl *S*-(4,6-dimethylpyrimidin-2-yl)thiocarbonate, rt, 6 h, 91%. (*h*) 1. THF, TBAF, rt, 1 h, 98%; 2. $(COCl)_2$, DMSO, NEt_3, CH_2Cl_2, 82%. (*i*) H_2O, SO_2, rt, 60 h; then CH_3OH, SO_2, 63%. (*j*) Dowex 1X2 (OH^-) resin, H_2O, 90%.

122. Catalytic hydrogenation of compound **122** followed by treatment with triethylamine in methanol furnished the protected 1-deoxynojirimycin **123**. Removal of the protecting groups from **123** led to **2** in 34% overall yield from **118**.

On the other hand, catalytic hydrogenation of **121** followed by protection of the resulting amine with *p*-methoxybenzyl *S*-(4,6-dimethylpyrimidin-2-yl)thiocarbonate afforded the carbamate **124**, which underwent desilylation with TBAF followed by Swern oxidation of the resulting primary hydroxyl group to provide the aldehyde **125**. Removal of the protecting groups from **125** with aqueous sulfurous acid at room temperature provided 1-deoxynojirimycin-1-sulfonic acid (**17**). Finally, (+)-nojirimycin (**1**) was generated, by treatment of **22** with Dowex 1X2 (OH^-) resin, in 36% overall yield from **121**.

3.1.1.4 *Synthesis from* D*-mannitol* A methodology has utilized a double nucleophilic opening of C_2-symmetric bisepoxides to synthesize 1-deoxynojirimycin (**2**) (Scheme 23).[239] The tetrol **126**,[240] prepared from D-mannitol, underwent selective silylation of the primary hydroxyl groups with TBSCl followed by mesylation of the secondary hydroxyl groups to afford the dimesylate **127**. Desilylation and subsequent alkaline treatment of the resulting diol afforded the diepoxide **128** in 26% overall yield from D-mannitol. Regiospecific opening of one epoxy function followed by spontaneous 6-exo ring closure was expected to be favored kinetically according to Baldwin's rules[241] over the competing 7-endo ring closure reaction, and to be favored also over the substitution at both C-1 and C-6 by 2 equiv. of amine. However, treatment of **128** with $BnNH_2$ in toluene at reflux for 11 days was found to afford a 55:45 separable mixture of **129** (49%) and **130** (39%), which were hydrogenated in the presence of palladium black to give **2** and azepane analogue **131**, respectively.

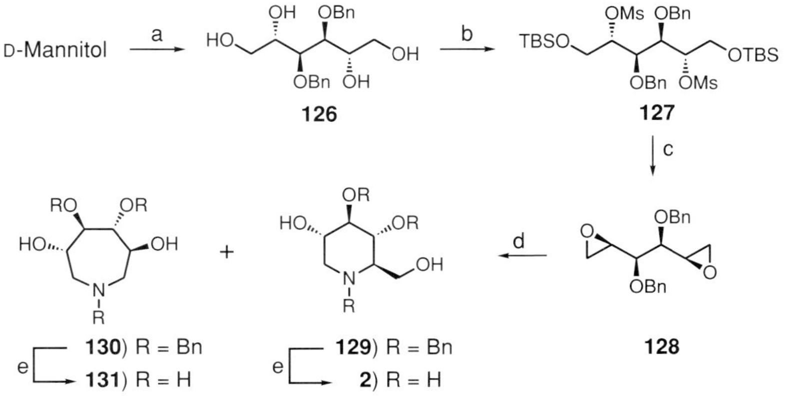

Scheme 23 (*a*) 1. DMP, $SnCl_2$, $(CH_2OCH_3)_2$, 50%; 2. NaH, BnBr, *n*-Bu_4NI, THF; 3. CH_3CO_2H, H_2O, 40°C, 87%. (*b*) 1. TBSCl, imidazole, DMF, 0°C, 80%; 2. MsCl, NEt_3, CH_2Cl_2, 0°C, 98%. (*c*) HCl, CH_3OH, 20°C; then NaOH, H_2O, 20°C, 75%. (*d*) $BnNH_2$ (5 equiv.), $PhCH_3$, reflux, 11 days. (*e*) H_2, Pd black, CH_3CO_2H, 15 h, 100%.

3.1.1.5 *Synthesis from gluconolactone* 1-Deoxynojirimycin (**2**) and the corresponding lactam **10** were stereoselectively synthesized[242] from tetra-*O*-benzyl-D-glucono-1,5-lactone (**132**)[243] (Scheme 24). Lactone **132** underwent amination to give the hydroxy amide

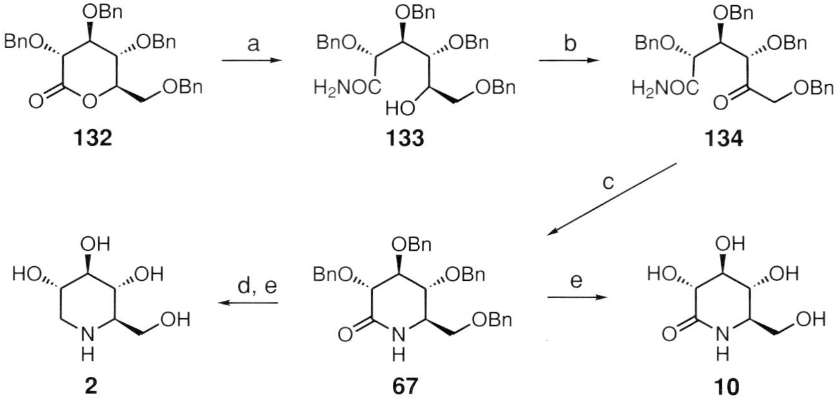

Scheme 24 (*a*) NH_3, 86%. (*b*) DMSO, Ac_2O. (*c*) $NaBH_3CN$, HCO_2H, 58% for two steps. (*d*) $LiAlH_4$. (*e*) H_2, Pd *on* C.

133 in 86% yield. Oxidation of **133** gave the corresponding keto amide **134**, which was treated with formic acid and sodium cyanoborohydride, in a one-pot reaction, to provide the lactam **67** in 58% overall yield from **133**. Debenzylation of **67** afforded gluconolactam **10**, whereas its reduction with $LiAlH_4$ afforded the tetra-*O*-benzyl-1-deoxynojirimycin, which was hydrogenated to produce **2**.

3.1.1.6 *Synthesis from glucuronolactone* Glucuronolactone was also used for the synthesis of 1-deoxynojirimycin (**2**) (Scheme 25).[244] Chlorination[245,246] of **135**[247] gave the chloride **136**, which upon reduction with sodium borohydride afforded 5-chloro-5-deoxy-1,2-*O*-isopropylidene-β-L-idofuranose (**137**). Selective protection of the primary hydroxyl group with DHP afforded **138**, which underwent S_N2 displacement with azide ion to give **139**. The respective bromo analogue of **136** produced the azide **139** in 55% overall yield via a similar pathway. Complete deprotection of **139** furnished **142**, which was finally hydrogenated to produce **2** in 33% overall yield from **135**.

On the other hand, reaction of **137** with sodium azide afforded **140** whose hydrogenation gave the amine **141**. Acid hydrolysis of **141** afforded **1** (Scheme 25).[244,245]

Alternatively, **1** and **2** were also prepared from **135** (Scheme 26)[248] by Swern oxidation,[249–251] followed by condensation with *O*-benzylhydroxylamine hydrochloride in refluxing benzene with azeotropic removal of water to afford the *E*-isomer of *O*-benzyloxime **143**. Reduction of **143** in the presence of Boc-anhydride furnished the Boc-amine **144** as a single diastereomer in 49% overall yield from **135**. Reduction of **144** with $LiAlH_4$ afforded **145** (92%), which was treated with saturated aqueous SO_2 to produce nojirimycin bisulfite (**17**) in 40% overall yield from **135**. Treatment of **17** with basic ion-exchange resin gave **1** in quantitative yield. Hydrogenation of **17** in the presence of Raney nickel and barium hydroxide produced **2**.

3.1.1.7 *Synthesis from inositols* Syntheses of **1** and **2** as well as the L-analogues starting from the seven-membered lactones **146** and **152**,[252] prepared from myo-inositol, have been

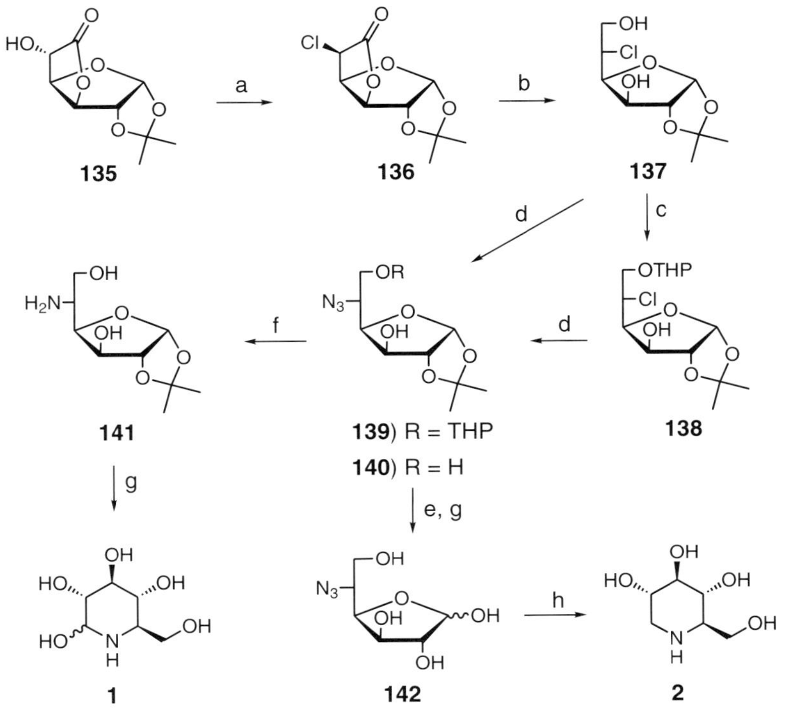

Scheme 25 (*a*) CH_2Cl_2, Py, 0°C, SO_2Cl_2, 1.5 h; then $NaHCO_3$, 94%; *or* $Cl_2C{=}N^+Me_2Cl^-$, CH_2Cl_2, 30 min, 87%. (*b*) CH_3OH, Amberlite IR-120 (H^+) resin, 0°C, $NaBH_4$, 89%; *or* $LiBH_4$, THF, 0°C, Dowex 50WX2 (H^+) resin, 92%. (*c*) CH_2Cl_2, DHP, PPTs, 12 h, rt. (*d*) DMF, NaN_3, 130°C, 4 days, 75%. (*e*) CH_2Cl_2, CH_3OH, *p*-TsOH, rt, overnight, 47% for four steps. (*f*) H_2, CH_3OH, 10% Pd *on* C, rt, 1 h, 95%. (*g*) EtOH–H_2O (2:5), TFA, 40°C, 92%. (*h*) CH_3OH–H_2O (5:6), H_2, 5% Pd *on* C, 3 days; then Amberlite CG-50 (NH_4^+) resin, 90%.

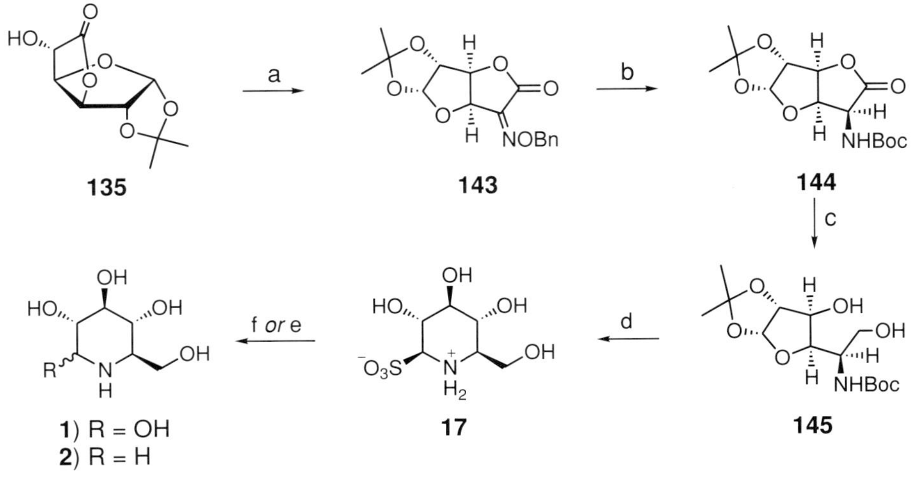

Scheme 26 (*a*) 1. $(COCl)_2$, DMSO, NEt_3, CH_2Cl_2, –70°C, 90%; 2. $BnONH_2{\cdot}HCl$, C_6H_6, reflux. (*b*) H_2, 10%, Pd *on* C, $(Boc)_2O$ (1.1 equiv.), EtOAc. (*c*) $LiAlH_4$, THF, 0°C, 92%. (*d*) Sat. aqueous SO_2, 35–40°C, 90%. (*e*) H_2, Raney nickel, $Ba(OH)_2{\cdot}8H_2O$, H_2O. (*f*) Dowex 1X2 basic ion-exchange resin.

reported (Schemes 27 and 28).[253,254] Treatment of **146** with trimethyl orthoformate in methanol in the presence of *p*-toluenesulfonic acid followed by $LiAlH_4$ reduction afforded **147** in 90% yield. Protection of the primary hydroxyl group in **147** with chloromethyl methyl ether followed by Mitsunobu reaction[255] using phthalimide furnished **148** in 46% yield and the two side products **149** (19%) and **150** (34%). Removal of the phthaloyl group from **148** by hydrazine hydrate followed by protection of the resulting amine and subsequent removal of the *O*-benzyl groups afforded the triol **151**. Treatment of **151** with sulfur dioxide produced (+)-nojirimycin bisulfite (**17**) (58%), which upon treatment with Dowex 1X2 (OH^-) resin afforded **1**. On the other hand, hydrogenolysis of **17** furnished **2** in 53% yield.

146 **147** **148**) X = Phth, Y = H **149**) X = H, Y = Phth **150** **151** **17** **1**) R = OH **2**) R = H

Scheme 27 (*a*) 1. $HC(OCH_3)_3$, *p*-TsOH·H_2O, CH_3OH, reflux, 1 h; 2. THF, $LiAlH_4$, 0–25°C, 2 h, 90% for two steps. (*b*) 1. DIPEA, CH_2Cl_2, $ClCH_2OCH_3$, 5°C, 4 h, 61%, **147** (33% recovery); 2. phthalimide, Ph_3P, benzene, diisopropyl azadicarboxylate, rt, 2 h, **148** (46%), **149** (19%), **150** (34%). (*c*) 1. $NH_2NH_2·H_2O$, CH_3OH, reflux, overnight; then CH_2Cl_2, $(Boc)_2O$, NEt_3, rt, 3 h, 95%; 2. EtOH, H_2, $Pd(OH)_2$ *on* C, rt, 4 h, 100%. (*d*) H_2O, SO_2, 40°C, 2 days, 58%. (*e*) H_2O, Dowex 1X2 (OH^-) resin, 40%. (*f*) 1. $Ba(OH)_2·8H_2O$, H_2O, Raney nickel (W-4), H_2, rt, 8 h; 2. Amberlite 1R-120B (H^+) resin, NH_4OH as eluent, 53%.

152 **153** **154**) R = OH **155**) R = H

Scheme 28

Likewise, (−)-nojirimycin (**154**) and (−)-1-deoxynojirimycin (**155**) were prepared starting from the lactone **152**.

References

1. Ishida, N.; Kumagai, K.; Niida, T.; Tsuruoka, T.; Yumoto, H. *J. Antibiot., Ser. A.* **20** (1967) 66.
2. Inouye, S.; Tsuruoka, T.; Ito, T.; Niida, T. *Tetrahedron* **24** (1968) 2125.
3. Nishikawa, T.; Ishida, N. *J. Antibiot., Ser. A* **18** (1965) 132.
4. Schimidt, D.D.; Frommer, W.; Muller, L.; Truscheit, E. *Naturwissenschaften* **66** (1979) 584.
5. Matsumura, S.; Enomoto, H.; Aoyagi, Y.; Yoshikuni, Y.; Yagi, M. Japanese Patent Application 78/99,208, Aug. 14, 1978; *Chem. Abstr.* **93** (1980) 66581s; Japanese Kokai Tokkyo Koho 8027,136, 1980; Japanese Patent Kokoku 59,27338, 1984.
6. Niwa, T.; Inouye, S.; Tsuruoka, T.; Koaze, Y.; Niida, T. *Agric. Biol. Chem.* **34** (1970) 966.
7. Reese, E.T.; Parrish, F.W.; Ettinger, M. *Carbohydr. Res.* **18** (1971) 381.
8. Hanozet, G.; Pircher, H.P.; Vanni, P.; Oesch, B.; Semenza, G. *J. Biol. Chem.* **256** (1981) 3703.
9. Truscheit, E.; Frommer, W.; Junge, B.; Muller, L.; Schmidt, D.D.; Wingender, W. *Angew. Chem., Int. Ed. Engl.* **20** (1981) 744.
10. Fellows, L.E. *Chem. Br.* **23** (1987) 842.
11. Fleet, G.W.J. *Spec. Publ. R. Soc. Chem.* **65** (Top. Med. Chem.) (1988) 149.
12. Ishida, N.; Kumagai, K.; Niida, T.; Hamamoto, K.; Shomura, T. *J. Antibiot., Ser. A* **20** (1967) 62.
13. Inouye, S.; Tsuruoka, T.; Niida, T. *J. Antibiot., Ser. A* **19** (1966) 291.
14. Kodama, Y.; Tsuruoka, T.; Niwa, T.; Inouye, S. *J. Antibiot.* **38** (1985) 116.
15. Hardick, D.J.; Hutchinson, D.W.; Trew, S.J.; Wellington, E.M.H. *Tetrahedron* **48** (1992) 6285.
16. Hughes, A.B.; Rudge, A.J. *Nat. Prod. Rep.* **11** (1994) 135.
17. Yagi, M.; Kauno, K.; Aoyagi, Y.; Murai, H. *Nippon Kagaku Kaishi* **50** (1976) 571; *Chem. Abstr.* **86** (1977) 167851r.
18. Paulsen, H.; Sangster, I.; Heyens, K. *Chem. Ber.* **100** (1967) 802.
19. Yoshikuni, Y. *Agric. Biol. Chem.* **52** (1988) 121.
20. Daigo, K.; Inamori, Y.; Takemoto, T. *Chem. Pharm. Bull.* **34** (1986) 2243.
21. Murao, S.; Miyata, S. *Agric. Biol. Chem.* **44** (1980) 219.
22. Scofield, A.M.; Fellows, L.E.; Nash, R.J.; Fleet, G.W.J. *Life Sci.* **39** (1986) 645.
23. Saeki, H.; Ohki, E. *Chem. Pharm. Bull.* **16** (1968) 2477.
24. Fuhrmann, U.; Bause, E.; Ploegh, H. *Biochim. Biophys. Acta* **825** (1985) 95.
25. Ezure, Y.; Moruo, S.; Miyazaki, K.; Kawamata, M. *Agric. Biol. Chem.* **49** (1985) 1119.
26. Winkler, D.A.; Holan, G. *J. Med. Chem.* **32** (1989) 2084.
27. Kang, M.S.; Elbein, A.D. *Plant Physiol.* **71** (1983) 551.
28. Truscheit, E.; Frommer, W.; Junge, B.; Muller, L.; Schmidt, D.D.; Wingender, W. *Angew. Chem., Int. Ed. Engl.* **20** (1981) 744.
29. Sinnott, M.L. *Chem. Rev.* **90** (1990) 1171.
30. Withers, S.G.; Street, I.P. *J. Am. Chem. Soc.* **110** (1988) 8551.
31. Schneider, M.J.; Ungemach, F.S.; Broquist, H.P.; Harris, T.M. *Tetrahedron* **39** (1983) 29.
32. Gupta, R.B.; Franck, R.W. *J. Am. Chem. Soc.* **109** (1987) 6554.
33. Withers, S.G.; Street, I.P.; Bird, P.; Dolphin, D.H. *J. Am. Chem. Soc.* **109** (1987) 7430.
34. Dorling, P.R.; Huxtable, C.R.; Colegate, S.M. *Biochem. J.* **191** (1980) 649.
35. Fuhrmann, U.; Bause, E.; Legler, G.; Ploegh, H. *Nature* **307** (1984) 755.
36. Gupta, R.B.; Franck, R.W. *J. Am. Chem. Soc.* **109** (1987) 6554.
37. Post, C.B.; Karplus, M. *J. Am. Chem. Soc.* **108** (1986) 1317.
38. Fleet, G.W.J. *Tetrahedron Lett.* **26** (1985) 5073.
39. Fleet, G.W.J.; Ramsden, N.G.; Nash, R.J.; Fellows, L.E.; Jacob, G.S.; Molyneux, R.J.; Cenci di Bello, I.; Winchester, B. *Carbohydr. Res.* **205** (1990) 269.
40. Segal, H.L.; Winkler, J.R. *Curr. Top. Cell. Regul.* **24** (1984) 229.
41. Battistini, L.; Zanardi, F.; Rassu, G.; Spanu, P.; Pelosi, G.; Fava, G.G.; Ferrari, M.B.; Casiraghi, G. *Tetrahedron: Asymmetry* **8** (1997) 2975.
42. Hudlicky, T.; Rouden, J.; Luna, H. *J. Org. Chem.* **58** (1993) 985.
43. Altenbach, H.J.; Himmeldirk, K. *Tetrahedron: Asymmetry* **6** (1995) 1077.
44. Heiker, F.R.; Schueller, A.M. *Carbohydr. Res.* **203** (1990) 314.
45. Dondoni, A.; Merino, P.; Perrone, D. *J. Chem. Soc., Chem. Commun.* (1991) 1578.
46. Meyers, A.I.; Price, D.A.; Andres, C.J. *Synlett* (1997) 533.
47. Bols, M.; Hazell, R.G.; Thomsen, I.B. *Chem. Eur. J.* **3** (1997) 940.

48. Johnson, C.R.; Golebiowski, A.; Schoffers, E.; Sundram, H.; Braun, M.P. *Synlett* (1995) 313.
49. Cook, G.R.; Beholz, L.G.; Stille, J.R. *J. Org. Chem.* **59** (1994) 3575.
50. Lindström, U.M.; Somfai, P. *Tetrahedron Lett.* **39** (1998) 7173.
51. Dondoni, A.; Merino, P.; Perrone, D. *Tetrahedron* **49** (1993) 2939.
52. Ishida, H.; Kitagawa, M.; Kiso, M.; Hasegawa, A. *Carbohydr. Res.* **208** (1990) 267.
53. Norris, P.; Horton, D.; Levine, B.R. *Tetrahedron Lett.* **36** (1995) 7811.
54. Meyers, A.I.; Andres, C.J.; Resek, J.E.; Woodall, C.C.; McLaughlin, M.A.; Lee, P.H.; Price, D.A. *Tetrahedron* **55** (1999) 8931.
55. Kajimoto, T.; Liu, K.K.C.; Pederson, R.L.; Zhong, Z.; Ichikawa, Y.; Proco, J.A., Jr.; Wong, C.H. *J. Am. Chem. Soc.* **113** (1991) 6187.
56. Battistini, L.; Zanardi, F.; Rassu, G.; Spanu, P.; Pelosi, G.; Gasparri Fava, G.; Ferrari, M.B.; Casiraghi, G. *Tetrahedron: Asymmetry* **8** (1997) 2975.
57. Yoshikuni, Y.; Ezure, Y.; Seto, T.; Mori, K; Watanabe, M.; Enomoto, H. *Chem. Pharm. Bull.* **37** (1989) 106.
58. Ezure, Y. *Agric. Biol. Chem.* **49** (1985) 2159.
59. Spohr, U.; Spiro, R.G.; Bach, M. *Can. J. Chem.* **71** (1993) 1919.
60. Spohr, U.; Bach, M.; Spiro, R.G. *Can. J. Chem.* **71** (1993) 1928.
61. Spohr, U.; Bach, M. *Can. J. Chem.* **71** (1993) 1943.
62. Marquis, C.; Picasso, S.; Vogel, P. *Synthesis* (1999) 1441.
63. Frérot, E.; Marquis, C.; Vogel, P. *Tetrahedron Lett.* **37** (1996) 2023.
64. Ardron, H.; Butters, T.D.; Platt, F.M.; Wormald, M.R.; Dwek, R.A.; Fleet, G.W.J.; Jacob, G.S. *Tetrahedron: Asymmetry* **4** (1993) 2011.
65. Suhara, Y.; Achiwa, K. *Chem. Pharm. Bull.* **43** (1995) 414.
66. Andersen, S.M.; Ekhart, C.; Lundt, I.; Stü, A.E. *Carbohydr. Res.* **326** (2000) 22.
67. Poupon, E.; Luong, B.-X.; Chiaroni, A.; Kunesch, N.; Husson, H.-P. *J. Org. Chem.* **65** (2000) 7208.
68. Bülow, A.; Plesner, I.W.; Bols, M. *J. Am. Chem. Soc.* **122** (2000) 8567.
69. Garcia-Moreno, M.I.; Diaz-Perez, P.; Mellet, C.O.; Fernández, J.M.G. *Chem. Commun.* (2002) 848.
70. Budzinska, A.; Sas, W. *Tetrahedron Lett.* **42** (2001) 105.
71. Shitara, E.; Nishimura, Y.; Kojima, F.; Takeuchi, T. *J. Antibiot.* **52** (1999) 348.
72. Koóš, M.; Steiner, B.; Mičová, J.; Langer, V.; Ďurik, M.; Gyepesová, D. *Carbohydr. Res.* **332** (2001) 351.
73. Shitara, E.; Nishimura, Y.; Kojima, F.; Takeuchi, T. *Bioorg. Med. Chem.* **7** (1999) 1241.
74. Nishimura, Y.; Shitara, E.; Takeuchi, T. *Tetrahedron Lett.* **40** (1999) 2351.
75. Shitara, E.; Nishimura, Y.; Nerome, K.; Hiramoto, Y.; Takeuchi, T. *Org. Lett.* **2** (2000) 3837.
76. Shitara, E.; Nishimura, Y.; Kojima, F.; Takeuchi, T. *Bioorg. Med. Chem.* **8** (2000) 343.
77. Nishimura, Y.; Shitara, E.; Adachi, H.; Toyoshima, M.; Nakajima, M.; Okami, Y.; Takeuchi, T. *J. Org. Chem.* **65** (2000) 2.
78. Kondo, K.-I.; Adachi, H.; Nishimura, Y.; Takeuchi, T. *Nat. Prod. Lett.* **15** (2001) 371.
79. Kondo, K.; Adachi, H.; Shitara, E.; Kojima, F.; Nishimura, Y. *Bioorg. Med. Chem.* **9** (2001) 1091.
80. Szolcsányi, P.; Gracza, T.; Koman, M.; Prónayová, N.; Liptaj, T. *Tetrahedron: Asymmetry* **11** (2000) 1.
81. Butters, T.D.; van den Broek, L.A.G.M.; Fleet, G.W.J.; Krulle, T.M.; Wormald, M.R.; Dwek, R.A.; Platt, F.M. *Tetrahedron: Asymmetry* **11** (2000) 113.
82. Monterde, M.I.; Brieva, R.; Gotor, V. *Tetrahedron: Asymmetry* **12** (2001) 525.
83. Comins, D.L.; Fulp, A.B. *Tetrahedron Lett.* **42** (2001) 6839.
84. Banba, Y.; Abe, C.; Nemoto, H.; Kato, A.; Adachi, I.; Takahata, H. *Tetrahedron: Asymmetry* **12** (2001) 817.
85. Fernández, J.M.G.; Mellet, C.Q.; Benito, J.M.; Fuentes, J. *Synlett* (1998) 316.
86. Szolcsányi, P.; Gracza, T.; Koman, M.; Prónayová, N.; Liptaj, T. *Tetrahedron: Asymmetry* **11** (2000) 2579.
87. Takahashi, S.; Kuzuhara, H.; Nakajima, M. *Tetrahedron* **57** (2001) 6915.
88. Mao, H.; Joly, G.J.; Peeters, K.; Hoornaert, G.J.; Compernolle, F. *Tetrahedron* **57** (2001) 6955.
89. Filichev, V.V.; Brandt, M.; Pedersen, E.B. *Carbohydr. Res.* **333** (2001) 115.
90. Pistia-Brueggeman, G.; Hollingsworth, R.I. *Tetrahedron* **57** (2001) 8773.
91. Subramanian, T.; Lin, C.-C.; Lin, C.-C. *Tetrahedron Lett.* **42** (2001) 4079.
92. Compain, P.; Martin, O.R. *Bioorg. Med. Chem.* **9** (2001) 3077.
93. Fouace, S.; Therisod, M. *Tetrahedron Lett.* **41** (2000) 7313.
94. Jourdant, A.; Zhu, J. *Tetrahedron Lett.* **41** (2000) 7033.
95. Scott, J.D.; Williams, R.M. *Tetrahedron Lett.* **41** (2000) 8413.
96. Langlois, N.; Clavez, O. *Tetrahedron Lett.* **41** (2000) 8285.
97. Carbonnel, S.; Fayet, C.; Gelas, J.; Troin, Y. *Tetrahedron Lett.* **41** (2000) 8293.
98. Saha, N.N.; Desai, V.N.; Dhavale, D.D. *Tetrahedron* **57** (2001) 39.
99. Joly, G.J.; Peeters, K.; Mao, H.; Brossette, T.; Hoornaert, G.J.; Compernolle, F. *Tetrahedron Lett.* **41** (2000) 2223.
100. Mehta, G.; Mohal, N. *Tetrahedron Lett.* **41** (2000) 5747.
101. Joseph, C.C.; Regeling, H.; Zwanenburg, B.; Chittenden, G.J.F. *Carbohydr. Res.* **337** (2002) 1083.

102. Mehta, G.; Lakshminath, S. *Tetrahedron Lett.* **43** (2002) 331.
103. Ruiz, M.; Ojiea, V.; Quintela, J.M. *Synlett* (1999) 204.
104. Godskesen, M.; Lundt, I.; Søtofte, I. *Tetrahedron: Asymmetry* **11** (2000) 567.
105. Pandey, G.; Kapur, M. *Tetrahedron Lett.* **41** (2000) 8821.
106. Jensen, H.H.; Bols, M. *J. Chem. Soc., Perkin Trans. 1* (2001) 905.
107. Hansen, S.U.; Bols, M. *J. Chem. Soc., Perkin Trans. 1* (2000) 911.
108. D'Andrea, F.; Catelani, G.; Mariani, M.; Vecchi, B. *Tetrahedron Lett.* **42** (2001) 1139.
109. Kojima, M.; Seto, T.; Kyotani, Y.; Ogawa, H.; Kitazawa, S.; Mori, K.; Maruo, S.; Ohgi, T.; Ezure, Y. *Biosci. Biotech. Biochem.* **60** (1996) 694.
110. Kiso, M.; Katagiri, H.; Furui, H.; Hasegawa, A. *J. Carbohydr. Chem.* **11** (1992) 627.
111. Davis, F.A.; Zhang, H.; Lee, S.H. *Org. Lett.* **3** (2001) 759.
112. Garcia-Moreno, M.I.; Benito, J.M.; Mellet, C.O.; Fernández, J.M.G. *J. Org. Chem.* **66** (2001) 7604.
113. Patil, N.T.; Tilekar, J.N.; Dhavale, D.D. *J. Org. Chem.* **66** (2001) 1065.
114. Szolcsányi, P.; Gracza, T.; Koman, M.; Prónayová, N.; Liptaj, T. *Chem. Commun.* (2000) 471.
115. Desai, V.N.; Saha, N.N.; Dhavale, D.D. *Chem. Commun.* (1999) 1719.
116. Cipolla, L.; Ferla, B.L.; Peri, F.; Nicotra, F. *Chem. Commun.* (2000) 1289.
117. Désiré, J.; Shipman, M. *Synlett* (2001) 1332.
118. Agami, C.; Comesse, S.; Kadouri-Puchot, C. *J. Org. Chem.* **65** (2000) 4435.
119. Langlois, N. *Org. Lett.* **4** (2002) 185.
120. Souers, A.J.; Ellman, J.A. *J. Org. Chem.* **65** (2000) 1222.
121. Haddad, M.; Larcheveque, M. *Tetrahedron Lett.* **42** (2001) 5223.
122. Brooks, C.A.; Comins, D.L. *Tetrahedron Lett.* **41** (2000) 3551.
123. Legler, G. *Adv. Carbohydr. Chem. Biochem.* **48** (1990) 319.
124. Ganem, B. *Acc. Chem. Res.* **29** (1996) 340.
125. Raadt, A.; Ekhart. C.W.; Ebner, M.; Stütz, A.E. *Top. Curr. Chem.* **187** (1997) 157.
126. Stick, R.V. *Top. Curr. Chem.* **187** (1997) 187.
127. Van de Broek, L.A.G.M. In *Carbohydrates in Drug Design*; Witczak, Z., Nieforth, K.A., Eds.; Marcel Dekker: New York, 1997; p. 471.
128. Van de Broek, L.A.G.M.; Vermaas, D.J.; Heskamp, B.M.; van Boeckel, C.A.A.; Tan, M.C.A.A.; Bolscher, J.G.M.; Ploegh, H.L.; van Kemenade, F.J.; de Goede, R.E.Y.; Miedema, F. *Recl. Trav. Chim. Pays-Bas* **112** (1993) 82.
129. Defoin, A.; Sarazin, H.; Streith, J. *Helv. Chem. Acta* **79** (1996) 560.
130. Sun, L.; Li, P.; Amankulor, N.; Tang, W.; Landry, D.W.; Zhao, K. *J. Org. Chem.* **63** (1998) 6472.
131. Sütz, A.E. *Iminosugars as Glycosidase Inhibitors*; Wiley-VCH: Weinheim, 1999.
132. Winchester, B.; Fleet, G.W.J. *J. Carbohydr. Chem.* **19** (2000) 471.
133. Jacob, G.S. *Curr. Opin. Struct. Biol.* **5** (1995) 605.
134. Bols, M. *Acc. Chem. Res.* **31** (1998) 1.
135. Heightman, T.D.; Vasella, A.T. *Angew. Chem., Int. Ed.* **38** (1999) 750.
136. Sears, P.; Wong, C.-H. *Angew. Chem., Int. Ed.* **38** (1999) 2300.
137. Ikeda, K.; Takahashi, M.; Nishida, M.; Miwa, M.; Kizu, H.; Kameda, Y.; Arisawa, M.; Watson, A.A.; Nash, R.J.; Fleet, G.W.J.; Asano, N. *Carbohydr. Res.* **323** (2000) 73.
138. Asano, N.; Tomioka, E.; Kizu, H.; Matsui, K. *Carbohydr. Res.* **253** (1994) 235.
139. Kato, A.; Asano, A.; Kizu, H.; Matsui, K. *J. Nat. Prod.* **60** (1997) 312.
140. Wong, C.-H.; Prosencher, L.; Porco, J.A.; Jung, S.-H.; Wang, Y.-F.; Chen, L.; Wang, R.; Steensma, D.H. *J. Org. Chem.* **60** (1995) 1492.
141. Rassu, G.; Pinna, L.; Spano, P.; Culeddu, N.; Casizaghi, G.; Fava, G.G.; Ferrari, M.B.; Pelosi, G. *Tetrahedron* **48** (1992) 727.
142. Herdeis, C.; Schiffer, T. *Tetrahedron* **52** (1996) 14745.
143. Compernolle, F.; Joly, G.; Peeters, K.; Toppel, S.; Hoomaert, G.; Kilonda, A.; Bila, B. *Tetrahedron* **53** (1997) 12739.
144. Asano, N.; Nash, R.J.; Molyneux, R.J.; Fleet, G.W.J. *Tetrahedron: Assmmetry* **11** (2000) 1645.
145. Winchester, B.; Barker, C.; Bathes, S.; Jacob, G.S.; Namgoong, S.K.; Fleet, G.W.J. *Biochem. J.* **265** (1990) 277.
146. Rhinehart, B.L.; Robinson, K.M.; Liu, P.S.; Payne, J.; Wheatley, M.E.; Wagner, S.R. *J. Pharmacol. Exp. Ther.* **241** (1987) 915.
147. Fleet, G.W.J.; Namgoong, S.K.; Barker, C.; Baines, S.; Jacob, G.S.; Winchester, B. *Tetrahedron Lett.* **30** (1989) 4439.
148. Szolesanyl, P.; Gracza, T.; Koman, M.; Prosayova, N.; Liptaj, T. *J. Chem. Soc., Chem. Commun.* (2000) 471.
149. Desai, V.N.; Saha, N.N.; Dhavale, D.D. *J. Chem. Soc., Chem. Commun.* (1999) 1719.
150. Dhavale, D.D.; Saha, N.N.; Desai, V.N. *J. Org. Chem.* **62** (1997) 7482.

151. Dhavale, D.D.; Desai, V.N.; Sindkhedkar, M.D.; Mali, R.S.; Castellari, C.; Trombini, C. *Tetrahedron: Asymmetry* **8** (1997) 1475.
152. Camiletti, C.; Dhavale, D.D.; Gentilucci, L.; Trombini, C. *J. Chem. Soc., Perkin Trans. 1* (1993) 3157.
153. Camiletti, C.; Dhavale, D.D.; Donati, F.; Trombini, C. *Tetrahedron Lett.* **36** (1995) 7293.
154. Dhavale, D.D.; Trombini, C. *Heterocycles* **34** (1992) 2253.
155. Dhavale, D.D.; Donati, F.; Trombini, C. *J. Chem. Soc., Chem. Commun.* (1992) 1268.
156. Chakraborty, T.K.; Jayaprakash, S. *Tetrahedron Lett.* **38** (1997) 8899.
157. Jotterand, N.; Vogel, P. *J. Org. Chem.* **64** (1999) 8973.
158. Nishimura, Y.; Adachi, H.; Satoh, T.; Shitara, E.; Nakamura, H.; Kojima, F.; Takeuchi, T. *J. Org. Chem.* **65** (2000) 4871.
159. Asano, N.; Oseki, K.; Kizu, H.; Matsui, K. *J. Med. Chem.* **37** (1994) 3701.
160. Schaller, C.; Vogel, P.; Jäger, V. *Carbohydr. Res.* **314** (1998) 25.
161. Von der Osten, C.H.; Sinskey, A.J.; Barbas, C.F., III; Pederson, R.L.; Wang, Y.F.; Wong, C.H. *J. Am. Chem. Soc.* **111** (1989) 3924.
162. Liao, L.X.; Wang, Z.M.; Zhang, H.X.; Zhou, W.S. *Tetrahedron: Asymmetry* **10** (1999) 3649.
163. Koulocheri, S.D.; Haroutounian, S.A. *Tetrahedron Lett.* **40** (1999) 6869.
164. Sharpless, K.B.; Amberg, W.; Bennani, Y.L.; Crispino, G.A.; Hartung, J.; Jeong, K.S.; Kwong, H.L.; Morikawa, K.; Wang, Z.M.; Xu, D.; Zhang, X.L. *J. Org. Chem.* **57** (1992) 2768.
165. Schmidt, U.; Werner, J. *Synthesis* (1986) 986.
166. Altenbach, H.J.; Wischant, R. *Tetrahedron Lett.* **36** (1995) 4983.
167. Xu, Y.M.; Zhou, W.S. *Tetrahedron Lett.* **37** (1996) 1461.
168. Xu, Y.M.; Zhou, W.S. *J. Chem. Soc., Perkin Trans. 1* (1997) 741.
169. Vonhoff, S.; Vasella, A. *Synth. Commun.* **29** (1999) 551.
170. Auberson, Y.; Vogel, P. *Angew Chem., Int. Ed. Engl.* **28** (1989) 1498.
171. Bordier, A.; Compain, P.; Martin, O.R.; Ikeda, K.; Asano, N. *Tetrahedron: Asymmetry* **14** (2003) 47.
172. Han, H. *Tetrahedron Lett.* **44** (2003) 1567.
173. Godin, G.; Compain, P.; Masson, G.; Martin, O.R. *J. Org. Chem.* **67** (2002) 6960.
174. Ginesta, X.; Pericás, M.A.; Riera, A. *Tetrahedron Lett.* **43** (2002) 779.
175. Goujon, J.-Y.; Gueyrard, D.; Compain, P.; Martin, O.R.; Asano, N. *Tetrahedron: Asymmetry* **14** (2003) 1969.
176. Takahata, H.; Banba, Y.; Ouchi, H.; Nemoto, H.; Kato, A.; Adachi, I. *J. Org. Chem.* **68** (2003) 3603.
177. Knight, J.G.; Tchabanenko, K. *Tetrahedron* **59** (2003) 281.
178. Holzgrabe, U.; Heller, E. *Tetrahedron* **59** (2003) 781.
179. Calmés, M.; Escale, F.; Rolland, M.; Martinez, J. *Tetrahedron: Asymmetry* **14** (2003) 1685.
180. Kinast, G.; Schedel, M. *Angew. Chem., Int. Ed. Engl.* **20** (1981) 805.
181. Long, J.W.; Bollenback, G.N. *Methods Carbohydr. Chem.* **2** (1963) 79.
182. Pistia-Brueggeman, G.; Hollingsworth, R.I. *Carbohydr. Res.* 328 (2000) 467.
183. Moutel, S.; Shipman, M. *J. Chem. Soc., Perkin Trans. 1* (1999) 1403.
184. Gent, P.A.; Gigg, R. *J. Chem. Soc., Perkin Trans. 1* (1974) 1446.
185. Ley, S.V.; Norman, J.; Griffith, W.P.; Marsden, S.P. *Synthesis* (1994) 639.
186. Mancuso, A.J.; Swern, D. *Synthesis* (1981) 165.
187. Vasella, A.; Voeffray, R. *Helv. Chem. Acta* **65** (1982) 1134.
188. Matos, C.R.R.; Lopes, R.S.C.; Lopes, C.C. *Synthesis* (1999) 571.
189. Kiguchi, T.; Tajiri, K.; Ninomiya, I.; Naito, T.; Hiramatus, H. *Tetrahedron Lett.* **36** (**1995**) 253.
190. Ikota, N. *Heterocycles* **29** (1989) 1469.
191. Gludemans, C.P.J.; Fletcher, H.G., Jr. *Methods Carbohydr. Chem.* **6** (1972) 373.
192. Ermert, P.; Vasella, A. *Helv. Chim. Acta* **74** (1991) 2043.
193. Garegg, P.J.; Samuelsson, B. *J. Chem. Soc., Perkin Trans. 1* (1980) 2866.
194. Omura, K.; Swern, D. *Tetrahedron* **34** (1978) 1651.
195. Mancuso, A.J.; Huang, S.L.; Swern, D. *J. Org. Chem.* **43** (1978) 2480.
196. Schmidt, R.R.; Michel, J.; Rücher, E. *Liebigs Ann. Chem.* (1989) 423.
197. Lubineau, A.; Thieffrey, A.; Veyrieres, A. *Carbohydr. Res.* **46** (1976) 143.
198. Bernet, B.; Vasella, A. *Helv. Chim. Acta* **62** (1979) 1990.
199. Bernotas, R.C.; Ganem, B. *Tetrahedron Lett.* **26** (1985) 1123.
200. Rajanikanth, B.; Seshadri, R. *Tetrahedron Lett.* **30** (1989) 755.
201. Fleet, G.W.J.; Ramsden, N.G.; Carpenter, N.M.; Petursson, S.; Aplin, R.T. *Tetrahedron Lett.* **31** (1990) 405.
202. Tsuda, Y.; Okuno, Y.; Kanemitsu, K. *Heterocycles* **27** (1988) 63.
203. Tsuda, Y.; Matsuhira, N.; Kanemitsu, K. *Chem. Pharm. Bull.* **33** (1985) 4095.
204. Tsuda, Y.; Okuno, Y.; Iwaki, M.; Kanemitsu, K. *Chem. Pharm. Bull.* **37** (1989) 2673.
205. Tsuda, Y.; Hanajima, M.; Yoshimoto, K. *Chem. Pharm. Bull.* **31** (1983) 3778.
206. Tsuda, Y.; Okuno, Y.; Kanemitsu, K. *Heterocycles* **27** (1988) 63.

207. Reitz, A.B.; Baxter, E.W. *Tetrahedron Lett.* **31** (1990) 6777.
208. Tsuda, Y.; Hanajima, M.; Matsuhira, N.; Okuno, Y.; Kanemitsu, K. *Chem. Pharm. Bull.* **37** (1989) 2344.
209. Helferich, B.; Bigelow, N.M. *Z. Physiol. Chem.* (1931) 200.
210. Kiely, D.E.; Fletcher, H.G., Jr., *J. Org. Chem.* **34** (1969) 1386.
211. Barnett, J.E.G.; Rasheed, A; Corina, D.L. *Biochem. J.* **131** (1973) 21.
212. Blattner, R.; Ferrier, R.J. *J. Chem. Soc., Perkin Trans. 1* (1980) 1523.
213. Ferrier, R.J.; Tyler, P.C. *J. Chem. Soc., Perkin Trans. 1* (1980) 1528.
214. Berger, A.; Ebner, M.; Stüz, A.E. *Tetrahedron Lett.* **36** (1995) 4989.
215. Dax, K.; Gaigg, B.; Grassberger, V.; Kolblinger, B.; Stütz, A.E. *J. Crabohydr. Chem.* **9** (1990) 479.
216. Fleet, G.W.J.; Carpenter, N.M.; Petursson, S.; Ramsden, N.G. *Tetrahedron Lett.* **31** (1990) 409.
217. Whistler, R.L.; Gramero, R.E. *J. Org. Chem.* **29** (1964) 2609.
218. Fleet, G.W.J.; Smith, P.W.; Nash, R.J.; Fellows, L.E.; Parekh, R.B.; Rademacher, T. *Chem. Lett.* (1986) 1051.
219. Fleet, G.W.J.; Smith, P.W. *Tetrahedron Lett.* **26** (1985) 1469.
220. Fleet, G.W.J.; Fellows, L.E.; Smith, P.W. *Tetrahedron* **43** (1987) 979.
221. Fleet, G.W.J.; Smith, P.W. *Tetrahedron* **43** (1987) 971.
222. Paulsen, H.; Sangster, I.; Heyns, K. *Chem. Ber.* **100** (1967) 802.
223. Slobodin, J.M. *Zh. Obshch. Khim. USSR* **17** (1947) 485.
224. Beaupere, D.; Stasik, B.; Demailly, R.U.G. *Carbohydr. Res.* **191** (1989) 163.
225. Behling, J.; Farid, P.; Medich, J.R.; Scaros, M.G.; Prunier, M.; Weier, R.M.; Khanna, I. *Synth. Commun.* **21** (1991) 1383.
226. Chen, C.; Whistler, R.L. *Carbohydr. Res.* **175** (1988) 265.
227. Vaultier, M.; Knioezi, N.; Carrie, R. *Tetrahedron Lett.* **24** (1983) 763.
228. Pederson, R.L.; Kim, M.J.; Wong, C.H. *Tetrahedron Lett.* **29** (1988) 4645.
229. Wong, C.H.; Whitesides, G.M. *J. Org. Chem.* **48** (1983) 3199.
230. Wong, C.H.; Mazenod, F.P.; Whitesides, G.M. *J. Org. Chem.* **48** (1983) 3493.
231. Durrwachter, J.R.; Drueckhammer, D.G.; Nozaki, K.; Sweers, H.M.; Wong, C.H. *J. Am. Chem. Soc.* **108** (1986) 7812.
232. Bednarski, M.D.; Waldmann, H.J.; Whitesides, G.M. *Tetrahedron Lett.* **27** (1986) 5807.
233. Ziegler, V.T.; Straub, A.; Effenberger, F. *Angew. Chem.* **100** (1988) 737; *Angew. Chem., Int. Ed. Engl.* **27** (1988) 842.
234. Straub, A.; Effenberger, F.; Fischer, P. *J. Org. Chem.* **55** (1990) 3926.
235. Liu, K.K.C.; Kajimoto, T.; Chen, L.; Zhong, Z.; Ichikawa, Y.; Wong, C.H. *J. Org. Chem.* **56** (1991) 6280.
236. Kajimoto, T.; Chen, L.; Liu, K.K.C.; Wong, C.H. *J. Am. Chem. Soc.* **113** (1991) 6678.
237. Iida, H.; Yamazaki, N.; Kibayashi, C. *J. Org. Chem.* **52** (1987) 3337.
238. Katsuki, T.; Sharpless, K.B. *J. Am. Chem. Soc.* **102** (1980) 5974.
239. Poitout, L.; Merrer, Y.L.; Depezay, J.C. *Tetrahedron Lett.* **35** (1994) 3293.
240. Jurczak, J.; Bauer, T.; Chmielewski, M. *Carbohydr. Res.* **164** (1987) 493.
241. Baldwin, J.E. *J. Chem. Soc., Chem. Commun.* (1976) 734.
242. Overkleeft, H.S.; van Wiltenburg, J.; Pandit, U.K. *Tetrahedron Lett.* **34** (1993) 2527.
243. Kuzuhara, H.; Fletcher, H.G. *J. Org. Chem.* **32** (1967) 2531.
244. Dax, K.; Gaigg, B.; Grassberger, V.; Kölblinger, B.; Stütz, A.E. *J. Carbohydr. Chem.* **9** (1990) 479.
245. Klemer, A.; Hofmeister, U.; Lemmes, R. *Carbohydr. Res.* **68** (1979) 391.
246. Parolis, H. *Carbohydr. Res.* **43** (1975) C1.
247. Sowden, J.C. *J. Chem. Soc.* **74** (1952) 4377.
248. Anzeveno, P.B.; Creemer, L.J. *Tetrahedron Lett.* **31** (1990) 2085.
249. Kitihara, T.; Ogawa, T.; Naganuma, T.; Matsui, M. *Agric. Biol. Chem.* **38** (1974) 2189.
250. Bashyal, B.P.; Chow, H.F.; Fleet, G.W.J. *Tetrahedron Lett.* **27** (1986) 3205.
251. Bashyal, B.P.; Chow, H.F.; Fellows, L.E.; Fleet, G.W.J. *Tetrahedron* **43** (1987) 415.
252. Chida, N.; Yamada, E.; Ogawa, S. *J. Carbohydr. Chem.* **7** (1988) 555.
253. Chida, N.; Furuno, Y.; Ikemoto, H.; Ogawa, S. *Carbohydr. Res.* **237** (1992) 185.
254. Chida, N.; Furuno, Y.; Ogawa, S. *J. Chem. Soc., Chem. Commun.* (1989) 1230.
255. Mitsunobu, O. *Synthesis* (1981) 1.

3.1.2 *Mannojirimycin*

Mannojirimycin (**1**) and 1-deoxymannojirimycin (1,5-dideoxy-1,5-imino-D-mannitol, **2**) were produced by *Streptomyces subrutilus* ATCC 27467 grown on medium containing glucose. Compound **1** was first produced and then underwent dehydration and reduction to give **2**.[1] Moreover, compound **2** was isolated from *Omphalea diandra* L.,[2] *Lonchocarpus sericeus* and *Lonchocarpus costaricensis*.[3] The legume *L. sericeus*, a native to West Indies and tropical America, was reported to have insecticidal and pesticidal properties, and its bark extracts are used to treat parasitic skin infections.[4] Compound **2** is an inhibitor of bovine α-L-fucosidase[5] and mannosidase I, a glycoprotein-processing enzyme. It is also a useful tool for the study of biochemical pathways.[6,7]

Microbiological oxidation of mannojirimycin with *Gluconobacter suboxydans* IAM 1829 gave D-mannono-δ-lactam (**3**).[3] It exhibited powerful inhibition of rat α-mannosidase and of apricot β-glucosidase.[3] The structures of **2** and **3** were determined on the basis of NMR spectroscopy and X-ray structural analysis.[4,5]

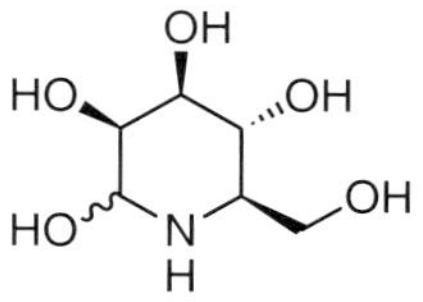

Mannojirimycin **1**   1-Deoxymannojirimycin **2**   D-Mannono-δ-lactam **3**

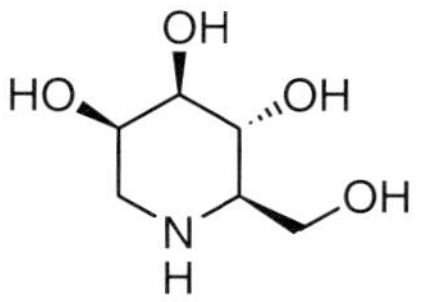

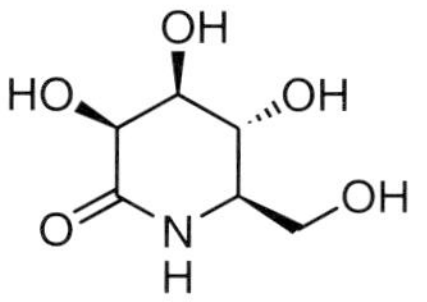

3.1.2.1 *Synthesis from D-glucose* 1-Deoxymannojirimycin (**2**) has been synthesized from diacetone D-glucose (**4**) (Scheme 1).[8–11] Compound **4** was benzylated quantitatively[12] and then underwent selective removal of the terminal isopropylidene group followed by reaction with dimethyl carbonate to afford the 5,6-carbonate **5**. Methanolysis of **5** afforded a mixture of β- and α-furanosides **6** in a ratio of 5:2 in 92% yield. Triflation of α-isomer **6** followed by displacement with azide ion led to azidomannofuranoside **7**, which underwent removal of the carbonate group by catalytic amount of methoxide ion in methanol to give the key intermediate **8**. The formation of **2** required an intramolecular nucleophilic displacement of a leaving group at C-6 by an amino group at C-2. Thus, selective tosylation of the primary hydroxyl group in **8**, followed by hydrogenation and subsequent cyclization of the resulting amine by treatment with sodium acetate in ethanol, and then protection of the resulting secondary amine with benzyl chloroformate afforded the bicyclic benzyl carbamate **9**. Acid hydrolysis of **9** followed by reduction with sodium borohydride gave **10**, whose deprotection furnished **2**.

Alternatively, an intramolecular nucleophilic displacement of a leaving group at C-2 by an amino group at C-6 has also been used for the synthesis of 1-deoxymannojirimycin (**2**) and (2*S*,3*R*,4*R*,5*R*)-3,4,5-trihydroxypipecolic acid (**14**) from diacetone D-glucose (**4**) (Scheme 2).[13] Selective removal of the terminal isopropylidene group in **4** followed by selective tosylation of the primary hydroxyl group and subsequent S_N2 displacement of the

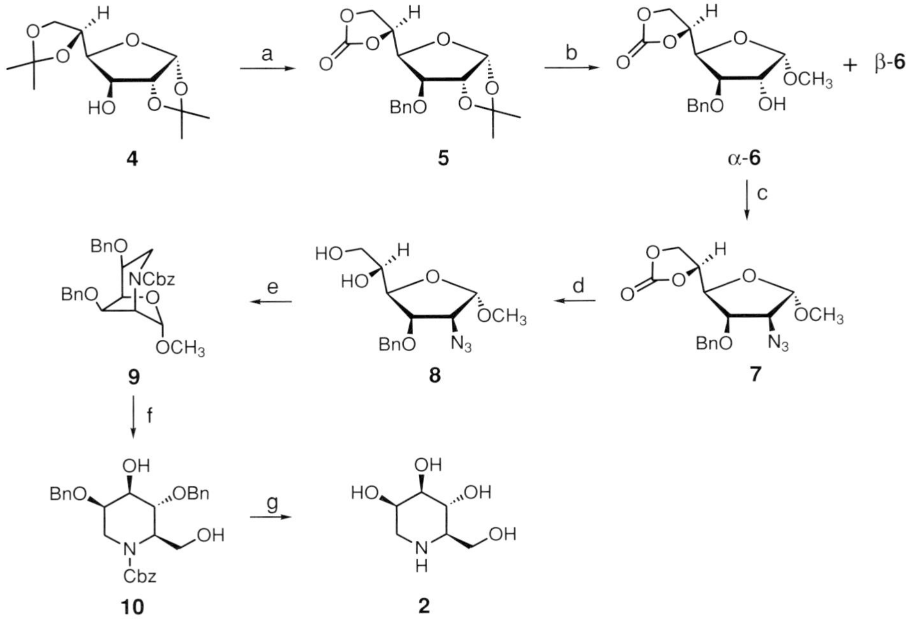

Scheme 1 (*a*) 1. BnBr, NaH, DMF, 100%; 2. 0.5% HCl, CH_3OH, rt, 12 h; 3. $(CH_3O)_2CO$, $NaOCH_3$, reflux, 79% for three steps. (*b*) Dowex 50WX8 (H^+) resin, CH_3OH, reflux, 92%, 5:2 β-**6**:α-**6**. (*c*) 1. Tf_2O, Py, CH_2Cl_2, –20°C, 20 min; 2. NaN_3, DMF, 50°C, 2 days. (*d*) CH_3OH, $NaOCH_3$, rt, 75% for three steps. (*e*) 1. *p*-TsCl, Py, rt, 6 h; 2. Pd black, H_2, EtOH, 30 min; then NaOAc, EtOH, 50°C; 3. CbzCl, ether, H_2O, $NaHCO_3$, 72% for three steps. (*f*) 1. TFA–H_2O (1:1), rt, 1 h; 2. $NaBH_4$, EtOH–H_2O, 81% for two steps. (*g*) $Pd(OH)_2$, H_2, EtOH.

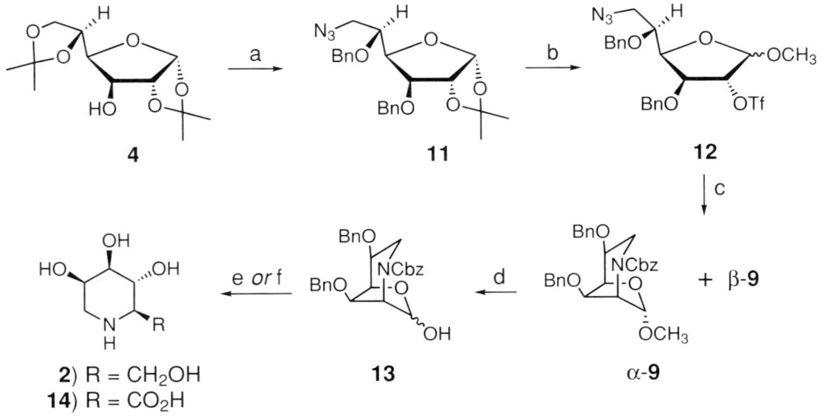

Scheme 2 (*a*) 1. AcOH–H_2O (2:1), rt, 6 h; then 4°C, 12 h, 93%; 2. Py, *p*-TsCl, –14°C, 12 h, 98%; 3. NaN_3, DMF, 40°C, 15 h, 92%; 4. THF, NaH, Bu_4NI, BnBr, 35°C, 18 h, 68%. (*b*) 1. CH_3OH, HCl, rt, 12 h, 95%; 2. CH_2Cl_2, Py, Tf_2O, –50 to 0°C, 90 min, 97%. (*c*) 1. CH_2Cl_2, Ph_3P, rt, 30 min, reflux, 2 h; then K_2CO_3, 48 h; 2. ether–aqueous $NaHCO_3$ (3:2), CbzCl, rt, 18 h, 87% for two steps. (*d*) H_2O–1,4-dioxane–TFA (1:2:1), rt, 24 h, 85%. (*e*) 1. EtOH, $NaBH_4$, 20 min, 94%; 2. AcOH, Pd black, H_2, 18 h, 95%. (*f*) 1. 1,4-dioxane–H_2O (3:1), Br_2, $BaCO_3$, rt, 36 h; 2. AcOH–H_2O (2:1), Pd black, H_2, 48 h, 84% for two steps.

tosyloxy group with azide ion and benzylation of the secondary hydroxyl groups afforded **11** in 57% overall yield from **4**. Treatment of **11** with methanolic hydrogen chloride followed by triflation of the resulting glucoside afforded the triflate **12**. Reaction of the azido group with triphenylphosphine followed by treatment with aqueous potassium carbonate and subsequent protection of the resulting secondary amine with benzyl chloroformate afforded the bicyclic carbamates α-**9** and β-**9**. Hydrolysis of the glycosidic bond in **9** with TFA gave **13**. Reduction of **13** with sodium borohydride followed by hydrogenation afforded **2** in 35% overall yield from **4**. Oxidation of **13** with bromine in water in the presence of barium carbonate followed by hydrogenation gave **14** in 33% overall yield from **4**.

The synthesis of **2** from 3-*O*-benzyl 1,2:5,6-di-*O*-isopropylidene-α-D-glucofuranose (**15**)[12] started by reaction with 5% HCl in methanol followed by benzaldehyde in the presence of zinc chloride to give methyl 3-*O*-benzyl-4,6-*O*-benzylidene-α-D-glucopyranoside (**16**) in 85% yield (Scheme 3).[14] Treatment of **16** with $LiAlH_4$ and aluminum chloride followed by triflation of the free hydroxyl groups at C-2 and C-6 and subsequent treatment with liquid ammonia produced compound **17**. Intramolecular nucleophilic displacement of the triflate by the 6-amino group occurred on standing in DMF at 35°C for several days to produce the bicyclic amine **18**. Alternatively, triflation of the hydroxyl group in **16** followed by treatment with sodium azide afforded the protected 2-azidomannose **20**. Subsequent debenzylidenation of **20** and selective mesylation of the primary hydroxyl group followed by benzylation afforded **21**, whose azido group was hydrogenated to the amine which upon intramolecular nucleophilic displacement of the mesyloxy group at C-6 afforded the bicyclic amine **18**, which was converted to the corresponding benzyl carbamate **19**. Hydrolysis of **19** by aqueous TFA followed by reduction with sodium borohydride and subsequent hydrogenolysis of the protecting groups gave **2**.

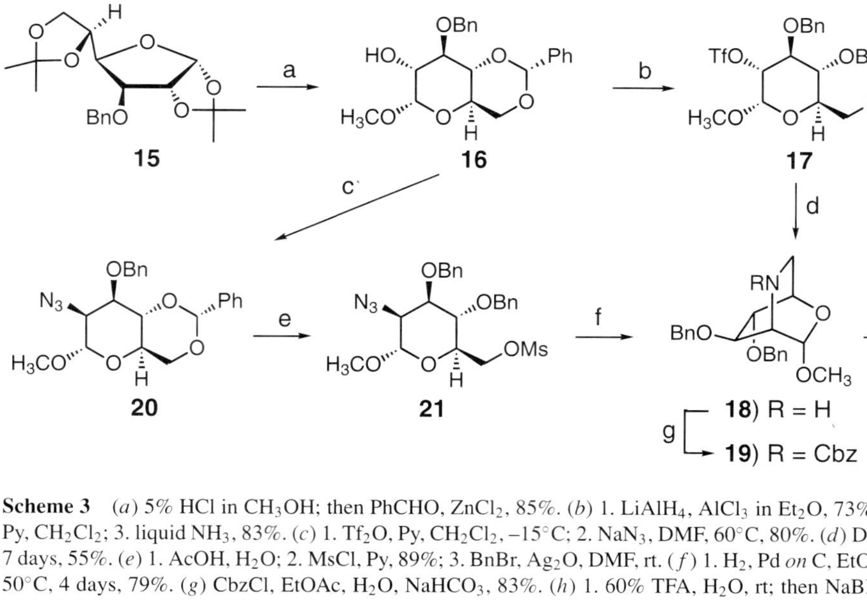

Scheme 3 (*a*) 5% HCl in CH_3OH; then PhCHO, $ZnCl_2$, 85%. (*b*) 1. $LiAlH_4$, $AlCl_3$ in Et_2O, 73%; 2. Tf_2O, Py, CH_2Cl_2; 3. liquid NH_3, 83%. (*c*) 1. Tf_2O, Py, CH_2Cl_2, −15°C; 2. NaN_3, DMF, 60°C, 80%. (*d*) DMF, 35°C, 7 days, 55%. (*e*) 1. AcOH, H_2O; 2. MsCl, Py, 89%; 3. BnBr, Ag_2O, DMF, rt. (*f*) 1. H_2, Pd *on* C, EtOAc; DMF, 50°C, 4 days, 79%. (*g*) CbzCl, EtOAc, H_2O, $NaHCO_3$, 83%. (*h*) 1. 60% TFA, H_2O, rt; then $NaBH_4$, EtOH, 87%; 2. Pd black, H_2, AcOH, 100%.

3.1.2.2 *Synthesis from* D-*mannose* The 2,3:5,6-di-*O*-isopropylidene-α-D-mannofuranose has proved to be a versatile precursor, used in a number of synthetic schemes, for the synthesis of 1-deoxymannojirimycin (**2**). In Scheme 4,[15] it was converted to **22**,[16] which was silylated with TBSCl followed by removal of the terminal acetonide group and subsequent selective silylation of the primary hydroxyl group to afford **23** in 11% overall yield from **22**. Oxidation of the hydroxyl group in **23** with Collins reagent afforded the ketone **24**, which was hydrogenated to afford directly the piperidine **25**. The reduction occurred selectively from the less hindered β-side of the cyclic imine intermediate. Removal of the protecting groups in **25** with 75% aqueous TFA afforded **2** in 7% overall yield from **22**.

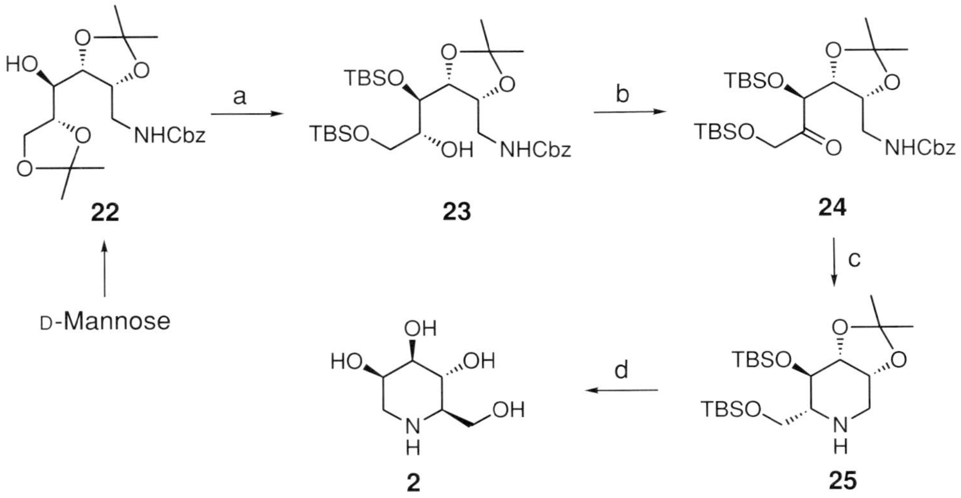

Scheme 4 (*a*) 1. TBSCl, imidazole, DMF, rt, 2 days, 77%; 2. *p*-TsOH, 90% aqueous acetone, rt, 30 h, 15 and 43% recovery; 3. TBSCl, imidazole, DMF, rt, 6 h, 94%. (*b*) Collins reagent, CH_2Cl_2, rt, 80%. (*c*) H_2, Pd black, EtOH, 74%. (*d*) 75% aqueous TFA, rt, overnight, 86%.

Mannojirimycin (**1**) and 1-deoxymannojirimycin (**2**) were prepared from 2,3:5,6-di-*O*-isopropylidene-α-D-mannofuranose (**26**) (Scheme 5).[7] The oxime **27**, prepared from **26**, underwent reduction by hydrogenation in the presence of Raney nickel, and subsequent removal of the trityl group afforded the amines **28** and **29**. Acid hydrolysis of **28** afforded **30** (21%), which was converted to **31** (28%) that upon treatment with Dowex 1 (OH^-) resin afforded **1**. On the other hand, hydrogenation of **28** gave **32**, which upon removal of the 2,3-*O*-isopropylidene group gave **31**. Further hydrogenation of **32** afforded **33**, which was subjected to acid hydrolysis to afford **2** in 67% from **28**.

1-Deoxymannojirimycin (**2**) has also been synthesized from benzyl 2,3-*O*-isopropylidene-α-D-mannofuranoside (**34**)[17] (Scheme 6).[18] Dimesylation of **34** followed by S_N2 displacement of the primary mesyloxy group with potassium acetate in the presence of 18-crown-6 ether afforded **35**. With potassium *tert*-butoxide, **35** gave the epoxide **36**. Regioselective opening of the 5,6-anhydro function followed by triflation of the C-5 hydroxyl group furnished the triflate **37**, which was subjected to S_N2 displacement either with lithium azide to afford **38** (50%) and the 5,6-*cis*-enolether derivative **40** (40%), or with benzylamine

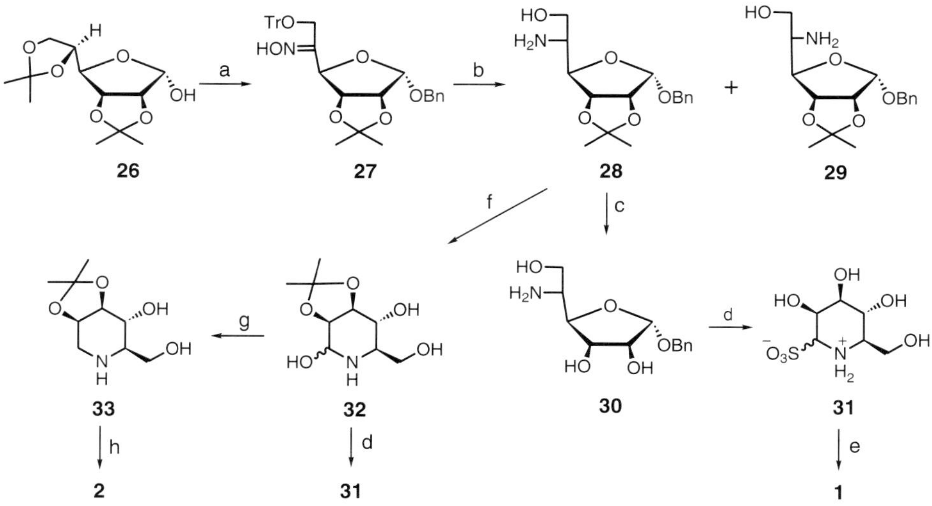

Scheme 5 (*a*) 1. BnCl, methyl tri-*n*-octylammonium chloride, benzene, 10 M NaOH, rt, 30 h, 78%; 2. CH_3OH, 12 M HCl, H_2O, 7 h, 94%; 3. TrCl, Py, 25°C, 18 h; 4. DMSO, Ac_2O, 25°C, 16 h, 65% for two steps; 5. NH_2OH. (*b*) 1. Raney nickel, H_2; 2. CH_3OH–H_2O (4:1), 0.5 M HCl, rt, 18 h; then aqueous Na_2CO_3, **28** (12% from **26**), **29** (11% from **26**). (*c*) *p*-TsOH, CH_3OH–H_2O (1:1). (*d*) *p*-TsOH, CH_3OH, H_2O, SO_2, 0–40°C for 3 days in a sealed tube. (*e*) Dowex 1 (OH^-) resin, 1 h. (*f*) $Pd(OH)_2$ *on* C, CH_3OH, H_2O, H_2, 30 min, AcOH, overnight. (*g*) H_2, Pd *on* C. (*h*) 1. 0.1 M HCl, rt, 2 days. 2. Same as (*e*).

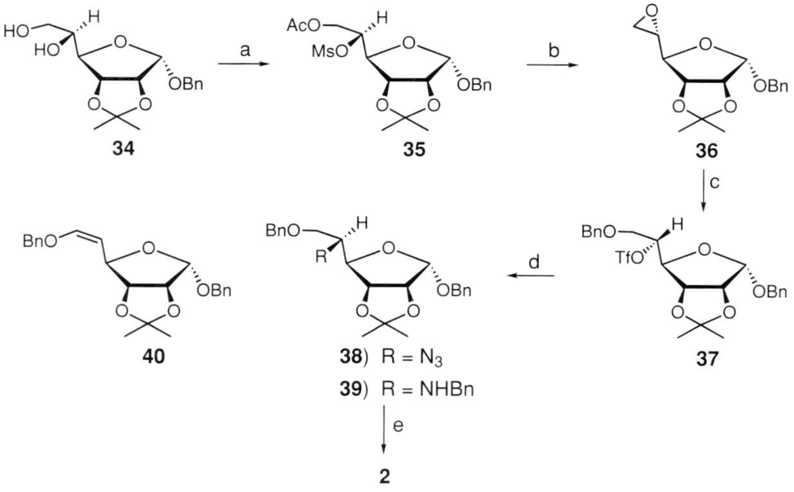

Scheme 6 (*a*) 1. Py, 0°C, MsCl, 16 h, 98%; 2. CH_3CN, KOAc, 18-crown-6, reflux, 20 h, 89%. (*b*) DMF, 0°C, *t*-BuOK, 30 min, 91%. (*c*) 1. DMF, NaH, BnOH, 0°C, 15 min; then rt, 16 h, 91%; 2. CH_2Cl_2, Py, –20°C, Tf_2O, MS, 1 h. (*d*) 1. DMF–toluene (1:3), LiN_3, Bu_4NN_3 *or* LiN_3, 12-crown-4 complex, 8 h, rt, **40** (40%), **38** (50%); *or* toluene, $BnNH_2$, 14 days, rt, **39** (56%). (*e*) 1. EtOH–H_2O–AcOH (5:1:1), 20% $Pd(OH)_2$ *on* C, H_2, 48 h, rt; 2. conc. HCl, 48 h, rt; then Amberlite IRA-400 (OH^-) resin, 92% from **39** and 95% from **38**.

in toluene to afford the 5-*N*-benzylmannofuranoside derivative **39** (56%). Hydrogenolysis of **38** or **39** over palladium hydroxide followed by acid hydrolysis afforded **2** in about 38% overall yield from **34**.

Methyl α-D-mannopyranoside was also used for the synthesis of 1-deoxymannojirimycin (**2**) by transformation into olefin **41**, which underwent intramolecular cyclization to give **42** and **43** (Scheme 7).[19] Compound **42** was hydrogenolyzed to produce **2**.

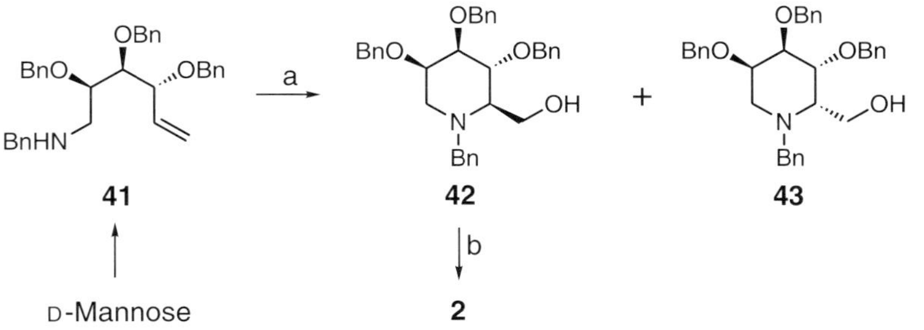

Scheme 7 (*a*) $Hg(CF_3CO_2)_2$, THF. (*b*) H_2, Pd *on* C.

3.1.2.3 *Synthesis from D-fructose* A facial synthesis of 1-deoxymannojirimycin (**2**) from D-fructose has been reported (Scheme 8).[20] Acetylation of D-fructose followed by bromination with triphenylphosphine dibromide afforded compound **44**. Removal of the acetyl groups from **44** afforded compound **45**, which was reacted with NaN_3 to give **46**. Hydrogenation of **46** afforded **2**.

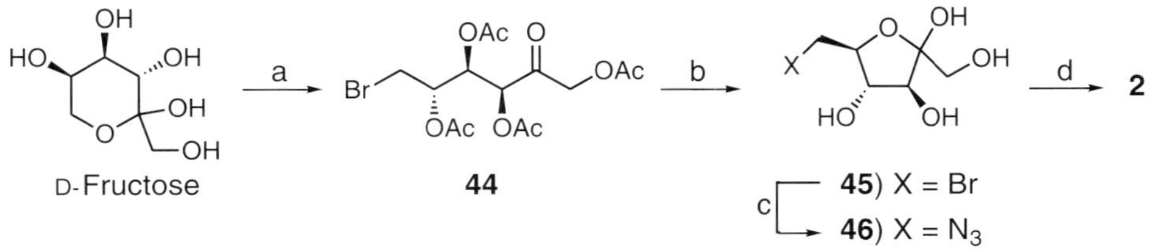

Scheme 8 (*a*) 1. Ac_2O, Py; 2. CH_2Cl_2, Py, Ph_3PBr_2, reflux, 3 h; then aqueous $NaHCO_3$, 89%. (*b*) CH_3OH, 1 M $NaOCH_3$, 0°C, pH 8, 5 h; then Amberlite IR-120 (H^+) resin, 70%. (*c*) NaN_3, DMF, rt, 7 days, 66%. (*d*) H_2, 5% Pd *on* C, CH_3OH, rt, 4 h, 60–70%.

3.1.2.4 *Synthesis from D-gluconolactone* D-Gluconolactone was converted to 1-deoxymannojirimycin (**2**) via introduction of an azide group at C-2, with retention of configuration as in **47**[21,22] (Scheme 9).[23] Reduction of **47** followed by protection of the resulting amine with $(Boc)_2O$ and then reduction of the ester group and subsequent acetylation afforded **51**. Selective removal of the terminal isopropylidene group followed by selective mesylation of the primary hydroxyl group and subsequent treatment with sodium hydroxide and silylation of the primary hydroxyl group afforded the epoxide **52**. Removal of the Boc group in **52** with Me_3SiCl followed by intramolecular nucleophilic cyclization and deprotection afforded **2** in 31% overall yield from D-gluconolactone.

Alternatively,[24] the manno azide **47** was hydrogenated, followed by protection of the resulting amine with benzyl chloroformate, removal of the terminal isopropylidene group and selective mesylation of the resulting primary hydroxyl group to furnish **48** in 69% overall yield from **47**. Hydrogenation of **48** was followed by intramolecular cyclization using sodium acetate to produce the ester **49**. Deprotection of **49** with Dowex 50W8X resin afforded 3,4,5-trihydroxypipecolic acid (**14**) in 58% overall yield from **47**. Reduction of **49** with $LiAlH_4$ gave the diol **50** in 93% yield, which underwent removal of the remaining isopropylidene group to furnish **2** in 59% overall yield from **47**. Treatment of **2** with conc. HCl afforded 1-deoxymannojirimycin hydrochloride (**2**·HCl).

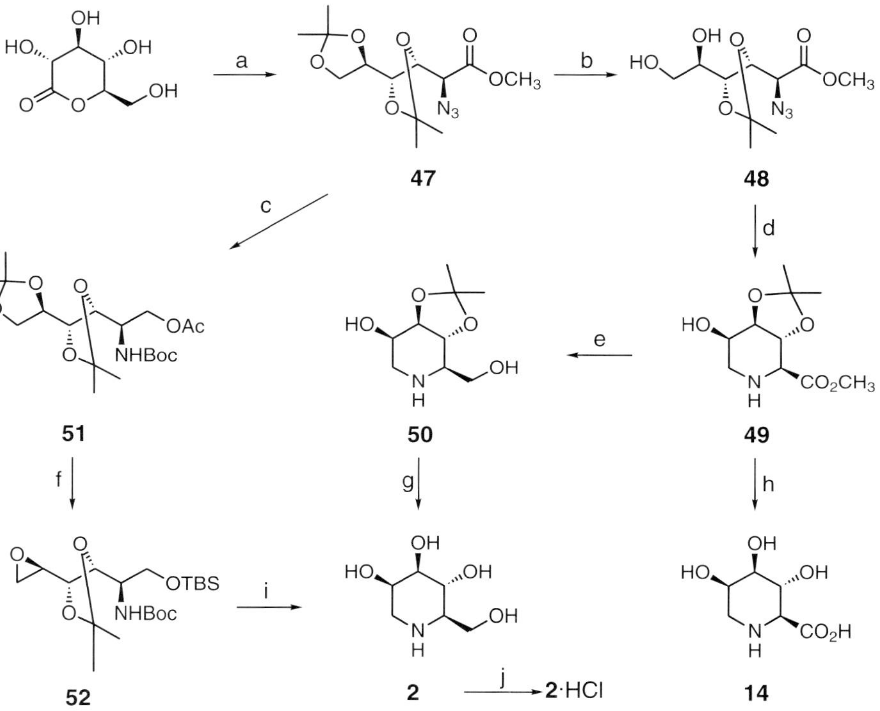

Scheme 9 (*a*) 95%, Refs. 21 and 22. (*b*) 1. 10% Pd *on* C, H_2, CbzCl, aqueous K_2CO_3, EtOAc, 6 h, CH_2Cl_2, 93%; 2. Dowex 50W8X resin, 90% CH_3OH, rt, 16 h, 95%; 3. MsCl, NEt_3, CH_2Cl_2, 0°C, 20 min, 78%. (*c*) 1. 10% Pd *on* C, H_2, EtOAc; 2. $(Boc)_2O$, CH_3OH, NEt_3, rt, 20 min, 93% for two steps; 3. $LiAlH_4$, THF, 0°C to rt, 13 h, 95%; 4. Ac_2O, Py, rt, 15 h, 93%. (*d*) 10% Pd *on* C, H_2, AcONa, CH_3OH, 10 h; then filtrate was refluxed for 1 h, 95%. (*e*) $LiAlH_4$, THF, rt, 3 h, 93%. (*f*) 1. Dowex 50WX8 resin, 90% CH_3OH, rt, 18 h, 98%; 2. MsCl, NEt_3, CH_2Cl_2, −10°C, 5 min, 97%; 3. NaOH, CH_3OH, rt, 5 min, 98%; 4. TBSCl, imidazole, DMF, rt, 12 h, 92%. (*g*) Dowex 50W8X resin, CH_3OH, reflux, 3 h, 97%. (*h*) Dowex 50W8X resin, THF–H_2O (3:1), reflux, overnight, 89%. (*i*) 1. TMSCl, PhOH, CH_2Cl_2, rt to reflux; 2. Dowex 50W resin, 90% CH_3OH, reflux, 55% for two steps. (*j*) Conc. HCl, recrystallized from CH_3OH.

3.1.2.5 *Synthesis from gulonolactone* Syntheses of 1-deoxymannojirimycin (**2**) and D-mannonolactam (**3**) have been done (Scheme 10)[25–27] starting from L-gulonolactone, commercially available or from hydrogenation of either D-glucuronolactone[28,29] or vitamin

C.[30–32] Treatment of L-gulonolactone with DMP in the presence of *p*-toluenesulfonic acid gave the corresponding di-*O*-isopropylidene derivative, which underwent selective hydrolysis with aqueous acetic acid followed by selective silylation of the primary hydroxyl group and then triflation to afford **53**. This was directly treated with sodium azide to furnish **54** in 43% overall yield from L-gulonolactone. Hydrogenation of **54** afforded the respective amine, which spontaneously underwent rearrangement to give the divergent δ-lactam **55** in 80% yield. Borane dimethylsulfide complex reduction of **55** afforded **56**, which underwent acid hydrolysis to give **2** in 18% overall yield from L-gulonolactone. On the other hand, removal of the protecting groups from **55** with aqueous TFA afforded **3** in 16% overall yield from L-gulonolactone.

D-Gulonolactone was similarly converted into 1-deoxy-L-mannojirimycin (**57**) and L-mannonolactam (**58**) in an overall yield 20 and 24%, respectively, from D-gulonolactone.

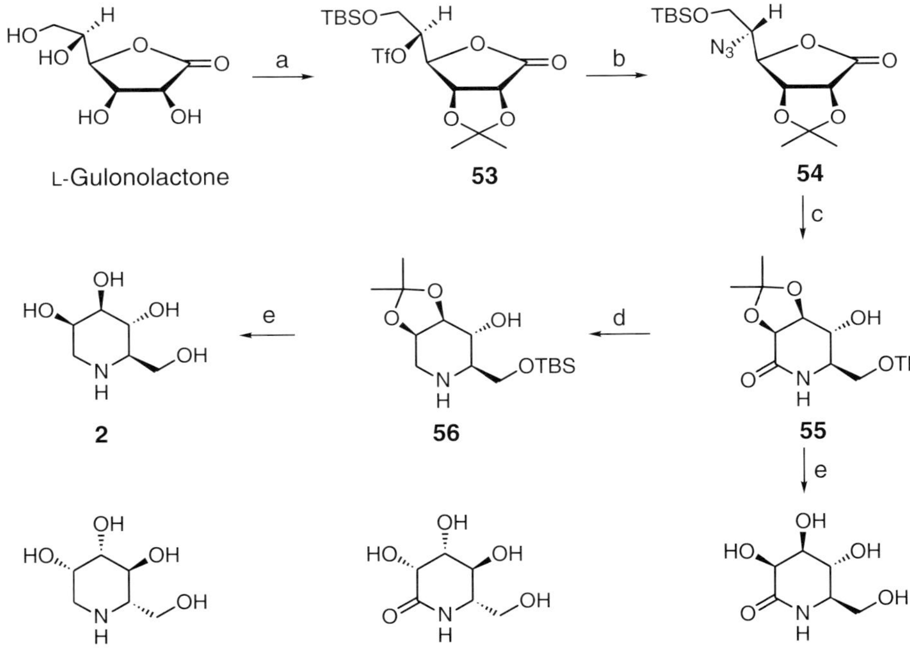

Scheme 10 (*a*) 1. Acetone, DMP, *p*-TsOH, 84%; 2. aqueous AcOH, 75%; 3. TBSCl, imidazole, DMF, –30°C, 82%; 4. Tf_2O, Py, CH_2Cl_2, –40°C. (*b*) NaN_3, DMF, 83% for two steps. (*c*) H_2, 10% Pd *on* C, CH_3OH, 80%. (*d*) BMS, THF, rt, 4 h. (*e*) Aqueous TFA, rt, 84% for two steps.

A facial synthesis of tetrazole **62** as an intermediate for the synthesis of 1-deoxymannojirimycin (**2**) has been reported (Scheme 11).[33] The azide **54**, obtained from L-gulonolactone **53**,[26,34] was treated with ammonia in methanol to produce **59**, which was reacted with trifluoroacetic anhydride in pyridine to give the nitrile **60**. When the azidonitrile **60** was heated in toluene for 3 days, an efficient 1,3-dipolar cycloaddition took place, resulting in the formation of the tetrazole **61**. Removal of both the silyl and acetonide groups in

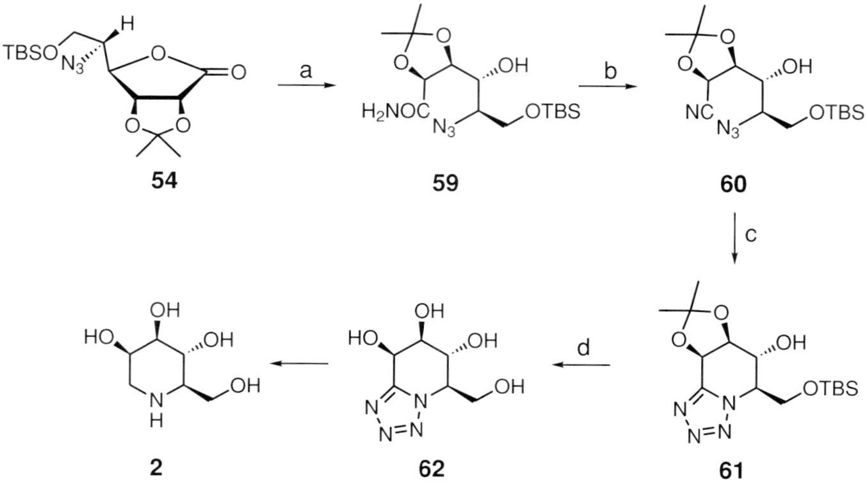

Scheme 11 (*a*) NH_3, CH_3OH. (*b*) TFAA, Py, −30°C, 75%. (*c*) Toluene, 100–105°C, 3 days, 91%. (*d*) TFA–H_2O (1:1), 55% from **54**.

61 by aqueous TFA afforded the target tetrazole **62** in 55% overall yield from **54**. Tetrazole **62** could be converted into **2**.[35]

3.1.2.6 *Synthesis from D-glucuronic acid* Synthesis of 1-deoxymannojirimycin (**2**) from 2-acetamido-2-deoxy-D-mannuronic acid, prepared from D-glucuronic acid, has been reported (Scheme 12).[36] 2-Acetamido-2-deoxy-D-mannofuranurono-6,3-lactone (**63**), prepared from 2-acetamido-2-deoxy-D-mannuronic acid, underwent N-deacetylation by treatment with HCl to give 2-amino-2-deoxy-D-mannofuranurono-6,3-lactone (**64**).[37,38] This was reduced at C-1 and C-6 by treatment with aqueous sodium borohydride whereby mannojirimycin (**1**) was produced, which upon further reduction gave 72% yield of **2**.

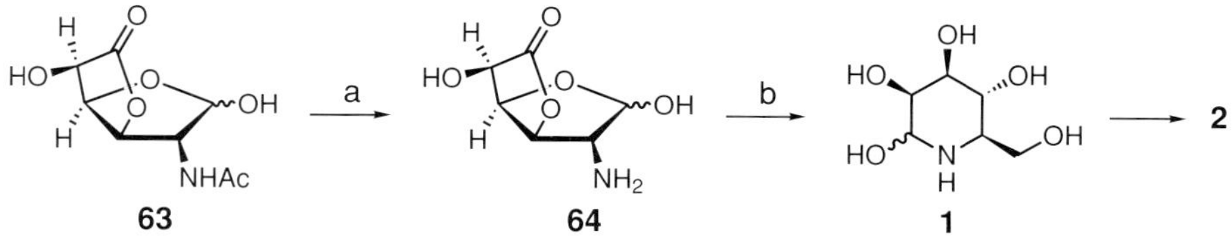

Scheme 12 (*a*) 4 M HCl, 100°C, 7 min. (*b*) $NaBH_4$, H_2O, overnight, rt.

3.1.2.7 *Synthesis from sucrose* 1-Deoxymannojirimycin (**2**) was synthesized[39] from sucrose by treatment with triphenylphosphine and tetrachloromethane followed by S_N2 displacement of the resulting chloride groups with azide ion to afford the 6,6′-diazido-6,6′-dideoxysucrose **65** in 57% overall yield from sucrose (Scheme 13). Hydrolysis of **65** with ion-exchange resin afforded a mixture of 6-azido-6-deoxy-D-glucose (**66**) and 6-azido-6-deoxy-D-fructofuranose (**67**). The azido derivative **66** was converted into **67** in 25% yield using glucose isomerase (SWEETZYMENT). Reductive cyclization of **67** afforded **2** in 78% yield.

The same intermediate **67** was also prepared[40] from methyl D-fructofuranoside (**68**), prepared from D-fructose (Scheme 13), by reaction with 2,4,6-triisopropylbenzensulfonyl chloride in pyridine, followed by acetylation and subsequent S_N2 displacement of the sulfonyloxy group with azide ion to afford methyl 1,3,4-tri-*O*-acetyl-6-azido-6-deoxy-D-fructofuranoside (**69**) in 50% overall yield from D-fructose. Zemplen deacetylation of **69** followed by acid hydrolysis afforded 6-azido-6-deoxy-D-fructose (**67**) in 82% yield, which underwent catalytic hydrogenation to furnish **2** in 25% overall yield from D-fructofuranose.

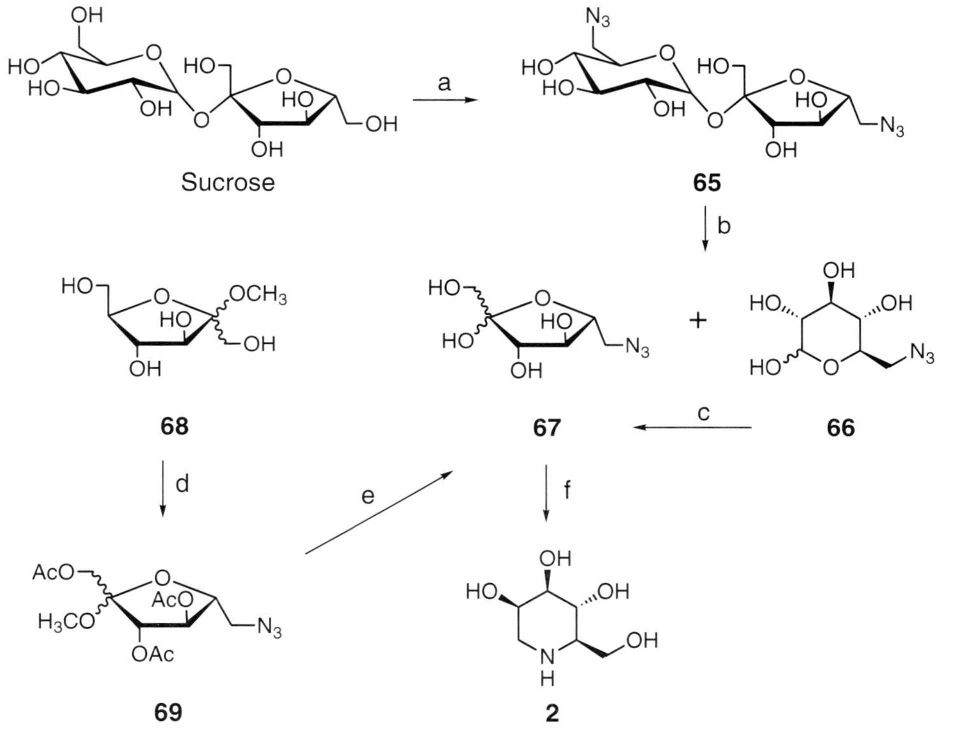

Scheme 13 (*a*) 1. Ph_3P, CCl_4, Py, 65–75%; 2. NaN_3, DMF, 81%, 57% for two steps. (*b*) Amberlite IR-120 (H^+) resin, H_2O. (*c*) H_2O, $MgSO_4$, pH 8.4, Na_2CO_3, 60°C, 60 h, polymer-supported glucose isomerase, 25%. (*d*) 1. 2,4,6-Triisopropylbenzenesulfonyl chloride, Py; then Ac_2O; 2. NaN_3, DMSO, 80°C; 50% from D-fructose. (*e*) 1. $NaOCH_3$, CH_3OH; 2. 50% aqueous TFA, 82%. (*f*) H_2, CH_3OH, H_2O, Pd *on* C, Amberlite CG-50 resin, 78%, *or* H_2, Pd *on* C, EtOH, 61%.

References

1. Hardick, D.J.; Hutchinson, D.W.; Trew, S.J.; Wellington, E.M.H. *Tetrahedron* **48** (1992) 6285.
2. Kite, G.C.; Fellows, L.E.; Fleet, G.W.J.; Liu, P.S.; Scofield, A.M.; Smith, N.G. *Tetrahedron Lett.* **29** (1988) 6483.
3. Niwa, T.; Tsuruoka, T.; Goi, H.; Kodama, Y.; Itoh, J.; Inouye, S.; Yamada, Y.; Niida, T.; Nobe, M.; Ogawa, Y. *J. Antibiot.* **37** (1984) 1579.

4. Irvine, F.R. *Woody Plants of Ghana*; Cambridge University Press: London, 1961.
5. Evans, S.V.; Fellows, L.E.; Shing, T.K.M.; Fleet, G.W.J. *Phytochemistry* **24** (1985) 1953.
6. Fuhrmann, U.; Bause, E.; Legler, G.; Ploegh, H. *Nature* **307** (1984) 755.
7. Legler, G; Jülich, E. *Carbohydr. Res.* **128** (1984) 61.
8. Fleet, G.W.J.; Smith, P.W.; Nash, R.J.; Fellows, L.E.; Parekh, R.B.; Rademacher, T. *Chem. Lett.* (1986) 1051.
9. Fleet, G.W.J.; Smith, P.W. *Tetrahedron Lett.* **26** (1985) 1469.
10. Fleet, G.W.J.; Fellows, L.E.; Smith, P.W. *Tetrahedron* **43** (1987) 979.
11. Fleet, G.W.J.; Smith, P.W. *Tetrahedron* **43** (1987) 971.
12. Czernechi, S.; Georgonlis, G.; Provelenghious, C. *Tetrahedron Lett.* (1976) 3535.
13. Fleet, G.W.J; Ramsden, N.G.; Witty, D.R. *Tetrahedron* **45** (1989) 327.
14. Fleet, G.W.J.; Gough, M.J.; Shing, T.K.M. *Tetrahedron. Lett.* **25** (1984) 4029.
15. Setoi, H.; Takeno, H.; Hashimoto, M. *Chem. Pharm. Bull.* **34** (1986) 2642.
16. Setoi, H.; Takeno, H.; Hashimoto, M. *J. Org. Chem.* **50** (1985) 3948.
17. Edge, A.S.B.; Spiro, R.G. *J. Biol. Chem.* **259** (1984) 4710.
18. Broxterman, H.J.G.; Neefjes, J.J.; van der Marel, G.A.; Ploegh, M.L.; van Boom, J.H. *J. Carbohydr. Chem.* **7** (1988) 593.
19. Bernotas, R.C.; Ganem, B. *Tetrahedron Lett.* **26** (1985) 1123.
20. Spreitz, J.; Stütz, A.E.; Wrodnigg, T.M. *Carbohydr. Res.* **337** (2002) 183.
21. Regeling, H.; Rouville, E.D.; Chittenden, G.J.F. *Recl. Trav. Chim. Pays-Bas* **106** (1987) 461.
22. Csuk, R.; Hugemer, M.; Vasella, A. *Helv. Chim. Acta* **71** (1988) 609.
23. Lee, S.G.; Yoon, Y.-J.; Shin, S.C.; Lee, B.Y.; Cho, S.-D.; Kim, S.K.; Lee, J.-H. *Heterocycles* **45** (1997) 701.
24. Park, K.H.; Yoon, Y.J.; Lee, S.G. *J. Chem. Soc., Perkin Trans. 1* (1994) 2621.
25. Fleet, G.W.J.; Ramsden, N.G.; Witty, D.R. *Tetrahedron Lett.* **29** (1988) 2871.
26. Fleet, G.W.J.; Ramsden, N.G.; Witty, D.R. *Tetrahedron* **45** (1989) 319.
27. Shing, T.K.M. *J. Chem. Soc., Chem. Commun.* (1988) 1221.
28. Ishidate, M.; Imai, Y.; Hirasaka, Y.; Umemoto, K. *Chem. Pharm. Bull.* **11** (1965) 173.
29. Berends, W.; Konings, J. *Recl. Trav. Chim. Pays-Bas* **74** (1955) 1365.
30. Vekemans, J.A.J.; Boerekamp, J.; Godefroi, E.F.; Chittenden, G.J.F. *Recl. Trav. Chim. Pays-Bas* **104** (1985) 266.
31. Vekemans, J.A.J.; Franken, G.A.M.; Dapperens, C.W.M.; Godefroi, E.F.; Chittenden, G.J.F. *J. Org. Chem.* **53** (1988) 627.
32. Andrews, G.C.; Crawford, T.C.; Bacon, B.E. *J. Org. Chem.* **46** (1981) 2976.
33. Brandstetter, T.W.; Davis, B.; Hyett, D.; Smith, C.; Hackett, L.; Winchester, B.G.; Fleet, G.W.J. *Tetrahedron Lett.* **36** (1995) 7511.
34. Fleet, G.W.J.; Ramsden, N.G.; Witty, D.R. *Tetrahedron Lett.* **29** (1988) 2871.
35. Ermert, P.; Vasella, A. *Helv. Chim. Acta* **74** (1991) 2043.
36. Leontein, K.; Lindberg, B.; Lönngren, J. *Acta Chem. Scand. B* **36** (1982) 515.
37. Leontein, K.; Lindberg, B.; Lönngren, J. *Can. J. Chem.* **59** (1981) 2081.
38. Darakas, E.; Hultberg, H.; Leontein, K.; Lönngren, J. *Carbohydr. Res.* **103** (1982) 176.
39. Raadt, A.D.; Stütz, A.E. *Tetrahedron Lett.* **33** (1992) 189.
40. Furneaux, R.H.; Tyler, P.C.; Whitehouse, L.A. *Tetrahedron Lett.* **34** (1993) 3613.

3.1.3 *Galactonojirimycin (galactostatin)*

Galactonojirimycin (5-amino-5-deoxy-D-galactopyranose, galactostatin, **1**) has been isolated as its bisulfite adducts **2** from the culture broth of *Streptomyces lydicus* PA-5726[1–3] collected from a soil sample in Nagasaki Prefecture, Japan. Its derivatives galactostatin lactam (**3**) and 1-deoxygalactostatin (**4**) have been prepared from galactostatin (Scheme 1).[2] Galactostatin (**1**) strongly inhibits β-galactosidase. Galactostatin lactam (**3**) and 1-deoxygalactostatin [(+)-1,5-dideoxy-1,5-imino-D-galactitol, **4**] are also competitive inhibitors with high affinities for *Penicillium multicolor* β-galactosidase, and their K_i values were 4.0×10^{-9} and 3.3×10^{-8} M at pH 6.0, respectively.[4–7] Galactostatin and its derivatives showed some antiviral activities, but no antimicrobial activity was observed.[3] The 50% inhibition values for plaque formation (ID_{50}) against coxsackie virus A9 were 200 μg/mL for galactostatin, 360 μg/mL for galactostatin bisulfite, 125 μg/mL for galactostatin lactam and 250 μg/mL for 1-deoxygalactostatin.

Galactostatin bisulfite **2** ← a — Galactostatin **1** — b → Galactostatin lactam **3**

Galactostatin **1** — c → 1-Deoxygalactostatin **4**

Scheme 1 (*a*) 6% H_2SO_3, 50%. (*b*) 1. 0.2 N I_2, 0.2 N NaOH, H_2O, rt, 2 h; 2. Dowex 50WX8 (H^+) resin; 3. Amberlite IRA-47 (OH^-) resin, 53%. (*c*) 1. H_2, AcOH, 50%, EtOH, 5 h; 2. Dowex 2X8 (OH^-) resin, 59%.

Fabry's disease leads to a storage of glycosphingolipids having a terminal α-D-galactosyl residues in most visceral tissues,[8] thus characterized by a deficiency of lysosomal α-D-galactosidase A. 1-Deoxygalactostatin (**4**) is a potent and selective α-D-galactosidase inhibitor that may be useful in developing a reversible effect that can be used in developing an animal model of Fabry's disease.

3.1.3.1 *Synthesis from D-galactose* Methyl α-D-galactopyranoside (**5**) has been converted to 1-deoxygalactonojirimycin (**4**) (Scheme 2).[6] Thus, methyl 2,3,4-tri-*O*-benzyl-6-bromo-6-deoxy-α-D-galactopyranoside (**6**),[9] prepared from **5**, was heated with zinc, benzylamine and $NaBH_3CN$ to afford the aminoalkene **7**, which on treatment with mercuric trifluoroacetate in THF afforded **9** and **8** in 75 and 10% yield, respectively; the equatorial isomer **9** is the favored one. This mixture was treated with sodium borohydride in DMF containing oxygen to afford **10**, which underwent catalytic debenzylation to give **4** in 26% overall yield from **5**.

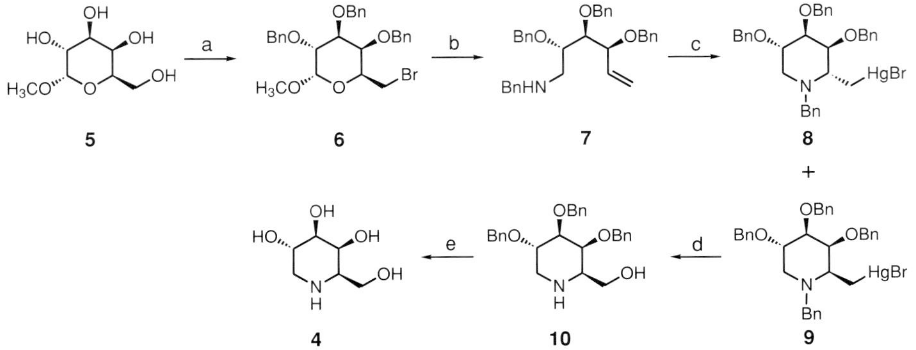

Scheme 2 (*a*) 1. Py, rt, TrCl, 90°C, 1.5 h; 2. DMF, NaH, 0°C; then BnBr, 3 h, rt, overnight; 3. H_2SO_4, CH_3OH, rt, 90 min; 4. MsCl, NEt_3, CH_2Cl_2, 0°C, 2 h; 5. LiBr, butanone, reflux, 2.3 h, 61% for five steps. (*b*) 1-Propanol–H_2O (19:1), zinc, $BnNH_2$, $NaBH_3CN$, reflux, 2 h, 90%. (*c*) $(F_3CCO_2)_2Hg$, THF, rt, 1 h; then saturated $NaHCO_3$, 10 min; then KBr, 2.5 h. (*d*) DMF, $NaBH_4$, O_2, 1 h; then 10% HCl, 30 min, 54% from **6**. (*e*) EtOH, H_2, 4 M HCl–CH_3OH, 10%, Pd *on* C, 28 h, 88%.

An approach to synthesize 1-deoxygalactostatin (**4**) has utilized the tetrazole derivative **17** (Scheme 3).[10] Thus, the D-galactose oxime derivative **11** was treated with Ph_3P and CBr_4 to give **12** and **13**. The inversion of the configuration at C-5 of **13** by oxidation–reduction processes was disappointing, leading to a 1:1 ratio of the galacto–altro derivatives **13** and **15** via **14**. Treatment of **15** with *p*-toluenesulfonyl chloride in pyridine afforded **16**, which was heated with sodium azide in dimethylsulfoxide to produce the tetrazole **17**. Reduction of **17** followed by catalytic hydrogenation gave the 1-deoxygalactostatin (**4**).

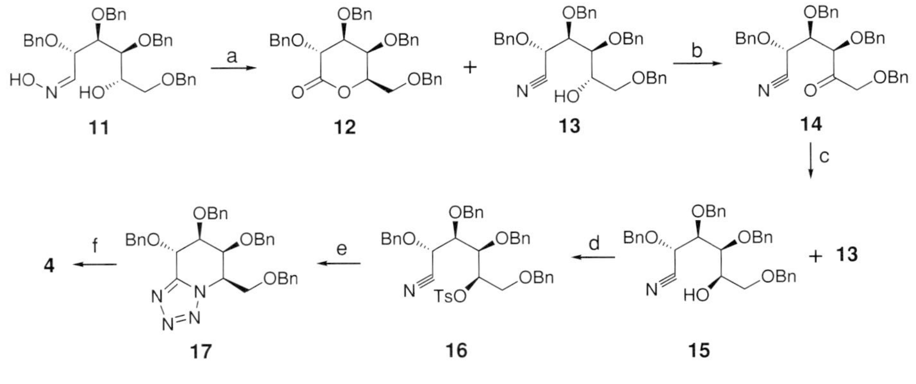

Scheme 3 (*a*) PPh_3, CBr_4, CH_3CN, rt, 20 min, 51% of **13** and 32% of **12**. (*b*) PCC, CH_2Cl_2, MS 3 Å , rt, 1.5 h, 92%. (*c*) $NaBH_4$, CH_3OH, –60°C, 30 min, 39% of **13** and 40% of **15**. (*d*) *p*-TsCl, Py, 50°C, 20 h, 67%. (*e*) NaN_3, DMSO, 100°C, 16 h, 71%. (*f*) 1. $LiAlH_4$, Et_2O, rt, 5 h, 78%; 2. H_2, 10%, Pd *on* C, CH_3OH, AcOH, rt.

An improved methodology for **17** has utilized instead of the nitrile **13** the nitrile **19**,[11] prepared from 4,6-*O*-benzylidene-2,3-di-*O*-benzyl-D-galactose (**18**) using PPh_3 and CBr_4 (Scheme 4).[10] Oxidation of **19** with PCC followed by reduction with sodium borohydride in THF gave 1:5 mixture of the diastereoisomeric alcohols **19** and **20**, which were tosylated to give the corresponding separable tosylates **21** and **22**. Treatment of **21** with sodium azide afforded the tetrazole **23**, in 77% yield, whose acetal cleavage followed by benzylation produced **17**.

Scheme 4 (*a*) 1. NH_2OH, CH_3OH, 55°C, 3 h, 99%; 2. PPh_3, CBr_4, CH_3CN, Py, rt, 20 min, 79%. (*b*) 1. PCC, CH_2Cl_2, MS 3 Å, rt, 1.5 h, 73%; 2. $NaBH_4$, THF, –78°C, 3 h, 81% of **20** and **22** in 1:5 ratio. (*c*) *p*-TsCl, Py, 75°C, 48 h, 69% of **21** and 13% of **22**. (*d*) NaN_3, DMSO, 120°C, 12 h, 77%. (*e*) 1. HCl, CH_3OH, 60°C; 2. BnBr, NaH, 60°C, 5 h, 80%.

1-Deoxygalactonojirimycin (**4**) and its L-altro analogue **27** have been synthesized from partially pivaloylated galactofuranoside derivative **24**[12] (Scheme 5).[13] Triflation of the

Scheme 5 (*a*) Pivaloyl imidazole, DMF, 60°C, 24 h. (*b*) Tf_2O, Py, CH_2Cl_2, 15°C, 40 min, 100%. (*c*) NaN_3, DMF, 2 h, 85%. (*d*) 1. $NaNO_2$, DMF, 55%; 2. Tf_2O, Py, CH_2Cl_2, 100%. (*e*) 1. $NaOCH_3$, CH_3OH, 12 h; 2. H_2, Pd *on* C, CH_3OH, 12 h, 100%. (*f*) 1. NaN_3, DMF, 75%, 12 h; 2. $NaOCH_3$, CH_3OH, 12 h, 97%. (*g*) H_2, Pd *on* C; CH_3OH, 12 h, 100%.

C-5-OH group of **24** afforded the divergent intermediate **25** in quantitative yield. The triflate **25** underwent S_N2 displacement with sodium azide to give the 5-azido-L-altrofuranoside **26**, which subsequently treated with sodium methoxide in methanol followed by hydrogenation to afford 1,5-dideoxy-1,5-imino-L-altritol (**27**) in 77% overall yield from **24**. Inversion of the configuration at C-5 in **25** was achieved by treatment with sodium nitrite in DMF, followed by triflation of the resulting L-altro derivative to afford the triflate **28**. Treatment of **28** with sodium azide in DMF followed by Zémplen deacetylation afforded the azide **29**, which on hydrogenation afforded **4** in 40% overall yield from **24**.

Oxidation of 1,6-anhydro-α-D-galactofuranose (**30**) with PtO_2 afforded the ketone **31**, which was treated with hydroxylamine to give a 3:1 mixture of isomeric oximes **32**. Hydrogenolysis of **32** in the presence of Raney nickel followed by treatment with CbzCl afforded the carbamates **33** and **34**. Acid hydrolysis of **33** afforded **35**, followed by hydrogenolysis to give **4** (Scheme 6).[14]

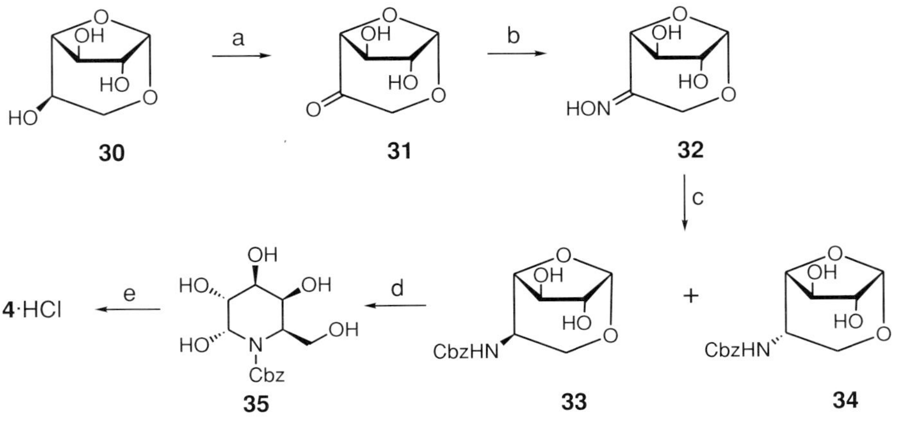

Scheme 6 (*a*) PtO_2, 45°C, 80%. (*b*) $NH_2OH{\cdot}HCl$, 90%. (*c*) 1. Raney nickel, H_2, CH_3OH; 2. $NaHCO_3$, CbzCl. (*d*) 1 N HCl, 100°C, 3 h, 10%. (*e*) Pd *on* C, H_2, HCl, 4.5 h.

3.1.3.2 *Synthesis from* D*-glucose* 1-Deoxynojirimycin, synthesized from D-glucose, has been converted to 1-deoxygalactonojirimycin (**4**) (Scheme 7).[15] Protection of the nitrogen of 1-deoxynojirimycin with CbzCl and subsequent isopropylidenation followed by benzylation afforded **30**. Acid hydrolysis of **30** followed by treatment with aqueous DMF containing potassium carbonate gave the cyclic carbamate **31**. Mesylation of **31** and then S_N2 displacement with lithium benzoate afforded the galacto derivative **32**. Saponification of **32** with aqueous NaOH in dichloromethane and methanol followed by treatment with barium hydroxide in boiling aqueous methanol and subsequent hydrogenolysis afforded **4** in 15% overall yield from 1-deoxynojirimycin.

Diacetone D-glucose (**33**) has been used for the synthesis of galactostatin (**1**) and 1-deoxygalactostatin (**4**) (Scheme 8).[5] Compound **33** was converted[16] to the D-galacto derivative **36** by oxidation of the hydroxyl group at C-3, followed by acetylation and then hydrogenation of the resulting enol acetate to afford **34**, which was deacetylated and tosylated to give **35**. Inversion of configuration at C-3 of **35** with tetrabutylammonium acetate in

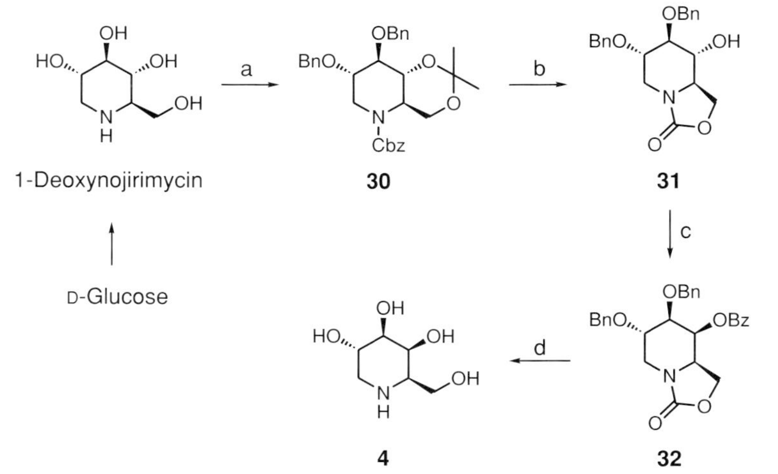

Scheme 7 (*a*) 1. DMF, $NaHCO_3$, 0°C, CbzCl, 1 h; 2. *p*-TsOH, DMP–2-methoxypropene (1:1), 90 min, 40°C, 77% for two steps; 3. DMF, NaH, 0°C, BnBr, rt, 16 h. (*b*) 60% AcOH, 60°C, 24 h; then aqueous K_2CO_3, 80°C, 62% for two steps. (*c*) 1. Acetone, NEt_3, –10°C, MsCl, CH_2Cl_2, 80%; 2. DMF, lithium benzoate, 100°C, 60 h, 80%. (*d*) 1. CH_3OH, CH_2Cl_2, 1 M NaOH, 40°C, 16 h, 80%; 2. $Ba(OH)_2{\cdot}8H_2O$, reflux, 6 h, 76%; 3. CH_3OH, 1 M HCl, H_2, 10% Pd *on* C, 6 h, 80%.

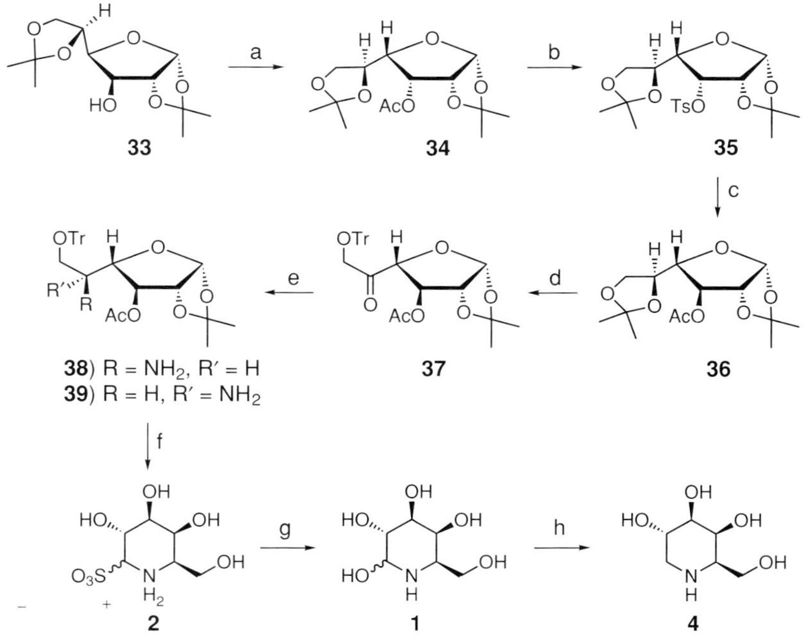

Scheme 8 (*a*) 1. PDC, Ac_2O, CH_2Cl_2, reflux, 1 h, 94%; 2. 20%, $Pd(OH)_2$ *on* C, –25 to 10°C, H_2, 7 h, 91%. (*b*) 1. $NaOCH_3$, CH_3OH; 2. *p*-TsCl, Py. (*c*) Bu_4NOAc, chlorobenzene, reflux, 5 h, 78%. (*d*) 1. 50% aqueous AcOH, 25°C, 5 h; 2. TrCl, Py, 70°C, 7 h, 98% for two steps; 3. PDC, Ac_2O, CH_2Cl_2, reflux, 1 h, 95%. (*e*) 1. $KHCO_3$, $NH_2OH{\cdot}HCl$, CH_3OH, reflux, 30 min, 96%; 2. $NaOCH_3$, CH_3OH; 3. Raney nickel, H_2, 70% for two steps, separation. (*f*) CH_3OH, SO_2, 40°C in sealed vessel, 4 h, 84%. (*g*) H_2O, 0.3 M $Ba(OH)_2$, rt, 1 h, pH ~8, 97%. (*h*) 1. H_2O, H_2, 2.5 h, PtO_2, AcOH; 2. Dowex 50 (H^+) resin, 98%.

chlorobenzene afforded **36**, which underwent removal of the terminal isopropylidene group followed by selective tritylation and subsequent oxidation of the secondary hydroxyl group at C-5 to produce the ketone **37**. Condensation of **37** with hydroxylamine followed by deacetylation and subsequent hydrogenation afforded a 1.7:1 mixture of the D-galacto and L-altro derivatives **38** and **39**. Treatment of **38** with methanol saturated with SO_2 in a sealed vessel afforded **2**, which was treated with aqueous barium hydroxide to afford **1** in 82% from **38**. Hydrogenation of **1** afforded **4** in 98% yield.

L-Arabino-hexos-5-ulose (**41**) has been used to synthesize 1-deoxygalactostatin (**4**) (Scheme 9).[17] Compound **41**,[18] obtained from methyl β-D-glucopyranoside (**40**), was subjected to reductive amination[19–21] with benzhydrylamine and sodium cyanoborohydride in a diastereospecific manner to give a moderate yield (36%) of **42**. The conversion of compound **42** into **4** was achieved by hydrogenation.[22]

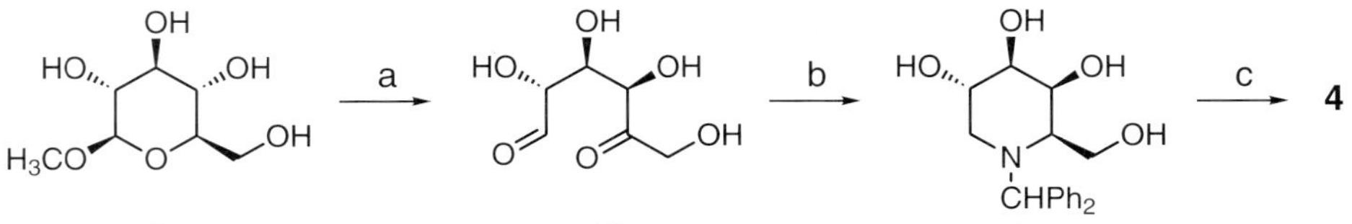

Scheme 9 (*a*) Ref. 18. (*b*) Ph_2CHNH_2, $NaBH_3CN$, CH_3OH, –78°C to rt, 36%. (*c*) H_2, $Pd(OH)_2$ *on* C, CH_3OH; then IRA-400 (OH^-), quantitative.

3.1.3.3 *Synthesis from* L-*sorbose* 1-Deoxygalactostatin has been prepared from L-sorbose stereoselectively by acetonation with 2,2-dimethoxypropane to give 1,2:4,6-di-*O*-isopropylidene-α-L-sorbofuranose (**43**) (Scheme 10).[23] Swern oxidation of the free hydroxyl group at C-3 followed by reduction with sodium borohydride produced **44** whose rearrangement gave 1,2:3,4-di-*O*-isopropylidene-α-L-tagatofuranose (**45**) in 85% yield, which

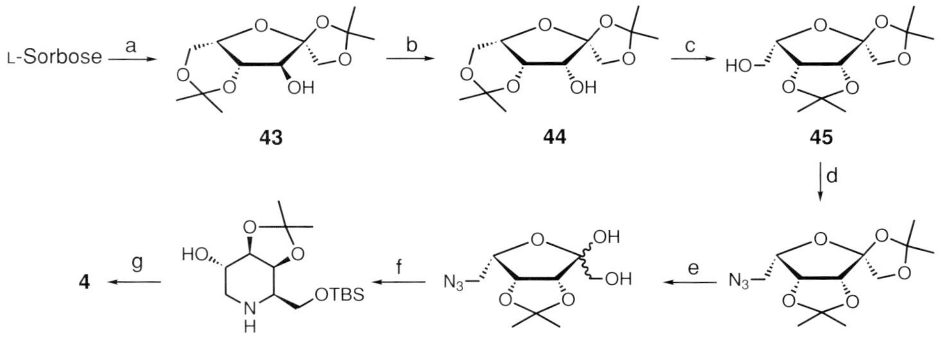

Scheme 10 (*a*) DMP, $SnCl_2$, $CH_3OCH_2CH_2OCH_3$. (*b*) 1. DMSO, Tf_2O, NEt_3, CH_2Cl_2; 2. $NaBH_4$, EtOH, 70% for two steps. (*c*) CSA, acetone, 85%. (*d*) 1. MsCl, NEt_3, CH_2Cl_2; 2. NaN_3, DMSO, 80°C, 82% for two steps. (*e*) 1. 0.5% $BF_3{\cdot}OEt_2$ *in* Ac_2O, 0°C; 2. CH_3ONa, CH_3OH, 90% for two steps. (*f*) 1. TBSCl, imidazole, DMF; 2. H_2, Pd *on* C, EtOH, 75% for two steps. (*g*) TFA–H_2O (3:7), rt, 100%.

is thermodynamically favored because of the presence of two five-membered ring ketals. Mesylation of the primary hydroxyl group in **45** followed by nucleophilic displacement with azide ion afforded the 6-azido-6-deoxy-derivative **46** in 82% yield. The attempted conversion of **46** directly to **4** resulted in a considerable decomposition of **46**. Moreover, removal of the 1,2-isopropylidene group from **46** gave **47**, which upon hydrogenation led to traces of the desired product. However, silylation of the primary hydroxyl group in **47** with TBSCl and subsequent hydrogenation afforded **48** in 68% yield from **46**. Removal of the protecting groups from **48** afforded **4** in 32% overall yield from **43**.

3.1.3.4 *Synthesis from* L*-threose* 4-*O*-(*tert*-Butyldiphenylsilyl)-2,3-*O*-isopropylidene-L-threose (**49**) has been used for the synthesis of 1-deoxygalactostatin (**4**) and the unnatural pipecolic acid derivative **56** (Scheme 11).[24] Compound **49**,[25,26] obtained from L-tartaric acid, was treated with tin(II) azaenolate **51**,[27] obtained from **50** by treatment with stannous chloride, to afford compound **52** in 79% yield with a diastereomeric excess of 90%. Benzylation of adduct **52** led to **53**. After removal of the silyl group, the hydroxyl group

Scheme 11 (*a*) 1. THF, stannous chloride, –78°C, 1 h; 2. aqueous NH_4Cl. (*b*) NaH, BnBr, Bu_4NI, rt, 24 h, 75%. (*c*) 1. TBAF, THF, rt, 4 h, 95%; 2. MsCl, NEt_3, DMAP, CH_2Cl_2, rt, 1 h, 100%; 3. 0.25 M HCl–EtOH (1:2), 9 h, 65%. (*d*) DMSO, NEt_3, 70°C, 2 h, 85%. (*e*) 1. 0.25 M HCl–THF (1:1), H_2, Pd *on* C, rt, 9 h; 2. Dowex (H^+) resin, 88%. (*f*) $LiEt_3BH$, THF, rt, 2 h, 84%.

was mesylated and then the pyrazino moiety was hydrolyzed to yield the amino ester **54**. Heating of **54** in dimethylsulfoxide with triethylamine gave the piperidine **55**. Reduction of **55** with lithium triethylborohydride followed by catalytic hydrogenation in acidic medium and subsequent ion-exchange chromatography furnished 1-deoxygalactostatin (**4**). On the other hand, deprotection of **55** gave rise to pipecolic acid **56**.

Again L-tartaric acid was used for the synthesis of galactostatin (**1**) and 1-deoxygalactostatin (**4**) by conversion to **57**,[28] whose epoxidation afforded **58a** and **58b** (Scheme 12).[29,30] Regio- and stereoselective epoxide opening of **58b** was effected by treatment with dilithium tetrabromonickelate in THF to give the bromohydrin **59** in 74% yield. This was converted to the corresponding diacetonide, followed by desilylation to produce **60**. Treatment of **60** with NaN_3 and subsequent reduction of the resulting azide and protection of the resulting amine with *p*-methoxybenzyl *S*-4,6-dimethylpyrimidin-2-yl thiocarbonate gave carbamate **63** in 93% yield. Oxidation of the primary hydroxyl group followed by treatment

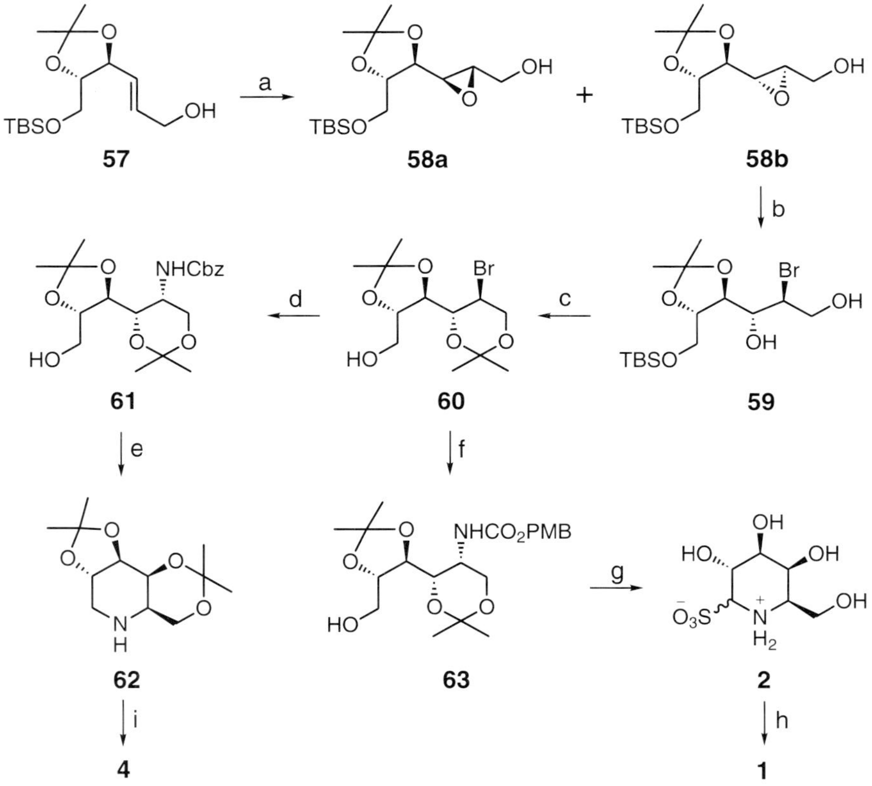

Scheme 12 (*a*) RCO_3H *or* *t*-BuO_2H, $VO(acac)_2$. (*b*) Li_2NiBr_4, THF. (*c*) 1. DMP, *p*-TsOH, Py; 2. TBAF, THF. (*d*) 1. NaN_3, DMSO; 2. H_2, Pd *on* C, CH_3OH, 82%; 3. CbzCl, aqueous Na_2CO_3, CH_2Cl_2, quantitative. (*e*) 1. MsCl, NEt_3, CH_2Cl_2, 96%; 2. H_2, Pd *on* C, CH_3OH; 3. NEt_3, CH_3OH, 57% for two steps. (*f*) 1. NaN_3, DMSO, 63% for three steps; 2. H_2, Pd *on* C, CH_3OH, 82%; 3. *p*-methoxybenzyl *S*-4,6-dimethylpyrimidin-2-yl thiocarbonate, NEt_3, dioxane, 93%. (*g*) 1. $(COCl)_2$, DMSO, NEt_3, 98%; 2. SO_2, H_2O, 47%. (*h*) Dowex 1X8 (OH^-) resin, elution with H_2O, 69%. (*i*) HCl, CH_3OH, 89%.

with SO_2 in water gave **2** in 47% yield that can readily furnish (+)-galactostatin (**1**) in 69% yield. On the other hand, a similar conversion of **60** but protecting the amine with CbzCl afforded the carbamate **61**. Mesylation of **61** followed by hydrogenation and subsequent cyclization with NEt_3 afforded the protected 1-deoxygalactonojirimycin **62**, whose deprotection with hydrochloric acid in methanol led to the formation of **4** in 89% yield.

3.1.3.5 *Synthesis from D-ribonolactone* An efficient synthesis of 1-deoxygalactostatin (**4**) has been carried out by addition of $LiCH_2OMOM$ to 5-azido-D-ribono-1,4-lactone (**64**)[31,32] to furnish the azidolactol **65** (64%), which underwent catalytic hydrogenation in the presence of palladium black to afford the protected D-galacto piperidine **66** as a single diastereoisomer in 94% yield (Scheme 13).[33,34] Acid hydrolysis effected the removal of the protecting groups from **66** to produce **4** in 54% overall yield from **64**.

Scheme 13 (*a*) $LiCH_2OMOM$, THF, –78°C, 64%. (*b*) H_2, 10% Pd black, EtOH, 72 h, 94%. (*c*) HCl, CH_3OH, rt, 24 h; Amberlite IR-120 (H^+) resin, 1 M NH_4OH, 89%.

References

1. Miyake, Y.; Ebata, M. *Agric. Biol. Chem.* **52** (1988) 153.
2. Miyake, Y.; Ebata, M. *Agric. Biol. Chem.* **52** (1988) 661.
3. Miyake, Y.; Ebata, M. *Agric. Biol. Chem.* **52** (1988) 1649.
4. Saunier, B.; Kilker, R.D., Jr.; Tkacz, J.S.; Quaroni, A.; Herscovics, A. *J. Biol. Chem.* **257** (1982) 14155.
5. Legler, G.; Pohl, S. *Carbohydr. Res.* **155** (1986) 119.
6. Bernotas, R.C.; Pezzone, M.A.; Ganem, B. *Carbohydr. Res.* **167** (1987) 305.
7. Miyake, Y.; Ebata, M. *J. Antibiot.* **40** (1987) 122.
8. Desnick, R.J.; Klionsky, B.; Sweeley, C.C. In *The Metabolic Basis of Inherited Disease*, 4th ed.; Stanbury, J.B., Wyngaarden, J.B., Fredrickson, D.S., Eds.; McGraw-Hill: New York, 1982; Chapt. 39, p. 810.
9. Bernet, B.; Vasella, A. *Helv. Chem. Acta* **62** (1979) 1990.
10. Heightman, T.D.; Ermert, P.; Klein, D.; Vasella, A. *Helv. Chim. Acta* **78** (1995) 514.
11. Rajanbabu, T.V.; Fukunaga, T.; Reddy, G.S. *J. Am. Chem. Soc.* **111** (1989) 1759.
12. Santoyo-González, F.; Uriel, C.; Calvo-Asin, J.A. *Synthesis* (1998) 1787.
13. Uriel, C.; Santoyo-González, F. *Synlett* (1999) 593.
14. Paulsen, H.; Hayauchi, Y.; Sinnwell, V. *Chem. Ber.* **113** (1980) 2601.
15. Heiker, F.R.; Schueller, A.M. *Carbohydr. Res.* **203** (1990) 314.
16. Meyer, W.; Reckendorf, Z. *Angew. Chem.* **79** (1967) 151.
17. Barili, P.L.; Berti, G.; Catelani, G.; D'Andrea, F.; Rensis, F.D.; Puccioni, L. *Tetrahedron* **53** (1997) 3407.
18. Barili, P.L.; Berti, G.; Catelani, G.; D'Andrea, F. *Gazz. Chim. Ital.* **122** (1992) 135.
19. Baxter, E.W.; Reitz, A.B. *Tetrahedron* **31** (1990) 6777.
20. Baxter, E.W.; Reitz, A.B. *Bioorg. Med. Chem. Lett.* **2** (1992) 1419.
21. Baxter, E.W.; Reitz, A.B. *J. Org. Chem.* **59** (1994) 3175.
22. Kondo, A.; Ando, K.; Ishida, H.; Kato, I.; Hasegawa, A.; Kiso, M. *J. Carbohydr. Chem.* **13** (1994) 554.
23. Furneaux, R.H.; Tyler, P.C.; Whitehouse, L.A. *Tetrahedron Lett.* **34** (1993) 3609.
24. Ruiz, M.; Ruanova, T.M.; Ojea, V.; Quintela, J.M. *Tetrahedron Lett.* **40** (1999) 2021.
25. Martin, S.F.; Chen, H.J.; Yang, C.P. *J. Org. Chem.* **58** (1993) 2867.
26. Mash, E.A.; Nelson, K.A.; Van Deusen, S.; Hemperly, S.B. *Org. Synth.* **8** (1993) 155.

27. Kobayashi, S.; Furuta, T.; Hayashi, T.; Nishijima, M.; Hanada, K. *J. Am. Chem. Soc.* **120** (1998) 908.
28. Iida, H.; Yamazaki, N.; Kibayashi, C. *J. Org. Chem.* **52** (1987) 3337.
29. Aoyagi, S.; Fujimaki, S.; Yamazaki, N.; Kibayashi, C. *Heterocycles* **30** (1990) 783.
30. Aoyagi, S.; Fujimaki, S.; Yamazaki, N.; Kibayashi, C. *J. Org. Chem.* **56** (1991) 815.
31. Fleet, G.W.J.; Ramsden, N.G.; Witty, D.R. *Tetrahedron* **45** (1989) 319.
32. Varela, O.; Zunsain, P.A. *J. Org. Chem.* **58** (1993) 7860.
33. Shilvock, J.P.; Fleet, G.W.J. *Synlett* (1998) 554.
34. Shilvock, J.P.; Nash, R.J.; Watson, A.A.; Winters, A.L.; Butters, T.D.; Dwek, R.A.; Winkler, D.A.; Fleet, G.W.J. *J. Chem. Soc., Perkin Trans. 1* (1999) 2747.

3.1.4 *Fagomine*

(+)-Fagomine (1,5-imino-1,2,5-trideoxy-D-arabino-hexitol, **1**) has been found as free base in Japanese buckwheat *Fagopyrum esculentums* Moench[1,2] and also as a glycoside **3** in *Xanthocercis zambesiaca*.[3] In addition, fagomine (**1**) and 4-*epi*-fagomine (**2**) were isolated from *Morus alba*.[4] Although compound **1** has no inhibitory effect on glycosidases from various sources, it inhibits α-glycosidase activity in mouse gut.[5]

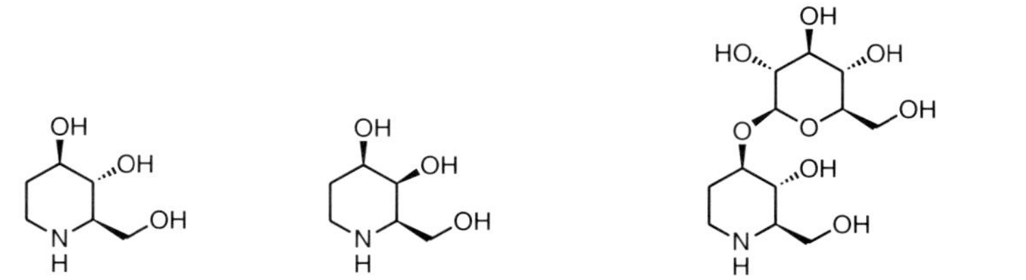

Fagomine **1** 4-*epi*-Fagomine **2** 3-*O*-(β-D-Glucopyranosyl)-fagomine **3**

3.1.4.1 *Synthesis from* D-*glucose* The synthesis of fagomine (**1**) from carbamate **5**, obtained from **4**, requires removal of the hydroxyl group at C-5 (Scheme 1).[6–8] Two deoxygenation methods were applied on **5** to give **6**: the first was Barton deoxygenation[9] of the phenyloxythiocarbonyl derivative of **5** and the second method was the reduction of the triflate of **5** with lithium triethylborohydride to afford **6** in 69 and 78% overall yield, respectively. Acid hydrolysis of **6** followed by borohydride reduction of the resulting lactol

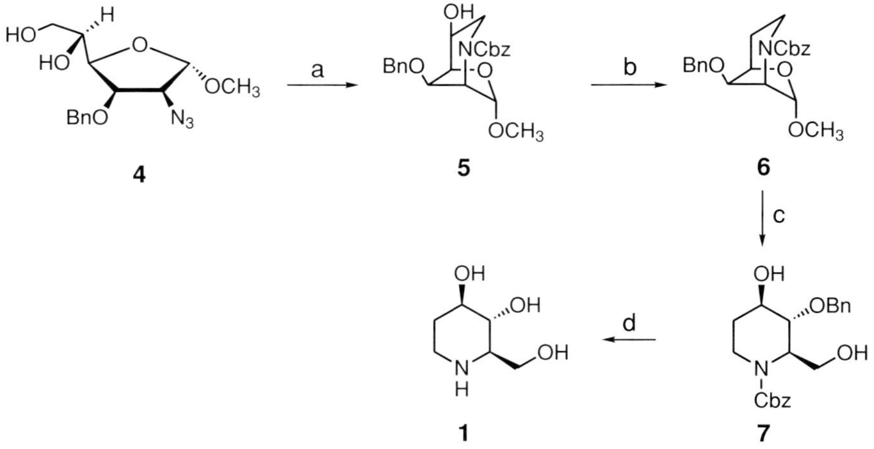

Scheme 1 (*a*) Refs. 6 and 7. (*b*) 1. DMAP, CH_3CN, rt, phenyl chlorothionocarbonate, 24 h, 91%; 2. toluene, AIBN, *n*-Bu_3SnH, 75°C, 24 h, 76% *or* 1. CH_2Cl_2, Py, Tf_2O, –30°C, 1 h; 2. THF, $LiB(C_2H_5)_3H$, rt, 6 h; then CbzCl, $NaHCO_3$, 30 min, 78%. (*c*) 50% aqueous TFA, rt, 30 min; then EtOH, $NaBH_4$, 1 h, 58%. (*d*) EtOH, H_2, 10% $Pd(OH)_2$ *on* C, 12 h; ion-exchange chromatography CG-120 (H^+), 74%.

afforded **7**, which underwent hydrogenolysis of the protecting groups to give **1** in 33% overall yield from **5**.

Syntheses of (+)-fagomine (**1**) and the unnatural 3,4-dihydroxypipecolic acid (**14**) from diacetone D-glucose have been achieved by conversion to the xylofuranose derivative **8**[10] in 74% overall yield (Scheme 2).[11] Triflation of **8** followed by treatment with potassium cyanide in *N,N*-dimethylformamide gave the nitrile **9**. Methanolysis of **9** gave the methyl furanosides as anomeric mixture whose triflation of the hydroxyl group at C-2 gave **10**. Reduction of **10** with BMS complex afforded the corresponding 6-amino sugar, which was treated with benzyl chloroformate to give the bicyclic piperidine **11**. Hydrolysis of **11** by aqueous TFA furnished the lactol **12**, whose reduction and removal of the protecting groups produced **1** in 34% overall yield from diacetone D-glucose. On the other hand, oxidation of **12** with bromine in aqueous dioxane containing barium carbonate gave the protected lactone **13** from which the free amino acid **14** was obtained as the monohydrate in 25% overall yield from diacetone glucose.

HO BnO O O O 8 a NC BnO O O O 9 b NC BnO O OCH3 OTf 10 c

OH OH N H COOH 14 f BnO NCbz O O 13 e BnO NCbz O OR

d 11) R = CH3 12) R = H g 1

Scheme 2 (*a*) 1. Py, CH_2Cl_2, –50°C, TfCl, –30°C, 1 h, 94%; 2. KCN, DMF, 30°C, 6 h, 96%. (*b*) 1. CH_3OH, AcCl, –5°C, 12 h, 81%; 2. CH_2Cl_2, Py, –5°C, Tf_2O, 0°C, 90 min, 95%. (*c*) 1. Cyclohexane, 40°C, BMS complex, rt, 24 h, 96%; 2. CbzCl, ether, aqueous $NaHCO_3$, 18 h, 79%. (*d*) TFA–H_2O (1:1), rt, 20 min, 87%. (*e*) H_2O–1,4-dioxane (1:3), $BaCO_3$, 0°C, Br_2, rt, 24 h, 84%. (*f*) AcOH–H_2O (2:1), H_2, Pd black, 48 h, ion-exchange chromatography, 89%. (*g*) 1. EtOH, $NaBH_4$, 20 min, 97%; 2. AcOH, H_2, Pd black, 18 h, 98%.

3.1.4.2 *Synthesis from D-glucal* Synthesis of (+)-fagomine (**1**) from D-glucal derivative **15** has also been reported (Scheme 3).[12] The tri-*O*-benzyl-D-glucal was converted to alkene **16**,[13,14] followed by oxidation of the free hydroxyl group in **16** to give the corresponding ketone, which was converted into the oxime **17**. This was reduced with $LiAlH_4$ to provide the primary amine, which was immediately transformed into the protected amines **18** and **19** in 1:3.5 ratio. Ozonolysis of **19** followed by treatment with triphenylphosphine provided **20**, which underwent dehydration to furnish imino glucal **21**. Hydrogenation of the double bond in **21** was achieved using 10% Pd in the presence of morpholine, followed by hydrogenation in the presence of hydrochloric acid to effect removal of the benzyl groups to furnish **1** as hydrochloride salt. Applying similar reaction sequence on **18** afforded 5-*epi*-fagomine (**22**).

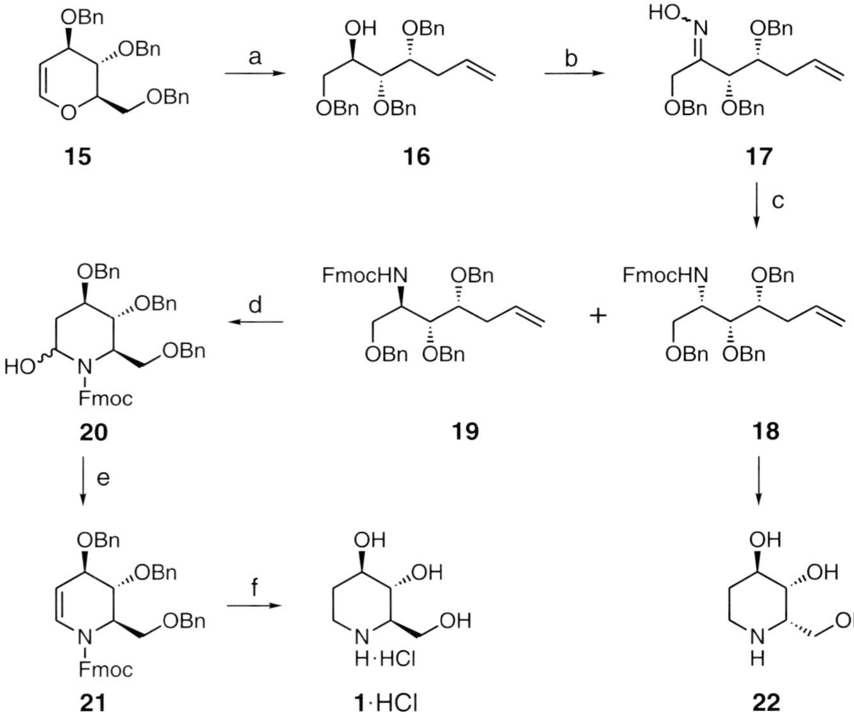

Scheme 3 (*a*) Refs. 13 and 14, 64%. (*b*) 1. TPAP, NMO, CH_2Cl_2, MS 4 Å, 83%; 2. $NH_2OH{\cdot}HCl$, Py, EtOH, 60°C, 98%. (*c*) 1. $LiAlH_4$, Et_2O; 2. Fmoc-Cl, K_2CO_3, THF, H_2O, 0°C, **19** (44%), **18** (13%). (*d*) 1. O_3, CH_2Cl_2, –78°C; 2. Ph_3P, CH_2Cl_2, 87% for two steps. (*e*) $(COCl)_2$, CH_2Cl_2, DMF, 95%. (*f*) 1. H_2, Pd *on* C, morpholine, EtOH, 70%; 2. H_2, Pd *on* C, HCl, 85%.

References

1. Koyama, M.; Sakamura, S. *Agric. Biol. Chem.* **38** (1974) 1111.
2. Koyama, M.; Aijima, T.; Sakamura, S. *Agric. Biol. Chem.* **38** (1974) 1467.
3. Evans, S.V.; Hayman, A.R.; Fellows, L.E.; Shing, T.K.M.; Derome, A.E.; Fleet, G.W.J. *Tetrahedron Lett.* **26** (1985) 1465.
4. Asano, N.; Oseki, K.; Tomioka, E.; Kizu, H.; Matsui, K.N. *Carbohydr. Res.* **259** (1994) 243.
5. Scofield, A.M.; Fellows, L.E.; Nash, R.J.; Fleet, G.W.J. *Life Sci.* **39** (1986) 645.
6. Fleet, G.W.J.; Smith, P.W. *Tetrahedron Lett.* **26** (1985) 1469.
7. Fleet, G.W.J.; Smith, P.W.; Nash, R.J.; Fellows, L.E.; Parekh, R.B.; Rademacher, T. *Chem. lett.* (1986) 1051.

8. Fleet, G.W.J.; Fellows, L.E.; Smith, P.W. *Tetrahedron* **43** (1987) 979.
9. Barton, D.H.R.; Motherwell, W.B. *Pure Appl. Chem.* **53** (1981) 15.
10. Anderson, R.C.; Nabinger, R.C. *Tetrahedron Lett.* **24** (1983) 2743.
11. Fleet, G.W.J.; Witty, D.R. *Tetrahedron: Asymmetry 1* (1990) 119.
12. Désiré, K.; Dransfield, P.J.; Gore, P.M.; Shipman, M. *Synlett* (2001) 1329.
13. Bettelli, M.; Cherubini, P.; D'Andrea, P.; Passacantilli, P.; Piancatelli, G. *Tetrahedron* **54** (1998) 6011.
14. Tius, M.A.; Busch-Petersen, J. *Tetrahedron Lett.* **35** (1994) 5181.

3.1.5 *Homonojirimycin analogues*

α-Homonojirimycin (2,6-dideoxy-2,6-imino-D-glycero-L-gulo-heptitol, HNJ, **1**) was isolated from the leaves of larval food plant *Omphalea diandra* L.[1] and from the moth *Urania fulgens*;[2] HNJ has been shown to be accumulated in moths feeding on plants *O. diandra* L.[3] It was the first example of a naturally occurring iminopyranose analogue of a heptose and it has been identified as a drug candidate for antidiabetic therapy.[4]

The nine homonojirimycins – α-homonojirimycin (**1**), β-homonojirimycin (**2**), α-homomannojirimycin (**3**), β-homomannojirimycin (**4**), α-3,4-di-*epi*-homonojirimycin (revised[5] to be α-4-*epi*-homonojirimycin or α-homoallonojirimycin, **5**), α-homogalactonojirimycin (α-homogalactostatin, **6**), β-homogalactonojirimycin (β-homogalactostatin, **7**), 7-*O*-β-D-glucopyranosyl-α-homonojirimycin (MDL25,637, **8**) and 5-*O*-α-D-glalactopyranosyl-α-homonojirimycin (**9**) – were isolated from 50% aqueous ethanol extract of *Aglaonema treubii* Engle.[6] α-Homomannojirimycin (**3**) and β-homomannojirimycin (**4**) have been shown to be popular in cultivated plants such as *Hyacinths*[7] and *Aglaonema*.[8] The syntheses of homonojirimycin and its analogues as well as the related pipecolic acid derivatives from carbohydrate precursors are presented herein.

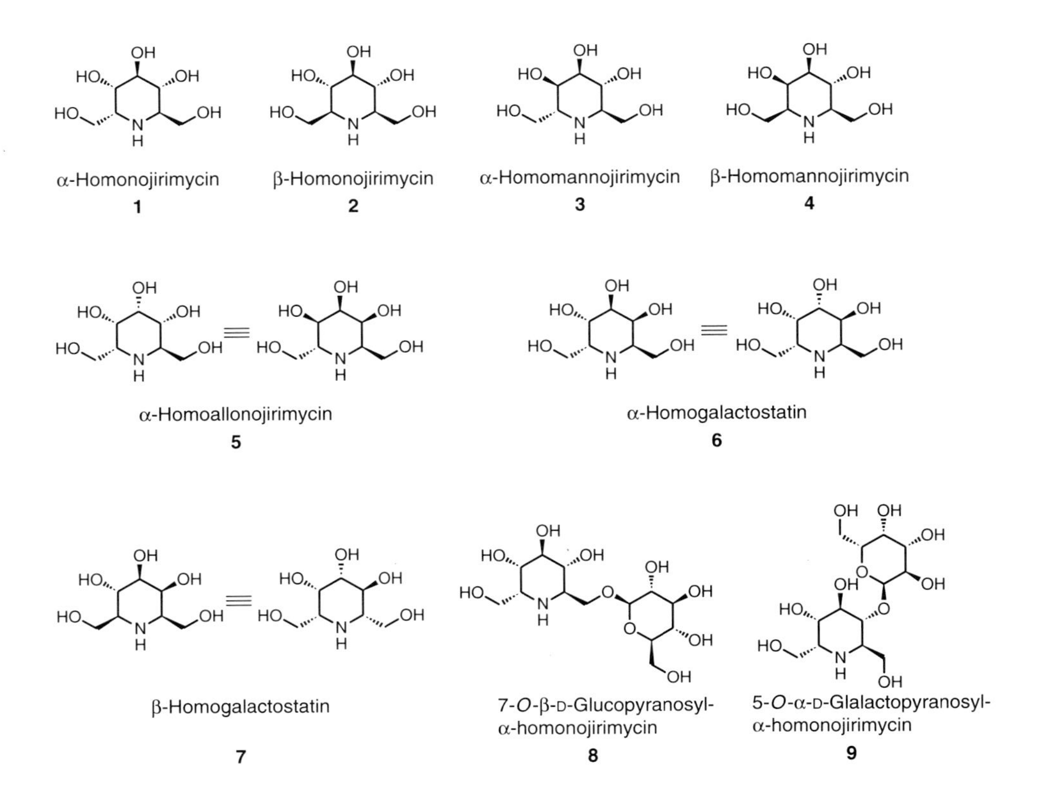

3.1.5.1 *Synthesis from D-galactose* Syntheses of α- (**6**) and β-homogalactostatin (**7**) from D-galactose have been done (Schemes 1 and 2)[9,10] by Wittig methylenation of the tetra-*O*-benzyl derivative **10** to give the heptenitol **11**[11] in 84% yield. Double inversion at C-6 under Mitsunobu's conditions gave the D-galacto amino heptenitol derivative **13**, via L-altro heptenitol **12**. Exchange of the protecting group on the amine function provided the benzyloxycarbonyl derivative **14**, which underwent internal amidomercuration followed by treatment with iodine to achieve iododemercuration that produced the cyclic carbamate **15**. Removal of the protecting groups from **15** gave **6** in 15% overall yield from **10**.

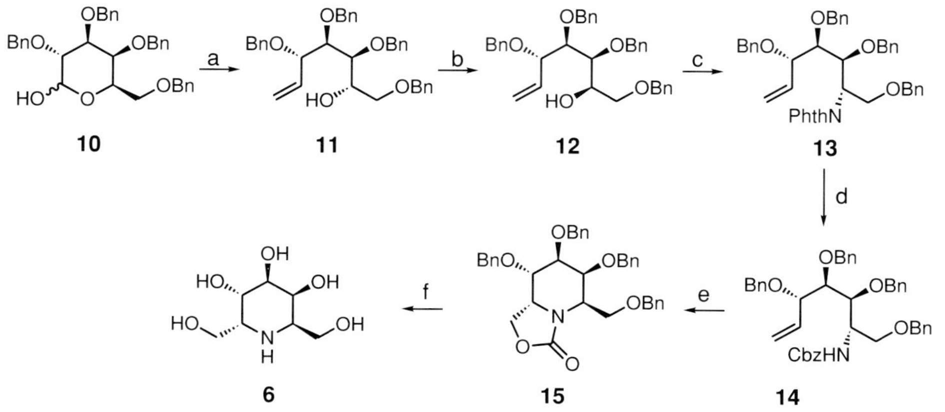

Scheme 1 (*a*) $Ph_3P{=}CH_2$, toluene, rt, 48 h, 84%. (*b*) 1. *p*-$O_2NC_6H_4CO_2H$, Ph_3P, DEAD, THF, 78%; 2. CH_3ONa, CH_3OH, rt, 5 h; then Amberlite IR-120 (H^+) resin. 77%. (*c*) Phthalimide, Ph_3P, DEAD, THF, 79%. (*d*) 1. $NH_2NH_2{\cdot}H_2O$, CH_3OH, 70°C, 1 h; 2. CbzCl, K_2CO_3, THF, 84% for two steps. (*e*) 1. $(CF_3CO_2)_2Hg$, THF, rt, 48 h; 2. I_2, THF, 0–25°C, 45 min, 71%. (*f*) 1. H_2, 10% Pd *on* C, AcOH, 50°C, 75%; 2. KOH, CH_3OH, H_2O, rt, overnight; then 60°C, 2 h, Amberlite IR-120 (H^+) resin, eluted with 10% aqueous NH_3, 81%.

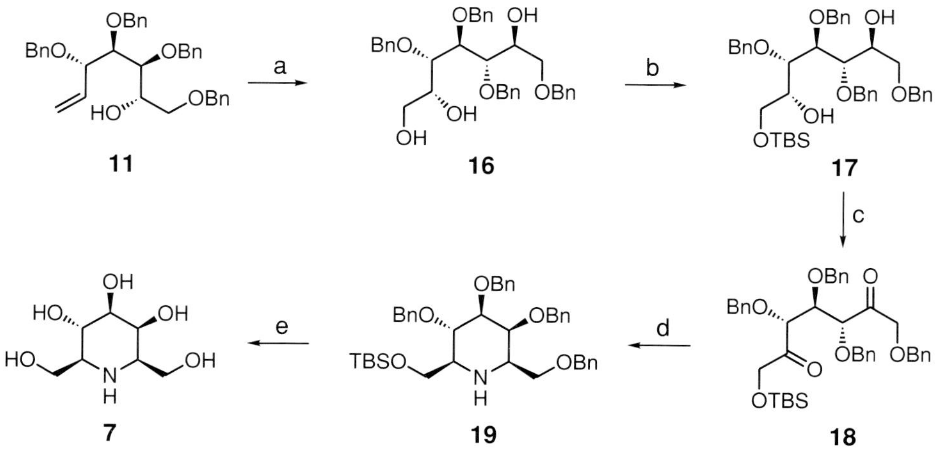

Scheme 2 (*a*) OsO_4, NMO, acetone, H_2O, rt, 14 h, 98% (de ~90%). (*b*) TBSCl, CH_2Cl_2, DMAP, NEt_3, rt, 1 h, 82%. (*c*) $(COCl)_2$, DMSO, CH_2Cl_2, –78°C, 1 h; then NEt_3, –78°C, 15 min. (*d*) NH_4HCO_3, $NaBH_3CN$, MS 3 Å, rt, 1 h, 44% for two steps. (*e*) 1. AcOH, H_2O, THF, 55°C, overnight, 78%; 2. TMSI, CH_2Cl_2, rt, 12 h; then Dowex 1X2-200 (OH^-) ion-exchange resin, eluted with H_2O.

On the other hand, synthesis of β-homogalactostatin (**7**) was also achieved from heptenitol **11** (Scheme 2)[9,10] by dihydroxylation of the double bond with catalytic OsO_4 to give the L-glycero-L-galacto heptitol derivative **16**, which underwent selective protection of the primary alcohol function as a TBS to give **17**. Oxidation of **17** to the diketone **18**, under Swern conditions, followed by reductive amination gave the single piperidine derivative **19**, which was subjected to complete deprotection to **7** in ~20% overall yield from **10**.

3.1.5.2 *Synthesis from D-glucose* The synthesis of α-homonojirimycin (**1**) from tetra-*O*-benzyl-D-glucopyranose (**20**) (Scheme 3)[12] was carried out by reaction with methylenetriphenylphosphorane to provide the alcohol **21** in 80% yield that underwent Moffatt[13,14] or Swern oxidation[15] to give the corresponding ketone, which was immediately converted to the oxime **22**. Reduction of **22** with lithium aluminum hydride followed by protection of the resulting amine as carbamate and then cyclization afforded exclusively the α-mercuriomethyl derivative **23**. Reductive oxygenation of **23** led to the protected α-homonojirimycin **24**, which was hydrogenated to give **1**.

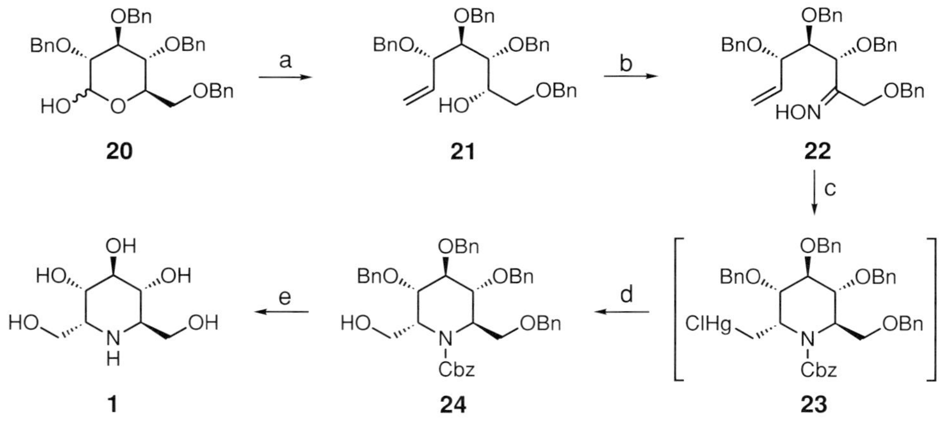

Scheme 3 (*a*) $Ph_3P{=}CH_2$, 80%. (*b*) 1. DCC, DMSO; 2. $NH_2OH{\cdot}HCl$, $KHCO_3$. (*c*) 1. $LiAlH_4$; 2. CbzCl, K_2CO_3; 3. $Hg(OAc)_2$; 4. KCl, H_2O. (*d*) O_2, $NaBH_4$. (*e*) Removal of protecting groups.

The synthesis of α-homonojirimycin (**1**) has been carried out from nojirimycin sulfite (**25**) (Scheme 4).[16] Conversion of **25** to the nitrile **26**,[17] followed by benzoylation, and protection of the NH with TFAA afforded the nitrile **27**. Hydrolysis of **27** in 90% TFA furnished the corresponding amide, whose subsequent reaction with N_2O_4 gave carboxylic acid **28**. Reduction of **28** afforded the alcohol **29**, which underwent removal of the protecting groups to produce **1**.

3.1.5.3 *Synthesis from D-mannose* Syntheses of α-homonojirimycin (**1**), 6-*epi*-α-homomannojirimycin (**39**) and the pipecolic acid derivative **36** from D-mannose have been reported (Scheme 5).[18,19] The azidolactone **31**,[20,21] prepared from diacetone mannose (**30**), underwent selective removal of the terminal isopropylidene group, followed by selective protection of the resulting primary hydroxyl group with *tert*-butyldimethylsilyl chloride to

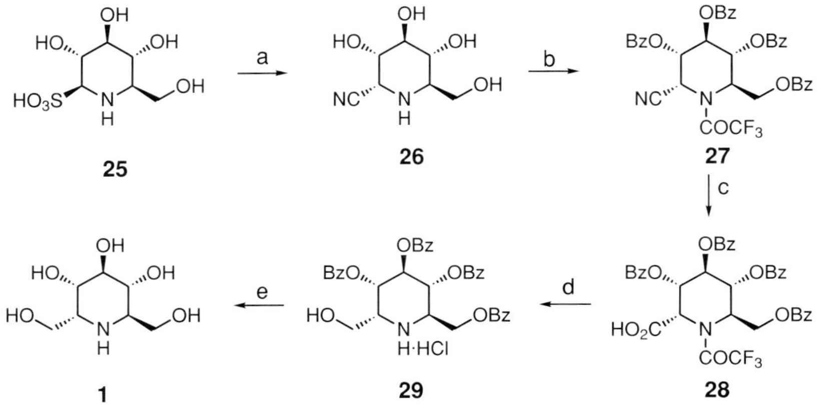

Scheme 4 (*a*) $Ba(CN)_2$, 85%. (*b*) 1. PhCOCl, NEt_3, 82%; 2. TFAA, NEt_3, 89%. (*c*) 1. TFA–H_2O (9:1), $(CF_3CO_2)_2Hg$; 2. N_2O_4, 97% for two steps. (*d*) 1. $NaBH_4$, $BF_3 \cdot Et_2O$, B_2H_6; 2. HCl, Et_2O, 75%. (*e*) EtOAc, $NaHCO_3$, aqueous NaCl, charcoal, 1 h; then methanolic NH_3, 69%.

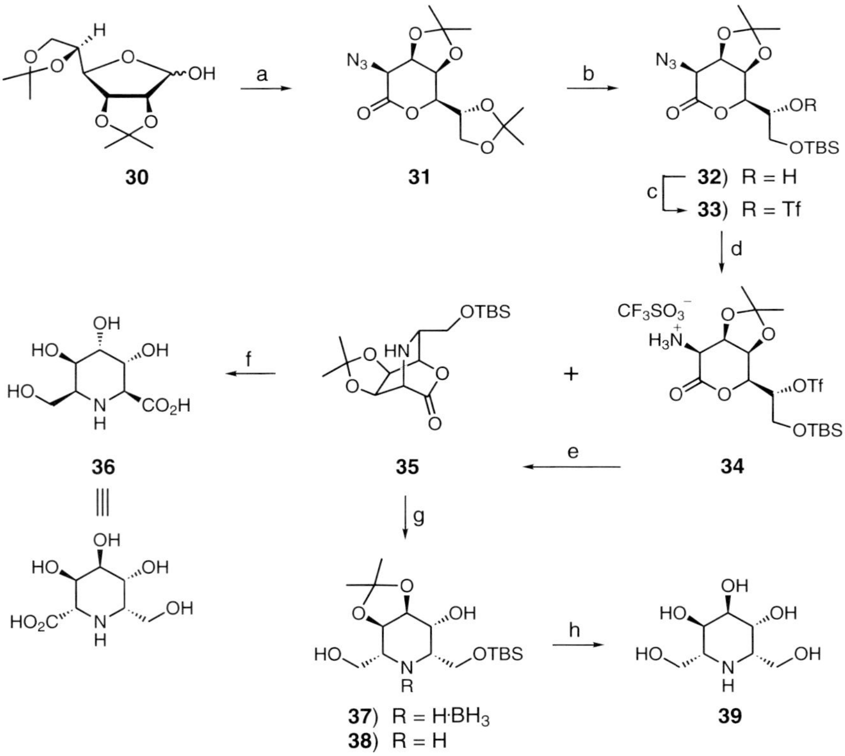

Scheme 5 (*a*) Refs. 20 and 21. (*b*) 1. 80% AcOH, 50°C, 3.5 h, 94%; 2. TBSCl, DMF, imidazole, –10°C, 15 min, 75%. (*c*) Tf_2O, CH_2Cl_2, Py, –20°C, 95%. (*d*) 1. EtOAc, H_2, 10% Pd *on* C, rt, 24 h, **35** (52%), **34** (43%); *or* H_2, EtOAc, 10% Pd *on* C, excess NaOAc, **35** (96%). (*e*) NaOAc, DMF, rt, 20 h, 86%; *or* Na_2CO_3, THF, 24 h, 79%. (*f*) 50% aqueous TFA, 20°C, 20 h; then Dowex 50X8-100 (H^+) resin, eluting with 0.8 M aqueous Py, 48%. (*g*) $LiAlH_4$, THF, 0°C, 2 h, 54%; *or* $LiBH_4$, THF, –20°C to rt, 2 h, 39%. (*h*) 50% aqueous TFA, rt, 20 h, Dowex 50X8-100 (H^+) resin, 0.5 M aqueous ammonia, 85% from **37** and 82% from **38**.

produce **32**, which was triflated to afford **33**. Hydrogenation of **33** afforded the aminotriflate salt **34** (43%) and the cyclized product **35** (52%). The former, **34**, was cyclized to **35** by treatment with either anhydrous sodium acetate or anhydrous sodium carbonate in DMF or THF. Alternatively, hydrogenation of **33** in the presence of excess of anhydrous sodium acetate furnished the bicyclic amine **35** in 96% yield. Treatment of **35** with 50% aqueous TFA gave the pipecolic acid derivative **36**. Reduction of the bicyclic lactone **35** with $LiAlH_4$ in THF afforded **37** (54%). On the other hand, lithium borohydride reduction of **35** gave both **37** and **38** in a combined yield of 58%, which were subjected to acid hydrolysis to give **39**.

On the other hand, oxidation of the secondary hydroxyl group in **32** using pyridinium chlorochromate gave the ketone **40** (74%), which was reduced with triethylphosphite[19,22] to give an intermediate iminophosphorane which spontaneously underwent an intramolecular aza-Wittig reaction[23,24] to give the bicyclic imine **41**. The imine **41** was reduced with $NaBH_3CN$ to produce the lactone **42** (70%). Treatment of **42** with sodium acetate or sodium carbonate in methanol led to the formation of **43** and **44**. Reduction of **44** gave **45**, which was also obtained from **41** by lithium borohydride reduction. Removal of the protecting groups produced the α-homonojirimycin (**3**) in 28% overall yield from **32**.

The bicyclic imine **41** was also used for the synthesis of β-homomannojirimycin (**4**) by conversion to **43**, which was reduced to give **46** that upon deprotection gave **4** (Scheme 6).[25] Toward the synthesis of **4**, compound **43** was also obtained from the epimerization of **44**.

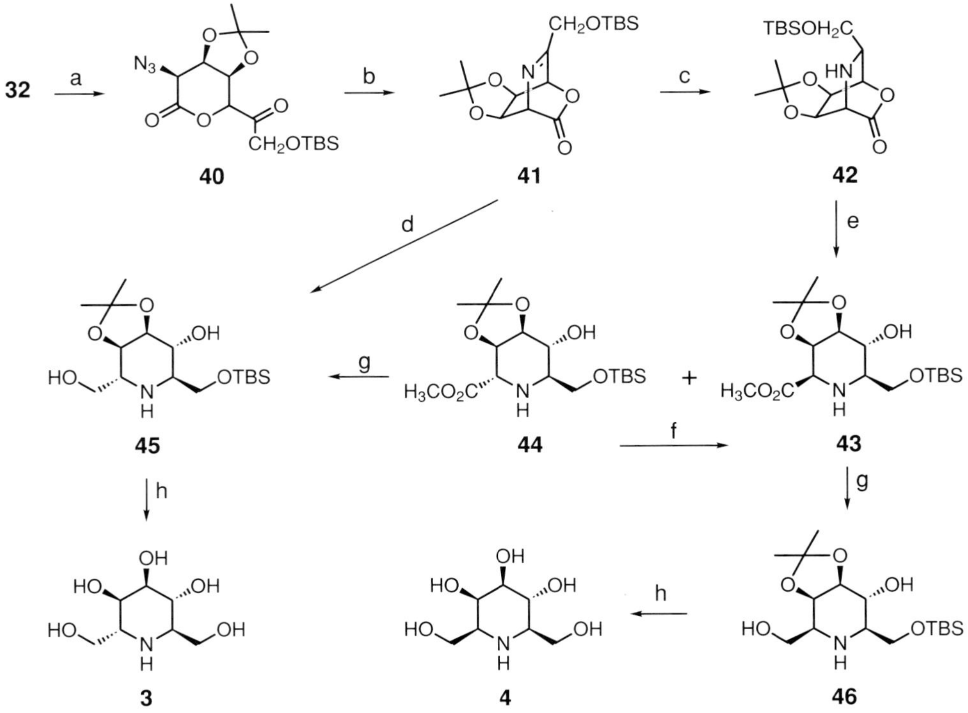

Scheme 6 (*a*) CrO_3·Py, MS 3 Å, CH_2Cl_2, rt, 18 h, 74%. (*b*) $(EtO)_3P$, THF, rt, 18 h, 89%. (*c*) $NaBH_3CN$, AcOH, 70%. (*d*) $LiBH_4$, THF, –78°C to rt, 5 h, 46%. (*e*) NaOAc, CH_3OH, reflux; *or* Na_2CO_3, CH_3OH, reflux, **44** (13%), **45** (59%). (*f*) Na_2CO_3, CH_3OH, reflux, 52%. (*g*) $LiBHEt_3$, THF, –60°C, 67%. (*h*) 50% aqueous TFA, *or* aqueous HCl, rt; Dowex 50X8-100 (H^+) resin, 0.5 M aqueous ammonia; then Amberlite CG-400 (OH^-) resin, H_2O, 79–92%.

3.1.5.4 *Synthesis from erythrose* α-Homonojirimycin (**1**) has been synthesized from the erythrose derivative **47**[26,27] by Sharpless asymmetric epoxidation to the *syn*-epoxide **48** (Scheme 7).[28] Regio- and stereoselective ring opening of the epoxide using dialkylaluminum benzylamine[29] and subsequent protection of the resulting secondary amine with benzyl chloroformate produced the carbamate **49**. This was protected with MOMCl, followed by desilylation with TBNF and subsequent oxidation using Swern oxidation to give **50**. Wittig reaction of the aldehyde **50** afforded the alkene **51** (84%), which was hydroxylated diastereoselectively to give **53** and **52** (2.5:1) in 90% total yield. Selective silylation of **53** and subsequent mesylation and then de-N-protection afforded **54**, which underwent an intramolecular cyclization in boiling methanol containing triethylamine to produce the secondary amine **55**, whose deprotection afforded **1**.

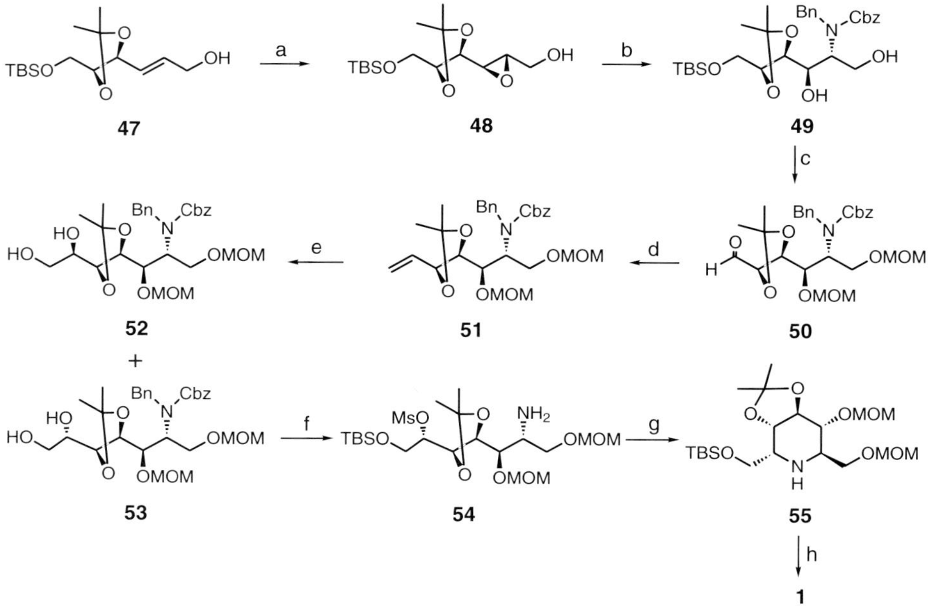

Scheme 7 (*a*) (+)-DET, $Ti(Oi\text{-}Pr)_4$, TBHP. (*b*) 1. $Et_2AlNHCH_2Ph$, CH_2Cl_2; 2. CbzCl, aqueous Na_2CO_3, CH_2Cl_2, 98%. (*c*) 1. CH_3OCH_2Cl, $(i\text{-}Pr)_2NEt$, $CHCl_3$; then TBAF, THF, 60% for three steps; 2. $(COCl)_2$, DMSO, CH_2Cl_2, −78°C; then NEt_3, 98%. (*d*) Ph_3PCH_3Br, *n*-BuLi, THF, 84%. (*e*) *N*-Methylmorpholine oxide, OsO_4, aqueous acetone, 90%. (*f*) 1. TBSCl, imidazole, DMF; then MsCl, NEt_3, CH_2Cl_2, 77%; 2. H_2, $Pd(OH)_2$, CH_3OH. (*g*) NEt_3, CH_3OH, reflux; 81% for two steps. (*h*) Conc. HCl, CH_3OH, reflux, 68%.

3.1.5.5 *Synthesis from aldonolactones* The synthesis of β-homonojirimycin (**2**) from tetra-*O*-benzyl-D-glucono-1,5-lactone (**56**) (Scheme 8)[30] was achieved by treatment of the latter with (methoxymethoxy)methyl lithium[31] to give the α-D-gluco-heptulose derivative **57**, which underwent reduction with $LiAlH_4$ to produce a mixture of heptitols **58** (1:1 ratio). Oxidation of **58** using Swern oxidation (DMSO–TFAA) gave the heptodiulose **59**. Compound **59** was immediately submitted to reductive amination using ammonium formate in the presence of sodium cyanoborohydride to produce **60** in 50% yield from **58**. Removal of the MOM protecting group from **60** afforded **61**, which underwent debenzylation with iodotrimethylsilane to furnish **2**.

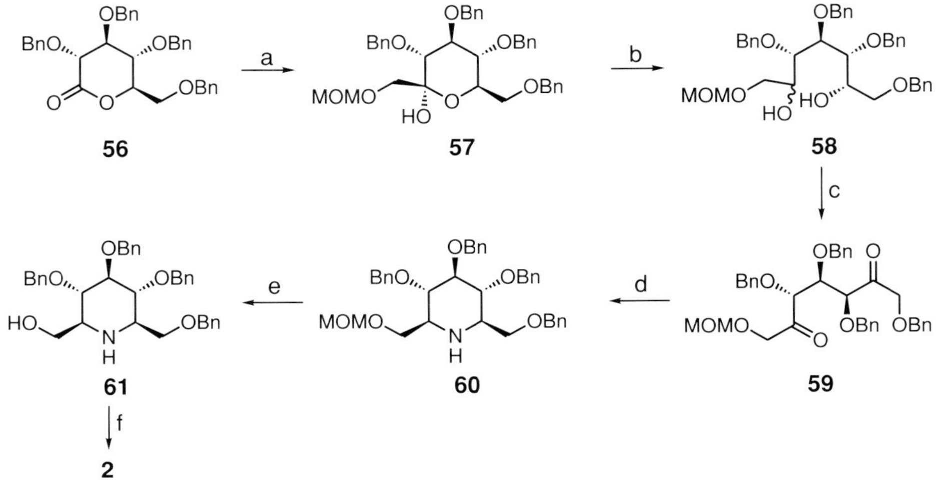

Scheme 8 (*a*) $LiCH_2OMOM$, THF, –78°C, 70%. (*b*) $LiAlH_4$, THF, rt, overnight, 98% (1:1). (*c*) TFAA, DMSO, CH_2Cl_2, NEt_3, –78°C, 90%. (*d*) $NH_4^+HCO_2^-$, $NaBH_3CN$, CH_3OH, rt, 30 min. (*e*) 6 N aqueous HCl, THF, 50°C, overnight, 93%. (*f*) 1. TMSI, CH_2Cl_2, 0°C to rt, 12 h; 2. Dowex 1X2-200 (OH^-) resin, H_2O.

Syntheses of a number of homogalactonojirimycins by addition of $LiCH_2OMOM$ to 5-azido-aldono-1,4-lactones followed by hydrogenation of the resulting azidolactol have been achieved (Scheme 9).[32,33] Thus, β-homogalactonojirimycin (**7**) was synthesized by the hydroxymethylation of the 5-azido-L-mannono-1,4-lactone **62**[34] via the intermediates **63** and **64**. Homojirimycins **6** and **65** were obtained from the *C*-5-epimer of **62** following similar steps.

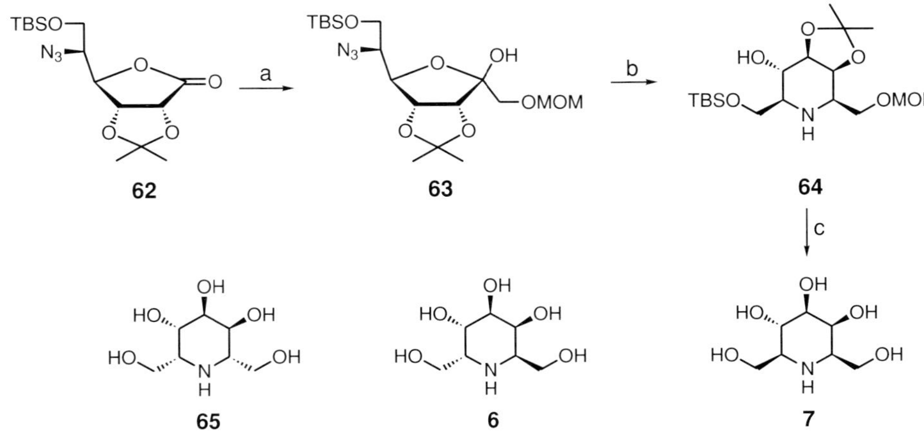

Scheme 9 (*a*) $LiCH_2OMOM$, THF, –78°C, 81%. (*b*) H_2, 10% Pd black, EtOH, 72 h, 94%. (*c*) HCl, CH_3OH, rt, 24 h; Amberlite IR-120 (H^+) resin, 1 M NH_4OH, 85%.

References

1. Kite, G.C.; Fellows, L.E.; Fleet, G.W.J.; Liu, P.S.; Scofield, A.M.; Smith, N.G. *Tetrahedron Lett.* **29** (1988) 6483.
2. Kite, G.C.; Horn, J.M.; Romeo, J.T.; Fellows, L.E.; Lees, D.C.; Scofield, A.M.; Smith, N.G. *Phytochemistry* **29** (1990) 103.
3. Kite, G.C.; Fellows, L.E.; Lees, D.C.; Kitchen, D.; Monteith, G.B. *Biochem. Syst. Ecol.* **19** (1991) 441.
4. Rhinehart, B.L.; Robinson, K.M.; Liu, P.S.; Payne, A.J.; Whealtley, M.E.; Wanger, S.R. *J. Pharmacol. Exp. Ther.* **241** (1987) 915.
5. Martin, O.R.; Compain, P.; Kizu, H.; Asano, N. *Bioorg. Med. Chem. Lett.* **9** (1999) 3171.
6. Asano, N.; Nishida, M.; Kizu, H.; Matsui, K.; Watson, A.A.; Nash, R.J. *J. Nat. Prod.* **60** (1997) 98.
7. Asano, N.; Kato, A.; Miyauchi, M.; Kizu, H.; Matsui, K.; Watson, A.A.; Nash, R.J.; Fleet, G.W.J. *J. Nat. Prod.* **61** (1998) 625.
8. Asano, N.; Nishida, M.; Kato, A.; Kizu, H.; Matsui, K.; Shimada, Y.; Itoh, T.; Baba, M.; Watson, A.A.; Nash, R.J.; Lilley, P.M.Q.; Watkin, D.J.; Fleet, G.W.J. *J. Med. Chem.* **41** (1998) 2565.
9. Martin, O.R.; Saavedra, O.M.; Xie, F.; Liu, L.; Picasso, S.; Vogel, P.; Kizu, H.; Asano, N. *Bioorg. Med. Chem.* **9** (2001) 1269.
10. Martin, O.R.; Xie, F.; Liu, L. *Tetrahedron Lett.* **36** (1995) 4027.
11. Martin, O.R.; Yang, F.; Xie, F. *Tetrahedron Lett.* **36** (1995) 47.
12. Liu, P.S. *J. Org. Chem.* **52** (1987) 4717.
13. Pfitzner, K.E.; Moffatt, J.G. *J. Am. Chem. Soc.* **87** (1965) 5661.
14. Pfitzner, K.E.; Moffatt, J.G. *J. Am. Chem. Soc.* **87** (1965) 5670.
15. Mancuso, A.J.; Huang, S.L.; Swern, D. *J. Org. Chem.* **43** (1978) 2480.
16. Anzeveno, P.B.; Greemer, L.J.; Daniel, J.K.; King, C.H.R.; Liu, P.S. *J. Org. Chem.* **54** (1989) 2539.
17. Böshagen, H.; Geiger, W.; Junge, B. *Angew. Chem., Int. Ed. Engl.* **20** (1981) 806.
18. Bruce, I.; Fleet, G.W.J.; Cenci de Bello, I.; Winchester, B. *Tetrahedron Lett.* **30** (1989) 7257.
19. Bruce, I.; Fleet, G.W.J.; Cenci de Bello, I.; Winchester, B. *Tetrahedron* **48** (1992) 10191.
20. Beacham, A.R.; Bruce, I.; Choi, S.; Fairbanks, A.J.; Fleet, G.W.J.; Skead, B.M.; Peach, J.M.; Saunders, J.; Watkin, D.J. *Tetrahedron: Asymmetry* **2** (1991) 883.
21. Fleet, G.W.J.; Bruce, I.; Girdhar, A.; Haraldsson, M.; Peach, J.M.; Watkin, D.J. *Tetrahedron* **46** (1990) 19.
22. Bruce, I.; Fleet, G.W.J.; Cenci di Bello, I.; Winchester, B. *Tetrahedron Lett.* **30** (1989) 7257.
23. Takeuchi, H.; Yanigada, S.; Ozaki, T.; Hagiwara, S.; Eguchi, S. *J. Org. Chem.* **54** (1989) 431.
24. Eguchi, S.; Takeuchi, H. *J. Chem. Soc., Chem. Commun.* (1989) 602.
25. Shilvock, J.P.; Nash, R.J.; Lloyd, J.D.; Winters, A.L.; Asano, N.; Fleet, G.W.J. *Tetrahedron: Asymmetry* **9** (1998) 3505.
26. Iida, H.; Yamazaki, N.; Kibayashi, C. *J. Org. Chem.* **52** (1987) 3337.
27. Aoyagi, S.; Fujimaki, S.; Yamazaki, N.; Kibayashi, C. *Heterocycles* **30** (1990) 783.
28. Aoyagi, S.; Fujimaki, S.; Kibayashi, C. *J. Chem. Soc., Chem. Commun.* (1990) 1457.
29. Overman, L.E.; Flippin, L.A. *Tetrahedron Lett.* **22** (1981) 195.
30. Saavedra, O.M.; Martin, O.R. *J. Org. Chem.* **61** (1996) 6987.
31. Shiozaki, M. *J. Org. Chem.* **56** (1991) 528.
32. Shilvock, J.P.; Fleet, G.W.J. *Synlett* (1998) 554.
33. Shilvock, J.P.; Nash, R.J.; Watson, A.A.; Winters, A.L.; Butters, T.D.; Dwek, R.A.; Winkler, D.A.; Fleet, G.W.J. *J. Chem. Soc., Perkin Trans. 1* (1999) 2747.
34. Fleet, G.W.J.; Ramsden, N.G.; Witty, D.R. *Tetrahedron* **45** (1989) 319.

3.2 Miscellaneous substituted piperidines

The second part of Chapter 3 discusses six groups of heterocycles, namely 2,6-disubstituted-3-hydroxypiperidines, hydroxylated pipecolic acid, sesbanimide, siastatin, meroquinene and pyridyl of pyridomycin.

3.2.1 *2,6-Disubstituted 3-hydroxypiperidines*

The 2,3,6-trisubstituted piperidine alkaloids are widely distributed in nature and have a common structure possessing 3-hydroxypiperidine ring with a side chain in position 6 and methyl or hydroxymethyl group in position 2. They are distinguished by the configuration of the substituents, the length and functionality of the side chain in position 6.

(–)-Prosophylline (**1**) and (–)-prosopinine (**2**) are naturally occurring alkaloids isolated from the leaves of the African mimosa *Prosopis africana* Taub,[1–5] which are used in indigenous medicine. These alkaloids possess a variety of antibiotic and anesthetic properties.[6–8] The racemic alkaloid of (–)-desoxoprosophylline (**3**), (–)-desoxoprosopinine (**4**)[9] and (+)-prosafrinine (**5**)[10] have been isolated from the same plant *Prosopis africana*.[10]

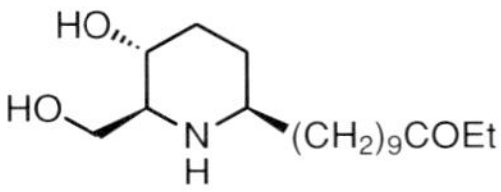

HO, HO, N, H, $(CH_2)_9COEt$ — (–)-Prosophylline **1**

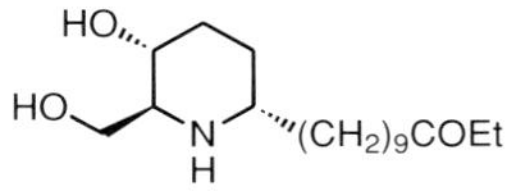

HO, HO, N, H, $(CH_2)_9COEt$ — (–)-Prosopinine **2**

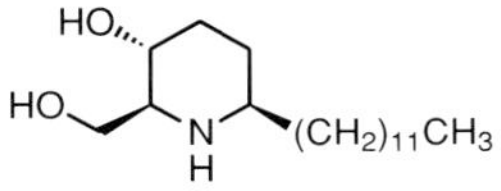

HO, HO, N, H, $(CH_2)_{11}CH_3$ — (–)-Desoxoprosophylline **3**

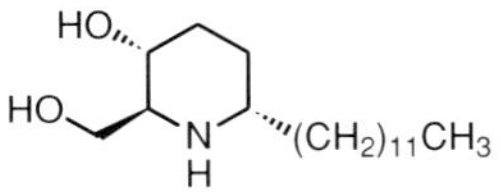

HO, HO, N, H, $(CH_2)_{11}CH_3$ — (–)-Desoxoprosopinine **4**

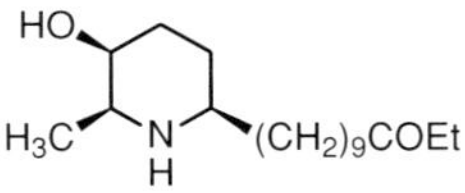

HO, H_3C, N, H, $(CH_2)_9COEt$ — (+)-Prosafrinine **5**

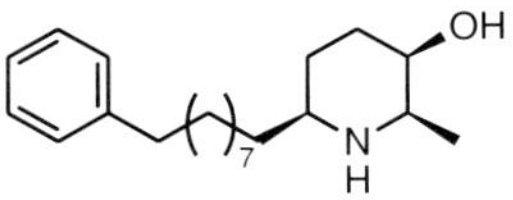

OH, 7, N, H — Irnigaine **6**

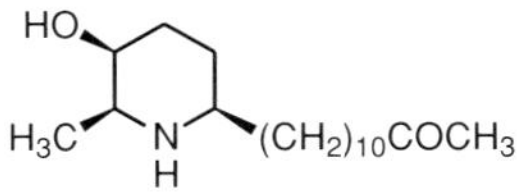

HO, H_3C, N, H, $(CH_2)_{10}COCH_3$ — (–)-Cassine **7**

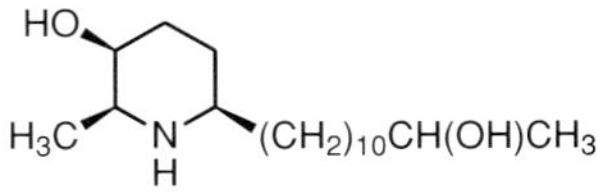

HO, H_3C, N, H, $(CH_2)_{10}CH(OH)CH_3$ — (–)-Iso-6-cassine **8**

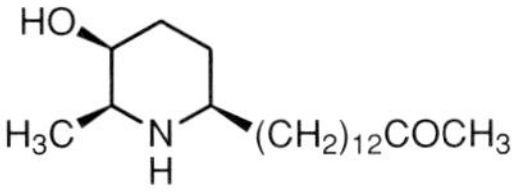

HO, H_3C, N, H, $(CH_2)_{12}COCH_3$ — (+)-Spectaline **9**

Irnigaine [(2*R*,3*R*,6*S*)-2-methyl-6-(9′-phenyl-nonyl)-piperidin-3-ol, **6**] was isolated in small amounts from the tubers of *Arisarum vulgare* (Araceae).[11] (–)-Cassine (**7**),[12] (–)-iso-6-cassine (**8**)[13] and (+)-spectaline (**9**)[14] have been isolated from *Cassia* species. The absolute

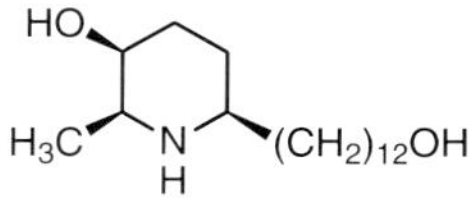

Julifloridine **10**

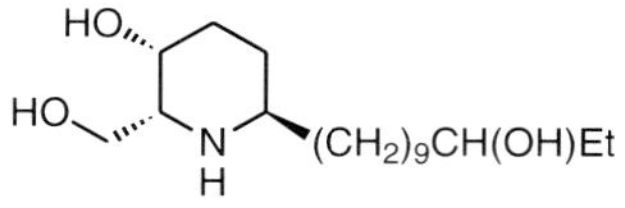

Prosopine **11**

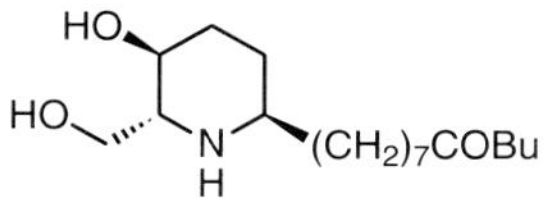

(±)-Isoprosopinine B **12**

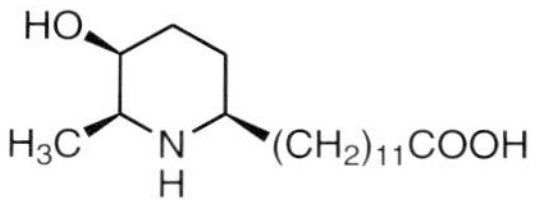

Spicigerine **13**

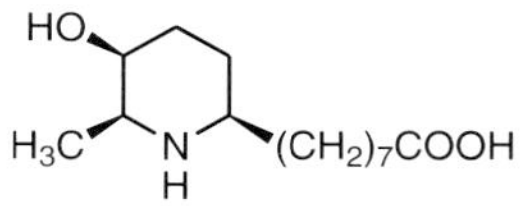

(+)-Carpamic acid **14**

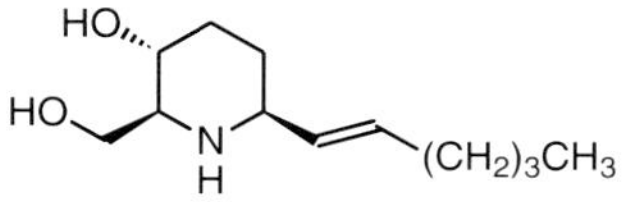

Micropine **15**

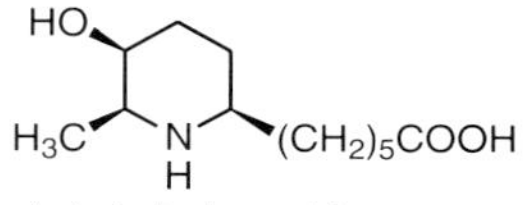

(+)-Azimic acid **16**

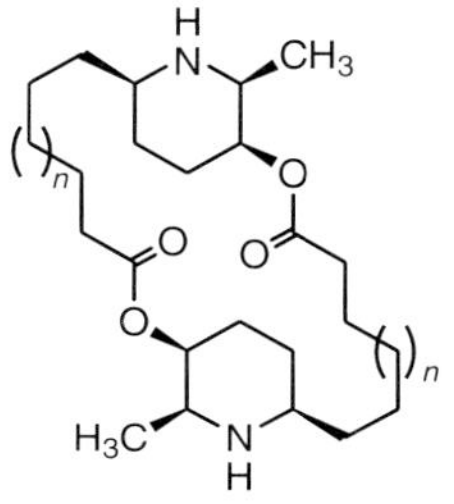

17) Azimine ($n = 1$)
18) Carpaine ($n = 3$)

configuration of cassine (**7**)[13,15] has been studied. (+)-Prosafrinine (**5**), julifloridine (**10**), prosopine (**11**), (±)-isoprosopinine B (**12**) and spicigerine (**13**) were isolated from *P. africana* and *Cassia* species.[16–21] (+)-Carpamic acid (**14**) derived from carpaine (**18**) and (+)-azimic acid (**16**) derived from azimine (**17**) were isolated from *Carica papaya*,[22] whose pharmacological properties are well documented.[23–25] Micropine (**15**) was isolated from *Microcos philippinensis*.[26]

Syntheses of the naturally occurring trisubstituted piperidines from noncarbohydrates and their analogues from carbohydrates and noncarbohydrates, as starting materials, have been reported.[27–68] The syntheses of the natural ones from carbohydrates are discussed here.

3.2.1.1 *Synthesis from D-glucose* The total syntheses of (+)-azimic acid and (+)-carpamic acid are described, based on the use of optically active precursors derived from D-glucose (Scheme 1).[69] The unsaturated derivative **20**,[70] obtained from methyl α-D-glucopyranoside (**19**), underwent selective reduction of the double bond followed by treatment with NBS to give **21**. Sequential debenzylation, reduction and O-benzylation gave **22**, which upon thiolysis gave **23**. The tosylate **24** was then treated with excess sodium azide to give the respective azido derivative, which was reacted with bromine to give the aldehyde derivative **25**. Treatment of **25** with Grignard reagent, prepared from 8-bromo-1-(tetrahydropyran-2-yloxy)octane, led to the formation of **26**, presumably as a mixture of epimers. Oxidation with pyridimium chlorochromate produced the corresponding azidoketone derivative, which upon hydrogenation, gave **27**. Conversion to the *N*-benzyloxycarbonyl derivative followed by deprotection and oxidation of the terminally protected primary alcohol led to *N*-benzyloxycarbonyl-3-*O*-benzyl carpamic acid, which upon reductive removal of the N,O-protecting groups gave carpamic acid (**14**). Following the same sequence described above, but by using the Grignard reagent[71] derived from 6-bromo-1-(tetrahydropyran-2-yloxy)hexane, the crystalline azimic acid (**16**) has been obtained.

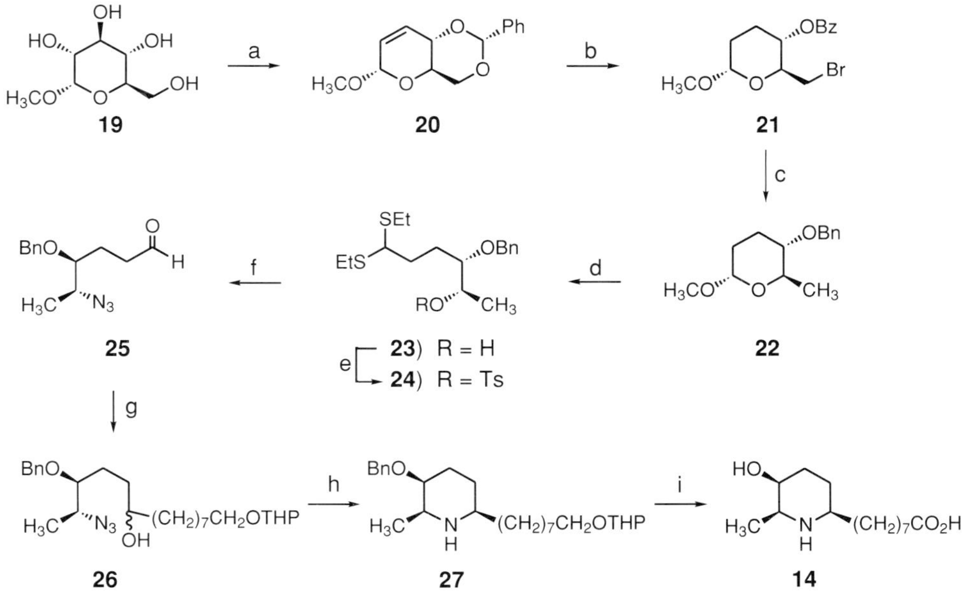

Scheme 1 (*a*) Ref. 70. (*b*) 1. Pd *on* C, H_2, CH_3OH, quantitative; 2. NBS, CCl_4, reflux, 30 min, 95%. (*c*) 1. $NaOCH_3$, CH_3OH, 2. $LiAlH_4$, BnBr, NaH, DMF, 86% for two steps. (*d*) EtSH, HCl, 71%. (*e*) *p*-TsCl, Py, 98%. (*f*) 1. DMF, 80°C, NaN_3, 98%; 2. Br_2, ether, H_2O, 63%. (*g*) $BrMg(CH_2)_7CH_2OTHP$, THF, –50°C, 30 min, 80%. (*h*) 1. CrO_3, Py, NaOAc, CH_2Cl_2, 83%; 2. 10% Pd *on* C, H_2, EtOAc, 83%. (*i*) 1. CbzCl, aqueous acetone, 88%; 2. CrO_3, aqueous H_2SO_4, acetone, 57%; 3. 10% Pd *on* C, H_2, CH_3OH, 8 h, 83%.

D-Glucose has also been used for the synthesis of (–)-desoxoprophylline (**3**) and (–)-desoxoprosopinine (**4**) by conversion to methyl 3,4-dideoxy-α-D-erythro-pyranoside (**28**),[72,73] whose reaction with EtSH in the presence of conc. HCl gave, after complete protection, the dithioacetal **29** (Schemes 2 and 3).[74] Removal of the isopropylidene group from **29** followed by selective tritylation of the primary hydroxyl group and subsequent mesylation of the secondary hydroxyl group afforded **30**. The latter underwent S_N2 displacement with azide ion followed by reduction of the azido group and protection of the resulting amine to afford **31**. Mercury(II) chloride oxidation of the dithioacetal in **31**, followed by Horner–Emmons olefination of the aldehyde **32**, afforded the (*E*,*Z*)-α,β-unsaturated esters, which were subjected to DIBAL-H reduction to give the *E*,*Z*-mixture of allylic alcohols **33** (83%); the *Z*-isomer is the minor (3%). The major one was treated with *p*-toluenesulfonyl chloride in the presence of DMAP to afford the chloride **34**, which was treated with $Pd(Ph_3P)_4$ to produce a mixture of the piperidine derivatives **35** and **36** (75%) in a diastereomeric ratio of 10:1, respectively. Ozonolysis of **35** and **36** followed by reduction and MOM protection of the resulting hydroxyl group afforded **37**, which was subjected to acid hydrolysis followed by Swern oxidation and olefination with $Ph_3P{=}CH(CH_2)_9CH_3$ to give the 2,6-*cis*-substituted piperidine **38**. Hydrogenation of **38** followed by complete deprotection furnished (–)-desoxoprophylline (**3**).

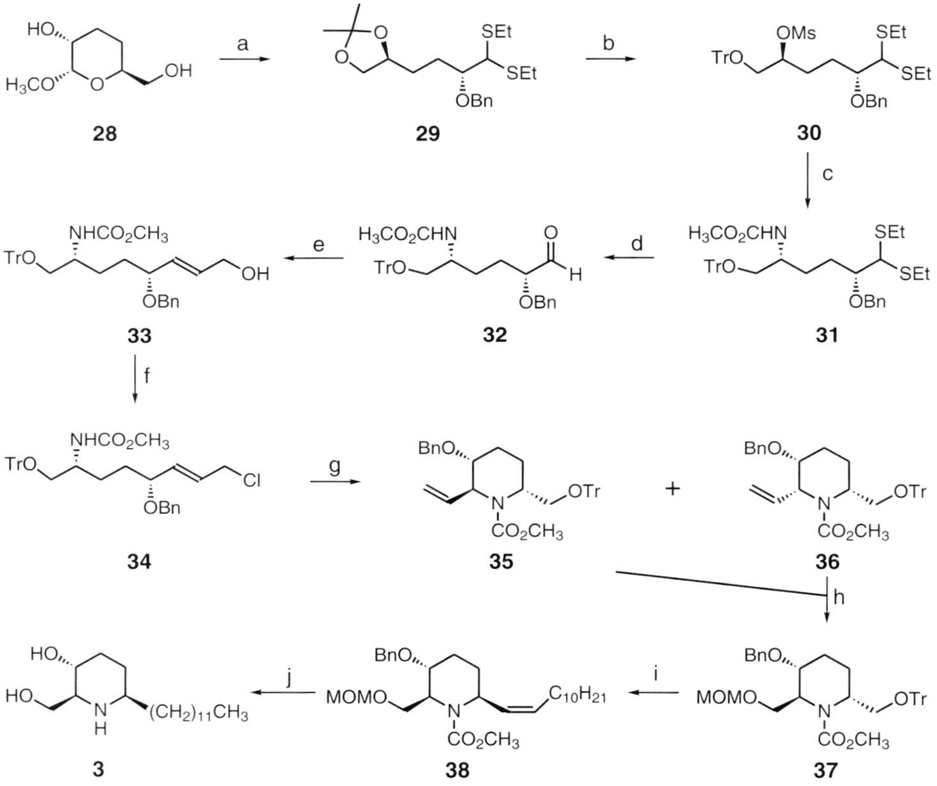

Scheme 2 (*a*) 1. EtSH, conc. HCl, –15°C, 5.5 h; 2. DMP, CSA, acetone, rt, 12 h, 84% for two steps; 3. BnBr, NaH, THF, rt, 13 h, 98%. (*b*) 1. 50% aqueous AcOH, rt, 15 h; 2. TrCl, DMAP, Py, 70°C, 2.5 h; 3. MsCl, Py, rt, 13 h, 86% for three steps. (*c*) 1. NaN_3, DMF, 70°C, 24 h, 91%; 2. H_2S, aqueous Py, rt, 2 days; 3. $ClCO_2CH_3$, K_2CO_3, aqueous acetone, rt, 2 h, 94% for two steps. (*d*) $HgCl_2$, $CaCO_3$, aqueous CH_3CN, rt, 30 min. (*e*) 1. $(EtO)_2P(O)CH_2CO_2Et$, NaH, THF, rt, 30 min, 99% for two steps; 2. DIBAL-H, CH_2Cl_2, –78°C, 90 min; then separation on SiO_2, *E* (83%), *Z* (3%). (*f*) *p*-TsCl, DMAP, CH_2Cl_2, rt, 2 days, 94%. (*g*) THF, NaH, $Pd(Ph_3P)_4$, *n*-Bu_4NI, rt, 4 days, 75%. (*h*) 1. O_3, CH_2Cl_2, CH_3OH, –78°C, 30 min, Ph_3P, –78°C, 15 min; then $NaBH_4$, 0°C, 2 h; 2. CH_3OCH_2Cl, (*i*-Pr)$_2$NEt, CH_2Cl_2, rt, 16 h; then separation, 76% for two steps. (*i*) 1. *p*-TsOH, CH_3OH, rt, 90 min, 94%; 2. $(COCl)_2$, DMSO, CH_2Cl_2, –78°C, 45 min; then NEt_3, rt, 1 h; 3. $Ph_3P{=}CH(CH_2)_9CH_3$, THF, 0°C, 10 min, separation, 59% for two steps. (*j*) 1. H_2, 10% Pd *on* C, EtOH, rt, 2 h, 79%; 2. 3 M KOH, $(CH_2OH)_2$, $NH_2NH_2 \cdot H_2O$, reflux, 2.5 h, 88%; 3. 4 M HCl, 1,4-dioxane, 100°C, 17 h, 81%.

Ozonolysis of the mixture of **35** and **36** followed by sodium borohydride reduction afforded **39** (86%) and **40** (6%) (Scheme 3).[74] Detritylation of **39** followed by Swern oxidation and Wittig olefination of the resulting aldehyde with $Ph_3P{=}CH(CH_2)_9CH_3$ furnished the Z-olefin **41** (49%) and the E-olefin **42** (3%). The Z-olefin was hydrogenated and the oxazolidinone ring was saponified to produce (–)-desoxoprosopinine (**4**).

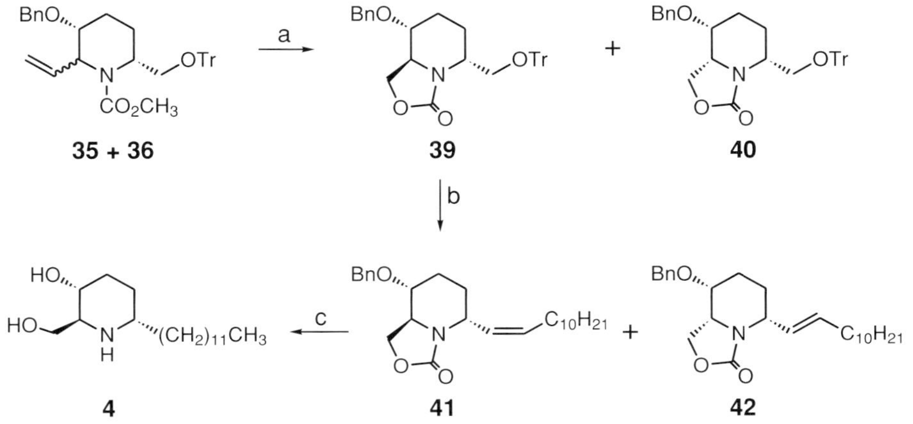

Scheme 3 (*a*) 1. O_3, CH_2Cl_2, CH_3OH, –78°C, 30 min; then Ph_3P, –78°C, 15 min; then $NaBH_4$, 0°C, 2 h; 2. NaH, THF, reflux, 1 h, **39** (86%), **40** (6%). (*b*) 1. *p*-TsOH, CH_3OH, rt, 90 min, 98%; 2. $(COCl)_2$, DMSO, CH_2Cl_2, –78°C, 45 min; then NEt_3, rt, 1 h; 3. $Ph_3P{=}CH(CH_2)_9CH_3$, THF, 0°C, 10 min, **41***E* (3%), **42***Z* (49%). (*c*) 1. H_2, 10% Pd *on* C, CH_3OH, conc. HCl, rt, 1 h; 2. 8 M KOH, EtOH, 100°C, 24 h, 80% for two steps.

3.2.1.2 *Synthesis from* D-*glucal* Synthesis of (+)-desoxoprosophylline (**49**) was based on D-glucal as a precursor (Scheme 4).[75] Protection of the hydroxy groups in D-glucal as *p*-methoxybenzyl ethers followed by hydration of the double bond afforded **43**, which underwent Wittig olefination with methylenetriphenylphosphorane and then TPAP oxidation of the resulting secondary alcohol to furnish **44**. Reduction of the corresponding oxime gave a 77:23 mixture of amine isomers of **45**. The PMB groups were changed to acetyl groups to give **46**. Ozonolytic cleavage of the terminal double bond of **46** and subsequent dehydration of the resulting hemiacetal using oxalyl chloride gave the imino glucal **47**, which was converted into **48** in 78% yield. After Fmoc deprotection, hydrogenation and removal of the acetyl groups, **49** was obtained.

On the other hand, D-glucal has been used for constructing such piperidines via another synthetic strategy (Schemes 5[76] and 6[77]). Thus, D-glucal was reacted with $HgSO_4$, followed by selective protection of the primary hydroxyl group with TBDPSCl to furnish **50**. Displacement of the secondary hydroxyl group with azide with inversion of configuration was achieved in high enantiomeric excess (>98%) on treating compound **50** with DBU and DPPA. Hydrogenation of the azido group and tosylation of the resulting amine afforded the sulfonamide **51**.[78] This was reacted with *m*-CPBA to afford hydroxymethyldihydropyridone **52**,[79] which was treated with $HC(OEt)_3$ followed by hydrogenation and subsequent reduction to afford (3*R*)-piperidinol **53**, as a single diastereomer. Benzylation of **53** followed by treatment with allyltrimethylsilane in the presence of $TiCl_4$ furnished

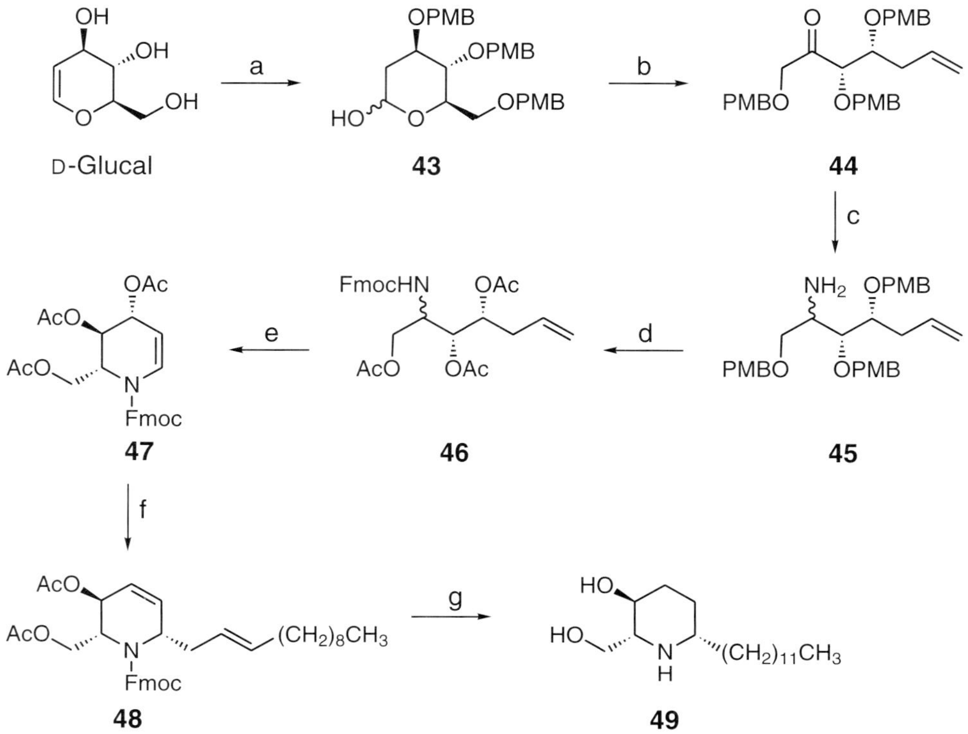

Scheme 4 (*a*) 1. NaH, PMBCl, DMF; 2. $Hg(OAc)_2$, THF, H_2O; then $NaBH_4$, 51%. (*b*) 1. $Ph_3P{=}CH_2$, toluene; 2. TPAP, NMO, MS 4 Å, CH_2Cl_2, 69% for two steps. (*c*) 1. $HONH_2{\cdot}HCl$, Py, EtOH, 60°C; 2. $LiAlH_4$, Et_2O, rt. (*d*) 1. Fmoc-Cl, K_2CO_3, THF–H_2O (3:1); 2. TFA, CH_2Cl_2; 3. Ac_2O, Py, rt, 54% for five steps. (*e*) 1. O_3, −78°C, CH_2Cl_2; then DMS, rt; 2. $(COCl)_2$, NEt_3, DMF, CH_2Cl_2, 53% for two steps. (*f*) 1. $BF_3{\cdot}Et_2O$, CH_2Cl_2, $H_2C{=}CHCH(TMS)(CH_2)_8CH_3$, −60 to 0°C, 3 h; 2. piperidine, CH_2Cl_2, rt, 1 h, 78% for two steps. (*g*) 1. H_2, Pt *on* C, EtOH, 1.5 h; 2. LiOH, THF·H_2O, 2.5 h, 51% for two steps.

54 in 87% yield. Dihydroxylation and subsequent periodate cleavage produced aldehyde **55**, whose elongation of C-6 chain was carried out through the introduction of the 8-oxo-*n*-decanyl side chain by Wittig reaction to give **56**. Finally, cleavage of the acetal group followed by hydrogenation gave **57**, whose removal of the protecting groups provided (−)-prosophylline (**1**). On the other hand, **53** gave the protected piperidinol **58**, which can be elaborated in the stereoselective synthesis of (−)-desoxoprosophylline (**3**).[80]

The isomer of **51** has been used to prepare (+)-prosophylline and (+)-desoxoprosophylline (**49**)[81] by applying methodology similar to that reported above.

Synthesis of (+)-desoxoprosopinine [(+)-**4**] through a key amidoalkylation reaction, which permits stereospecific generation of protected *trans*-2,6-disubstituted piperidines, has been achieved (Scheme 6).[77] Allylation of **60**,[82] obtained from compound **59**, occurred rapidly upon treatment with allyltrimethylsilane[83–88] in the presence of $TiCl_4$ to provide **61** in 88% yield, which underwent deacetylation and benzylation to furnish **62**. Ozonolysis of the allyl group in **62** followed by olefination of the resulting aldehyde and subsequent

OH
OH
OH
O
D-Glucal
a
O
OTPS
OH
50
b
O
OTPS
NHTs
51
c
O
TPSO
N
OH
Ts
52
d
RO
TPSO
N
OEt
Ts
53a) R = H
53b) R = Bz
e
BnO
TPSO
N
Ts
54
f
BnO
TPSO
N
CHO
Ts
55
g
BnO
TPSO
N
Ts
H
H
O
O
56
h
HO
TPSO
N
$(CH_2)_9COCH_2CH_3$
Ts
57
j
BzO
TPSO
N
$(CH_2)_{11}CH_3$
Ts
58
i
HO
HO
N
H
$(CH_2)_9COCH_2CH_3$
1
i
HO
HO
N
H
$(CH_2)_{11}CH_3$
3

Scheme 5 (*a*) Ref. 78. (*b*) 1. DPPA, DBU, toluene, 0°C to rt, 84%; 2. H_2, Pd *on* C, CH_3OH; 3. *p*-TsCl, NEt_3, CH_2Cl_2, 87%. (*c*) 1. *m*-CPBA, CH_2Cl_2, 91%. (*d*) 1. $HC(OEt)_3$, $BF_3{\cdot}OEt_2$, MS 4 Å, THF, 0°C, 95%; 2. H_2, Pd *on* C, AcOEt, 91%; 3. $NaBH_3CN$, AcOH, CH_3OH, 0°C to rt, 85% (**53a**); *or* 2. $NaBH_4$, $CeCl_3{\cdot}7H_2O$, CH_3OH, –30°C; 3. H_2, Pd *on* C, AcOEt, 75% for two steps; 4. DEAD, Ph_3P, BzOH, THF, rt, 91% (**53b**). (*e*) 1. NaH, BnBr, Bu_4NI, THF, 91%; 2. allyltrimethylsilane, $TiCl_4$, CH_2Cl_2, –78°C, 87%. (*f*) 1. $K_3Fe(CN)_6$, K_2CO_3, $K_2OsO_2(OH)_2$, Na_2SO_3, *t*-BuOH–H_2O (1:1); 2. $NaIO_4$, H_2O–EtOH (1:1), 96% for two steps. (*g*) PPh_3, $CH_3CH_2C(OCH_2)_2C_7H_{14}Br$, *n*-BuLi, 68%. (*h*) 1. HCl, H_2O; 2. H_2, Pd *on* C, EtOH, 88% for two steps. (*i*) 1. TBAF, THF, 91%; 2. Na, naphthalene, 64%. (*j*) Ph_3P, $CH_3(CH_2)_{10}CH_2Br$, *n*-BuLi.

O
HN
O
O
59
a
AcO
H
N
O
O
OCH_3
60
b
AcO
H
N
O
O
61
c
BnO
H
N
O
O
62
d
BnO
H
N
O
$H_3C(H_2C)_{11}$
O
63
e
HO
HO
N
H
$(CH_2)_{11}CH_3$
(+)-**4**

Scheme 6 (*a*) Ref. 82. (*b*) $(CH_3)_3SiCH_2CH{=}CH_2$, $TiCl_4$, CH_2Cl_2, 25°C, 88%. (*c*) 1. K_2CO_3, CH_3OH, 98%; 2. NaH, BnBr, THF, 95%. (*d*) 1. O_3, CH_2Cl_2, CH_3OH; then DMS, 90%; 2. *n*-$C_9H_{18}CH{=}PPh_3$, THF, reflux, 35%; 3. H_2, Pd *on* C, 100%. (*e*) 1. Aqueous NaOH, EtOH, reflux, 97%; 2. Li, liquid NH_3, 77%.

hydrogenation of the resulting olefin afforded compound **63**. Removal of the benzyl group from **63** with Li in liquid NH_3 afforded (+)-**4**.

3.2.1.3 *Synthesis from* D*-glyceraldehyde* Synthesis of (–)-prosophylline (**1**) from D-glyceraldehyde acetonide (**64**) has been reported (Scheme 7).[89] The enantioselective allylation of aldehyde **64**[90] with (*S*,*S*)-**75**[91] afforded the homoallyl alcohol **65** in 86% yield. Protection of **65**, as the benzyl derivative, followed by hydroboration and transformation to

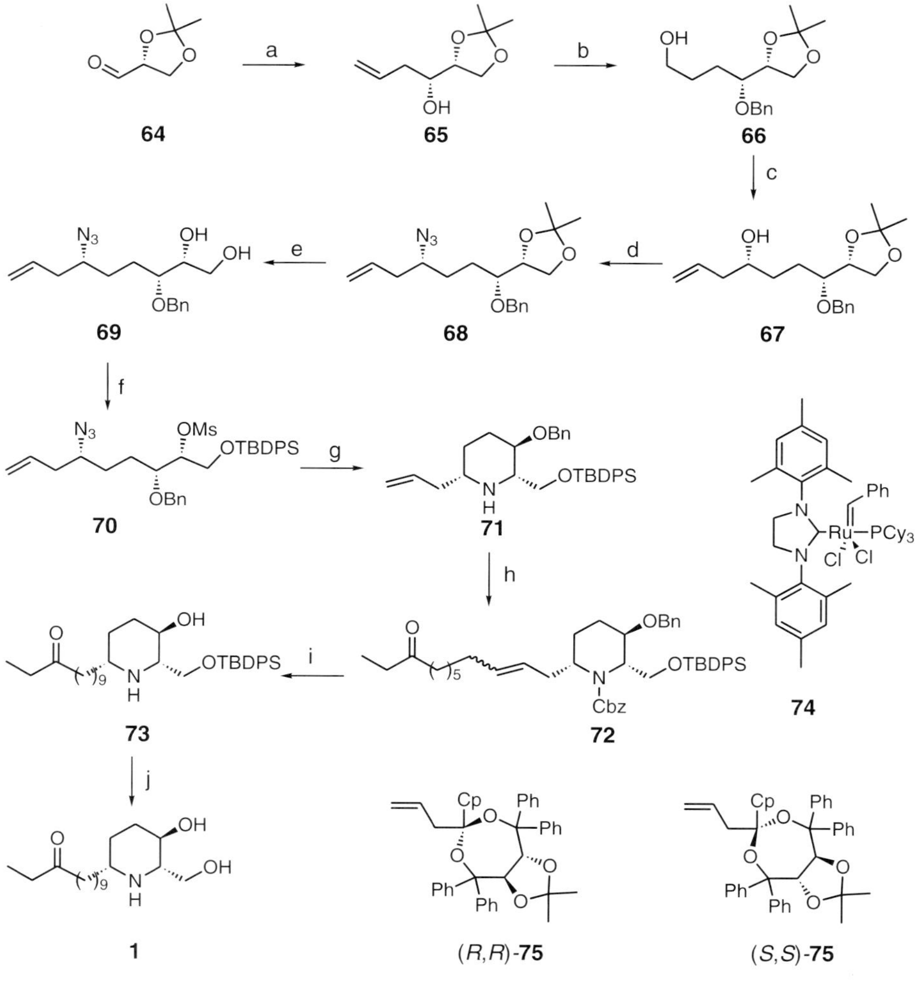

Scheme 7 (*a*) 1. (*S*,*S*)-**75**, Et_2O, –78°C, 86%. (*b*) 1. *t*-BuOK, BnBr, THF, 91%; 2. BH_3–THF; 3. H_2O_2, NaOH, H_2O, 83%. (*c*) 1. DMSO, $(COCl)_2$, NEt_3, CH_2Cl_2, quantitative; 2. (*R*,*R*)-**75**, Et_2O, –78°C, 81%. (*d*) Ph_3P, DEAD, DPPA, THF, 0°C to rt. (*e*) AcOH–H_2O (80:20), 76% for two steps. (*f*) 1. TBDPSCl, imidazole, CH_2Cl_2, 96%; 2. MsCl, DMPA, Py. (*g*) 1. Ph_3P, THF, H_2O, 89%; 2. NEt_3, CH_3OH, reflux, 88%. (*h*) 1. CbzCl, Na_2CO_3, CH_2Cl_2, H_2O, quantitative; 2. $CH_3CH_2C(O)(CH_2)_5CH_2CHCH_2$, **74**, CH_2Cl_2, reflux, 58%. (*i*) H_2, 10%, Pd *on* C, CH_3OH, HCl, 60%. (*j*) TBAF, THF, 90%.

the primary alcohol gave **66** in 83% yield. Swern oxidation of **66** and then treatment with the allyltitanium complex (*R*,*R*)-**75** gave **67** in 98:2 diastereomeric ratio and 81% yield. The homoallylic alcohol **67** was treated with DPPA using Mitsunobu reaction to give the azide **68**. Compound **68** was deisopropylidenated to give **69** in 76% yield. Protection of the primary hydroxyl group of **69** as TBDPS (96%) followed by mesylation gave **70** in 98% yield. Reduction of the azide group in **70** afforded the respective amine in 89% yield, which upon treatment with NEt_3 in boiling methanol furnished **71** in 88% yield. Piperidine **71** was N-protected with CbzCl and treated with the Grubbs' catalyst **74** to give **72**, which was hydrogenated to produce **73**. This was then treated with TBAF in THF to afford **1** in 54% yield.

Synthesis of (–)-desoxoprosopinine (**4**) from the acetonide **76** has been reported by O-silylation with TBDPSCl and imidazole, followed by ring cleavage of the acetonide to afford the diol **77**, which was converted to the epoxide **78** (Scheme 8).[92] Reaction of **78** with allylmagnesium bromide gave alcohol **79**, which was condensed with oxazolidine-2,4-dione

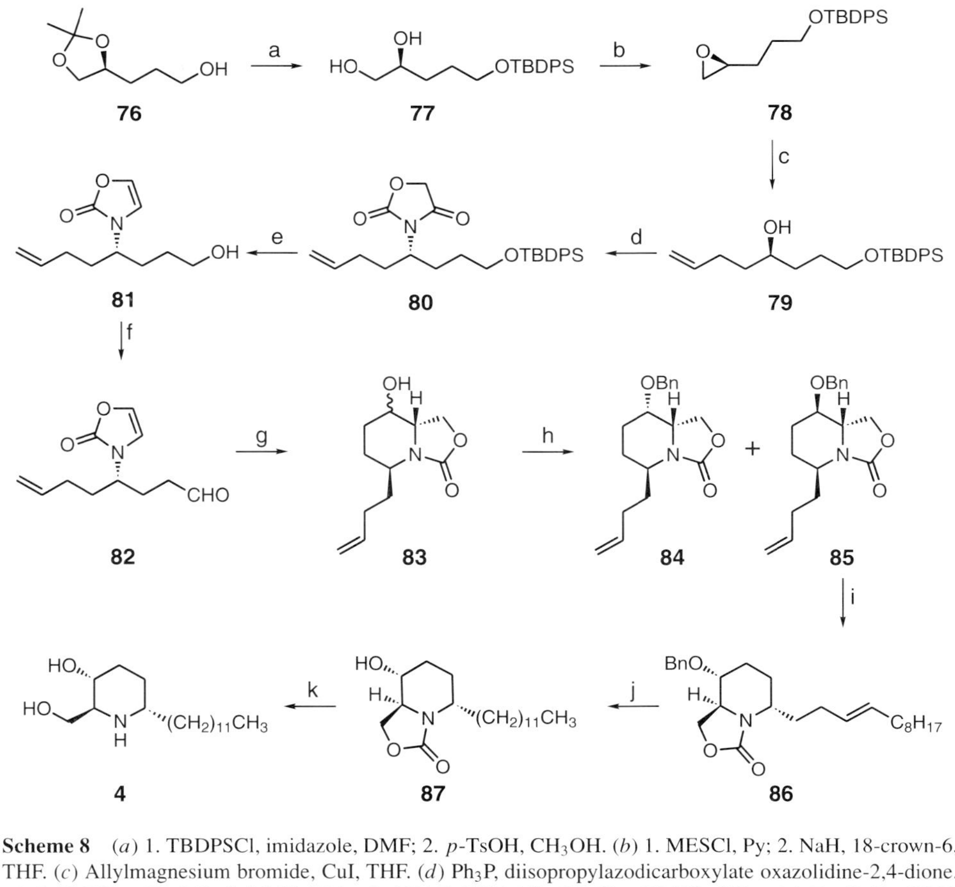

Scheme 8 (*a*) 1. TBDPSCl, imidazole, DMF; 2. *p*-TsOH, CH_3OH. (*b*) 1. MESCl, Py; 2. NaH, 18-crown-6, THF. (*c*) Allylmagnesium bromide, CuI, THF. (*d*) Ph_3P, diisopropylazodicarboxylate oxazolidine-2,4-dione. (*e*) 1. $NaBH_4$, CH_3OH; 2. MsCl, NEt_3; 3. TBAF, THF. (*f*) $(COCl)_2$, DMSO, NEt_3. (*g*) Bu_3SnH, AIBN, benzene. (*h*) NaH, BnBr, Bu_4NBr, THF, **85** (50%), **84** (25%). (*i*) 1. O_3, CH_3OH, CH_2Cl_2; then $(CH_3)_2S$; 2. *n*-$C_9H_{19}Ph_3PBr$, *n*-BuLi. (*j*) H_2, 10% Pd *on* C, CH_3OH, conc. HCl. (*k*) 8 M KOH, EtOH, 100°C, 24 h.

via Mitsunobu reaction to afford **80**. Reduction of **80** followed by mesylation and desilylation produced **81**. Swern oxidation of **81** afforded aldehyde **82**, which was converted into the 8-hydroxyoxazolopiperidine **83**, as a diastereomeric mixture. Treatment of **83** with BnBr afforded **84** and **85** in 25 and 50% yield, respectively. Ozonolysis of **85** followed by olefination afforded **86**. Hydrogenation of **86** afforded **87**, which was converted into **4**.[93]

3.2.1.4 *Synthesis from L-gulonolactone* Synthesis of (+)-desoxoprosophylline (**49**), using L-gulonolactone (**88**) as a chiral starting material and tandem Wittig [2+3]-cycloaddition reaction to form the heterocyclic core unit, has been reported (Scheme 9).[94] Thus, compound **88** was transformed to 5,6-*O*-isopropylidene-L-gulonolactone (**89**),[95,96] and then to the α-mesylated lactone **90**,[96,97] which underwent S_N2 displacement of the α-mesyloxy group by iodine followed by catalytic hydrogenation to afford the lactone **91**. Removal of the isopropylidene group from **91** afforded the lactone **92**, which was treated with excess

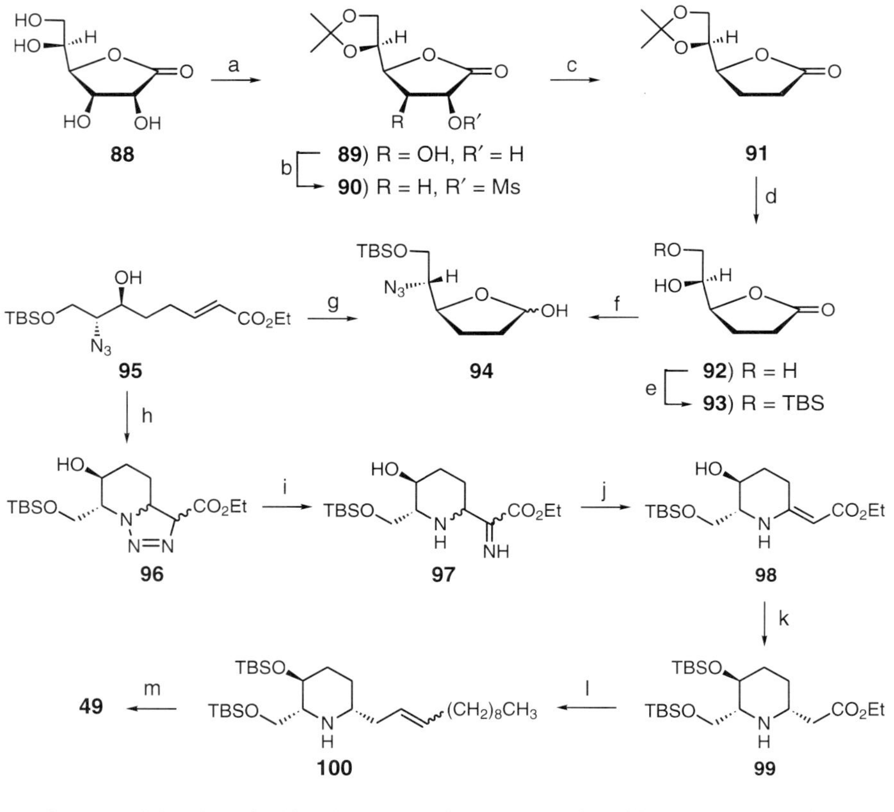

Scheme 9 (*a*) Ref. 96. (*b*) Refs. 96 and 97. (*c*) 1. NaI, acetone, reflux, 92%; 2. H_2, NEt_3, 10% Pd *on* C, 20–30 h, 82%. (*d*) Conc. HCl, *i*-PrOH, 48 h, 96%. (*e*) Excess TBSCl, NEt_3, DMAP, DMF, 12 min, 99%. (*f*) 1. MsCl, NEt_3, CH_2Cl_2, 20 min, 79%; 2. NaN_3, DMPU, 70°C, 24 h; 3. DIBAL-H, THF, –78°C, 4–6 h, 69%. (*g*) Ph_3PCHCO_2Et, toluene, rt, 1 day. (*h*) Toluene, rt, 4 days, 98%. (*i*) NEt_3, CH_2Cl_2, 12 h, 96%. (*j*) $Rh_2(OAc)_4$, 12 h, 97%. (*k*) 1. H_2, 10% Pd *on* C, EtOH, 48 h, 71%; 2. TBSCl, imidazole, DMF, 84%. (*l*) 1. DIBAL-H, *n*-pentane, –78°C, 25 min, 66%; 2. $Ph_3P(CH_2)_9CH_3Br$, $NaN[Si(CH_3)_3]_2$, THF, –40°C, 6 min, rt, 160 min, 79%. (*m*) 1. H_2, 10% Pd *on* C, EtOH, 12 h, 93%; 2. HCl, EtOH, 15 min; then 6 N KOH, 87%.

TBSCl to give the silylated lactone **93**. Mesylation of the free secondary hydroxyl group followed by nucleophilic substitution with azide ion and reduction with DIBAL-H afforded the lactol **94**. Reaction of **94** with Ph_3PCHCO_2Et gave **95**, which was transformed into the triazolines **96** in 98% overall yield from **94**.[94] Rearrangement of **96** had taken place in nearly quantitative yield by the action of triethylamine to give **97**. Treatment of **97** with $Rh_2(OAc)_4$ afforded **98**, which was hydrogenated followed by silylation with TBSCl to give the ester **99**. Reduction of **99** followed by Wittig reaction with decylphosphonium bromide under salt-free conditions afforded the olefin **100**, which was then hydrogenated and finally deprotected to give **49** in 23% overall yield from **90**.

3.2.1.5 *Synthesis from calcium* D*-gluconate* The nonracemic intermediates for (–)-cassine (**7**) were synthesized from calcium D-gluconate (Scheme 10).[98] The stereoisomeric

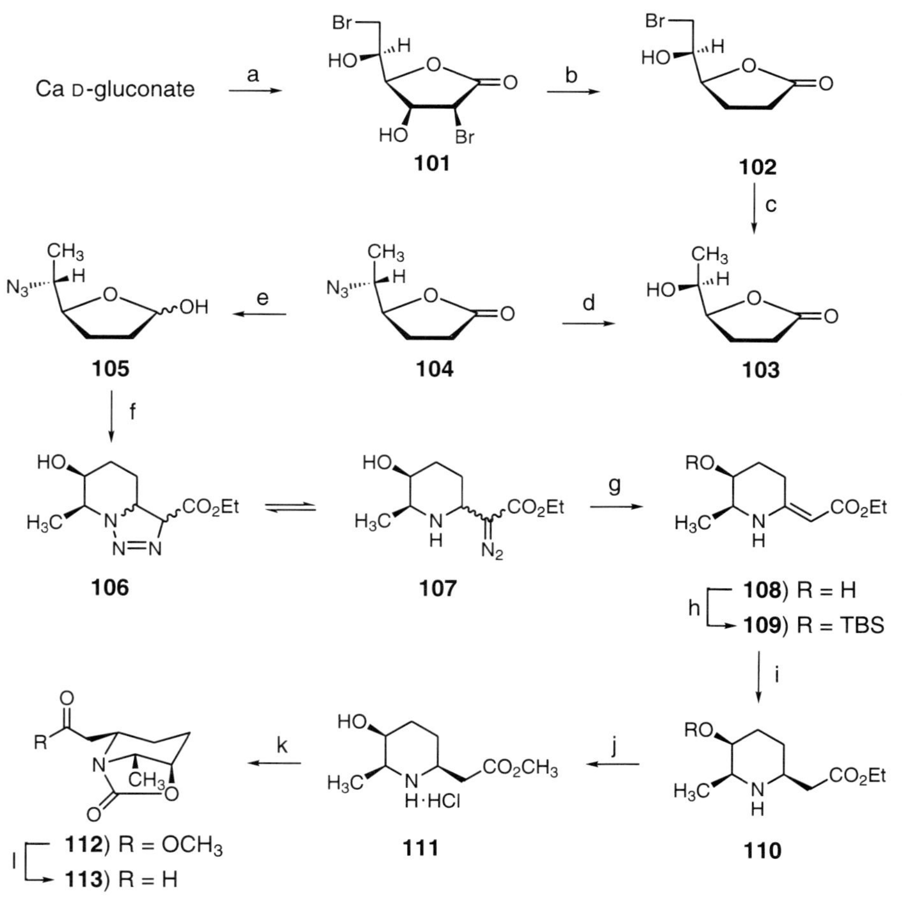

Scheme 10 (*a*) 1. HBr, HOAc; 2. CH_3OH, 42% for two steps. (*b*) H_2, Pd *on* C, EtOH, 71%. (*c*) H_2, Pd *on* C, NEt_3, EtOAc, 78%. (*d*) 1. CH_2Cl_2, NEt_3, –40°C, MsCl, –30°C to rt, 2 h, 98%; 2. DMF, LiN_3, 60°C, 18 h, 77%. (*e*) THF, DIBAL-H, –78°C, 45 min, 81%. (*f*) $Ph_3P{=}CHCO_2Et$, toluene, 20°C, 4 days, 87%, **106** (53%), **107** (34%). (*g*) Toluene, 14 h, 90–100°C, 68%. (*h*) TBSCl, DMAP, imidazole, rt, 4 days, 76%. (*i*) Pd *on* C, H_2, EtOH, 40°C, 36 h, 99%. (*j*) CH_3OH, HCl, 65°C, 1 h, 87%. (*k*) $(Boc)_2O$, DABCO, THF, rt, 14 h, 68%. (*l*) 1. DIBAL-H, THF, –30°C, 20 min; then 4 h at –20°C, 52%; 2. Tf_2O, DMSO, CH_2Cl_2, –50°C, NEt_3, 59%.

lactone **103** was obtained from calcium D-gluconate, via the intermediates **101** and **102**,[99,100]. After converting the hydroxyl group in **103** to the corresponding mesylate and on subsequent azidolysis, **104** was obtained. Treatment of **104** with DIBAL-H produced **105** in 14% overall yield from calcium D-gluconate. When **105** was treated with ethoxycarbonyl methylenetriphenylphosphorane, the corresponding olefin intermediate could not be isolated, because an intramolecular 1,3-dipolar cycloaddition took place immediately to provide the diastereomeric triazolines **106** and the diazoamines **107** in a ratio of 2:1. The mixture of **106** and **107** gave a stereochemically homogeneous product **108**, when they were heated in toluene at 90–100°C. Thereby, elimination of nitrogen took place with concomitant 1,2-H shift to provide the *Z*-olefin as the only product. Treatment of **108** with TBSCl provided **109** whose reduction occurred exclusively from the less shielded β-face to give **110** as the only product. Desilylation of **110** afforded **111**, which was treated with $(Boc)_2O$ and diazabicyclooctane to provide **112**, in which both functional groups were protected. Reduction of the ester group in **112** with DIBAL-H gave the corresponding alcohol, whose Moffat oxidation produced the aldehyde **113**, which is a precursor for the synthesis of *Cassia* and *Prosopis* alkaloids.

References

1. Ratle, G.; Monseur, X.; Das, B.; Yassi, J.; Khuong-Huu, Q.; Goutarel, R. *Bull. Soc. Chim. Fr.* (1966) 2945; *Chem. Abstr.* **66** (1966) 18779h.
2. Bourrinet, P.; Quevauviller, A. *C. R. Soc. Biol.* **162** (1968) 1138; *Chem. Abstr.* **70** (1969) 95233k.
3. Bourrinet, P.; Quevauviller, A. *Ann. Pharm. Fr.* **26** (1968) 787; *Chem. Abstr.* **71** (1969) 29012g.
4. Khuong-Huu, Q.; Ratle, G.; Monseur, X.; Goutarel, R. *Bull. Soc. Chim. Belg.* **81** (1972) 425.
5. Khuong-Huu, Q.; Ratle, G.; Monseur, X.; Goutarel, R. *Bull. Soc. Chim. Belg.* **81** (1972) 423.
6. Pinder, A.R. *Nat. Prod. Rep.* **9** (1992) 941.
7. Numata, A.; Ibuka, T. In *The Alkaloids*, Vol. 31; Brossi, A., Ed.; Academic Press: New York, 1985; Chapt. 6.
8. Foder, G.B.; Colasanti, B. In *The Alkaloids: Chemical and Biological Respectives*, Vol. 30; Peppetier, S.W., Ed.; Wiley: New York, 1985; Chapt. 1.
9. Agami, C.; Couty, F.; Lequesne, C. *Tetrahedron* **51** (1995) 4043.
10. Khuong-Huu, Q.; Ratle, G.; Monseur, X.; Goutarel, R. *Bull. Soc. Chim. Belg.* **81** (1972) 4432.
11. Agami, C.; Couty, F.; Mathieu, H. *Tetrahedron Lett.* **37** (1996) 4000.
12. Melhaoui, A.; Bodo, B. *Nat. Prod. Lett.* **7** (1995) 101.
13. Christofidis, I.; Welter, A.; Jadot, J. *Tetrahedron* **33** (1977) 977.
14. Hight, R.J. *J. Org. Chem.* **29** (1964) 471.
15. Highet, R.J.; Highet, P.F. *J. Org. Chem.* **31** (1966) 1275.
16. Rice, W.Y., Jr.; Coke, J.L. *J. Org. Chem.* **31** (1966) 1010.
17. Strunz, G.M.; Findlay, J.A. In *The Alkaloids*, Vol. 26; Brossi, A., Ed.; Academic Press: New York, 1985; p. 89.
18. Wang, C.-L.J.; Wuonola, M.A. *Org. Prep. Proc. Int.* **24** (1992) 585.
19. Pinder, A.R. *Nat. Prod. Rep.* **9** (1992) 491.
20. Pinder, A.R. *Nat. Prod. Rep.* **9** (1992) 17.
21. Pinder, A.R. *Nat. Prod. Rep.* **7** (1990) 447.
22. Spiteller-Friedmann, M.; Spiteller, G. *Monatsh. Chem.* **95** (1964) 1234.
23. Watt, J.M.; Breyer-Brandwijk, M.G. *The Medicinal and Poisonous Plants of Southern and Eastern Africa*, 2nd ed.; E & S Livingstone: London, 1962.
24. Coke, J.L.; Rice, W.Y., Jr. *J. Org. Chem.* **30** (1965) 3420.
25. Smalberger, T.M.; Rall, G.J.H.; de Wall, H.L.; Arndt, R.R. *Tetrahedron* **24** (1968) 6417.
26. Aguinaldo, A.M.; Read, R.W. *Phytochemistry* **29** (1990) 2309.
27. Brown, E.; Dhal, R.; Lavoue, J. *Tetrahedron Lett.* **12** (1971) 1055.
28. Brown, E.; Dhal, R. *Bull. Soc. Chim. Fr.* (1972) 4292.
29. Brown, E.; Dhal, R.; Casals, P.F. *Tetrahedron* **28** (1972) 5607.

30. Fodor, G.; Funmeaux, J.-P.; Sankaran, V. *Synthesis* (1972) 464.
31. Brown, E.; Bourgouin, A. *Chem. Lett.* (1974) 109.
32. Brown, E.; Dhal, R. *Tetrahedron Lett.* **15** (1974) 1029.
33. Brown, E.; Bourgouin, A. *Tetrahedron* **31** (1975) 1047.
34. Brown, E.; Bonte, A. *Tetrahedron Lett.* **16** (1975) 2881.
35. Baxter, A.J.G.; Holmes, A.B. *J. Chem. Soc., Perkin Trans. 1* (1977) 2343.
36. Saitoh, Y.; Moriyama, Y.; Hirota, H.; Takahashi, T.; Khuong-Huu, Q. *Bull. Chem. Soc. Jpn.* **54** (1981) 488.
37. Bonte, A. *Bull. Soc. Chim. Fr.* (1981) 281.
38. Saitoh, Y.; Moriyama, Y.; Takahashi, T. *Tetrahedron Lett.* **21** (1980) 75.
39. Natsume, M.; Ogawa, M. *Heterocycles* **14** (1980) 615.
40. Natsume, M.; Ogawa, M. *Heterocycles* **16** (1981) 973.
41. Natsume, M.; Ogawa, M. *Heterocycles* **20** (1983) 601.
42. Holmes, A.B.; Thompson, J.; Baxter, A.J.G.; Dixon, J. *J. Chem. Soc.; Chem. Commun.* (1985) 37.
43. Singh, R. Ghosh, S.K. *Tetrahedron Lett.* **43** (2002) 7711.
44. Paterne, M.; Brown, E. *J. Chem. Res.* (1985) 278.
45. Birkinshaw, T.N.; Holmes, A.B. *Tetrahedron Lett.* **28** (1987) 813.
46. Paterne, M.; Dhal, R.; Brown, E. *Bull. Chem. Soc. Jpn.* **62** (1989) 1321.
47. Hsseberg, H.-A.; Gerlach, H. *Ann. Chem.* (1989) 255.
48. Momose, T.; Toyooka, N. *Tetrahedron Lett.* **34** (1993) 5785.
49. Lu, Z.-H.; Zhou, W.-S. *Tetrahedron* **49** (1993) 4659.
50. Lu, Z.-H.; Zhou, W.-S. *J. Chem. Soc., Perkin Trans. 1* (1993) 593.
51. Cook, G.R.; Beholz, L.G.; Stille, J.R. *J. Org. Chem.* **59** (1994) 3575.
52. Takao, K.; Nigawara, Y.; Nishino, E.; Takagi, I.; Maeda, K.; Tadano, K.; Ogawa, S. *Tetrahedron* **50** (1994) 5681.
53. Terakado, M.; Murai, K.; Miyazawa, M.; Yamamoto, K. *Tetrahedron* **50** (1994) 5705.
54. Cook, G.R.; Beholz, L.G.; Stille, J.R. *Tetrahedron Lett.* **35** (1994) 1669.
55. Toyooka, N.; Yoshida, Y.; Momose, T. *Tetrahedron Lett.* **36** (1995) 3715.
56. Yuasa, Y.; Shibuya, S. *Tetrahedron: Asymmetry* **6** (1995) 1525.
57. Yuasa, Y.; Ando, J.; Shibuya, S. *J. Chem. Soc., Perkin Trans. 1* (1996) 793.
58. Luker, T.; Hiemstra, H.; Speckamp, W.N. *J. Org. Chem.* **62** (1997) 3592.
59. Kadota, I.; Kawada, M.; Muramatsu, Y.; Yamamoto, Y. *Tetrahedron Lett.* **38** (1997) 7469.
60. Hirai, Y.; Watanabe, J.; Nozaki, T.; Yokoyama, H.; Yamaguchi, S. *J. Org. Chem.* **62** (1997) 776.
61. Kadota, I.; Kawada, M.; Muramatsu, Y.; Yamamoto, Y. *Tetrahedron: Asymmetry* **8** (1997) 3887.
62. Pahl, A.; Wartchow, R.; Meyer, H.H. *Tetrahedron Lett.* **39** (1998) 2095.
63. Agami, C.; Couty, F.; Lam, H.; Mathieu, H. *Tetrahedron* **54** (1998) 8783.
64. Ojima, I.; Vidal, E.S. *J. Org. Chem.* **63** (1998) 7999.
65. Agami, C.; Couty, F.; Mathieu, H. *Tetrahedron Lett.* **39** (1998) 3505.
66. Toyooka, N.; Yoshida, Y.; Yotsui, Y.; Momose, T. *J. Org. Chem.* **64** (1999) 4914.
67. Bailey, P.D.; Smith, P.D.; Morgan, K.M.; Rosair, G.M. *Tetrahedron Lett.* **43** (2002) 1071.
68. Jourdant, A.; Zhu, J. *Tetrahedron Lett.* **42** (2001) 3431.
69. Hanessian, S.; Frenette, R. *Tetrahedron Lett.* **20** (1979) 3391.
70. Horton, D.; Thompson, J.K.; Tindall, C.G., Jr. *Methods Carbohydr. Chem.* **6** (1972) 297.
71. Corey, E.J.; Nicolaou, K.C.; Melvin, L.S. *J. Am. Chem. Soc.* **97** (1975) 654.
72. Holder, N.L.; Fraser-Reid, B. *Can. J. Chem.* **51** (1973) 3357.
73. Umezawa, S.; Okazaki, Y.; Tsuchiya, T. *Bull. Chem. Soc. Jpn.* **45** (1972) 3619.
74. Tadano, K.; Takao, K.; Nigawara, Y.; Nishino, E.; Takagi, I.; Maeda, K.; Ogawa, S. *Synlett* (1993) 565.
75. Dransfield, P.J.; Gore, P.M.; Shipman, M.; Slawin, A.M.Z. *Chem. Commun.* (2002) 150.
76. Koulocheri, S.D.; Haroutounian, S.A. *Tetrahedron Lett.* **40** (1999) 6869.
77. Ciufolini, M.A.; Hermann, C.W.; Whitmire, K.H.; Byrne, N.E. *J. Am. Chem. Soc.* **111** (1989) 3473.
78. Koulocheri, S.D.; Haroutounian, S.A. *Synthesis* (1999) 1889.
79. Koulocheri, S.D.; Magiatis, P.; Skaltsounis, A.L.; Haroutounian, S.A. *Tetrahedron* **56** (2000) 6135.
80. Luker, T.; Hiemstra, H.; Speckamp, W.N. *J. Org. Chem.* **62** (1997) 3592.
81. Yang, C.; Liao, L.; Xu, Y.; Zhang, H.; Xia, P.; Zhou, W. *Tetrahedron: Asymmetry* **10** (1999) 2311.
82. Ciufolini, M.A.; Wood, C.Y. *Tetrahedron Lett.* **27** (1986) 5085.
83. Hart, D.J.; Tsai, Y.M. *Tetrahedron Lett.* **22** (1981) 1567.
84. Kraus, G.A.; Neuenschwanser, K. *J. Chem. Soc., Chem. Commun.* (1982) 134.
85. Speckamp, W.N.; Hiemstra, H. *Tetrahedron* **41** (1985) 4367.
86. Shono, T. *Tetrahedron* **40** (1984) 811.
87. Zaugg, H.E. *Synthesis* **85** (1984) 181.
88. Yang, C.-F.; Xu, Y.-M.; Liao, L.-X.; Zhou, W.-S. *Tetrahedron Lett.* **39** (1998) 9227.
89. Cossy, J.; Willis, C.; Bellosta, V.; BouzBouz, S. *J. Org. Chem.* **67** (2002) 1982.

90. Schmid, C.R.; Bryant, J.D. *Org. Synth.* **72** (1995) 6.
91. Hafner, A.; Duthaler, R.O.; Marti, R.; Rihs, G.; Rothe-Streit, P.; Swarzenbach, F. *J. Am. Chem. Soc.* **114** (1992) 2321.
92. Yuasa, Y.; Ando, J.; Shibuya, S. *Tetrahedron: Asymmetry* **6** (1995) 1525.
93. Takao, K.; Nigawara, Y.; Nishino, E.; Takagi, I.; Maeda, K.; Tadano, K.; Ogawa, S. *Tetrahedron* **50** (1994) 5681.
94. Herdeis, C.; Telser, J. *Eur. J. Org. Chem.* (1999) 1407.
95. Andrews, G.C.; Crawford, T.C.; Bacon, B.E. *J. Org. Chem.* **46** (1981) 2976.
96. Vekemans, J.A.J.M.; Boerekamp, J.; Godefroi, E.F.; Chittenden, G.I. *Recl. Trav. Chim. Pays-Bas* **104** (1985) 266.
97. Kalwinsh, I.; Metten, K.-H.; Brückner, R. *Heterocycles* **40** (1993) 939.
98. Herdeis, C.; Schiffer, T. *Tetrahedron* **55** (1999) 1043.
99. Bock, K.; Lundt, I.; Pedersen, C. *Carbohydr. Res.* **68** (1979) 313.
100. Lundt, I.; Pedersen, C. *Synthesis* (1986) 1052.

3.2.2 *Hydroxylated pipecolic acids*

(2*S*,3*R*,4*R*,5*S*)-3,4,5-Trihydroxypipecolic acid (**1**) was isolated from the seeds of legume *Baphia racemosa* Bak.[1] It is a specific inhibitor of human liver β-D-glucuronidase and idouronidase but it has no effect on α- and β-glucosidases or mannosidases.[2] (2*S*,4*S*,5*S*)-4,5-Dihydroxypipecolic acid (**2**) was isolated from the leaves of *Derris eliptica*.[3]

(2*S*,4*R*)-4-Hydroxypipecolic acid [(−)-*cis*-4-hydroxy-2-piperidine carboxylic acid, **3**] was isolated from the leaves of *Calliandra pittieri* and *Strophantus scandeus*[4,5] and it was identified as a constituent of cyclopeptide antibiotics, such as virginiamycin S_2.[6] It was also employed as a precursor in the preparation of selective *N*-methyl-D-aspartate receptor antagonists.[7] Furthermore, (−)-**3** has served as a building block in a recent synthesis of palinavir, a potent peptidomimetic-based HIV protease inhibitor.[8,9]

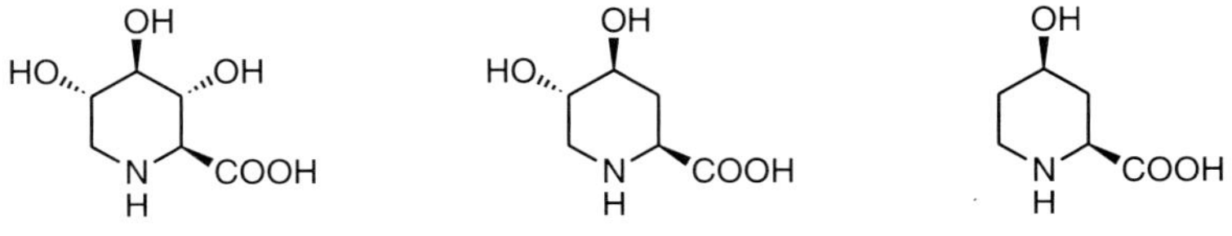

3.2.2.1 *Synthesis from D-glucose*

(2*S*,3*R*,4*R*,5*S*)-3,4,5-Trihydroxypipecolic acid (**1**) was prepared from methyl α-D-glucopyranoside (**4**) by conversion to **5** (Scheme 1).[10] The Hg^{2+}-mediated cyclization of the aminoalkene **5** gave the bromomercurial **6** as a major product, which was reductively oxygenated to the alcohol **7** in 72% yield. Swern oxidation of **7** furnished the sensitive aldehyde **8** in 90% yield, which was immediately oxidized to give **9** in 40% yield. Catalytic hydrogenation of **9** led to complete removal of the protecting groups to give **1**.

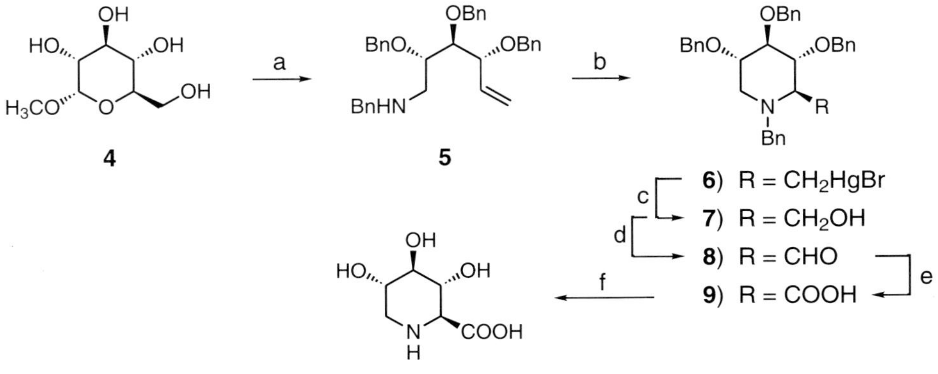

Scheme 1 (*a*) See Scheme 9 in Section 3.1.1. (*b*) $HgBr_2$. (*c*) $NaBH_4$, DMF, O_2. (*d*) $(COCl)_2$, DMSO, CH_2Cl_2; then NEt_3, 90%. (*e*) $KMnO_4$, acetone, H_2O, −10°C, 40%. (*f*) H_2, Pd *on* C, EtOH, 96%.

Synthesis of pipecolic acid **1** and its epimer **14** from D-glucose has been achieved from the carbamate **10** (Scheme 2).[11–13] Oxidation of the C-5-OH in **10** with pyridinium chlorochromate followed by $NaBH_4$ reduction afforded **11**. Hydrolysis of **11** followed by bromine water oxidation afforded the lactone **12**, which underwent hydrogenation to furnish **1** in 72% yield. On the other hand, acid hydrolysis of **10** followed by oxidation with Br_2 afforded the lactone **13**. Removal of the protecting groups and subsequent ion-exchange chromatography furnished **14**.

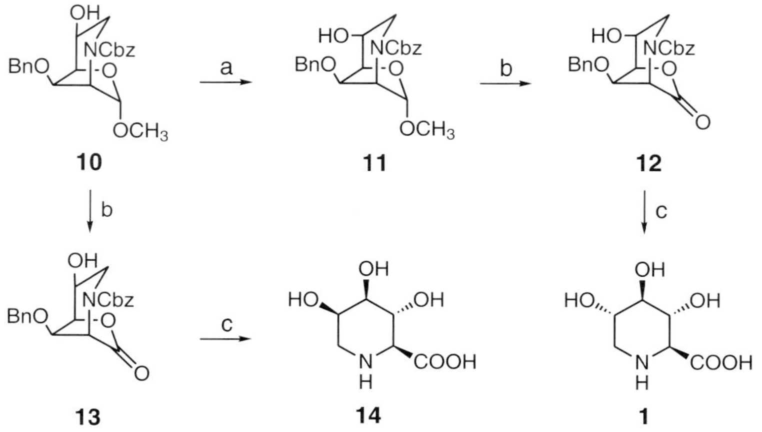

Scheme 2 (*a*) CH_2Cl_2, rt, PCC, MS 3 Å, 2 h; then EtOH, $NaBH_4$, 0°C, 1 h, 78%. (*b*) 50% aqueous TFA, rt, 15 min; then dioxane–H_2O (3:1), $BaCO_3$, 0°C, Br_2, rt, 24 h; then $Na_2S_2O_3$, 93%. (*c*) AcOH–H_2O (2:1), H_2, Pd black, 24 h, ion-exchange chromatography, Dowex 50X8-100 (H^+) resin, **1** (72%), **14** (100%).

3.2.2.2 *Synthesis from* D-*glucosamine* Synthesis of *cis*-4-hydroxypipecolic acid (**3**) from D-glucosamine has been reported by oxidizing it with HgO to give 2-amino-2-deoxy-D-gluconic acid (D-glucosaminic acid, **15**[14]), which underwent β-elimination with acetic anhydride and sodium acetate to produce the furanone **16** (Scheme 3).[15] Hydrogenation of **16** afforded the respective 3,5-dideoxylactone in 90% yield, whose deprotection in acidic medium led to **17** (87%), which was N-protected with benzyl chloroformate or 2-(*tert*-butoxycarbonyloxyimino)-2-phenylacetonitrile to furnish after O-mesylation the corresponding mesylates **18** or **19**, respectively. Removal of the Boc group from **19** with iodotrimethylsilane in chloroform gave poor yield. However, hydrogenation of the carbobenzyloxy group in **18** followed by cyclization with 2 M aqueous KOH furnished **3** in 55% overall yield from **18**.

3.2.2.3 *Synthesis from* D-*glucuronolactone* Pipecolic acids **1** and **2** were synthesized from the D-glucuronolactone derivative **20** via introducing a nitrogen at C-5 with overall retention of configuration, followed by connecting it with the aldehydic group (Scheme 4).[16,17] Compound **20** was converted to the ido compound **22**[18] in 78% yield. Triflation of **22** followed by displacement with azide ion gave **23**. Hydrogenation and subsequent protection with benzyl chloroformate gave the carbamate **24** in 44% overall yield from **22**. Removal of the isopropylidene group in **24** followed by catalytic hydrogenation afforded **1**.

Scheme 3 (*a*) HgO, Ref. 14. (*b*) Ac_2O, NaOAc, 100°C, 1 min. (*c*) 1. H_2, Pd *on* C (15 psi), 90% for three steps; 2. 5 N HCl, 65°C, 18 h, 87%. (*d*) 1. CbzCl, CH_2Cl_2, NEt_3, 0°C to rt, 21 h, 68%; *or* Boc-ONHCH(Ph)CN, H_2O–dioxane (1:1), NEt_3, rt, 24 h, 54%; 2. MsCl, Py, $CHCl_3$, –10°C, 1 h to rt, 10 h, ~80%. (*e*) 1. H_2, 10% Pd *on* C, 16 h, 98%; *or* $(CH_3)_3SiI$, 1.5 h, 40%; 2. 2 M KOH, rt, 1 h, Dowex 50 (H^+) resin, 55% for two steps.

Scheme 4 (*a*) Ref. 18, 78%. (*b*) 1. Tf_2O, Py, CH_2Cl_2, –20°C; 2. NaN_3, DMF, –10 to –20°C, 1 h. (*c*) Tf_2O, Py, CH_2Cl_2, –20°C. (*d*) 1. H_2, 10% Pd *on* C, EtOAc; 2. CbzCl, $NaHCO_3$, EtOAc, H_2O, 44% for four steps. (*e*) 1. TFA, H_2O, rt; 2. H_2, Pd black, H_2O–AcOH (9:1), 4 days, 60%. (*f*) NaN_3, DMF, –20°C, 5 h, 84%. (*g*) 1. H_2, Pd *on* C, 20°C, 3 h; 2. CbzCl, $NaHCO_3$, EtOAc, H_2O, 0°C, 10 min, 72%. (*h*) 1. CH_3ONa, CH_3OH, 0°C, 1 min; 2. $NaBH_4$, CH_3OH, 0°C, 5 min, 91%. (*i*) 1. MsCl (1.1 equiv.), Py, –20°C, 30 h, 80%; 2. H_2, Pd black, EtOAc, Py, 20°C, 11 h, 100%. (*j*) 0.1 M KOH *in* EtOH–H_2O (1:1), 20°C, 5 min, 82%.

On the other hand, triflation of **20** afforded **21**, which underwent nucleophilic substitution with azide ion to give **25** (84% yield). Reduction of **25** and subsequent protection of the resulting amine afforded the carbamate **26** (72% yield). Treatment of **26** with base and then with sodium borohydride gave the unsaturated diol **27** in 55% overall yield from **20**. Selective mesylation of the primary hydroxyl group in **27** followed by hydrogenolysis afforded the single diastereomeric aminomesylate **28**, which was treated with KOH to give **2** in 37% yield from **20**.

3.2.2.4 *Synthesis from heptono-1,4-lactone* D-Glycero-D-guloheptono-1,4-lactone (**29**) has been used as a chiral template for the syntheses of the pipecolic acid and its analogues (Scheme 5).[19] Thus, treatment of per-*O*-benzoyl-heptonolactone **30**, obtained from **29**,[20]

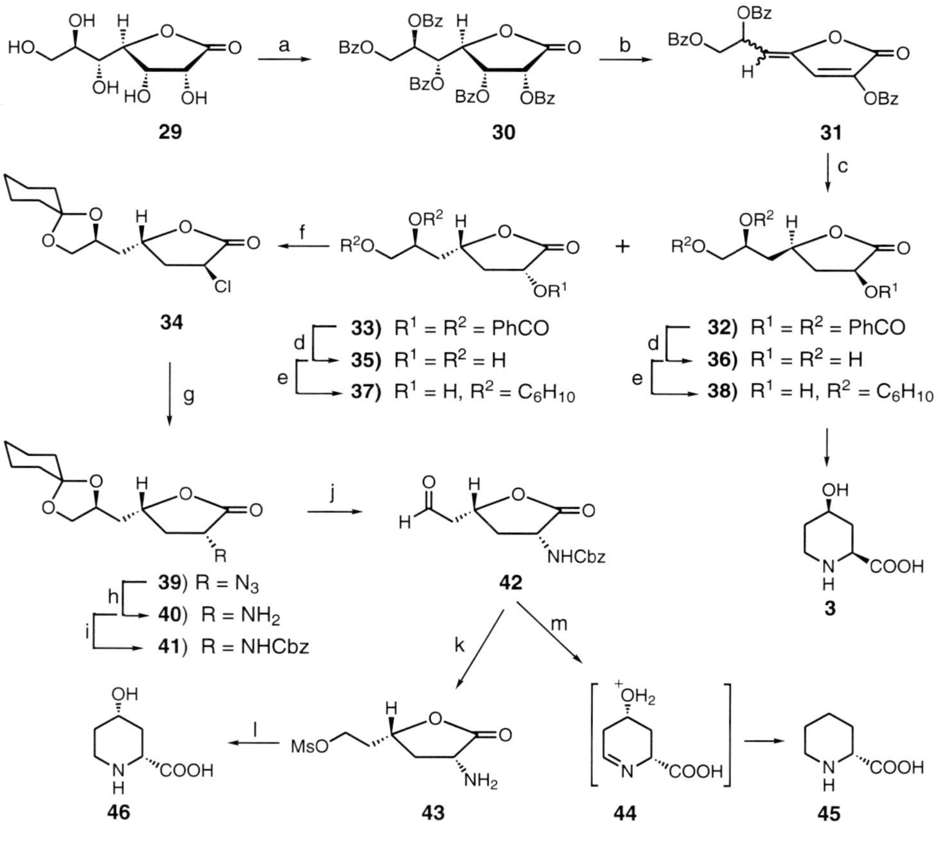

Scheme 5 (*a*) BzCl, Py. (*b*) NEt_3–$CHCl_3$ (1:10), **31***E*:**31***Z* in 1:1 ratio in 90%. (*c*) H_2, Pd *on* C, **33** (40%). (*d*) $NaOCH_3$, CH_3OH. (*e*) Cyclohexanone, $CuSO_4$, *p*-TsOH, rt, 20 h, **37** (32%), **38** (35%). (*f*) *p*-TsCl, Py, 0°C to rt, 22 h; then aqueous HCl, CH_2Cl_2, 82%. (*g*) NaN_3, DMF, rt, 24 h, 89%. (*h*) H_2, 10% Pd *on* C, EtOAc, 85%. (*i*) CbzCl, $NaHCO_3$, EtOAc, 0°C, 70%. (*j*) 1. CH_3OH, 0.5 N aqueous HCl, 94%; 2. $NaIO_4$, CH_3OH, 3 h, 84%. (*k*) 1. $NaBH_3CN$, CH_3OH, rt, 89%; 2. MsCl, Py, rt, 24 h, 85%; 3. H_2, 10% Pd *on* C, EtOH, rt, 2 h, 96%. (*l*) 0.1 M aqueous KOH, rt, 30 min, Dowex 50W (H^+) resin, 81%. (*m*) 10% Pd *on* C, H_2, EtOAc, AcOH, Dowex 50W (H^+) resin.

with triethylamine in chloroform gave a mixture of *E* and *Z* 2-furanone **31** (1:1 ratio) in 90% yield via a double β-elimination process. Hydrogenation of this mixture gave the 3,5-dideoxylactone derivative **32** and **33**, which could not be separated. However, debenzoylation of the mixture to give **35** and **36** followed by treatment with cyclohexanone and $CuSO_4$ in the presence of *p*-toluenesulfonic acid afforded a mixture of **37** (32%) and **38** (35%), which were successfully separated by column chromatography. Treatment of **37** with *p*-toluenesulfonyl chloride in pyridine afforded the 2-chloro derivative **34** in 82% yield, which underwent S_N2 displacement with sodium azide to produce **39**. Catalytic hydrogenation of **39** followed by protection of the resulting amine **40** with benzyl chloroformate afforded the benzyl carbamate **41**, which was decyclohexylidenated followed by periodate oxidation of the resulting diol to produce the aldehyde **42**. Hydrogenolysis of **42** afforded **44** (20%) as a minor product and **45** as the major product. On the other hand, reduction of the aldehyde **42** with $NaBH_3CN$ in methanol followed by mesylation of the resulting primary hydroxyl group and subsequent deprotection of the amine afforded **43**, which was subjected to cyclization using KOH to furnish **46**. Similarly, the lactone **38** was converted into **3**.

References

1. Manning, K.S.; Lynn, D.G.; Shabanowitz, J.; Fellows, L.E.; Singh, M.; Schrire, B.D. *J. Chem. Soc., Chem. Commun.* (1985) 127.
2. Cence di Bello, I.; Dorling, P.; Fellowa, L.E.; Winchester, B. *FEBS Lett.* **176** (1984) 61.
3. Marlier, M.; Dardenne, G.; Casimir, J. *Phytochemistry* **15** (1976) 183.
4. Romeo, J.T.; Swain, L.A.; Bleecker, A.B. *Phytochemistry* **22** (1983) 1615.
5. Schenk, V.W.; Schutte, H.R. *Flora* **153** (1963) 426.
6. Vanderhaeghe, H.; Janssen, G.; Compernolle, F. *Tetrahedron Lett.* **28** (1971) 2687.
7. Hays, S.J.; Malone, T.C.; Johnson, G.J. *J. Org. Chem.* **56** (1991) 4084, and references therein.
8. Beaulieu, P.L.; Lavallée, P.; Abraham, A.; Anderson, P.C.; Boucher, C.; Bousquet, Y.; Duceppe, J.S.; Gillard, J.; Gorys, V.; Grand-Maitre, C.; Grenier, L.; Guindon, Y.; Guse, I.; Plamondom, L.; Soucy, F.; Valois, S.; Wernic, D.; Yoakim, C. *J. Org. Chem.* **62** (1997) 3440.
9. Anderson, P.C.; Soucy, F.; Yoakim, C.; Lavallée, P.; Beaulieu, P.L. U.S. Patent 5 614 533, 1997; *Chem. Abstr.* **126** (1997) 305785v.
10. Bernotas, R.C.; Ganem, B. *Tetrahedron Lett.* **26** (1985) 4981.
11. Fleet, G.W.J.; Smith, P.W.; Nash, R.J.; Fellows, L.E.; Parekh, R.B.; Rademacher, T. *Chem. Lett.* (1986) 1051.
12. Fleet, G.W.J.; Smith, P.W. *Tetrahedron Lett.* **26** (1985) 1469.
13. Fleet, G.W.J.; Fellows, L.E.; Smith, P.W. *Tetrahedron* **43** (1987) 979.
14. Wolform, M.L.; Cron, M.J. *J. Am. Chem. Soc.* **74** (1952) 1715.
15. Nin, A.P.; Varela, O.; De Lederkremer, R.M. *Tetrahedron* **49** (1993) 9459.
16. Bashyal, B.P.; Chow, H.F.; Fleet, G.W.J. *Tetrahedron Lett.* **27** (1986) 3205.
17. Bashyal, B.P.; Chow, H.F.; Fellows, L.E.; Fleet, G.W.J. *Tetrahedron* **43** (1987) 415.
18. Cauk, R.; Honig, H.; Nimpt, J.; Weidmann, H. *Tetrahedron Lett.* **21** (1980) 2135.
19. di Nardo, C.; Varela, O. *J. Org. Chem.* **64** (1999) 6119.
20. di Nardo, C.; Jeroncic, L.O.; de Lederkremer, R.M.; Varela, O. *J. Org. Chem.* **61** (1996) 4007.

3.2.3 *Sesbanimide*

(+)-Sesbanimide A (**1**) and its isomer sesbanimide B (**3**) were isolated from *Sesbania drummodii* seeds[1,2] and from *Sesbania punicea*.[3] The latter is a deciduous shrub found throughout the southern United States and South America. It is an introduced noxious weed in southern Africa, with a history of toxicity to livestock and fowl.[4,5] Sesbanimide A (**1**) was found to be the most active component of the *Sesbania* alkaloids as evaluated in screening in experimental leukemias. It has IC_{50} value of 7.7×10^{-3} μg/mL against KB cells *in vitro* and T/C values of 140–181% in 8–12 μg/kg dose level against P388 murine leukemia *in vivo*. Sesbanimide B (**3**) also shows considerable antitumor activity, but it is less than that of **1**.[6] The high activity of **1** presumably originates from a combination of the structural features of the three rings; a number of AB- and BC-ring systems of sesbanimide have been tested for antitumor activity, but none of these exhibited a cytotoxicity comparable to that of **1**.[7] The structure of sesbanimides, consisting of three rings linked by single bonds, was confirmed by a single-crystal X-ray crystallographic study.[1] However, the total synthesis of (−)-sesbanimide A (**2**), the antipode of the natural (+)-sesbanimide A (**1**), establishes the absolute configuration of **1** as the 7*S*,8*R*,9*S*,10*R*,11*R* compound.

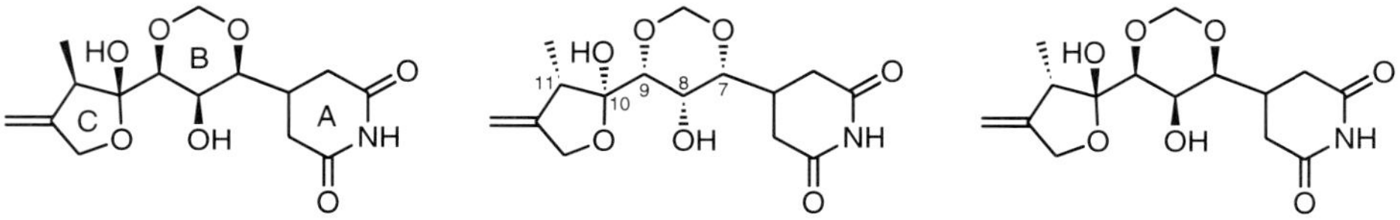

(+)-Sesbanimide A **1** (−)-Sesbanimide A **2** Sesbanimide B **3**

3.2.3.1 *Synthesis from* D*-glucose* Stereospecific syntheses of A, AB and ABC rings of sesbanimides have been achieved from D-glucose by constructing the A ring, followed by the B ring (Scheme 1).[8] Thus, diacetone D-glucose was transformed to the aldehyde **4**,[9] which upon treatment with Meldrum's acid afforded compound **5**, which underwent Michael addition with $LiCH_2CO_2Et$ in THF to produce **6**. This was subjected to decarboxylation, esterification with *p*-nitrophenol in the presence of Cu powder, followed by subsequent treatment of the resulting *p*-nitrophenyl ester with benzylamine in a one pot to give **7**. Hydrolysis of **7** followed by thermal dehydration gave the glutarimide derivative **8**, which underwent dithioacetalization with ethanedithiol in the presence of zinc chloride to afford **9**. Reaction of **9** with paraformaldehyde afforded the 1,3-dioxane **10**, which was subjected to oxidative hydrolysis of dithioacetal group with mercuric perchlorate to form **11**, the AB ring moiety of sesbanimides **1** and **3**, in 35% overall yield from **4**.

Alternatively, the aldehyde **4** was used for constructing the glutarimide ring A of sesbanimide via the α,β-unsubstituted ester **12** (Scheme 2).[10,11] The ester **12** was treated with the potassium salt of dimethyl malonate, and the resulting triester **13** was decarbomethoxylated to give **14**,[12–14] which was also obtained by radical addition of methyl bromoacetate to **12**.[15,16] Hydrolysis[17] of **14** with lithium hydroxide afforded the diacid, which upon heating with urea or by treatment with sodium amide and liquid ammonia gave the glutarimide **15**. This glutarimide has been converted to the AB rings of sesbanimide.[18]

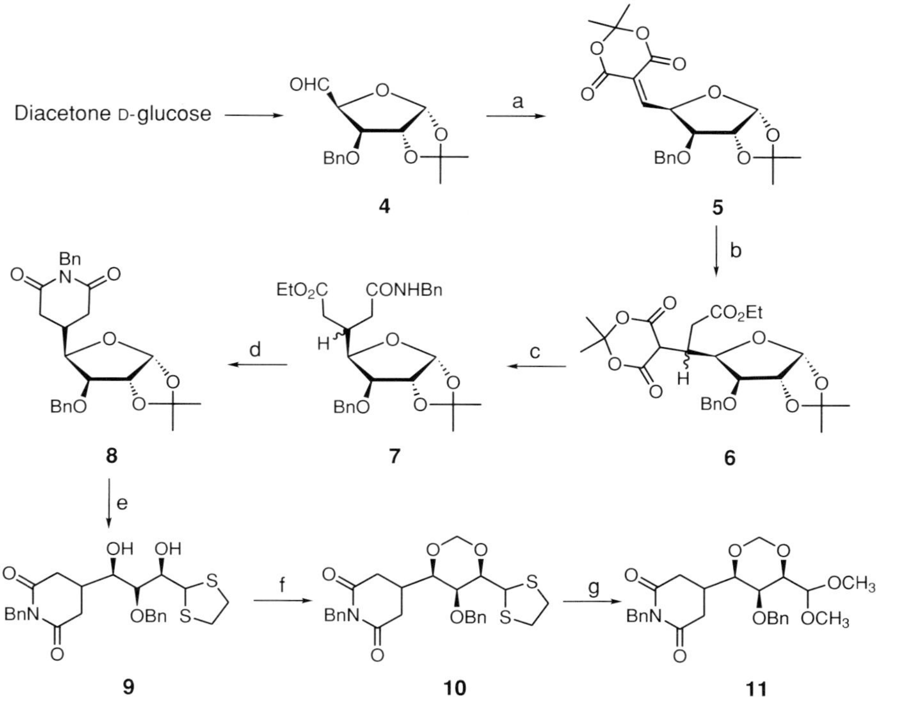

Diacetone D-glucose

Scheme 1 (*a*) Meldrum's acid, MS 4 Å, piperidine, AcOH, CH_2Cl_2, rt, 88%. (*b*) $LiCH_2CO_2Et$ (1.2 equiv.), THF, –78°C, 93%. (*c*) 1. *p*-Nitrophenol, Cu powder, CH_3CN, reflux; 2. $PhCH_2NH_2$, NEt_3, rt, 91%. (*d*) 1. 1 N NaOH, EtOH, 70°C; 2. 210°C, 20 mm Hg, 87%. (*e*) $HSCH_2CH_2SH$, $ZnCl_2$, 0°C, 83%. (*f*) Paraformaldehyde, toluene, *p*-TsOH, 100°C, 85%. (*g*) $Hg(ClO_4)_2 \cdot 3H_2O$, $CH_3OH–CHCl_3$ (1:2), rt, 77%.

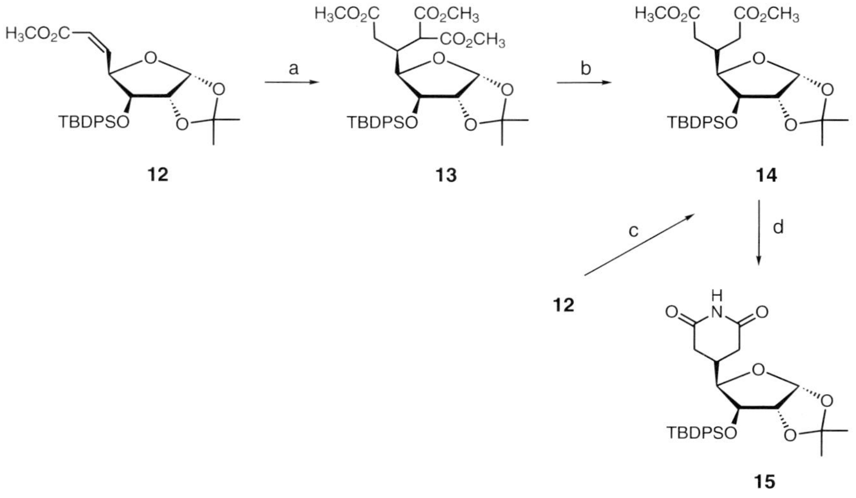

Scheme 2 (*a*) $KCH(CO_2CH_3)_2$. (*b*) Decarbomethoxylated. (*c*) Methyl bromoacetate (20 equiv.), 80°C, AIBN, Bu_3SnH, 15 h, 30–56%, 60% recovery of start material. (*d*) 1. LiOH; 2. urea, 165°C, 80%; *or* $NaNH_2$, liquid NH_3, –33°C, 3 h, 45–87%.

On the other hand, diacetone D-glucose was transformed to (–)- and (+)-sesbanimide A via the open chain derivative 3-*O*-benzyl-D-glucose diethyldithioacetal **17**, which was prepared by hydrolysis of 3-*O*-benzyl-1,2:5,6-di-*O*-isopropylidene-D-glucofuranose (**16**) and subsequent treatment with ethanethiol in the presence of acid (Scheme 3).[18] Acetonation of **17** under kinetic control followed by methylenation with dibromomethane afforded the key intermediate 1,3-dioxane **18**. Deprotection of the dithioacetal unit in **18** followed by

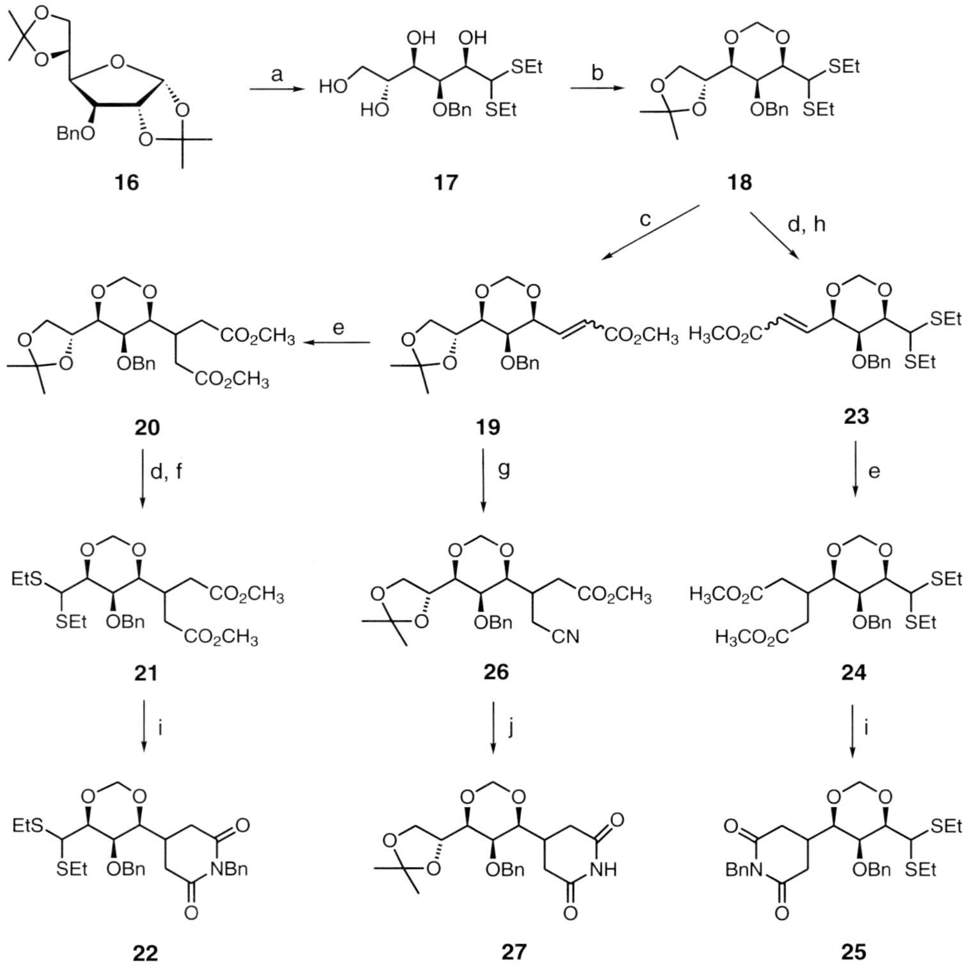

Scheme 3 (*a*) Dowex (H^+) resin; then EtSH, conc. HCl, 92%. (*b*) 1. Acetone, anhydrous $CuSO_4$, 75%; 2. CH_2Br_2, NaOH, Bu_4NI, dioxane, H_2O, 84%. (*c*) 1. $HgCl_2$, HgO, acetone; 2. $Ph_3P{=}CHCO_2CH_3$, 88%. (*d*) CH_3OH, HCl; then $NaIO_4$, aqueous CH_3OH. (*e*) $CH_2(CO_2CH_3)_2$, $NaOCH_3$, CH_3OH; then aqueous NaCl, DMSO. (*f*) EtSH, $Et_2O{\cdot}BF_3$, CH_2Cl_2, 73%. (*g*) 1. Li(TMS)CHCN, THF; 2. CsF, CH_3CN, 84%. (*h*) $Ph_3P{=}CHCO_2CH_3$, acetone. (*i*) $PhCH_2NHLi$, THF, –78°C. (*j*) 1. H_2O_2, KOH, aqueous EtOH; 2. NaOEt, THF, 41% for two steps.

treatment with [(methoxycarbonyl)methylene]triphenylphosphorane gave a mixture of the diastereoisomeric esters **19**. Michael addition of dimethyl malonate anion to **19** followed by demethoxycarbonylation yielded the dimethyl glutarate **20**. Removal of the isopropylidene group in **20** followed by periodate oxidation and subsequent treatment with ethanethiol gave **21**, whose reaction with lithium benzylamine gave the benzylglutarimide **22**. Alternatively, methanolysis of the isopropylidene ring of **18** followed by periodate oxidation and then treatment with the [(methoxycarbonyl)methylene]triphenylphosphorane gave the diastereoisomeric α,β-unsaturated esters **23** (72%) (Z:E 4:1). Michael addition of dimethyl malonate to **23** followed by demethoxycarbonylation gave the dimethyl glutarate **24**. Treatment of **24** with lithium benzylamine gave the required benzyl glutarimide **25**. Ring B has also been constructed[19] from **19**, by addition of the lithiated trimethylsilylacetonitrile, to afford **26**. Selective hydrolysis of the cyano group to the amide followed by cyclization with sodium ethoxide afforded **27**.

Toward the total synthesis of (+)-sesbanimide A (**1**), compound **27**[18,19] was used as a precursor (Scheme 4).[20] The isopropylidene group in **27** was hydrolyzed, followed by oxidative cleavage of the resulting glycol and subsequent sodium borohydride reduction to afford the alcohol **28**. Protection of the hydroxyl group gave **29**, which was followed by changing the benzyl group to *tert*-butyldiphenylsilyl group to give **30**. Selective deprotection of **30** with DIBAL-H followed by Collins oxidation of the resulting hydroxyl group in **31** and subsequent coupling with **32** afforded a mixture of **33** and **34**. Collins oxidation of the adduct **33** followed by deprotection of the TBDPS afforded **1**.

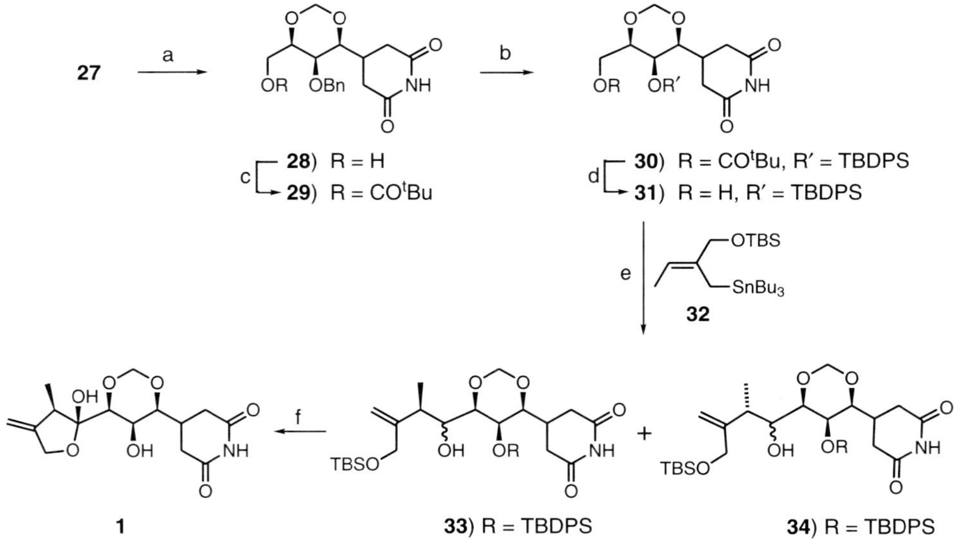

Scheme 4 (*a*) 1. Aqueous HCl, THF, quantitative; 2. $NaIO_4$, aqueous CH_3OH, quantitative; 3. $NaBH_4$, EtOH, 89%. (*b*) 1. H_2, 10% Pd *on* C, EtOH, 92%; 2. TBDPSOTf, 2,6-lutidine, CH_2Cl_2, 97%. (*c*) *t*-BuCOCl, Py, 92%. (*d*) DIBAL-H, CH_2Cl_2, 88%. (*e*) 1. $CrO_3{\cdot}2Py$, CH_2Cl_2, 95%; 2. $BF_3{\cdot}OEt_2$, CH_2Cl_2, **33** (18%), **34** (17%). (*f*) 1. $CrO_3{\cdot}2Py$, CH_2Cl_2; 2. AcOH, aqueous THF, 91% for two steps.

3.2.3.2 *Synthesis from* D*-xylose* A construction of the AB ring has also been achieved from D-xylose (Scheme 5).[21] D-Xylose was reacted with ethanethiol followed by treatment with formaldehyde to afford the dithioacetal **35**, which underwent partial acetolysis of the 3,5-*O*-methylene group to afford the diacetate **36**. Deacetylation followed by benzylation of the resulting diol furnished **37**. Mercuric perchlorate hydrolysis of **37** followed by Wittig reaction afforded **38**. Michael reaction of **38** with diethyl malonate and subsequent removal of the ethoxycarbonyl group afforded the diester **39**. This was treated with benzylamine to afford the benzylglutarimide derivative **40** in 22% overall yield from D-xylose.

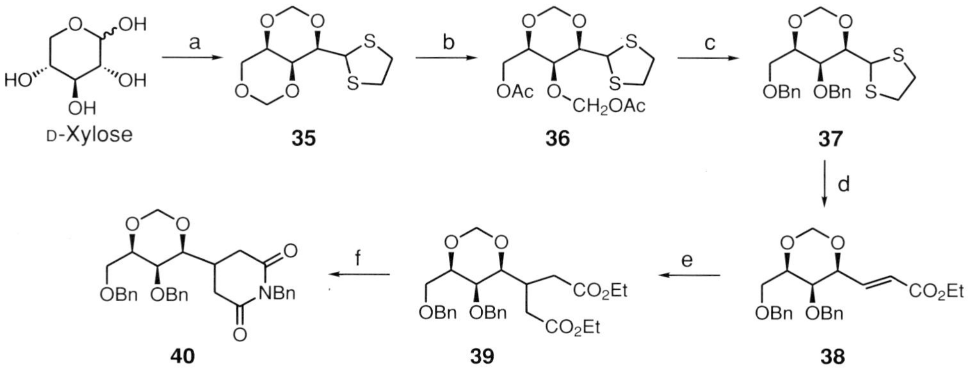

Scheme 5 (*a*) $HSCH_2CH_2SH$, HCl, aqueous HCHO, 70%. (*b*) Ac_2O, AcOH, H_2SO_4, 76%. (*c*) 1. $NaOCH_3$, CH_3OH, $CHCl_3$; 2. BnBr, NaH, 95%. (*d*) 1. $Hg(ClO_4)_2{\cdot}3H_2O$, $CHCl_3$, THF, 87%; $Ph_3P{=}CHCO_2Et$, 82%. (*e*) 1. $CH_2(CO_2Et)_2$, $NaOCH_3$; 2. DMSO, NaCl, H_2O, 70%. (*f*) $C_6H_5CH_2NH_2$, DMF, 170°C, 3 days in sealed tube, 87%.

Alternatively, the readily accessible 1,2-*O*-isopropylidene-α-D-xylofuranose (**41**) was used for the total syntheses of the natural (+)-sesbanimide A (**1**) and the unnatural (–)-sesbanimide B (**3**) (Scheme 6).[22–24] Benzylation of **41** afforded **42** whose acetonide group was removed by treating with conc. HCl, followed by Wittig reaction with [(methoxycarbonyl)methylene]triphenylphosphorane. Subsequent reaction with methylsilyl trifluoromethanesulfonate in dimethoxymethane afforded **43**, the B ring of **1** and **3**. Michael addition to the C-4 position for constructing the carbon framework of the A-ring system was done by addition of the sodium salt of dimethyl malonate to give **44**, which was followed by demethoxycarbonylation of the resulting adduct to give the corresponding diester. Hydrolysis of the two ester groups followed by ammonolysis afforded the amide **44** as a mixture of two diastereomers. Dehydration of **44** with acetic anhydride smoothly produced the corresponding glutarimide **45**, which underwent catalytic hydrogenation to effect removal of the benzyl groups, followed by conversion into **46** as shown before, and then subjected to regioselective Reformatsky reaction employing (*E*)-ethyl 2-(bromomethyl)crotonate to give the *exo*-methylene-γ-lactone **47**. Treatment of **47** with diisobutylaluminum hydride yielded the hemiacetal, which without isolation was further reduced with sodium borohydride in the presence of cerium(III) chloride to afford **48** in 73% yield. Selective protection of the primary hydroxyl group of **48**, as a *tert*-butyldiphenylsilyl ether **49**, followed by

Collins oxidation of the remaining secondary hydroxyl group and subsequent removal of the two silyl groups afforded a 1:1 mixture of **1** and **3**, which could be separated.

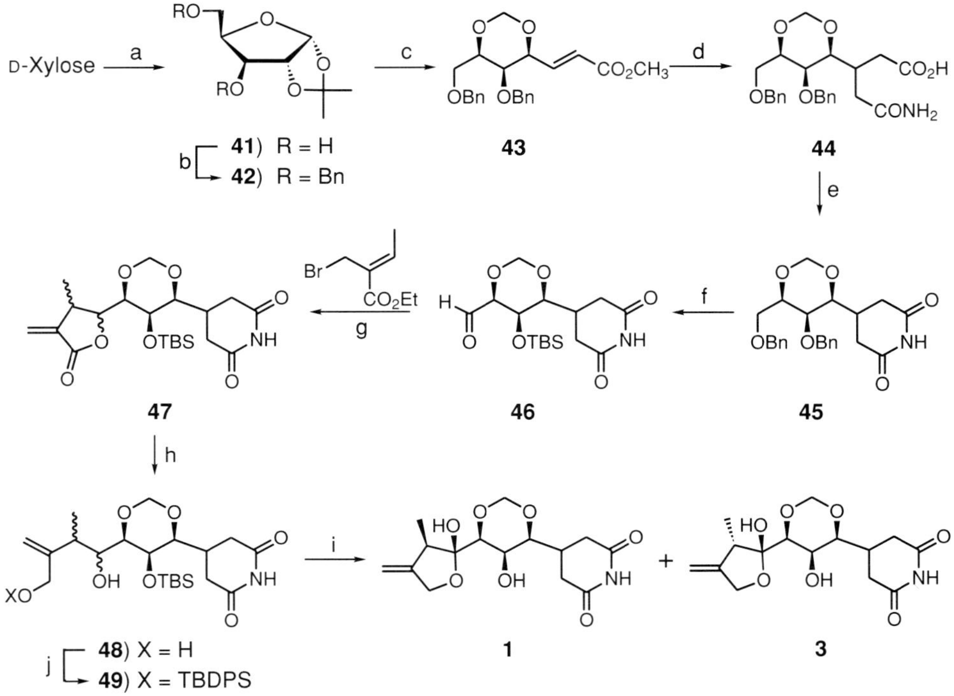

Scheme 6 (*a*) 1. 18 M H_2SO_4, $CuSO_4$, acetone, rt, 25 h, 74%; 2. 0.12 M HCl, rt, 1 h, 96%. (*b*) NaH, THF, reflux, 15 min; then BnCl, *n*-Bu_4NBr, reflux, 5 min, 92%. (*c*) 1. 12 M HCl, AcOH, rt, 5 min, 73%; 2. $Ph_3P{=}CHCO_2CH_3$, toluene, reflux, 30 s, 92%; 3. TMSOTf, 2,6-lutidine, $(CH_3O)_2CH_2$, 0°C, 15 min, 79%. (*d*) 1. $NaCH{=}C(CO_2CH_3)_2$, *n*-Bu_4NBr, rt, 12 h; 2. NaCl, H_2O–DMSO, 160°C, 1 h, 89%; 3. 1 M KOH, rt, 48 h; 4. CH_3OCOCl, NEt_3, THF, –20°C, 3 h; 5. NH_3 gas, 0°C, 30 min. (*e*) 1. NaOAc, Ac_2O, 100°C, 20 min, 51%; 2. H_2, Pd *on* C, AcOH, CH_3OH, rt, 2 h, 95%. (*f*) 1. *t*-BuCOCl, Py, 0°C, 2.5 h, 91%; 2. *t*-$Bu(CH_3)_2SiOTf$, 2,6-lutidine, CH_2Cl_2, rt, 10 min, 86%; 3. *i*-Bu_2AlH, CH_2Cl_2, –78°C, 1 h, 87%; 4. $CrO_3{\cdot}2Py$, CH_2Cl_2, rt, 10 min, 84%. (*g*) Zn, THF, reflux, 6 min, 73%. (*h*) 1. *i*-Bu_2AlH, CH_2Cl_2, –78°C, 1 h; 2. $NaBH_4$, $CeCl_3{\cdot}7H_2O$, CH_3OH, 0°C, 10 min, 73%. (*i*) TBDPSCl, imidazole, DMF, rt, 40 min. (*j*) $CrO_3{\cdot}2Py$, CH_2Cl_2, rt, 30 min, TBAF, THF, rt, 10 min, 16–19% .

3.2.3.3 *Synthesis from* D*-threose* The synthesis of ring A in sesbanimide A (**1**) was constructed by treatment of **51**,[25] obtained from D-threose derivative **50**, with lithium trimethylsilylacetonitrile[26,27] to afford exclusively **52** in quantitative yield (Scheme 7).[28] The latter underwent desilylation with CsF to produce the ester **53** in 94% yield. Hydrolysis of the cyanide group in **53** followed by imide formation afforded **54** (47%).

3.2.3.4 *Synthesis from* D*-mannitol* The AB ring of sesbanimide was also synthesized from cyclohexylidene D-glyceraldehyde **57**,[29] readily available from D-mannitol (Scheme 8).[30] The addition of (*Z*)-γ-alkoxyallylboronate **56**, prepared from **55**, to **57**

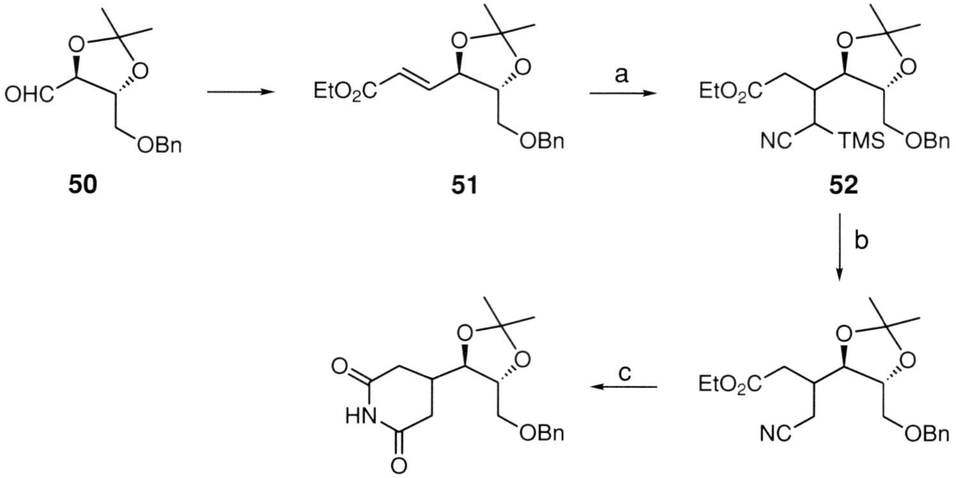

Scheme 7 (*a*) LiTMSCHCN, THF, –78°C, 1 h; then aqueous NH_4Cl, quantitative. (*b*) CsF, 10% aqueous $NaHCO_3$, 94%. (*c*) H_2O_2, NaOH, aqueous EtOH, 50°C; then *t*-BuOK at 200°C; 47%.

afforded homoallyl alcohol **58**. Epoxidation of **58** using the $VO(acac)_2$–TBHP provided **59** as the sole product. The epoxide **59** was treated with PhSNa in THF, followed by methylenation to provide **60**. Hydrolysis of **60** with 2% aqueous TFA followed by periodate cleavage and subsequent Wittig olefination afforded the unsaturated ester **61**. The glutarimide ring system was formed by the Michael addition of *tert*-butyl cyanoacetate to the α,β-unsaturated ester **61** to afford **62** as a mixture of diastereomers. Decarboxylation and deprotection of the MOM ether afforded a mixture (~1:1) of cyanoester **63** and cyanolactone **64**. Ammonolysis of **64** provided the amide **65**. This was transformed into the succinimide **66** (17–21% overall yield from **57**).

3.2.3.5 *Synthesis from* D*-sorbitol* Two approaches for the synthesis of (–)-sesbanimide A (**2**) from D-sorbitol have been achieved (Scheme 9).[31] The aldehyde **67**,[32,33] obtained from D-(–)-sorbitol, was reacted with the sodium salt of diisopropyl(carboethoxy)methyl phosphonate to afford the olefin **68**, which was treated with magnesium monoethyl malonate to afford the diester **69**. Treatment of **69** with ammonium hydroxide followed by pyrolysis gave the imide **70**. Selective hydrolysis of the terminal acetal using a mixture of trifluoroacetic anhydride and acetic acid led to the primary acetate **71**, which was treated with *tert*-butyldiphenylsilyl triflate to afford the silyl **72**. Deprotection of the acetyl residue of **72** with diisobutylaluminum hydride gave the alcohol **73**, which underwent Swern oxidation[34] to give the aldehyde **74**.

Treatment of **74** with **75** in the presence of borontrifluoride etherate followed by hydrolysis afforded **76** and **77**. Swern oxidation of **77** gave **78**, which on treatment with acetic acid afforded (–)-sesbanimide A (**2**) in 17% overall yield from the aldehyde **67**. The analogue **80** was similarly prepared from **76**, via compound **79**.

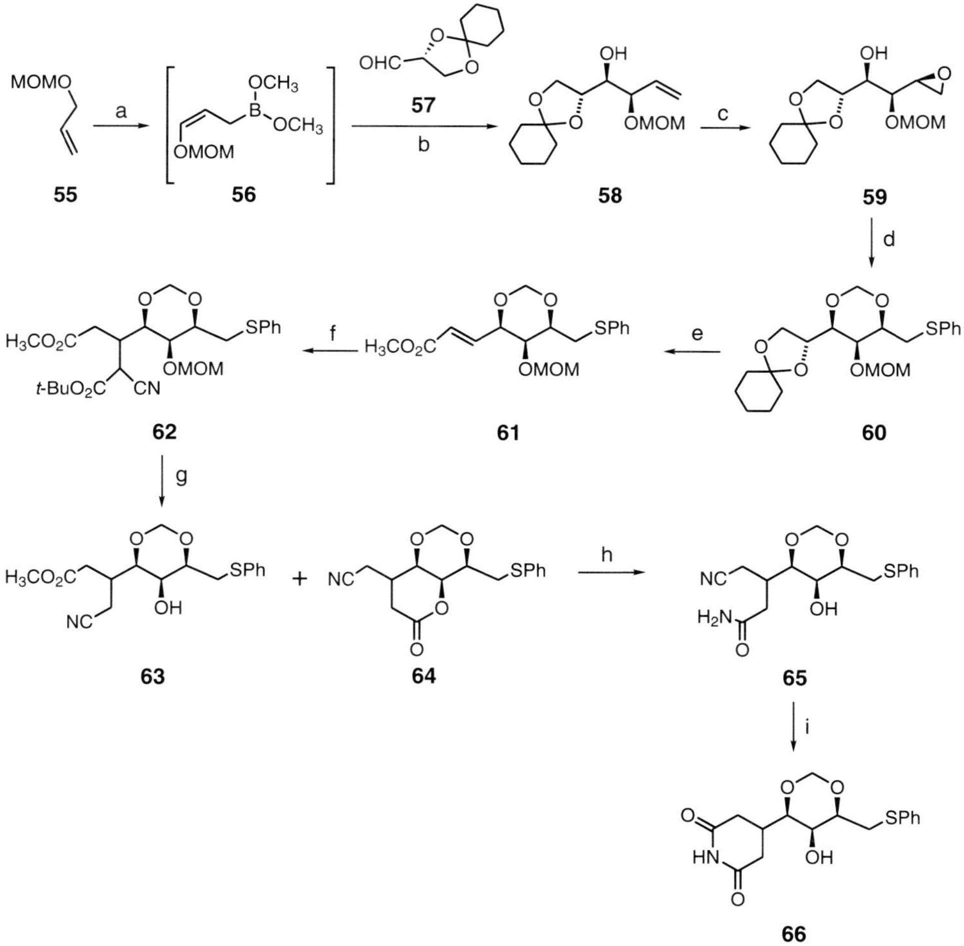

Scheme 8 (*a*) 1. *n*-BuLi, THF, –50°C; 2. $FB(OCH_3)_2$, –78°C. (*b*) 23°C, 75–80%. (*c*) $VO(acac)_2$, TBHP, CH_2Cl_2. (*d*) 1. PhSNa, THF; 2. CH_2Br_2, NaOH, Bu_4NI, dioxane. (*e*) 1. TFA–H_2O (1:50); 2. $NaIO_4$; 3. $Ph_3P{=}CHCO_2CH_3$, 66–70%. (*f*) *t*-BuO_2CCH_2CN, *t*-BuOK, THF, 23°C, 93%. (*g*) DMSO, NaCl, H_2O, 165°C. (*h*) NH_3–CH_3OH, 93%. (*i*) 1. NaH, *i*-PrOH; 2. HCO_2H, H_2O, 85%.

The second approach has also utilized the aldehyde **67** (Scheme 10).[35–37] Wittig olefination of **67** with $Ph_3P{=}CHCO_2CH_3$ afforded 1:9 ratio of *Z*/*E* isomers **81** in 69% yield. The glutarimide moiety was constructed from the *E*-isomer of **81** by reaction with *tert*-butyl carbamoyl acetate to produce isomeric mixture of glutarimides **82**. Hydrolysis of **82** followed by decarboxylation furnished **83** in 45% overall yield from **67**. Acetolysis of **83** led to opening of one of the dioxolane rings, giving the diacetate **84**. Compound **84** represents the synthon containing the AB rings with the correct absolute configuration of the sugar moiety and possessing a functionalization at C-9 for further elaboration to the sesbanimide molecule.

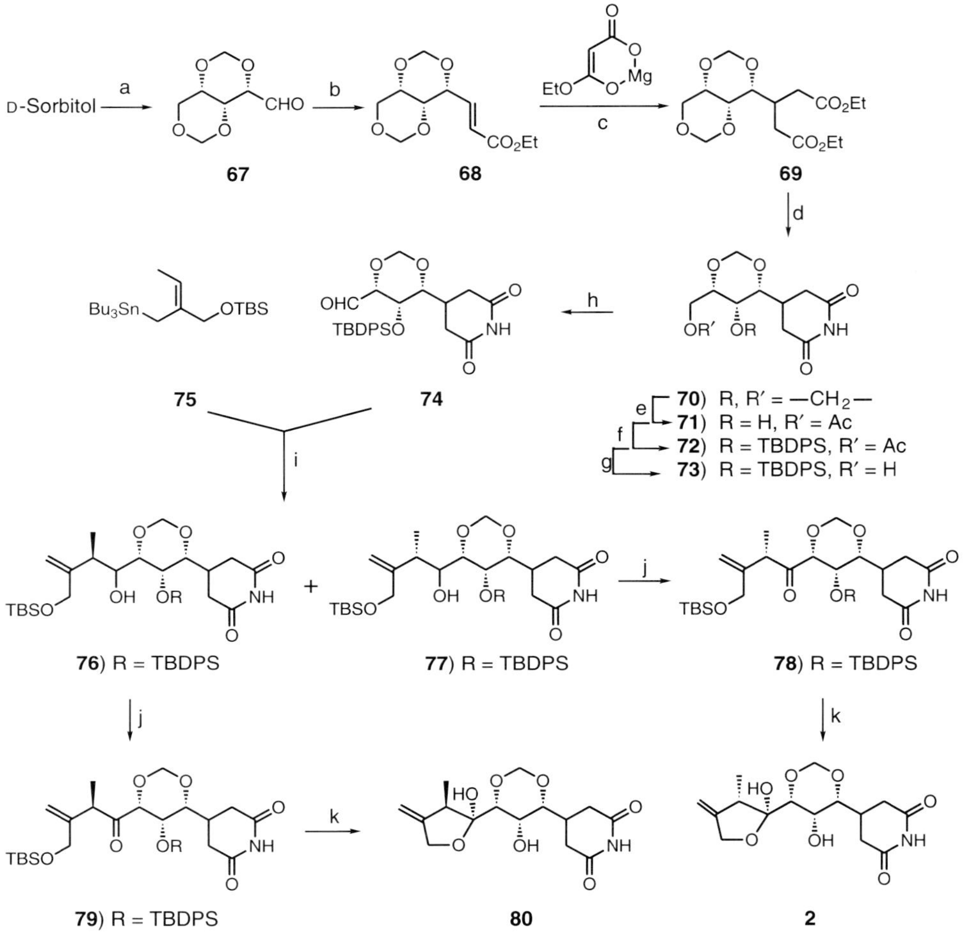

Scheme 9 (*a*) 1. HCHO, HCl; 2. HIO_4. (*b*) Sodium salt of diisopropyl(carboethoxy)methyl phosphonate, ether, 87%. (*c*) Ether, 70%. (*d*) NH_4OH, 155–210°C, 68%. (*e*) TFAA, AcOH, 22°C, 6 h, 87%. (*f*) TBDPSOTf, CH_2Cl_2, 2,6-lutidine. (*g*) DIBAL-H, –78 to 0°C, 2 h, 97%. (*h*) Swern oxidation, 97%. (*i*) $BF_3 \cdot OEt_2$; then aqueous $NaHCO_3$, –78°C. (*j*) Same as (*h*), 89%. (*k*) HOAc, THF, H_2O, 22°C, 4.3 h, 100%.

The diacetate **84** underwent deacetylation followed by acetal formation to afford **85**. Selective opening of the cyclic acetal ring followed by Collins oxidation afforded the aldehyde intermediate **86**. Treatment of **86** with allylsilane derivative **87** afforded a mixture of diastereoisomeric alcohols **88** in a 1.7:1 ratio. The major product **88** was oxidized to the corresponding ketone **89**, which upon removal of the protecting groups led to the formation of (–)-sesbanimide A (**2**).

Synthesis of the natural (+)-sesbanimide A (**1**) from D-xylose, employing a similar method to that used above for **2**, has been achieved[22,35–37] starting with **43**.

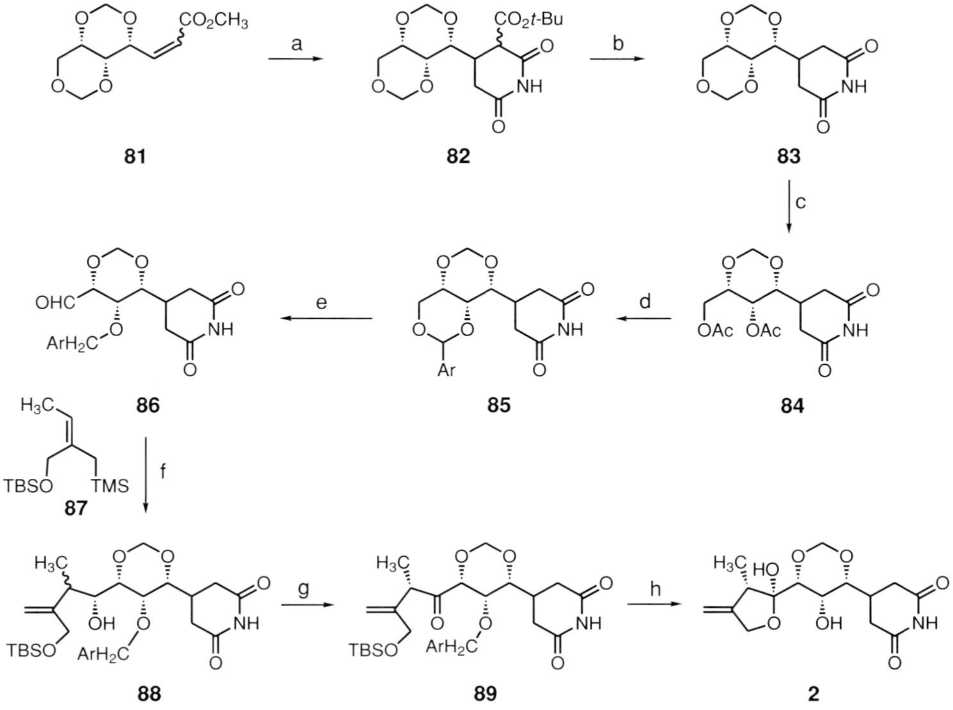

Scheme 10 (*a*) $H_2NCOCH_2CO_2t$-Bu, *t*-BuOK. (*b*) TFA, 1.5 h at rt; then DMF, reflux, 78%. (*c*) Ac_2O, AcOH, H_2SO_4, 2 h, 0°C, 91.5%. (*d*) 1. $NaOCH_3$, CH_3OH, –15°C, 86%; 2. *m*,*p*-$(CH_3O)_2C_6H_3CHO$, *p*-TsOH, 5 h, reflux, 98.7%. (*e*) 1. Et_2AlCl, CH_2Cl_2, Et_3SiH, –60°C, 90 min; then 2 h at 0°C; 2. CrO_3·Py, CH_2Cl_2–DMF (4:1), rt, 65%. (*f*) BF_3·OEt_2, CH_2Cl_2, $CH_3CH{=}C[Si(CH_3)_3]OSi(CH_3)_2t$-Bu, –78°C, 4 h, 50%. (*g*) CrO_3·Py, CH_2Cl_2, DMF, 15 min, rt; then Ac_2O, rt, 8 min; 65%. (*h*) DDQ, CH_2Cl_2, H_2O, rt, 2 h, 69%; THF–AcOH–H_2O (1:1:1), rt, 86%.

References

1. Powell, R.G.; Smith, C.R.; Weisleder, D., Jr.; Matsumoto, G.K.; Clardy, J.; Kozlowski, J. *J. Am. Chem. Soc.* **105** (1983) 3739.
2. Powell, R.G.; Smith, C.R.; Weisleder, D., Jr. *Phytochemistry* **23** (1984) 2789.
3. Gorst-Allman, C.P.; Steyn, P.S.; Vleggaar, R. *J. Chem. Soc., Perkin Trans. 1* (1984) 1311.
4. Powell, R.G.; Smith, C.R.; Madrigal, R.V. *Planta Med.* **30** (1976) 1.
5. Terblanche, M.; Deklerk, W.A.; Smit, J.D.; Adelaar, T.F. *J. S. Afr. Vet. Med. Assoc.* **37** (1966) 191.
6. Powell, R.G.; Smith, C.R., Jr. U.S. Patent 4,534,327, 1985.
7. Matsuda, F.; Ohsaki, M.; Yamada, Y.; Tereshima, S. *Bull. Chem. Soc. Jpn.* **61** (1988) 2123.
8. Shibuya, M. *Heterocycles* **23** (1985) 61.
9. Wolfrom, M.L.; Hanessian, S. *J. Org. Chem.* **27** (1962) 1800.
10. Sacripante, G.; Tan, C.; Just, G. *Tetrahedron Lett.* **26** (1985) 5643.
11. Kinoshita, T.; Okamoto, K. *Synthesis* (1985) 402.
12. Krapcho, A.P.; Lovey, A.J. *Tetrahedron Lett.* **14** (1973) 957.
13. Krapcho, A.P. *Synthesis* (1982) 805.
14. Krapcho, A.P. *Synthesis* (1982) 893.

15. Maercker, A. *Org. React.* **14** (1965) 270.
16. Bergmann, E.D.; Ginsburg, D. Pappo, R. *Org. React.* **10** (1959) 179.
17. Corey, E.J.; Szekely, I.; Shiner, S. *Tetrahedron Lett.* **18** (1977) 3529.
18. Fleet, G.W.J.; Shing, T.K.M. *J. Chem. Soc., Chem. Commun.* (1984) 835.
19. Valverde, S.; Garcia-Ochoa, S.; Martin-Lomas, M. *Carbohydr. Res.* **147** (1986) C1.
20. Tomioka, K.; Hagiwara, A.; Koga, K. *Tetrahedron Lett.* **29** (1988) 3095.
21. Rama Roa, A.V.; Yadav, J.S.; Naik, A.M.; Chaudhary, A.G. *Indian J. Chem., Sect. B* **25** (1986) 579.
22. Matsuda, F.; Kawasaki, M.; Terashima, S. *Tetrahedron Lett.* **26** (1985) 4639.
23. Matsuda, F.; Terashima, S. *Tetrahedron Lett.* **27** (1986) 3407.
24. Matsuda, F.; Terashima, S. *Tetrahedron* **44** (1988) 4721.
25. Mukaiyama, T.; Suzuki, K.; Yamada, T. *Chem. Lett.* (1982) 929.
26. Matsuda, I.; Murata, S.; Ishii, Y. *J. Chem. Soc., Perkin Trans. 1* (1979) 26.
27. Haruta, R.; Ishiguro, M.; Furuta, K.; Mori, A.; Ikeda, N.; Yamamoto, H. *Chem. Lett.* (1982) 1093.
28. Tomioka, K.; Koga, K. *Tetrahedron Lett.* **25** (1984) 1599.
29. Sugiyama, T.; Sugawara, H.; Watanabe, M.; Yamashita, K. *Agric. Biol. Chem.* **48** (1984) 1841.
30. Roush, W.R.; Michaelides, M.R. *Tetrahedron Lett.* **27** (1986) 3353.
31. Schlessinger, R.H.; Wood, J.L. *J. Org. Chem.* **51** (1986) 2621.
32. Ness, A.T.; Hann, R.M.; Hudson, C.S. *J. Am. Chem. Soc.* **66** (1944) 665.
33. Bourne, E.J.; Wiggins, L.F. *J. Chem. Soc.* (1944) 517.
34. Mancuso, A.J.; Huang, S.-L.; Swern, D. *J. Org. Chem.* **43** (1978) 2480.
35. Wanner, M.J.; Koomen, G.-J.; Pandit, U.K. *Heterocycles* **22** (1984) 1483.
36. Wanner, M.J.; Willard, N.P.; Koomen, G.-J.; Pandit, U.K. *J. Chem. Soc., Chem. Commun.* (1986) 396.
37. Wanner, M.J.; Willard, N.P.; Koomen, G.-J.; Pandit, U.K. *Tetrahedron* **43** (1987) 2549.

3.2.4 *Siastatin*

Siastatin A and B were isolated by Umezawa *et al.*[1] in 1974 from a *Streptomyces* culture (*Streptomyces verticillus* var. quintum MB695-A4). Siastatin B was found to inhibit neuraminidases.[2] It is involved in various biological functions such as immune response,[3,4] oncogenesis,[5,6] metastasis of tumors,[7–10] sperm penetration[11] and viral infection.[12–15] Siastatin A is more effective than siastatin B in the inhibition of sialidases prepared from *Cl. perfringens* and chicken chorioallantoic membrane. However, siastatin B is a stronger inhibitor of sialidases prepared from *Streptomyces* and rat organs than is siastatin A. The relative configuration of siastatin B was determined as 2(*S*/*R*)-acetamido-3(*S*/*R*)-4(*R*/*S*)-dihydroxypiperidine-5(*R*/*S*)-carboxylic acid by ^{1}H NMR and X-ray crystallographic studies. The structure of siastatin A is still unknown.[16] Although many efforts have been made to develop appropriate synthetic methods for siastatin B analogues **2–12**,[17–31] only few of them utilized carbohydrates as chiral precursors.[30,31]

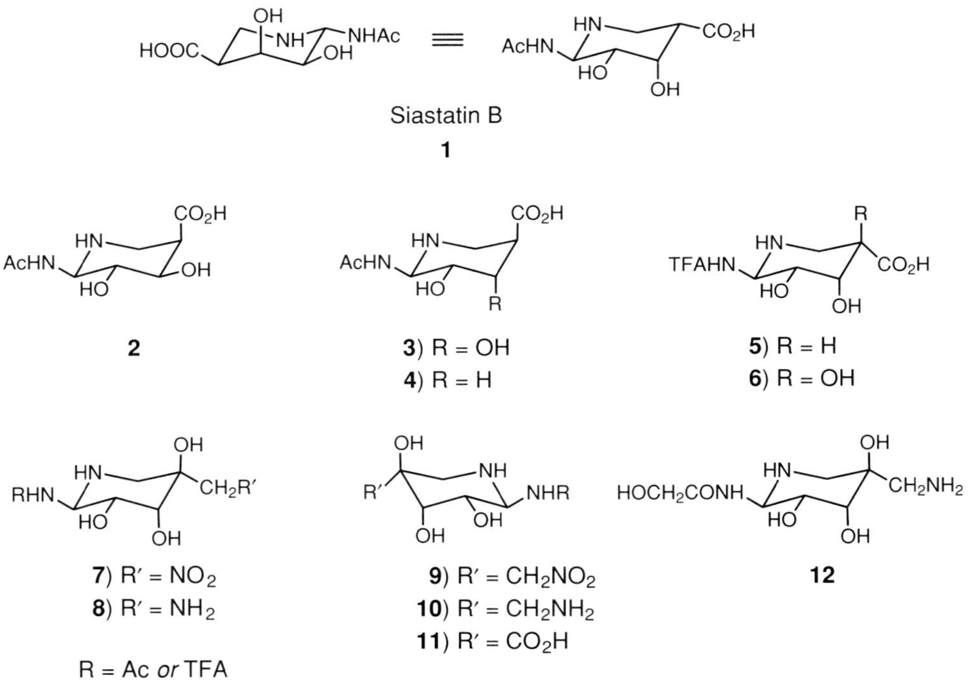

Siastatin B
1

2

3) R = OH
4) R = H

5) R = H
6) R = OH

7) R′ = NO_2
8) R′ = NH_2

R = Ac *or* TFA

9) R′ = CH_2NO_2
10) R′ = CH_2NH_2
11) R′ = CO_2H

12

The first total synthesis of siastatin B (**1**) was achieved from L-ribose by protection of the 2,3-diol, followed by introduction of an azide group on C-5 and oxidation of the anomeric hydroxyl group to give 5-azido-5-deoxy-2,3-*O*-isopropylidene-L-ribonolactone (**13**) (Scheme 1).[30,31] Hydrogenation of **13** in the presence of Raney nickel afforded the corresponding amine, which underwent ring expansion followed by complete protection to furnish lactam **14**. Hydride reduction of **14** gave **15**, which was followed by Swern oxidation to give **16**. Mitsunobu reaction converted the axial hydroxyl group in **16** to the equatorial phthalimido group in **17**. Removal of the *tert*-butyldimethylsilyl group at C-5 followed by oxidation and then condensation with nitromethane gave **18** quantitatively as a single stereoisomer. Acetylation of **18** was followed by elimination of the acetoxy group and then transformation to the carboxylate **19**, which was converted to the α,β-saturated hydroxylmethyl derivative **20** using sodium borohydride. The carboxylic acid formed from oxidation of **20** was converted, upon removal of the protecting groups, to **1**. The enantiomer of **1** was also synthesized from D-ribose by the same method.

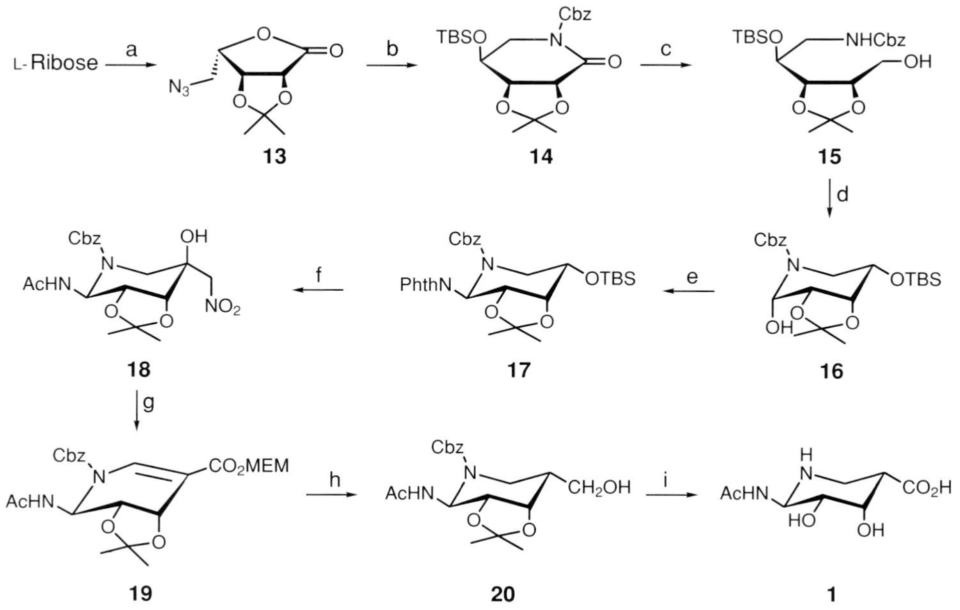

Scheme 1 (*a*) 1. *p*-TsOH, acetone; 2. MsCl, Py; 3. NaN_3, DMSO; 4. $CrO_3 \cdot Py$, CH_2Cl_2, 89% for four steps. (*b*) 1. H_2, Raney nickel, CH_3OH, 88%; 2. TBSCl, imidazole, DMF; 3. CbzCl, NaH, DMF, 99% for two steps. (*c*) $NaBH_4$, EtOH, 70%. (*d*) $(COCl)_2$, DMSO, NEt_3, CH_2Cl_2, 88%. (*e*) Phthalimide, Ph_3P, DEAD, DMF, 100%. (*f*) 1. NH_2NH_2, CH_3OH; 2. Ac_2O, Py; 3. TBAF, THF, 100% for three steps; 4. RuO_4, CH_2Cl_2, CCl_4, 99%; 5. CH_3NO_2, NaH, DMF, 100%. (*g*) 1. *p*-TsOH, Ac_2O, K_2CO_3, benzene, 100%; 2. Py, 88°C, 80%; 3. $CH_3CH{=}C(CH_3)_2$, *t*-BuOH, $NaOCl_2$, NaH_2PO_4, H_2O; 4. MEMCl, $(i\text{-Pr})_2NEt$, CH_2Cl_2, 55%. (*h*) $NaBH_4$, CF_3CH_2OH–THF (1:10), 75%. (*i*) 1. PDC, DMF; 2. H_2, 5% Pd *on* C, CH_3OH; 3. 1 M aqueous HCl; then Dowex 50WX4 (H^+) resin, eluted with NH_4OH, 66%.

References

1. Umezawa, H.; Aoyagi, T.; Komiyama, T.; Morishima, H.; Hamada, M.; Takeuchi, T. *J. Antibiot.* **27** (1974) 963.
2. Roseman, S. *Chem. Phys. Lipids* **5** (1970) 270.
3. Simons, R.L.; Rios, A. *J. Immunol.* **111** (1973) 1820.
4. Rott, R.; Becht, H.; Orlich, M. *J. Gen. Virol.* **22** (1974) 35.
5. Bekesi, J.G.; St-Arneault, G.; Holland, J.F. *Cancer Res.* **31** (1971) 2130.
6. Bernacki, R.J. *Science (Washington, D.C.)* **195** (1977) 577.
7. Gasic, G.; Gasic, T.; *Proc. Nat. Acd. Sci. USA* **48** (1962) 1172.

8. Yogeeswaran, G.; Salk, P.L. *Science (Washington, D.C.)* **212** (1981) 1514.
9. Dennis, J.; Waller, C.; Timpl, R.; Schirrmacher, V. *Nature (London)* **300** (1982) 274.
10. Fogel, M.; Aptevogt, P.; Schirrmacher, V. *J. Exp. Med.* **157** (1983) 371.
11. Souport, P.S.; Clewe, T.H. *Fert. Steril.* **16** (1965) 677.
12. Ackemann, W.W.; Ishida, N.; Massab, H.F. *Exptl. Med.* 102 (1955) 545.
13. Haff, R.F.; Stewart, R.C. *Immunol. J.* **94** (1965) 842.
14. Lipkind, M.A.; Tsvetkova, I.V. *Arch. Ges. Virusforsch.* **29** (1970) 370.
15. Huang, R.T.C.; Rott, R.; Wahn, K.; Klenk, H.D.; Kohama, T. *Virology* **107** (1980) 313.
16. Nishimura, Y. Institute of Microbial Chemistry, Tokyo, Japan. Personal communication, 2004.
17. Nishimura, Y.; Kudo, T.; Kondo, S.; Takeuchi, T.; Tsuruoka, T.; Fukuyasu, H.; Shibahara, S. *J. Antibiot.* **47** (1994) 101.
18. Satoh, T.; Nishimura, Y.; Kondo, S.; Takeuchi, T. *Carbohydr. Res.* 286 (1996) 173.
19. Clinch, K.; Vasella, A. *Tetrahedron Lett.* **28** (1987) 6428.
20. Kudo, T.; Nishimura, Y.; Kondo, S.; Takeuchi, T. *J. Antibiot.* **45** (1992) 954.
21. Nishimura, Y.; Kudo, T.; Kondo, S.; Takeuchi, T. *J. Antibiot.* **45** (1992) 963.
22. Kudo, T.; Nishimura, Y.; Kondo, S.; Takeuchi, T. *J. Antibiot.* **45** (1992) 1662.
23. Nishimura, Y.; Umezawa, Y.; Kondo, S.; Takeuchi, T.; Mori, K.; Kijima-Suda, I.; Tomita, K.; Sugawara, K.; Nakamura, K. *J. Antibiot.* **46** (1993) 1883.
24. Kudo, T.; Nishimura, Y.; Kondo, S.; Takeuchi, T. *J. Antibiot.* **46** (1993) 300.
25. Nishimura, Y.; Satoh, T.; Kudo, T.; Kondo, S.; Takeuchi, T. *Bioorg. Med. Chem.* **4** (1996) 91.
26. Nishimura, Y.; Kudo, T.; Umezawa, Y.; Kondo, S.; Takeuchi, T. *Nat. Prod. Lett.* *1* (1992) 39.
27. Nishimura, Y.; Kudo, T.; Umezawa, Y.; Kondo, S.; Takeuchi, T. *Nat. Prod. Lett.* *1* (1992) 33.
28. Nishimura, Y.; Satoh, S.; Kondo, S.; Takeuchi, T.; Azetaka, M.; Fukuyasu, H.; Iizuka, Y.; Shibahara, S. *J. Antibiot.* **47** (1994) 840.
29. Nishimura, Y. In *Studies in Natural Products Chemistry*, Vol. 16; Atta-ur-Rahamn, Ed.; Elsevier: Amsterdam, 1995; p. 75.
30. Nishimura, Y.; Wang, W.M.; Kondo, S.; Aoyagi, T.; Umezawa, H. *J. Am. Chem. Soc.* **110** (1988) 7249.
31. Nishimura, Y.; Wang, W.M.; Kudo, T.; Kondo, S. *Bull. Chem. Soc. Jpn.* **65** (1992) 978.

3.2.5 *Meroquinene*

(+)-Meroquinene (**1**) is a key synthetic precursor of a number of medicinally important alkaloids such as quinine (**2**) and cinchonamine (**3**).[1,2] Meroquinene and homomeroquinene, (3*R*)-vinyl-(4*S*)-piperidine propionic acid, are also degradation products of cinchonine.[3,4]

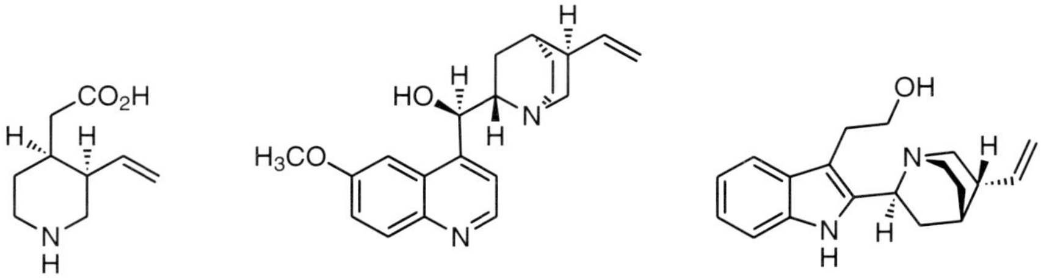

(+)-Meroquinene
1

Quinine
2

Cinchonamine
3

Synthesis of meroquinene (**1**) from D-glucose has been achieved (Scheme 1).[5] Treatment of 2-acetoxy-D-glucal triacetate (**4**),[6,7] obtained from D-glucose, with *tert*-butyl alcohol in the presence of boron trifluoride etherate afforded **5** (78%). Treatment of **5** with potassium carbonate in methanol followed by acetylation afforded **6** (70%). This was treated with bromomagnesium cyano divinylcuprate followed by addition of methyl bromoacetate to produce the *trans*-disubstituted glycoside **7**, which underwent epimerization to the *cis*-isomer **8** by treatment with NEt_3 in DMF. Deoxygenation of the carbonyl group in **8** was achieved via its tosyl hydrazone **9**, which was converted into the corresponding tosyl hydrazine derivative **10**, by treatment with $NaBH_3CN$, and then converted to **11** by reaction with sodium acetate trihydrate at 72°C. Acid hydrolysis of **11** followed by removal of the acetyl group afforded the lactol **12**, which underwent periodate oxidation to produce the dialdehyde **13**. Subsequent reductive Borch-type amination[8,9] furnished the *N*-benzyl meroquinene methyl ester **14** in 30% overall yield from **11**. Finally, N-debenzylation of **14** was accomplished by treatment[10] with ethyl chloroformate to give the *N*-ethoxycarbonyl derivative **15**, which underwent acid hydrolysis to give the hydrochloride of **1**.

Scheme 1 (*a*) *t*-BuOH, toluene, $BF_3 \cdot OEt_2$, 78%. (*b*) $(CH_3O)_2POCH_2Li$, THF, rt, 3 h, 63%, *or* K_2CO_3, CH_3OH; then Ac_2O, Py, DMAP, CH_2Cl_2, 70%. (*c*) $(CH_2{=}CH)_2CuCN(MgBr)_2$, THF, –78°C, 30 min; then $BrCH_2CO_2CH_3$, 25°C, 8 h. (*d*) 1. NEt_3, DMF, 0°C to rt, overnight, **8** (61%); 2. $TsNHNH_2$, EtOH, 72–74°C, 1.5 h, **9** (81%). (*e*) 1. $NaBH_3CN$, THF, CH_3OH, pH 3.8, 97%; 2. $NaOAc \cdot 3H_2O$, EtOH, 72°C, 72%. (*f*) 1. AcOH, 10%, THF, H_2O, 35°C, 24 h, 89%; 2. $NaOCH_3$, CH_3OH, 100%. (*g*) $NaIO_4$, H_2O, acetone, 48 h. (*h*) $BnNH_2HCl$, CH_3CN, $NaCNBH_3$, 24 h, pH 4.3, 34%. (*i*) EtOCOCl, benzene, reflux, 81%. (*j*) 10% aqueous HCl, reflux, 3 h, 93%.

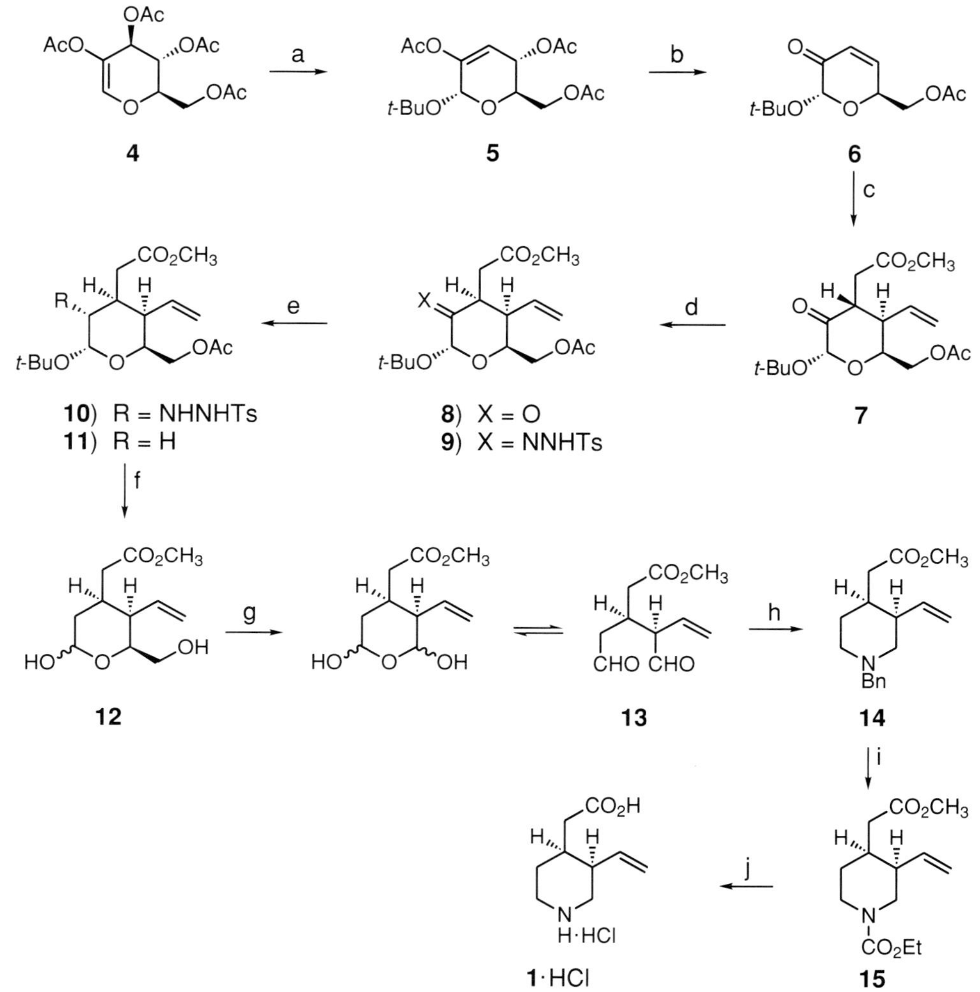

References

1. Dalton, D.R. *The Alkaloids*; Marcel Dekker: New York, 1979; p. 509.
2. Uskokovic, M.R.; Grethe, G. In *The Alkaloids*, Vol. 24; Manske, R.H.F., Ed.; Academic Press: New York, 1973.
3. Prostenils, M.; Prelog, V. *Helv. Chim. Acta* **26** (1943) 1965.
4. Doering, W.E.; Chanley, J.-D. *J. Am. Chem. Soc.* **68** (1946) 586.
5. Hanessian, S.; Faucher, A.-M.; Léger, S. *Tetrahedron* **46** (1990) 231.
6. Roth, W.; Pigman, W. *Methods Carbohydr. Chem. 2* (1963) 405.
7. Rao, D.R.; Lerner, L. *Carbohydr. Res.* **22** (1972) 345.
8. Borch, R.F.; Bernstein, M.; Durst, H.D. *J. Am. Chem. Soc.* **93** (1971) 2897.
9. Borch, R.F. *Org. Synth.* **52** (1972) 124.
10. Ziegler, F.E.; Bennett, G.B. *J. Am. Chem. Soc.* **95** (1973) 7458.

3.2.6 *Pyridyl fragment of pyridomycin*

(2*R*,3*S*,4*S*)-4-Amino-3-hydroxy-2-methyl-5-(3-pyridyl)pentanoic acid (**8**) was isolated as a degradation product from the antimycobacterial antibiotic pyridomycin.[1–3] The absolute configuration of the antibiotic was determined by X-ray crystallographic analysis.[4]

Synthesis of the fragment **8** starting with D-glucose has been reported (Scheme 1).[5] 3-Deoxy-1,2:5,6-di-*O*-isopropylidene-3-*C*-methyl-α-D-allofuranose (**1**),[6] prepared from D-glucose, underwent removal of the terminal isopropylidene group with aqueous acetic acid followed by selective tosylation of the primary hydroxyl group to afford **2**, which upon treatment with sodium methoxide afforded the epoxide **3**. Regiospecific opening of the epoxide ring of **3** with 3-pyridyllithium afforded **4**, which was mesylated and subjected to S_N2 displacement with sodium azide to give **6** in addition to the elimination reaction product **5**. Acid hydrolysis of **6** followed by sodium periodate oxidation and subsequent oxidation with bromine in aqueous acetic acid afforded the pentanoic acid derivative **7**. Removal of

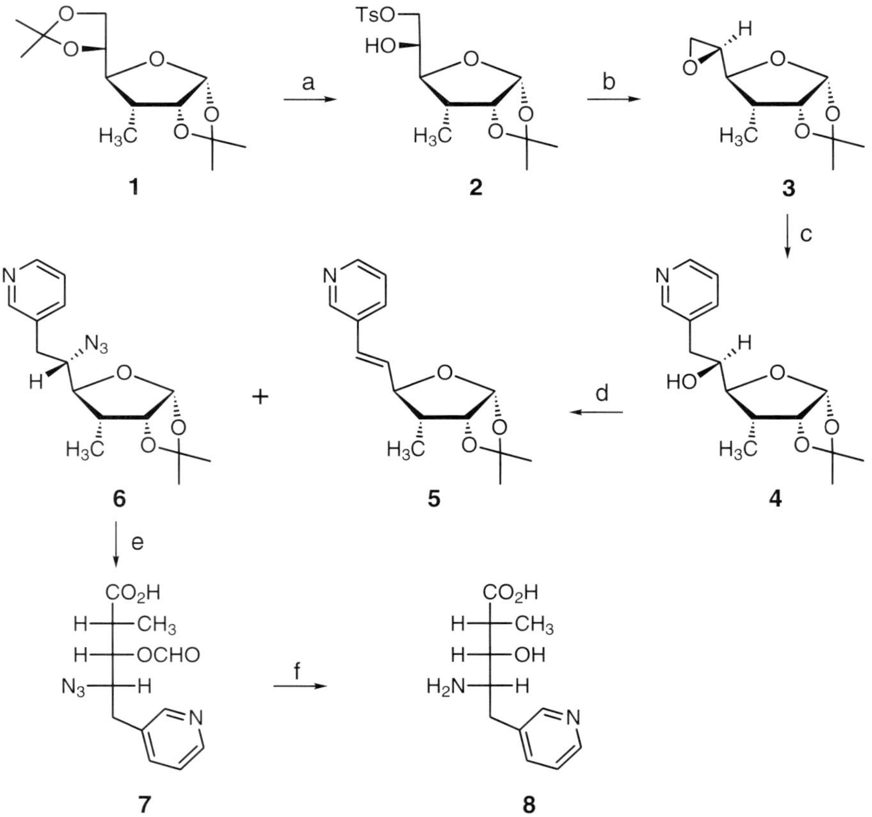

Scheme 1 (*a*) 1. 91% aqueous AcOH, 30 min, 90.5%; 2. *p*-TsCl, Py, 72%. (*b*) $NaOCH_3$, CH_3OH, rt, 2 h, 76%. (*c*) BuLi, –35°C, 3-bromopyridine, –10°C, 1.5 h, ether, 88%. (*d*) 1. MsCl, Py, rt, 75%; 2. NaN_3, DMSO, 85°C, 2 h, **5** (23%), **6** (70%). (*e*) 1. 20% aqueous AcOH, reflux, 5 h; 2. 33% aqueous AcOH, $NaIO_4$, 5°C, 20 min; 3. Br_2, rt, overnight, 59%. (*f*) 1. 2% HCl–dioxane (1:1), rt, 1 h, 81%; 2. H_2, CH_3OH, Pd black, 1 h, 97%.

the *O*-formyl group in **7** with dil. HCl in aqueous dioxane followed by reduction of the azido group afforded **8** in 10.6% overall yield from **1**.

References

1. Maeda, K.; Kosaka, H.; Okami, Y.; Umezawa, H. *J. Antibiot. (Tokyo)* **6A** (1953) 140.
2. Okami, Y.; Maeda, K.; Umezawa, H. *J. Antibiot. (Tokyo)* **7A** (1954) 55.
3. Ogawara, H.; Koyama, G.; Naganawa, H.; Maeda, K.; Umezawa, H. *Chem. Pharm. Bull.* **16** (1968) 679.
4. Koyama, G.; Iitaka, Y.; Maeda, K.; Umezawa, H. *Tetrahedron Lett.* **8** (1967) 3587.
5. Kinoshita, M.; Mariyama, S. *Bull. Chem. Soc. Jpn.* **48** (1975) 2081.
6. Rosenthal, A.; Sprinze, M. *Can. J. Chem.* **47** (1969) 3941.

4 Seven-membered nitrogen heterocycles

Only two types of heterocycles that belong to this group have been obtained from sugars; they are the antihelminthic and anti-infectious bengamides as well as the lipid-containing nucleoside antibiotics liposidomycins. This is why this chapter is the shortest one in the book.

4.1 Bengamides

Bengamides **1–7**[1–3] were isolated from *Jaspidage* and *Choristid* marine sponges collected from the Benga lagoon of the Fiji Islands. Bengamides A (**1**) and B (**2**) showed significant antihelminthic and anti-infectious activities as well as cytotoxicity. Bengamides have a unique structure having a cyclo-L-lysine [(*S*)-α-aminocaprolactam] and a C_{10} side chain with four contiguous hydroxyl groups as well as an *E*-olefin. The absolute configuration of the side chain of the bengamides has been tentatively assigned as (2*R*,3*R*,4*S*,5*R*,6*E*)-3,4,5-trihydroxy-2-methoxy-8-methylnon-6-enyl as a common structural feature by ^{1}H NMR study of its *O*-methyl mandlate derivative.[3] Total syntheses of bengamides A, B and C from noncarbohydrates have been reported.[4,5] Moreover, carbohydrates have also been used for their synthesis.

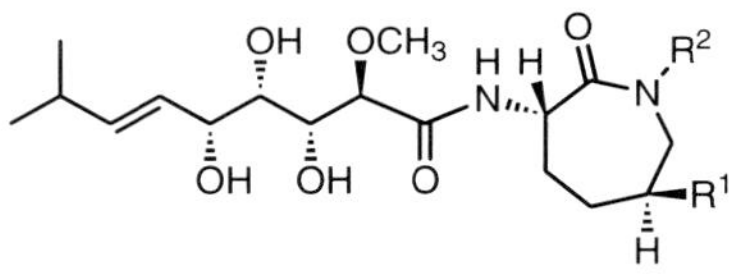

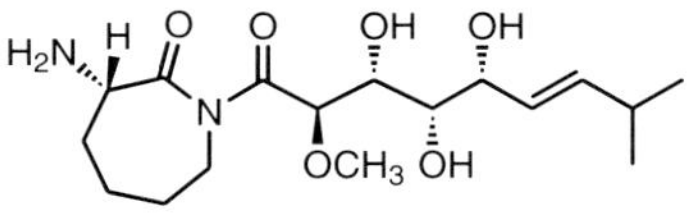

Bengamide	R^1	R^2
(**1**) A	n-$C_{13}H_{27}CO_2$	H
(**2**) B	n-$C_{13}H_{27}CO_2$	CH_3
(**3**) E	H	H
(**4**) F	H	CH_3
(**5**) C	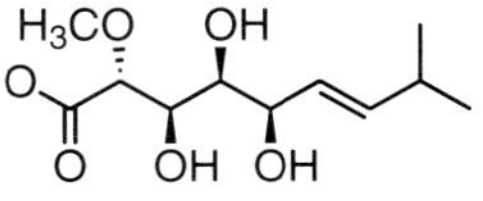	H
(**6**) D	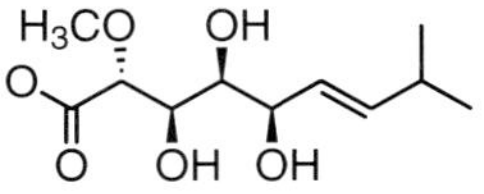	CH_3

(**7**) Isobengamide E

4.1.1 *Synthesis from* D*-glucose*

Syntheses of bengamide E (**3**) and *Z*-bengamide E (**21**) have been achieved utilizing diacetone glucose **8** as a starting material, which is converted into 3-*O*-acetyl-1,2:5,6-di-*O*-isopropylidene-α-D-gulofuranose[6–10] (**9**) in 50% overall yield from D-glucose (Scheme 1).[11] Replacement of the acetyl group in **9** with benzyl group followed by methanolysis afforded the β-D-gulopyranoside **10** (65%) together with its α-anomer **11** (9%) and the β-furanoside **12** (26%). Treatment of **10** with dimethoxybenzaldehyde followed by methylation of the hydroxyl group at C-2 afforded **13** (89%), which was subjected to a reductive ring opening of the 4,6-*O*-benzylidene group with DIBAL-H to afford **14** (75%). Swern oxidation[12] of **14** gave **15**, which upon Wittig olefination gave exclusively the *Z*-isomer.

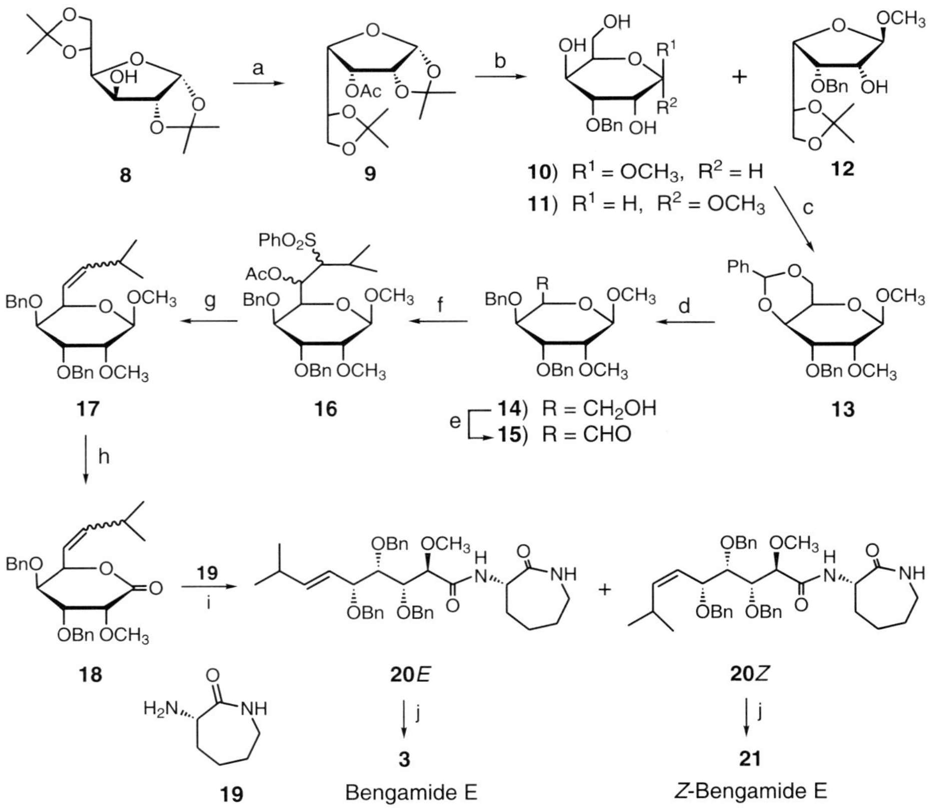

Scheme 1 (*a*) 1. RuO_2, KIO_4; 2. Ac_2O, Py; H_2, Pd *on* C, 50% for three steps. (*b*) 1. KOH, BnCl, 140°C, 2 h, 83%; 2. Dowex 50W resin, CH_3OH, reflux, 20 h, **10** (65%), **11** (9%), **12** (26%). (*c*) 1. $PhCH(OCH_3)_2$, *p*-TsOH, CH_2Cl_2, rt, 1 h, 87%; 2. NaH, CH_3I, DMF, 0°C to rt, 5 h, 89%. (*d*) DIBAL-H, CH_2Cl_2, 0°C to rt, 10 h, 75%. (*e*) DMSO, $(COCl)_2$, –70°C, 30 min; then NEt_3, rt, quantitative. (*f*) 1. *n*-BuLi, $(CH_3)_2CHCH_2SO_2Ph$, THF, 0°C, 30 min; 2. Ac_2O, Py, DMAP, rt, 12 h, 50% from **14**. (*g*) 1. 5% Na(Hg); Na_2HPO_4, 0°C, 4 h, 88%, 3:1 *E*/*Z*. (*h*) 1. 50% AcOH, 110°C, 20 h; 2. DMSO, Ac_2O, rt, 24 h, 52% from **17**. (*i*) NEt_3, dioxane, 3 days, *Z* (26%), *E* (71%). (*j*) Na, NH_3, THF, –78°C, 42%, 30 min.

Julia's protocol[13] for stereocontrolled elimination of acetoxy sulfone to yield *E*-olefins was done by condensation of isobutyl phenyl sulfone with **15** and subsequent acetylation to give a mixture of two diastereomers of **16** whose treatment with Na(Hg) gave a mixture of *E*- and *Z*-olefins **17** in a ratio of 3:1 in 88% yield. Hydrolysis of the glycosidic linkage of **17** with 50% acetic acid gave the corresponding anomeric mixture, which was oxidized to give the lactone **18**. This was condensed with the commercially available (*S*)-α-aminocaprolactam[14] (**19**) in the presence of triethylamine to give *E*- and *Z*-condensates **20**. Birch reduction of **20***E* and **20***Z* afforded **3** and the *Z*-isomer of **21**, respectively.

4.1.2 *Synthesis from* L*-glucose*

Total syntheses of bengamide B (**2**) and bengamide E (**3**) have been achieved utilizing L-glucose as a starting material (Scheme 2).[15] 2,3,4,6-Tetra-*O*-benzyl-L-glucopyranose[16] (**22**) was reacted with the monoanion generated from isobutylphenylsulfone to give **23** as a diastereomeric mixture, which was treated with Na(Hg) to give the respective *E*-olefin, followed by methylation of the free hydroxyl group to afford **24**. Removal of the benzyl

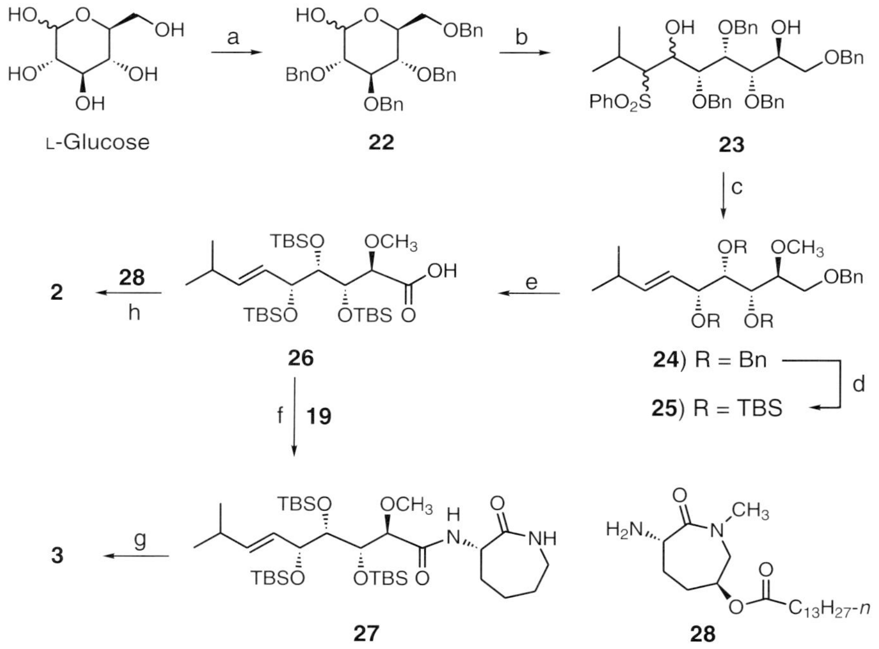

Scheme 2 (*a*) Ref. 16. (*b*) Isobutylphenylsulfone (3 equiv.); *n*-BuLi, THF, −78°C, 15 min; then added **22**, −78 to 20°C, 90 min, 80–85%. (*c*) 1. 6% Na(Hg), Na_2HPO_4, CH_3OH, 20°C, 2 h; 2. CH_3I (excess), KH, THF, 20°C, 90 min, 59–63%. (*d*) 1. Na/NH_3 (liquid) (5 min); 2. Ac_2O, Py, DMAP (20°C, 2 h); 3. CH_3OH, KH (cat.), 20°C, 2 h; 4. pivaloyl chloride, Py, 0°C, 4 h; 5. TBSOTf (5 equiv.), $NEt(i\text{-}Pr)_2$, CH_2Cl_2, 0°C, 60 min; 6. CH_3Li, CH_2Cl_2, −78°C, 4 h, 50% for six steps. (*e*) 1. $(COCl)_2$, DMSO, CH_2Cl_2, NEt_3, −78 to −40°C; 2. $NaClO_2$, NaH_2PO_4, *t*-BuOH, 2-methyl-2-butene, 0–20°C, 30 min, 90%. (*f*) HOBT, 1-(3-dimethylaminopropyl)-ethylcarbodiimide hydrochloride, CH_2Cl_2, 0–20°C, 8 h, 74%. (*g*) TBAF (excess), THF, 20°C, 90 min, 91%. (*h*) Same as (*f*), 10 h, 61%; then (*g*), 54%.

group with sodium in liquid ammonia followed by several protection and deprotection steps afforded **25**, which was oxidized to give the acid **26** in 21–24% overall yield from **22**. The coupling between **26** and (*S*)-α-aminocaprolactam (**19**) afforded **27**, which upon desilylation produced **3**. Similarly, compound **28**[17–19] was coupled with compound **26** to produce **2**.

4.1.3 *Synthesis from* L*-mannose*

Synthesis of bengamide E (**3**) was also achieved using 2,3-*O*-isopropylidene-5-*O*-methyl-L-mannofuranose[20,21] (**29**), obtained from quebrachitol [(1*R*,2*S*,3*S*,4*S*,5*R*,6*R*)-6-methoxy-cyclohexane-1,2,3,4,5-pentaol)], as a starting material (Scheme 3).[22] Protection of the primary hydroxyl group with TBSCl followed by Wittig olefination produced the olefin **30**. Mild acid treatment of **30** caused a migration of the *O*-isopropylidene group, followed by protection of the primary hydroxyl group with TBS to produce **31** in 59% yield. Oxidation of the hydroxyl group at the C-5 position in **31** followed by reduction with $Zn(BH_4)_2$ afforded

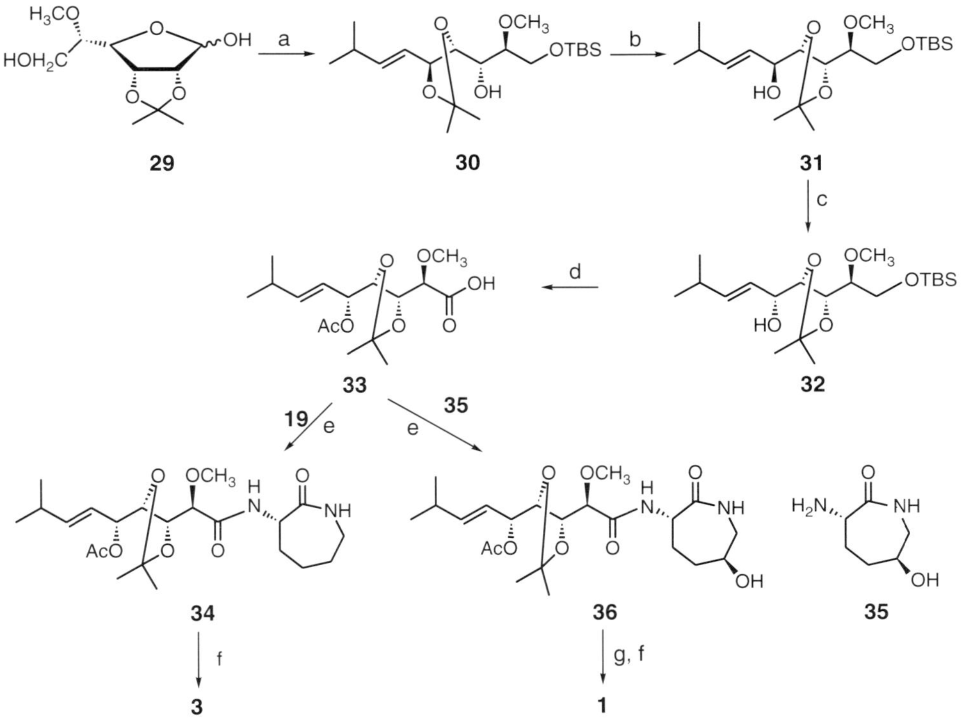

Scheme 3 (*a*) 1. TBSCl (1.9 mol. equiv.), NEt_3 (2.8 mol. equiv.), DMAP, CH_2Cl_2, rt, 20 h, 73%; 2. $(CH_3)_2CHCH_2P^+Ph_3Br^-$ (10 mol. equiv.), *n*-BuLi (9 mol. equiv.), benzene, rt, 4 h, *E* (76%), *Z* (14%). (*b*) 1. *p*-TsOH (0.05 mol. equiv.), acetone; 2. TBSCl, NEt_3 (2.8 mol. equiv.), DMAP, CH_2Cl_2, rt, 20 h. (*c*) 1. MnO_2 (30 mol. equiv.), CH_2Cl_2, rt, 16 h, 74%; 2. $Zn(BH_4)_2$ (7 mol. equiv), ether–toluene (1:1), −78 to 0°C, 1 h, 66%. (*d*) 1. Ac_2O, Py, rt, 15 h; 2. TBAF (10 mol. equiv.), AcOH (20 mol. equiv.), THF, 0°C to rt, 12 h, 90%; 3. Jones reagent (3 mol. equiv.), acetone, 0°C, 2 h, 85%. (*e*) $(EtO)_2P(O)CN$ (1.3 mol. equiv.), NEt_3, DMF, 0°C, 2 h, 88%. (*f*) 1. CH_3ONa, CH_3OH–THF (5:1), 5°C, 14 h; 2. TFA–THF–H_2O (3:3:2), 0°C to rt. (*g*) Myristic acid (2.5 equiv.), EDAC (2.5 equiv.), DMAP (1 equiv.), CH_2Cl_2, −15°C to rt, 62%.

the inverted alcohol **32** in 49% yield. Acetylation of **32** followed by removal of the silyl group and oxidation of the resulting primary alcohol with Jones reagent afforded **33** (85%), which was treated with cyclo-L-lysine (**19**)[14] to afford **34** whose deprotection gave **3** in 50% yield.

Similarily, synthesis of bengamide A (**1**) was achieved by coupling of **33** and hexahydro-2-azepinone **35**[5,23] to afford **36** (82%). Removal of the *O*-acetyl group followed by reaction with myristic acid, and finally, removal of the isopropylidene group provided **1** in 26% overall yield from **35**.

4.1.4 *Synthesis from D-threose*

Synthesis of bengamide E (**3**) from the D-threose derivative **37** has been reported (Scheme 4).[24] Corey and Fuchs dibromoolefination[25] of **37** produced the olefin **38**, which was treated with butyllithium to give the acetylene derivative **39**. The acetylide anion of **39** was generated

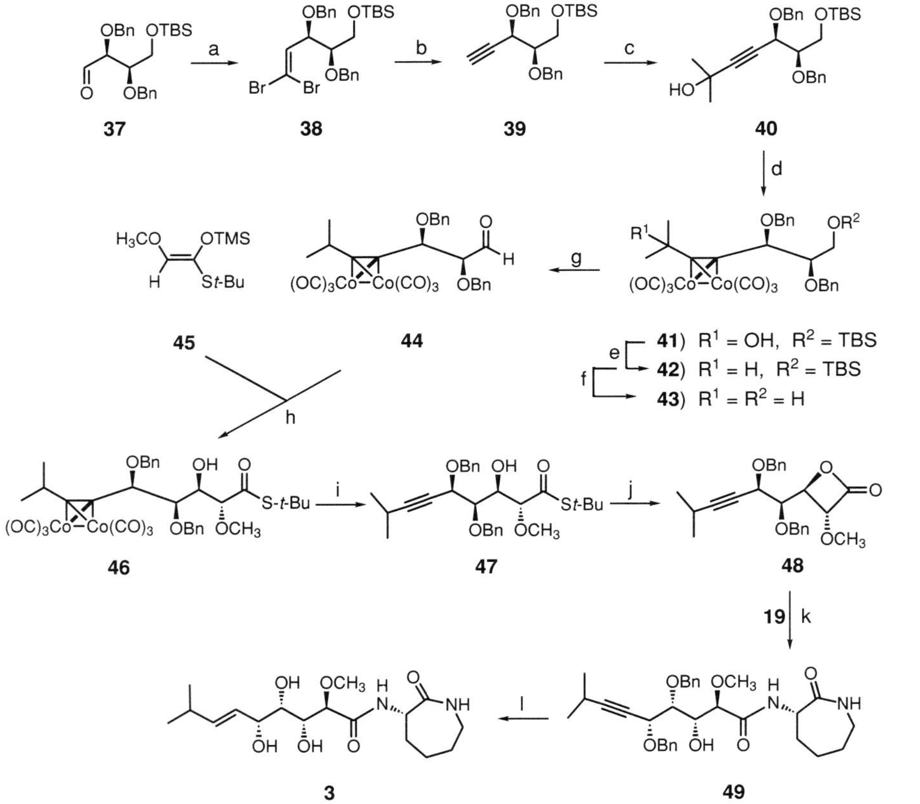

Scheme 4 (*a*) CBr_4, CH_2Cl_2, PPh_3, 15 min, NEt_3, 65%. (*b*) Ether, *n*-BuLi, 0°C, 10 min, 96%. (*c*) 1. THF, *n*-BuLi, −78°C, 1 h, acetone, 3 h, 84%; 2. CH_2Cl_2, octacarbonyldicobalt, rt, 3 h, 85%; 3. CH_2Cl_2, ZnI_2, $NaBH_3CN$, rt, 7 h; 4. CH_3OH, 0°C, CAN, 1 h, 20% HCl, rt, 2 h, 75%. (*d*) CH_2Cl_2, octacarbonyldicobalt, rt, 3 h, 85%. (*e*) CH_2Cl_2, ZnI_2, $NaBH_3CN$, rt, 7 h. (*f*) CH_3OH, 0°C, CAN, 1 h, 20% HCl, 2 h, rt, 75%. (*g*) $(COCl)_2$, DMSO, CH_2Cl_2, −78°C; then NEt_3. (*h*) CH_2Cl_2, $SnCl_4$, −78 to 0°C. (*i*) CAN, 0°C, 45 min, 47% from **43**. (*j*) CH_3OH, rt, CF_3CO_2Ag, 45°C, 5 h. (*k*) CH_2Cl_2, $(CH_3)_3Al$, 0°C to rt, 15 min; then 45°C, 5 h, 75%. (*l*) THF, Na *in* liquid NH_3, −78°C, 30 min, 65%.

with BuLi and subsequently trapped with acetone to furnish **40**, which was treated with octacarbonyldicobalt to afford the cobalt-complex derivative **41**. Reductive dehydration of **41** with $NaBH_3CN$ in the presence of zinc iodide[26] furnished the deoxygenated product **42**, which was demetallated with CAN and desilylated to afford **43**. Removal of the tertiary hydroxyl group of **41** with $NaBH_3CN$ gave the deoxygenated product **42**, which was hydrolyzed with hydrochloric acid to afford **43** in 78% yield. The alcohol **43** was oxidized[27] to produce the aldehyde **44**, which was subsequently reacted with **45** to give the cobalt-complexed aldol product **46**. Decomplexation of **46** by treatment with CAN in methanol gave **47** in 47% overall yield from **43**. Compound **47** underwent lactonization to furnish the β-lactone **48**, which was treated with **19** to give the amide **49**. Finally, reduction of **49** using Na/NH_3 effected debenzylation and reduction of the triple bond to the *trans*-double bond to provide **3**.

4.1.5 *Synthesis from (R)-glyceraldehyde*

Synthesis of bengamide E (**3**) was achieved utilizing (*R*)-glyceraldehyde acetonide (Scheme 5).[28] Thus, the furan adduct of (*R*)-glyceraldehyde acetonide[29] **50** was methylated to afford

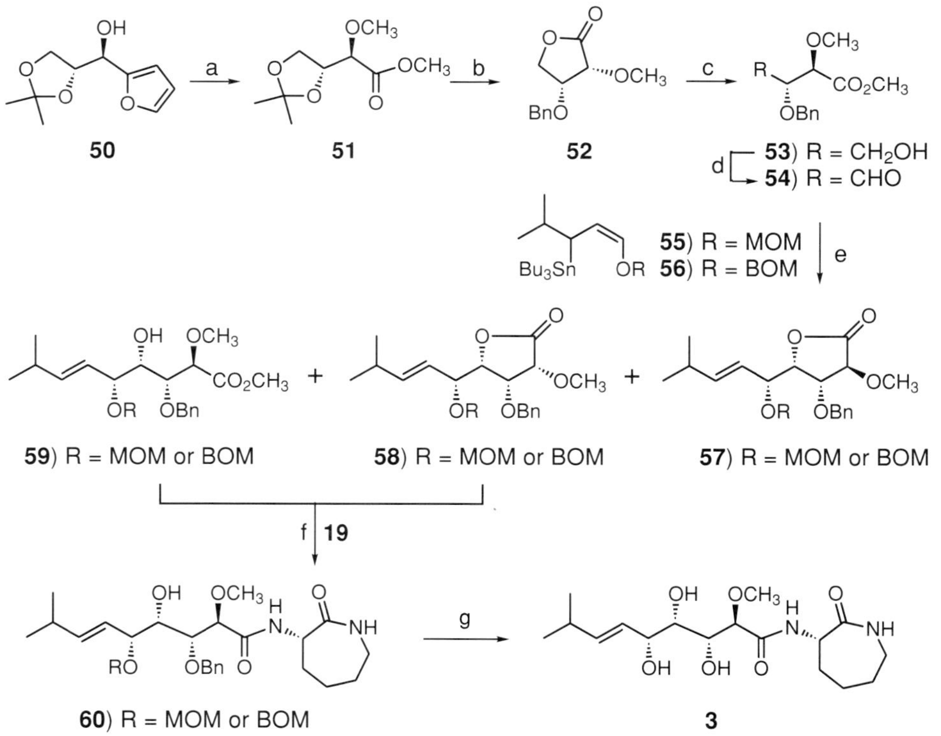

Scheme 5 (*a*) 1. NaH, CH_3I, THF, 0°C, 87%; 2. RuO_4, $CH_3CN–CCl_4–H_2O$ (3:2:3), $NaIO_4$, CH_2N_2, Et_2O, 0°C, 69%. (*b*) 1. TFA, H_2O, THF, 0°C, 100%; 2. $BnOC(NH)CCl_3$, CH_2Cl_2, cyclohexane, rt, TFA, 2 h, 63%. (*c*) K_2CO_3, CH_3OH, 0°C, 25 min. (*d*) Dess–Martin periodinate reagent, CH_2Cl_2, rt, 25 min, 85%. (*e*) CH_2Cl_2, −20°C, $MgBr_2{\cdot}OEt_2$, 45 min; then rt, 48 h, 90%. (*f*) CH_2Cl_2, 0°C, $(CH_3)_3Al$ in hexane, 78%. (*g*) THF, −78°C, Li *in* liquid NH_3, 76%.

the methyl ether derivative, followed by furan cleavage[30] and esterification to afford **51**. Removal of the acetonide group followed by benzylation afforded the lactone **52**. Treatment of **52** with K_2CO_3 in methanol afforded **53**, which upon oxidation by Dess–Martin periodinane reagent[31] afforded **54**. Addition of aldehyde **54** to the (*S*)-enantiomer of stannane **55** or **56** under chelation-controlled[32] condition afforded three products **59** (30%), **58** (35%) and **57** (25%). A mixture of **59** and **58** underwent aminolysis[33] with (*S*)-2-aminocaprolactam (**19**) to afford the protected bengamide E derivative **60** (78%). Deprotection with lithium in liquid ammonia afforded **3** in 76% yield.

The (*R*)-glyceraldehyde derivative **61** has also been used for the synthesis of bengamide E (**3**) (Scheme 6).[34] Silylation of enoate **61** followed by reduction with DIBAL-H afforded **62**, which underwent Sharpless epoxidation[35] followed by iodination of the primary hydroxyl group to afford **63**. Reaction of **63** with *tert*-butyllithium and dimethyl sulfate and subsequent *in situ* methylation gave the methyl ether **64** in a 95:5 mixture of *anti* and *syn* products. Ozonolysis of **64** followed by hydrolysis of the silyl ether gave **53**, which under similar sequence of reactions used in the former scheme gave **3**.

Scheme 6 (*a*) 1. TBSCl, imidazole, DMF, 99%; 2. DIBAL-H, −78°C, 76%. (*b*) 1. D-(−)-DIPT, TBHP, TIP, 98%; 2. Ph_3P, imidazole, I_2, 90%. (*c*) *t*-BuLi, $(CH_3O)_2SO_2$, 90%. (*d*) 1. O_3, CH_3OH, NaOH, CH_2Cl_2, 72%; 2. AcOH, H_2O, THF, 88%.

4.1.6 *Synthesis from* D*-glucoheptonolactone*

The first synthesis of the side chain (2-methoxy-3,4,5-trihydroxy-8-methylnon-6*E*-enoyl) of bengamides has been achieved utilizing α-D-glucoheptonic γ-lactone (Scheme 7).[36] 3,5:6,7-Di-*O*-isopropylidene-α-D-glucoheptonic γ-lactone (**65**)[37] underwent methylation of the free hydroxyl group at C-2 followed by $LiAlH_4$ reduction to afford the respective diol. Selective protection of the resulting primary hydroxyl group with MPMBr followed by protection of the secondary hydroxyl group with benzyl bromide afforded compound **66**, which was subjected to selective hydrolysis of the terminal isopropylidene group to give **67** as the major product, in addition to a minor product resulting from a cleavage of the nonterminal one. Benzylation of **67** followed by removal of the terminal acetonide group and then $NaIO_4$ oxidation afforded the aldehyde **68**. Treatment of **68** with isobutylsulfone followed by acetylation of the resulting hydroxyl group afforded the acetoxysulfone **69**, which underwent elimination[13] to afford **70** (50%). Cleavage of MPM group in **70** afforded the alcohol **24**, which was oxidized using Jones reagent, followed by esterification with diazomethane in ether to yield **71**.

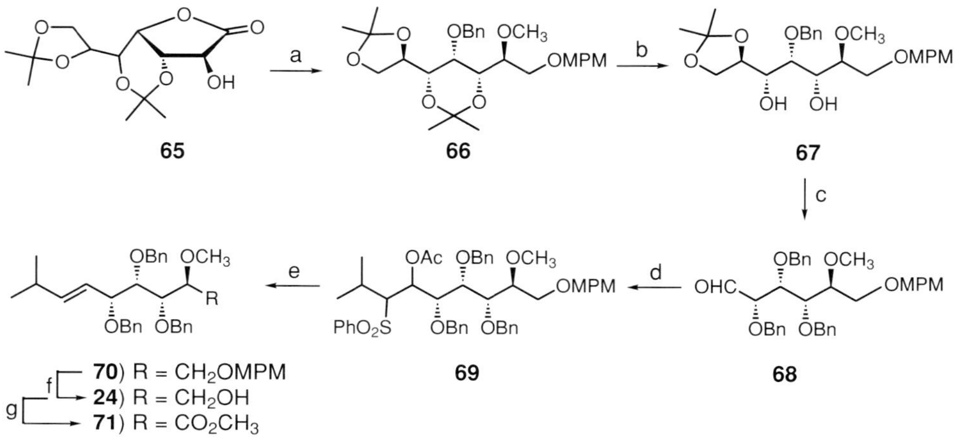

Scheme 7 (*a*) 1. Moist Ag_2O, CH_3I, CH_2Cl_2, rt, 6 h, 85%; 2. $LiAlH_4$, THF, reflux, 2 h; 3. NaH (1 equiv.), MPMBr (1 equiv.), THF, 0°C to rt, 18 h, 70%. (*b*) 1. NaH, BnBr, THF; 2. 0.8% H_2SO_4, CH_3OH, rt, 48 h. (*c*) 1. NaH, BnBr, THF, rt, 18 h; 2. *p*-TsOH, CH_3OH, rt, 4 h; 3. $NaIO_4$, EtOH, rt, 1 h. (*d*) 1. $PhSO_2CH_2CH(CH_3)_2$; then *n*-BuLi, THF, −30°C, 1 h; 2. Ac_2O, Py, DMAP, rt, 3 h. (*e*) 6% Na/Hg, Na_2HPO_4, CH_3OH, 0°C, 2 h, 50%. (*f*) DDQ, CH_2Cl_2, 1 h. (*g*) 1. Jones reagent, EtOEt, 0°C, 1 h; 2. CH_2N_2, EtOEt.

References

1. Guiñoă, E.; Adamezeski, M.; Bakus, G. J. *J. Org. Chem.* **51** (1986) 4494.
2. Adamezeski, M.; Guiñoă, E.; Crews, P. *J. Am. Chem. Soc.* **111** (1989) 647.
3. Adamezeski, M.; Guiñoă, E.; Crews, P. *J. Org. Chem.* **55** (1990) 240.
4. Chida, N.; Tobe, T.; Yamazaki, K.; Ogawa, S. *Heterocycles* **38** (1994) 2383.
5. Chida, N.; Tobe, T.; Okada, S.; Ogawa, S. *J. Chem. Soc., Chem. Commun.* (1992) 1064.
6. Stevens, J.D. *Methods Carbohydr. Chem.* **6** (1972) 123.
7. Baker, D.C.; Horton, D.; Tindall, C.G. *J. Carbohyr. Res.* **24** (1972) 192.
8. Meyerzu Reckendorf, W. *Methods Carbohydr. Chem.* **6** (1972) 129.
9. Slessor, K.N.; Tracey, A.S. *Can. J. Chem.* **47** (1969) 3989.
10. Lemiuex, R.U.; Stick, R.V. *Aust. J. Chem.* **28** (1975) 1799.
11. Kishimoto, H.; Ohrui, H.; Meguro, H. *J. Org. Chem.* **57** (1992) 5042.
12. Mancuso, A.J.; Huang, S.L.; Swern, D.J. *J. Org. Chem.* **43** (1978) 2480.
13. Julia, M.; Badet, B. *Bull. Chem. Soc. Fr.* (1975) 1363.
14. Pellegate, R.; Pinza, M.; Pifferi, G. *Synthesis* (1978) 614.
15. Broka, C.A.; Ehrler, J. *Tetrahedron Lett.* **32** (1991) 5907.
16. Glaudemans, C.P.J.; Fletcher, H.G., Jr. *Methods Carbohydr. Chem.* **6** (1972) 373.
17. Mulzer, J.; DeLasalle, P. *J. Chem. Res. (S)* (1983) 10.
18. Kende, A.S.; Kawamura, K.; Orwat, M.J. *Tetrahedron Lett.* **30** (1989) 5821.
19. Evans, D.A.; Chapman, K.T.; Bisaha, J. *J. Am. Chem. Soc.* **110** (1988) 1238.
20. Van Alphen, J. *Ind. Eng. Chem.* **43** (1951) 141.
21. Chida, N.; Suzuki, M.; Suwama, M.; Ogawa, S. *J. Carbohydr. Chem.* **8** (1989) 319.
22. Chida, N.; Tobe, T.; Ogawa, S. *Tetrahedron Lett.* **32** (1991) 1063.
23. Yamada, Y.; Kasai, Y.; Shioiri, T. *Tetrahedron Lett.* **14** (1973) 1595.
24. Mukai, C.; Moharram, S.M.; Kataoka, O.; Hanaoka, M. *J. Chem. Soc., Perkin Trans. 1* (1995) 2849.

25. Corey, E.J.; Fuchs, P.L. *Tetrahedron Lett.* **13** (1972) 3769.
26. Lau, C.K.; Dufresne, C.; Bélanger, P.C.; Piétré, S.; Scheigetz, J. *J. Org. Chem.* **51** (1986) 3038.
27. Jeong, N.; Lee, B.Y.; Lee, S.M.; Chung, Y.K.; Lee, S.-G. *Tetrahedron Lett.* **34** (1993) 4023.
28. Marshall, J.A.; Luke, G.P. *J. Org. Chem.* **58** (1993) 6229.
29. Suzuki, K.; Yuki, Y.; Mukaiyama, T. *Chem. Lett.* (1981) 1529.
30. Danishefsky, S.J.; DeNinno, M.P.; Chen, S.-H. *J. Am. Chem. Soc.* **110** (1988) 3929.
31. Dess, D.B.; Martin, J.C. *J. Org. Chem.* **48** (1983) 4155.
32. Marshall, J.A.; Luke, G.P. *J. Org. Chem.* **56** (1991) 483.
33. Basha, A.; Lipton, M.; Weinreb, S.M. *Tetrahedron Lett.* **18** (1977) 4171.
34. Marshall, J.A.; Luke, G.P. *Synlett* (1992) 1007.
35. Hanson, R.M.; Sharpless, K.B. *J. Org. Chem.* **51** (1986) 1922.
36. Gurjar, M.K.; Srinivas, N.R. *Tetrahedron Lett.* **32** (1991) 3409.
37. Brimacombe, J.S.; Tucker, L.C.N. *Carbohydr. Res.* *1* (1965) 332.

4.2 Liposidomycins

The liposidomycins are a family of novel lipid-containing nucleoside antibiotics that were found in the culture filtrate and mycelia of *Streptomyces griseoporeus*.[1] These antibiotics, which have unique biological activity and structures, inhibit the formation of the lipid intermediate in bacterial peptidoglycan synthesis three times more than does tunicamycin and have extremely high specificity.[2,3] The structures of liposidomycins A,[4] B (**1**)[2] and C were proposed on the basis of degradation and spectroscopic studies. They are identical except for slight variations in the lipid portion.

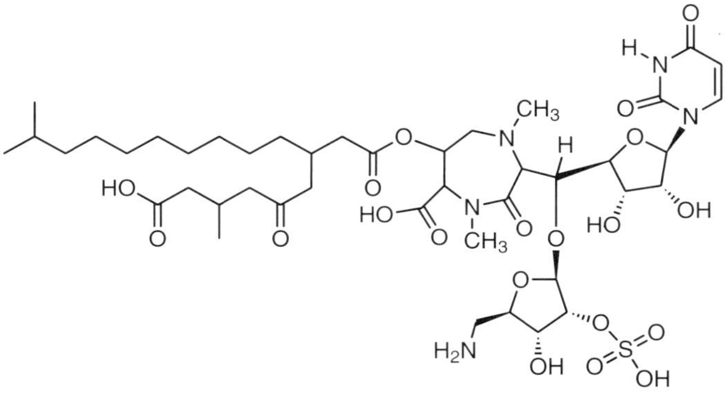

Liposidomycin B
1

Partial synthesis of the diazepanone part has been achieved from carbohydrate derivatives, where L-ascorbic acid was the selected precursor (Schemes 1 and 2).[5] The methyl threonate **2**,[6–8] obtained from L-ascorbic acid in 65% yield, was reduced, followed by selective benzylation of the primary hydroxyl group to furnish the monobenzyl ether **3**. Tosylation of **3** followed by S_N2 displacement of the resulting tosyloxy group with sodium azide afforded **4**. Removal of the isopropylidene group from **4** afforded a diol, which upon selective tosylation gave **5**. Silylation of **5** with TBSCl followed by catalytic hydrogenation and subsequent protection of the resulting amine with Cbz-glycine furnished the peptide **6**, which upon cyclization with potassium carbonate provided the unwanted four-membered ring azetidine **7**. However, the tosylate **5** was converted into the epoxide **8**, which underwent a nucleophilic opening of the epoxide ring with sarcosine (*N*-methylglycine) in refluxing methanol to furnish the corresponding azidocarboxylic acid, via the expected attack at the primary carbon, whose subsequent catalytic hydrogenation afforded the amino acid **9**. Cyclization of **9** with DCC in methylene chloride furnished the desired 1,4-diazepan-2-one **10**. Protection of the hydroxyl group in **10** and subsequent N-methylation with methyl iodide provided 1,4-dimethyl-1,4-diazepanone (**11**) in 6% overall yield from L-ascorbic acid.

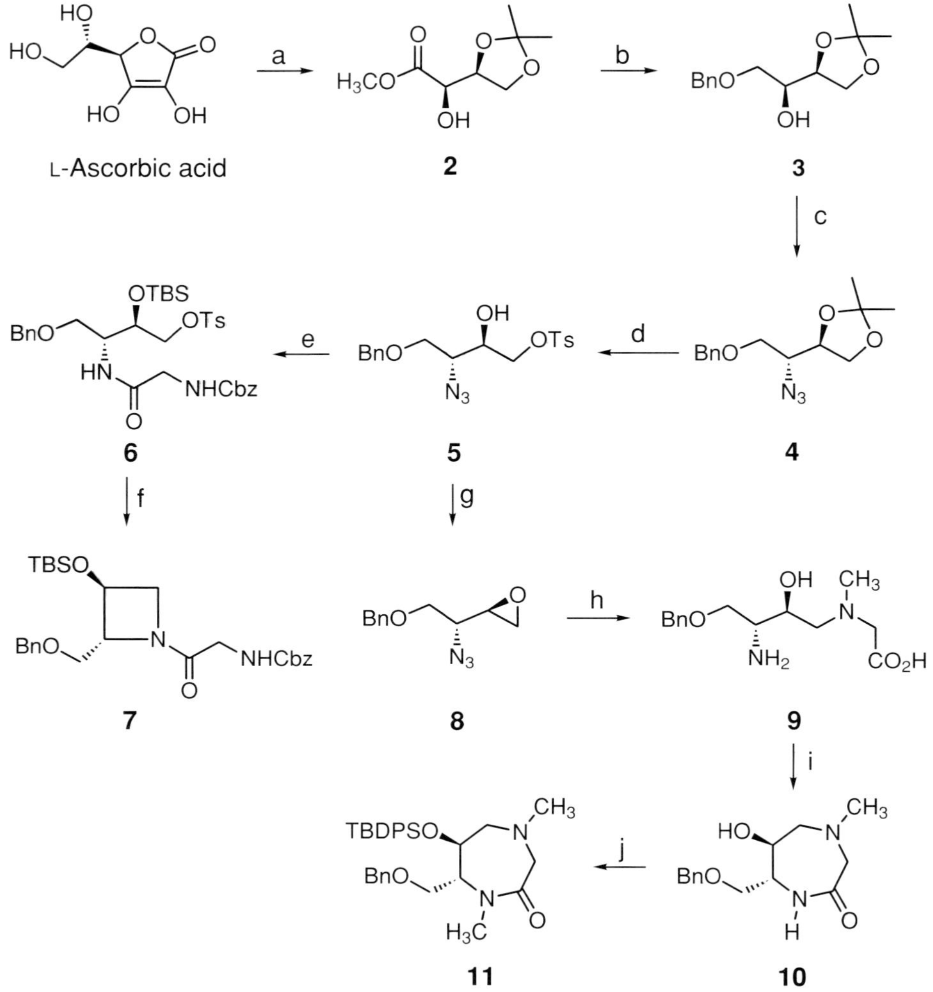

Scheme 1 (*a*) 1. Acetone, AcCl, rt, 3 h; 2. 35% H_2O_2, $CaCO_3$, H_2O, 0°C to rt, 3 h; 3. CH_3I, $NaHCO_3$, $AcN(CH_3)_2$, rt, 2 days, 65% for three steps. (*b*) 1. $NaBH_4$, EtOH, 0°C to rt, 83%; 2. $(Bu)_2SnO$, CH_3OH, reflux, 5 h; 3. BnBr, DMF, 70–80°C, 90% for two steps. (*c*) 1. *p*-TsCl, Py, 0°C, 12 h, 95%; 2. NaN_3, DMF, 70–80°C, 12 h, 90%. (*d*) 1. 1 M HCl, CH_3CN, 87%; 2. *p*-TsCl, Py, 0°C, 15 h, 75%. (*e*) 1. TBSCl, imidazole, DMF, 89%; 2. H_2, Pd *on* C, EtOAc, 92%; 3. $CbzNHCH_2CO_2H$, DCC, CH_2Cl_2, 0°C, 2 h, 88%. (*f*) K_2CO_3, DMF, 40°C, 10 h, 50%. (*g*) K_2CO_3, CH_3OH, rt, 85%. (*h*) 1. $CH_3NHCH_2CO_2H$, NEt_3, CH_3OH, reflux, 10 h, 83%; 2. H_2, Pd *on* C, CH_3OH, 72%. (*i*) DCC, CH_2Cl_2, 0°C, 15 h, 71%. (*j*) 1. TBDPSCl, imidazole, DMF, 83%; 2. CH_3I, NaH, DMF, 76%.

On the other hand, the corresponding isomer **17** was prepared from **2** by tosylation to give **12**, which was converted to the epoxide **13** that underwent ring opening with BnONa to afford the benzyl ether **14**. This was converted into the 1,4-dimethyl-1,4-diazepanone derivative **17** in 16% overall yield from **14** via the intermediates **15** and **16** (Scheme 2).[5]

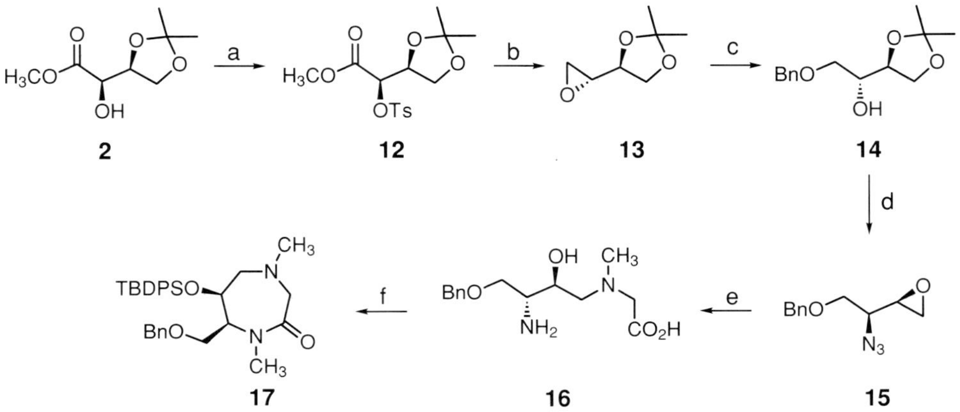

Scheme 2 (*a*) *p*-TsCl, Py, 0°C, 90%. (*b*) 1. $NaBH_4$, EtOH, 0°C, 30 min, then rt, 12 h; 2. K_2CO_3, CH_3OH, rt, 63% for two steps. (*c*) BnONa, DMF, 50°C, 3 h, 88%. (*d*) 1. *p*-TsCl, Py, 0°C, 12 h, 95%; 2. NaN_3, DMF, 70–80°C, 12 h, 90%; 3. 1 M HCl, CH_3CN, 87%; 4. *p*-TsCl, Py, 0°C, 15 h, 75%; 5. K_2CO_3, CH_3OH, rt, 85%. (*e*) 1. $CH_3NHCH_2CO_2H$, NEt_3, CH_3OH, reflux, 10 h, 83%; 2. H_2, Pd *on* C, CH_3OH, 72%. (*f*) 1. DCC, CH_2Cl_2, 0°C, 15 h, 71%; 2. TBDPSCl, imidazole, DMF, 83%; 3. CH_3I, NaH, DMF, 76%.

References

1. Isono, K.; Uramoto, M.; Kusakabe, H.; Kimura, K.; Izaki, K.; Nelson, C.C.; McCloskey, J.A. *J. Antibiot.* **38** (1985) 1617.
2. Ubukata, M.; Isono, K.; Kimura, K.; Nelson, C.C.; McCloskey, J.A. *J. Am. Chem. Soc.* **110** (1985) 4416.
3. Kimura, K.; Miyata, N.; Kawanishi, G.; Kamio, Y.; Izaki, K.; Isono, K. *Agric. Biol. Chem.* **53** (1989) 1811.
4. Ubukata, M.; Kimura, K.; Isono, K.; Nelson, C.C.; Gregson, J.M.; McCloskey, J.A. *J. Org. Chem.* **57** (1992) 6392.
5. Kim, K.S.; Cho, I.H.; Ahn, Y.H.; Park, J.I. *J. Chem. Soc., Perkin Trans 1* (1995) 1783.
6. Jackson, K.; Jones, J. *Can. J. Chem.* **47** (1969) 2498.
7. Isbell, H.S.; Frush, H.L. *Carbohydr. Res.* **72** (1979) 301.
8. Wei, C.C.; Bernardo, S.D.; Tengi, J.P.; Borgese, J.; Weigele, M. *J. Org. Chem.* **50** (1985) 3462.

5 Fused nitrogen heterocycles

This is the longest chapter of the book. It discusses the different types of naturally occurring fused nitrogen heterocycles synthesized from sugars. The chapter is divided into six parts arranged according to the size of the fused rings. The first one deals with 3:5-fused heterocycles and contains the azinomycins. The second discusses 4:5-fused heterocycles and includes β-lactams, noted for their antibiotic activity. The third part presents the 5:5-fused heterocycles, which comprise four groups, namely polyhydroxy-pyrrolizidines, trehazolin, allosamidin and biotin, all of which have a wide range of biological activities. The fourth part contains two groups: the bioactive indolizidine alkaloids isolated from fungal and higher plant sources. The first group includes castanospermines, swainsonine, lentiginosines and slaframine, and the second group comprises various miscellaneous natural 5:6-fused rings. The fifth part of the chapter includes kifunensine, nagastatin, calystegines, mesembrine and streptolidine, and the sixth part of the chapter contains 6:6-fused rings such as hydroxylated quinuclidines, biopterins and isoquinolines, which includes calycotomine, decumbenosine, laudanosine and glaucine.

5.1 3:5-Fused heterocycles

5.1.1 *Azinomycins*

In 1986, azinomycins A (**1**) and B (**2**) were isolated from the culture broth of *Streptomyces griseofuscus* S42227.[1,2] They exhibit potent antitumor activity against a number of different

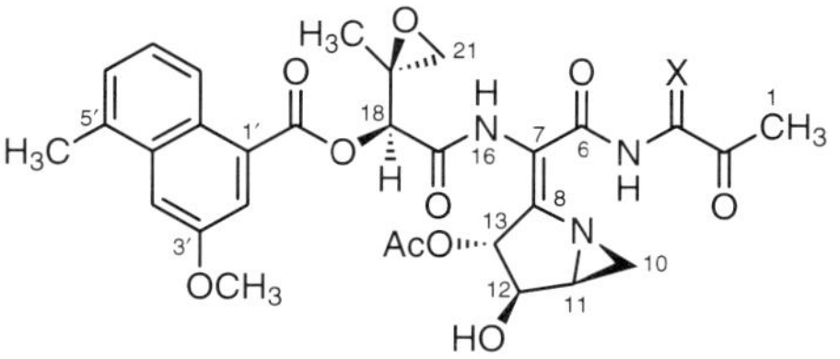

1) Azinomycin A, X = H_2

2) Azinomycin B, X = CHOH

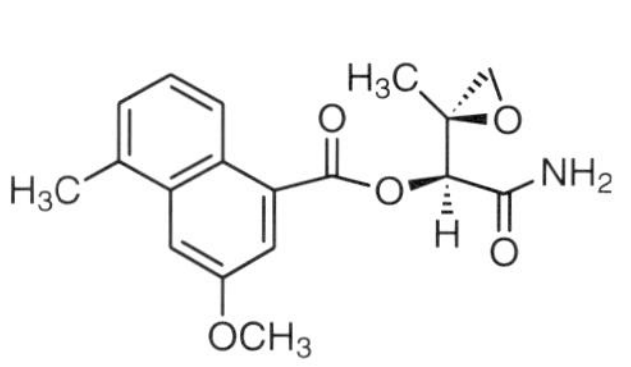

3

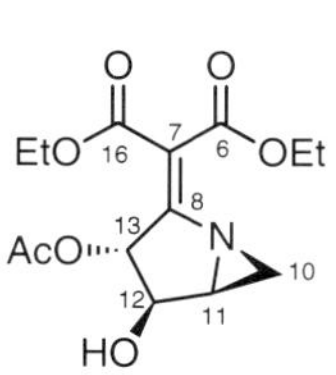

4

tumor cell lines[1,2] such as P388 leukemia.[3] Azinomycin B was found to be identical to carzinophilin,[4–16] isolated from *Streptomyces sahachiroi*, based on detailed comparison of ^{1}H and ^{13}C NMR spectra and reinvestigation of the FAB mass spectrum (FAB-MS).[17,18] In addition, the epoxy amide **3**, devoid of the 1-azabicyclo[3.1.0]hexane ring system **4**, was isolated from *Str. griseofuscus* S42227 and it was found to have a significant cytotoxic activity.[19] It has been established that azinomycins act by interstrand cross-linking of DNA,[20,21] a process associated with many clinically important antitumor agents such as mitomycin C[22] and are well known as a bisalkylating agent for DNA.[20,23] Syntheses of the fragments of azinomycins from noncarbohydrates have been reported,[24–58] and herein those synthesized from carbohydrates are discussed.

5.1.1.1 *Synthesis from* D*-fructose* The epoxide subunit of azinomycins has been synthesized from D-fructose (Scheme 1).[59] D-Fructose was initially converted to the lactone **5**, whose reduction and subsequent isopropylidenation gave **6**. Tosylation of the free hydroxyl group followed by selective hydrolysis of the terminal isopropylidene group afforded **7**, which underwent periodate oxidative cleavage of the diol followed by further oxidation to furnish the carboxylic acid, which was esterified to give **8**. Hydrolysis of the remaining isopropylidene group followed by base-induced ring-closure afforded **9**, which was coupled with 3-methoxy-5-methyl-1-naphthoic acid (**10**) to give **11**. Selective removal of the benzyl group by hydrogenation gave the carboxylic acid **12**, which is suitable for coupling with the dehydroamino acid fragment of azinomycins.

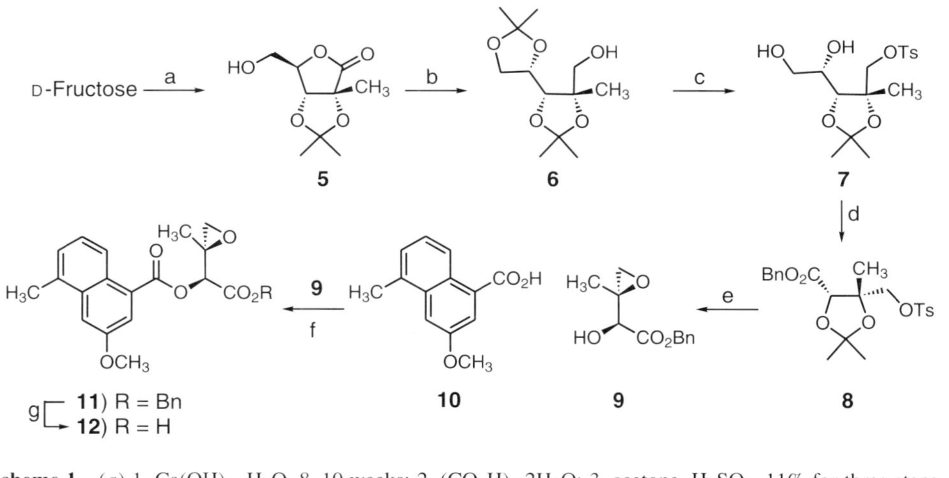

D-Fructose —a→ **5** —b→ **6** —c→ **7** —d→ **8** —e→ **9**; **9** + **10** —f→ **11**) R = Bn —g→ **12**) R = H

Scheme 1 (*a*) 1. $Ca(OH)_2$, H_2O, 8–10 weeks; 2. $(CO_2H)_2{\cdot}2H_2O$; 3. acetone, H_2SO_4, 11% for three steps. (*b*) 1. $LiAlH_4$, Et_2O; 2. acetone, *p*-TsOH, 67% for two steps. (*c*) 1. *p*-TsCl, DMAP, pyridine; 2. 70% aqueous AcOH, 67% for two steps. (*d*) 1. $NaIO_4$, CH_3OH–H_2O (1:1); 2. Jones reagent, acetone; 3. BnOH, DCC, DMAP, THF, 70% for three steps. (*e*) 1. 80% aqueous AcOH; 2. K_2CO_3, acetone, 82% for two steps. (*f*) DCC, DMAP, 76%. (*g*) H_2, 10% Pd *on* C, CH_3OH.

5.1.1.2 *Synthesis from* D*-glucose* A synthesis of the fragment **3** of azinomycin has been reported from D-glucose (Scheme 2).[60] The aldehyde **13**,[61] obtained from diacetone D-glucose in 70% yield, was reduced with sodium borohydride, followed by reflux with catalytic amount of *p*-TsOH under azeotropic removal of EtOH to give the γ-lactone **14** in 92% yield. Treatment of **14** with the acid chloride **15** afforded the ester **16** in 84% yield. Aminolysis of **16** was performed with methanolic ammonia at 0°C to afford the amide **17** in 91% yield. Removal of the benzyl group by hydrogenolysis gave the diol **18** (73% yield), which underwent selective mesylation followed by epoxide formation with K_2CO_3 in refluxing acetone to afford the epoxide **3** in 32% overall yield from diacetone D-glucose.

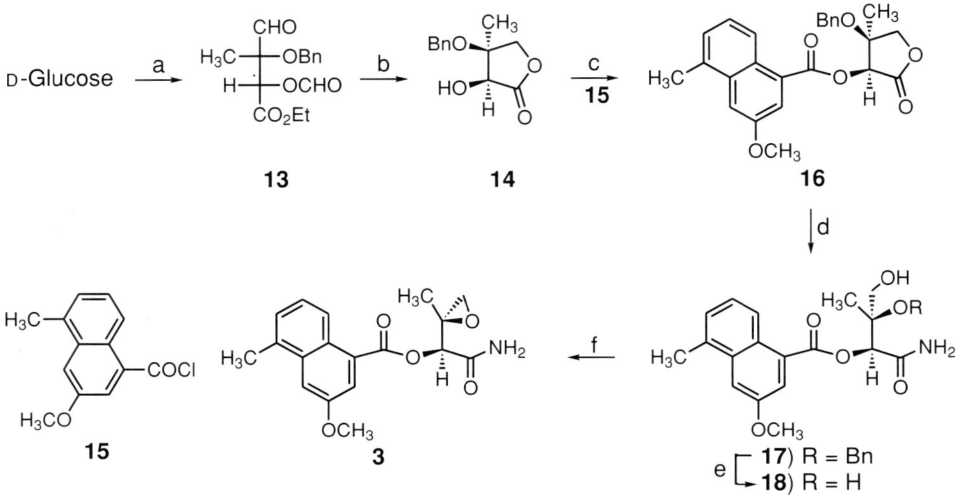

Scheme 2 (*a*) Ref. 61, 70%. (*b*) 1. $NaBH_4$, EtOH–THF (1:1), 0°C; 2. *p*-TsOH, benzene, reflux, 92%. (*c*) $(i\text{-Pr})_2NEt$, DMAP, CH_2Cl_2, 0°C, 84%. (*d*) 15% NH_3, CH_3OH, 0°C, 91%. (*e*) H_2, 10% Pd *on* C, AcOH, 73%. (*f*) 1. MsCl, $(i\text{-Pr})_2NEt$, CH_2Cl_2, rt; 2. K_2CO_3, acetone, reflux, 89%.

5.1.1.3 *Synthesis from D-glucosamine* Aziridino[1,2-*a*]pyrrolidine (**26**), a substructure of azinomycins A and B, has been synthesized from D-glucosamine (Scheme 3).[62] The *p*-methoxybenzylidene acetal **19**,[63–65] derived from D-glucosamine in overall yield 40–50%, was protected as *tert*-butyldimethylsilyl ether. Subsequent cleavage of the acetal and selective iodination of the primary hydroxyl group followed by acetylation of the secondary hydroxyl group provided **20**. The iodide **20** underwent fragmentation[66] followed by immediate reduction of the generated aldehyde to afford **21**. This underwent a Mitsunobu cyclization followed by ozonolysis of the olefin function to furnish the aldehyde **22**, which was reacted with the phosphonate **23** to afford the dehydroamino acid **24** with 10:1 *Z*/*E* diastereoselectivity. Treatment of **24** with excess NBS in the presence of 1,4-diazobicyclo[2.2.2]octane afforded **25**, which underwent removal of the *N*-benzyloxycarbonyl group followed by Michael addition to produce the aziridino[1,2-*a*]pyrrolidine **26** with retention of configuration of the olefin.

5.1.1.4 *Synthesis from D-arabinose* The lactams (3*S*,4*R*,5*S*)-1-benzoyl- and 1-benzyloxycarbonyl-5-benzyloxymethyl-3,4-dibenzyloxypyrrolidin-2-ones (**34** and **35**), important intermediates for fragment **36**, have been synthesized from D-arabinose (Scheme 4).[67] Treatment of D-arabinose with methanol–HCl afforded the methyl arabinofuranoside,

Scheme 3 (*a*) 1. TBSOTf, 2,6-lutidine, CH_2Cl_2, 0°C, 90%; 2. $(NH_4)_2Ce(NO_3)_6$, CH_3CN, H_2O, 25°C, 1 h, 73%; 3. I_2, Ph_3P, Py, toluene, 70°C, 88%; 4. Ac_2O, Py, CH_2Cl_2, DMAP, 24°C, 98%. (*b*) 1. Zn, 95% EtOH, reflux, 1 h; 2. $NaBH_4$, THF, H_2O, −43°C, 84%. (*c*) 1. Ph_3P, THF, 23°C, 12 h, DEAD, 85%; 2. O_3, CH_2Cl_2, −78°C, DMS, 23°C, 100%. (*d*) $(H_3CO)_2P(O)CH(CO_2CH_3)NHCO_2CH_3$, *t*-BuOK, CH_2Cl_2, −78 to 24°C. (*e*) 1. Excess NBS, 1,4-diazobicyclo[2.2.2]octane, 24°C, 8 h; 2. $Na_2S_2O_4$, 0°C, 15%. (*f*) 1. Et_3SiH, $PdCl_2$, NEt_3, 25°C, 30 min, 100%; 2. 1,4-diazobicyclo[2.2.2]octane, $CDCl_3$, 50°C, 1 h.

Scheme 4 (*a*) 1. CH_3OH, HCl; 2. BnBr, NaH, DMF; 3. AcOH, HCl. (*b*) 1. $NaBH_4$, EtOH, 98%; 2. Ph_3CCl, NEt_3, DMAP, CH_2Cl_2, 81%; 3. MsCl, Py, 93%. (*c*) 1. NaN_3, 15-crown-5, HMPA, 83%; 2. $LiAlH_4$, ether, 96%. (*d*) BzCl, Py, CH_2Cl_2, 71% or CbzCl, NEt_3, CH_2Cl_2, 72%. (*e*) Aqueous HBF_4, CH_3CN. (*f*) PCC, DMF.

which was benzylated and subsequently hydrolyzed to give **27**[68] in 29% overall yield from D-arabinose. Compound **27** was converted to the tri-*O*-benzyl arabinitol by reduction and the resulting primary hydroxyl group was selectively protected using trityl chloride. Subsequent mesylation gave **28** in 74% overall yield from **27**. Treatment of **28** with sodium azide in the presence of 15-crown-5 produced the corresponding azide, which was reduced to the amine **29**. The amino group of **29** was protected by treatment with benzoyl chloride or benzyloxycarbonyl chloride to give **30** or **31**, respectively, whose detritylation afforded **32** and **33**. Oxidation of each of **32** and **33** with PCC afforded the five-membered lactams **34** and **35** in 34 and 70% yield, respectively, via spontaneous ring closure.

The D-arabinose derivative **27** has also been served as a precursor for the synthesis of fragment **43** (Scheme 5).[69] It was converted into **37**, which was treated with vinylmagnesium bromide to produce the olefin **38**, which upon oxidative degradation with pyridinium chlorochromate gave **39**.[70] Ozonolysis of the vinyl function in **39** followed by reduction afforded, after selective removal of the PMB group with CAN, the lactam **40**. Silylation of the primary hydroxyl group in **40** followed by treatment with Lawesson's reagent afforded the pyrrolidine-2-thione **41**, which was treated with diethylbromomalonate, DBU and triphenylphosphine to furnish after desilylation with fluoride ion the 2-methylidenepyrrolidine **42**. Mesylation of **42** followed by heating in the presence of KHMDS in THF afforded the aziridine **43** in 12% overall yield from **27**.

Scheme 5 (*a*) $PMBNH_2$, toluene, reflux. (*b*) $CH_2{=}CHMgBr$, THF, 71% for two steps. (*c*) PCC, MS 4 Å, CH_2Cl_2, 71%. (*d*) 1. O_3, EtOH, −20°C; then $NaBH_4$, 96%; 2. CAN, CH_3CN, H_2O, 96%. (*e*) 1. TBSOTf, 2,6-lutidine, CH_2Cl_2, 79%; 2. Lawesson's reagent, toluene, reflux, 83%. (*f*) 1. $(EtO_2C)_2CHBr$, CH_2Cl_2; then DBU, Ph_3P, 82%; 2. TBAF, THF, 88%. (*g*) 1. MsCl, NEt_3, CH_2Cl_2, 88%; 2. KHMDS, THF, 60°C, 63%.

The fragments **50** and **51** have been synthesized from the thiolactam **41** (Scheme 6).[71] Methylation of **41** gave the 2-methylthiopyrroline **44**, which underwent condensation with 2-phenyl-Δ^2-oxazolin-5-one to furnish the adduct **45** as an inseparable 6:4 mixture of the E- and Z-isomers. The stable Δ^2-oxazolin-5-one derivative **45** can be activated by acylation of the amino group of the pyrrolidine ring as the corresponding N-allyloxycarbonyl (N-Alloc) derivative. Its treatment with isopropylamine gave the separable (pyrrolidin-2-ylidene)glycine amide **47***E* and **47***Z* in 20 and 45% yield, respectively, which were readily transformed into mesylates **49** by a combination of desilylation, mesylation and removal of the N-Alloc group. Finally, construction of the aziridine ring was achieved by treating **49** with KHMDS in THF to furnish **50** as a single isomer in 15–31% overall yield from **41**. On

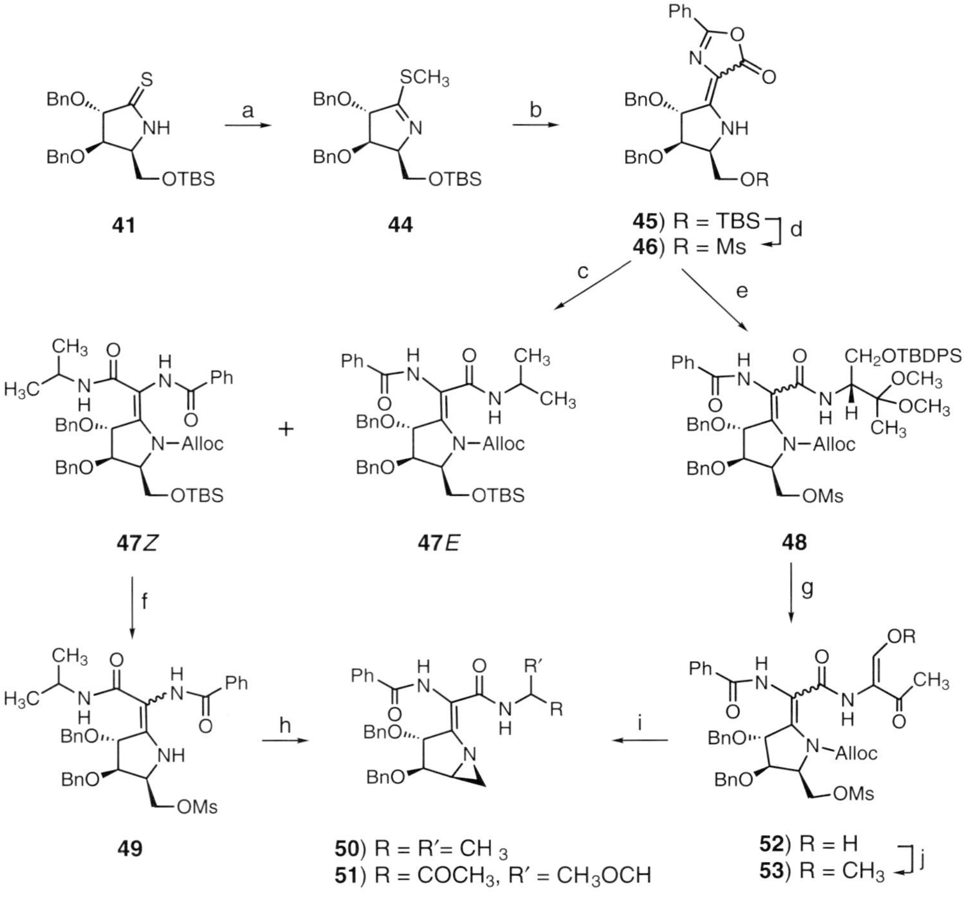

Scheme 6 (*a*) CH_3I, CH_2Cl_2, 99%. (*b*) 2-Phenyl-Δ^2-5-oxazolinone, toluene, 80°C, 82%. (*c*) $Alloc_2O$, DMAP, THF, then isopropylamine, 20% (**47***E*), 45% (**47***Z*). (*d*) 1. TBAF, THF, 90%; 2. MsCl, NEt_3, CH_2Cl_2, −78°C, 90%. (*e*) 1. $Alloc_2O$, DMAP, THF; 2. $TBDPSOCH_2CH(NH_2)CH(OCH_3)_2$, toluene, 68% (**48***E*), 23% (**48***Z*) for two steps. (*f*) 1. HCl, CH_3OH, 98%; 2. MsCl, NEt_3, CH_2Cl_2, −78°C, 65–96%; 3. $Pd(Ph_3P)_4$, Ph_3P, dimedone, THF, 87–97%. (*g*) 1. HF, Py, *E* (93%), *Z* (79%); 2. PCC, MS 4 Å, CH_2Cl_2, *E* (68%), *Z* (83%); 3. *p*-TsOH, THF, H_2O. (*h*) KHMDS, THF, 64%. (*i*) TBAF, MS 4 Å, THF, 73%. (*j*) 1. CH_2N_2, THF, Et_2O, *E* (67%), *Z* (78%); 2. $Pd(Ph_3P)_4$, Ph_3P, AcOH, THF, *E* (57%), *Z* (88%).

the other hand, compound **45** was transformed into mesylates **46**, which were acylated and transformed to the amide **48** (91%) as a mixture of *E*- and *Z*-isomers. These were converted into the 2-methylidenepyrrolidine **53** via the β-ketoaldehyde **52**. Finally, TBAF was found to be effective for the aziridine ring formation in **53**, to give **51** as a single isomer.

The core structure **63** of azinomycins A and B has been synthesized from the 1,1-bis(ethylthio)-4,6-isopropylidene D-arabinose derivative **54**[63] by treatment with *p*-methoxybenzyl chloride, followed by removal of the dithioacetal groups with *N*-bromosuccinimide and subsequent reduction of the resulting aldehyde to afford **55** (Schemes 7 and 8).[72–74]

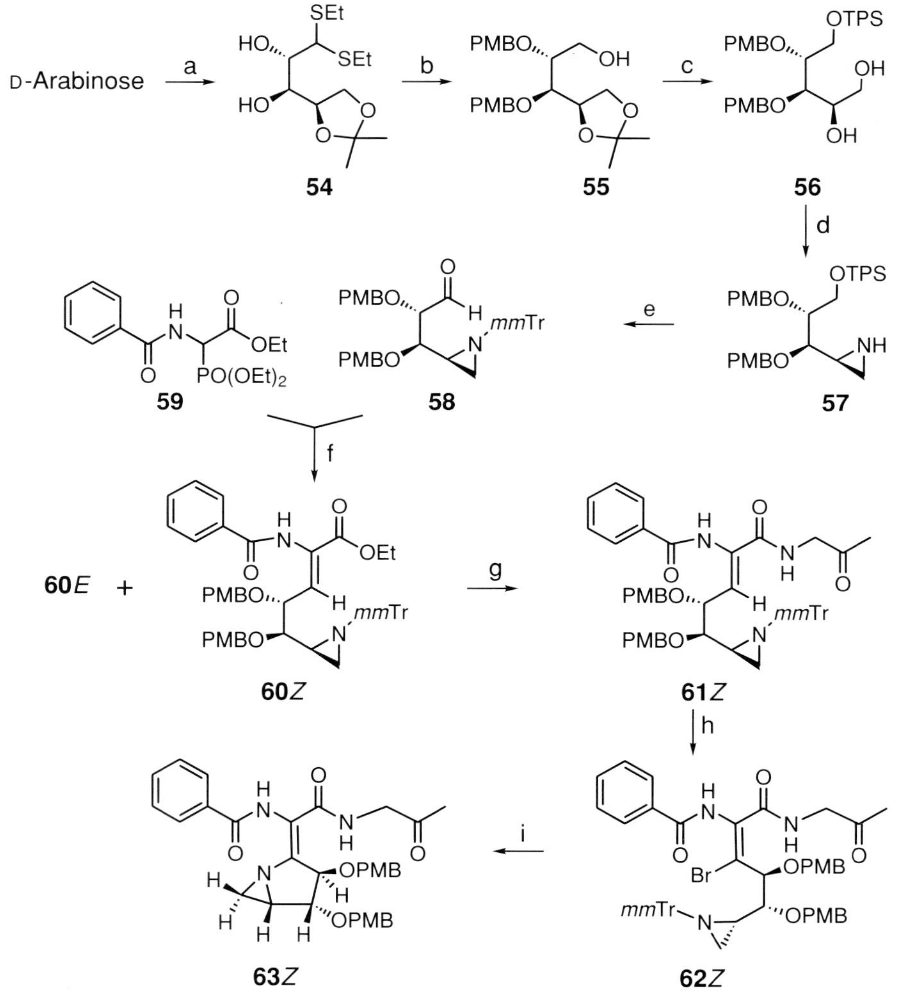

Scheme 7 (*a*) Ref. 63. (*b*) 1. PMBCl, NaH, DMF, 25°C, 76%; 2. NBS, CH_3CN, H_2O, 2,6-lutidine; 3. $NaBH_4$, EtOH, 74%. (*c*) 1. TPSCl, imidazole, DMF, rt, 95%; 2. PPTS, CH_3OH, 50°C, 66%. (*d*) 1. MsCl, Py, CH_2Cl_2; 2. NaN_3, DMF, 50°C; 3. $(Ph)_3P$, toluene, 40°C, 76%. (*e*) 1. *mm*TrCl, Py, NEt_3, −78 to 0°C, 64%; 2. TBAF, THF, rt, 92%; 3. $(COCl)_2$, DMSO, NEt_3, CH_2Cl_2, −78°C. (*f*) LDA, THF, −78 to 0°C, 3.7:1 *Z*/*E*, 70%. (*g*) 1. LiOH, THF, 50°C; 2. 2-propanol amine, DCC, HOBt, CH_2Cl_2, rt; 3. $(COCl)_2$, DMSO, NEt_3, CH_2Cl_2, −78°C, 59%. (*h*) Br_2, −78°C, DABCO, CH_2Cl_2, 71%. (*i*) 1. TCA, $CDCl_3$, rt; 2. NEt_3, $CDCl_3$, 50°C, 65%.

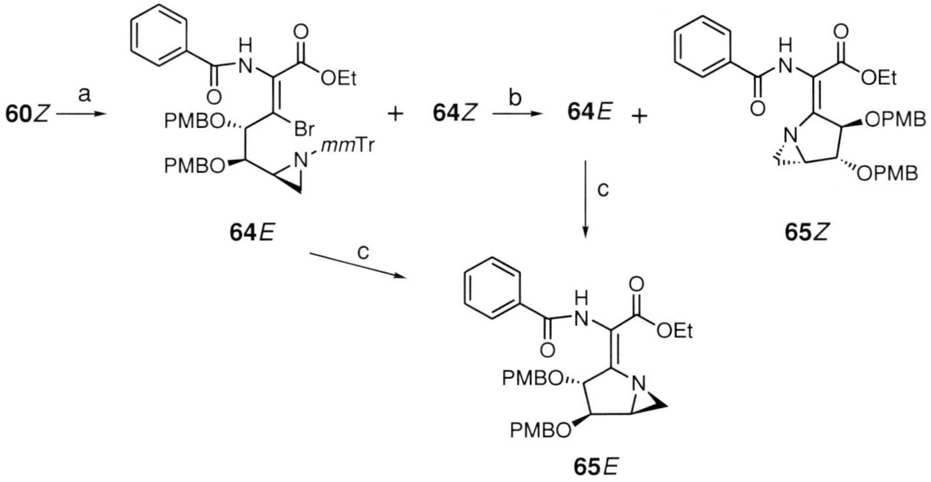

Scheme 8 (*a*) Br_2, 2,6-lutidine, −78 to 0°C, 1.5 h, DABCO, 77%, 1:1.5 *E*/*Z*; *or* NBS, $CHCl_3$, rt, 76%, *E* only. (*b*) 1. TCA, CD_3CN, rt; 2. NEt_3, 50°C, 4 h, **64***E* (32%), **65***Z* (65%). (*c*) 1. TCA, CD_3CN, rt, 45 min; 2. NEt_3, 50°C, 16 h, 91%.

Silylation of **55** followed by removal of the isopropylidene group afforded the diol **56**. Selective mesylation followed by azidolysis and subsequent transformation, via an aza-ylide intermediate,[75] afforded the aziridine **57**. Protection of the secondary amine with *mm*Tr group followed by desilylation and subsequent Swern oxidation of the generated primary hydroxyl group afforded the aldehyde **58**. Condensation of **58** with ethyl *N*-benzoyl-α-(diethylphosphono)glycinate (**59**) using LDA afforded the dehydroamino acid ester products **60***E* and **60***Z*. Hydrolysis of **60***Z* to the corresponding carboxylic acid followed by coupling with 1-amino-2-propanol and subsequent oxidation afforded **61***Z*. However, under the same conditions the *E*-isomer decomposed without conversion to its corresponding amide. Bromination of **61***Z* followed by treatment with DABCO afforded a single β-bromo dehydroamino acid amide isomer **62***Z*. Removal of the *mm*Tr group with trichloroacetic acid (TCA) followed by formation of the azabicyclo[3.1.0]hex-2-ylidene ring system by heating with NEt_3 provided **63***Z* in 3% overall yield from **54**.

On the other hand, treatment of the dehydroamino ester **60***Z* with Br_2 in the presence of 2,6-lutidine afforded a mixture of the respective *Z* and *E* vinyl bromides **64** (1.5:1). Treatment of this mixture with TCA led to deprotection of the *E*-isomer more rapidly than the *Z*-isomer, followed by heating in NEt_3 to afford a single azabicyclo[3.1.0] **65***Z*. The vinyl bromide **64***E* was deprotected by TCA in CD_3CN and then cyclized with NEt_3 to give [3.1.0] product **65***E* in 91% yield (Scheme 8).[72–74]

References

1. Nagaoka, K.; Matsumoto, M.; Ono, J.; Yokoi, K.; Ishizeki, S.; Nakashima, T. *J. Antibiot.* **39** (1986) 1527.
2. Yokoi, K.; Nagaoka, M.; Nakashima, T. *Chem. Pharm. Bull.* **34** (1986) 4554.
3. Ishizeki, S.; Ohtsuka, M.; Irinoda, K.; Kukita, K.-I.; Nagaoka, K.; Nakashima, T. *J. Antibiot.* **40** (1987) 60.
4. Tanaka, M.; Kishi, T.; Maruta, Y. *J. Antibiot., Ser. B* **12** (1959) 361.

5. Tanaka, M.; Kishi, T.; Maruta, Y. *J. Antibiot., Ser. B* **13** (1960) 177.
6. Onda, M.; Konda, Y.; Noguchi, A.; Omura, S.; Hata, T. *J. Antibiot., Ser. B* **22** (1969) 42.
7. Onda, M.; Konda, Y.; Hatano, A.; Hata, T.; Ōmura, S. *Chem. Pharm. Bull.* **32** (1984) 2995.
8. Hata, T.; Koga, F.; Sano, T.; Kanamori, K.; Matsumae, A.; Sugawara, R.; Shima, T.; Ito, S.; Tomizawa, S. *J. Antibiot., Ser. A (Tokyo)* **7** (1954) 107.
9. Hata, T.; *Tanpakushitu Kakusan Koso* **4** (1959) 96.
10. Onda, M.; Konda, Y.; Omura, S.; Hata, T. *Chem. Pharm. Bull.* **19** (1971) 2013.
11. Lown, J.W.; Hanstock, C.C. *J. Am. Chem. Soc.* **104** (1982) 3213.
12. Onda, M.; Konda, Y.; Hatano, A.; Hata, T.; Ōmura, S. *J. Am. Chem. Soc.* **105** (1983) 6311.
13. Shibuya, M. *Tetrahedron Lett.* **24** (1983) 1175.
14. Garner, P.; Park, J.J.; Rotello, V. *Tetrahedron Lett.* **26** (1985) 3299.
15. Moran, E.J.; Armstrong, R.W. *Tetrahedron Lett.* **32** (1991) 3807.
16. Kondo, Y.; Machida, T.; Sasaki, T.; Takeda, K.; Takayanagi, H.; Harigaya, Y. *Chem. Pharm. Bull.* **42** (1994) 285.
17. Armstrong, R.W.; Tellew, J.E.; Moran, E.J. *J. Org. Chem.* **57** (1992) 2208.
18. England, P.; Chun, K.H.; Moran, E.J.; Armstrong, R.W. *Tetrahedron Lett.* **31** (1990) 2669.
19. Hashimoto, M.; Matsumoto, M.; Yamada, K.; Terashima, S. *Tetrahedron Lett.* **35** (1994) 2207.
20. Armstrong, R.W.; Salvati, M.E.; Naguyen, M. *J. Am. Chem. Soc.* **114** (1992) 3144.
21. Lown, J.W.; Majumdar, K.C. *Can. J. Biochem.* **55** (1977) 630.
22. Partt, W.B.; Ruddon, R.W.; Ensminger, W.D.; Maybaum, J. *The Anticancer Drugs*, 2nd ed.; Oxford University Press: Oxford, 1994.
23. Terawaki, A.; Greenberg, J. *Nature* **209** (1966) 481.
24. Bryant, H.J.; Dardonville, C.Y.; Hodgkinson, T.J.; Hursthouse, M.B.; Abdul Malik, K.M.; Shipman, M. *J. Chem. Soc., Perkin Trans.* **1** (1998) 1249.
25. Bryant, H.J.; Dardonville, C.Y.; Hodgkinson, T.J.; Shipman, M.; Slawin, A.M.S. *Synlett* (1996) 973.
26. Coleman, R.S.; Carpenter, A.J. *Tetrahedron* **53** (1997) 16313.
27. Armstrong, R.W.; Tellew, J.E.; Moran, E.J. *Tetrahedron Lett.* **37** (1996) 447.
28. Hashimoto, M.; Terashima, S. *Heterocycles* **47** (1998) 59.
29. Coleman, R.S.; Kong, J.-S. *J. Am. Chem. Soc.* **120** (1998) 3538.
30. Coleman, R.S.; Sarko, C.R.; Gittinger, J.P. *Tetrahedron Lett.* **38** (1997) 5917.
31. Coleman, R.S.; Carpenter, A.J. *J. Org. Chem.* **58** (1993) 4452.
32. Coleman, R.S.; Mckinley, J.D. *Tetrahedron Lett.* **39** (1998) 3433.
33. Shishido, K.; Omodai, T.; Shibuya, M. *J. Chem. Soc., Perkin Trans. 1* (1992) 2053.
34. Shibuya, M.; Terauchi, H. *Tetrahedron Lett.* **28** (1987) 2619.
35. England, P.; Chun, K.H.; Moran, E.J.; Armstrong, R.W. *Tetrahedron Lett.* **31** (1990) 2669.
36. Hashimoto, M.; Matsumoto, M.; Yamada, K.; Terashima, S. *Tetrahedron Lett.* **35** (1994) 2207.
37. Coleman, R.S.; Li, J.; Navarro, A. *Angew. Chem., Int. Ed.* **40** (2001) 1736.
38. Hodghinson, T.J.; Shipman, M. *Tetrahedron* **57** (2001) 4467.
39. Miyashita, K.; Park, M.; Adachi, S.; Seki, S.; Obika, S.; Imanishi, T. *Bioorg. Med. Chem. Lett.* **12** (2002) 1075.
40. Hartley, J.A.; Hazrati, A.; Kelland, L.R.; Khanim, R.; Shipman, M.; Suzenet, F.; Walker, L.F. *Angew. Chem., Int. Ed.* **39** (2000) 3467.
41. Coleman, R.S.; Kong, J.-S.; Richardson, T.E. *J. Am. Chem. Soc.* **121** (1999) 9088.
42. Hodgkinson, T.J.; Kelland, L.R.; Shipman, M.; Suzenet, F. *Bioorg. Med. Chem. Lett.* **10** (2000) 239.
43. Hartley, J.A.; Hazrati, A.; Hodgkinson, T.J.; Kelland, L.R.; Khanim, R.; Shipman, M.; Suzenet, F. *Chem. Commun.* (2000) 2325.
44. Fujiwara, T.; Saito, I.; Sugiyama, H. *Tetrahedron Lett.* **40** (1999) 315.
45. Coleman, R.S. *Synlett* (1998) 1031.
46. Shibuya, M.; Terauchi, H. *J. Pharm. Sci.* **76** (1987) 213.
47. Yoshimura, J.; Hara, K.; Yamaura, M.; Mikami, K.; Hashimoto, H. *Bull. Chem. Soc. Jpn.* **55** (1982) 933.
48. Schreiber, S.L.; Schreiber, T.S.; Smith. D.B. *J. Am. Chem. Soc.* **109** (1987) 1525.
49. Armstrong, R.W.; Combs, A.P.; Tempest, P.A.; Brown, S.D.; Keating, T.A. *Acc. Chem. Res.* **29** (1996) 123.
50. Kim, S.W.; Bauer, S.M.; Armstrong, R.W. *Tetrahedron Lett.* **39** (1998) 7031.
51. Coleman, R.S.; Richardson, T.E.; Carpenter, A.J. *J. Org. Chem.* **63** (1998) 5738.
52. Coleman, R.S.; Richardson, T.E.; Kong, J.-S. *Kim. Geterotsikl. Soedin.* (1998) 1632.
53. Hashimoto, M.; Yamada, K.; Terashima, S. *Chem. Lett.* (1992) 975.
54. Hashimoto, M.; Terashima, S. *Chem. Lett.* (1994) 1001.
55. Hodgkinson, T.J.; Kelland, L.R.; Shipman, M.; Vile, J. *Tetrahedron* **54** (1998) 6029.
56. Goujon, J.-Y.; Shipman, M. *Tetrahedron Lett.* **43** (2002) 9573.
57. Coleman, R.S.; Perez, R.J.; Burk, C.H.; Navarro, A. *J. Am. Chem. Soc.* **124** (2002) 13008.
58. Coleman, R.S.; Burk, C.H.; Navarro, A.; Brueggemeier, R.W.; Diaz-Cruz, E.S. *Org. Lett.* **4** (2002) 3545.

59. Ando, K.; Yamada, T.; Shibuya, M. *Heterocycles* **29** (1989) 2209.
60. Shibuya, M.; Terauchi, H. *Tetrahedron Lett.* **28** (1987) 2619.
61. Shibuya, M. *Tetrahedron Lett.* **24** (1983) 1175.
62. Coleman, R.S.; Carpenter, A.J. *J. Org. Chem.* **57** (1992) 5813.
63. Coleman, R.S.; Dong, Y.; Carpenter, A.J. *J. Org. Chem.* **57** (1992) 3732.
64. Chargaff, E.; Bovarnick, M. *J. Biol. Chem.* **118** (1937) 421.
65. Neuberger, A.; Rivers, R.P. *J. Chem. Soc.* (1939) 132.
66. Bernet, B.; Vasella, A.; *Helv. Chim. Acta* **67** (1984) 1328.
67. Konda, Y.; Machida, T.; Akaiwa, M.; Takeda, K; Harigaya, Y. *Heterocycles* **43** (1996) 555.
68. Tejima, S.; Fleetcher, H.G., Jr., *J. Org. Chem.* **28** (1963) 2999.
69. Hashimoto, M.; Terashima, S. *Chem. Lett.* (1994) 1001.
70. Lay, L.; Nicotra, F.; Paganini, A.; Pangrazio, C.; Panza, L. *Tetrahedron Lett.* **34** (1993) 4555.
71. Hashimoto, M.; Terasima, S. *Tetrahedron Lett.* **35** (1994) 9409.
72. Armstrong, R.W.; Moran, E.J. *J. Am. Chem. Soc.* **114** (1992) 371.
73. Armstrong, R.W.; Tellew, J.E.; Moran, E.J. *J. Org. Chem.* **57** (1992) 2208.
74. Moran, E.J.; Tellew, J.E.; Zhao, Z.; Armstrong, R.W. *J. Org. Chem.* **58** (1993) 7848.
75. Staudinger, H.; Meyer, M. *Helv. Chem. Acta* **2** (1919) 519.

5.2 4:5-Fused heterocycles

5.2.1 *β-Lactams*

The naturally occurring β-lactam antibiotics are a well-known group of compounds. (2*R*,5*S*)-2-(Hydroxymethyl)clavam {(3*R*,5*S*)-3-hydroxymethyl-4-oxa-1-azabicyclo-[3.2.0]heptan-7-one, **1**} has been isolated from culture fluids of *Streptomyces clavuligerus*.[1] It exhibits antifungal activity against a number of fungi. Clavalanine {Ro 22-5417, 3-[(3*S*,5*S*)-7-oxo-1-aza-4-oxabicyclo[3.2.0]hept-3-yl]-L-alanine, **2**} was isolated from *Str. clavuligerus*.[2–4] It is an antimetabolite of *O*-succinylhomoserine and intervenes in the biosynthesis of methionine, whereas most β-lactams are peptidoglycan biosynthesis inhibitors.[2–4]

Clavulanic acid (**3**) was isolated from the fermentation of the microorganism *Str. clavuligerus*.[5] It is a highly potent, broad spectrum and irreversible β-lactams inhibitor[6] with clinical applications.[7] The β-lactam **4**, structurally related to clavulanic acid, has been isolated from *Str. clavuligerus*.[2–4]

Carbapenem SQ 27860 {(5*R*)-7-oxo-1-aza-bicyclo[3.2.0]hept-2-ene-2-carboxylic acid, **5**} was the first carbapenem antibiotic produced by bacteria.[8] It shows a wide antibacterial spectrum and it is unstable, but it can be characterized as its *p*-nitrobenzyl ester.[8]

(+)-Thienamycin (**6**),[9–11] 8-*epi*-thienamycin (**7**),[12] olivanic acids[13–16] and (+)-PS-5 (**8**)[17,18] were isolated from the fermentation broths of the soil microorganism *Streptomyces cattleya* and exhibit unique activities as broad spectra antibiotics.[19–23] Thienamycin (**6**) has two peerless functional and structural features compared to the traditional penicillins (**9**):

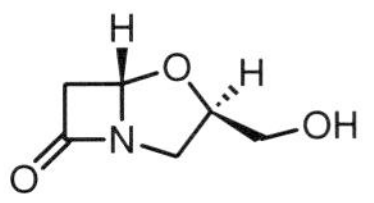

2-(Hydroxymethyl)clavam **1**

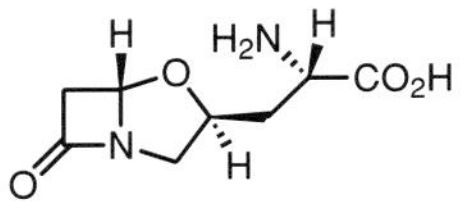

Clavalanine (Ro 22-5417) **2**

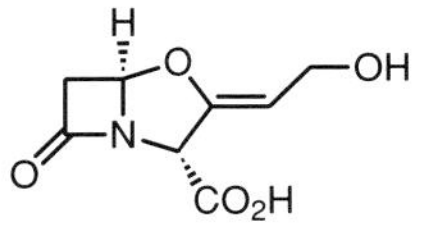

Clavulanic acid **3**

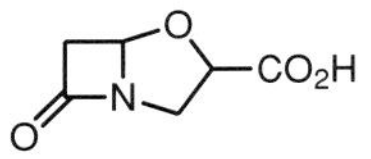

4

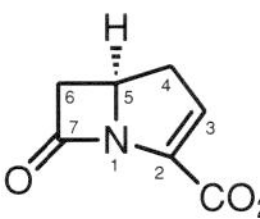

Carbapenem SQ 27860 **5**

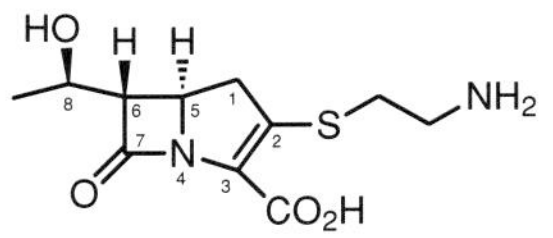

(+)-Thienamycin **6**

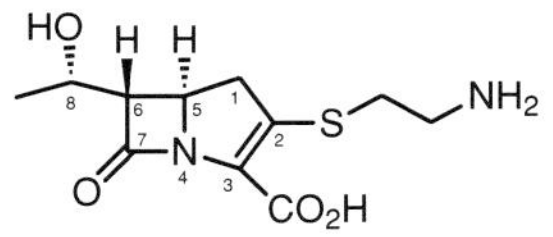

8-*epi*-Thienamycin **7**

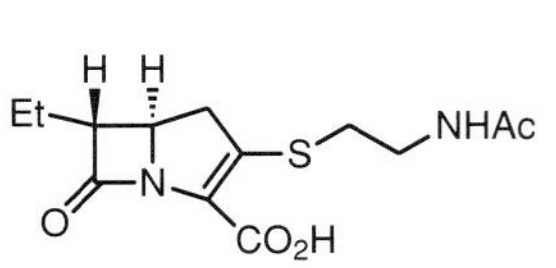

(+)-PS-5 **8**

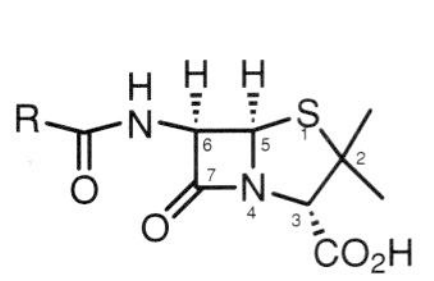

Penicillins **9**

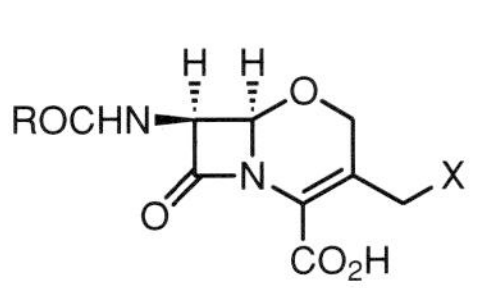

Oxacephalosporins **10**

the α-hydroxyethyl side chain, which replaces the traditional 6-acylamino group; and the unsaturated substituted pyrrolidine ring instead of the thiazine ring.

1-Oxacephems, exemplified by 1-oxacephalosporins (**10**), exhibit a higher antibacterial activity than that of 1-thia congeners.[24,25]

Syntheses of the β-lactams and their analogues from noncarbohydrates have been frequently reported.[26–63] Those syntheses from carbohydrates are presented here.

5.2.1.1 *Synthesis from* D-*allose* A synthesis of the β-lactam **15** has been reported from the D-allose derivative **11** (Scheme 1).[64] Silylation of **11** followed by treatment with trichloroacetyl isocyanate afforded after crystallization the β-lactam **12** (50%). Benzylation of **12** afforded **13**, which underwent removal of the protecting groups to give **14**. Cleavage of the vicinal diol group in **14**, with sodium metaperiodate under standard conditions, led to the formation of a dialdehyde, which without isolation was reduced to give the β-lactam derivative **15**.

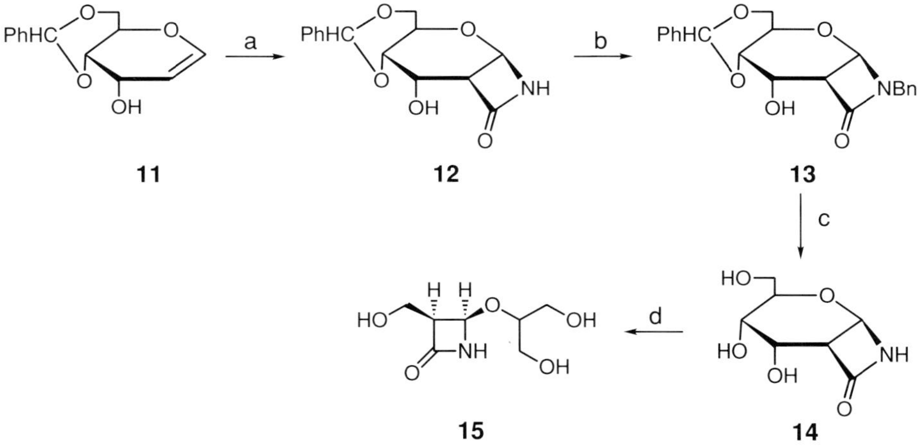

Scheme 1 (*a*) 1. TMSCl, Py; 2. CH_3NO_2, $Cl_3CCONCO$, rt, 4 days, crystallization from CH_3OH, 50%. (*b*) Benzene, K_2CO_3, Bu_4NBr, BnBr, reflux, 2 h, 68%. (*c*) CH_3OH, rt, 10% Pd *on* C, H_2, 4 h, 80%. (*d*) $NaIO_4$, CH_3OH, H_2O, $(NH_4)_2SO_4$, −5°C, 5 min; then $NaBH_4$, 90%.

5.2.1.2 *Synthesis from* D-*galactose* The β-lactam isomers **17** and **18** have been prepared-from the D-galactose derivative **16** (Scheme 2).[65,66] Isomerization of **16** with $NaHCO_3$ followed by treatment with sodium metaperiodate and then reduction afforded **17**. On the

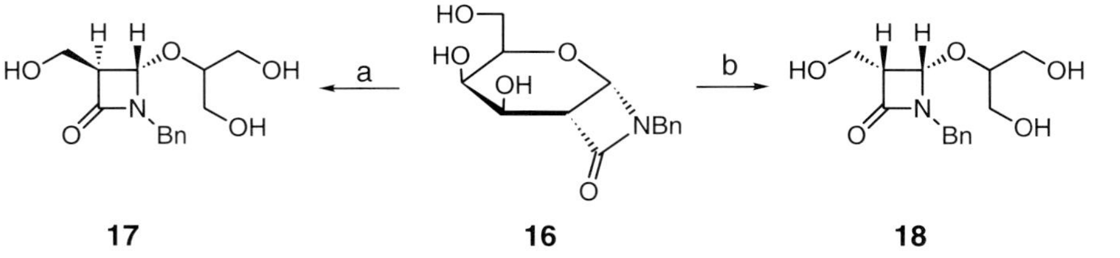

Scheme 2 (*a*) 1. CH_3OH, H_2O, $NaHCO_3$, −5 to 16°C; 2. $NaIO_4$, 30 min, then $NaBH_4$, 29%. (*b*) CH_3OH, H_2O, $(NH_4)_2SO_4$, −5°C, $NaIO_4$, 15 min; then $NaBH_4$, H_2O, 78%.

other hand, treatment of **16** with sodium metaperiodate in the presence of ammonium sulfate followed by sodium borohydride reduction led to the formation of the β-lactam **18**.

5.2.1.3 *Synthesis from D-glucose* Synthesis of 6-*epi*-thienamycin (**34**) has been achieved from D-glucose by transforming it to the epoxide **20**, via **19**,[67] in 30% overall yield (Scheme 3).[68] Reaction of diethylaluminum cyanide with **20** gave regioselectively the 4-cyano compound **21**, whose hydrolysis to **22** followed by mesylation gave **23**. Subsequent cyclization with *t*-BuOK afforded 45% yield of the azetidinone **25**, together with 20% of the unsaturated amide **24**. Mild hydrolysis of **25** furnished the hemiacetal **26**, which underwent Wittig reaction to give the unsaturated ester **27**. Oxidation of the protected ester **28** afforded

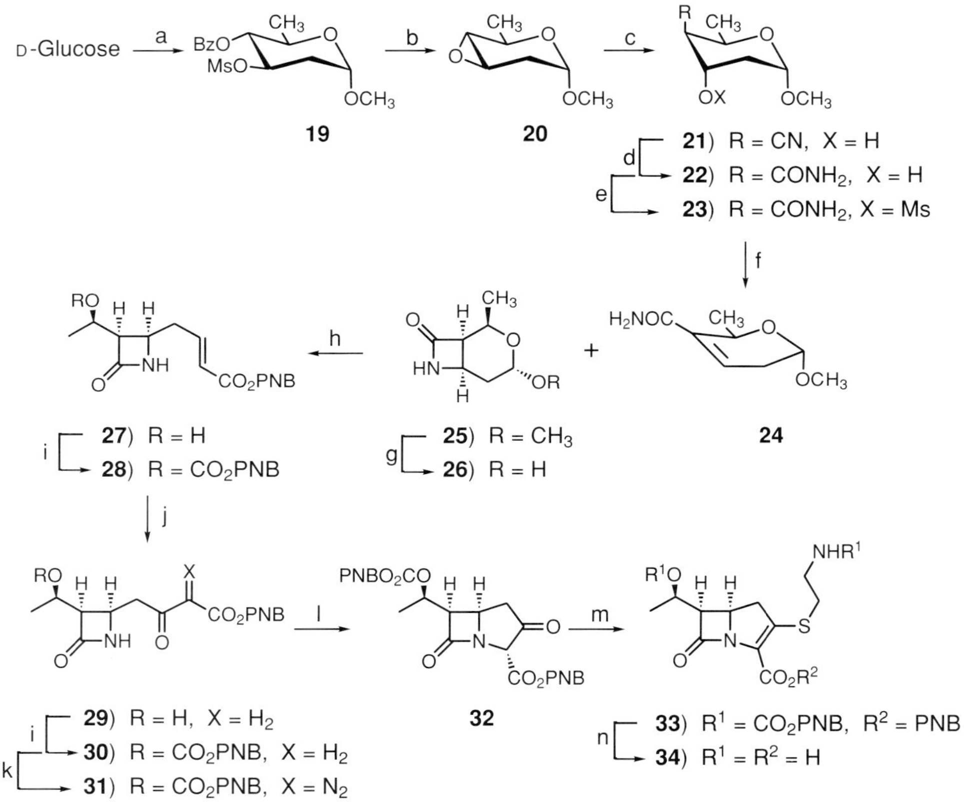

Scheme 3 (*a*) Ref. 67. (*b*) $NaOCH_3$, CH_3OH, 0°C to rt, 5 h, 90%. (*c*) Et_2AlCN, Et_2O, −40°C, 3 h, 65%. (*d*) H_2O_2, 1 M K_2CO_3, H_2O, rt, 12 h, 90%. (*e*) MsCl, Py, 0°C to rt, 12 h, 77%. (*f*) *t*-BuOK, 18-crown-6, DMF, 0°C, 3 h, 45%. (*g*) 70% aqueous HCO_2H, rt, 2.5 h, 95%. (*h*) $Ph_3P{=}CHCO_2PNB$, CH_3CN, 80°C, 3 h, 35%. (*i*) $ClCO_2PNB$, 4-*N*,*N*′-dimethylaminopyridine, CH_2Cl_2, −10°C, 1 h, heat up to 0°C, 4 h, 76%. (*j*) *t*-BuO_2H, $PdCl_2$, Na_2PdCl_4, 50% aqueous AcOH, 60°C, 70 min, 67%. (*k*) $CH_3C_6H_4SO_2N_3$, NEt_3, CH_3CN, 0°C, 2 h, 85%. (*l*) $Rh_2(OAc)_4$, benzene, 80°C, 2 h, 91%. (*m*) 1. $(PhO)_2P(O)Cl$, $(i\text{-}Pr)_2EtN$, CH_3CN, 0°C, 1 h; 2. $(i\text{-}Pr)_2EtN$, $HS(CH_2)_2NH{-}CO_2PNB$, CH_3CN, 0°C, 1 h; then −25°C, 24 h, 82%. (*n*) H_2, Pd *on* C, THF, sodium morpholine propane sulfonate, rt, 1 h, 60%.

the ketoester **29**, whose hydroxyl group was protected to give **30**.[69] The synthesis was completed[70] by conversion to the diazo compound **31**, followed by heating with $Rh_2(OAc)_4$ to provide the bicyclic lactam **32** as a single diastereoisomer. The cysteamine side chain was then introduced to furnish **33**, whose hydrogenolysis gave **34**.

The lactone **41**, as intermediate for the synthesis of thienamycin, has been prepared from D-glucose (Scheme 4).[71] The glucoside derivative **35**,[72] obtained from D-glucose, was treated with NBS to give the bromodeoxy derivative **36**, which underwent hydrogenation followed by S_N2 displacement of the mesyloxy group with azide ion and subsequent debenzoylation with sodium methoxide to furnish **37**. Oxidation of the OH group in **37** with pyridinium chlorochromate and subsequent condensation of the resulting ketone with *O*,*O*-dimethyl formylphosphonate *S*,*S*-dimethyl thioacetal[73] afforded the ketenedithioacetal derivative **38**, which was reduced with $LiAlH_4$, followed by treatment with trifluoroacetic anhydride to produce the *N*-trifluoroacetyl derivative **39**. Oxidation of the dithioacetal functional group in **39** with $HgCl_2$ and HgO in aqueous acetone followed by treatment with CrO_3 and subsequent diazotization afforded the ester **40**, which was subjected to selective hydrolysis, oxidation and deprotection to give lactone **41**. Treatment of **41** with benzyl alcohol followed by cyclization of the resulting open chain **42** gave the β-lactam **43**. Compound **43** was prepared from noncarbohydrate precursors and used in a total synthesis of (±)-thienamycin.[74,75]

Scheme 4 (*a*) Ref. 72. (*b*) NBS, CCl_4, $BaCO_3$, reflux, 95%. (*c*) 1. Pd *on* C, H_2, EtOAc, 90%; 2. *n*-Bu_4NN_3, benzene, reflux, 85%; 3. $NaOCH_3$, CH_3OH, 93%. (*d*) 1. PCC, CH_2Cl_2, MS 4 Å, 95%; 2. $(CH_3O)_2POCH(SCH_3)_2$, *n*-BuLi, THF, −78 to 25°C, 60%. (*e*) 1. $LiAlH_4$, THF, 84%; 2. TFAA, Py, CH_2Cl_2, 87%. (*f*) 1. $HgCl_2$, HgO, aqueous acetone, 76%; 2. CrO_3, acetone, H_2SO_4; then CH_2N_2, 90%. (*g*) 1. 12 N HCl–THF (2:5), 68%; 2. aqueous Br_2, CH_3CN, $CaCO_3$, 80%; 3. 12 N HCl, reflux; then azeotropic evaporation from toluene at 50°C, 73%; evaporation at low temperature. (*h*) BnOH, 7 mL/g, 70°C, 4.5 h. (*i*) 1. NEt_3 (1.0 equiv.), DCC (1.0 equiv.), BnOH, 55°C, 4.5 h; 2. 40 psi H_2, Pd *on* C.

Alternatively, (+)-thienamycin (**6**) has been synthesized from D-glucose by conversion into **44**[76,77] (Scheme 5).[78,79] Hydrogenation of the azido group in **44** and protection of the resulting amino group gave **45**. Swern oxidation of **45** followed by Horner–Wittig reaction using methoxymethyldiphenylphosphine oxide and subsequent treatment with potassium hydride afforded **46** as a mixture of isomers in a ratio of 2.8:1. The major isomer was converted into the ester **47** in two steps in 21% yield. The glucosidic bond in **47** was hydrolyzed with aqueous HCl, followed by Jones oxidation to provide the δ-lactone **48**. Deprotection of the amino group and then acid hydrolysis of the ester group gave **49**, which was heated in benzyl alcohol and subsequently cyclized with *N*,*N*′-dicyclohexylcarbodiimide to give the β-lactam **50**. Removal of the benzyl group from **50** followed by treatment with *N*,*N*′-carbonyldiimidazole and subsequent addition of the magnesium salt of *p*-nitrobenzyl hydrogen malonate afforded the β-ketoester **51**. This was treated with $Rh_2(OAc)_4$ to afford the bicyclic ketone **52**, whose enol phosphate was reacted

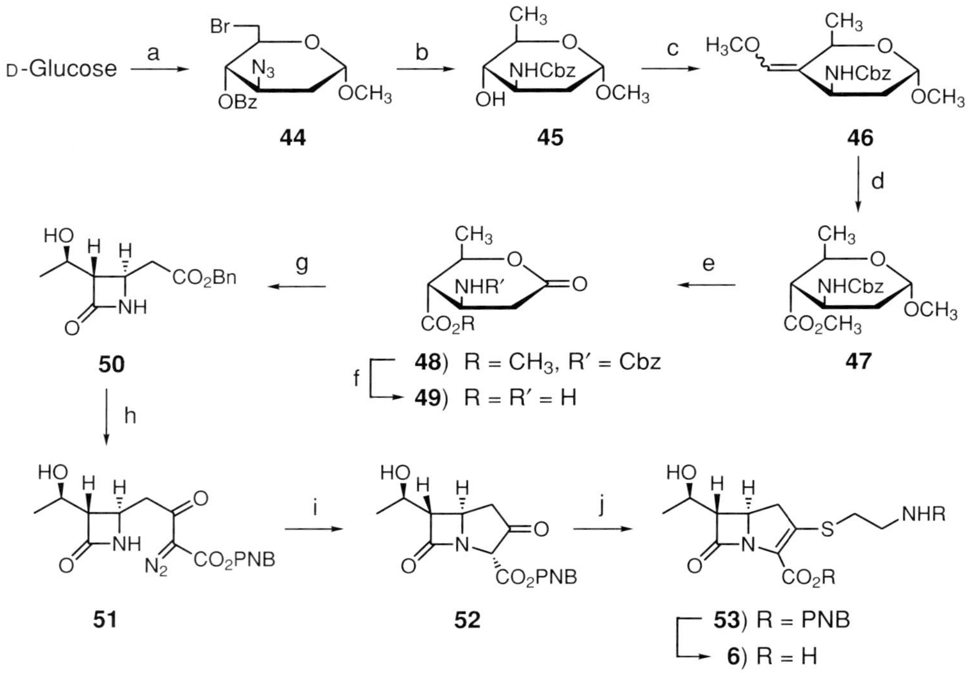

Scheme 5 (*a*) Refs. 76 and 77. (*b*) 1. H_2, Raney nickel (W-4), NEt_3, CH_3OH, 8 h; 2. NEt_3, CH_2Cl_2, CbzCl, rt, 4 h, 55% for two steps. (*c*) 1. TFAA, DMSO, CH_2Cl_2, −78°C; then NEt_3, 95%; 2. $Ph_2P(O)CH_2OCH_3$, THF, LDA, $(i\text{-Pr})_2NH$, BuLi, −65°C, 1.5 h, 92%; 3. DMF, KH, −10 to 0°C, 1 h; 75%. (*d*) 1. 5 N H_2SO_4, THF, rt, 8 h, 40%; 2. PCC, CH_2Cl_2, rt, 24 h, 30%; *or* $PdCl_2$, CuCl, DMF, H_2O, rt, 1 h, O_2, 70°C, 19 h, 53%. (*e*) 1. 0.6 N HCl, THF, reflux, 3 h; 2. Jones reagent, acetone, 0°C, 40 min, 65% for two steps. (*f*) H_2, 5% Pd *on* C, CH_3OH, 1 h; then conc. HCl, reflux, 40 min, 100%. (*g*) BnOH, DCC, NEt_3, 55°C, 5.5 h, 64%. (*h*) 1. CH_3OH, H_2, 5% Pd *on* C, rt, 1 h; 2. DMF, CH_3CN, *N*,*N*′-carbonyldiimidazole, rt, 1.5 h, magnesium salt of *p*-nitrobenzyl hydrogen malonate, rt, overnight, 74%; 3. *p*-carboxybenzenesulfonyl azide, NEt_3, rt, 30 min, 74%. (*i*) $Rh_2(OAc)_4$, benzene, reflux, 3 min, 74%. (*j*) $(i\text{-Pr})_2NEt$, CH_3CN, −20°C, 2 min, diphenyl phosphorochloridate, −20°C, 1.5 h; then *N*-*p*-nitrobenzyloxycarbonylamino ethanethiol, −20°C, 2.5 h, 65%.

with *N*-*p*-nitrobenzyloxycarbonylamino ethanethiol to afford the protected thienamycin **53**, which upon catalytic hydrogenation gave **6**.

3,5-Di-*epi*-clavalanine (**62**) was synthesized from diacetone D-glucose by conversion to 1,2-*O*-isopropylidene-3-*O*-tosyl-α-D-xylofuranose (**54**) (Scheme 6).[4] Oxidation of **54** with Jones reagent followed by esterification afforded **55**, which was transformed into the 3-deoxyaraburonic acid derivative **57** via the unstable α,β-unsaturated ester **56**. Condensation of **57** with 4-acetoxy-2-azetidinone[80] in the presence of palladium acetate afforded a 4:1 mixture of the respective epimers. After separation, the major one was debenzylated and esterified with diazomethane to afford the methyl ester **58**. Reduction of **58** with sodium borohydride followed by tosylation afforded **59**, which was treated with lithium azide at room temperature followed by lithium bromide to give the α-azido ester **60**. Cyclization of **60** to the clavam **61** was achieved with lithium *tert*-butoxide. Hydrogenation of **61** gave **62**.

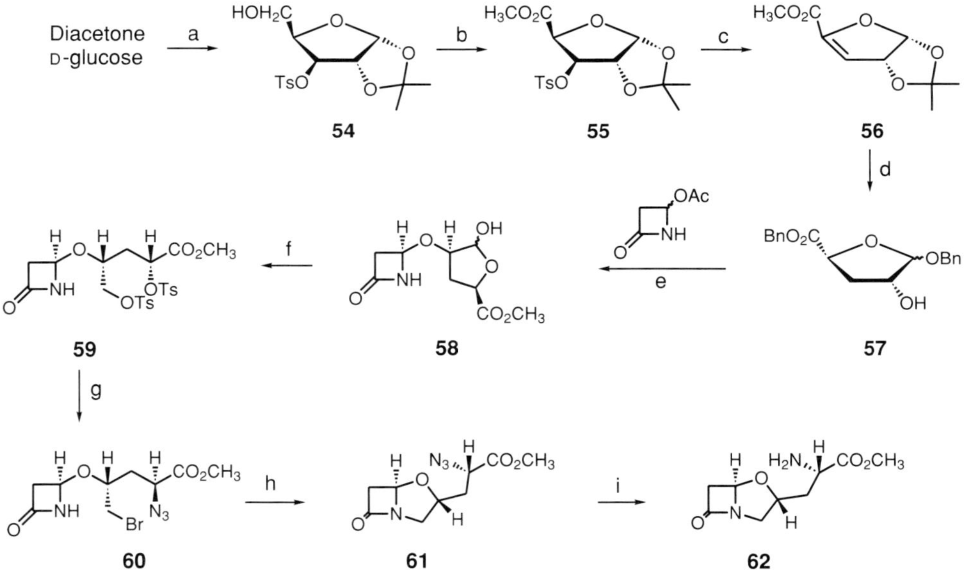

Scheme 6 (*a*) 1. *p*-TsCl, Py; 2. acid hydrolysis; 3. $NaIO_4$; 4. reduction, 70% for four steps. (*b*) 1. CrO_3, H_2SO_4, 0°C, 8 h; 71%; 2. CH_3OH, BF_3, rt, overnight, 89%. (*c*) DBU, $CHCl_3$, −10°C, 1 h; then rt, overnight. (*d*) 1. H_2, 10% Pd *on* C, EtOH, 95 min; 2. BnOH, 5% HCl gas, rt, overnight, 73%. (*e*) 4-Acetoxy-2-azetidinone, $Pd(OAc)_2$, benzene, NEt_3, 30 h, 71%; 2. H_2, 10% Pd *on* C, EtOH, rt, 36 h; 3. CH_2N_2, CH_3OH, 75% for two steps. (*f*) 1. $NaBH_4$, CH_3OH, 0°C, 75 min, 83%; 2. *p*-TsCl, CH_2Cl_2–acetone (4:1), DMAP, 0°C to rt, overnight, 44%. (*g*) 1. LiN_3, DMF, rt, 6 h, 93%; 2. LiBr, THF, reflux, 6 h, 82%. (*h*) *t*-BuOLi, DMF, −20°C, 60 min, 7%. (*i*) EtOAc, H_2, PtO_2, 45 min.

5.2.1.4 *Synthesis from D-glucosamine* The β-amino acid **63**,[81] derived from D-glucosamine, was used for the synthesis of *p*-nitrobenzyl ester of carbapenem SQ 27860 (**5**). Protection of the amino group in **63** followed by esterification with diazomethane produced **64** (90%) (Scheme 7).[82] Mesylation of **64** followed by treatment with *N*,*N*-diisopropylethylamine in DMF gave the pyrrolidine derivative **65** in 88% yield. Removal of the benzyl groups followed by selective benzoylation of the resulting primary hydroxyl

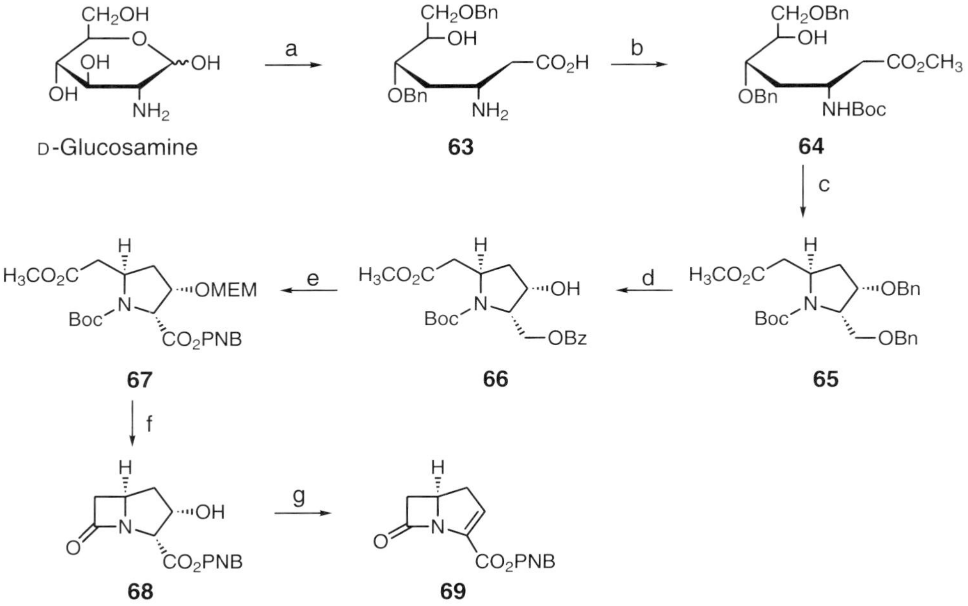

Scheme 7 (*a*) Ref. 81. (*b*) 1. $(Boc)_2O$, 2 M NaOH; 2. CH_2N_2, 90% for two steps. (*c*) 1. CH_3SO_2Cl, Py, 2. $(i\text{-}Pr)_2NEt$, DMF, 90–100°C, 88% for two steps. (*d*) 1. H_2, $Pd(OH)_2$ *on* C, EtOH, 100%; 2. BzCl, Py, 84%. (*e*) 1. $CH_3O(CH_2)_2OCH_2Cl$, $(i\text{-}Pr)_2NEt$; 2. $NaOCH_3$, CH_3OH; 3. Jones oxidation; then $p\text{-}NO_2C6H_4CH_2Br$, $NaHCO_3$, DMF, 81% for three steps. (*f*) 1. LiOH, aqueous CH_3OH, rt; then 2 M HCl; then CH_2Cl_2, NEt_3; 2. $CH_3CN–H_2O$, 2,2′-dipyridyl disulfide and Ph_3P, 85% for two steps. (*g*) MsCl, Py; then NEt_3, CH_2Cl_2; then Florisil column chromatography, 80% yield.

group afforded the benzoate **66** (84% yield). Protection of the secondary hydroxyl group with MEMCl followed by removal of the benzoyl group and subsequent oxidation of the resulting primary hydroxyl group and then esterification produced the nitrobenzyl ester **67** in 81% yield. Compound **67** was treated with lithium hydroxide and the product was hydrolyzed to give the corresponding β-amino acid, which was then converted into the β-lactam **68**. The β-lactam **68** was mesylated and then treated with triethylamine to furnish the SQ 27860 *p*-nitrobenzyl ester **69** in 36.6% yield from **63**.

Methyl 2-deoxy-2-methoxycarbonylamino-α-D-glucopyranoside (**70**),[83] readily accessible from D-glucosamine, has been used for the synthesis of a (+)-thienamycin intermediate (Scheme 8).[81] The acetonide of **70** was acetylated to give **71**, whose acetoxy group was then removed photochemically. Deisopropylidenation of **72** followed by benzylation gave **73**. Acid hydrolysis of **73** and then thioacetalization and acetylation gave **74**. Hydrolysis of **74** gave an unstable aldehyde, which was reacted with methoxymethylenetriphenylphosphorane to give **75**. The vinyl ether **75** was hydrolyzed by acid to the aldehyde **76**, which underwent oxidation followed by alkaline hydrolysis to give the β-amino acid **77**. Treatment of **77** with 2,2′-dipyridyl disulfide and Ph_3P afforded the β-lactam **78**, which was silylated and then converted into **79**, which upon submission to trialkylstannane reduction yielded **80**. Reaction of the lithium enolate of **80** with acetaldehyde gave a mixture of diastereoisomers,

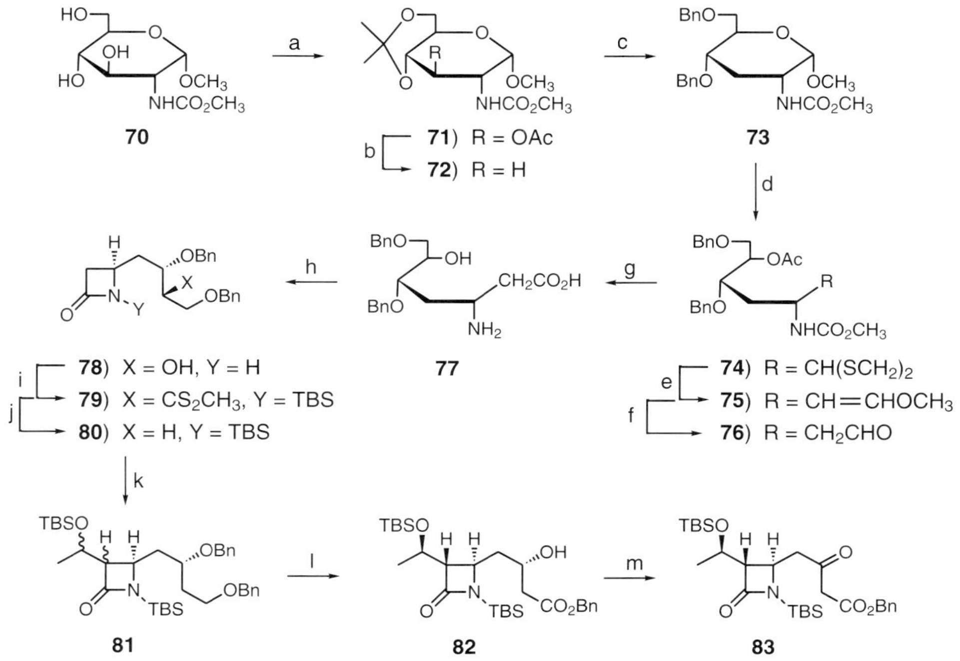

Scheme 8 (*a*) 1. DMP, *p*-TsOH, DMF; 2. Ac_2O, Py, 86%. (*b*) $h\nu$, aqueous hexamethylphosphoric triamide, 93%. (*c*) 1. Aqueous AcOH, 85%; 2. BnBr, NaH, dimethoxyethane, 80%. (*d*) 1. HCl, aqueous AcOH; 2. $(CH_2SH)_2$, $BF_3{\cdot}OEt_2$; 3. Ac_2O, Py, 78%. (*e*) 1. CH_3I, aqueous CH_3CN, 2. $Ph_3PCH_2OCH_3Cl$, $Et(CH_3)_2CONa$, benzene, 70%. (*f*) 1. Aqueous AcOH, 74%; 2. Jones reagent. (*g*) Aqueous $Ba(OH)_2$, 85%. (*h*) $(C_5H_4NS)_2$, Ph_3P, CH_3CN, 89%. (*i*) TBSCl, NEt_3, DMF. (*j*) 1. CS_2, NaH, THF; then CH_3I, 90%; 2. Bu_3SnH, azoisobutyro nitrile, 100%, toluene. (*k*) 1. Lithium diisopropylamide, CH_3CHO, THF; 2. TBSCl, imidazole, DMF, 80%. (*l*) 1. Cyclohexene, $Pd(OH)_2$, EtOH, 90%; 2. O_2, Pt, aqueous dioxane; then BnBr, DBU, CH_3CN, 79%. (*m*) CrO_3, Py.

which upon silylation and subsequent chromatographic separation furnished one of the pure diastereoisomer of **81** in 39% yield. The diastereoisomer (6*S*,8*R*) of **81** was debenzylated to give the respective diol, whose selective oxidation of the primary hydroxyl group was achieved by Pt-catalyzed autoxidation to yield a hydroxy acid, which was then esterified to give **82**. Collins oxidation of **82** provided the intermediate **83**.

5.2.1.5 *Synthesis from* D-*arabinose* D-Arabinose has been used for the synthesis of β-lactam and 1-oxacephem antibiotics (Scheme 9).[84] Thus, the bicyclic β-lactam **85**, obtained from 3,4-di-*O*-trimethylsilyl-D-arabinal (**84**),[85,86] was treated with *tert*-butyl glyoxylate to produce **86**, which underwent chlorination using thionyl chloride to afford **87**. Treatment of **87** with triphenylphosphine followed by removal of the silyl protecting groups afforded the diol **88**. Periodate oxidation of **88** in the presence of ammonium sulfate afforded **90** via **89**. Subsequent reduction of **90** with sodium borohydride followed by acetylation afforded the lactams **91** and **92**.

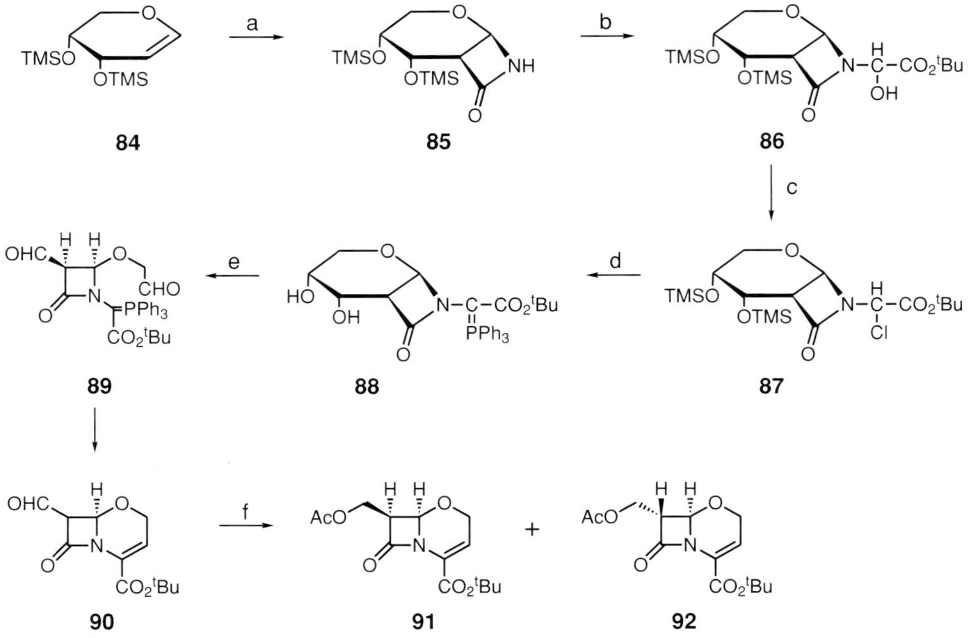

Scheme 9 (*a*) Refs. 85 and 86. (*b*) *t*-Butyl glyoxylate, MS 3 Å, toluene, DMF, 40°C, 70%. (*c*) THF, Py, Cl_2SO, −20°C, 30 min, 70%. (*d*) 1. THF, Ph_3P, 40°C, 16 h, 53%; 2. Et_2O, −70°C, HF, Py, −20°C, 2 days, 85%. (*e*) CH_3OH, 3% aqueous $(NH_4)_2SO_4$, −4°C, $NaIO_4$; then $NaHCO_3$, 10°C. (*f*) 1. $NaBH_4$; 2. Ac_2O, Py, 16% from **88**.

Syntheses of β-lactams **94** and **97** from the arabino derivative **93** have also been reported (Scheme 10).[65,66] Treatment of **93** with sodium metaperiodate in the presence of ammonium sulfate followed by sodium borohydride reduction led to the formation of β-lactam **94**. On the other hand, similar treatment but in the presence of sodium bicarbonate led to a mixture of **95** (25%), **96** (9%) and **97** (38%).

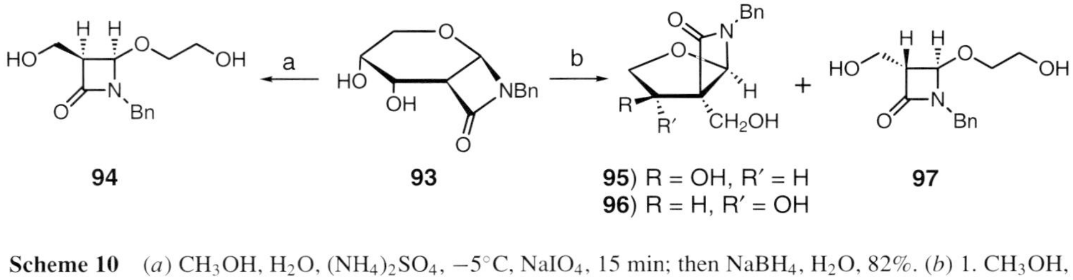

Scheme 10 (*a*) CH_3OH, H_2O, $(NH_4)_2SO_4$, −5°C, $NaIO_4$, 15 min; then $NaBH_4$, H_2O, 82%. (*b*) 1. CH_3OH, H_2O, $NaHCO_3$, −5 to 16°C; $NaIO_4$, 30 min; 2. $NaBH_4$, **95** (25%), **96** (9%), **97** (38%).

5.2.1.6 *Synthesis from* D*-xylose* Clavalanine (**2**) was synthesized from D-xylose by conversion to 1,2-*O*-isopropylidene-D-xylofuranose (**98**), which was selectively acetylated followed by treatment with 1,1′-thiocarbonyldimidazole (TCDI) to furnish **99** in 85% overall yield (Scheme 11).[87] Radical reduction of **99** followed by acid hydrolysis of the

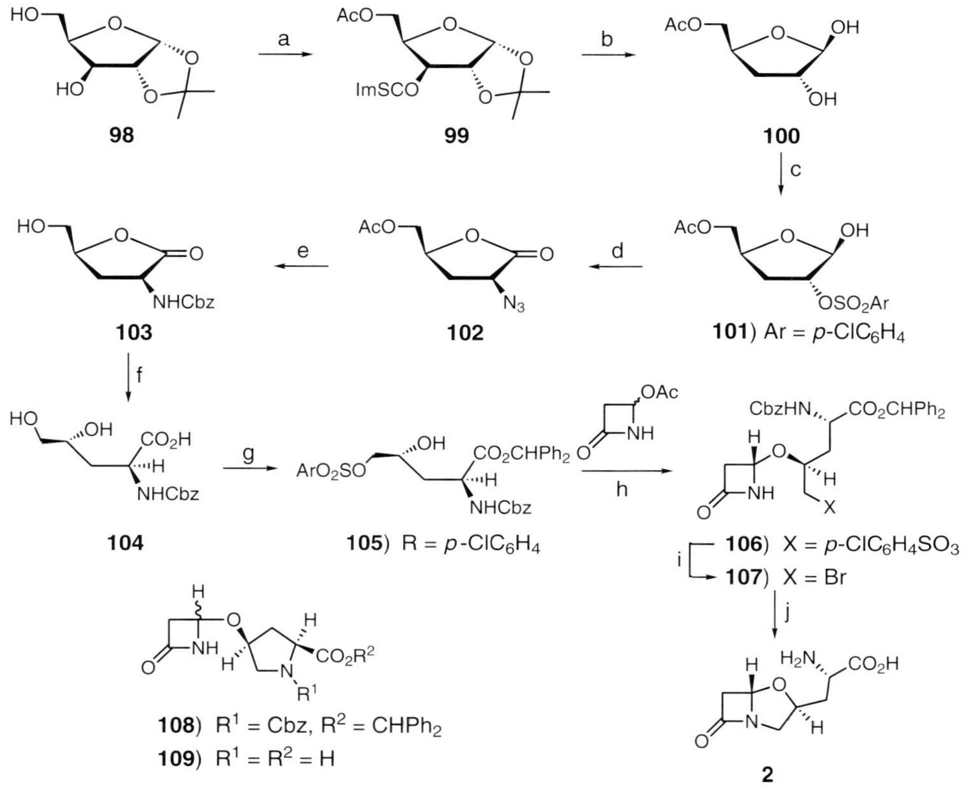

Scheme 11 (*a*) 1. CH_2Cl_2, Py, AcCl, $-10°C$ to rt, overnight, 85%; 2. TCDI, $ClCH_2CH_2Cl$, reflux, 1 h, quantitative. (*b*) 1. Bu_3SnH, toluene, reflux, AIBN, 95%; 2. 50%, TFA, 78%. (*c*) 1. $(Bu_2SnO)_x$, CH_3OH, reflux; 2. p-$ClC_6H_4SO_2Cl$, NEt_3, 60.5%. (*d*) 1. $NaIO_4$, CCl_4, RuO_4, $C_2H_4Cl_2$, 98%; 2. LiN_3, DMF, rt, overnight, 90%. (*e*) 1. H_2, Pd *on* C, 51%; 2. CbzCl, 70%; 3. 1,4-dioxane, H_2O, Dowex AG 50WX4 (H^+) resin. 93%. (*f*) 1. KOH, H_2O; 2. Dowex (H^+) resin. (*g*) 1. $Ph_2C{=}N_2$, $AcCH_3$, 78%; 2. p-$ClC_6H_4SO_2Cl$, DMAP, 63%. (*h*) $Pd(OAc)_2$, NEt_3, benzene, 52%. (*i*) LiBr, THF, 95%. (*j*) 1. 2,2-Dimethyl-6,6,7,7,8,8,8-heptafluoro-3,5-octanedionate, DMF, 70°C, 10 h, 50%; 2. H_2, 10% Pd *on* C, CH_3OH, 97%.

isopropylidene group afforded the 3-deoxy-*erythro*-furanopentose **100** (78%). Reaction of **100** with dibutyltin oxide in boiling methanol afforded the dibutylstannylene derivative, which upon treatment with p-$ClC_6H_4SO_2Cl$ afforded **101**. Oxidation of **101** with ruthenium tetraoxide followed by treatment of the resulting lactone with LiN_3 afforded **102** (90%), which underwent catalytic hydrogenation followed by protection of the resulting amine with benzyl chloroformate and subsequent deacetylation to furnish **103** (65%). The lactone ring in **103** was readily transformed to the acid **104**, which upon treatment with diphenyldiazomethane followed by p-chlorobenzene sulfonyl chloride afforded the (2*S*,4*S*)-2-amino-4,5-dihydroxypentanoic acid derivative **105** (49%). Condensation of **105** with racemic 4-acetoxy-2-azetidinone[80] was catalyzed with palladium acetate to produce **106**, which underwent solvolysis with lithium bromide to produce the bromide **107**. Treatment of **107** with silver 2,2-dimethyl-6,6,7,7,8,8,8-heptafluoro-3,5-octanedionate followed by hydrogenation afforded a mixture of products from which clavalanine (Ro 22-5417, **2**) was readily separated in pure form in addition to the hydroxyproline derivatives **108** and **109**.

5.2.1.7 *Synthesis from L-glyceraldehyde* Condensation of **110a** and **110b** with the glyceraldehyde derivative **111** gave the β-lactams **112** and **113**, respectively (Scheme 12).[88] Compound **112** was converted to β-lactam **114** in 68% yield, which underwent hydrogenation to afford β-lactam **115**. Similar reactions converted **113** to β-lactam **117** via intermediate **116**.

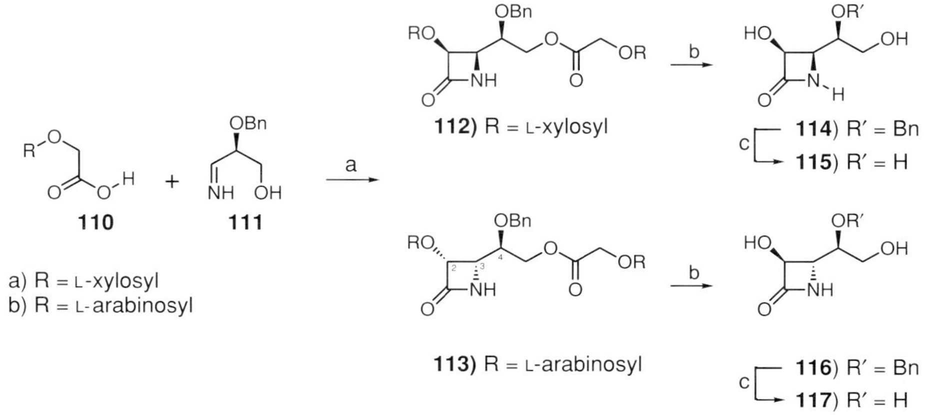

Scheme 12 (*a*) NEt_3, CH_2Cl_2, 0–25°C. (*b*) $(NH_4)_2Ce(NO_3)_6$, $CH_3CN–H_2O$ (1:1), −5 to 0°C, 45 min, 75%; then NaOH, THF–CH_3OH (5:1), 0°C, 30 min, 90%; overall 68%. (*c*) H_2, 10% Pd *on* C; CH_3OH, 25°C, 90%.

5.2.1.8 *Synthesis from D-glyceraldehyde* Synthesis of (2*R*,5*S*)-2-(hydroxymethyl)-clavam (**1**) from D-glyceraldehyde, prepared from D-mannitol, has been accomplished (Scheme 13).[89] Periodate oxidation of 1,2:5,6-diisopropylidene-D-mannitol gave **118**, a good chiral precursor for β-lactams. Sodium borohydride reduction of **118** afforded **119**. Silylation of **119** and subsequent removal[90] of the isopropylidene group gave the diol **120**. Selective tosylation of **120** gave compound **121** in 78% yield. Condensation of **121** with β-lactam **122**[91] in the presence of zinc acetate dihydrate gave the *trans*-lactam **123** in 63% yield. Treatment of **123** with hydrazine hydrate followed by addition of acetic acid afforded the unstable intermediate **124**, which was directly treated with conc. HCl and then potassium nitrite to afford the chlorinated lactam **125** in 86% overall yield from **123**. Radical dechlorination of **125** gave **126**. Iodination of **126** followed by treatment of the resulting iodide **127** with powdered potassium carbonate afforded the oxapenam derivative **128**, which underwent desilylation to give **1** in 15.6% overall yield from the starting diisopropylidene.

D-Glyceraldehyde acetonide (**118**)[89,92] has also been used for the synthesis of other β-lactams (Scheme 14).[93–95] It was converted into the Schiff base **129**, followed by reaction with potassium azidoacetate, cyanuric chloride and triethylamine to give the β-lactam **130** as a single *cis*-isomer in 55% yield. The reaction of **130** with methoxyacetyl chloride in the presence of triethylamine afforded the *cis*-β-lactam **131**. Similarly, *cis*-β-lactams **131**–**135** were also prepared as single isomers in comparable yields. Acid hydrolysis of **130** afforded **136**, which underwent oxidation of the diol side chain and then removal of the *p*-methoxybenzyl group with CAN to afford **137**. The allyl derivative **135** was oxidized to

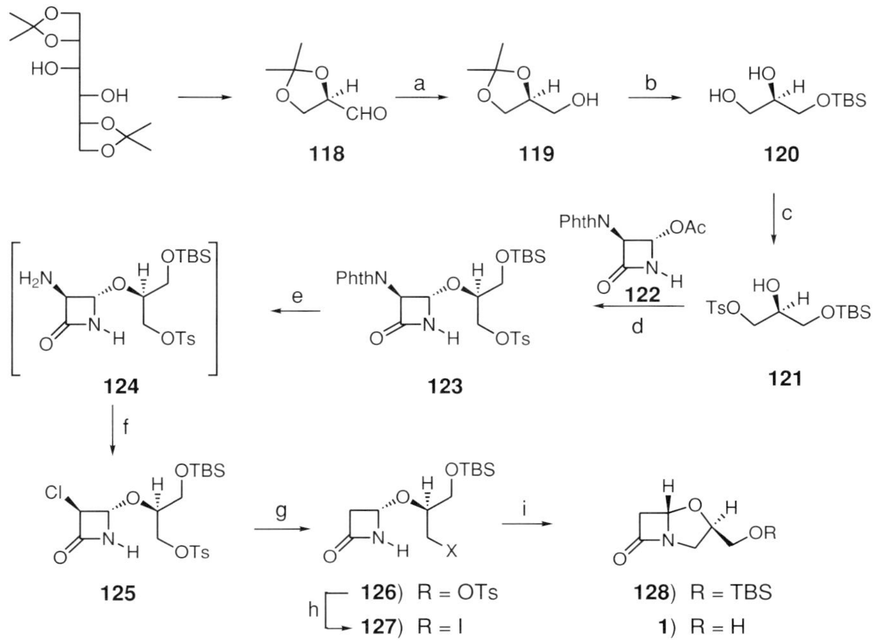

Scheme 13 (*a*) $NaBH_4$. (*b*) 1. TBSCl, imidazole, 96% for two steps; 2. CH_2Cl_2, 0°C, $BF_3{\cdot}OEt_2$, excess 1,3-propanedithiol, 89%. (*c*) *p*-TsCl, Py, 0°C, 78%. (*d*) Zinc acetate dihydrate, benzene, toluene with azeotropic removal of water, reflux, 63%. (*e*) $NH_2NH_2{\cdot}H_2O$, EtOH, 0°C, 0.5 h; then AcOH. (*f*) 1. Conc. HCl, KNO_2, below 0°C, 86%; 2. TBSCl, imidazole, 78%. (*g*) Bu_3SnH, AIBN, benzene, reflux, 78%. (*h*) Iodination, 86% for two steps. (*i*) 1. K_2CO_3, DMF, rt, 90%; 2. THF, TBAF, AcOH, 94%.

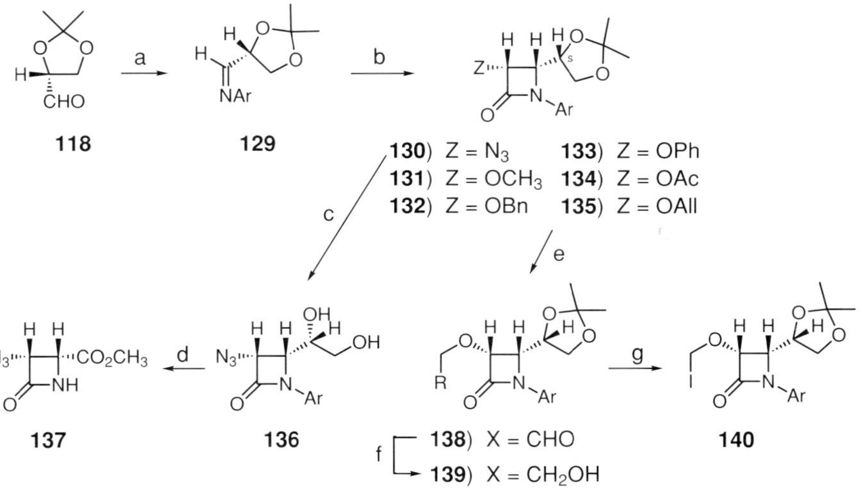

Scheme 14 (*a*) H_2NAr, Et_2O, 0°C, 1 h. (*b*) $N_3CH_2CO_2K$, NEt_3, cyanuric chloride; *or* CH_3OCH_2COCl, CbzCl, *or* $AcOCH_2COCl$, NEt_3, CH_2Cl_2, −20°C, then rt, overnight; *or* $CH_2{=}CHCH_2OCH_2COCl$, NEt_3, MS 3 Å, CH_2Cl_2. (*c*) 80% AcOH, 60°C, 6 h. (*d*) 1. RuO_2, $NaIO_4$, $(CH_3)_2CO$, H_2O, rt, overnight; 2. CH_2N_2, Et_2O; 3. CAN, CH_3CN, −5°C, 2 h. (*e*) O_3, then DMS, CH_2Cl_2. (*f*) $NaBH_4$, EtOH. (*g*) 1. *p*-TsCl, NEt_3, DMAP, CH_2Cl_2; 2. NaI, acetone.

give **138**, which was reduced with sodium borohydride to furnish **139**. Tosylation of **139** followed by treatment with sodium iodide gave **140**.

5.2.1.9 *Synthesis from* L*-ascorbic acid* L-Ascorbic acid has been used as a chiral starting material for the synthesis of β-lactams (Scheme 15).[96] It was converted into 5,6-*O*-isopropylidene-L-ascorbic acid (**141**), which underwent Ca_2CO_3–H_2O_2 oxidation[97] followed by esterification to give **143** via **142**. This was treated with NH_3 in the presence of NH_4OH to produce the key intermediate **144**, which can also be obtained from L-threonic acid (**147**) by conversion to the L-threonamide **145**, via the lactone **146**, in 47% yield from **147**. Treatment of **144** with *p*-chlorobenzenesulfonyl chloride followed by S_N2 displacement of the resulting sulfonate with azide ion gave the azide **148** in 52% yield from **144**. Alternatively, triflation of **144** followed by reaction with lithium azide afforded **148** in 72% yield. Hydrogenolysis of **148** followed by protection of the resulting amine with

Scheme 15 (*a*) Acetone, DMP, HCl, 1 h, 77%. (*b*) 1. Ca_2CO_3, 0°C; then 30% H_2O_2, 20 to 30–40°C, 30 min; then charcoal, 10% Pd *on* C, 100°C, 30 min, 78%; 2. dimethylacetamide, $NaHCO_3$, CH_3I, rt, 2 days, 95%; *or* H_2O, $NaHCO_3$, $(CH_3)_2SO_4$, 40°C, 6 h, 72.3%. (*c*) THF, 29% NH_4OH, NH_3, rt, overnight, 89%. (*d*) H_2O, 100°C, Bio-Rad AG 50WX4 (H^+) resin, 30 min; then CH_3CN, *p*-TsOH, reflux 1 h, 76%. (*e*) CH_3OH, NH_3, 0°C to rt, 48 h, 95.7%. (*f*) DMF, DMP, *p*-TsOH·H_2O, rt, 4.5 h; then Dowex AG 1X4 (OH^-) resin, 2 min, 65%. (*g*) NEt_3, $ClCH_2CH_2Cl$, rt, *p*-$ClC_6H_4SO_3Cl$, rt, 30 h, 74%; *or* $ClCH_2CH_2Cl$, Py, Tf_2O, −10°C, 30 min to 0°C, 30 min; then LiN_3, DMF, rt, 15 h, 72% or NaN_3, 60°C, 48 h, 70%. (*h*) 1. EtOH, H_2, 10% Pd *on* C, 25°C, 2 h; then K_2CO_3, CbzCl, 0°C, 2 h, 91%; 2. HCl, CH_3CN, rt, 2 h, 91%; 3. DMF, 2,6-lutidine, −10°C; then $ClCH_2COCl$, CH_2Cl_2, −10°C, 1 h to rt, 76%; then 1,2-dimethoxyethane, NEt_3, −20°C, MsCl, 1 h, 86%. (*i*) 1. 2-Picoline, $ClCH_2CH_2Cl$, −10°C, $ClSO_3H$, then 0°C, 1 h; then $KHCO_3$, $ClCH_2CH_2Cl$, reflux, 15 min, 87% for two steps. (*j*) CH_2Cl_2, 0°C, chloroacetyl isocyanate, 1 h; then sodium *N*-methyldithiocarbamate, rt, 1 h, evaporation; then EtOH, H_2O, AG 50WX4 (Na^+) resin, 50%. (*k*) H_2, 10% Pd *on* C, CH_3OH, rt, 1 h, 95%.

(benzyloxy)carbonyl chloride and subsequent removal of the isopropylidene group afforded **149**. Protection of the primary hydroxyl group of **149** was followed by mesylation to give **150**, which was sulfonated with 2-picoline–SO_3 complex, followed by boiling in a two-phase system consisting of 1,2-dichloroethane and aqueous potassium bicarbonate to produce stereospecifically the β-lactam **151**. Treatment of **151** with chloroacetyl isocyanate followed by removal of the resulting chloroacetyl group with sodium *N*-methyldithiocarbamate furnished the β-lactam **152**. This underwent catalytic hydrogenation to form the zwitter ion **153**.

The calcium derivative of L-threonic acid **147** was treated with *O*-benzylhydroxylamine hydrochloride in the presence of 1-[3-(dimethylamino)propyl]-3-ethylcarbodiimide hydrochloride as the condensing agent, followed by selective protection of the 2- and 4-hydroxyl groups with TBSCl to afford **154**. Cyclization of **154** gave the β-lactam **155**, which underwent selective removal of the protecting groups, either by hydrogenolysis to afford **156** or by acid hydrolysis to give **157** (Scheme 16).[96]

147 → (a) **154** → (b) **155**) R = Bn → (c) **156**) R = H; **155** → (d) **157**

Scheme 16 (*a*) 1. H_2O, $BnONH_2{\cdot}HCl$, $NaHCO_3$; then 1-[3-(dimethylamino)propyl]-3-ethylcarbodiimide hydrochloride, pH 4.5–5.25, 72.4%; 2. Py, $ClCH_2CH_2Cl$, $-10^\circ C$, TBSCl, rt, overnight, 73.5%; *or* $ClCH_2CH_2Cl$, Py, $0^\circ C$, TBSCl, rt, overnight, then Dowex AG 50WX4 (H^+, 100–200 mesh) resin, 92.6%. (*b*) Ph_3P, CH_3CN, rt, 15 min; then NEt_3, CCl_4, rt, overnight, 77%. (*c*) CH_3OH, H_2, 10% Pd *on* C, rt, 45 min, 100%. (*d*) 90% TFA, rt, 2 h, 64.4%.

5.2.1.10 *Synthesis from* D-*ribonolactone* The required intermediates **164** and **165** for the synthesis of clavalanine (**2**) were prepared from D-ribonolactone (Scheme 17).[98]

D-Ribonolactone → (a) **158** → (b) **159**) R = N_3 → (c) **160**) R = NHCbz → (d) **161** → (e) **162**) R = H → (g) **163**) R = H·HCl; (h) **164**) R = Cbz; **162** → (f) **165**

Scheme 17 (*a*) PhCHO, HCl. (*b*) 1. Tf_2O, Py, $0^\circ C$; 2. NaN_3, DMF, rt. (*c*) 1. H_2, 10% Pd *on* C, EtOAc; 2. CbzCl, $NaHCO_3$, $0^\circ C$, H_2O, THF. (*d*) NaH, THF, $-20^\circ C$; 1 N HCl. (*e*) H_2, Raney nickel, 2 atm, EtOH. (*f*) Dowex 50WX2 (H^+) resin. (*g*) HCl. (*h*) CbzCl.

D-Ribonolactone was converted into the azide **159** (57%) via **158** with retention of configuration.[99] Catalytic hydrogenation of the azide group followed by protection of the resulting amine gave the carbamate **160** (65%). Base-induced elimination of benzaldehyde and subsequent ring contraction afforded the 1,4-lactone **161**. Hydrogenation of the olefin in the presence of Raney nickel afforded diastereospecifically the amino lactone **162** as a single isomer. The lactone **162** underwent acid hydrolysis to give the amino acid **165** in 20% overall yield from D-ribonolactone. Treatment of the lactone **162** with hydrogen chloride gave the hydrochloride **163**, while treatment with benzyloxycarbonyl chloride yielded the carbamate **164**, a precursor of **2**.

References

1. Brown, D.B.; Evans, J.R. *J. Chem. Soc., Chem. Commun.* (1979) 282.
2. Pruess, D.L.; Kellet, M.J. *J. Antibiot.* **36** (1983) 208.
3. Evans, R.H.; Ax, H.; Jacoby, A.; Williams, T.H.; Jenkins, E.; Scannel, J. *J. Antibiot.* **36** (1983) 213.
4. Müller, J.C.; Yoome, V.; Pruess, D.L.; Blount, J.F.; Weigele, M. *J. Antibiot.* **36** (1983) 217.
5. Howarth, T.T.; Brown, A.G.; King, T.J. *J. Chem. Soc., Chem. Commun.* (1976) 266.
6. Reading, C.; Cole, M. *Antimicrob. Agents Chemother.* **11** (1977) 852.
7. Rolinson, G.N.; Watson, A., Eds. *Excerpta Medica*; Elsevier: Amsterdam, 1980.
8. Parker, W.L.; Rathnum, M.L.; Wells, J.S., Jr.; Trejo, W.H.; Principe, P.A.; Sykes, R.B. *J. Antibiot.* **35** (1982) 653.
9. Kahan, J.S.; Kahan, F.M.; Goegelman, R.; Currie, S.A.; Jackson, M.; Stapley, E.O.; Miller, T.W.; Miller, A.K.; Hendlin, D.; Mochales, S.; Hernandez, S.; Woodruff, H.B.; Birnbaum, J. *J. Antibiot.* **32** (1979) 1.
10. Albers-Schönberg, G.; Arison, B.H.; Hensens, O.D.; Hirshfield, J.; Hoogsteen, K.; Kaczka, E.A.; Rhodes, R.F.; Kahan, J.S.; Kahan, F.M.; Rateliffe, R.W.; Walton, E.; Russwinkle, L.J.; Morin, R.B.; Christensen, B.G. *J. Am. Chem. Soc.* **100** (1978) 6491.
11. Tufariello, J.J.; Lee, G.E.; Senaratne, P.A.; Al-Nuri, M. *Tetrahedron Lett.* **20** (1979) 4359.
12. Salzmann, T.N.; Ratcliffe, R.W.; Christensen, B.G.; Bouffard, F.A. *J. Am. Chem. Soc.* **102** (1980) 6161.
13. Brown, A.G.; Corbett, D.F.; Eglington, A.J.; Howarth, T.T. *J. Chem. Soc., Chem. Commun.* (1977) 523.
14. Corbett, D.F.; Eglington, A.J.; Howarth, T.T. *J. Chem. Soc., Chem. Commun.* (1977) 953.
15. Butterworth, D.; Cole, M.; Hornseomb, G.; Robinson, G.N. *J. Antibiot.* **32** (1979) 287.
16. Hood, J.D.; Box, S.J.; Verrall, M.S. *J. Antibiot.* **32** (1979) 295.
17. Okamura, K.; Hirata, S.; Koki, A.; Hori, K.; Shibamoto, N.; Okumura, Y.; Okabe, M.; Okamoto, R.; Kouno, K.; Fukagawa, Y.; Shimanehi, Y.; Ishikura, T.; Lein, J. *J. Antibiot.* **32** (1979) 262.
18. Okamura, K.; Hirata, S.; Okamura, Y.; Fukagawa, Y.; Shimanchi, Y.; Kouno, K.; Ishikura, T.; Lein, J. *J. Antibiot.* **31** (1978) 480.
19. Sakamoto, M.; Iguchi, H.; Okamura, K.; Hori, S.; Fukagawa. Y.; Ishikura, T.; Lein, J. *J. Antibiot.* **32** (1979) 280.
20. Kropp, H.; Kahan, J.S.; Kahan, F.M.; Sundelof, J.; Darland, G.; Birnbaum, J. In *16th Interscience Conference on Antimicrobial Agents and Chemotherapy*, Chicago, IL, 1976; Abstract 226.
21. Tally, F.P.; Jacobus, N.V.; Gorbach, S.L. *Antimicrob. Agents Chemother.* **14** (1978) 436.
22. Weaver, S.S.; Bodey, G.P.; LeBlane, B.M. *Antimicrob. Agents Chemother.* **15** (1979) 518.
23. Nagahara, T.; Kametani, T. *Heterocycles* **25** (1987) 729.
24. Cama, L.D.; Christensen, B.G. *J. Am. Chem. Soc.* **96** (1974) 7582.
25. Firestone, R.A.; Fahey, J.L.; Maciejewicz, N.S.; Palet, G.S.; Christensen, B.G. *J. Med. Chem.* **20** (1977) 551.
26. Shih, D.H.; Hannah, J.; Christensen, B.G. *J. Am. Chem. Soc.* **100** (1978) 8004.
27. Johnston, D.B.R.; Schmitt, S.M.; Bouffard, F.A.; Christensen, E.G. *J. Am. Chem. Soc.* **100** (1978) 313.
28. Cama, L.D.; Christensen, B.G. *J. Am. Chem. Soc.* **100** (1978) 8006.
29. Banik, B.K.; Samajdar, S.; Banik, I. *Tetrahedron Lett.* **44** (2003) 1699.
30. Kametani, T.; Huang, S.-P.; Yokohama, S.; Suzuki, Y.; Ihara, M. *J. Am. Chem. Soc.* **102** (1980) 2060.
31. Melillo, D.G.; Shinkai, I.; Liu, T.; Rayan, K.; Sletzinger, M. *Tetrahedron Lett.* **21** (1980) 2783.
32. Schmitt, S.M.; Johnston, D.B.R.; Christensen, B.G. *J. Am. Chem. Soc.* **45** (1980) 1142.
33. Melillo, D.G.; Liu, T.; Ryan, K.; Sletzinger, M.; Shinkai, I. *Tetrahedron Lett.* **22** (1981) 913.
34. Shiozaki, M.; Ishida, N.; Hiraoka, T.; Yanagisawa, H. *Tetrahedron Lett.* **22** (1981) 5205.
35. Hatanaka, M. *Tetrahedron Lett.* **28** (1987) 83.
36. Koskinen, A.M.P.; Ghiaci, M. *Tetrahedron Lett.* **31** (1990) 3209.

37. Somfai, P.; He, H.M.; Tanner, D. *Tetrahedron Lett.* **32** (1991) 283.
38. Honda, T.; Ishizone, H.; Naito, K.; Mori, W.; Suzuki, Y. *Chem. Pharm. Bull.* **40** (1992) 2031.
39. Konosu, T.; Furukawa, Y.; Hata, T.; Oida, S. *Chem. Pharm. Bull.* **39** (1991) 2813.
40. Ishiguro, M.; Tanaka, R.; Namikawa, K.; Nasu, T.; Inoue, H.; Nakatsuka, T.; Oyama, Y.; Imajo, S. *J. Med. Chem.* **40** (1997) 2126.
41. Hanessian, S.; Reddy, B. *Tetrahedron* **55** (1999) 3427.
42. Furman, B.; Thürmer, R.; Katuza, Z.; Voelter, W.; Chmielewski, M. *Tetrahedron Lett.* **40** (1999) 5909.
43. Udodong, U.E.; Fraser-Reid, B. *J. Org. Chem.* **53** (1988) 2132.
44. Shih, D.H.; Baker, F.; Cama, L.; Christensen, B.G. *Heterocycles* **21** (1984) 29.
45. Arribas, E.; Carreiro, C.; Valdeolmillos, A.M. *Tetrahedron Lett.* **29** (1988) 1609.
46. Kaluza, Z.; Lysek, R. *Tetrahedron: Asymmetry* **8** (1997) 2553.
47. Ohtake, N.; Jona, H.; Okada, S.; Okamoto, O.; Imai, Y.; Ushijima, R.; Nakagawa, S. *Tetrahedron: Asymmetry* **8** (1997) 2939.
48. Berks, A.H. *Tetrahedron* **52** (1996) 331.
49. Aoyama, Y.; Uenaka, M.; Konoike, T.; Iso, Y.; Nishitani, Y.; Kanda, A.; Naya, N.; Nakajima, M. *Bioorg. Med. Chem. Lett.* **10** (2000) 2403.
50. Aoyama, Y.; Uenaka, M.; Kii, M.; Tanaka, M.; Konoike, T.; Hayasaki-Kajiwara, Y.; Naya, N.; Nakajima, M. *Bioorg. Med. Chem.* **9** (2001) 3065.
51. Imamura, H.; Shimizu, A.; Sato, H.; Sugimoto, Y.; Sakuraba, S.; Nakajima, S.; Abe, S.; Miura, K.; Nishimura, I.; Yamada, K.; Morishima, H. *Tetrahedron* **56** (2000) 7705.
52. Izquierdo, I.; Plaza, M.T.; Robles, R.; Mota, A.J. *Tetrahedron: Asymmetry* **11** (2000) 4509.
53. Charpentier, E.; Tuomanen, E. *Microbes Infect.* **2** (2000) 1855.
54. Aoyama, Y.; Uenaka, M.; Konoike, T.; Iso, Y.; Nishitani, Y.; Kanda, A.; Naya, N.; Nakajima, M. *Bioorg. Med. Chem. Lett.* **10** (2000) 2397.
55. Aoyama, Y. *Expert. Opin. Ther. Patents* **11** (2001) 1424.
56. Aoyama, Y.; Uenaka, M.; Konoike, T.; Hayasaki-Kajiwara, Y.; Naya, N.; Nakajima, M. *Bioorg. Med. Chem. Lett.* **11** (2001) 1691.
57. Imamura, H.; Ohtake, N.; Jona, H.; Shimizu, A.; Moriya, M.; Sato, H.; Sugimoto, Y.; Ikeura, C.; Kiyonaga, H.; Nakano, M.; Nagano, R.; Abe, S.; Yamada, K.; Hashizume, T.; Morishima, H. *Bioorg. Med. Chem.* **9** (2001) 1571.
58. Aoyama, Y.; Konoike, T.; Kanda, A.; Naya, N.; Nakajima, M. *Bioorg. Med. Chem. Lett.* **11** (2001) 1695.
59. Sunagawa, M.; Sasaki, A. *Heterocycles* **54** (2001) 497.
60. Panfil, I.; Maciejewski, S.; Belzecki, C.; Chmielewski, M. *Tetrahedron Lett.* **30** (1989) 1527.
61. Maciejewski, S.; Panfil, I.; Belzecki, C.; Chmielewski, M. *Tetrahedron Lett.* **31** (1990) 1901.
62. Maciejewski, S.; Panfil, I.; Belzecki, C.; Chmielewski, M. *Tetrahedron* **48** (1992) 10363.
63. Chmielewski, M.; Grodner, J.; Fudong, W. *Tetrahedron* **48** (1992) 2935.
64. Kaluza, Z.; Chmielewski, M. *Tetrahedron* **45** (1989) 7195.
65. Chmielewski, M.; Kaluza, Z.; Abramski, W.; Grodner, J.; Belzecki, C.; Sedmera, P. *Tetrahedron* **45** (1989) 227.
66. Chmielewski, M.; Kaluza, Z.; Abramski, W.; Belzecki, C. *Tetrahedron Lett.* **28** (1987) 3035.
67. Bartner, P.; Boxler, D.L.; Brambilla, R.; Mallams, A.K.; Morton, J.B.; Reichert, P.; Sancilio, F.C.; Suprenant, U.; Tomaleski, G.; Lukacs, G.; Olesker, A.; Thang, T.T.; Valente, L.; Omura, S. *J. Chem. Soc., Perkin Trans. 1* (1979) 1600.
68. Knierzinger, A.; Vasella, A. *J. Chem. Soc., Chem. Commun.* (1984) 9.
69. Tsuji, T.; Nagashima, U.; Hori, K. *Chem. Lett.* (1980) 257.
70. Ratcliffe, R.W.; Salzman, T.N.; Christensen, B.G. *Tetrahedron Lett.* **21** (1980) 31.
71. Hanessian, S.; Desilets, D.; Rancourt, G.; Fortin, R. *Can. J. Chem.* **60** (1982) 2292.
72. Kovar, J.; Dienstbierova, V.; Jary, J. *Collect. Czech. Chem. Commun.* **32** (1967) 2498.
73. Mikolajczyk, M.; Grzejszczak, D.; Zatorski, M. *Tetrahedron Lett.* **17** (1976) 2731.
74. Melillo, D.G.; Liu, T.; Ryan, K.; Sletzinger, M.; Shinkai, I. *Tetrahedron Lett.* **21** (1980) 2783.
75. Melillo, D.G.; Shinkai, I.; Liu, T.; Ryan, K.; Sletzinger, M. *Tetrahedron Lett.* **22** (1981) 913.
76. Hanessian, S.; Plessas, N.R. *J. Org. Chem.* **34** (1969) 1045.
77. Richardson, A.C. *Carbohydr. Res.* **4** (1967) 422.
78. Ikota, N.; Yoshino, O.; Koga, K. *Chem. Pharm. Bull.* **30** (1982) 1929.
79. Ikota, N.; Yoshino, O.; Koga, K. *Chem. Pharm. Bull.* **39** (1991) 2201.
80. Clauss, K.; Grimm, D.; Prossel, G. *Liebigs Ann. Chem.* (1974) 539.
81. Miyashita, M.; Chida, N.; Yoshikoshi, A. *J. Chem. Soc., Chem. Commun.* (1982) 1354.
82. Miyashita, M.; Chida, N.; Yoshikoshi, A. *J. Chem. Soc., Chem. Commun.* (1984) 195.
83. Hanessian, S.; Liak, T.J.; Vanasse, B. *Synthesis* (1981) 396.
84. Grodner, J.; Chmielewski, M. *Tetrahedron* **51** (1995) 829.
85. Chmielewski, M.; Kaluza, Z. *J. Org. Chem.* **51** (1986) 2395.

86. Chmielewski, M.; Kaluza, Z. *Carbohydr. Res.* **167** (1987) 143.
87. De Bernardo, S.; Tengi, J.P.; Sasso, G.J.; Weigele, M. *J. Org. Chem.* **50** (1985) 3457.
88. Krämer, B.; Franz, T.; Picasso, S.; Pruschek, P.; Jäger, V. *Synlett* (1997) 295.
89. Konosu, T.; Oida, S. *Chem. Pharm. Bull.* **39** (1991) 2212.
90. Kelly, D.R.; Roberts, S.M.; Newton, R.F. *Synth. Commun.* **9** (1979) 295.
91. Sheehan, J.C.; Henery-Logan, K.R. *J. Am. Chem. Soc.* **84** (1962) 2983.
92. Baer, R.; Fischer, H.O.C. *J. Biol. Chem.* **128** (1939) 463.
93. Banik, B.K.; Manhas, M.S.; Kaluza, Z.; Barakat, K.J.; Bose, A.K. *Tetrahedron Lett.* **33** (1992) 3603.
94. Wagle, D.R.; Monteleone, M.G.; Krishnan, L.; Manhas, M.S.; Bose, A.K. *J. Chem. Soc., Chem. Commun.* (1989) 915.
95. Bose, A.K.; Hegde, V.R.; Wagle, D.R.; Bari, S.S.; Manhas, M.S. *J. Chem. Soc., Chem. Commun.* (1986) 161.
96. Wei, C.C.; De Bernardo, S.; Tengi, T.P.; Borgese, J.; Weigele, M. *J. Org. Chem.* **50** (1985) 3462.
97. Isbell, H.S.; Frush, H.L. *Carbohydr. Res.* **72** (1979) 301.
98. Ariza, J.; Font, J.; Ortuno, R.M. *Tetrahedron Lett.* **32** (1991) 1979.
99. Dho, J.C.; Fleet, G.W.J.; Peach, J.M.; Prout, K.; Smith, P.W. *Tetrahedron Lett.* **27** (1986) 3203.

5.3 5:5-Fused heterocycles

5.3.1 *Polyhydroxypyrrolizidines*

There are a large number of natural products including a pyrrolizidine ring as a basic skeleton and hydroxyl groups as substituents. The location and configuration of the hydroxyl and hydroxymethyl groups on the ring as well as their incorporation in macrocycles may lead to subgroups. Such alkaloids **1–10** may be named as stereoisomers of either alexine (**1**) or australine (**5**), based on the configuration at the bridgehead position C-7a, *S* configuration for alexines and *R* configuration for australines.[1–15]

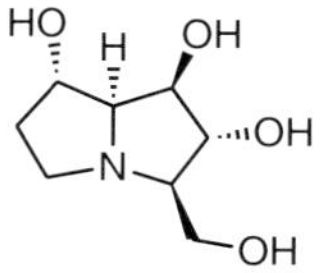

(+)-Alexine
1

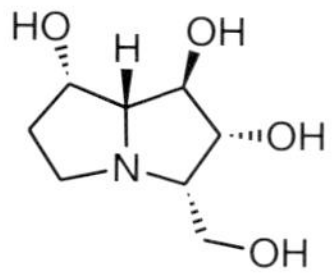

3,7a-Di-*epi*-alexine
[3-*epi*-australine]
2

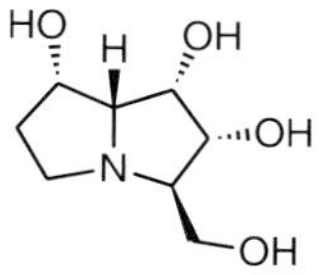

1,7a-Di-*epi*-alexine
[(+)-1-*epi*-australine]
3

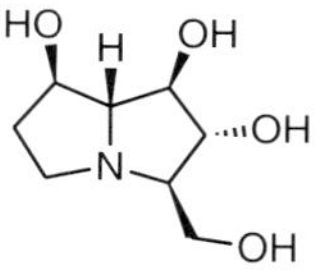

7,7a-Di-*epi*-alexine
[(+)-7-*epi*-australine]
4

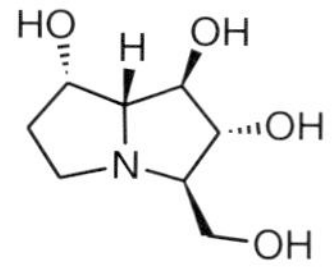

7a-*epi*-Alexine
[(+)-australine]
5

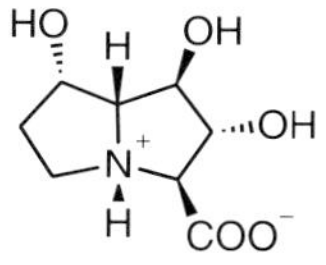

7a-*epi*-Alexaflorine
6

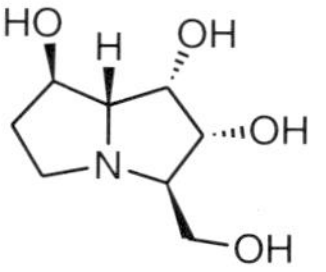

1,7,7a-Tri-*epi*-alexine
[1,7-di-*epi*-australine]
7

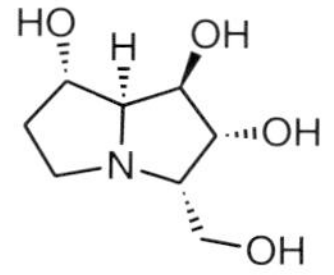

3-*epi*-Alexine
8

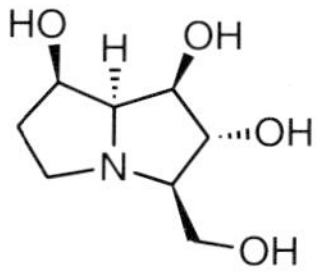

7-*epi*-Alexine
9

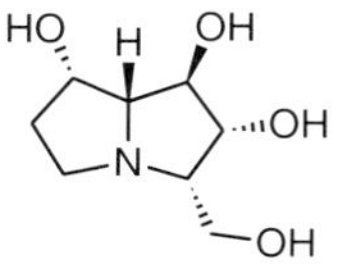

3-*epi*-Australine
10

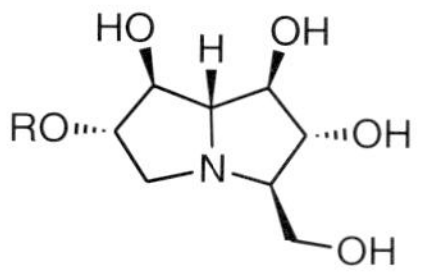

11) Casuarine, R = H
12) Casuarine-6-α-D-glucopyranoside, R = α-D-glucopyranosyl

Alexine [(1*R*,2*R*,3*R*,7*S*,8*S*)-3-hydroxymethyl-1,2,7-trihydroxypyrrolizidine, **1**] has been isolated from *Alexa leiopetala*.[16] Similar alkaloids 3,7a-di-*epi*-alexine (3-*epi*-australine, **2**),[17] (+)-australine (7a-*epi*-alexine, **5**),[18] 1-*epi*-australine (1,7a-di-*epi*-alexine, **3**)[19] and 7,7a-di-*epi*-alexine (7-*epi*-australine, **4**)[20] have been isolated from *Castanospermum australe* A. Cunn. (Leguminosae).

Alexine (**1**) and 3-*epi*-australine (**10**) are generally poor inhibitors of glucosidases and galactosidases,[16,17] but they display amyloglucosidase inhibition comparable with that of castanospermine,[20] while **1** is an effective thioglucosidase inhibitor.[21] The configurational and conformational analyses of alexines using NMR data and X-ray have been well documented.[22]

1-*epi*-Australine (**3**), 7-*epi*-australine (**4**) and australine (**5**) are also good amyloglucosidase inhibitors.[20,23–26] Compound **5** inhibits glucosidase I, but not glucosidase II,[23] and has recently been shown to exhibit antiviral activity.[27] Modest glucosidase I, β-glucosidase and α-mannosidase inhibitions have been observed for **3**,[19] which displayed good activity in a mouse gut digestive α-glucosidase assay, as did **4**.[19,28] All the three compounds inhibit HIV.[28]

7a-*epi*-Alexaflorine (**6**) is the first example of an amino acid with a carboxyl group substituent at C-3 of the pyrrolizidine nucleus having a stereochemistry corresponding to that of 7a-*epi*-alexine (**5**), and it was isolated from the leaves of *Alexa grandiflora*.[29]

Casuarine (**11**) and its 6-glucoside **12** occur in the bark of *Casuarinas equisetifolia* (Casuarinaceae),[30] which has been used for the treatment of cancer in Western Samoa. Casuarine also occurs as the major alkaloid in both the leaves and bark of *Eugenia jambolona* Lam. (Myrtaceae) and in an unidentified African plant, reported to be beneficial in treating AIDS patients.[31] It was traditionally used for treating diabetes in India. *Eugenia jambolana* is a tree in India well-known for the therapeutic value of its seeds, leaves and fruit against diabetes and bacterial infections.[32,33] Its fruit has been shown to reduce blood sugar levels in humans[34] and the aqueous extracts of the bark are claimed to affect glycogenolysis and glycogen storage in animals.[35] Casuarine is a potent inhibitor of glucosidase I (72% inhibition at 5 μg/mL), being only slightly less active than castanospermine (82% inhibition at 5 μg/mL).[36]

Hyacinthacines A_1 (**13**), A_2 (**14**), A_3 (**15**), B_3 (**18**) and C_1 (**19**) have been isolated from *Muscari armeniacum*,[37] and B_1 (**16**), B_2 (**17**) and C_1 (**19**) were isolated from *Hyacinthoides nonscripta* and *Scilla campanulata*.[38] Compounds **13** and **17** are potent inhibitors of rat intestinal lactase enzyme. Compound **13** is a moderate inhibitor of α-L-fucosidase and amyloglucosidase. The inversion of the hydroxyl group at C-l in **13** as in **14** caused an enhancement of its inhibitory potential toward amyloglucosidase but abolished the inhibition of α-L-fucosidase. Compound **15** is a less effective inhibitor of rat intestinal lactase and amyloglucosidase than **14**. Compound **18** has proved to be a moderate inhibitor of lactase and amyloglucosidase, but had no significant activity toward other glycosidases.

Compounds **20–68** were isolated from various plants and butterflies.[39–103] Rosmarinecine (**20**) has been isolated from various plants in the Compositae family, including *Senecio pleitocephalus*,[39] *Senecio triangularis*,[40] *Senecio taiwanesis* Hayata,[41] *Senecio pterophorus*,[42] *Senecio hygrophilus*,[43] *Senecio adnatus* D.C.,[44] *Senecio angulatus* L.,[45] *Senecio hadiensis* and *Senecio syringifolius*.[46] (+)-Crotanecine (**21**) was isolated from the leaves and twigs gathered from *Crotalaria agatiflora* grown in Australia.[47–49] The X-ray

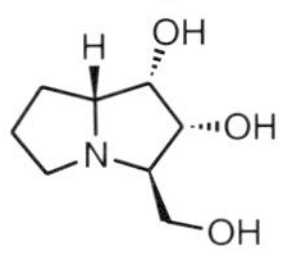

Hyacinthacine A_1
13

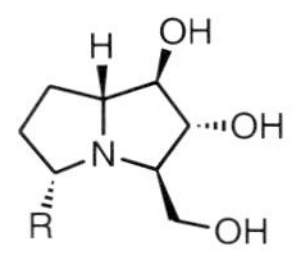

14) Hyacinthacine A_2, R = H
15) Hyacinthacine A_3, R = CH_3

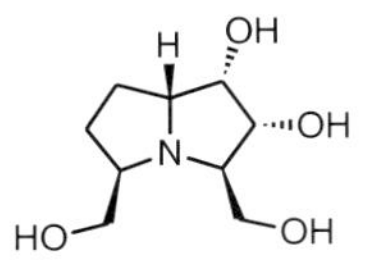

Hyacinthacine B_1
16

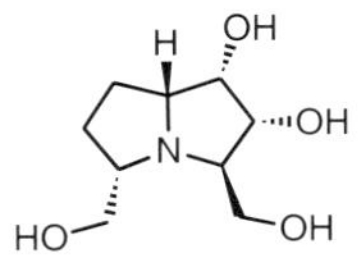

Hyacinthacine B_2
17

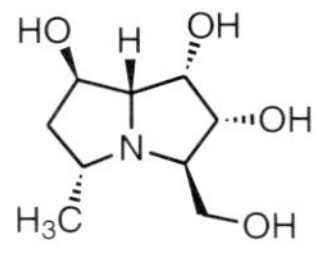

Hyacinthacine B_3
18

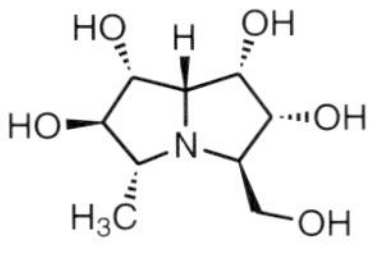

Hyacinthacine C_1
19

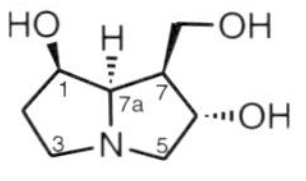

(–)-Rosmarinecine
20

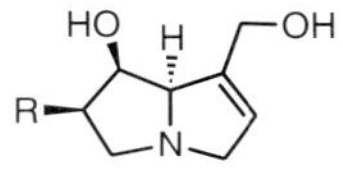

21) (–)-Crotanecine, R = OH
22) (+)-Retronecine, R = H

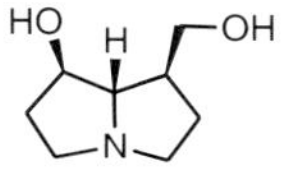

(–)-Hastanecine
23

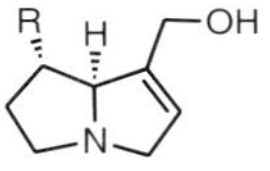

24) (+)-Heliotridine, R = OH
25) (–)-Supinidine, R = H

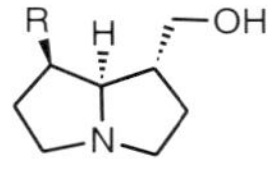

26) (–)-Turneforcidine, R = OH
27) (–)-Trachelanthamidine, R = H

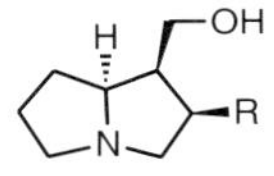

28) Petasinecine, R = OH
29) (–)-Isoretronecanol, R = H

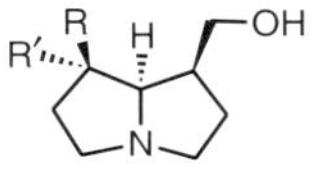

30) (–)-Platynecine, R = OH, R′ = H
31) (–)-Dihydroxyheliotridane, R = H, R′ = OH

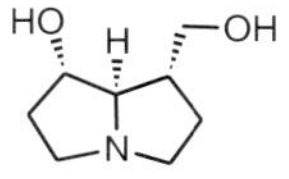

(+)-Hastanecine
32

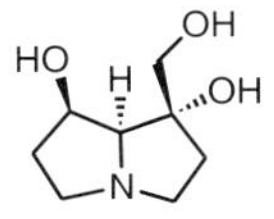

(–)-Hadinecine
33

crystal structure of **21** has been determined.[94] (+)-Retronecine (**22**), (+)-heliotridine (**24**) and (−)-supinidine (**25**) were isolated from the seeds of *Crotalaria spectabilis*.[51,52] (–)-Platynecine (**30**) was isolated from *Senecio platyphyllusis*[53–55]; it is the base portion of several pyrrolizidine alkaloids, including platyphyllin and neoplatyphilline.[55]

Tussilagine (**36**) and isotussilagine (**37**) are two nontoxic pyrrolizidines bearing a methyl group at the C-2 position. Both compounds exist in *Tussilago farara*, *Echinacea purpurea*, *Arnica* and *Echiracea angustifolia* and their structures were identified by X-ray analysis.[57–59] Otonecine (**38**) experiences a strong interaction between its basic nitrogen

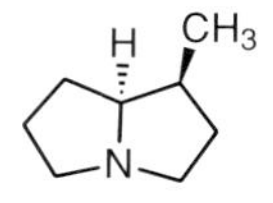

Heliotridane **34**

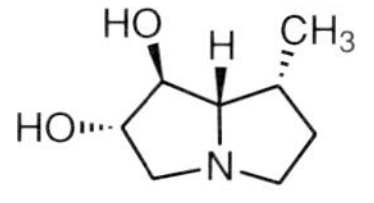

6,7-Dihydroxyheliotridane **35**

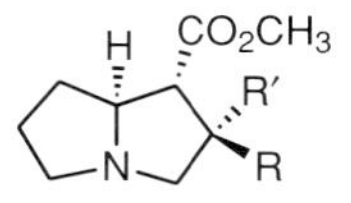

36) Tussilagine, R = OH, R′ = CH_3
37) Isotussilagine, R = CH_3, R′ = OH

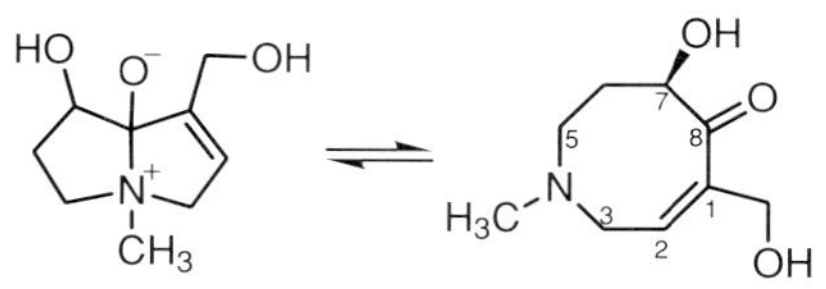

Otonecine **38**

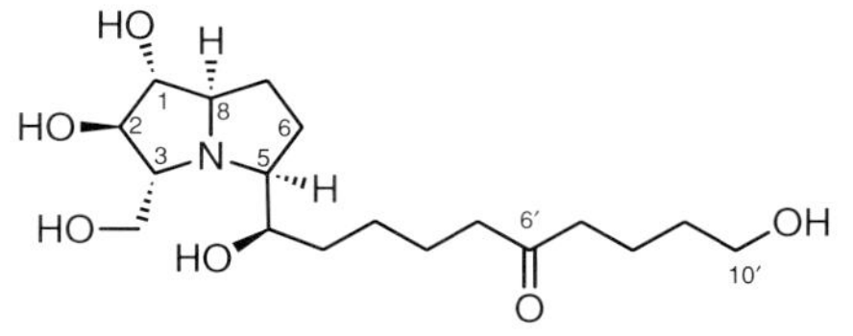

Broussonetine N **39**

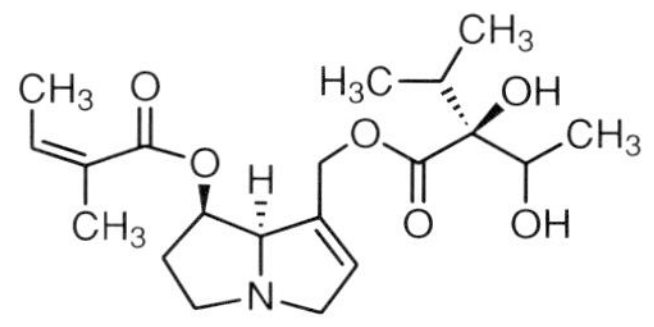

Punctanecine **40**

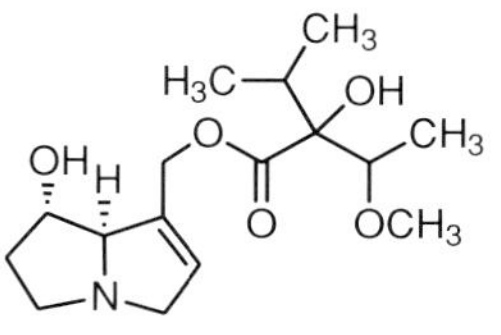

Heliotrine **41**

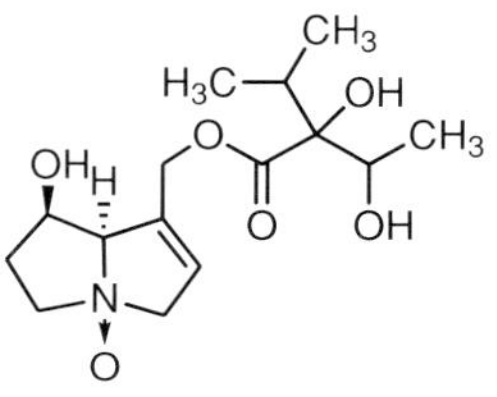

Indicine *N*-oxide **42**

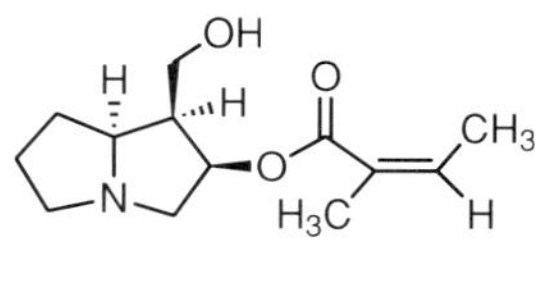

Petasinine **43**

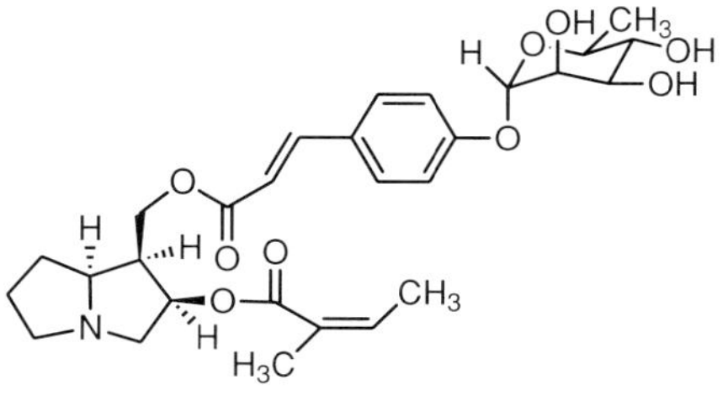

Petasinoside **44**

and the ketone carbonyl groups, thus leading to the equilibrium between the valence bond tautomers.[60–63] Broussonetine N (**39**) was isolated from *Broussonetia kazinoki* Sieb (Moraceae).[64–67] Indicine *N*-oxide[68,69] (**42**) shows marked antitumor activity reaching to the clinical trials,[70] while heliotrine (**41**) was an established carcinogen.[71–73]

Petasinine (**43**) and petasinoside (**44**) were isolated from *Petasites japonicus* Maxim.[74]

Jacoline (**48**) was isolated from *Cirsium wallichii* D.C. collected from India and identified as *O*-acetyljacoline.[75] Yamataimine (**49**) was isolated from *Cacalia yatabei* Maxim.[76]

45) Acetylmadurensine, $R = CH_3$, $R' = H$
46) Acetyl-*cis*-madurensine, $R = H$, $R' = CH_3$

47) Crotaflorine

48) Jacoline

49) Yamataimine

50) (+)-Dicrotaline

51) Usaramine, $R = H$, $R' = R'' = CH_3$
52) Retrosine, $R = CH_3$, $R' = R'' = H$
53) Isatidine *N*-oxide, $R = CH_3$, $R' = R'' = H$

54) Rosmarinine, $R = H$, $R^1 = R^2 = CH_3$, $R^3 = OH$
55) Neorosmarinine, $R = R^2 = CH_3$, $R^1 = H$, $R^3 = OH$
56) Petitianine, $R = H$, $R^1 = CH_3$, $R^2 = CH_2OH$, $R^3 = OH$
57) Hastacine, $R = R^2 = CH_3$, $R^1 = R^3 = H$

58) Angularine, $R = H$, $R' = CH_2$
59) 12-*O*-Acetylrosmarinine, $R = Ac$, $R' = CH_3$

60) Anacrotine, $R = H$, $R' = CH_3$, $R'' = OH$
61) Acetylanacrotine, $R = H$, $R' = CH_3$, $R'' = OAc$
62) Acetyl-*trans*-anacrotine, $R = CH_3$, $R' = H$, $R'' = OAc$

63) Senecivernine, $R = R' = H$, $R'' = CH_3$
64) Integerrimine, $R = CH_3$, $R' = R'' = H$
65) Senecionine, $R = R'' = H$, $R' = CH_3$

66) Sceleratine, $R = OH$
67) Sceleratine *N*-oxide, $R = OH$
68) Merenskine, $R = Cl$
69) Merenskine *N*-oxide, $R = Cl$

Rosmarinine (**54**) was isolated from *S. rosmarinifolius* Linn (Compositae family) and upon hydrolysis afforded the necine base (–)-rosmarinecine (**20**);[77–79] biosynthesis of **20** is well understood.[80–86] The X-ray crystal structure of rosmarinine has been reported.[87] The necine portion of **54** has been found in other pyrrolizidine alkaloids such as neorosmarinine (**55**), petitianine (**56**), angularine (**58**) and 12-*O*-acetylrosmarinine (**59**).[46,88,89] In addition, **54**, among other alkaloids, has been isolated from butterflies of the *Danaus plexippus* L. and *Danaus chrysippus* L. species found in Southern Australia.[89] It was demonstrated that **54** was mainly obtained from the butterflies that consumed *S. pterophorus*. The butterflies can store the alkaloids for an extended period of time and it is speculated to make the insects distasteful to predators.

Anacrotine (**60**) was isolated from the seeds of *Crotalaria amagyroids*, obtained from Sri Lanka,[47,90] and from *Crotalaria laburnifolia*[91,92] as well as *Crotalaria incana* shrub, grown in South Africa.[48] The structure of **60** was unambiguously proven by X-ray crystallographic analysis, confirming the relative and absolute stereochemistry and the size of the macrolactone.[93] (–)-Senecionine (**65**) is the best-known hepatotoxic pyrrolizidine alkaloid isolated from senecio plants as a principal livestock poisoning of these plants.[95,96] Sceleratine (**66**) and its *N*-oxide (**67**) have been isolated from *Senecio latifolius* D.C. and their structures were determined by spectroscopic and chemical methods as well as X-ray crystallography.[97]

Senecivernine (**63**) was isolated from *Senecio vernalis*.[98] The stereochemistry in **63**, merenskine (**68**)[99] and sceleratine (**66**)[100] was determined with X-ray analysis.[101] Integerrimine (**64**) was isolated from *S. integerrinus*[102] and *Crotalaria incana*.[103] The structure of **64** has been determined as the isomer of senecionine (**65**), differing only in the configuration of the ethylidene group.[104,105]

Many synthetic approaches of the naturally occurring hydroxypyrrolizidines and their analogues from noncarbohydrates and carbohydrates have been reported.[106–239] Herein, those derived only from carbohydrates shall be discussed.

5.3.1.1 *Synthesis from D-glucose* A strategy for the syntheses of alexine (**1**) and 7-*epi*-alexine (**9**) has utilized the formation of the pyrrolidine ring in D-glucose by joining C-2 and C-5 by nitrogen (Scheme 1).[240] Thus, methyl 2-azido-3-*O*-benzyl-2-deoxy-α-D-mannofuranoside (**70**),[241] obtained from D-glucose, was reacted with *tert*-butyldimethylsilyl chloride to give the respective silyl ether. Triflation of the free hydroxyl group followed by hydrogenation and subsequent benzylation of the resulting cyclized secondary amine afforded the fully protected pyrrolidine **71**. Removal of the silyl ether from **71** followed by Swern oxidation afforded the corresponding aldehyde, which upon subsequent treatment with vinylmagnesium bromide gave a mixture of the epimeric allylic alcohols **73** (38%) and **72** (37%). The more polar alcohol **73** was converted into **1**. Thus, protection of **73** with TBSCl followed by treatment with BMS in THF and then alkaline hydrogen peroxide afforded **74**. Tosylation of **74** gave the salt **75**, which was hydrogenated to give **76**. This was treated with aqueous TFA, followed by reduction of the resulting lactol with sodium borohydride and then by ion-exchange chromatography to give **1**. A minor epimerization of the open chain form of the intermediate lactol has taken place to give a 7% of 3-*epi*-alexine (**8**). The less polar alcohol **72** was transformed into **9** in 20% overall yield from **72** by a similar sequence of reactions to those used above.

The first enantioselective synthesis of retronecine (**22**) and its enantiomer (–)-**22** has been reported by conversion of D-glucose, via diacetone glucose, to 3-azido-1,2-*O*-isopropylidene-5,6-di-*O*-mesyl-α-D-glucofuranose (**77**)[242] (Scheme 2).[243] Reductive cyclization of **77** by hydrogenation followed by protection of the resulting secondary amine afforded the pyrrolidine derivative **78**. Compound **78** was treated with lithium chloride to give **79**, which underwent reduction followed by methanolysis of the isopropylidene

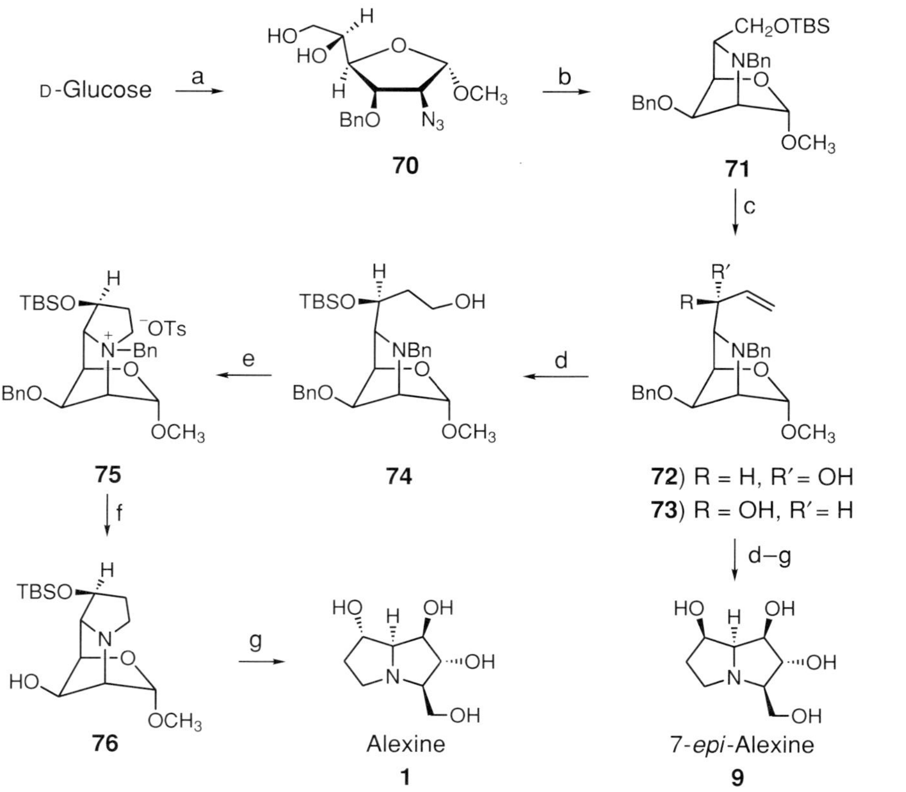

Scheme 1 (*a*) Ref. 241. (*b*) 1. TBSCl, DMF, 0°C, 95%; 2. Tf_2O, Py, −30°C; 3. H_2, EtOAc, Pd; 4. BnBr, DMF, NaOH, 77%. (*c*) 1. TBAF, THF, 88%; 66% from **70**; 2. $(COCl)_2$, DMSO, CH_2Cl_2, −78°C; then NEt_3; 3. vinylmagnesium bromide, THF, **72** (37%), **73** (38%). (*d*) 1. TBSCl, 89%; 2. borane, DMS, THF, then H_2O_2, base 67%. (*e*) *p*-TsCl, Py–CH_2Cl_2, 77%. (*f*) 10% Pd *on* C, H_2, AcOH, 72%. (*g*) 1. Aqueous TFA, 36 h; 2. $NaBH_4$, EtOH, ion-exchange chromatography, Amberlite CG-120 (NH_4^+) resin, 54%.

group and subsequent benzylation of the resulting secondary hydroxyl group to furnish **80**. Acid hydrolysis of **80** followed by olefination with methylenetriphenylphosphorane gave the vinyl alcohol **81**, whose treatment with MEMCl in the presence of imidazole followed by hydroboration and oxidation afforded **82**. Mesylation of **82** followed by reductive cyclization afforded **83**. Selective removal of the MEM group in **83** by acid hydrolysis gave the 7-hydroxy derivative **84**, while catalytic hydrogenolysis of **83** afforded the 1-hydroxy derivative **86**. The transformation of **84** and **86** to (–)-**22** and its enantiomer **22**, respectively, involved hydroxymethylation at C-7 and C-1. Thus, mesylation of **86** followed by treatment with sodium thiophenolate afforded the sulfide **87**. This was oxidized to the corresponding sulfoxide followed by benzyloxymethylation of the sulfoxide to afford the (benzyloxy)methyl phenyl sulfoxide, which upon subsequent *syn* elimination furnished the olefin **88**. Acid hydrolysis of **88** followed by benzyl ether cleavage with lithium in liquid ammonia afforded (+)-**22** in 5% overall yield from **77**. Similarly, conversion of **84** to (–)-**22** via **85** and **89** was achieved.

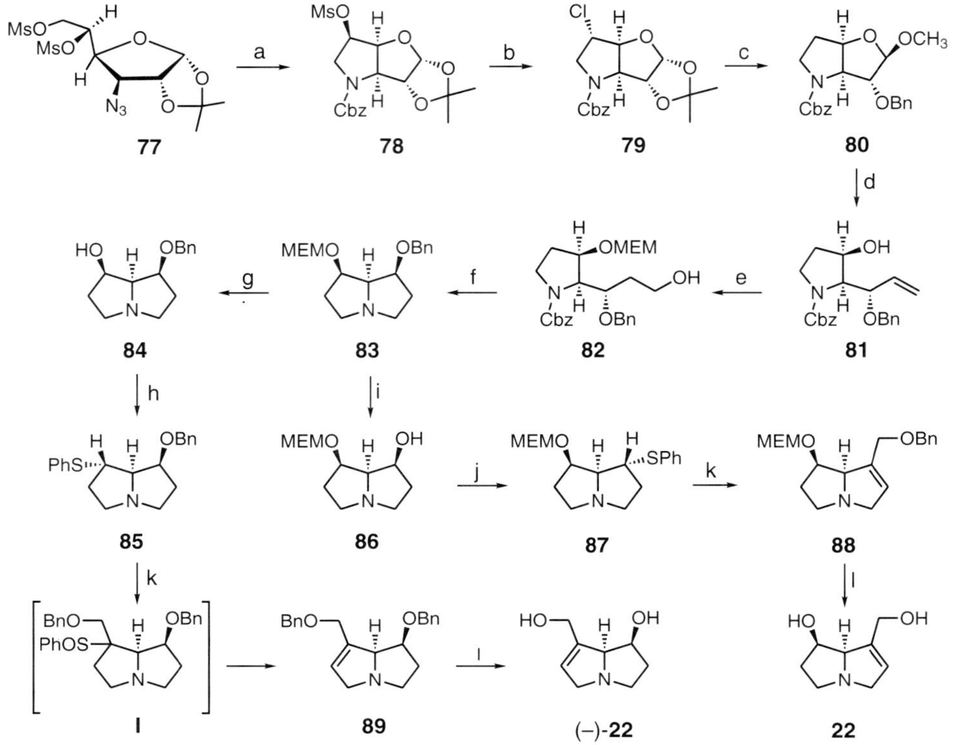

Scheme 2 (*a*) 1. EtOAc–CH_3OH (5:2), Raney nickel (W-4), H_2, 2 h, 100%; 2. CH_3OH, NEt_3, benzyl *S*-4, 6-dimethylpyrimid-2-yl thiocarbonate, rt, 2 h, 100%. (*b*) DMF, LiCl, 130°C, 17 h, 85%. (*c*) 1. Bu_3Sn, toluene, reflux, 1 day; 2. 10% HCl, CH_3OH, 50°C, 1 h, 61%; 3. DMF, NaH, rt, 30 min; then BnBr, rt, 1 h, 94%. (*d*) 1. AcOH–3 M HCl (3:1), 70°C, 2 h, 92%; 2. $Ph_3P^+CH_3Br^-$, THF, BuLi, rt, 30 min, 50°C, 5 h, 71%. (*e*) 1. CH_2Cl_2, *N*,*N*-diisopropylethylamine, (2-methoxyethoxy)methyl chloride, 45°C, 5 h, 93%; 2. THF, 9-borabicyclo[3.3.1]nonane, rt to 60°C, 1.5 h; then 2 M NaOH, 30% H_2O_2, 45°C, 2 h, 100%. (*f*) 1. Py, MsCl, −40°C to rt, 1 h, 92%; 2. Raney nickel (W-4), H_2, EtOAc, rt, 1 day, 100%. (*g*) 3 M HCl, H_2O, rt, overnight; then $NaHCO_3$, 100%. (*h*) 1. Py, MsCl, rt, 1 h, 88%; 2. PhSH, NaOH, DMF, 50°C, 2 h, 69%. (*i*) EtOH, H_2, Raney nickel (W-4), reflux, 20 h, 100%. (*j*) Same as (*h*); 1. 86%; 2. 65%. (*k*) 1. HCl–Et_2O (1.0 M solution), CH_3OH, 0°C, CH_2Cl_2, 85%, *m*-CPBA, −30°C, 30 min; then 10% aqueous KOH, 84%; 2. lithium diisopropylamine, $BnOCH_2Cl$, THF, HMPA, −78 to −15°C to rt, 77%; 3. xylene, reflux, 10 min, 69%. (*l*) 3 M HCl, rt, 1 day; then liquid NH_3, THF, lithium, −33°C, 5 h, isoprene, NH_4Cl, 56%.

Alternatively, the mesylate **77** was treated with sodium iodide in refluxing 2-butanone to produce the olefin **90** (Scheme 3).[244] Oxymercuration–demercuration reaction of **90** using mercuric acetate–sodium borohydride afforded a 1:1 mixture of **91** and **92**. Compound **92** was mesylated, reduced and cyclized in boiling ethanol, in the presence of sodium acetate, to give **93**. This was converted by standard steps into **94**, and subsequent hydrolysis of **94** followed by oxidation with pyridinium chlorochromate afforded the lactone **95** (28% overall yield from **92**), which was used in the synthesis of retronecine (**22**).[245]

The intermediate **95** and its analogue **103** could be otherwise prepared from **96** by selective tosylation of the primary hydroxyl group, followed by azide reduction and subsequent cyclization to afford the pyrrolidine **97** (Scheme 4).[246] Compound **97** was heated with dry

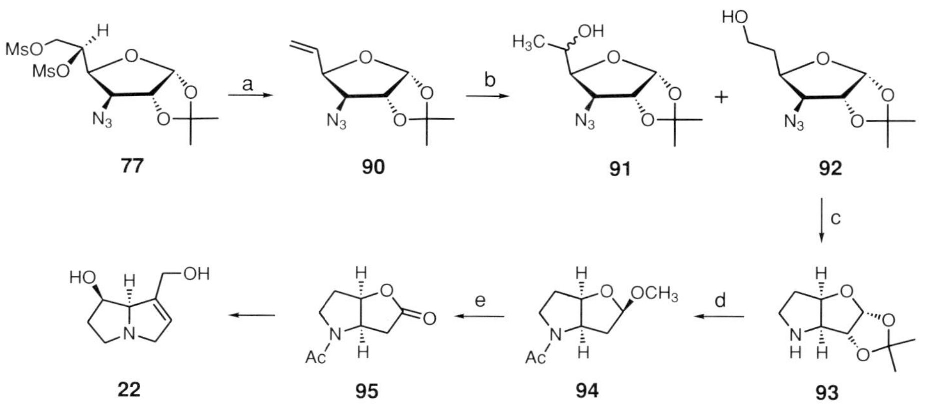

MsO MsO H N3 O O O → a → 77 90 → b → H3C OH 91 + HO 92 → c → 93 → d → OCH3 94 → e → 95 → 22 HO H OH N Ac H

Scheme 3 (*a*) NaI, $CH_3COCH_2CH_3$. (*b*) $Hg(OAc)_2$, $NaBH_4$, THF, H_2O. (*c*) 1. MsCl, Py, 0°C to rt, 1.5 h, 95%; 2. H_2, 10% Pd *on* C, CH_3OH, 18 h, rt, 90%; 3. NaOAc, EtOH, reflux, 6 h, 85%. (*d*) 1. Ac_2O, Py, rt, 2 h, 81%; 2. CH_3OH, HCl, reflux; *or* CH_3OH, Amberlite IR-120 resin, reflux, 3 h, 85%; 3. NaH, CS_2, CH_3I, THF, 4 h, 91%; 4. Bu_3SnH, toluene, reflux, AIBN, 8 h, 82%. (*e*) 1. AcOH–H_2O (1:1), reflux, 3 h, 86%; 2. PCC, CH_2Cl_2, rt, 20 h, 86%.

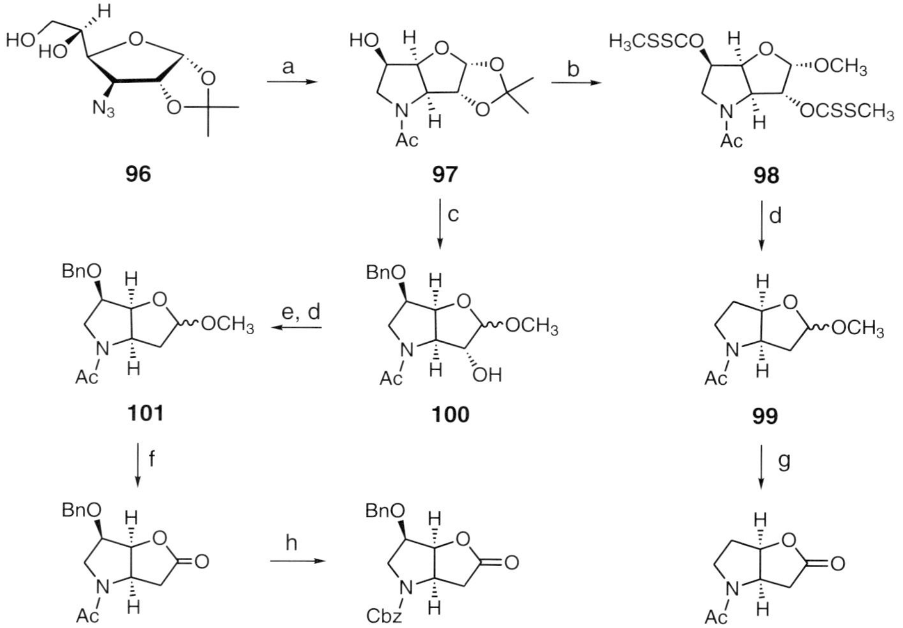

HO HO H N3 96 → a → HO H N Ac H 97 → b → H3CSSCO H OCH3 OCSSCH3 N Ac H 98 → d → 99 OCH3 → g → 95; 97 → c → BnO H OCH3 OH 100 → e, d → BnO OCH3 101 → f → BnO O 102 → h → BnO N Cbz H 103

Scheme 4 (*a*) 1. Py, *p*-TsCl, rt, overnight, 72%; 2. CH_3OH, 10% Pd *on* C, H_2, rt, 10 h; 3. EtOH, NaOAc, reflux, 4 h; Ac_2O, CH_3OH, rt, 2 h, 42% for three steps. (*b*) 1. CH_3OH, Amberlite IR-120 (H^+) resin, reflux, 6 h, 76%; 2. THF, NaH, CS_2, 20 min; then CH_3I, overnight, 81%. (*c*) 1. THF, NaH, 1.5 h, rt; then BnBr, rt, overnight, 81%; 2. CH_3OH, HCl, reflux, 4 h, 86%. (*d*) Toluene, *t*-BuSnH, AIBN, reflux, 8 h, 82%. (*e*) 1. THF, NaH, 1.5 h, CS_2; then CH_3I, rt, 3.5 h, 88%; 2. toluene, *n*-Bu_3SnH, AIBN, reflux, 6 h, 82%. (*f*) 1. AcOH–H_2O (1:1), 100°C, 2.5 h, 74%; 2. CH_2Cl_2, PCC, rt, 16 h, 80%. (*g*) Same as (*h*); 1. 86%; 2. 86%. (*h*) 1. H_2O, $Ba(OH)_2$, reflux, 17 h; 2. EtOH–$NaHCO_3$ (1:1), H_2O, CbzCl, rt, 3 h, 31%.

methanol and Amberlite IR-120 (H^+) resin, followed by conversion to the dixanthate **98**, whose deoxygenation with n-Bu_3SnH afforded **99**. Acid hydrolysis of **99** and oxidation afforded the lactone **95**. On the other hand, benzylation of **97** followed by methanolysis afforded **100**, which was deoxygenated to produce **101**. Acid hydrolysis of **101** followed by oxidation with PCC afforded the lactone **102**, which was converted into the *N*-carbobenzyloxy derivative **103** by hydrolysis of the *N*-acetyl group followed by carbobenzyloxylation.

A synthesis of (–)-platynecine (**30**) has been reported by conversion of diacetone D-glucose to **104**,[247] whose reaction with triflic anhydride followed by reduction with sodium borohydride and subsequent replacement of the carbamate group with trifluoroacetyl group afforded **105** in 86% overall yield from **104** (Scheme 5).[248] Methanolysis of the acetonide **105** gave the corresponding methyl furanoside (α:β 1:13). Oxidation at C-2 of the major β-anomer followed by Wittig reaction afforded the unsaturated ester **106**. Hydrogenation of **106** proceeded with high stereoselectivity to give a single isomer, whose treatment with sodium methoxide caused removal of the trifluoroacetyl group and intramolecular cyclization of the resulting amine with the ester group to afford the tricyclic amide **107**. Hydrolysis of **107** with aqueous TFA followed by reduction of the resulting lactol afforded **30** in 20% overall yield from **104**.

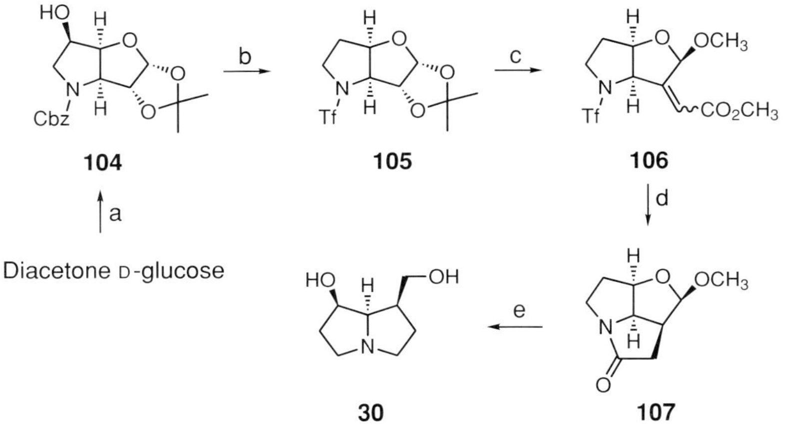

Scheme 5 (*a*) Ref. 247. (*b*) 1. Tf_2O, Py–CH_2Cl_2 (1:30), −30°C, 1 h and 10–15°C, 2 h, 95%; 2. $NaBH_4$, CH_3CN, under N_2, 3 days, 96%; 3. EtOH–EtOAc (10:7), H_2, Pd black, 12 h; 4. CH_2Cl_2, Py, Tf_2O, 0°C, 3 h, 94% for two steps. (*c*) 1. CH_3OH, AcCl, 50°C, 3 h, β (81%), α (6%); 2. PCC, MS 3 Å, CH_2Cl_2, rt, overnight, 78%; 3. $Ph_3P{=}CHCO_2CH_3$, benzene, reflux, 3 h, 86%. (*d*) 1. H_2, 10%, Pd *on* C, EtOAc, 12 h, 99%; 2. $NaOCH_3$, CH_3OH, reflux, 48 h, under N_2, 52%. (*e*) Aqueous TFA, rt, 4 h; then THF, $LiAlH_4$, reflux under N_2 for 3.5 h, H_2O, NaOH, 78%.

5.3.1.2 *Synthesis from* D-*glucosamine* The first asymmetric syntheses of (–)-rosmarinecine (**20**) and (–)-7-deoxyrosmarinecine (**118**) have been carried out using the carbamate **108**,[249] prepared from methyl α-D-glucosaminide (Schemes 6 and 7).[250] Silylation of **108** with TBSCl followed by condensation with allylmagnesium bromide afforded the threo isomer **109**. The *R* configuration of the newly formed asymmetric carbon atom was rationalized by a chelation-controlled approach.[251,252] Oxidation of the olefin **109** with sodium

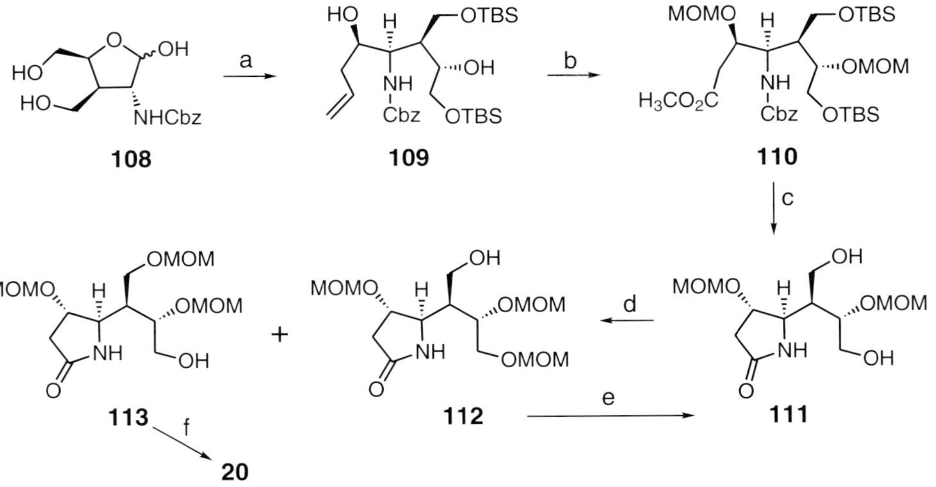

Scheme 6 (*a*) 1. TBSCl, Py, 87%; 2. $CH_2{=}CHCH_2MgBr$, Et_2O, 5°C, 30 min; 20°C, 30 min, reflux, 3 h, 92%. (*b*) 1. $NaIO_4$, $KMnO_4$, aqueous *t*-BuOH; then 5% K_2CO_3, 15 h; 2. CH_2N_2, Et_2O, 30 min; 3. MOMCl, $(i\text{-Pr})_2NEt$, $CHCl_3$, 60°C, 5 h, 92%. (*c*) 1. H_2, 5% Pd *on* C, THF, AcOH, 3 h; 2. TBAF, THF, 5°C, 30 min, 93%. (*d*) MOMCl, $(i\text{-Pr})_2NEt$, THF, 6 h, **112** (34%), **113** (51%). (*e*) Cat. CSA, CH_3OH, 80%. (*f*) 1. MsCl, Py; 2. BMS, THF, 60°C; 3. 0.5 N HCl, dioxane, 80°C, 6 h, 50%.

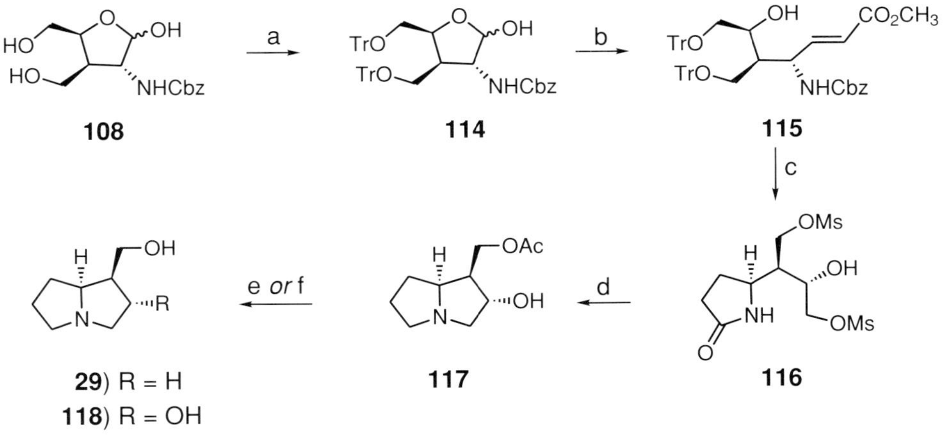

Scheme 7 (*a*) *p*-TrCl, Py, 70°C, 26 h. (*b*) $Ph_3P{=}CHCO_2CH_3$, toluene, 60°C, 62 h, 85%. (*c*) 1. H_2, 5% Pd *on* C, THF, AcOH, 5 h; 2. Amberlyst 15, CH_3OH, 60°C, 5 h; 3. MsCl, Py, 0°C, 2 h, 80% for three steps. (*d*) 1. BMS, THF, 60°C, 12 h; 2. KOAc, DMSO, 80°C, 4 h, 90%. (*e*) 1. $SOCl_2$, reflux, 3 h; 2. H_2, Raney nickel, EtOH, 15 h; 3. NH_3, CH_3OH, 40 h, 74% for three steps. (*f*) NH_3, CH_3OH, 40 h, 84%.

metaperiodate and potassium permanganate followed by esterification with diazomethane and subsequent methoxymethylation of the secondary hydroxyl groups afforded the protected ester **110**. Removal of the N-blocking group afforded the corresponding γ-lactam, which upon desilylation with fluoride ion afforded the diol **111**. Selective methoxymethylation of the two hydroxyl groups in **111** gave a mixture of **113** (51%) and **112** (34%); the minor product **112** could be recycled to **111** by selective deprotection. Mesylation, reduction and deprotection of **113** afforded **20** in 17.5% overall yield from **108**.

The same intermediate **108** was also used for the synthesis of (–)-7-deoxyrosmarinecine (**118**) and (–)-isoretronecanol (**29**). The ditrityl derivative **114**, obtained from **108**, was subjected to Wittig reaction to give the unsaturated ester **115**, which upon hydrogenation, lactamization and mesylation gave the lactam **116**. Its reductive cyclization followed by displacement of the mesylate with acetate anion afforded the acetate **117**, which was deacetylated to form **118** in 51% yield from **114**. On the other hand, the acetate **117** was converted into **29**[253] in 45% overall yield from **114** (Scheme 7).[250]

5.3.1.3 *Synthesis from* D*-mannose* A divergent approach to the synthesis of 1-*epi*-australine (**3**) and 1,7-di-*epi*-australine (**7**) is to use **120** as a starting substrate containing five stereocenters; no other stereocenters need to be created to achieve the goal (Scheme 8).[254] Thus, a mixture of **119**[255] and its epimer, obtained from D-mannose, were triflated followed by S_N2 displacement with sodium azide to afford the same azide **120**, which underwent reduction with DIBAL-H and $NaBH_4$ followed by selective silylation of the primary hydroxyl group to afford **121**. Mesylation of **121** followed by removal of the terminal isopropylidene group and subsequent epoxide formation afforded **122**. Triflation of the primary hydroxyl group in **122** followed by treatment with lithium cyanide afforded the nitrile **123**. Hydrogenation of **123** gave **124** in 16% overall yield from **119**. Disappointingly, treatment of the aminonitrile **124** with aqueous ammonia in ethanol in the presence of ammonium chloride gave very low yield (about 5%) of the required bicyclic lactam **125**. Reduction of the bicyclic lactam **125** gave the epimeric borane adduct, which on acid hydrolysis gave **3**.

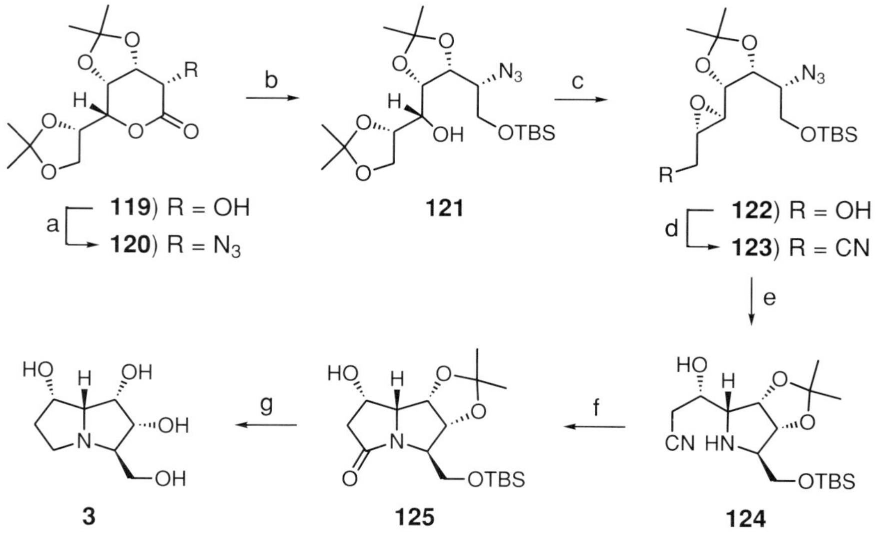

Scheme 8 (*a*) 1. Tf_2O, CH_2Cl_2, Py, 84%, 89%; 2. NaN_3, DMF, rt, 90% yield. (*b*) 1. DIBAL-H, THF, 2. $NaBH_4$, EtOH, 89%; 3. TBSCl, THF, imidazole, 79%. (*c*) 1. MsCl, Py, DMAP, 83%; 2. aqueous AcOH, dioxane, 92%; 3. CH_3OH, $Ba(OH)_2$, 89%. (*d*) 1. Tf_2O, CH_2Cl_2, Py, $-50°C$, 2. LiCN, CH_2Cl_2, THF, 60% for two steps. (*e*) H_2, Pd black, EtOAc, 82%, 16% from **119**. (*f*) aqueous NH_3, EtOH, NH_4Cl, very low yields 5%. (*g*) 1. BMS, THF; 2. 50% aqueous TFA.

On the other hand, similar sequence of transformation was carried out on **124** to give the bicyclic lactam **126**. Oxidation of **126** by pyridinium chlorochromate followed by sodium borohydride reduction afforded the lactam **127**, which was converted into **3** in 13% overall yield from **119** (Scheme 9).[254] Reduction of **126** followed by acid hydrolysis produced **7** in 17% overall yield from **119**.

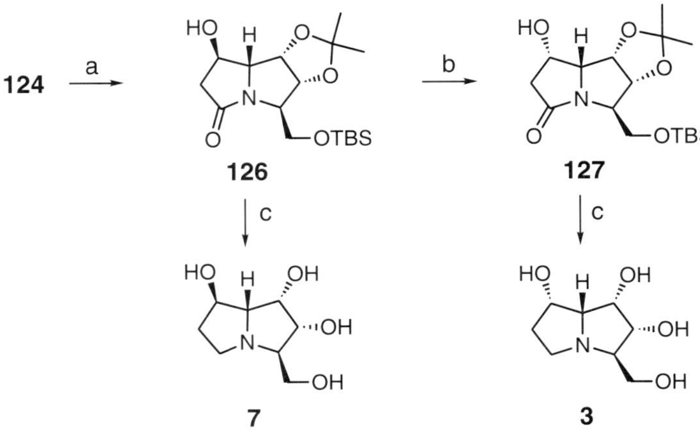

Scheme 9 (*a*) Aqueous NH_3, EtOH, NH_4Cl, 100°C, 20 h, 60%. (*b*) 1. PCC, CH_2Cl_2; 2. $NaBH_4$, EtOH, 0°C. (*c*) 1. THF, BMS, 94%; 2. 50% aqueous TFA, 100%.

5.3.1.4 *Synthesis from* D-*fructose* Syntheses of 7a-*epi*-hyacinthacine A_2 (7-deoxyalexine, **139**) and 5,7a-di-*epi*-hyacinthacine A_3 (**143**) from D-fructose have been reported (Scheme 10).[256,257] Mesylation of 3-*O*-benzoyl-4-*O*-benzyl-1,2-*O*-isopropylidene-β-D-fructopyranose (**128**)[258] followed by azidolysis gave 5-azido-3-*O*-benzoyl-4-*O*-benzyl-5-deoxy-1,2-*O*-isopropylidene-α-L-sorbopyranose (**129**). The removal of the 1,2-*O*-isopropylidene group followed by protection of the primary hydroxyl group as silyl ether afforded **130**. The azide in **130** underwent catalytic hydrogenation to form the corresponding amine that rearranged in a fast process to its cyclic imine intermediate **131**, which was finally hydrogenated in a highly stereocontrolled manner to afford **132**. Oxidation of **132** using tetra-*n*-propylammonium perruthenate yielded the aldehyde **133**, which was directly treated with [(methoxycarbonyl)methylene]triphenylphosphorane to afford **134**. Catalytic hydrogenation of **134** afforded **135**, which was heated with methanolic sodium methoxide followed by treatment with TBAF to produce lactam **138**. This was reduced with BMS in THF to produce the partially protected 7a-*epi*-hyacinthacine A_2 (**137**). On the other hand, compound **133** was treated with (triphenylphosphoranylidene) acetaldehyde to give the (*E*)-α,β-unsaturated aldehyde **136**, which underwent catalytic hydrogenation followed by partial deprotections and cyclization to give the pyrrolizidine **137**. Catalytic hydrogenation of **137** afforded **139**.

For the synthesis of 5,7a-di-*epi*-hyacinthacine A_3 (**143**), compound **133** was treated with 1-triphenylphosphoranylidene-2-propanone to give **140**, which underwent catalytic hydrogenation to afford **142** via the intermediate **141**. Finally, removal of the protecting groups in **142** gave **143** (Scheme 11).[256,257]

Scheme 10 (*a*) Ref. 258. (*b*) 1. *p*-TsCl, NEt_3, CH_2Cl_2, rt, overnight; 70%; 2. NaN_3, DMF, 80°C, 15 h, 84%. (*c*) 1. 70% aqueous TFA, rt, 5 h, quantitative; 2. TBDPSCl, DMF, imidazole, rt, 5 h, 96%. (*d*) H_2, Raney nickel, CH_3OH, 4 h. (*e*) 1. H_2, Raney nickel, CH_3OH, 15 h, 97% from **130**; 2. CbzCl, NEt_3, CH_3OH, rt, 2 h, 68%. (*f*) TPAP, NMO, CH_2Cl_2, MS 4 Å, 4 h, 95%. (*g*) $Ph_3P{=}CHCO_2CH_3$, CH_2Cl_2, rt, overnight, 89%. (*h*) H_2, 10% Pd *on* C, CH_3OH, 8 h, 66%. (*i*) $Ph_3P{=}CHCHO$, CH_2Cl_2, rt, 36 h, 44%. (*j*) 1. $NaOCH_3$, CH_3OH, reflux, 10 h, 71%; 2. TBAF, CH_3OH, 6 h, 81%. (*l*) BMS, THF, 0°C, 30 min, rt, 6 h, 26%. (*k*) 1. H_2, 10% Pd *on* C, CH_3OH, HCl, 7 h, 45%; 2. TBAF, CH_3OH, 6 h, 59%. (*m*) 1. H_2, 10% Pd *on* C, CH_3OH, HCl, 93%; 2. Amberlite IRA-400 (OH^-) resin, CH_3OH, 76%.

5.3.1.5 *Synthesis from* D*-arabinose* (+)-Alexine (**1**) has been synthesized from D-arabinose via 2,3,5-tri-*O*-benzyl-D-arabinofuranose (**144**)[259,260] (Scheme 12).[261] The functionalized lactam **146** was obtained by the nucleophilic addition of vinylmagnesium bromide to the furanosylamine **145**, obtained from **144**, followed by oxidative degradation with PCC. The olefinic part in **146** was then cleaved to give the corresponding aldehyde intermediate, which was in turn subjected to allylation to give **147**. Removal of the *N*-MPM moiety from **147** followed by protection of the free hydroxyl group with MOMCl and subsequent oxidative cleavage of the double bond gave an aldehyde compound, which underwent

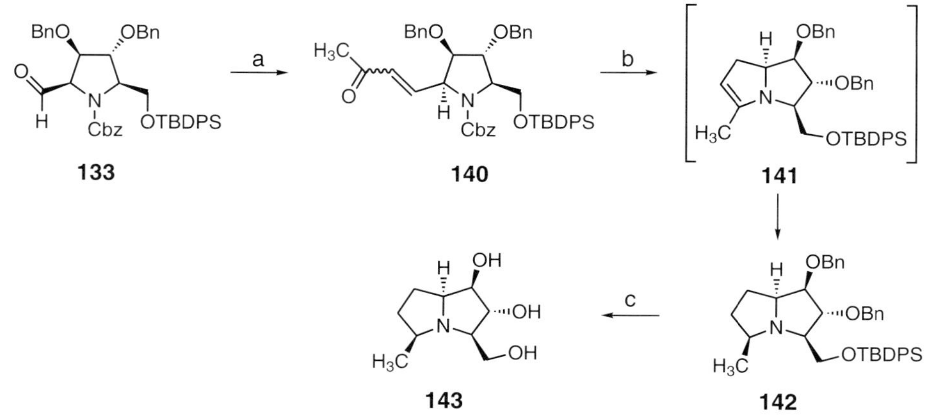

Scheme 11 (*a*) $Ph_3P{=}CHCOCH_3$, CH_2Cl_2, reflux, 81%. (*b*) H_2, 10% Pd *on* C, 24 h, 60%. (*c*) 1. TBAF, CH_3OH, 6 h, 85%; 2. H_2, 10% Pd *on* C, CH_3OH, HCl, 3. Amberlite IRA-400 (OH^-) resin, CH_3OH, 83%.

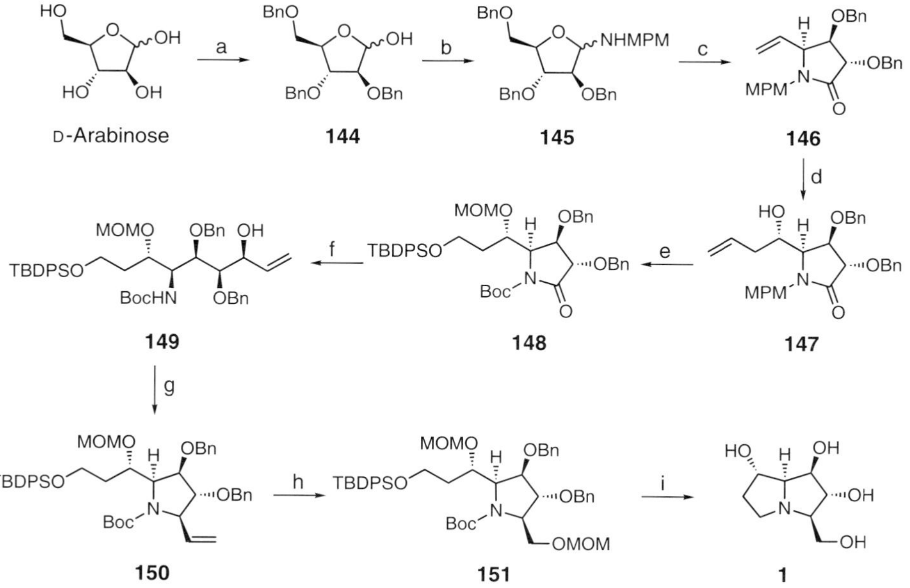

Scheme 12 (*a*) Refs. 260 and 261, **144** is commercially available. (*b*) $MPMNH_2$, benzene–$CHCl_3$ (1:1), MS 4 Å, reflux; quantitative. (*c*) 1. $CH_2{=}CHMgBr$, THF, −78 to −40°C, 70%; 2. PCC, MS 4 Å, CH_2Cl_2, 68%. (*d*) 1. OsO_4 NMO, acetone–H_2O (1:1), 98%; 2. $NaIO_4$, Et_2O–H_2O (2:1); 3. allyltrimethylsilane, $BF_3{\cdot}OEt_2$, CH_2Cl_2, −78 to −20°C; 82% for two steps. (*e*) 1. CAN, CH_3CN–H_2O (9:1), 71%; 2. MOMCl, $(i\text{-Pr})_2NEt$, CH_2Cl_2, 75%; 3. OsO_4, NMO, acetone–H_2O (1:1), 91%; 4. $NaIO_4$, Et_2O–H_2O (2:1); 5. $NaBH_4$, EtOH, 90% for two steps; 6. TBDPSCl, imidazole, DMF, quantitative; 7. $(Boc)_2O$, DMAP, NEt_3, CH_2Cl_2, 99%. (*f*) 1. $CH_2{=}CHMgBr$, THF, −78°C; 2. $NaBH_4$, CH_3OH, −45°C, 66% for two steps. (*g*) 1. MsCl, NEt_3, CH_2Cl_2; 2. *t*-BuOK, THF, 84% for two steps. (*h*) 1. OsO_4, NMO, acetone–H_2O (1:1), 92%; 2. $NaIO_4$, Et_2O–H_2O (2:1); 3. $NaBH_4$, EtOH, 74% for two steps; 4. MOMCl, $(i\text{-Pr})_2NEt$, CH_2Cl_2, 99%. (*i*) 1. TBAF, THF, quantitative; 2. *p*-TsCl, Py, 92%; 3. conc. HCl, CH_3OH; 4. K_2CO_3, CH_3OH, 94% for two steps; 5. H_2, 10% Pd *on* C, EtOH, 70%.

reduction to give the corresponding alcohol compound whose protection gave the lactam **148**. Grignard addition to **148** afforded a labile quaternary α-hydroxypyrrolidine intermediate, which was subsequently reduced to provide **149**. Cyclization of **149** gave the pyrrolidine derivative **150** in 84% yield. Oxidative cleavage of **150** followed by reduction and MOM protection afforded **151**, which was subjected to mild basic conditions to construct the bicyclic pyrrolizidine ring, followed by partial deprotection with HCl, leading to the dibenzyl derivative of alexine that upon debenzylation gave **1**.

The synthesis of hyacinthacine A_2 (**14**) has been achieved by addition of divinylzinc to **144** to give the heptenitol **152** in 95% yield (Scheme 13).[262] Regioselective benzoylation of the allylic hydroxyl group using benzoyl chloride in a two-phase system afforded a 3.5:1 mixture of **153** and **154** whose Swern oxidation provided the keto-benzoate **155**, readily separable from other products. Compound **155** was converted under reductive amination into isomers of **156**. Ring-closing metathesis of the epimeric mixture using Grubb's catalyst afforded **157**, which was obtained as the main product in 30% yield. Finally, the removal of the benzyl groups and reduction of the double bond in **157** was conducted with catalytic hydrogenation to give **14**.

153) R^1 = Bz, R^2 = H
154) R^1 = H, R^2 = Bz

Scheme 13 (*a*) $(CH_2{=}CH)_2Zn$, 95%. (*b*) BzCl, *n*-Bu_4NI, CH_2Cl_2, 1 N NaOH, 0°C, 3 h. (*c*) TFAA, DMSO, NEt_3, CH_2Cl_2, −78°C to rt, 63% for two steps. (*d*) Allylamine, AcOH, $NaBH_3CN$, MS 3 Å, CH_3OH, 0–40°C, 6 days 78%, (3:1). (*e*) Bis(tricyclohexylphosphine) benzylidine ruthenium(IV) dichloride (Grubb's catalyst), toluene, 60°C, 72 h, 30%. (*f*) H_2, Pd *on* C, CH_3OH–THF–6 N HCl (4:1:0.25), rt, 20 h, 82%.

On the other hand, attempted synthesis of 3-*epi*-australine (**10**) from **144** has failed (Scheme 14).[263] The lactol **144** was subjected to a Wittig reaction to produce the alkene **158**, which was triflated followed by azidolysis to give the alkene **159**. Ozonolysis of **159** followed by olefination of the resulting aldehyde **160** with the allylic borane reagent produced the diene **161** in 50% overall yield from the alkene **159**. Heating of **161** in chloroform produced the two cyclized products **162** and **163** in equal amounts. Subjection of **163** to different conditions has failed to produce the ketone **164** required for the synthesis of **10**, where decomposition had taken place.

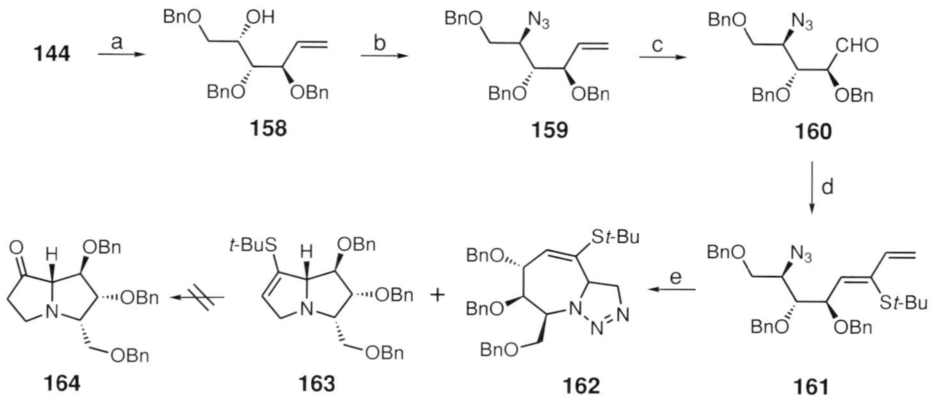

Scheme 14 (*a*) $Ph_3P^+CH_3Br^-$, *n*-BuLi, THF, −78°C to rt, 24 h, 80%. (*b*) 1. Tf_2O, CH_2Cl_2, Py, −40°C to rt, 30 min; 2. *n*-Bu_4NN_3, benzene, −10°C to rt, 1.5 h, 71%. (*c*) 1. O_3, CH_3OH; 2. DMS. (*d*) 1. 9-BBN, $H_2C{=}C{=}C(S\textit{t}\text{-Bu})TMS$; 2. NaOH, 50% from **159**. (*e*) $CHCl_3$, 75°C, 18 h, **163** (25%), **162** (25%).

5.3.1.6 *Synthesis from* L*-xylose* L-Xylose has been utilized for the synthesis of australine (**5**) and the unnatural (–)-7-*epi*-alexine (**9**) (Scheme 15).[263,264] L-Xylose was converted into 2,3,5-tri-*O*-benzyl-L-xylofuranose (**165**).[265] Wittig olefination of **165** followed by triflation of the resulting secondary hydroxyl group and subsequent displacement of the generated

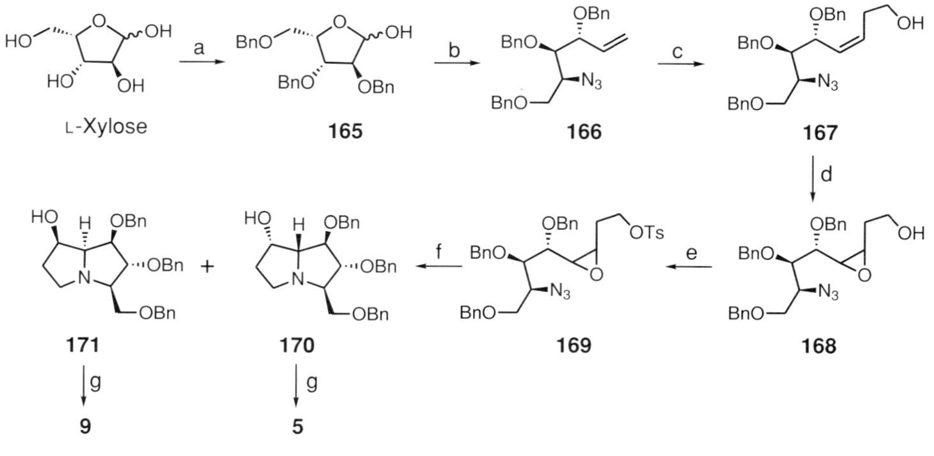

Scheme 15 (*a*) Ref. 265. (*b*) 1. $Ph_3P^+CH_3Br^-$ (2.2 equiv.), *n*-BuLi (2.3 equiv.), THF, 0°C, 15 min to 23°C over 3 h, 66%; 2. Tf_2O (1.2 equiv.), Py (1.4 equiv.), CH_2Cl_2, −40 to 0°C, 2.5 h; Bu_4NN_3 (5 equiv.), benzene, 23°C, 1 h, 75%. (*c*) 1. O_3, CH_2Cl_2–CH_3OH (6:1), −78°C; then DMS (3 equiv.), −78 to 23°C, 3.5 h; 2. $Ph_3P^+(CH_2)_3OHBr^-$ (1.05 equiv.), $KN(TMS)_2$ (2.1 equiv.), THF, 0°C, 1 h; 23°C, 1 h; TMSCl (1.08 equiv.), 0°C, 10 min; −78°C, 1 h; 23°C, 1 h; 1 M HCl, 23°C, 1 h, 35% from **166**. (*d*) *m*-CPBA (1.5 equiv.), CH_2Cl_2, 0°C; 23°C, 24 h, 65%. (*e*) *p*-TsCl (2 equiv.), Py (3 equiv.), DMAP (0.1 equiv.), CH_2Cl_2, −15°C, 48 h, based on 12% recovered **168**, 77%. (*f*) 5 wt% of 10% Pd *on* C, ether–EtOH (2:1), 1 atm H_2, 23°C, 15 h; K_2CO_3 (6 equiv.), EtOH, reflux, 20 h; flash chromatography, 71%. (*g*) 300 wt% of 10% Pd *on* C, EtOH, 1 atm H_2, 23°C, 48 h, **5** (87%), **9** (87%).

triflate with azide ion afforded the unstable azide **166**. Ozonolysis of **166** gave the corresponding aldehyde, which was directly elongated by three carbons to create the eight-carbon skeleton of the *Z*-alkene **167**. The unstable azidoalkene **167** was treated with *m*-CPBA to afford a 1:1 mixture of *cis*-epoxides **168**, which were directly tosylated to produce **169** as a 2:1 mixture of isomers, since one of the isomers of **168** underwent tosylation faster than the other and the reaction could not be driven to completion. Reduction of the azide group of **169** followed by boiling of the resulting amine in ethanol containing potassium carbonate afforded a 2:1 mixture of the two pyrrolizidines **170** and **171**. Separation and then debenzylation led to the two tetrahydroxypyrrolizidines australine (**5**) and (–)-7-*epi*-alexine (**9**) in 3.6 and 1.8% overall yield, respectively, from **165**.

5.3.1.7 *Synthesis from erythrose* An efficient approach for the synthesis of (+)-trihydroxyheliotridane (**180**) via a chiral erythrose derivative has been reported (Scheme 16).[266] Wittig reaction of 2,3-*O*-isopropylidene-L-erythrose (**172**) with $Ph_3P{=}CHCH{=}CHCO_2Et$ produced a 1:5 mixture of the (*E*,*E*)-**173** and (*Z*,*E*)-**174** isomeric dienes, respectively. The diene **173** could be quantitatively obtained by isomerization of **174** with I_2. The diene **174** was converted to the azide **177**, which upon boiling in benzene gave the vinyl aziridine **176**. Pyrolysis of **176** furnished the pyrrolizidine **178**. On the other hand, the diene **173** was

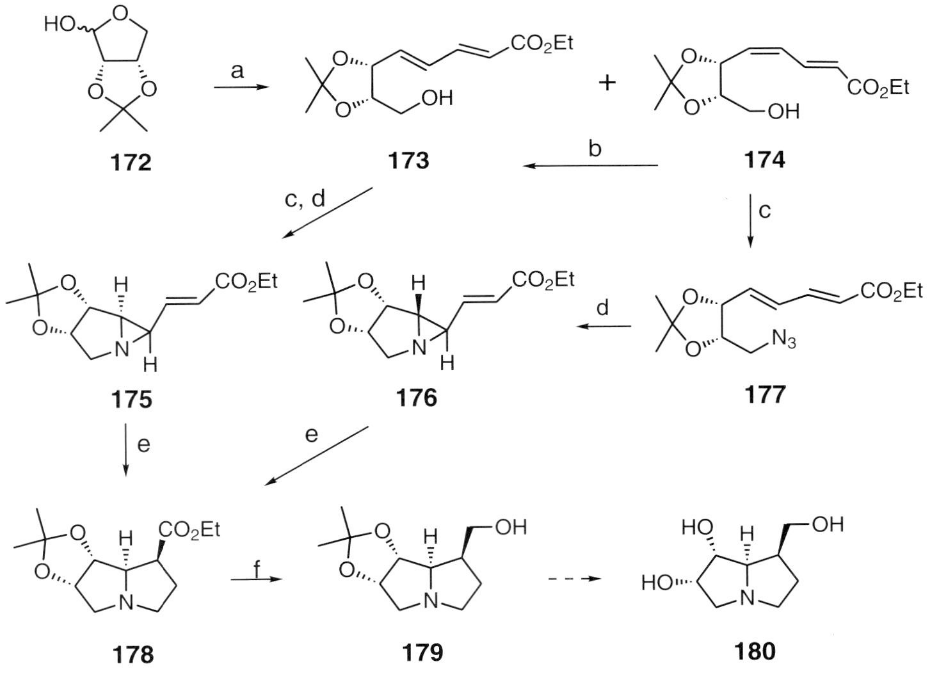

Scheme 16 (*a*) $Ph_3P{=}CHCH{=}CHCO_2Et$, CH_2Cl_2, 84%, **173** (14%), **174** (70%). (*b*) I_2, CH_2Cl_2, 94%. (*c*) 1. Tf_2O, Py, CH_2Cl_2; 2. NaN_3, 18-crown-6, CH_2Cl_2, 69% for two steps. (*d*) Benzene, reflux, 44%. (*e*) 1. FVP, 520°C, ca. 10^{-4} Torr; 2. H_2, Pd *on* C, CH_3OH. (*f*) $LiAlH_4$, THF, 34% for three steps.

similarly converted into **178**, via **175**, which underwent $LiAlH_4$ reduction to furnish **179**, whose deprotection would give **180**. In an identical fashion, (–)-trihydroxyheliotridane was prepared from 2,3-*O*-isopropylidene-D-erythrose.

2,3-*O*-Isopropylidene-D-erythrose (**181**) could be obtained from a number of readily available carbohydrates, and a particularly efficient route was carried out from D-araboascorbic acid (D-isoascorbic acid).[267] It has been used as a precursor for various natural products such as crotanecine (**21**) (Scheme 17).[268,269] Thus, its oxime can be converted upon mesylation to the cyano derivative **182**, which was allowed to react with methyl bromoacetate in the presence of activated zinc dust to yield the enamino esters **183** (*Z*/*E* 30:1). Each isomer of **183** could be cyclized in high yield to the respective pyrrolidine **184** on treatment with DBU. Both isomers gave the same saturated pyrrolidine derivative **185** on reduction with sodium cyanoborohydride in acidified methanol. Alkylation of **185** with ethyl bromoacetate in the presence of triethylamine gave the corresponding diester, which was converted by acid hydrolysis into the corresponding lactone whose silylation afforded **186**. When compound **186** was treated with potassium ethoxide, the intermediate keto ester **187** was formed, which was directly reduced with borohydride and subsequently acetylated to give a diastereoisomeric mixture of diacetates **188**. Elimination of acetic acid from **188**

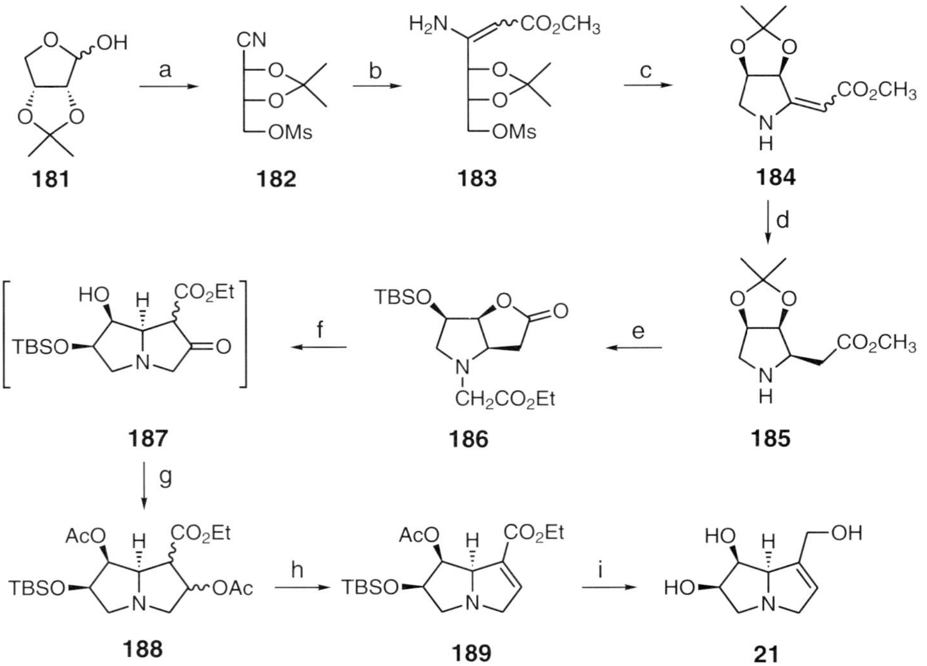

Scheme 17 (*a*) 1. $NH_2OH{\cdot}HCl$ (10 equiv.), Py, rt, 96%; 2. MsCl (12 equiv.), Py, −23°C. (*b*) Activated Zn, $BrCH_2CO_2CH_3$ (5 equiv.), THF, reflux, *Z* (78%), *E* (2.2%). (*c*) DBU (3 equiv.), CH_2Cl_2, rt, 24 h, *Z* gave 98%; *E* gave 80%. (*d*) $NaBH_3CN$, CH_3OH, HCl, 2 h, 90%. (*e*) 1. $BrCH_2CO_2Et$, NEt_3, THF, 89%; 2. 80% aqueous TFA, rt, 77%; then TBSCl, imidazole, DMF, 95%. (*f*) KOEt, benzene, rt; then AcOH. (*g*) $NaBH_4$, EtOH; then Ac_2O, Py, 39% for two steps. (*h*) DBU, CH_2Cl_2, rt, 90%. (*i*) Ref. 270.

occurred smoothly on treatment with DBU to give the unsaturated ester **189** in 70% yield. Conversion of compound **189** into (–)-**21** took place upon reduction of the ester function and deprotection.[270]

Alternatively, synthesis of (–)-crotanecine (**21**) has been accomplished from **181**[267,271,272] by the Wittig olefination followed by tosylation to give **190** (Scheme 18).[273] Nucleophilic substitution of the tosyloxy group in **190** with sodium azide was accompanied by an intramolecular [2+3] dipolar cycloaddition to provide the imine **191**. Carbomethoxylation of **191** by sequential treatment with LDA and Mander's reagent[274,275] or methyl chloroformate followed by removal of the tetrahydropyranyl group afforded **192**, which underwent mesylation[276] to afford the cyclopropyl imine **193**. This was treated with aqueous HCl to give **194** whose cyclization afforded a mixture of amino ester **195** and its diastereomer. Subsequent reaction of **195** with diphenyl diselenide proceeded to give α-seleno ester **196**. Protonation of the nitrogen of **196** followed by oxidation and thermal elimination afforded **197**. Reduction of the ester group of **197** and subsequent acid hydrolysis afforded **21**.

Scheme 18 (*a*) 1. $THPOCH_2CH_2CH{=}PPh_3$; 2. *p*-TsCl, NEt_3, CH_2Cl_2. (*b*) NaN_3, DMF, 65% from **181**. (*c*) 1. LDA, −78°C; $CNCO_2CH_3$ *or* $ClCO_2CH_3$, 57%; 2. PPTs, CH_3OH, 73%. (*d*) MsCl, NEt_3, DMAP, CH_2Cl_2, 91%. (*e*) Aqueous HCl, CH_2Cl_2. (*f*) 1. $NaBH_3CN$, HCl, CH_3OH; 2. pH 9. (*g*) LDA, PhSeSePh, THF, −78°C, 56%. (*h*) 1. H_2SO_4; 2. *m*-CPBA, CCl_4, reflux, 47% (95% based *on* recovered **196**). (*i*) 1. DIBAL-H; 2. 1 N HCl, THF.

The pyrrolidines **201** and **202**, which are important intermediates in the synthesis of (+)-retronecine (**22**) and (–)-crotonecine (**21**),[277,278] were prepared from the respective alcohols *Z*-**198** and *E*-**198** by mesylation and subsequent treatment with saturated ethanolic ammonia

(Scheme 19).[279] The *Z*-alkene **198** gave the pyrrolidine **201** as a sole product in 88% yield, but the *E*-alkene **198** gave a 9:1 mixture of the pyrrolidines **201** and **202** in 81% yield.

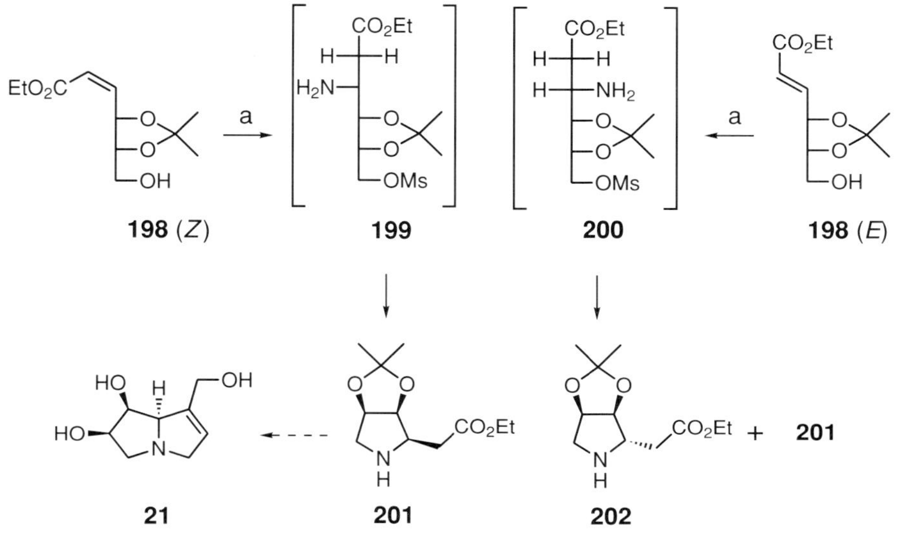

Scheme 19 (*a*) 1. MsCl, Py, CH_2Cl_2, 0°C, 20 min, rt, overnight, 88%; 2. EtOH, NH_3, rt, 96 h, **201**:**202** 9:1, 81%.

5.3.1.8 *Synthesis from* D-*mannitol* Syntheses of (+)-**30**, **31** and (+)-**31** from chiral *O*-benzylglycidol (**203**), readily available from D-mannitol, have been achieved by conversion[280] to the aziridines **206** and **207** via **204**[281] and **205**[282] (Schemes 20 and 21).[283] Thermolysis of **206** in diphenyl ether afforded the separable isomeric products **208** (8%) and **209** (70%). Partial reduction of **209** gave the lactol **210**, which upon subjection to Horner–Emmons reaction gave the bicyclic ester **211** (76%) which underwent selective N-debenzylation, saponification and cyclization to furnish the tricyclic lactam **212**. Removal of the *O*-benzyl group of **212** followed by iodination and olefination afforded the enol ether **213**, which underwent acid hydrolysis to produce the bicyclic ketone **214**. Acetylation of **114** followed by oxidation under Baeyer–Villiger conditions gave **215**, whose reduction with $LiAlH_4$ gave (–)-dihydroxyheliotridane (**31**).

Similarily, compound **207** was thermally cyclized to give **216** and **217**. The latter, **217**, was converted into **218** and **219** in 72 and 11% yield, respectively (Scheme 21).[280] The major one was sequentially debenzylated and diacylated to give **220**, which was converted to the iodo derivative **221**, followed by exposure to zinc in boiling ethanol to allow concurrent reductive ring cleavage, N-deprotection and cyclization to afford the vinyl lactam **222** in a diastereoisomerically pure state. Ozonolysis of **222** followed by reduction with sodium borohydride and then further reduction with $LiAlH_4$ afforded (+)-**31**. On the other hand, the minor isomer **219** afforded the (+)-platynecine [(+)-**30**] by following similar steps.

The open chain triol derivative **223**,[284–286] readily available from D-mannitol, has been used for the synthesis of (–)-hastanecine (**23**) and (–)-dihydroxyheliotridane (**31**)

Scheme 20 (*a*) 1. $CH_2{=}CHMgBr$, CuI, THF, −20°C, **204** (95%); *or* 1. $NaCH_2SOCH_3$, DMSO; 2. $CaCO_3$, 1,2-$Cl_2C_6H_4$, reflux, **205** (70%). (*b*) 2,3-Dibromopropionyl chloride, NEt_3, then $BnNH_2$, **206** (87%), **207** (77%). (*c*) PhOPh, 260°C, 5 min, **209** (70%), 2,3-*epi*-**208** (8%). (*d*) DIBAL-H, THF, −40°C. (*e*) $(EtO)_2P(O)CH_2CO_2Et$, NaH, THF, 76%. (*f*) 1. H_2, $Pd(OH)_2$, 97%; 2. LiOH, aqueous THF; 3. $(PhO)_2P(O)N_3$, NEt_3, DMF, 70% for two steps. (*g*) 1. EtSH, $BF_3{\cdot}OEt_2$, CH_2Cl_2, 95%; 2. I_2, PPh_3, imidazole; 3. DBU, THF, 70% for two steps. (*h*) 1% HCl, 94%. (*i*) Ac_2O, NEt_3, DMAP, then urea, H_2O_2, TFAA, 54%. (*j*) $LiAlH_4$, THF, 80%.

(Scheme 22).[287] Ozonolysis of **223** followed by sodium borohydride reduction afforded **224** (85%). The latter was benzylated, followed by removal of the isopropylidene group and subsequently treated with $Pb(OAc)_4$ in dichloromethane to afford the aldehyde **225** in 69% yield from **223**. Treatment of **225** with $(EtO)_2PCH_2CO_2Et$ in the presence of sodium hydride afforded **226** in 90% yield, which underwent reduction with DIBAL-H and the resulting hydroxyl group was protected with MOMCl, followed by debenzylation with sodium in liquid ammonia to afford **227** in 72% yield. Selective tritylation of the primary hydroxyl group in **227** afforded **228** (95%). The required branching was stereoselectively

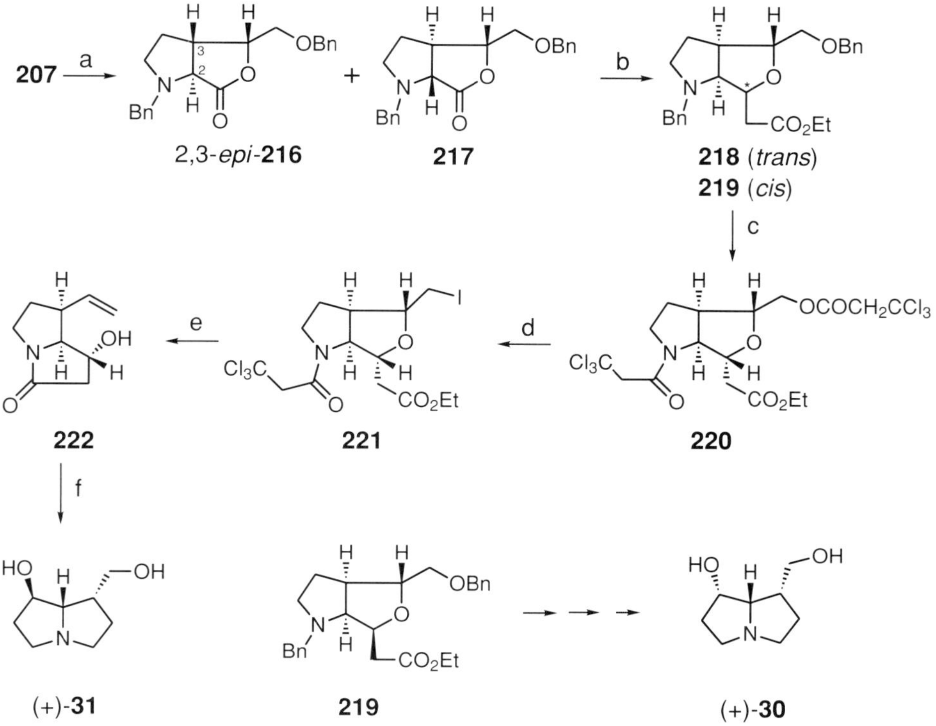

Scheme 21 (*a*) PhOPh, 260°C, 13 min, 70% inseparable 3:1 mixture of **217** and **216**. (*b*) 1. DIBAL-H, THF; 2. $(EtO)_2P(O)CH_2CO_2Et$, NaH, THF, *trans* 72% and *cis* 11% overall from **207**. (*c*) 1. H_2, Pd *on* C, conc. HCl, CH_3OH; 2. CCl_3CH_2COCl, Py, CH_2Cl_2–DMF (1:2), 87% from **218**. (*d*) 1. K_2CO_3, CH_3OH, 94%; 2. I_2, PPh_3, imidazole, THF, 99%. (*e*) Zn, EtOH, reflux, 82%. (*f*) 1. O_3, CH_3OH, −78°C; then $NaBH_4$, 77%; 2. $LiAlH_4$, THF, 79%.

introduced via an ortho ester rearrangement[288] using $(EtO)_3CCH_3$ in the presence of propionic acid to afford the ester **229** in 90% yield. Reduction of **229** with DIBAL-H followed by Mistunobu reaction using PhthNH gave **230**. The epoxidation of the double bond of **230** with *m*-CPBA afforded a 3–4:1 mixture of the *syn*-diastereomers **231** and **232**. Epoxide **231** was cyclized with hydrazine hydrate in ethanol to give a 7:1 mixture of the pyrrolidine **235** and the piperidine **234**. Compound **235** was detritylated and monomesylated at the primary position to give **236**, which was deprotected and cyclized to furnish **23**. The epoxide **232** was similarly transformed to **31** via intermediate **233**.

On the other hand, the acyclic epoxy alcohol **224** was also used for the synthesis of (+)-australine (**5**) (Scheme 23).[289,290] Reaction of **224** with 4-butenylisocyanate, prepared from 4-pentenoic acid via Curtius rearrangement of the corresponding azide *in situ*, gave the urethane **237**. Exposure of **237** to potassium *tert*-butoxide afforded the oxazolidinone **238**, which readily underwent acetonide migration to give the ketal **239**. Swern oxidation of the hydroxyl group in **239** followed by a Wittig reaction of the resultant aldehyde **240** with methylenetriphenylphosphorane furnished a diene, whose ring-closing metathesis with Grubbs catalyst produced the azacyclooctene derivative **241** in virtually quantitative yield.

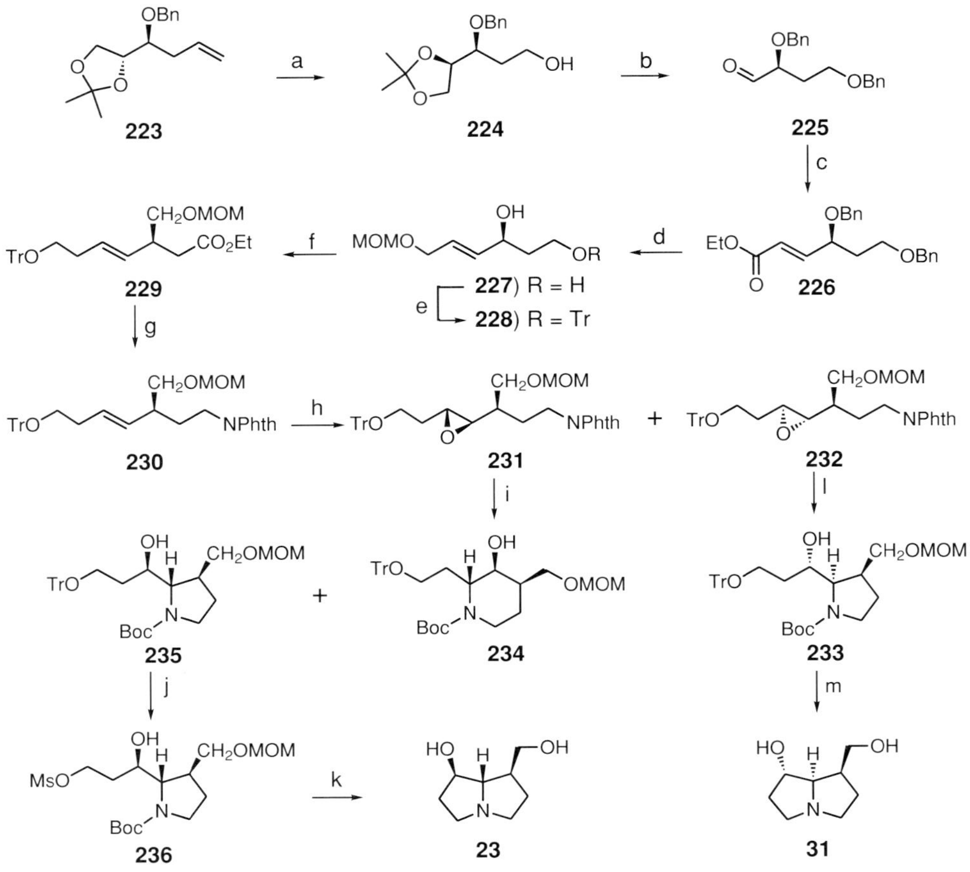

Scheme 22 (*a*) 1. O_3, CH_3OH, −78°C; 2. $NaBH_4$, CH_3OH, −78°C to rt, 10 h, 85% for two steps. (*b*) 1. BnBr, NaH; 2. DMF, −5°C to rt, 4–6 h, 90%; 3. $Pb(OAc)_4$, CH_2Cl_2, rt, 30 min, 90%. (*c*) $(EtO)_2PCH_2CO_2Et$, NaH, THF, −20°C, 30 min; then rt, 2 h, 90%. (*d*) 1. DIBAL-H, THF, −20°C, 2 h, 95%; 2. MOMCl, $(i\text{-Pr})_2NEt$, CH_2Cl_2, 0°C, 1 day, 95%; 3. Na/NH_3, −40°C, 80%. (*e*) TrCl, DMAP, Py, rt, 2 days, 95%. (*f*) $(EtO)_3CCH_3$, $EtCO_2H$, 100°C, 15 h, 90%. (*g*) 1. DIBAL-H, THF, −20°C, 2 h, 95%; 2. PPh_3, PhthNH, DEAD, THF, rt, 1 h, 95%. (*h*) *m*-CPBA, CH_2Cl_2, $NaHCO_3$, 0°C, 4 days, 80%. (*i*) 1. N_2H_4, EtOH, rt, 2 days; 2. $(Boc)_2O$, THF, $(i\text{-Pr})_2NH$, rt, 24 h, 82% for two steps. (*j*) 1. Pd *on* C, H_2, CH_3OH, conc. HCl (cat.), rt, 6 h, 95%; 2. MsCl, Py, CH_2Cl_2, rt, 14 h, 80%. (*k*) TFA, CH_3OH, rt, 12 h, 90%. (*l*) Same as (*i*), 75%. (*m*) Same as (*j*); 1. 90%; 2. 25%; *k*, 70%.

Removal of the isopropylidene group with HBr afforded the diol **242**, which upon benzylation furnished the dibenzyl ether **243**. This was treated with *m*-CPBA to produce the epoxide **244**. Treatment of **244** with lithium hydroxide gave **170**, which upon deprotection furnished **5** in 35.5% overall yield from **224**.

5.3.1.9 *Synthesis from aldonolactone* Syntheses of casuarines **249**, **253**, **261** and **262** from lactone **245** have been reported (Schemes 24 and 25).[291] The triflate in **245**[292] was displaced with azide ion to give the inverted azide, which on reduction with lithium borohydride

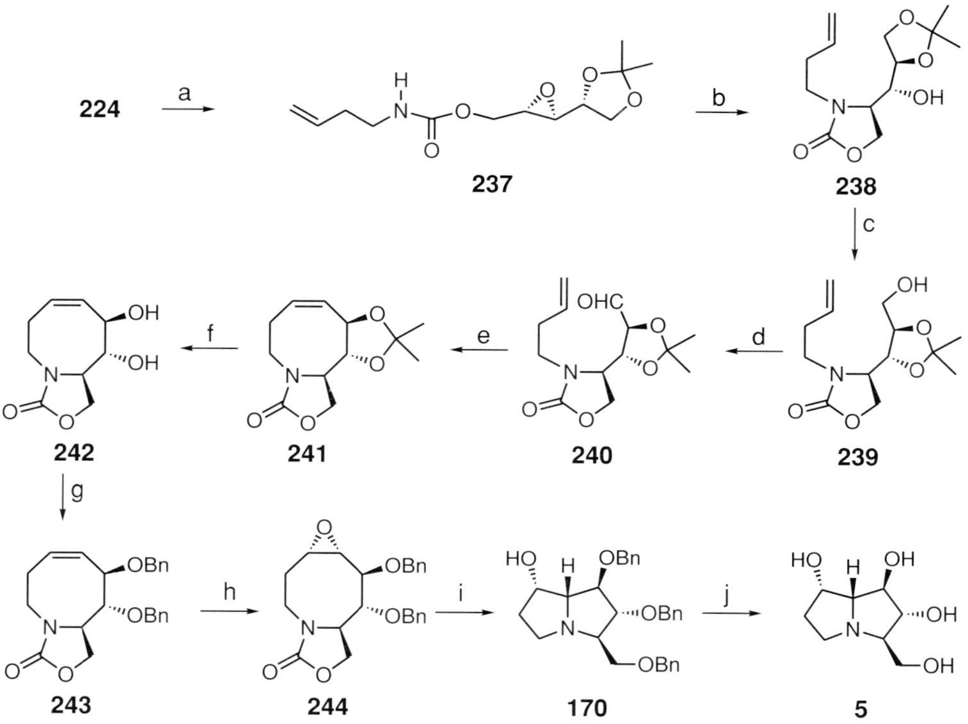

Scheme 23 (*a*) $CH_2{=}CH(CH_2)_2NCO$, $(i\text{-}Pr)_2NEt$, C_6H_6, reflux, 93%. (*b*) *t*-BuOK, THF, 0°C, 96%. (*c*) Amberlyst 15, acetone, rt, 62%, 98% based on recovered **238**. (*d*) $(COCl)_2$, DMSO, NEt_3, CH_2Cl_2, −78°C, 90%. (*e*) 1. $Ph_3P^+MeBr^-$, KHMDS, THF, −78°C to rt, 76%; 2. $(PCy_3)_2Ru(Cl)_2CHPh$, CH_2Cl_2, rt, 97%. (*f*) HBr, CH_3CN, rt, 99%. (*g*) NaH, BnBr, Bu_4NI, THF, 60°C, 84%. (*h*) *m*-CPBA, CH_2Cl_2, rt, 75%. (*i*) LiOH, $EtOH–H_2O$ (1:1), 95°C, 99%. (*j*) H_2, 20% $Pd(OH)_2/C$, CH_3OH, rt, 99%.

followed by mesylation of the resulting diol afforded **246**. The intramolecular double cyclization of **246** after azide reduction was not possible, since the resulting pyrrolidine ring that was firstly formed would contain a *trans*-acetonide group. On the other hand, removal of the ketal group followed by reduction of the azide function led to impure **249**. However, removal of the isopropylidene group followed by treatment with chlorotriethylsilane afforded **247** whose reduction, followed by treatment with sodium acetate in ethyl acetate, gave the pyrrolizidine **248**. Subsequent removal of the protecting groups led to 3,7-di-*epi*-casuarine (**249**) in 26% overall yield from **245**.

Introducing azide with retention of configuration via double inversion at C-7 of the triflate **245** led to the synthesis of 7-*epi*-casuarine (**253**). Thus, treatment of **245** with cesium trifluoroacetate in butanone followed by potassium carbonate in methanol afforded the inverted alcohol **250**. Triflation of the free hydroxyl group of **250** followed by S_N2 displacement using azide ion furnished the azide **251**. Reduction of the lactone **251** followed by mesylation afforded **252**. This was converted into **253** in 17% overall yield from **245** by a series of steps analogous to that used in the synthesis of **249**.

Scheme 24 (*a*) 1. NaN_3, DMF, 96%; 2. $LiBH_4$, THF; then MsCl, Py, DMAP, 80%. (*b*) 1. TFA–H_2O (1:1), 90%; 2. Et_3SiCl, imidazole, DMF, 45%. (*c*) CF_3CO_2Cs, $CH_3COCH_2CH_3$; then CH_3OH, K_2CO_3, 66%. (*d*) Te, $NaBH_4$, EtOH; then NaOAc, 84%. (*e*) TFA–H_2O (1:1), 100%. (*f*) 1. Tf_2O, Py, CH_2Cl_2; 2. NaN_3, DMF, 71% for two steps. (*g*) $LiBH_4$, THF; then MsCl, Py, DMAP, 69% for two steps. (*h*) 1. TFA–H_2O (1:1); 2. Et_3SiCl, imidazole, DMF, 56% for two steps; 3. Te, $NaBH_4$, EtOH; then NaOAc, 92%; 4. TFA–H_2O (1:1), 100%.

On the other hand, the triflate **254** was treated with sodium azide, followed by reduction of the lactone carbonyl function, and subsequent mesylation, complete removal of the protecting groups and reprotection by Et_3SiCl, to afford **259** (Scheme 25).[291] Sodium hydrogen telluride reduction of the azide function in **259** followed by intramolecular double cyclization with sodium acetate afforded the bicycle **260**, which on deprotection furnished **261** in 45% overall yield from the triflate **254**.

For the synthesis of **262**, the triflate **254** was treated with cesium trifluoroacetate in butanone to give the alcohol **255**. Triflation of **255** followed by triflate displacemet by azide ion furnished **256**. This was converted into the azidomesylate **257** by applying steps similar to those described for the epimer **259**. Similarly, compound **257** was converted to **262** via **258** in 11% overall yield from the triflate **254**.

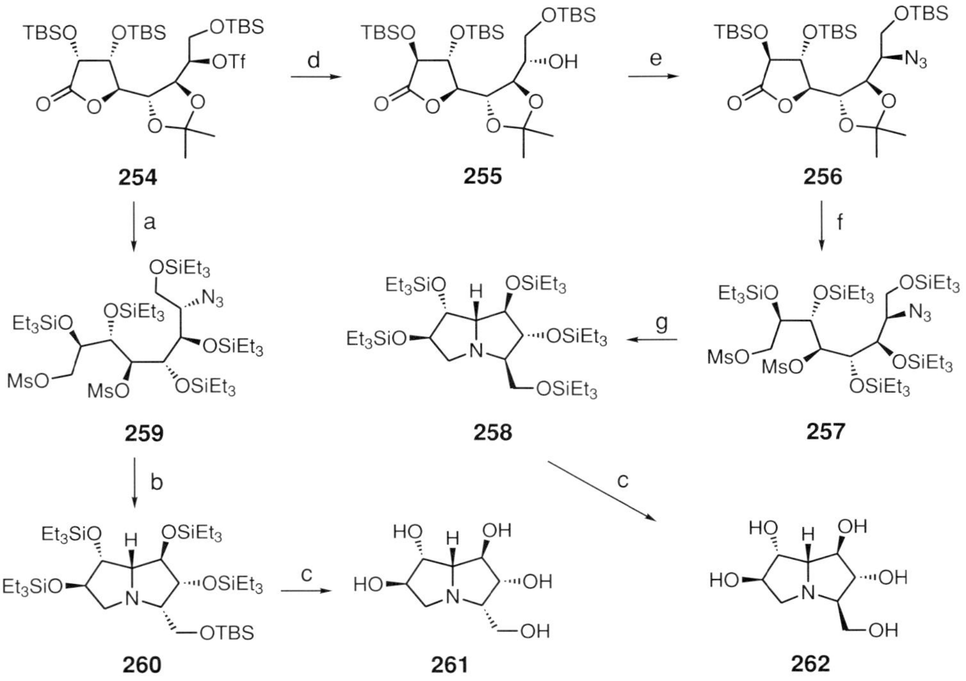

Scheme 25 (*a*) 1. NaN_3, DMF; 2. $LiBH_4$, THF; then MsCl, Py, DMAP; then TFA–H_2O (1:1); then Et_3SiCl, imidazole, DMF, 50%. (*b*) Te, $NaBH_4$, EtOH; then NaOAc, 89%. (*c*) TFA–H_2O (1:1), 100%. (*d*) CF_3CO_2Cs, $CH_3COCH_2CH_3$; then CH_3OH, K_2CO_3, 76%. (*e*) 1. Tf_2O, Py, CH_2Cl_2; 2. NaN_3, DMF, 63%. (*f*) $LiBH_4$, THF; then MsCl, Py, DMAP; then TFA–H_2O (1:1); then Et_3SiCl, imidazole, DMF. (*g*) Te, $NaBH_4$, EtOH; then NaOAc, 83%.

References

1. Rizk, A.-F.M. *Naturally Occurring Pyrrolizidine Alkaloids*; CRC Press: Boca Raton, FL, 1991.
2. Mattocks, A.R. *Chemistry and Toxicology of Pyrrolizidine Alkaloids*; Academic Press: London, 1986.
3. Robins, D.J. *Nat. Prod. Rep.* **8** (1991) 213.
4. Robins, D.J. *Nat. Prod. Rep.* **9** (1992) 313.
5. Robins, D.J. *Nat. Prod. Rep.* **10** (1993) 487.
6. Robins, D.J. *Nat. Prod. Rep.* **11** (1994) 613.
7. Liddell, J.R. *Nat. Prod. Rep.* **13** (1996) 187.
8. Liddell, J.R. *Nat. Prod. Rep.* **14** (1997) 653.
9. Liddell, J.R. *Nat. Prod.Rep.* **15** (1998) 363.
10. Liddell, J.R. *Nat. Prod. Rep.* **16** (1999) 499.
11. Hudlicky, T.; Seoane, G.; Price, J.D.; Gadamasetti, K. *Synlett* (1990) 433.
12. Dai, W.-M.; Nagao, Y.; Fugita, E. *Heterocycles* **30** (1990) 1231.
13. Ibuka, T. In *The Alkaloids*, Vol. 31; Brossi, A., Ed.; Academic Press: San Diego, 1987; p. 193.
14. Broggini, G.; Zecchi, G. *Synthesis* (1999) 905.
15. Casiraghi, G.; Zanardi, F.; Rassu, G.; Pinna, L. *Org. Prep. Proced. Int.* **28** (1996) 641.
16. Nash, R.J.; Fellows, L.E.; Dring, J.V.; Fleet, G.W.J.; Derome, A.E.; Hamor, T.A.; Scofield, A.M.; Watkin, D.J. *Tetrahedron Lett.* **29** (1988) 2487.
17. Nash, R.J.; Fellows, L.E.; Plant, A.C.; Fleet, G.W.J.; Derome, A.E.; Baird, P.D.; Hegarty, M.P.; Scofield, A.M. *Tetrahedron* **44** (1988) 5959.
18. Molyneux, R.J.; Benson, M.J.; Wong, R.Y. *J. Nat. Prod.* **51** (1988) 1198.

19. Harris, C.M.; Harris, T.M.; Molyneux, R.J.; Tropea, J.E.; Elbein, A.D. *Tetrahedron Lett.* **30** (1989) 5685.
20. Nash, R.J.; Fellows, L.E.; Dring, J.V.; Fleet, G.W.J.; Girdhar, A.; Ramsden, N.G.; Peach, J.M.; Hegarty, M.P.; Scofield, A.M. *Phytochemistry* **29** (1990) 111.
21. Scofield, A.M.; Rossiter, J.T.; Witham, P.; Kite, G.C.; Nash, R.J.; Fellows, L.E. *Phytochemistry* **29** (1990) 107.
22. Wormald, M.R.; Nash, R.J.; Hrnciar, P.; White, J.D.; Molyneux, R.J.; Fleet, G.W.J. *Tetrahedron: Asymmetry* **9** (1998) 2549.
23. Molyneux, R.J.; Benson, M.; Wong, R.Y.; Tropea, I.E.; Elbein, A.D. *J. Nat. Prod.* **51** (1988) 1206.
24. Tropea, J.E.; Molyneux, R.J.; Kaushal, G.P.; Pan, Y.T.; Mitchell, M.; Elbein, A.D. *Biochemistry* **28** (1989) 2027.
25. Robins, D.J. *J. Nat. Prod. Rep.* **7** (1990) 377.
26. Nash, R.J.; Fellows, L.E.; Dring, J.V.; Fleet, G.W.J.; Girdhar, A.; Ramsden, N.G.; Peach, J.V.; Hegarty, M.P.; Scofield, A.M. *Phytochemistry* **29** (1990) 114.
27. Elbein, A.D.; Tropea, J.E.; Molyneux, R.J. U.S. Pat. Appl. 289,907, 1989; *Chem. Abstr.* **113** (1990) 91444p.
28. Fellows, L.; Nash, R. PCT Int. Appl. WO GB Appl. 89/7,951, 1989; *Chem. Abstr.* **114** (1990) 143777f.
29. Pereira, A.C. de S.; Kaplan, M.A.C.; Maia, J.G.S.; Gottlieb, O.R.; Nash, R.J.; Fleet, G.W.J.; Pearce, L.; Watkin, D.J.; Scofield, A.M. *Tetrahedron* **47** (1991) 5637.
30. Nash, R.J.; Thomas, P.I.; Waigh, R.D.; Fleet, G.W.J.; Wormald, M.R.; Lilley, P.M.D.; Watkin, D.J. *Tetrahedron Lett.* **35** (1994) 7849.
31. Wormald, M.R.; Nash, R.J.; Watson, A.A.; Bhadoria, B.K.; Langford, R.; Sims, M.; Fleet, G.W.J. *Carbohydr. Lett.* **2** (1996) 169.
32. Nair, R.B.; Santhakumari, G. *Ancient Sci. Life* **6** (1986) 80.
33. Maiti, A.P.; Pal, S.C.; Chattopadhyay, D.; De, S.; Nandy, A. *Ancient Sci. Life* **5** (1985) 113.
34. Nande, C.V.; Kale, P.M.; Wagh, S.Y.; Antarkar, D.S.; Vaidya, A.B. *J. Res. Ayur. Siddha* **4** (1983) 1.
35. Chadha, Y.R., Ed. *The Wealth of India*; CSIR Publication and Information Directorate: New Dehli, India, 1976; p. 93.
36. Pan, Y.T.; Nori, H.; Saul, R.; Sanford, B.A.; Molyneux, R.; Elbein, A.D. *Biochemistry* **22** (1983) 3975.
37. Asano, N.; Kuroi, H.; Ikeda, K.; Kizu, H.; Kameda, Y.; Kato, A.; Adachi, I.; Watson, A.A.; Nash, R.J.; Fleet, G.W.J. *Tetrahedron: Asymmetry* **11** (2000) 1.
38. Kato, A.; Adachi, I.; Miyauchi, M.; Ikeda, K.; Komae, T.; Kizu, H.; Kameda, Y.; Watson, A.A.; Nash, R.J.; Wormald, M.R.; Fleet, G.W.J.; Asano, N. *Carbohydr. Res.* **316** (1999) 95.
39. Kunec, E.K.; Robins, D.J. *J. Chem. Soc., Chem. Commun.* (1986) 250.
40. Roitman, J.N. *Aust. J. Chem.* **36** (1983) 1203.
41. Lu, S.-T.; Lin, C.-N.; Wu, T.-S.; Shieh, D.-C. *J. Chin. Chem. Soc.* **19** (1972) 127.
42. Edga, J.A.; Cockrum, P.A.; Frahn, J.L. *Experientia* **15** (1976) 1535.
43. Richardson, M.F.; Warren, F.L. *J. Chem. Soc.* (1943) 452.
44. Hughes, C.A.; Gordon-Gray, C.G.; Schlosser, F.D.; Warren, F.L. *J. Chem. Soc.* (1965) 2370.
45. Porter, L.A.; Geissman, T.A. *J. Org. Chem.* **27** (1962) 4132.
46. Were, O.; Benn, M.; Munavu, R.M. *J. Nat. Prod.* **54** (1991) 491.
47. Atal, C.K.; Kapur, K.K.; Culvenor, C.C.J.; Smith, L.W. *Tetrahedron Lett.* **7** (1966) 537.
48. Mattocks, A.R. *J. Chem. Soc. (C)* (1968) 235.
49. Culvenor, C.C.J.; Smith, L.W. *Anal. Quim.* **68** (1972) 883.
50. Hartmann, T.; Witte, L. In *Alkaloids: Chemical and Biological Perspectives*, Vol. 9; Pelletier, S.W., Ed.; Pergamon: Oxford, 1995; p. 155.
51. Gelbaum, L.T.; Gordon, M.M.; Miles, M.; Zalkow, L.H. *J. Org. Chem.* **47** (1982) 2501.
52. Glinski, J.A.; Zalkow, L.H. *Tetrahedron Lett.* **26** (1985) 2857.
53. Orecholf, A. *Ber. Dtsch. Chem. Ges.* **68** (1935) 650.
54. Orecholf, A.; Konowalowa, R. *Ber. Dtsch. Chem. Ges.* **68** (1935) 1886.
55. Adams, R.; Rogers, E.F. *J. Am. Chem. Soc.* **63** (1941) 537.
56. Danilova, A.V.; Utkin, L.M.; Kozyreva, G.V.; Syreneva, Y.I. *J. Gen. Chem. USSR* **29** (1958) 2396.
57. Roder, E.; Wiedenfeld, H.; Jost, E.J. *Plant Med.* **43** (1981) 99.
58. Passreiter, C.M.; Willuhn, G.; Roder, E. *Planta Med.* **57** (1991) A101.
59. PaBreiter, C.M. *Phytochemistry* **31** (1992) 4135.
60. Wunderlich, J.A. *Acta Crystallogr.* **23** (1967) 846.
61. Culvenor, C.C.J.; O'Donovan, G.M.; Smith, L.W. *Aust. J. Chem.* **20** (1967) 801.
62. Bernbaum, G.I. *J. Am. Chem. Soc.* **96** (1974) 6165.
63. Perez-Salazar, A.; Cano, F.H.; Fayos, J.; Martinez-Carrera, S.; Garica-Blanco, S. *Acta Crystallogr. B* **33** (1977) 3525.
64. Shibano, M.; Kitagawa, S.; Kusano, G. *Chem. Pharm. Bull.* **45** (1997) 505.
65. Shibano, M.; Kitagawa, S.; Nakamura, S.; Akazawa, N.; Kusano, G. *Chem. Pharm. Bull.* **45** (1997) 700.

66. Shibano, M.; Nakamura, S.; Kubori, M.; Minoura, K.; Kusano, G. *Chem. Pharm. Bull.* **46** (1998) 1048.
67. Shibano, M.; Nakamura, S.; Kubori, M.; Minoura, K.; Kusano, G. *Chem. Pharm. Bull.* **46** (1998) 1416.
68. Kugelman, M.; Liu, W.C.; Axelrod, M.; McBride, J.J.; Roa, K.V. *J. Nat. Prod.* **39** (1976) 125.
69. Kovach, J.S.; Ames, M.M.; Dowis, G.; Moertel, C.G.; Hahn, R.G.; Cregan, E.T. *Cancer Res.* **39** (1979) 4540.
70. Zalkow, L.H.; Glinski, J.A.; Gelbaum, L.T.; Fleischmann, T.J.; McGowan, L.S.; Gordon, M.M. *J. Med. Chem.* **28** (1985) 687.
71. Men'shikov, G.P. *Ber. Dtsch. Chem. Ges. B* **65** (1932) 974.
72. Zalkow, L.H.; Bonett, S.; Gelbaum, L.; Gordon, M.M.; Patil, B.B.; Shani, A.; Van Derveer, D. *J. Nat. Prod.* **42** (1979) 603.
73. Atal, C.K. *J. Nat. Prod.* **41** (1978) 312.
74. Yamada, K.; Tatematsu, H.; Unno, R.; Hirata, Y.; Hirono, I. *Tetrahedron Lett.* **19** (1978) 4543.
75. Negi, P.K.S.; Fakhir, T.M.; Rajagopalan, T.R. *Indian J. Chem.* **28B** (1989) 524.
76. Hikichi, M.; Furuya, T. *Tetrahedron Lett.* **19** (1978) 767.
77. Wall, H.L. *Nature* **146** (1940) 777.
78. Wall, H.L. *Onderst. J. Vet. Sci. An. Ind.* **15** (1940) 241.
79. Wall, H.L. *Onderst. J. Vet. Sci. An. Ind.* **16** (1941) 149.
80. Kelly, H.A.; Robins, D.J. *J. Chem. Soc., Perkin Trans. 1* (1987) 177.
81. Kelly, H.A.; Robins, D.J. *J. Chem. Soc., Perkin Trans. 1* (1987) 2195.
82. Kelly, H.A.; Robins, D.J. *J. Chem. Soc., Chem. Commun.* (1988) 329.
83. Kunec, E.K.; Robins, D.J. *J. Chem. Soc., Perkin Trans. 1* (1989) 1437.
84. Denholm, A.A.; Kelly, H.A.; Robins, D.J. *J. Chem. Soc., Perkin Trans. 1* (1991) 2003.
85. Robins, D. *J. Chem. Soc. Rev.* **18** (1989) 375.
86. Robins, D.J. In *The Alkaloids*, Vol. 46; Cordell, G.A., Ed.; Academic Press: New York, 1995; Chapt. 1.
87. Freer, A.A.; Kelly, H.A.; Robins, D.J. *J. Acta Crystallogr. C* **42** (1986) 1348.
88. Porter, L.A.; Geissman, T.A. *J. Org. Chem.* **27** (1992) 4132.
89. Were, O.; Benn, M.; Munavu, R.M. *Phytochemistry* **32** (1993) 1595.
90. Culvenor, C.C.J.; Smith, L.W.; Willing, R.I. *Chem. Commun.* (1970) 65.
91. Sawhey, R.S.; Girotra, R.N.; Atal, C.K.; Culvenor, C.C.J.; Smith, L.W. *Indian J. Chem.* **5** (1967) 655.
92. Crout, D.H.G. *J. Chem. Soc., Perkin Trans. 1* (1972) 1602.
93. Mackay, M.F.; Sadek, M. Culvenor, C.C. *J. Acta Crystallogr. C* **40** (1984) 1073.
94. Richardson, J.F.; Culvenor, C.C. *J. Acta Crystallogr. C* **41** (1985) 1475.
95. Mattocks, A.R. *Chemistry and Toxicology of Pyrrolizidine Alkaloids*; Academic Press: London, 1986; Chapts. 1, 3, 7 and 11.
96. Barger, G.; Machle, J.J. *J. Chem. Soc.* (1936) 743.
97. Bredenkamp, M.W.; Wiechers, A.; van Rooyen, P.H. *Tetrahedron Lett.* **26** (1985) 5721.
98. Roder, E.; Wiedenfeld, H.; Pastewka, U. *Planta Med.* **37** (1979) 131.
99. Bredenkamp, M.W.; Wiechers, A.; van Rooyen, P.H. *Tetrahedron Lett.* **26** (1985) 929.
100. Bredenkamp, M.W.; Wiechers, A. *Tetrahedron Lett.* **26** (1985) 5721.
101. Parvez, M.; Benn, M.H. *Acta Crystallogr. C* **51** (1995) 1202.
102. Manske, R.H.F. *Can J. Res., Sect. B* **17B** (1939) 1; *Chem. Abstr.* **33** (1939) 6321.
103. Adams, R.; van Duuren, B.L. *J. Am. Chem. Soc.* **75** (1953) 4631.
104. Kropman, M.; Warren, F.L. *J. Chem. Soc.* (1950) 700.
105. Nair, M.D.; Adams, R. *J. Am. Chem. Soc.* **82** (1960) 3787.
106. Denmark, S.E.; Martinborough, E.A. *J. Am. Chem. Soc.* **121** (1999) 3046.
107. Ikota, N. *Tetrahedron Lett.* **33** (1992) 2553.
108. Ikota, N.; Nakagawa, H.; Ohno, S.; Noguchi, K.; Okuyama, K. *Tetrahedron* **54** (1998) 8985.
109. Hudlicky, T.; Seoane, G.; Lovelace, T.C. *J. Org. Chem.* **53** (1988) 2094.
110. Ohsawa, T.; Ihara, M.; Fukumoto, K.; Kametani, T. *J. Org. Chem.* **48** (1983) 3644.
111. Tufariello, J.J.; Lee, G.E. *J. Am. Chem. Soc.* **102** (1980) 373.
112. Robins, D.J.; Sakdart, S. *J. Chem. Soc., Perkin Trans. 1* (1981) 909.
113. Chamberlin, A.R.; Chung, T.Y.L. *J. Am. Chem. Soc.* **105** (1983) 3653.
114. Knight, D.W.; Share, A.C.; Gallagher, P.T. *J. Chem. Soc., Perkin Trans. 1* (1991) 1615.
115. Denmark, S.E.; Thorarensen, A.; Middleton, D.S. *J. Org. Chem.* **60** (1995) 3574.
116. Gruszecka-Kowalik, E.; Zalkow, L.H. *J. Org. Chem.* **55** (1990) 3398.
117. Kame-tani, T.; Yukawa, H.; Honda, T. *J. Chem. Soc., Perkin Trans. 1* (1988) 833.
118. Hudlicky, T.; Frazier, J.O.; Seoane, G.; Tiedje, M.; Seoane, A.; Kwart, L.D.; Beal, C. *J. Am. Chem. Soc.* **108** (1986) 3755.
119. Kametani, T.; Yukawa, H.; Honda, T. *J. Chem. Soc., Chem. Commun.* (1986) 651.
120. Kametani, T.; Higashiyama, K.; Otomasu, H.; Honda, T. *Isr. J. Chem.* **27** (1986) 57.

121. Chamberlin, A.R.; Nguyen, H.D.; Chung, J.Y.L. *J. Org. Chem.* **49** (1984) 1682.
122. Kametani, T.; Higashiyama, K.; Otomasu, H.; Honda, T. *Heterocycles* **22** (1984) 729.
123. Burnett, D.A.; Choi, J.-K.; Hart, D.J.; Tsai, Y.-M. *J. Am. Chem. Soc.* **106** (1984) 8201.
124. McDonald, T.L.; Narayanan, B.A. *J. Org. Chem.* **48** (1983) 1129.
125. Hart, D.J.; Yang, T.K. *Tetrahedron Lett.* **23** (1982) 2761.
126. Chamberlin, A.R.; Chung, J.Y.L. *Tetrahedron Lett.* **23** (1982) 2619.
127. Terao, Y.; Imai, N.; Achiwa, K.; Sekiya, M. *Chem. Pham. Bull.* **30** (1982) 3167.
128. Robins, D.J.; Sakdarat, S. *J. Chem. Soc., Perkin Trans. 1* (1979) 1734.
129. Tufariello, J.J.; Tette, J.P. *J. Org. Chem.* **40** (1975) 3866.
130. Tufariello, J.J.; Tette, J.P. *J. Chem. Soc., Perkin Trans. 1* (1971) 469.
131. Rueger, H.; Benn, M. *Heterocycles* **19** (1982) 1677.
132. Kraus, G.A.; Neuenschwander, K. *Tetrahedron Lett.* **21** (1980) 3841.
133. Robins, D.J.; Sakdarat, S. *J. Chem. Soc., Chem. Commun.* (1979) 1181.
134. Casiraghi, G.; Spanu, P.; Rassu, G.; Pinna, L.; Ulgheri, F. *J. Org. Chem.* **59** (1994) 2906.
135. Furneaux, R.H.; Gainsford, G.J.; Mason, J.M.; Tyler, P.C. *Tetrahedron* **50** (1994) 2131.
136. Takahata, H.; Banba, Y.; Momose, T. *Tetrahedron* **47** (1991) 7635.
137. Robertson, J.; Peplow, M.A.; Pillai, J. *Tetrahedron Lett.* **37** (1996) 5825.
138. Hart, D.J.; Yang, T.-K. *J. Chem. Soc., Chem. Commun.* (1983) 135.
139. Hart, D.J.; Yang, T.-K. *J. Org. Chem.* **50** (1985) 235.
140. Kano, S.; Yuasa, S.; Shibuya, S. *Heterocycles* **27** (1988) 253.
141. Röder, E.; Bourauel, T.; Wiedenfeld, H. *Liebigs Ann. Chem.* (1990) 607.
142. Pandely, G.; Lakshmaiah, G. *Synlett* (1994) 277.
143. Brand, M.; Drewes, S.E.; Loizou, G.; Roos, G.H.P. *Synth. Commun.* **17** (1987) 795.
144. Kelly, H.A.; Kunec, E.K.; Rodgers, M.; Robins, D.J. *J. Chem. Res. (S)* (1989) 358.
145. Kang, S.H.; Kim, G.T.; Yoo, Y.S. *Tetrahedron Lett.* **38** (1997) 603.
146. Ma, D.; Zhang, J. *J. Chem. Soc., Perkin Trans. 1* (1999) 1703.
147. Vedejs, E.; Galante, R.J.; Goekjian, P.G. *J. Am. Chem. Soc.* **120** (1998) 3613.
148. Nagao, Y.; Dai, W.-M.; Ochiai, M. *Tetrahedron Lett.* **29** (1988) 6133.
149. Tsai, Y.-M.; Ke, B.-W.; Yang, C.-T.; Lin, C.-H. *Tetrahedron Lett.* **33** (1992) 7895.
150. Denmark, S.E.; Thorarensen, A. *J. Am. Chem. Soc.* **119** (1997) 125.
151. Niwa, H.; Sakata, T.; Yamada, K. *Bull. Chem. Soc. Jpn.* **67** (1994) 1990.
152. Niwa, H.; Okamoto, O.; Ishiwata, H.; Kuroda, A.; Uosaki, Y.; Yamada, K. *Bull. Chem. Soc. Jpn.* **61** (1988) 3017.
153. Keck, G.E.; Nickell, D.G. *J. Am. Chem. Soc.* **102** (1980) 3632.
154. Goti, A.; Fedi, V.; Nannelli, L.; DeSarlo, F.; Brandi, A. *Synlett* (1997) 577.
155. Niwa, H.; Kunitani, K.; Nagoya, T.; Yamada, K. *Bull. Chem. Soc. Jpn.* **67** (1994) 3094.
156. Liu, Z.-Y.; Zhao, L.-Y. *Tetrahedron Lett.* **40** (1999) 5593.
157. Denmark, S.E.; Herbert, B. *J. Am. Chem. Soc.* **120** (1998) 7357.
158. Narasaka, K.; Sakakura, T.; Uchimaru, T.; Guédin-Vuong, D. *J. Am. Chem. Soc.* **106** (1984) 2954.
159. Narasaka, K.; Sakakura, T.; Uchimaru, T.; Morimoto, K.; Mukaiyama, T. *Chem. Lett.* (1982) 455.
160. Narasaka, K.; Uchimaru, T.; *Chem. Lett.* (1982) 57.
161. Niwa, H.; Kuroda, A.; Sakata, T.; Yamada, K. *Bull. Chem. Soc. Jpn.* **70** (1997) 2541.
162. Niwa, H.; Kuroda, A.; Yamada, K. *Chem. Lett.* (1983) 125.
163. Niwa, H.; Uosaki, Y.; Yamada, K. *Tetrahedron Lett.* **24** (1983) 5731.
164. Mulzer, J.; Shanyoor, M. *Tetrahedron Lett.* **34** (1993) 6545.
165. Denmark, S.E.; Parker, D.L., Jr.; Dixon, J.A. *J. Org. Chem.* **62** (1997) 435.
166. Shibano, M.; Tsukamoto, D.; Kusano, G. *Chem. Pharm. Bull.* **47** (1999) 907.
167. Gallos, J.K.; Sarli, V.C.; Koftis, T.V.; Coutouli-Argyropoulou, E. *Tetrahedron Lett.* **41** (2000) 4819.
168. Casiraghi, G.; Rassu, G.; Spanu, P. *Chemtracts Org. Chem.* **5** (1992) 201.
169. Denmark, S.E.; Hurd, A.R. *Org. Lett. 1* (1999) 1311.
170. de Vicente, J.; Arrayas, R.G.; Carretero, J.C. *Tetrahedron Lett.* **40** (1999) 6083.
171. Pearson, W.H.; Hembre, E.J. *J. Org. Chem.* **61** (1996) 5546.
172. Hall. A.; Meldrum, K.P.; Therond, P.R.; Wightman, R.H. *Synlett* (1997) 123.
173. Denmark, S.E.; Hurd, A.R. *J. Org. Chem.* **65** (2000) 2875.
174. Denmark, S.E.; Herbert, B. *J. Org. Chem.* **65** (2000) 2887.
175. Yoda, H.; Asai, F.; Takabe, K. *Synlett* (2000) 1001.
176. Romero, A.; Wong, C.H. *J. Org. Chem.* **65** (2000) 8264.
177. de Faria; A.R.; Salvador, E.L.; Correia, C.R.D. *J. Org. Chem.* **67** (2002) 3651.
178. Rabiczko, J.; Urbańczyk-Lipkowska, Z.; Chmielewski, M. *Tetrahedron* **58** (2002) 1433.

179. Gallos, J.K.; Sarli, V.C.; Stathakis, C.I.; Koftis, T.V.; Nachmia, V.R.; Coutouli-Argyropoulou, E. *Tetrahedron* **58** (2002) 9351.
180. Kato, A.; Kano, E.; Adachi, I.; Molyneux, R.J.; Watson, A.A.; Nash, R.J.; Fleet, G.W.J.; Wormald, M.R.; Kizu, H.; Ikeda, K.; Asano, N. *Tetrahedron: Asymmetry* **14** (2003) 325.
181. Robins, D.J. *Nat. Prod. Rep.* **12** (1995) 413.
182. Adams, R.; van Duuren, B.L. *J. Am. Chem. Soc.* **76** (1954) 6379.
183. Dener, J.M.; Hart, D.J. *Tetrahedron* **44** (1988) 7037.
184. Munchowski, J.M.; Nelson, P.H. *Tetrahedron Lett.* **21** (1980) 4585.
185. Flitsch, W.; Wernsmann, P. *Tetrahedron Lett.* **22** (1981) 719.
186. Robins, D.J. *J. Chem. Soc., Chem. Commun.* (1982) 1289.
187. Macdonald, T.L.; Narayanan, B.A. *J. Org. Chem.* **48** (1983) 1131.
188. Keck, G.E.; Nickell, D.G. *J. Am. Chem. Soc.* **102** (1980) 3634.
189. Vedejs, E.; Martinez, G.R. *J. Am. Chem. Soc.* **102** (1980) 7993.
190. Ohsawa, T.; Ihara, M.; Fukumoto, K.; Kametani, T. *J. Org. Chem.* **48** (1983) 4644.
191. Rüeger, H.; Benn, M. *Heterocycles* **20** (1980) 1331.
192. Pearson, W.H.; Hembre, E.J. *J. Org. Chem.* **61** (1996) 5545.
193. Vedejs, E.; Larsen, S.; West, F.G. *J. Org. Chem.* **50** (1985) 2170.
194. Chamberlin, A.R.; Chung, J.Y.L. *J. Org. Chem.* **50** (1985) 4425.
195. Tatsuta, K.; Takahashi, H.; Anemiya, Y.; Kinoshita, M. *J. Am. Chem. Soc.* **105** (1983) 4086.
196. Shishido, K.; Sukegawa, Y.; Fukumoto, K.; Kametani, T. *Heterocycles* **23** (1985) 1629.
197. Niwa, H.; Miyachi, Y.; Okamoto, O.; Uosaki, Y.; Yamada, K. *Tetrahedron Lett.* **27** (1986) 4605.
198. Ohsawa, T.; Ihara, M.; Fukumoto, K.; Kametani, T. *Heterocycles* **19** (1982) 1605.
199. Vedejs, E.; Martinez, G.R. *J. Am. Chem. Soc.* **102** (1980) 7994.
200. Kametani, T.; Ohsawa, T.; Ihara, M.; Fukumoto, K. *Heterocycles* **19** (1982) 2075.
201. White, J.D.; Ohira, S. *J. Org. Chem.* **51** (1986) 5492.
202. Niwa, H.; Okamoto, O.; Miyachi, Y.; Uosaki, Y.; Yamada, K. *J. Org. Chem.* **52** (1987) 2941.
203. Hart, D.J.; Choi, J.-K. *Tetrahedron* **41** (1985) 3959.
204. Danishefsky, S.; McKee, R.; Singh, R.K. *J. Am. Chem. Soc.* **99** (1977) 7711.
205. Aasen, A.J.; Culvenor, C.C.J.; Smith, L.W. *J. Org. Chem.* **34** (1969) 4137.
206. Aasen, A.J.; Culvenor, C.C.J. *Aust. J. Chem.* **22** (1969) 2657.
207. Keusenkothen, P.F.; Smith, M.B. *J. Chem. Soc., Perkin Trans. 1* (1994) 2485.
208. Haviari, G.; Célérier, J.P.; Petit, H.; Lhommet, G. *Tetrahedron Lett.* **24** (1993) 1599.
209. Knight, J.G.; Ley, S.V. *Tetrahedron Lett.* **32** (1991) 7119.
210. Le Coz, S.; Mann, A.; Thareau, F.; Taddei, M. *Heterocycles* **36** (1993) 2073.
211. Kelly, H.A.; Robins, D.J. *J. Chem. Soc., Perkin Trans. 1* (1989) 1339.
212. Seijas, J.A.; Vázquez-Tato, M.P.; Castedo, L.; Estévez, R.J.; Ónega, M.G.; Ruíz, M. *Tetrahedron* **48** (1992) 1637.
213. Sato, T.; Matsubayashi, K.-I.; Tsujimoto, K.; Ikeda, M. *Heterocycles* **36** (1993) 1205.
214. Sato, T.; Tsujimoto, K.; Matsubayashi, K.-I.; Ishibashi, H.; Ikeda, M. *Chem. Pharm. Bull. Jpn.* **40** (1992) 2308.
215. Ishibashi, H.; Uemura, N.; Nakatani, H.; Okazaki, M.; Sato, T.; Nakamura, N.; Ikeda, M. *J. Org. Chem.* **58** (1993) 2360.
216. Knight, D.W.; Share, A.C.; Gallagher, P.T. *Gazz. Chim. Ital.* **49** (1991) 49.
217. Ito, H.; Ikeuchi, Y.; Taguchi, T.; Hanzawa, Y.; Shiro, M. *J. Am. Chem. Soc.* **116** (1994) 5469.
218. Keck, G.E.; Cressman, E.N.K.; Enholm, E.J. *J. Org. Chem.* **54** (1989) 4345.
219. Hanselmann, R.; Benn, M. *Tetrahedron Lett.* **34** (1993) 3511.
220. Arai, Y.; Kontani, T.; Koizumi, T. *Chem. Lett.* (1991) 2135.
221. Nagao, Y.; Dai, W.-M.; Ochiai, M.; Tsukagoshi, S.; Fujita, E. *J. Org. Chem.* **55** (1990) 1148.
222. Nagao, Y.; Dai, W.-M.; Ochiai, M.; Shiro, M. *Tetrahedron* **46** (1990) 6361.
223. Denmark, S.E.; Thorarensen, A. *J. Org. Chem.* **59** (1994) 5672.
224. Denmark, S.E.; Schnute, M.E.; Marcin, L.R.; Thorarensen, A. *J. Org. Chem.* **60** (1995) 3205.
225. Ley, S.V.; Gutteridge, C.E. *Chemtracts Org. Chem.* **8** (1995) 222.
226. Horni, von A.; Hubácek, I.; Hesse, M. *Helv. Chim. Acta* **77** (1994) 579.
227. Murray, A.; Proctor, G.R.; Murray, P.J. *Tetrahedron Lett.* **36** (1995) 291.
228. Pandey, G.; Lakshmaiah, G. *Tetrahedron Lett.* **34** (1993) 4861.
229. Eguchi, M.; Zeng, Q.; Korda, A.; Ojima, I. *Tetrahedron Lett.* **34** (1993) 915.
230. Ojima, I.; Donovan, R.J.; Eguchi, M.; Shai, W.R.; Ingallina, P.; Korda, A.; Zeng, Q. *Tetrahedron* **49** (1993) 5431.
231. Cabezas, N.; Thierry, J.; Potier, P. *Heterocycles* **28** (1989) 607.
232. Correia, C.R.D.; de Faria, A.R.; Carvalho, E.S. *Tetrahedron Lett.* **36** (1995) 5109.

233. Gramain, J.-C.; Remuson, R.; Valle-Goyet, D.; Guilhem, J.; Lavaud, C. *J. Nat. Prod.* **54** (1991) 1062.
234. Bureau, R.; Mortier, J.; Joucla, M. *Tetrahedron* **48** (1992) 8947.
235. Burgess, K.; Henderson, I. *Tetrahedron Lett.* **31** (1990) 6949.
236. McCaig, A.E.; Wightman, R.H. *Tetrahedron Lett.* **34** (1993) 3939.
237. Fairbanks, A.J.; Fleet, G.W.J.; Jones, A.H.; Bruce, I.; Al Daher, S.; Cenci di Bello, I.; Winchester, B. *Tetrahedron* **47** (1991) 131.
238. Collin, W.F.; Fleet, G.W.J.; Haraldsson, M.; Cenci di Bello, I.; Winchester, B. *Carbohydr. Res.* **202** (1990) 105.
239. Marek, D.; Wadouachi, A.; Uzan, R.; Beaupere, D.; Nowogrocki, G.; Laplace, G. *Tetrahedron Lett.* **37** (1996) 49.
240. Fleet, G.W.J.; Haraldsson, M.; Nash, R.J.; Fellows, L.E. *Tetrahedron Lett.* **29** (1988) 5441.
241. Fleet, G.W.J.; Smith, P.W. *Tetrahedron* **43** (1987) 971.
242. Kovar, J.; Jary, J. *Collect. Czech. Chem. Commun.* **34** (1969) 2619.
243. Nishimura, Y.; Kodo, S.; Umezawa, H. *J. Org. Chem.* **50** (1985) 5210.
244. Gurjar, M.K.; Patil, V.J. *Indian J. Chem., Sect. B.* **24B** (1985) 1282.
245. Geissman, T.A.; Waiss, A.C., Jr., *J. Org. Chem.* **27** (1962) 139.
246. Gurjar, M.K.; Patil, V.J.; Pawar, S.M. *Indian J. Chem.* **26B** (1987) 1115.
247. Austin, G.N.; Baird, P.D.; Fleet, G.W.J.; Peach, J.M.; Smith, P.W.; Watkin, D.J. *Tetrahedron* **43** (1987) 3095.
248. Fleet, G.W.J.; Seijas, J.A.; Vàzquez-Tato, M.P. *Tetrahedron* **47** (1991) 525.
249. Tatsuta, K.; Miyashita, S.; Akimoto, K.; Kinoshita, M. *Bull. Chem. Soc. Jpn.* **55** (1982) 3254.
250. Tatsuda, K.; Takahashi, H.; Amemiya, Y.; Kinoshita, M. *J. Am. Chem. Soc.* **105** (1983) 4096.
251. Tsuda, Y.; Nunozawa, T.; Yoshimoto, K. *Chem. Pharm. Bull.* **28** (1980) 3223.
252. Kinoshita, M.; Ohsawa, N.; Gomi, S. *Carbohydr. Res.* **109** (1982) 5.
253. Aasen, A.J.; Culvenor, C.C.J. *J. Org. Chem.* **34** (1969) 4143.
254. Bruce, I.; Fleet, Girdhar, A.; Haraldsson, P.J.M.; Watkin, D.J. *Tetrahedron* **46** (1990) 19.
255. Choi, S.; Bruce, I.; Fairbanks, A.J.; Fleet, G.W.J.; Jones, A.H.; Nash, R.J.; Fellows, L.E. *Tetrahedron Lett.* **32** (1991) 5517.
256. Cubero, I.I.; López-Espinosa, M.T.P.; Dìaz, R.R.; Montalbán, F.F. *Carbohydr. Res.* **330** (2001) 401.
257. Izquierdo, I.; Plaza, M.T.; Robles, R.; Franco, F. *Tetrahedron: Asymmetry* **12** (2001) 2481.
258. Izquierdo, I.; Plaza, M.-T.; Tornel, P.L. *An. Quim.* **84C** (1988) 340.
259. Barker, R.; Fletcher, H.G. *J. Org. Chem.* **26** (1961) 4605.
260. Tejima, S.; Fletcher, H.G. *J. Org. Chem.* **28** (1963) 4605.
261. Yoda, H.; Katoh, H. Takabe, K. *Tetrahedron Lett.* **41** (2000) 7661.
262. Ramboud, L.; Copain, P.; Martin, O.R. *Tetrahedron: Asymmetry* **12** (2001) 1807.
263. Pearson, W.H.; Hines, J.V. *J. Org. Chem.* **65** (2000) 5785.
264. Pearson, W.H.; Hines, J.V. *Tetrahedron Lett.* **32** (1991) 5513.
265. MacCoss, M.; Chen, A.; Tolman, R.L. *Tetrahedron Lett.* **26** (1985) 4287.
266. Hudlicky, T.; Luna, H.; Price, J.D.; Rulin, F. *J. Org. Chem.* **55** (1990) 4683.
267. Cohen, N.; Banner, B.L.; Lopresti, R.J.; Wong, F.; Rosenberger, M.; Liu, Y.-Y.; Thom, E.; Liebman, A.A. *J. Am. Chem. Soc.* **105** (1983) 3661.
268. Buchanan, J.G.; Jigajinni, B.; Singh, G.; Wightman, R.H. *J. Chem. Soc., Perkin Trans. 1* (1987) 2377.
269. Buchanan, J.G.; Singh, G.; Wightman, R.H. *J. Chem. Soc., Chem. Commun.* (1984) 1299.
270. Yadav, V.K.; Rüeger, H.; Benn, M. *Heterocycles* **22** (1984) 2735.
271. Cohen, N.; Banner, B.L.; Laurenzano, A.J.; Carozza, L. *Synthesis* **63** (1984) 127.
272. Ballou, C.E. *J. Am. Chem. Soc.* **79** (1957) 165.
273. Bennett, R.B., III; Cha, J.K. *Tetrahedron Lett.* **31** (1990) 5437.
274. Mander, L.N.; Sethi, P. *Tetrahedron Lett.* **24** (1983) 5425.
275. Ziegler, F.E.; Wang, T.-F. *Tetrahedron Lett.* **26** (1985) 2291.
276. Crossland, R.K.; Servis, K.L. *J. Org. Chem.* **35** (1970) 3195.
277. Buchanan, J.G.; Edgar, A.R., Hewitt, B.D. *J. Chem. Soc., Perkin Trans. 1* (1987) 2371.
278. Buchanan, J.G.; Jigajinni, V.B.; Singh, G.; Wightman, R.H. *J. Chem. Soc., Perkin Trans. 1* (1987) 2377.
279. Robina, I.; Gearing, R.P.; Buchanan, J.G.; Wightman, R.H. *J. Chem. Soc., Perkin Trans. 1* (1990) 2622.
280. Deshong, P.; Kell, D.A.; Sidler, D.R. *J. Org. Chem.* **50** (1985) 2309.
281. Takano, S.; Tomita, S.; Iwabuchi, Y.; Ogasawara, K. *Synthesis* (1988) 610.
282. Takano, S.; Iwabuchi, Y.; Ogasawara, K. *J. Chem. Soc., Chem. Commun.* (1988) 1527.
283. Hashimura, K.; Tomita, S.; Hiroya, K.; Ogasawara, K. *J. Chem. Soc., Chem. Commun.* (1995) 2291.
284. Mulzer, J.; Angermann, A.; Münch, W. *Liebigs. Ann. Chem.* (1986) 825.
285. Mulzer, J.; Angermann, A. *Tetrahedron Lett.* **24** (1983) 2843.
286. Hoffmann, R.W.; Zeiss, H.J.; Endesfelder, A. *Carbohydr. Res.* **123** (1983) 320.

287. Mulzer, J.; Scharp, M. *Synthesis* (1993) 615.
288. Johnson, W.S.; Brockson, T.J.; Loew, P.; Rich, D.H.; Werthemann, R.A.; Arnold, R.A.; Li, T.; Faulkner, D.J. *J. Am. Chem. Soc.* **92** (1970) 4463.
289. White, J.D.; Hrnciar, P.; Yokochi, A.F.T. *J. Am. Chem. Soc.* **120** (1998) 7359.
290. White, J.D.; Hrnciar, P. *J. Org. Chem.* **65** (2000) 9129.
291. Bell, A.A.; Pickering, L.; Watson, A.A.; Nash, R.J.; Pan, Y.T.; Elbein, A.D.; Fleet, G.W.J. *Tetrahedron Lett.* **38** (1997) 5869.
292. Bell, A.A.; Pickering, L.; Watson, A.A.; Nash, R.J.; Griffiths, R.C.; Jones, M.G.; Fleet, G.W.J. *Tetrahedron Lett.* **37** (1996) 8561.

5.3.2 *Trehazolin*

Trehazolin (**1**), a pseudodisaccharide, is obtained from a culture broth of *Micromonospora* strain SANK 62390[1,2] and it is widely distributed in microorganisms, insects, plants and animals. It is a strong inhibitor of trehalase enzyme that specifically hydrolyzes α,α-trehalose (**5**). It probably acts as a close mimic of **5** or more likely to the postulated glycopyranosyl cation intermediate involved in the hydrolytic step of glycosides or a transition state leading to it. Compound **5** is ubiquitously found in insects such as insect flight, and it is the principal blood sugar used to support various energy-requiring functions.[3,4] Trehalose and trehalase enzyme have been reported to also participate in germination of ascospores in fungi[5–8] and in glucose transport in mammalian kidney and intestine.[9] Trehazolin and its analogues have important implications in immunology, virology and oncology.[10] Its structure was elucidated as a pseudodisaccharide consisting of an α-D-glucopyranose moiety through a cyclic isourea group bonded to a unique aminocyclopentitol, trehazolamine (**3**). This was deduced from degradation and ^{1}H NMR analysis[1,2] and confirmed through synthetic studies, that established its absolute configuration. The structure of the inhibitor, isolated from the culture broth of *Amycotlalopsis trehalostatica*,[11–13] was wrongly assigned as 5-*epi*-trehazolin, named trehalostatin (**2**). The structure of **2**[13–17] has been postulated to be the same as **1** through comparison of their physical data. Analogues of trehazolin have also been synthesized.[18–36]

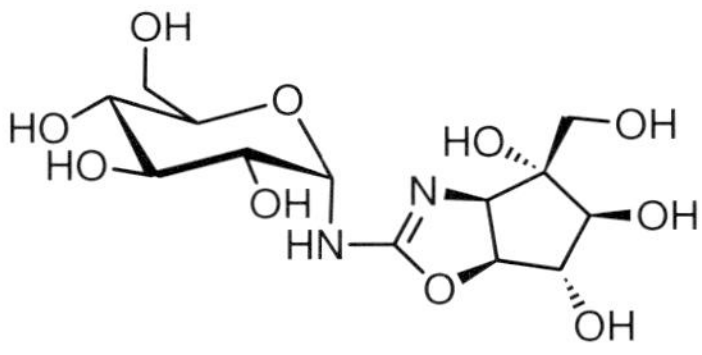

Trehazolin
1

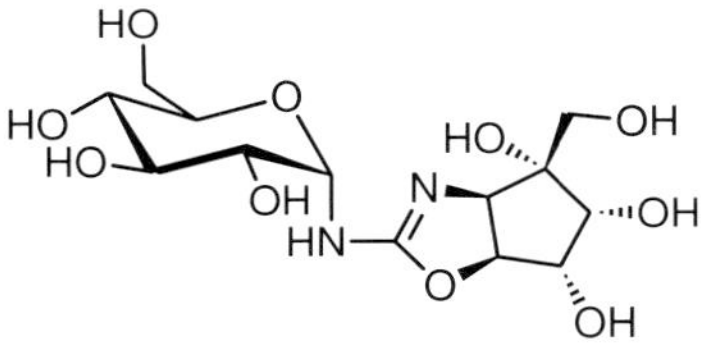

Trehalostatin
2

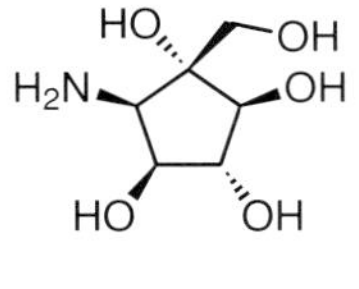

Trehazolamine
3

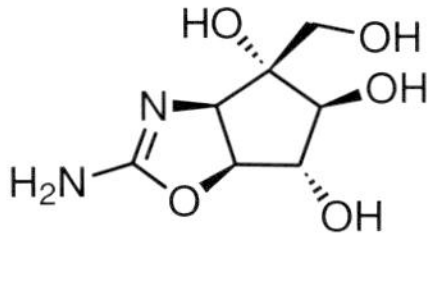

Trehalamine
4

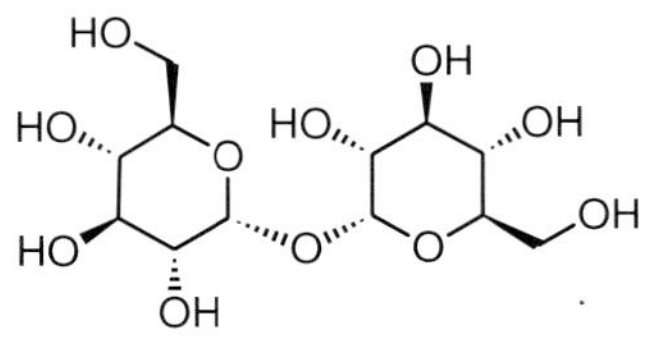

α,α-Trehalose
5

A biosynthetic pathway for trehazolin (**1**) could be outlined[37] as shown in Scheme 1. Two molecules of glucosylamine (**6**) were reacted with carbon dioxide to give the carbodiimide **7**, which could give **8**. Subsequent regioselective oxidation to **9** and stereoselective pinacol-type coupling afforded trehazolin (**1**).

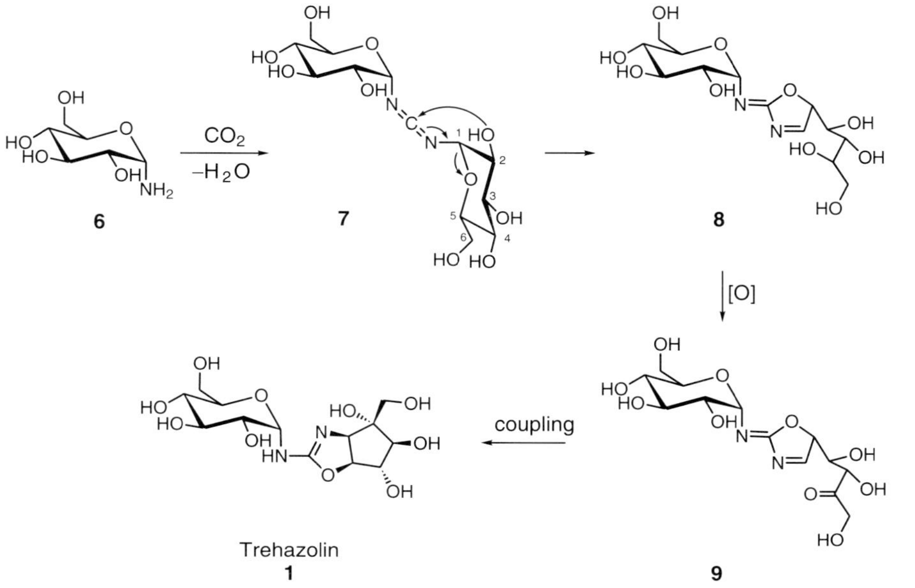

Scheme 1

5.3.2.1 *Synthesis from D-glucose* Various methods have been reported for the synthesis of trehazolamine (**3**) from carbohydrate precursors such as D-glucose. Thus, the 4,6-benzylidene derivative **10**, easily prepared[38] from D-glucose, was converted quantitatively to the respective open chain *O*-methyloxime derivative (Scheme 2).[37] Subsequent oxidation[39]

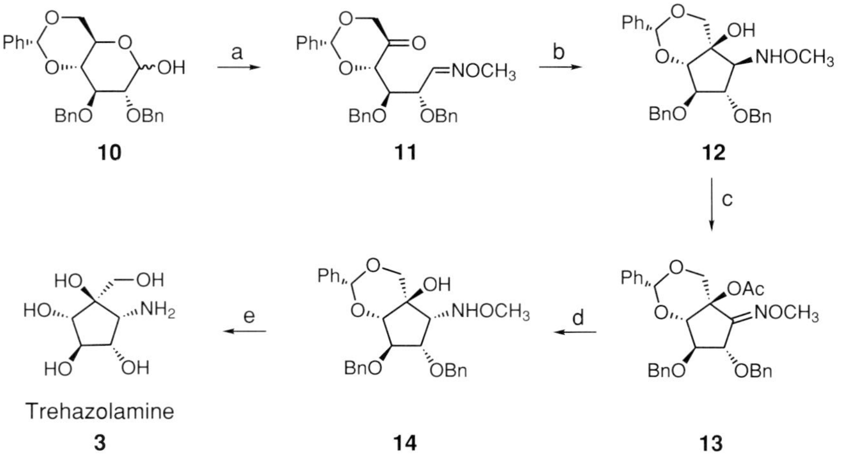

Scheme 2 (*a*) 1. $CH_3ONH_2 \cdot HCl$, Py, 40°C, quantitative; 2. Dess–Martin periodinane, CH_2Cl_2, quantitative. (*b*) SmI_2 (5 equiv.), *t*-BuOH (2.5 equiv.), THF, −78°C to rt, 84%. (*c*) 1. Ac_2O, Py, DMAP, 2. $Pb(OAc)_4$, benzene, 40°C, 44% for two steps. (*d*) 1. K_2CO_3, CH_3OH; 2. $LiAlH_4$, CH_3ONa, THF, −78°C, 67% for two steps. (*e*) Na, NH_3 (liquid), −78°C, 90%.

afforded ketone **11**, which underwent intramolecular coupling by using SmI_2 to give exclusively the diastereoisomer **12** in 84% yield. The only oxidizing agent for the conversion of the acetylated *O*-methoxyamine to the oxime ether **13** was found to be lead tetraacetate,[40,41] but with a modest yield (44%). Deacetylation of **13** followed by reduction with $LiAlH_4$ afforded **14**, which underwent full deprotection to afford **3** in 22% overall yield from **10**.

An alternative route for the synthesis of trehazolamine (**3**) has started with 2,3,4,6-tetra-*O*-benzyl-D-glucopyranose (**15**) (Schemes 3–5).[42] Sodium borohydride reduction of **15**[43] afforded quantitatively the D-glucitol derivative **16**. Swern oxidation of **16** gave **17**, whose cyclization[44] with SmI_2 afforded a 1:1 mixture of **18** and **19** in 90% yield. These were chromatographically inseparable, but they were converted into the separable cyclic thionocarbonates **20** and **21** using 1,1′-thiocarbonyldiimidazole. However, the mixture of thionocarbonates **22** and **23** gave the same product **24** upon heating with triethylphosphite. Direct epoxidation of **24** afforded an inseparable mixture of epoxides. This problem was solved by doing the epoxidation on the deacetylated derivative **25**. Sharpless epoxidation of **25** using diisopropyl L-tartrate yielded **26** (93%). Opening of **26** with LiN_3 yielded the azide **27** (89%), which underwent hydrogenolysis to give **3** in 39% overall yield from **15**.

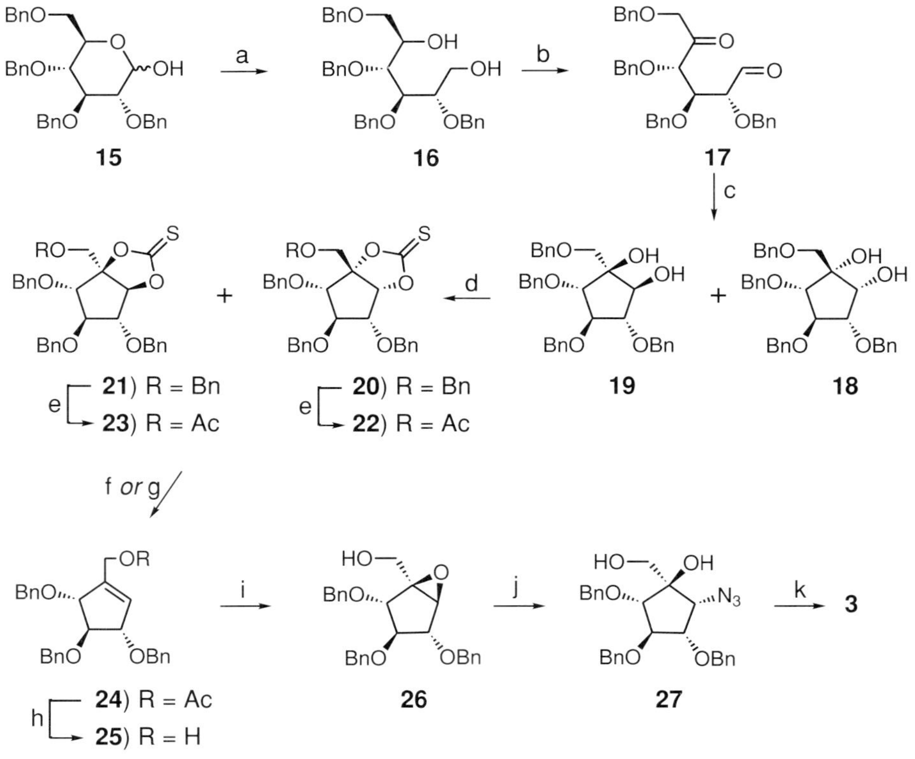

Scheme 3 (*a*) $NaBH_4$, EtOH–CH_2Cl_2 (1:1), rt, 98%. (*b*) $(COCl)_2$, DMSO, THF, −65°C; then NEt_3, −65°C to rt. (*c*) SmI_2, THF, *t*-BuOH, −50°C to rt, 90% for two steps. (*d*) 1,1′-Thiocarbonyldiimidazole, toluene, 110°C, 97%. (*e*) Ac_2O, TMSOTf, rt. (*f*) $(EtO)_3P$, reflux, 97%. (*g*) 1. $(EtO)_3P$, reflux, 93%; 2. Ac_2O, TMSOTf, −65°C, 50%. (*h*) $NaOCH_3$, CH_2Cl_2, CH_3OH, 96%. (*i*) L-DIPT, $Ti(O\textit{i}\text{-}Pr)_4$, *t*-$BuO_2H$, CH_2Cl_2, −30°C, 93%. (*j*) LiN_3, NH_4Cl, DMF, 125°C, 89%. (*k*) H_2, $Pd(OH)_2$, EtOH, THF, TFA, 67%.

On the other hand, epoxidation of **25** using diisopropyl D-tartrate furnished the other epoxide **28** as a single diastereoisomer. An analogous sequence of reactions on **28** produced, via the azide **29**, trehazolamine diastereoisomer **30** (Scheme 4).[42]

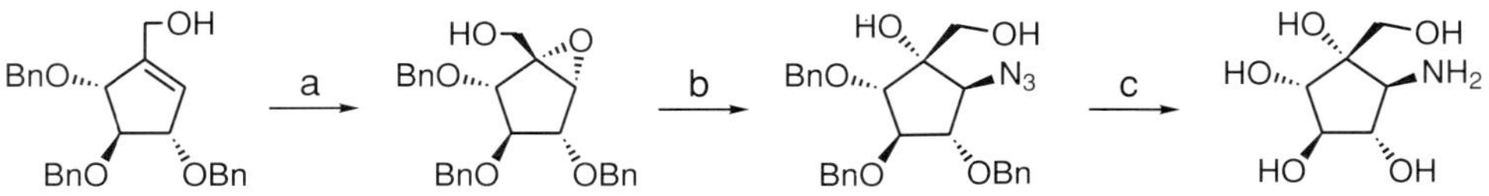

Scheme 4 (*a*) D-DIPT, Ti(O*i*-Pr)$_4$, *t*-BuO$_2$H, CH$_2$Cl$_2$, −30°C, 86%. (*b*) LiN$_3$, NH$_4$Cl, DMF, 125°C, 92%. (*c*) H$_2$, Pd(OH)$_2$, EtOH, THF, TFA, 63%.

A more direct route to **30** was also developed from the D-glucose derivative **15** (Scheme 5).[42] Reductive carbocyclization of the keto oxime derivative **31**, obtained[45,46] from **15**, using an excess of SmI$_2$, took place with subsequent N–O reductive cleavage to afford the aminocyclopentitol **32** in 88% yield. Hydrogenolysis of **32** afforded trehazolamine analogue **30** in 57% overall yield from **15**.

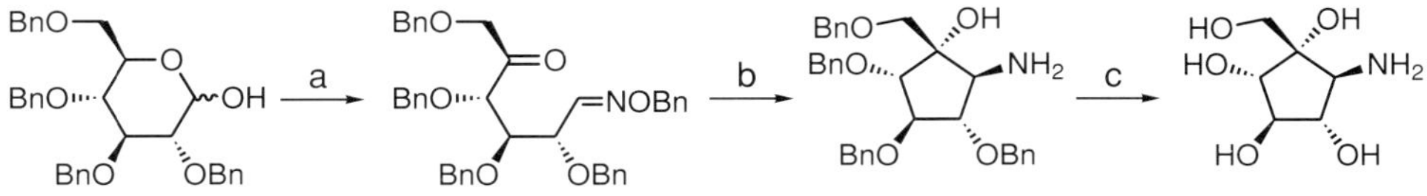

Scheme 5 (*a*) 1. BnONH$_2$·HCl, Py; 2. oxidation, 81% for two steps. (*b*) 0.1 M, SmI$_2$, THF, *t*-BuOH; then H$_2$O, −30°C to rt, 1 h, 88%. (*c*) H$_2$, Pd(OH)$_2$, *on* C, EtOH, THF, TFA, 80%.

Trehazolin has been synthesized from D-glucose via conversion[47] to the aldehyde **33**, which upon treatment with hydroxylamine hydrochloride afforded a 4:1 *anti*/*syn* mixture of **34** (Schemes 6–8).[15,17] Subsequent [2+3] cycloaddition of the oxime **34** with 5% aqueous sodium hypochlorite furnished the corresponding isoxazoline **35**, which was hydrogenolyzed to give the enone **36**. Silylation of **36** with TBSCl furnished the corresponding silyl ether whose subsequent reduction with sodium borohydride in the presence of CeCl$_3$·7H$_2$O afforded a 1:2.5 mixture of **37** and **38** without affecting the double bond. Benzylation of **38**, which possesses the desired configuration, afforded **39**. Removal of the TBS with TBAF afforded the corresponding allyl alcohol **40**. Sharpless epoxidation of **40** with diisopropyl L-tartrate, titanium tetraisopropoxide and *tert*-butyl hydroperoxide furnished the epoxide **41** as a single isomer. After benzylation of **41** with BnBr and NaH, the corresponding benzyl ether was treated with NaN$_3$ and NH$_4$Cl to afford azido alcohol **42** regiospecifically. Reduction of compound **42** with LiAlH$_4$ and subsequent treatment of the corresponding amino alcohol with benzyl isothiocyanate furnished the thiourea derivative **43**. Reduction of **42** with LiAlH$_4$ and subsequent cleavage of the two MOM groups with 5% methanolic hydrogen chloride gave **46**.

Compound **43** was hydrogenolyzed to cleave the two MOM groups and the resulting product was treated with 2-chloro-3-ethylbenzoxazolium tetrafluoroborate and triethylamine to afford the corresponding aminooxazoline **45** via **44**. Finally, compound **45** was hydrogenolyzed to give trehalamine (**4**).

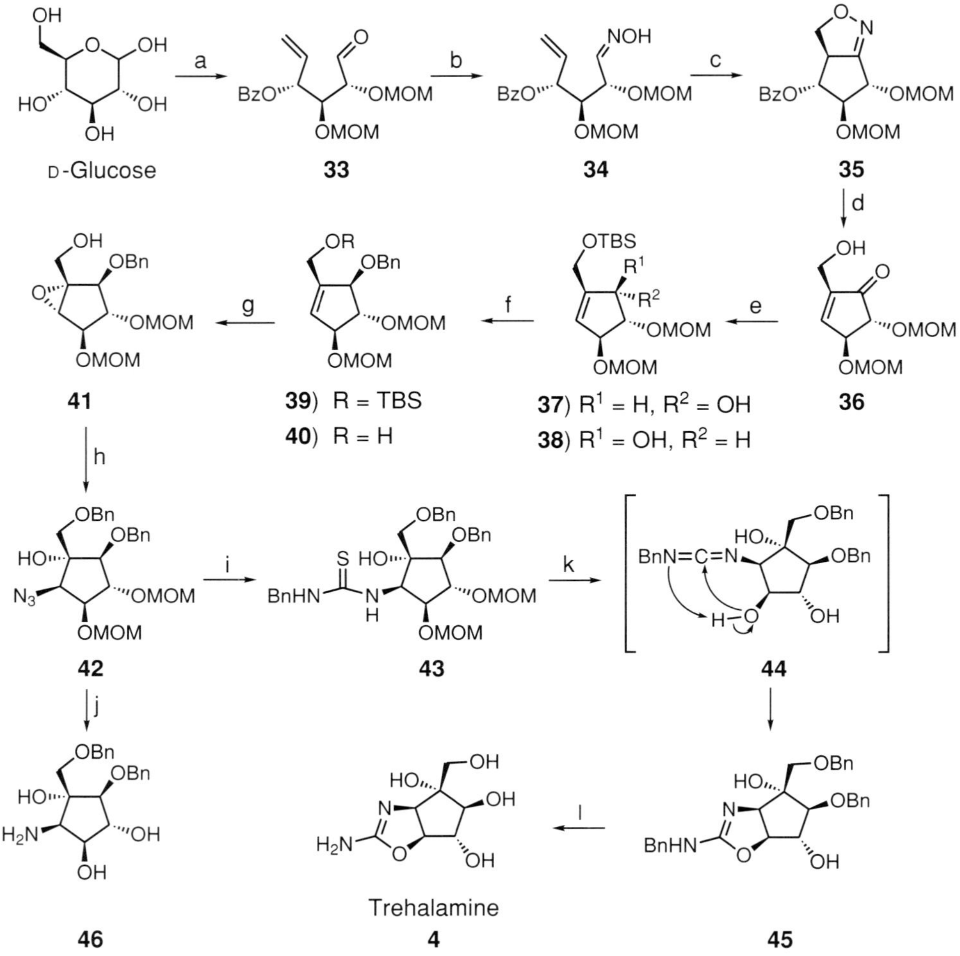

Scheme 6 (*a*) Ref. 50. (*b*) $NH_2OH{\cdot}HCl$, Na_2CO_3, 74%. (*c*) Aqueous NaOCl, cat. NEt_3, 66%. (*d*) H_2, Raney nickel, $B(OH)_3$, 72%. (*e*) 1. TBSCl, imidazole, 88%; 2. $NaBH_4$, $CeCl_3{\cdot}7H_2O$. (*f*) 1. BnBr, NaH; 2. TBAF, 58% for two steps. (*g*) L-DIPT, Ti(O*i*-Pr)$_4$, *t*-BuOOH, CH_2Cl_2, −25°C, 5 h. (*h*) 1. BnBr, NaH, 98%; 2. NaN_3, NH_4Cl, ethylene glycol, DMF, 78%. (*i*) 1. $LiAlH_4$; 2. BnNCS, 83%. (*j*) 1. $LiAlH_4$; 2. 5% HCl in CH_3OH. (*k*) 0.5 M aqueous HCl, 2-chloro-3-ethylbenzoxazolium tetrafluoroborate, NEt_3, 74%. (*l*) H_2, $Pd(OH)_2$ *on* C, 71%.

Hydrogenation of the azido group in **42** and then acetylation gave **47**. Cleavage of the MOM groups in **47** with 5% methanolic hydrogen chloride gave **48**. Complete acetylation of **48** furnished **49**, which upon hydrogenation and subsequent acetylation gave **50**. Hydrolysis of **50** followed by purification using an ion-exchange resin afforded the corresponding trehazolamine **3** (Scheme 7).[15,17]

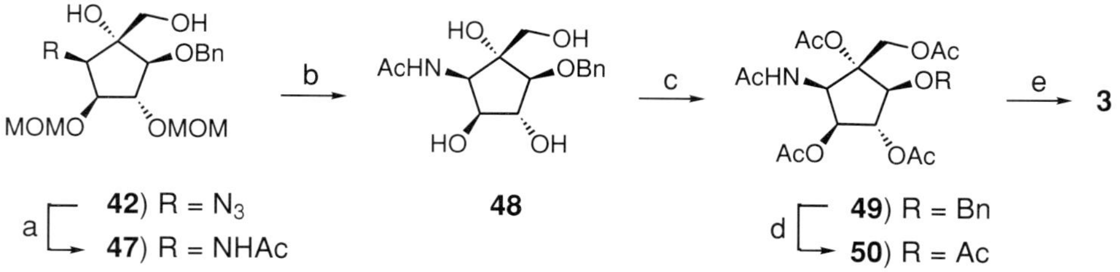

Scheme 7 (*a*) 1. H_2, 10% Pd *on* C; 2. Ac_2O, CH_3OH, 76%. (*b*) 5% HCl–CH_3OH. (*c*) Ac_2O, DMAP, 74%. (*d*) 1. H_2, $Pd(OH)_2$ *on* C, 2. Ac_2O, DMAP, 61%. (*e*) 1. 2 M HCl; 2. Amberlite CG-50 (NH_4^+) resin, 89%.

Toward a complete synthesis[15,17] of trehazolin (**1**), 2,3,4,6-tetra-*O*-benzyl-1-deoxy-α-D-glucopyranosyl isothiocyanate (**51**)[48] was reacted with the amines **3** and **46** in the presence of triethylamine to afford the α-D-glucopyranosylthiourea derivative **52** or **53**, respectively. Subsequent treatment with 2-chloro-3-ethylbenzoxazolium tetrafluoroborate and triethylamine afforded the respective amino oxazoline derivatives **54** and **55**. Finally, hydrogenation over $Pd(OH)_2$ on carbon afforded **1** (Scheme 8).

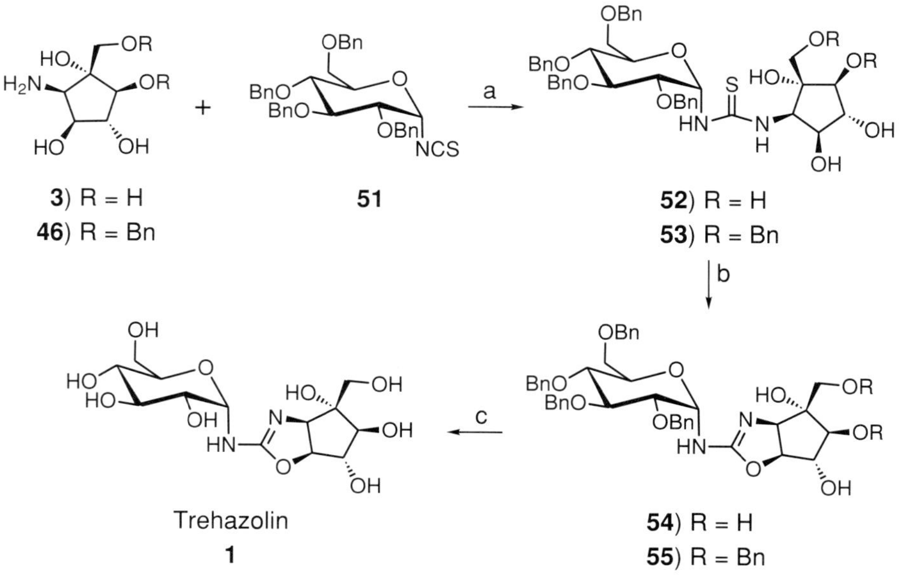

Scheme 8 (*a*) NEt_3, 69%. (*b*) $EtBF_4$, NEt_3, 68%. (*c*) H_2, $Pd(OH)_2$ *on* C, 44%.

5.3.2.2 *Synthesis from* D-*arabinose* The tri-*O*-benzyl trehazolamine **62** has been synthesized from D-arabinose (Scheme 9).[49] The precursors **57** and **58** were prepared from 2,3,5-tri-*O*-benzyl D-arabinose **56** in 47% overall yield.[50] Removal of the *p*-methoxybenzyl group from **57** (84%) followed by inversion of configuration under Mitsunobu conditions afforded **59**, which was also obtained from **58** by removal of the *p*-methoxybenzyl group. The combined **59** was then treated with *m*-CPBA to afford **60**, whose epoxide ring was opened with NaN_3 to give **61**. The azido group in **61** was reduced[52] using Ph_3P to give the aminocyclopentitol unit **62**. The amine **62** could be converted[52] to trehazolin (**1**). The pseudo anomeric center C-4 was inverted using triflic anhydride in the presence of pyridine at low temperature to give the corresponding aminooxazoline, which was then subjected to hydrogenolysis to afford **1**.

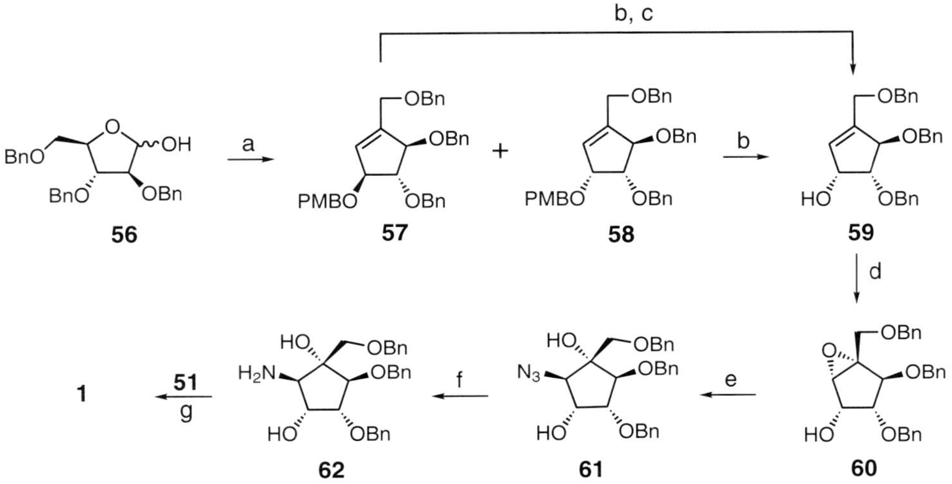

Scheme 9 (*a*) Ref. 50. (*b*) DDQ, CH_2Cl_2, H_2O, 84%. (*c*) PPh_3, DEAD, $C_6H_5CO_2H$, $NaOCH_3$, 95%. (*d*) *m*-CPBA, CH_2Cl_2, 89%. (*e*) NaN_3, DMF, 97%. (*f*) PPh_3, THF, 98%. (*g*) Ref. 51.

5.3.2.3 *Synthesis from D-mannitol* D-Mannitol was also used for providing trehazolamine via the conversion to (*R*)-(–)-epichlorohydrin (**63**),[53,54] which gave the optically active 1-(hydroxymethyl)spiro[2,4]cyclohepta-4,6-diene (**64**) in 60% yield upon treatment with lithium cyclopentadienide (Scheme 10).[55] Conversion of **64** into the corresponding trichloroacetimidate **65** was effected[56] by treatment with sodium hydride and Cl_3CCN. Reaction of **65** with I(*Sym*-Collidine)$_2$ClO$_4$ afforded **66** in 61% yield, which underwent silylation of the secondary carbinol to produce **67** in 95% yield. Treatment of **67** with Li_2NiBr_4 followed by treatment of the resulting cyclopropylcarbinyl bromide with a solution of dimethyldioxirane in acetone afforded the epoxide **68**. Epoxide ring opening by the vicinal trichloroacetamido group, upon treatment with $BF_3{\cdot}OEt_2$ in toluene, followed by free-radical reduction produced the oxazoline **69**. Treatment with aqueous PPTs followed by acetylation and subsequent hydroboration of the terminal alkene and then oxidation of the resulting primary alcohol provided the aldehyde **70**. Conversion of **70** to the corresponding phenylketone **71** took place by reaction with PhMgBr. Norrish-type II cleavage, upon irradiation in benzene, gave the alkene **72**, which, without purification, was reacted with catalytic OsO_4 to yield **73** as a single diastereomer.

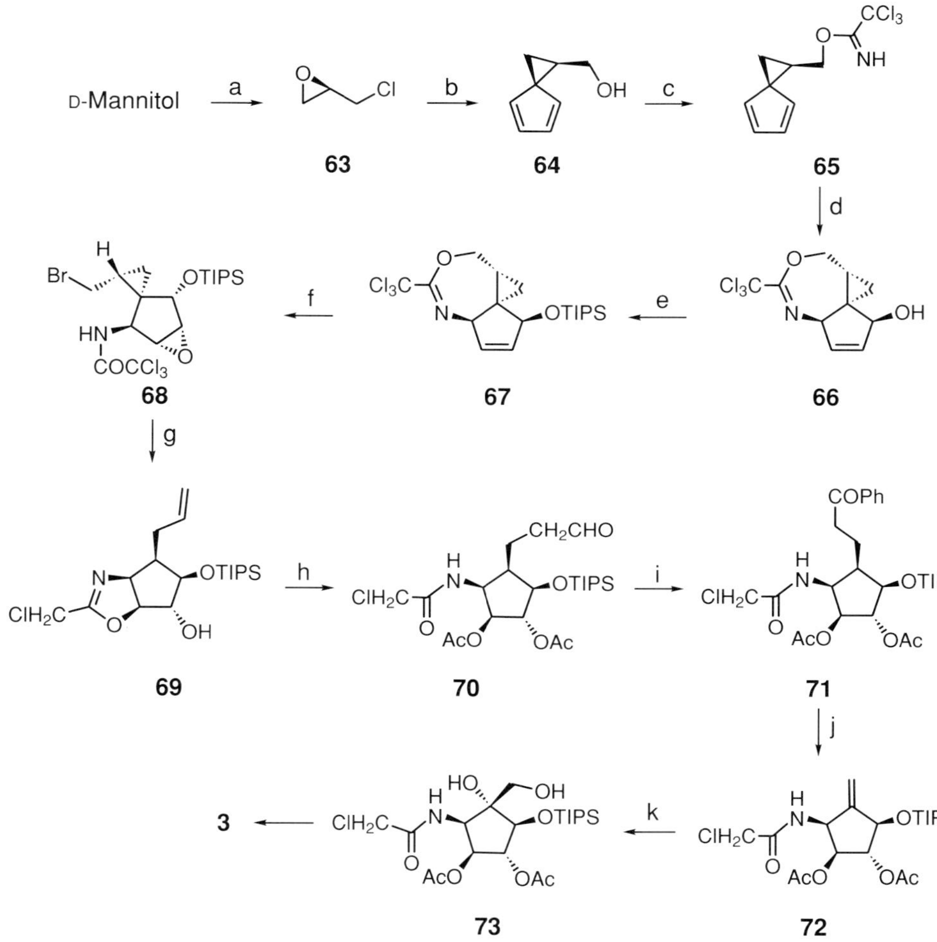

Scheme 10 (*a*) Refs. 53 and 54. (*b*) LiC_5H_5, NaH, THF, 60%. (*c*) NaH, Cl_3CCN, THF, 95%. (*d*) $I(Sym\text{-Collidine})_2ClO_4$, $NaHCO_3$, aqueous CH_3CN, 61%. (*e*) *i*-Pr_3SiOTf, 2,6-lutidine, CH_2Cl_2, 95%. (*f*) 1. Li_2NiBr_4, THF, 80%; 2. $(CH_3)_2CO_2$, acetone, 65%. (*g*) 1. $BF_3{\cdot}OEt_2$, 87%; 2. Bu_3SnH, Et_3B, $NaBH_4$, EtOH, 75%. (*h*) 1. PPTs, aqueous CH_3CN; then Ac_2O, DMAP, 77%; 2. CHX_2BH, H_2O_2, 83%; 3. $(COCl)_2$, DMSO, CH_2Cl_2; then NEt_3, 83%. (*i*) PhMgBr, LiBr, THF, 60%. (*j*) 1. $h\nu$ and then OsO_4, NMO; 2. $(COCl)_2$, DMSO, CH_2Cl_2; then NEt_3, 100% for two steps. (*k*) $h\nu$ and then OsO_4, NMO, 75%.

5.3.2.4 *Synthesis from myo-inositol* Total syntheses of trehazolin (**1**) and its isomers have established both its structure and absolute configuration (Schemes 11–13).[57,58] Thus, base-catalyzed nitromethane condensation[59,60] of the dialdehyde generated by periodate oxidation of (±)-1,2-*O*-cyclohexylidene-myo-inositol (**74**)[61] gave a mixture of the nitrodiols, which was hydrogenated in the presence of Raney nickel, followed by acetylation to afford the three diastereoisomeric 2,3-*O*-cyclohexylidene derivatives **75** (40%), (±)-**76** (5%) and **77** (5%) of 5-acetamido-1,4-*O*-acetylcyclopentane-1,2,3,4-tetraol. The minor racemic mixture **76** was de-O-acetylated, N,O-isopropylidenated, and then resolved by

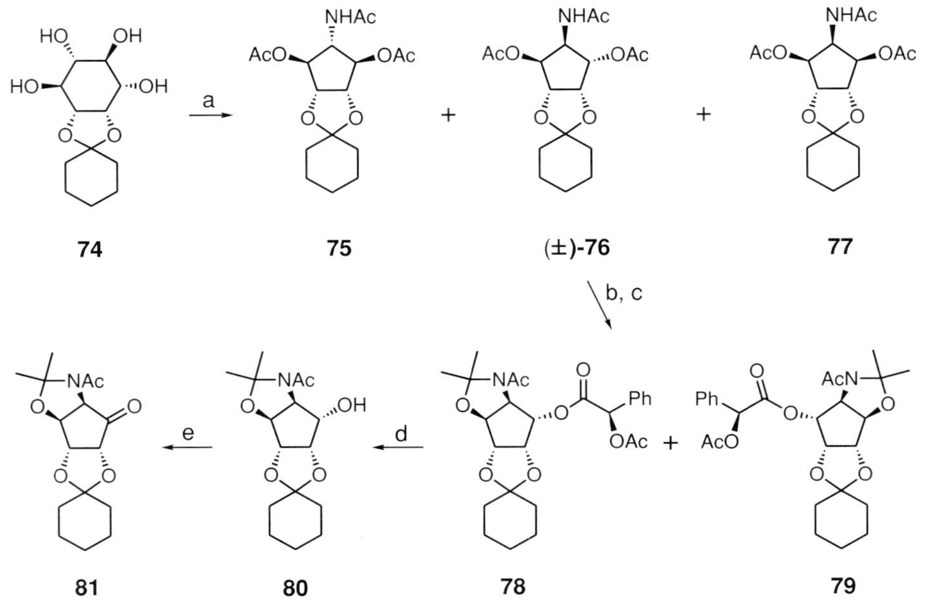

Scheme 11 (*a*) 1. $NaIO_4$; 2. CH_3NO_2, base, H_2, Raney nickel; then Ac_2O, Py, **75** (40%), (±)-**76** (5%), **77** (5%). (*b*) $NaOCH_3$, CH_3OH; then 2,2-dimethoxypropane, *p*-TsOH, DMF, 4 h at 50°C; then AcOH, CH_3OH, 48 h, 75%. (*c*) (*S*)-*O*-Acetylmandelic acid, DMAP, CH_2Cl_2, DCC, CH_2Cl_2, 0°C, 0.5 h, **78** (50%), **79** (48%). (*d*) $NaOCH_3$, CH_3OH, rt, quantitative. (*e*) PCC, MS 4 Å, rt, 2 h, 98%.

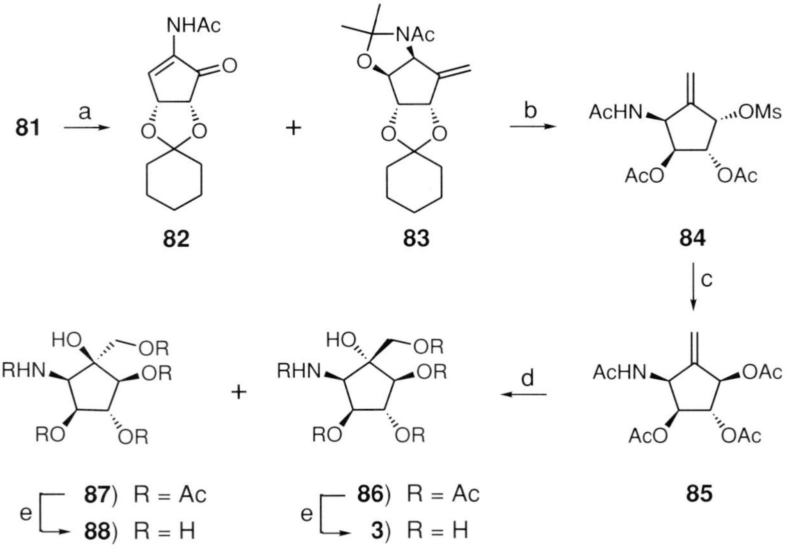

Scheme 12 (*a*) CH_2N_2, DMSO, ether, $P(OCH_3)_3$, 130°C, **83** (45%), **82** (11%). (*b*) Ref. 62. (*c*) NaOAc, DMF, 88%. (*d*) 1. OsO_4, acetylation, **86** (87%), **87** (13%). (*e*) 2 M HCl, 4.5 h at 80°C; Dowex 50WX2 (H^+) resin, elution with aqueous 5% NH_3, 94%.

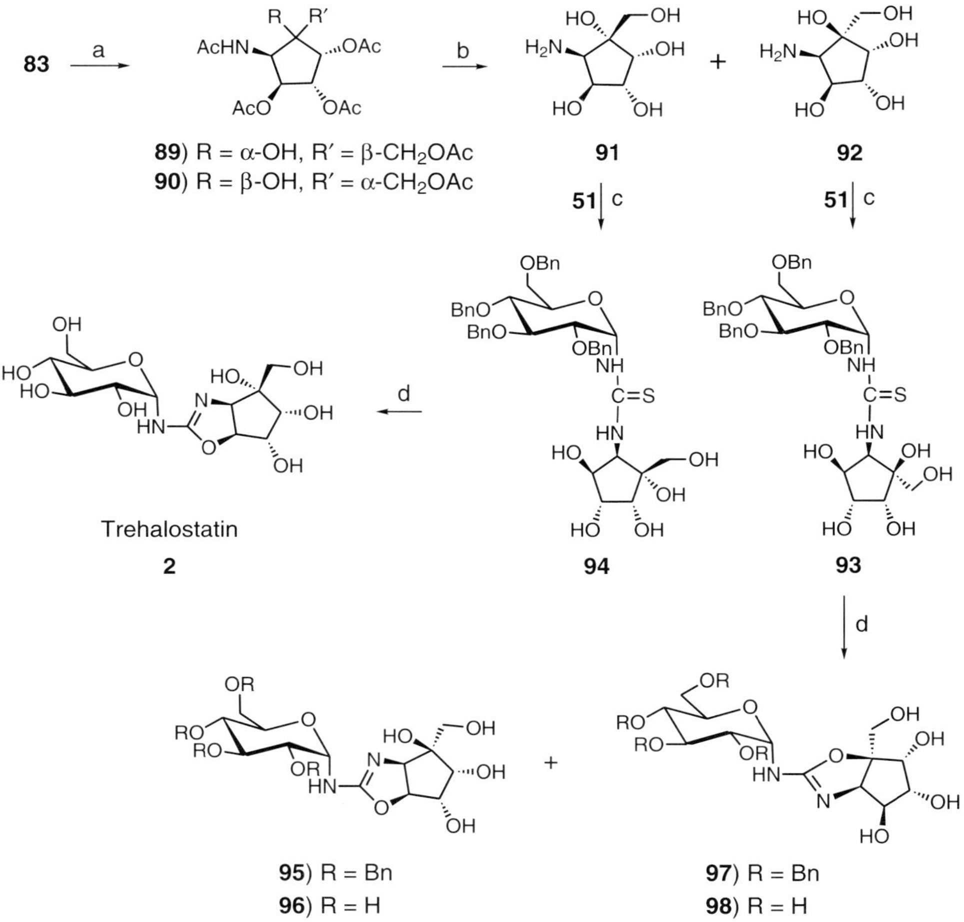

Scheme 13 (*a*) 1. OsO_4, 2. acetylation. (*b*) 2 M HCl, 4.5 h at 80°C; Dowex 50WX2 (H^+) resin, elution with aqueous 5% NH_3. (*c*) Aqueous 75% DMF, 4 h, rt, 92%. (*d*) 1. Ether, HgO, 3 h, 100%; 2. Na, liquid NH_3, 94%.

chromatographic separation of its (*S*)-acetylmandelates to give **78** and **79**. Deacylation of **78** gave **80**, which upon PCC oxidation furnished **81** (Scheme 11). Likewise, **81** was synthesized from **77** by a similar sequence of reactions.[62–64]

Compound **81** was transformed into the exo-olefin **83** via the respective spiro epoxide; the enone **82** (11%) was obtained as a side product (Scheme 12). Compound **83** was deprotected and the obtained triol was selectively mesylated at the allylic position to give after acetylation, compound **84** (68%). Treatment of **84** with sodium acetate resulted in the inversion of the configuration of C-1 to give the tetra-*N*,*O*-acetyl derivative **85**. Oxidation of **85** with OsO_4 in aqueous acetone followed by acetylation afforded **86** (87%) and **87** (13%) whose acid hydrolysis provided the free base **3** and **88**, respectively.

Treatment of **83** with OsO_4 followed by conventional decyclohexylidenation, deisopropylidenation and acetylation gave two branched amino cyclitols **89** (49%) and **90** (51%), which afforded the respective free amino alcohols **91** and **92** almost quantitatively by acid hydrolysis, followed by purification over Dowex 50WX2 (H^+) resin (Scheme 13). Coupling of **91** and **51** afforded **2** via **94**. Also coupling of **92** and **51** afforded **93**, which was converted into **96** and **98** via the intermediates **95** and **97**, respectively.

5.3.2.5 *Synthesis from* D*-ribonolactone* D-Ribonolactone has been converted to trehazolamine derivatives via the allylic alcohol **99**,[65] whose condensation with *p*-methoxybenzylisothiocyanate followed by anti-Markovnikov iodo cyclization with iodine afforded the iodo oxazolidinone **100** (82%) (Scheme 14).[66] The latter was treated with a mixture of acetic anhydride and sulfuric acid followed by activated zinc to furnish the allylic acetate **101** (90%), which underwent inversion at C-2′ under Mitsunobu conditions and the resulting alcohol was epoxidized to produce **102**. Hydrolysis of the epoxide **102** followed by acetylation of the resulting triol **103** afforded **104**, which was treated with CAN to furnish the triacetate **105**. Finally, **105** was converted into hexaacetate **86** in three steps.

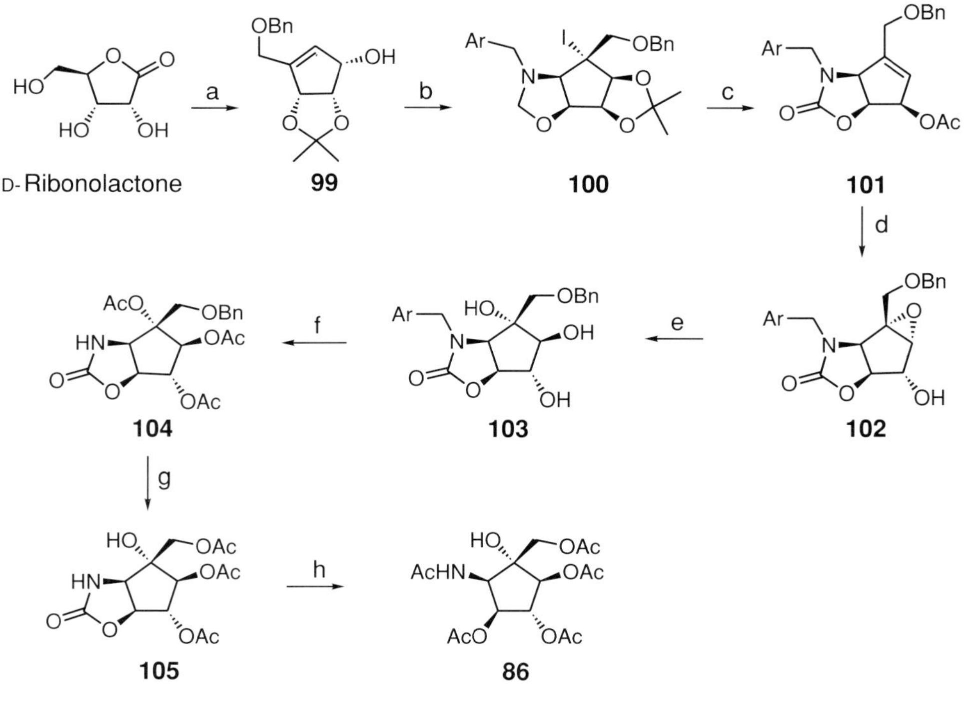

Scheme 14 (*a*) Six steps, Ref. 65. (*b*) 1. NaH, *p*-$(OCH_3)C_6H_4CH_2NCS$, CH_3I; 2. I_2, THF, Na_2CO_3, Na_2SO_3, 82% overall. (*c*) Ac_2O, H_2SO_4, Zn, THF, 90%. (*d*) 1. K_2CO_3, aqueous CH_3OH; 2. $PhCO_2H$, DEAD, Ph_3P, toluene; 3. Na_2CO_3, aqueous CH_3OH, 83% for three steps; 4. CF_3CO_3H, CH_2Cl_2, Na_2CO_3, −20°C, 90%. (*e*) $PhCO_2Na$, aqueous DMF, 100°C, 12 h, 89%. (*f*) 1. Ac_2O, Py, CH_2Cl_2, DMAP; 2. CAN, aqueous CH_3CN, 87% for two steps. (*g*) H_2, 10% Pd *on* C; CH_3OH, 98%. (*h*) 1. 2 N aqueous KOH, EtOH, reflux, 12 h; 2. Ac_2O, Py, DMAP, 70%.

References

1. Ando, O.; Satake, H.; Itoi, K.; Sato, A.; Nakajima, M.; Takahashi, S.; Haruyama, H. *J. Antibiot.* **44** (1991) 1165.
2. Ando, O.; Nakajima, M.; Hamano, K.; Itoi, K.; Takahashi, S.; Takamatsu, Y.; Sato, A.; Enokita, R.; Haruyama, H.; Kinoshita, T. *J. Antibiot.* **46** (1993) 1116.
3. Clegg, J.S.; Evans, D.R. *J. Exp. Biol.* **38** (1961) 771.
4. Sacktor, B.S.; Wormser-Shavit, E. *J. Biol. Chem.* **241** (1966) 634.
5. Hecker, L.I.; Sussman, A.S. *J. Bacteriol.* **115** (1973) 592.
6. Inoue, I.; Shimoda, C. *Mol. Gen. Gent.* **183** (1981) 32.
7. Thevelein, J.M.; Den-Hollander, J.A.; Shulman, R.G. *Proc. Natl. Acad. Sci.* **79** (1982) 3503.
8. Thevelein, J.M.; Jones, K.A. *Eur. J. Biochem.* **136** (1983) 583.
9. Sacktor, B. *Proc. Natl. Acad. Sci.* **60** (1968) 1007.
10. Elbein, A. *Annu. Rev. Biochem.* **56** (1987) 497.
11. Murao, S.; Sakai, T.; Gibo, H.; Shin, T.; Nakayama, T.; Komura, H.; Nomoto, K.; Amachi, T. Presented at the *XVTh International Carbohydrayte Symposium*, Yokohama, 1990.
12. Murao, S.; Sakai, T.; Gibo, H.; Nakayama, T.; Shin, T. *Agric. Biol. Chem.* **55** (1991) 895.
13. Nakayama, T.; Amachi, T.; Murao, S.; Sakai, T.; Shin, T.; Kenny, P.T.M.; Iwashita, T.; Zagorski, M.; Komura, H.; Nomoto, K. *J. Chem. Soc., Chem. Commun.* (1991) 919.
14. Ogawa, S.; Uchida, C.; Yuming, Y. *J. Chem. Soc., Chem. Commun.* (1992) 886.
15. Kobayashi, Y.; Miyazaki, H.; Shiozaki, M. *J. Am. Chem. Soc.* **114** (1992) 10065.
16. Kobayashi, Y.; Miyazaki, H.; Shiozaki, M. *Tetrahedron Lett.* **34** (1993) 1505.
17. Kobayashi, Y.; Miyazaki, H.; Shiozaki, M. *J. Org. Chem.* **59** (1994) 813.
18. Uchida, C.; Kitahashi, H.; Watanabe, S.; Ogawa, S. *J. Chem. Soc., Perkin Trans 1* (1995) 1707.
19. Kobayashi, Y.; Shiozaki, M. *J. Org. Chem.* **60** (1995) 2570.
20. Marco-Contelles, J.; Destable, C.; Gallego, P.; Chiara, J.L.; Bernabe, M. *J. Org. Chem.* **61** (1996) 1354.
21. Uchida, C.; Kitahashi, H.; Yamagishi, Y.; Iwaisaki, Y.; Ogawa, S. *J. Chem. Soc., Perkin Trans 1* (1994) 2775.
22. Goering, B.K.; Li, J.; Ganem, B. *Tetrahedron Lett.* **36** (1995) 8905.
23. Trost, B.M.; Van Vrankon, D.L. *J. Am. Chem. Soc.* **115** (1993) 444.
24. Kobayashi, Y.; Shiozaki, M. *J. Antibiot.* **47** (1994) 243.
25. Kobayashi, Y.; Miyazaki, H.; Shiozaki, M.; Haruyama, H. *J. Antibiot.* **47** (1994) 932.
26. Ogawa, S.; Uchida, C. *Chem. Lett.* (1993) 173.
27. Uchida, C.; Ogawa, S. *Bioorg. Med. Chem.* **4** (1996) 275.
28. Uchida, C.; Ogawa, S. *Carbohydr. Lett. 1* (1994) 77.
29. Elliot, R.P.; Hui, A.; Fairbanks, A.J.; Nash, R.J.; Winchester, B.G.; Fleet, W.J. *Tetrahedron Lett.* **34** (1993) 7949.
30. Shiozaki, M.; Mochizuki, T.; Hanzawa, H.; Haruyama, H. *Carbohydr. Res.* **288** (1996) 99.
31. Li, J.; Lang, F.; Ganem, B. *J. Org. Chem.* **63** (1998) 5877.
32. Kobayashi, Y. *Carbohydr. Res.* **315** (1999) 3.
33. Schoenfeld, R.C.; Lumb, J.-P.; Ganem, B. *Tetrahedron Lett.* **42** (2001) 6447.
34. Berecibar, A.; Granjean, C.; Siriwardena, A. *Chem. Rev.* **99** (1999) 779.
35. Kassab, D.J.; Ganem, B. *J. Org. Chem.* **64** (1999) 1782.
36. Clark, M.A.; Goering, B.K.; Li, J.; Ganem, B. *J. Org. Chem.* **65** (2000) 4058.
37. Boiron, A.; Zillig, P.; Faber, D.; Giese, B. *J. Org. Chem.* **63** (1998) 5877.
38. Qiao, L.; Veredas, J.C. *J. Org. Chem.* **58** (1993) 3480.
39. Ireland, R.E.; Liu, L. *J. Org. Chem.* **58** (1993) 2899.
40. Norman, R.O.C.; Purchase, R.; Thomas, C.B. *J. Org. Chem. Soc., Perkin Trans. 1* (1972) 1701.
41. Weiss, R.H.; Furfine, E.; Hausleden, E.; Dixon, D.W.J. *J. Org. Chem.* **49** (1984) 4969.
42. Storch de Gracia, I.; Dietrich, H.; Bobo, S.; Chiara, J.L. *J. Org. Chem.* **63** (1998) 5883.
43. Decoster, E.; Lacombe, J.-M.; Streber, J.-L.; Ferrari, B.; Pavia, A.A. *J. Carbohydr. Chem.* **2** (1983) 329.
44. Girard, P.; Namy, J.L.; Kagan, H.B. *J. Am. Chem. Soc.* **102** (1980) 2693.
45. Chiara, J.L.; Marco-Conteller, J.; Khira, N.; Gallego, P.; Destabel, C.; Bernabe, M. *J. Org. Chem.* **60** (1995) 6010.
46. Marco-Contelles, J.L.; Gallego, P.; Rodriguez-Fernandez, M.; Khiar, N.; Destabel, C.; Bernanbe, M.; Marinez-Grau, A.; Chiara, J.L. *J. Org. Chem.* **62** (1997) 7397.
47. Bernet, B.; Vasella, A. *Helv. Chim. Acta* **62** (1979) 1990.
48. Camarasa, M.J.; Fernández-Resa, P.A.; García-López, M.T.; de las Heras, F.G.; Mendez-Castrillón, P.P.; San Felix, A. *Synthesis* (1984) 509.
49. Seepersaud, M.; Al-Abed, Y. *Tetrahedron Lett.* **42** (2001) 1471.
50. Seepersaud, M.; Al-Abed, Y. *Tetrahedron Lett.* **41** (2000) 7801.

51. Storch de Gracia, I.; Bobo, S.; Martin-Ortega, M.D.; Chiara, J.L. *Org. Lett. 1* (1999) 1705.
52. Shiozaki, M.; Arai, M.; Kobayashi, Y.; Kasuya, A.; Miyamoto, S.; Furukawa, Y.; Takayama, T.; Haruyama, H. *J. Org. Chem.* **59** (1994) 4450.
53. Baldwin, J.J.; Raab., A.W.; Mensler, K.; Arison, B.H.; McClure, D.E. *J. Org. Chem.* **43** (1978) 4876.
54. Klunder, J.M.; Onami, T.; Sharpless, K.B. *J. Org. Chem.* **54** (1989) 1295.
55. Ledford, B.E.; Carreira, E.M. *J. Am. Chem. Soc.* **117** (1995) 11811.
56. Overman, L.E. *J. Am. Chem. Soc.* **96** (1974) 597.
57. Uchida, C.; Yamagishi, Y.; Ogawa, S. *Chem. Lett.* (1993) 971.
58. Uchida, C.; Yamagishi, Y.; Ogawa, S. *J. Chem. Soc., Perkin Trans. 1* (1994) 589.
59. Angyal, S.J.; Gero, S.D. *Aust. J. Chem.* **18** (1965) 1973.
60. Ahluwalia, R.; Angyal, S.J.; Luttrel, B.M. *Aust. J. Chem.* **23** (1970) 1819.
61. Angyal, S.J.; Tate, M.E.; Gero, S.D. *J. Chem. Soc.* (1961) 4116.
62. Suami, T.; Tadano, K.; Nishiyama, S.; Lichtenthaler, F.W. *J. Org. Chem.* **38** (1973) 3691.
63. Ogawa, S.; Uchida, C.; Yuming, Y. *J. Chem. Soc., Chem. Commun.* (1992) 886.
64. Ogawa, S.; Uchida, C. *J. Chem. Soc., Perkin Trans. 1* (1992) 1939.
65. Marquez, V.E.; Lim, M.; Khan, M.S.; Kaskar, B. *Nucleic Acid Chem.* **4** (1991) 27.
66. Knapp, S.; Purandare, A.; Rupitz, K.; Withers, S.G. *J. Am. Chem. Soc.* **116** (1994) 7461.

5.3.3 *Allosamidin*

Chitin (**1**), the β-1,4-linked polymer of *N*-acetylglucosamine, is widely known as one of the main skeletal components of insect cuticles[1–5] and microbial cell walls[6,7] as well as the principal macromolecule in crustacean shells, a major waste product of the seafood processing industry.[8] The metamorphosis of insects is controlled by two different types of chitinases and is an essential step for regulating their life cycles. Consequently, much attention has been focused on discovering substances that interact with its biosynthesis and metabolism.[9–11] The metabolism of chitin is controlled by the activity of synthetases, which transfer *N*-acetyl-D-glucosamine to the growing chitin chain, whereas exo- and endochitinases degrade the polymer to chitobiose.

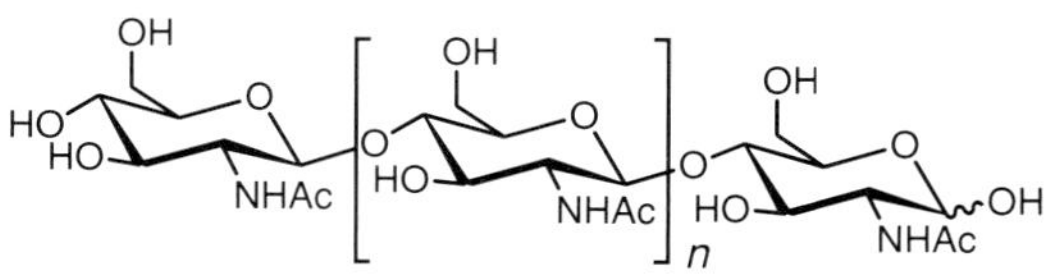

Chitin
1

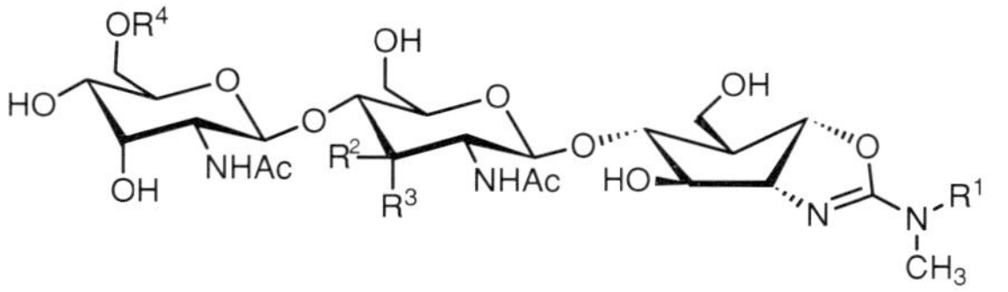

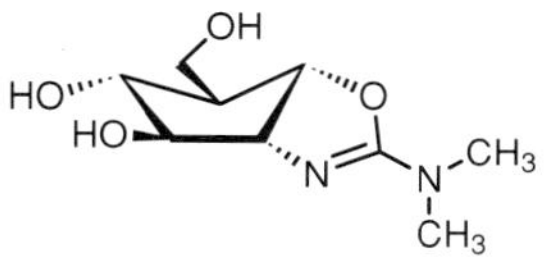

2) Allosamidin, $R^1 = CH_3$, $R^2 = R^4 = H$, $R^3 = OH$
3) Demethylallosamidin, $R^1 = CH_3$, $R^2 = R^4 = H$, $R^3 = OH$
4) Methylallosamidin, $R^1 = R^4 = CH_3$, $R^2 = H$, $R^3 = OH$
5) Methyl *N*-demethylallosamidin, $R^1 = R^2 = H$, $R^3 = OH$, $R^4 = CH_3$
6) Glucoallosamidin A, $R^1 = R^4 = CH_3$, $R^2 = OH$, $R^3 = H$
7) Glucoallosamidin B, $R^1 = R^3 = H$, $R^2 = OH$, $R^4 = CH_3$

Allosamizoline
8

Allosamidin (**2**) and its congeners demethylallosamidin (**3**), methylallosamidin (**4**), methyl *N*-demethylallosamidin (**5**), glucoallosamidin A (**6**) and glucoallosamidin B (**7**) are the first examples of endochitinase inhibitors. They were isolated from the mycelial extract of *Streptomyces* sp. 1713 and related *actinomycete* SA-684 and A82516. They exhibit the inhibitory activity against the chitinases of the silkworm *Bombyx mori in vitro* and prevent its larval ecdysis *in vivo*.[12–15] It has been thought that the chitinase inhibitor would be the good models for insect growth regulators.[16]

Allosamidin (**2**) has a unique pseudotrisaccharide structure consisting of two *N*-acetyl-D-allosamine units and a novel five-membered aminocyclitol, named allosamizoline[17–20] (**8**). This is the first example, in nature, having allosamine derivatives. The relative configuration of **8** was initially suggested to have a 3,4-*cis* diol[17] configuration and later it was revised to the 3,4-*trans*[18] one. The absolute configuration was then elucidated by studying its 3,4-bis(*p*-dimethoxyamino)-6-trityl derivative.[19]

The mechanism of cyclopentane ring formation of allosamizoline[21,22] may take place via pathway A or B during inositol biosynthesis, whereas via pathway C during shikimic acid biosynthesis (Scheme 1). This was based on a study using [3-^{2}H]-, [4-^{2}H]-, [5-^{2}H]- and [6-$^2H_2$]-D-glucosamine feeding in experiments which indicated that the cyclization to form the cyclopentanoid moiety of allosamizoline is presumed to proceed via a 4-keto or 6-aldehyde glucosamine derivative or their enol equivalents, which would undergo an aldol condensation of C-5 with C-1.

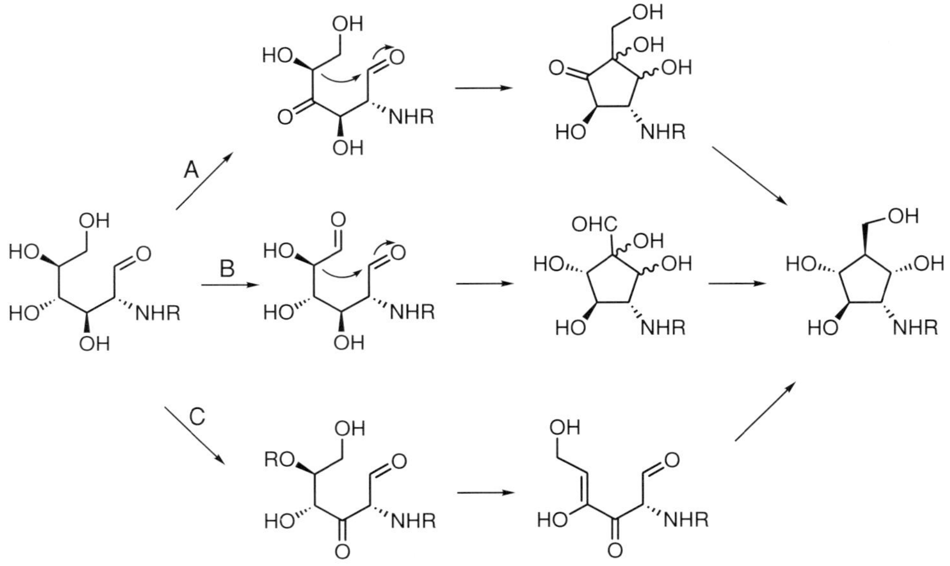

Scheme 1 Plausible mechanism of formation of the cyclopentane ring of allosamidin (**2**).

Many syntheses of allosamizoline and its analogues from noncarbohydrate have been reported.[23–43] Carbohydrates have been also used for the synthesis of allosamizoline, which upon suitable protection to be a glycosyl acceptor that can be coupled with the required oligosaccharide donor would led to the total synthesis of allosamidin.

5.3.3.1 *Synthesis from D-glucose* The synthesis of allosamizoline has been achieved by starting with methyl α-D-glucopyranoside (**9**) (Scheme 2).[44,45] Selective tosylation of **9** gave methyl 2,6-bis-*O*-(toluene-*p*-sulfonyl)-α-D-glucopyranoside (**10**),[46] whose selective benzoylation at O-3 followed by acetylation at O-4 gave compound **11**, which on treatment with sodium iodide afforded **12** (83%). Treatment of **12** with zinc in ethanol gave **13**, which was reacted with *N*-methyl hydroxylamine to furnish the isoxazolidine **14** (57% from **12**). Reduction with hydrogen over Raney nickel gave the aziridine **15**, whose treatment with peracid gave the cyclopentene **16**. Silylation of **16** followed by deacetylation and then benzoylation gave **17**. Oxyamination of **17** afforded adducts **18** and **19** together with the

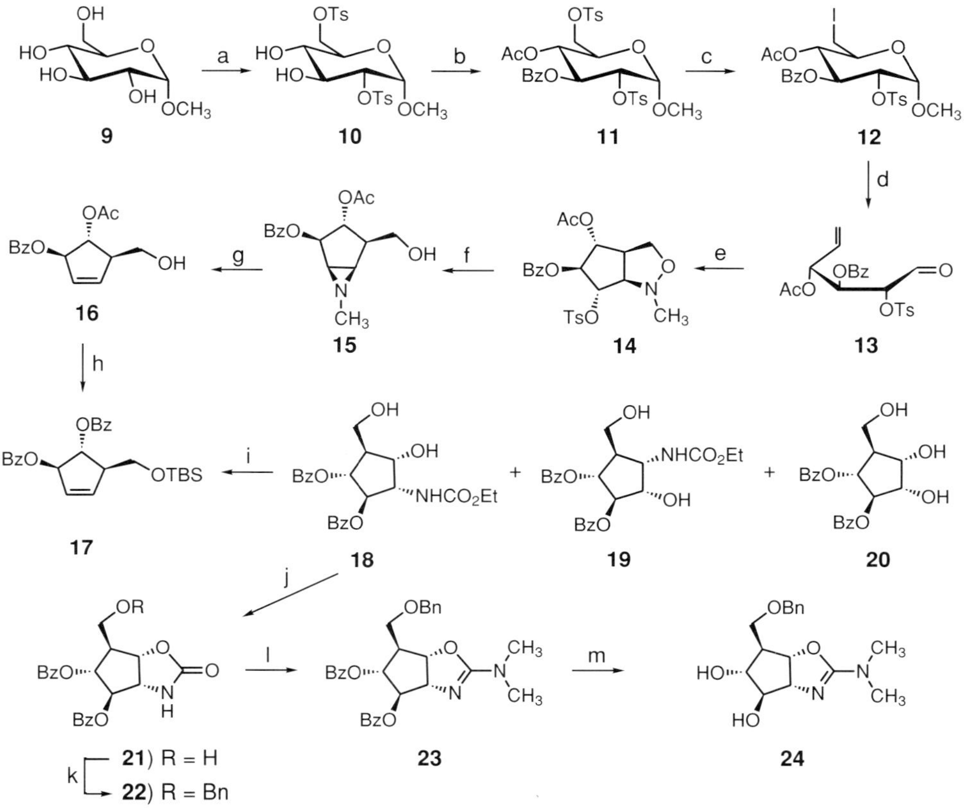

Scheme 2 (*a*) *p*-TsCl, Py. (*b*) 1. BzCl, Py, CH_2Cl_2, 55%; 2. Ac_2O, Py, 86%. (*c*) NaI, Ac_2O, reflux, 1 h, 83%. (*d*) EtOH, Zn powder, reflux, 40 min. (*e*) *N*-Methylhydroxylamine hydrochloride, EtOH, Py, 45°C, 57%. (*f*) Raney nickel (W-2), H_2, 68%. (*g*) Oxidation with magnesium monoperoxyphthalate hexahydrate propan-2-ol, **15** (81%). (*h*) 1. TBSCl, Py; 2. $NaOCH_3$, CH_3OH; 3. BzCl, Py. (*i*) Oxyamination, *t*-BuOH, Chloramine-T trihydrate, OsO_4, **20** (14%). (*j*) Bis(tributyltin)oxide, toluene, reflux, 72%. (*k*) BnBr, NaH, DMF. (*l*) CH_2Cl_2, MS 4 Å, trimethyloxonium tetrafluoroboranuide, 20°C, 24 h, $(CH_3)_2NH$, 81%. (*m*) CH_3OH, $NaOCH_3$, 20°C, 5 h, 97%.

diol **20**. Boiling of **18** with bis(tributyltin)oxide in toluene effected the removal of ethanol and promoted the cyclization involving the hydroxyl group to give **21**. Benzylation of **21** gave **22**, which was converted into the *N,N*-dimethylamino derivative **23**, followed by debenzoylation to afford **24**.

5.3.3.2 *Synthesis from* D*-glucosamine* Allosamizoline (**8**) has also been synthesized from D-glucosamine (Scheme 3).[47] Methyl 2-amino-4,6-*O*-benzylidene-3-*O*-benzyl-2-deoxy-α-D-glucopyranoside[48] (**26**), obtained from D-glucosamine hydrochloride (**25**), was converted into the corresponding *N,N*-dimethylurea derivative, which underwent hydrolysis with aqueous acetic acid to give the diol **27** (91%). Selective iodination of the primary position and subsequent protection of the secondary hydroxyl group with

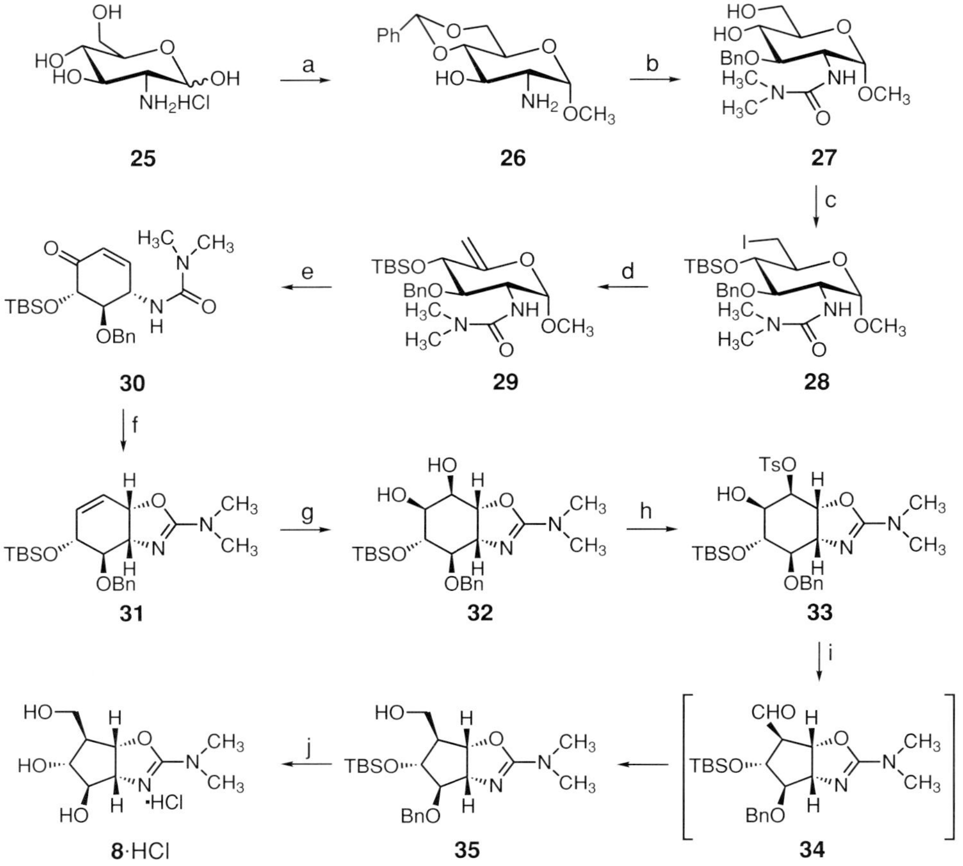

Scheme 3 (*a*) Ref. 48. (*b*) 1. $(CH_3)_2NCOCl$, NEt_3, CH_2Cl_2, 2. aqueous AcOH, 91% for two steps. (*c*) 1. *N*-Iodosuccinimide, Ph_3P, THF; 2. TBSOTf, 2,6-lutidine, CH_2Cl_2, 71% for two steps. (*d*) *t*-BuOK, THF, 96%. (*e*) 1. $HgSO_4$, 5 mM H_2SO_4, acetone; 2. MsCl, Py, 65% for two steps. (*f*) 1. $NaBH_4$, $CeCl_3 \cdot 7H_2O$, CH_3OH; 2. Ms_2O, NEt_3, CH_2Cl_2, 86% for two steps. (*g*) OsO_4, $(CH_3)_3NO$, *t*-BuOH, H_2O, 92%. (*h*) *p*-TsCl, DMAP, Py, CH_2Cl_2, 86%, 91% based on **32**. (*i*) L-Selectride, THF, 65°C, 86%. (*j*) 1. 1 M HCl, aqueous THF; 2. H_2, 10% Pd *on* C, 0.1 M HCl, H_2O, 90% for two steps.

tert-butyldimethylsilyl trifluoromethanesulfonate gave **28** (71%). The *t*-BuOK effected the dehydroiodination of **28** to give **29** (96%), which underwent Ferrier reaction[49] and subsequent β-elimination to afford **30** (65%). The enone **30** was stereoselectively reduced, followed by treatment of the resulting alcohol with methanesulfonic anhydride to furnish the oxazoline derivative **31** (86%). Dihydroxylation of **31** using OsO_4 occurred exclusively from the convex face to produce the *cis* diol **32** (92%). Selective tosylation of **32** gave **33** (86%), which underwent ring contraction with L-Selectride in THF to furnish **35** (86%) via the unstable aldehyde **34**. This contraction has been explained to be due to the presence of the tosyloxy group in nearly antiperiplanar relationship to the C-4–C-5 bond in the half-chair-like conformation of **33**. Finally, removal of the protecting groups from **35** afforded (−)- **8** as the hydrochloride salt in 21% overall yield from **26**.

Alternatively, the synthesis of allosamizoline (**8**) from D-glucosamine hydrochloride (**25**) using a free radical cyclization as a key step has also been reported (Scheme 4).[50,51] Compound **25** was converted to the *N*-Cbz tri-*O*-acetyl derivative **36** (69% overall yield),[52,53] which was treated with the *O*-benzyl ether of hydroxylamine followed by Im_2CS to produce the adduct thiocarbonylimidazolide **37**. Free radical cyclization of **37** using Bu_3SnH and AIBN afforded a mixture of diastereomeric products **38** and **39** in 2:9 ratio. Oxidation of the mixture of benzyloxyamines with *m*-CPBA afforded the oxime **40** in 79% yield,[54] which was treated with ozone followed by sodium borohydride reduction to furnish the alcohol **41**. Treatment of **41** with thionyl chloride gave the oxazolidinone **42**, which upon treatment with Et_3OBF_4 followed by $(CH_3)_2NH$ gave the allosamizoline triacetate **43**. Subsequent saponification of **43** followed by treatment with HCl gave allosamizoline hydrochloride (**8**·HCl).

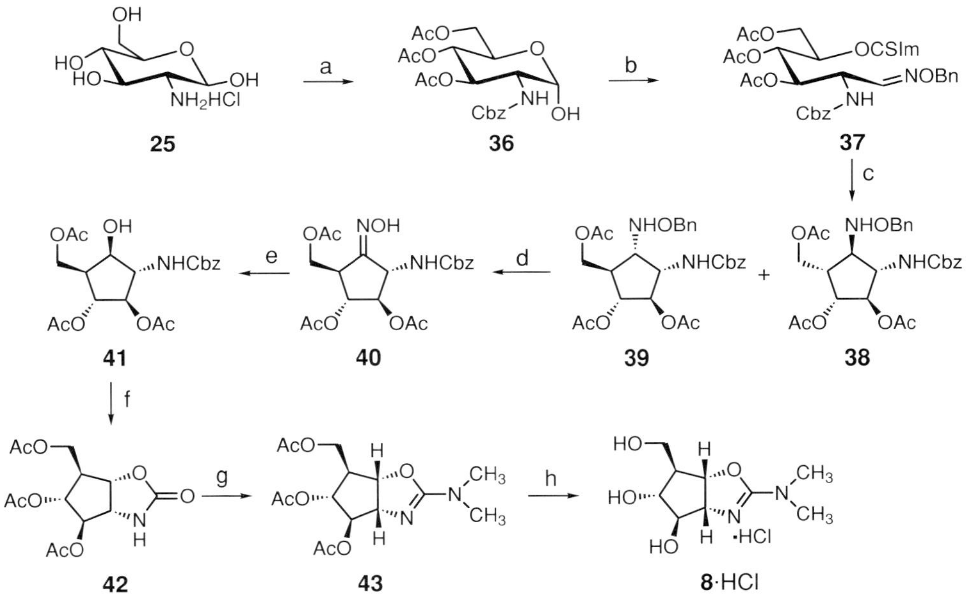

Scheme 4 (*a*) 1. CbzCl, $NaHCO_3$, H_2O, 96%; 2. Ac_2O, NEt_3, DMAP, THF, 82%; 3. $(NH_4)_2CO_3$, CH_3OH, THF, 88%. (*b*) 1. $NH_2OBn{\cdot}HCl$, Py, CH_2Cl_2, 88%; 2. $Im_2C{=}S$, benzene, 82%. (*c*) Bu_3SnH, AIBN, benzene, **38** (12%), **39** (54%). (*d*) *m*-CPBA, Na_2CO_3, EtOAc, 79%. (*e*) O_3, CH_2Cl_2, −40°C, CH_3OH, $NaBH_4$, −40°C to rt. (*f*) $SOCl_2$, 82%. (*g*) Et_3OBF_4, CH_2Cl_2, $(CH_3)_2NH$, CH_2Cl_2, 80%. (*h*) $NaOCH_3$, CH_3OH, HCl, 98%.

Another synthesis of (−)-allosamizoline (**8**) was also carried out using the glucoseamine derivative **44**[55–58] by an intramolecular cycloaddition of a nitrile oxide to an olefin as a key step (Scheme 5).[59,60] Iodination of **44** followed by reductive β-elimination using zinc in THF afforded the 5-enofuranose **46**, whose reaction with ethanethiol in conc. HCl followed by silylation with TBSOTf afforded **47**. Dethioacetalization of **47** with $HgCl_2$–$CaCO_3$ followed by treatment of the resulting aldehyde with NH_2OH afforded the oxime **50**, which underwent intramolecular cycloaddition to produce the isoxazoline **51**. Alternatively, treatment

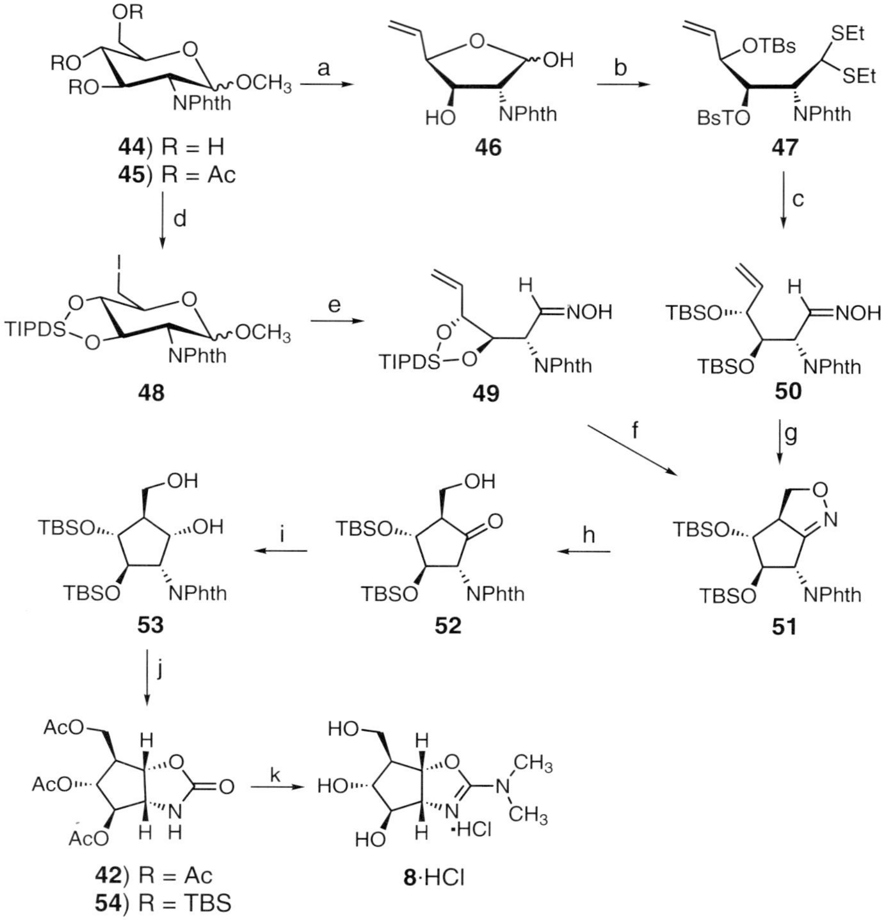

Scheme 5 (*a*) 1. I_2, Ph_3P, imidazole, CH_2Cl_2, 35°C, 4 days, 90%; 2. Zn, THF, 25°C, 1.5 h. (*b*) 1. Ethanethiol, conc. HCl, 0°C, 18 h, 61% for two steps; 2. TBSOTf, 2,6-lutidine, CH_2Cl_2, 0°C, 2 h, 90%. (*c*) $HgCl_2$–$CaCO_3$, 80%, aqueous acetone, 25°C, 12 h; then $NH_2OH{\cdot}HCl$, Py, 25°C, 18 h, 81% from **7**. (*d*) 1. *p*-TsCl, Py, 0–5°C, 20 h, 73% for two steps; 2. $TIPDSCl_2$, imidazole, DMF, 45°C, 18 h; 3. NaI, $NaHCO_3$, DMF, 70°C, 18 h, 69%. (*e*) 1. Zn, THF, reflux, 40 min; 2. $NH_2OH{\cdot}HCl$, NaOAc, CH_3OH, rt, 18 h. (*f*) 1. 2.5% NaClO, CH_2Cl_2, 0°C to rt, 10 h, 91%; 2. 1 M *n*-Bu_4NF, THF, rt, 20 min; 3. TBSCl, imidazole, DMF, 40°C, 4 days, 60% for two steps. (*g*) 0.7 M aqueous NaOCl, CH_2Cl_2, 0°C, 18 h, 91%. (*h*) O_3, O_2, CH_2Cl_2–CH_3OH (10:1), −78 to −30°C, 24 h, 60%; *or* O_3, CH_2Cl_2, CH_3OH, −30°C, 12 h; then DMS, rt. (*i*) $Zn(BH_4)_2$, THF–ether, (1:1), 0°C, 3 h, 100%. (*j*) 1. $NH_2NH_2{\cdot}H_2O$, 95% aqueous EtOH, 70°C, 4 h, 60%; 2. CbzCl, Na_2CO_3, H_2O–CH_2Cl_2 (1:2), 0°C, 1.5 h; 3. NaH, THF, 25°C; 4. 1% HCl–CH_3OH, 25°C, 2.5 h; 5. Ac_2O, Py, 25°C, 18 h; *or* TBSCl, imidazole. (*k*) 1. CH_3OTf (4 equiv.), CH_2Cl_2, 25°C, 5.5 h; 2. $(CH_3)_2NH{\cdot}HCl$, NEt_3, CH_2Cl_2, 25°C, 2 h; 3. 1 M aqueous HCl, 50°C, 4 h, 80% for seven steps; or 1. Et_3OBF_4, CH_2Cl_2, rt, 20 h; 2. $(CH_3)_2NH$, rt, 24 h, CH_2Cl_2, 46%; 3. aqueous HCl, 50°C, 4.5 h.

of methyl 3,4,6-tri-*O*-acetyl-2-deoxy-2-phthalimido-D-glucopyranoside (**45**) with sodium methoxide followed by selective monotosylation of the primary hydroxyl group, protection of the secondary hydroxyl group with dichlorotetraisopropyldisiloxane and then S_N2 displacement of the tosyl group with iodide ion afforded **48**. Reductive ring cleavage of **48** using freshly activated zinc powder in aqueous THF followed by condensation of the resulting vinyl aldehyde with hydroxylamine afforded the oxime **49** (69%). The oxime **49** was treated with sodium hypochlorite to afford the corresponding cyclized isoxazoline whose desilylation using TBAF and resilylation with TBSCl afforded the bis-TBS ether **51** in 60% yield from **49**. Ozonolytic cleavage of **51** furnished the β-hydroxy ketone **52**, which was treated with $Zn(BH_4)_2$ to afford a single isomer **53**. Compound **53** was converted to **8** via **42** or **54** through the steps shown in Scheme 5.

5.3.3.3 *Total synthesis of allosamidin* The cyclopentene diol **56**,[61] prepared from cyclopentadienylthallium **55**, was converted into the imidate **57**, which was subsequently cyclized via stereoselective Hg(II) mediated ring closure to provide oxazoline **58** as a single diastereomer (Scheme 6).[62] Subjection of **58** to radical oxygenation conditions led to demercuration by introducing a hydroxyl from the convex face of the ring system to provide

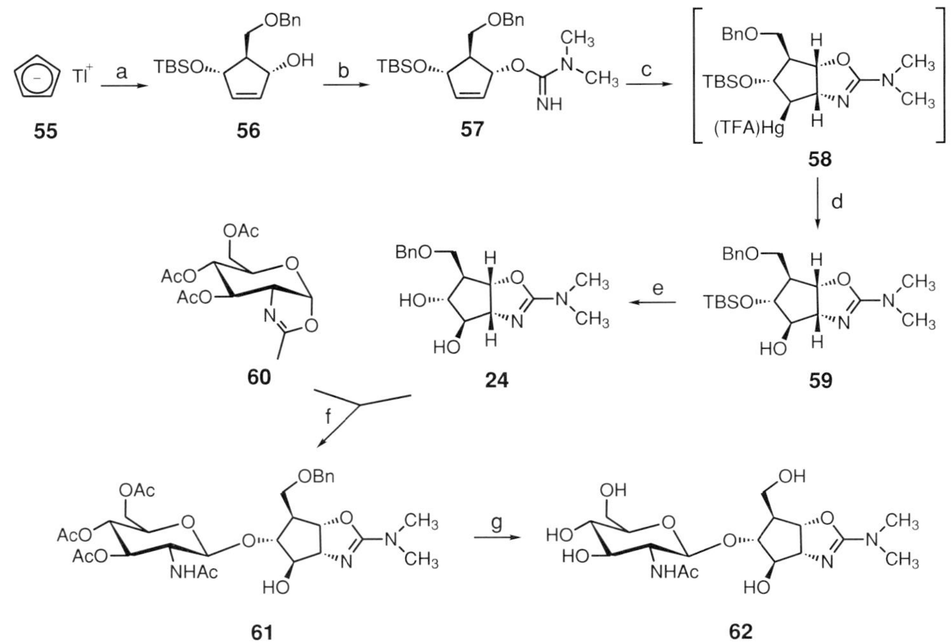

Scheme 6 (*a*) 1. Benzylchloromethylether (1.05 equiv.), Et_2O, −20°C, 3.5 h; 2. O_2, *hν*, methylene blue, thiourea, CH_3OH, 0°C, 1 h, followed by stirring at 0°C, 18 h, 60% for two steps; 3. Ac_2O (2 equiv.), DMAP (0.1 equiv.), CH_2Cl_2, 0°C, 1 h, 95%; 4. electric eel acetylcholinesterase, 0.2 M KH_2PO_4, 5%, CH_3OH, pH 7, rt, 4 days, 90%; 5. TBSCl, imidazole, CH_2Cl_2, 30 min, rt, 95%; 6. anhydrous NH_3, CH_3OH, −10°C to rt, 18 h, 95%. (*b*) Neat dimethylcyanamide, NaH (1.1 equiv.), −78°C to rt, 5 h, 93%. (*c*) Mercury(II) trifluoroacetate (1.5 equiv.), THF, rt, 36 h. (*d*) Vigorous O_2 flush, 1 M $NaBH_4$ in 2 M NaOH, 1,4-dioxane, rt, 2 h, 69% for two steps. (*e*) 1 N HF, CH_3CN, rt, 28 h, 90%. (*f*) TfOH (1 equiv.), CH_3NO_2, toluene, 2 h, 60°C, 50%. (*g*) 1. $Pd(OH)_2$, H_2, CH_3OH, 18 h; 2. anhydrous NH_3, CH_3OH, 36 h, rt, 95%.

the protected allosamizoline **59** in 69% yield from **57**. Desilylation of **59** afforded 6-*O*-benzyl-allosamizoline (**24**) in 56% yield from **56**.

For the synthesis of allosamidin, the glycosylation of **24** with oxazoline **60**, obtained by treatment of 2-acetamido-2-deoxy-β-D-glycopyranose-1,3,4,6-tetraacetate with $BF_3{\cdot}OEt_2$, using triflic acid as catalyst gave the β-anomer of the pseudodisaccharide **61** (50%), which underwent deprotection to produce **62**.

A related synthesis of **24** has been carried out from the di-*O*-acetyl derivative **63** (Scheme 7).[61,63] Selective deacetylation[64] of **63**[65] provided **64** in 95% yield. Protection of **64** with TBSCl followed by deacetylation with methanolic ammonia gave **65**. The resulting carbamate **66** (82%) was desilylated with aqueous HF to produce **67** (94%). This was subsequently converted to **24**, via **68**, whose selective benzylation gave **69** as a suitable glycosyl acceptor. Finally, hydrogenolysis of **24** provided (−)-allosamizoline (**8**).

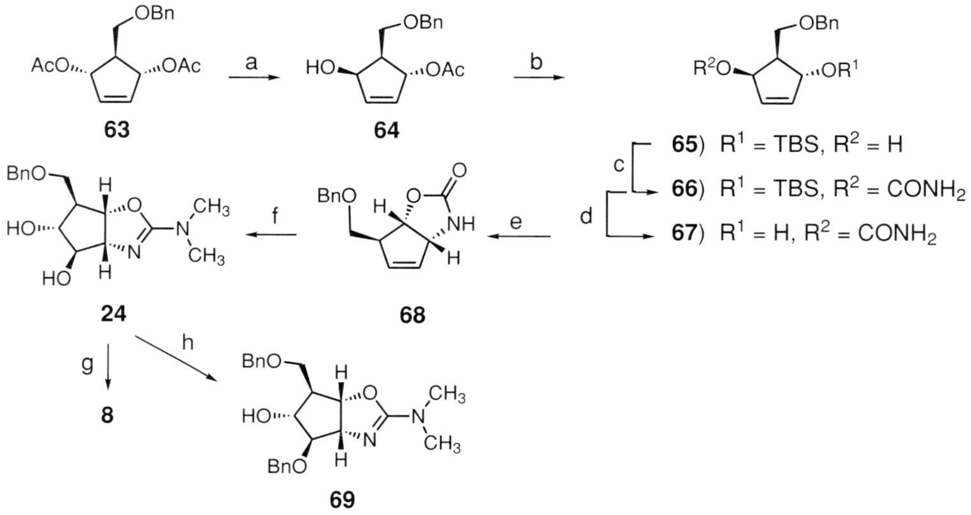

Scheme 7 (*a*) 1.45 M, NaH_2PO_4, buffer (pH 6.9), NaN_3, acetylcholinesterase, 6 days, 95%. (*b*) TBSCl, imidazole; NH_3, CH_3OH. (*c*) $ClCO_2Ph$, Py, NH_3, CH_3CN. (*d*) HF, CH_3CN. (*e*) TFAA, NEt_3, THF, −78°C to rt. (*f*) 1. CH_3OTf; 2. $(CH_3)_2NH$; 3. CF_3CO_3H, CF_3CO_2H; 4. TFA, H_2O, 38%. (*g*) H_2, 10% Pd *on* C, CH_3OH, AcOH, 84%. (*h*) Bu_2SnO, CH_3OH, reflux; BnBr, CsF, DMF, 35%.

For the synthesis of allosamidine (**2**), the required disaccharide as a glycosyl donor was prepared and coupled with the acceptor allosamizoline (Scheme 8). Thus, the glycosyl donor was prepared from the peracetylglucal **70** which was converted by Ferrier rearrangement[66,67] to **71**, whose deacetylation followed by treatment with benzaldehyde dimethylacetal afforded **72** in 70% yield. Treatment of **72** with 3,3-dimethyldioxirane in the presence of diethylamine underwent [2,3]-sigmatropic rearrangement to provide **73** in 96% yield. This was treated with [2-(trimethylsilyl)ethoxy]methyl chloride (SEMCl) to give **74** in quantitative yield. Debenzylidenation of **74** afforded the respective diol (99%), whose subsequent selective benzylation at C-6 via its stannylene derivative delivered **75** (69%). Coupling of **75** with bromosulfonamide **76** occurred under standard conditions to provide the disaccharide **77** (81%), which was treated with *N*,*N*-dibromobenzenesulfonamide to afford the bromosulfonamide **78** (57%). Glycosylation of **61** with **78** under basic conditions provided the pseudotrisaccharide **79** (42%). The SEM and benzylidene groups were cleaved by treatment with 5% HCl in methanol, followed by deblocking of the sulfonamide and benzyl groups and then acetylation to give the allosamidin heptaacetate (**80**) in 36% yield from **79**. Finally, cleavage of the acetyl esters with methanolic ammonia afforded **2**.

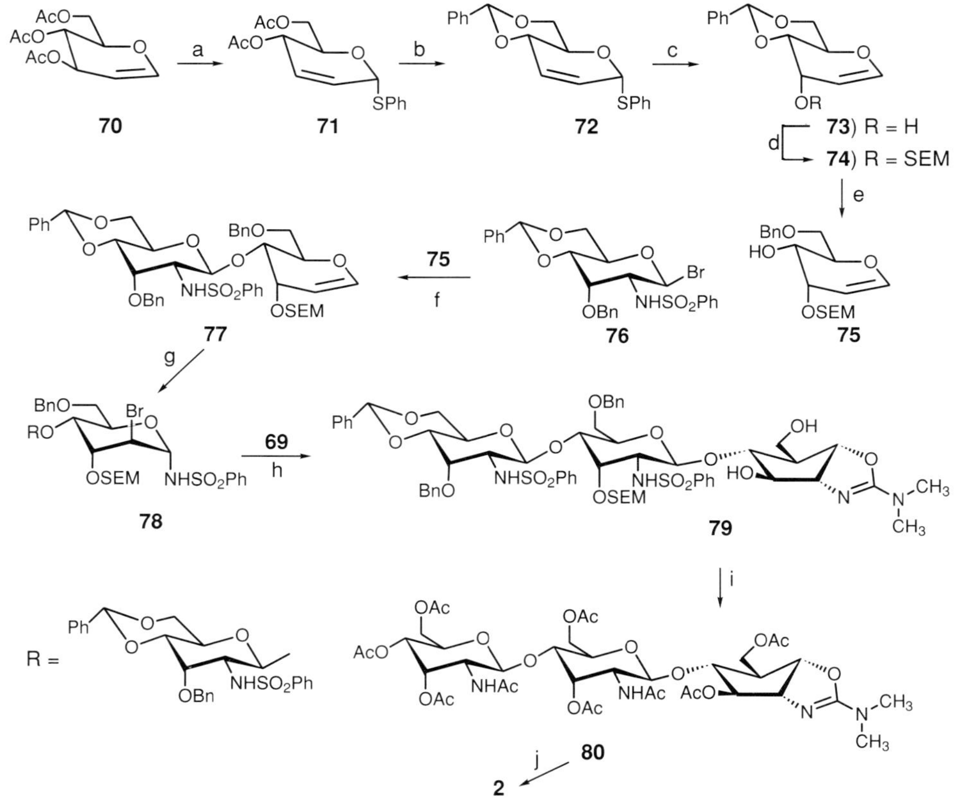

Scheme 8 (*a*) PhSH, $BF_3 \cdot OEt_2$. (*b*) 1. CH_3ONa, CH_3OH; 2. $PhCH(OCH_3)_2$, *p*-TsOH. (*c*) 1. Dimethyl dioxirane, Et_2NH, THF, 96%. (*d*) SEMCl, $(i\text{-Pr})_2NEt$. (*e*) 1. Na, NH_3, $-78°C$; 2. Bu_2SnO; BnBr, CsF. (*f*) KHMDS, DMF, $-40°C$ to rt. (*g*) $PhSO_2NBr_2$, NH_4I, EtOH, 57%. (*h*) KHMDS, DMF, $-40°C$ to rt, 42%. (*i*) 1. HCl, CH_3OH; 2. Na, NH_3, $-78°C$; 3. Ac_2O, Py, 36%. (*j*) NH_3, CH_3OH, 79%.

The glycosyl donor was also prepared from allyl 2-acetamido-4,6-*O*-benzylidene-2-deoxy-β-D-glucopyranoside (**81**)[68] whose mesylation gave compound **82**, which on solvolysis in wet 2-methoxyethanol gave the 2-acetamido-2-deoxy-D-allose derivative **85** in 91% yield. Benzylation of **85** afforded **86**, which on N-deacetylation with potassium hydroxide in methanol followed by N-phthaloylation gave the respective glycoside **87** in 71% yield. Removal of the allyl group, via its isomerization, gave **88**, which was further converted to the glycosyl donor **89**. Alternatively, the triflate **84**, derived from 2-phthalimidoglucoside **83**,[69] underwent displacement of the triflate group with inversion of configuration to give the respective alloside **91** in 62% yield. This approach avoided the time-consuming

N-deacetylation of compound **86** and gave access to both the glycosyl donor **89** and glycosyl acceptor **92,** which was otherwise obtained by N-deacetylation of **86** with potassium hydroxide followed by N-phthaloylation to give the fully substituted product **90** (61%). Subsequent selective reductive ring opening of the benzylidene acetal gave **92**. Coupling of the acceptor **92** with the trichloroacetimidate **89** gave the allyl glycoside **93** (85%), which on deallylation gave **94**. Compound **94** was converted to the trichloroacetimidate **95**, a disaccharide glycosylating agent (Scheme 9).[44]

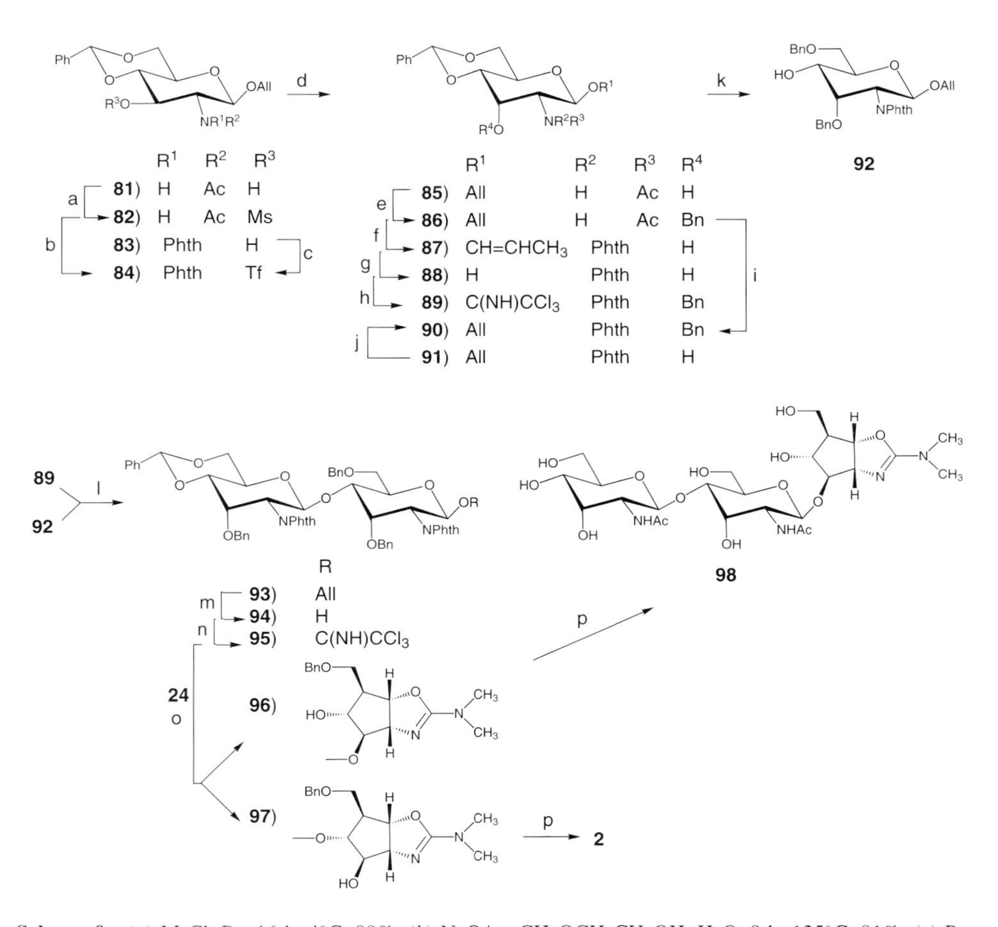

Scheme 9 (*a*) MsCl, Py, 16 h, 4°C, 89%. (*b*) NaOAc, $CH_3OCH_2CH_2OH$, H_2O, 9 h, 125°C, 91%. (*c*) Py, CH_2Cl_2, −30°C, Tf_2O, rt, 30 min, **84** (84%). (*d*) MsCl, Py, 16 h, 4°C, 89%; NaOAc, $CH_3OCH_2CH_2OH$, H_2O, 9 h, 125°C, **85** (91%). (*e*) DMF, BnBr, BaO, $Ba(OH)_2{\cdot}8H_2O$, 20°C, 16 h, $NaHCO_3$, **86** (94%). (*f*) KOH, CH_3OH, sealed tube, 125°C, 48 h; then phthalic anhydride, 20°C, 20 min; then Py, Ac_2O, 5 h, 125°C, **87** (71%). (*g*) Acetone, H_2O, $HgCl_2$, 20°C, 3 h, **88** (91%). (*h*) **88** to **89**, Cl_3CCN, CH_2Cl_2, MS 4 Å, DBU, 0°C, 15 min; **84** to **91**, BuNCN, DMF, 55°C, 3 h, **91** (62%). (*i*) KOH, CH_3OH, H_2O, sealed tube, 125°C, 5 days, **90** (61%). (*j*) DMF, NaH, BnBr, rt, 24 h, **90** (82%). (*k*) THF, sodium cyanoboranuide, MS 4 Å, HCl, ether, 30 min at 0°C, **92** (78%). (*l*) CH_2Cl_2, MS 4 Å, −30°C, TMSOTf, 15 min, **93** (85% from **88**). (*m*) DBU, EtOH–benzene–H_2O (7:3:1), tris(triphenylphosphine)rhodium(I) chloride, reflux, 24 h; then $HgCl_2$, acetone, H_2O, **94** (73%). (*n*) Cl_3CCN, CH_2Cl_2, 0°C, MS 4 Å, DBU, CH_2Cl_2, 15 min, **95** (84%). (*o*) 1. CH_2Cl_2, MS 4 Å, 0°C, TMSOTf, 30 min, at 20°C, 68%, **96**:**97** 5:1; 2. CH_3NH_2, EtOH, rt, 48 h; then Ac_2O, CH_3OH, −10°C, 16 h, NEt_3, 79%. (*p*) H_2, 10% Pd *on* C, 48 h, 93%.

Coupling of **95** with **24** gave the β-linked products **96** and **97** in the ratio 5:1 in 68% yield. Dephthaloylation of compounds **96** and **97** was effected by use of aqueous methylamine. The resulting mixture of diamines was acetylated and subsequently deprotected to give allosamidin **2** and its isomeric product **98**.

The glycosyl donor was also prepared from the D-allosamine derivative **99** (Scheme 10).[45,70,71] Deacetylation of **99** gave **100**,[72] which was treated with phthalic anhydride in the presence of NEt_3 to afford the phthalamide **101** (95%), which was benzylated to give the 3-*O*-benzylphthalimide **103** (52%) and the 3-*O*-benzylphthalamide **102** (44%). Hydrolysis of the benzyloxycarbonyl group of **102** followed by dehydration yielded 75% of **103**. Reductive ring opening[73–75] of the benzylidene group in **103** gave 84% of **107**, accompanied by only 5% of the regioisomer **108**, while reductive opening with $NaBH_3CN$ reagent[76] yielded only 59% of **107** and 30% of **108**. Removal of the allyloxy group from **103**

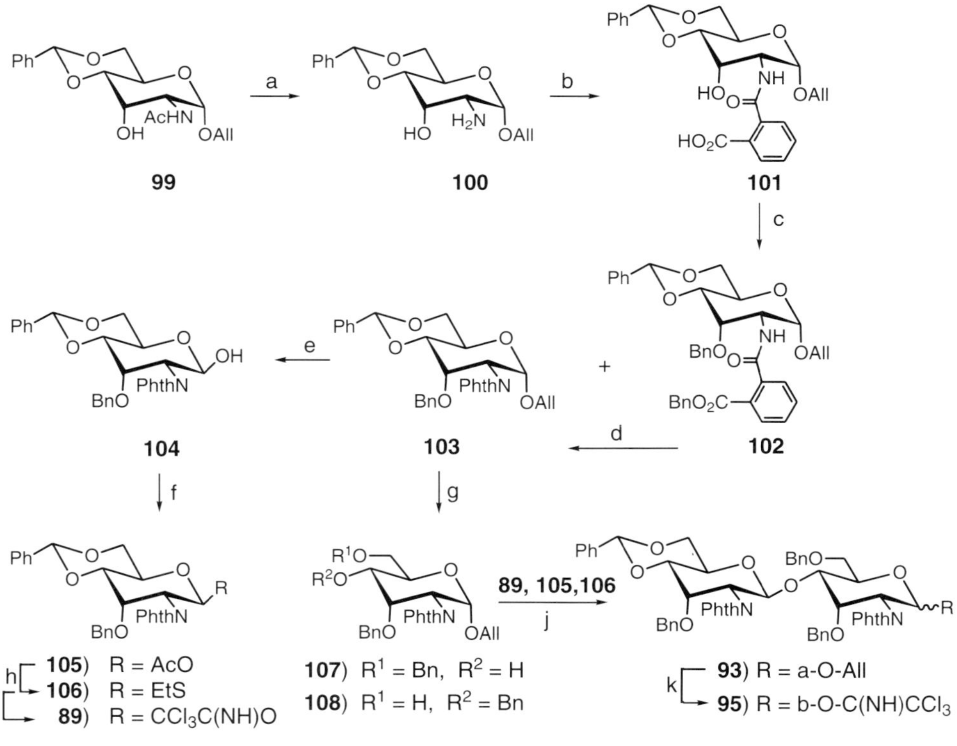

Scheme 10 (*a*) 1 M NaOH, 110°C, 6 days, 98%. (*b*) Phthalic anhydride, NEt_3, CH_3OH, 30 min, rt, 95%. (*c*) BnBr, NaH, DMF, 24 h, rt, **103** (52%), **102** (44%). (*d*) 1. 1 M NaOH, dioxane, 5 h, rt; 2. Py, Ac_2O, 48 h, rt, 75%. (*e*) 1. (Cycloocta-1,5-diene)bis(methyldiphenylphosphine)iridium hexafluorophosphate, H_2, THF, 3 h, rt; 2. HgO, $HgCl_2$, acetone–H_2O (9:1), 1 h, rt, 76%. (*f*) Py, Ac_2O, 12 h, rt, 97%. (*g*) 1. $NaBH_3CN$, THF, 2 h, 0°C; 2. HCl soln. in Et_2O, **107** (59%), **108** (30%); *or* $(CH_3)_3NBH_3$, $AlCl_3$, THF, 14 h, **107** (84%), **108** (5%). (*h*) $(CH_3)_3SiSEt$, TMSOTf, CH_2Cl_2, 12 h, rt, 51%. (*i*) CCl_3CN, K_2CO_3, CH_2Cl_2, 6 h, 77%. (*j*) $(CH_3)_3NBH_3$, CH_2Cl_2, 0°C, 20 min, 80%. (*k*) 1. H_2, THF, rt, 3 h, (cycloocta-1,5-diene)bis(methyldiphenylphosphine)iridium hexafluorophosphate; 2. HgO, $HgCl_2$, acetone–H_2O (9:1), 1 h, rt, 75%; 3. CCl_3CN, K_2CO_3, CH_2Cl_2, 16 h, rt, 90%.

afforded **104** which was converted to the glycosyl donors **89**, **105** and **106**. The β-acetate **105** was obtained almost quantitatively and treated with $(CH_3)_3SiSEt$ and trimethylsilyl triflate to form the β-thioglycoside **106** (51%). Reaction of **104** with Cl_3CCN afforded the β-D-imidate **89** (77%).[77] The glycosidation of the acceptor **107** with the donors **89**, **105** and **106** gave in each case the expected β-configurated disaccharide **93**, besides elimination product. The acetate **105** gave the lowest yield of **93** but the best yield (80%) was obtained with the imidate **89**. Removal of the allyl protecting group from glycoside **93** followed by treatment with CCl_3CN afforded the glycosyl donor **95** (67.5%).

The disaccharide donor was also synthesized by enzymatic degradation of chitin followed by condensation with suitable allosamizoline acceptor (Scheme 11).[78] Enzymatic degradation of chitin using chitinase (EC 3.2.1.14, *Streptomyces griseus*) followed by acetylation afforded the peracetate **109**. Treatment of **109** with PhSTMS afforded **110** in 87% yield,

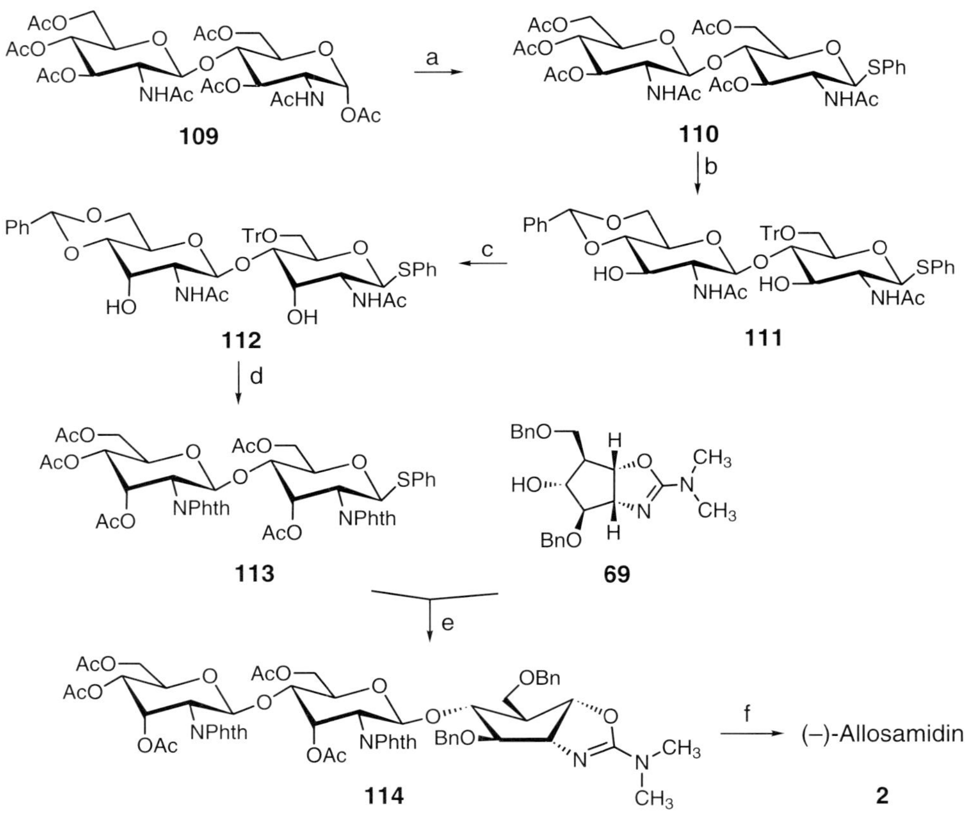

Scheme 11 (*a*) PhSTMS (4 equiv.), ZnI_2 (6–8 equiv.), dichloroethane, 50°C, 87%. (*b*) 1. $NaOCH_3$, CH_3OH, CH_2Cl_2; 2. $PhCH(OCH_3)_2$, *p*-TsCl, DMF; 3. TrCl, Py, DMF, 71% for three steps. (*c*) 1. MsCl, Py, 92%; 2. NaOAc, H_2O, 2-methoxyethanol, 82%. (*d*) 1. 80% AcOH; 2. 1 M NaOH; 3. phthalic anhydride, NEt_3, CH_3OH; 4. Ac_2O, Py, 41% for four steps. (*e*) 1. Benzyl 2,2,2-trichloroacetimidate, TfOH, CH_2Cl_2, hexane; 2. 1 M HCl, THF, 61% for two steps. (*f*) NBS, TfOH, MS 4 Å, CH_2Cl_2, 0°C, 40%. (*g*) 1. Aqueous CH_3NH_2, EtOH; 2. Ac_2O, CH_3OH; 3. H_2, 10% Pd *on* C, AcOH, CH_3OH, H_2O, 66%.

which underwent removal of the *O*-acetyl groups followed by successive benzylidenation and tritylation to produce **112** (71%). Mesylation of the free hydroxyl groups in **111**, followed by S_N2 displacement with inversion of configuration of the mesyloxy groups with sodium acetate afforded the allo-isomer **112** (75.5%), which underwent removal of the protecting groups followed by treatment with phthalic anhydride in methanol and subsequent acetylation to afford **113** (41%).

The donor **113** was coupled with the acceptor **69** to give the fully protected allosamidin derivative **114** (40%). Finally, removal of all protecting groups from **114** furnished (−)-allosamidin (**2**).

Glycosidation of the partially protected racemate **115**[27] by **95**, promoted by TMSOTf, afforded the four pseudotrisaccharides **116**, **117**, **118** and **119** in 61% overall yield and in a ratio of 40:44:9:7, and the main by-product was the aminoglycal. Hydrogenation of **119** over palladium on carbon afforded **120**. Dephthaloylation of **117** by treatment with excess of hydrazine hydrate followed by acetylation led to opening of the oxazoline ring to give **121** in 61%. Using only 2 equiv. of hydrazine hydrate and shorter reaction time led, after acetylation, to low yields of **121** and **123**. The latter **123** was obtained (73%) by effecting the dephthaloylation with aqueous methylamine followed by acetylation. Hydrogenolysis of **124** under acidic conditions yielded allosamidin (**2**) in 95% yield (Scheme 12).[45,70,71]

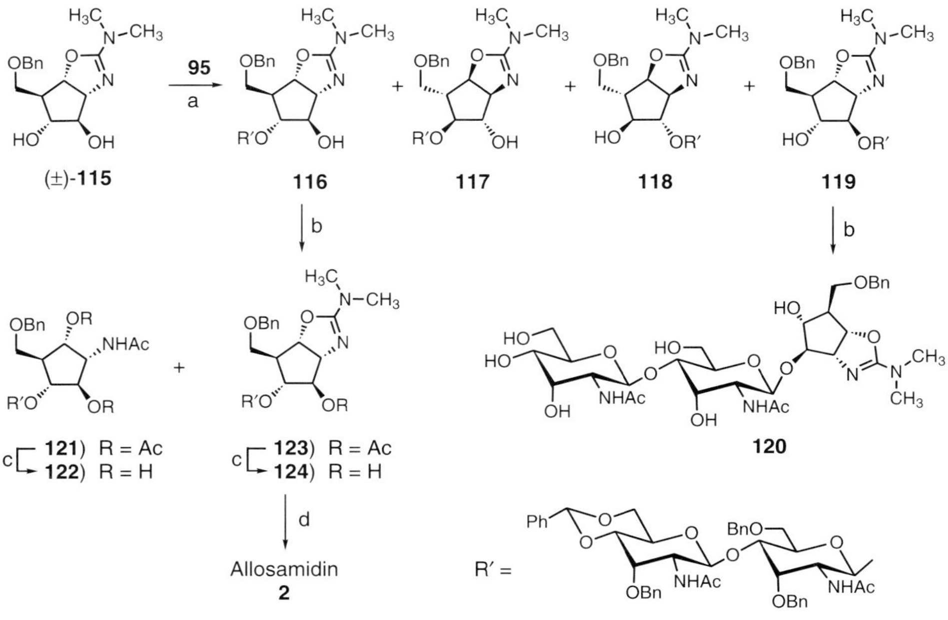

Scheme 12 (*a*) TMSOTf, MS 4 Å, CH_2Cl_2, 0°C, 20 min, **116** (24.5%), **117** (27%), **118** and **120** (5.5 and 4.3%) (*or* vice versa). (*b*) 1. $NH_2NH_2{\cdot}H_2O$, EtOH, 4 h, reflux; 2. Py, Ac_2O, 4-$(Me_2N)C_5H_4N$, 10 h, rt, **121** (61%); *or* 1. $NH_2NH_2{\cdot}H_2O$, EtOH, 45 min, reflux; 2. Py, Ac_2O, 4-$(Me_2N)C_5H_4N$, 10 h, rt, **121** (12%) and **123** (17%); *or* 1. 40% aqueous CH_3NH_2, CH_3OH, 48 h, rt; 2. Py, Ac_2O, 4-$(Me_2N)C_5H_4N$, 10 h, rt, **123** (73%). (*c*) $NaOCH_3$, CH_3OH, 10 h, rt, 97%. (*d*) H_2 (7 bar), 10% Pd *on* C, CH_3OH, AcOH, 36 h, rt, 95%.

References

1. Pariser, E.R. *Chitin Source Book: A Guide to the Research Literature*; Wiley: New York, 1989.
2. Flach, J. Jolles, P.-E. *Experientia* **48** (1992) 701.
3. Kramer, K.J.; Koga, D. *Insect Biochem.* **16** (1986) 851.
4. Anderson, S.D. *Annu. Rev. Entomol.* **24** (1979) 29.
5. Pont Lezica, R.; Quesada-Allue, L. *Methods Plant Biochem.* **2** (1990) 443.
6. Barrett-Bee, K.; Hamilton, M.J. *J. Gen. Microbiol.* **130** (1984) 1857.
7. Bartnichi-Garcia, S. *Annu. Rev. Microbiol.* **22** (1968) 87.
8. Knorr, D. *Food Technol.* **45** (1991) 114.
9. Muzzarelli, R.A.A. In *The Insect Itegument*; Hepburn, H.R., Ed.; Elsevier: New York, 1976.
10. Cabib, E.; Subrlati, A.; Bowers, B.; Silverman, S.J. *J. Cell. Biol.* **108** (1989) 1665.
11. Kato, T.; Shizuri, Y.; Izumida, H.; Yokoyama, A.; Endo, M. *Tetrahedron Lett.* **36** (1995) 2133.
12. Sakuda, S.; Isogai, A.; Matsumoto, S.; Suzuki, A. *J. Antibiot.* **40** (1987) 296.
13. Nishimoto, Y.; Sakuda, S.; Takayama, S.; Yamada, Y. *J. Antibiot.* **44** (1991) 716.
14. Somers, P.J.B.; Yao, R.C.; Doolin, L.E.; McGowan, M.J.; Fukuda, D.S.; Mynders, J.S. *J. Antibiot.* **40** (1987) 1751.
15. Koga, D.; Isogai, A.; Sakuda, S.; Matsumoto, S.; Suzuki, A.; Kimura, S.; Ide, A. *Agric. Biol. Chem.* **51** (1987) 471.
16. Kramer, K.J.; Dziadik-Turner, C.; Koga, D. In *Comprehensive Insect Physiology Biochemistry and Pharmacology*, Vol. 3; Kerkut, G.A., Gilbert, L.I., Eds.; Pergamon Press: Oxford, 1985; p. 75.
17. Sakuda, S.; Isogai, A.; Matsumoto, S.; Suzuki, A.; Koseki, K. *Tetrahedron Lett.* **27** (1986) 2475.
18. Sakuda, S.; Isogai, A.; Makita, T.; Matsumoto, S.; Koseki, K.; Kodama, H.; Suzuki, A. *Agric. Biol. Chem.* **51** (1987) 3251.
19. Sakuda, S.; Isogai, A.; Matsumoto, S.; Suzuki, A.; Koseki, K.; Kodama, H.; Yamada, Y. *Agric. Biol. Chem.*, **52** (1988) 1615.
20. Isogai, A.; Sato, M.; Sakuda, S.; Nakayama, J.; Suzuki, A. *Agric. Biol. Chem.* **53** (1989) 2825.
21. Sakuda, S.; Zhou, Z.-Y.; Takao, H.; Yamada, Y. *Tetrahedron Lett.* **37** (1996) 5711.
22. Sakuda, S.; Sugiyama, Y.; Zhou, Z.-Y.; Takao, H.; Ikeda, H.; Kakinuma, K.; Yamada, Y.; Nagasawa, H. *J. Org. Chem.* **66** (2001) 3356.
23. Knapp, S.; Kirk, B.A.; Vocadlo, D.; Withers, S.G. *Synlett* (1997) 435.
24. Elliott, R.P.; Fleet, G.W.J.; Pearce, L.; Smith, C.; Watkin, D.J. *Tetrahedron Lett.* **32** (1991) 6227.
25. Corbett, D.F.; Dean, D.K.; Robinson, S.R. *Tetrahedron Lett.* **35** (1994) 459.
26. Corbett, D.F.; Dean, D.K.; Robinson, S.R. *Tetrahedron Lett.* **34** (1993) 1525.
27. Trost, B.M.; Van Vranken, D.L. *J. Am. Chem. Soc.* **112** (1990) 1261.
28. Ling, R.; Mariano, P.S. *J. Org. Chem.* **63** (1998) 6072.
29. Li, J.; Lang, F.; Ganem, B. *J. Org. Chem.* **63** (1998) 3403.
30. Lu, H.; Mariano, P.S.; Lam, Y.-F. *Tetrahedron Lett.* **42** (2001) 4755.
31. Wakabayashi, T.; Saito, H.; Shiozaki, M. *Tetrahedron: Asymmetry* **11** (2000) 2083.
32. Mehta, G.; Mohal, N. *Tetrahedron Lett.* **42** (2001) 4227.
33. Terayama, H.; Kuzuhara, H.; Takahashi, S.; Sakuda, Y. *Biosci. Biotech. Biochem.* **57** (1993) 2067.
34. Takahashi, S.; Terayama, H.; Kuzuhara, H. *Tetrahedron Lett.* **35** (1994) 4149.
35. Goering, B.K.; Ganem, B. *Tetrahedron Lett.* **35** (1994) 6997.
36. Takahashi, S.; Terayama, H.; Kuzuhara, H.; Sakuda, S.; Yamada, Y. *Biosci. Biotech. Biochem.* **58** (1994) 2301.
37. Blattner, R.; Furneaux, R.H.; Lynch, G.P. *Carbohydr. Res.* **294** (1996) 29.
38. Berecibar, A.; Grandjean, C.; Siriwardena, A. *Chem. Rev.* **99** (1999) 779.
39. Arsequell, G.; Valencia, G. *Tetrahedron: Asymmetry* **10** (1999) 3045.
40. Kassab, D.J.; Ganem, B. *J. Org. Chem.* **64** (1999) 1782.
41. Trost, B.M.; Van Vranken, D.L. *J. Am. Chem. Soc.* **115** (1993) 444.
42. Maezaki, N.; Sakamoto, A.; Tanaka, T.; Iwata, C. *Tetrahedron: Asymmetry* **9** (1998) 179.
43. Felpin, F.-X.; Lebreton, J. *Tetrahedron Lett.* **43** (2002) 225.
44. Blattner, R.; Furneaux, R.H.; Kemmitt, T.; Tyler, P.C.; Ferrier, R.J.; Tiden, A.-K. *J. Chem. Soc., Perkin Trans. 1* (1994) 3411.
45. Maloisel, J.-L.; Vasella, A.; Trost, B.M.; Van Vanken, D.L. *Helv. Chim. Acta* **75** (1992) 1515.
46. Ferrier, R.J. *J. Chem. Soc., Perkin Trans. 1* (1979) 1455.
47. Takahashi, S.; Terayama, H.; Kuzuhara, H. *Tetrahedron Lett.* **32** (1991) 5123.
48. Ikeda, D.; Tsuchiya, T.; Umezawa, S. *Bull. Chem. Soc. Jpn.* **44** (1971) 2529.
49. Barton, D.H.; Augy-Dorey, S.; Camara, J.; Dalko, P.; Delaumeny, J.M.; Gero, S.D.; Quiclet-Sire, B.; Stutz, P. *Tetrahedron* **46** (1990) 215.

50. Simpkins, N.S; Stokes, S.; Whittle, A.J. *Tetrahedron Lett.* **33** (1992) 793.
51. Simpkins, N.S; Stokes, S.; Whittle, A.J. *J. Chem. Soc., Perkin Trans. 1* (1992) 2471.
52. Mikamo, M. *Carbohydr. Res.* **191** (1989) 150.
53. Chargoff, E.; Bovarnick, M. *J. Biol. Chem.* **118** (1937) 422.
54. Corey, E.J.; Pyne, S.G. *Tetrahedron Lett.* **24** (1983) 2821.
55. Nakata, M.; Akazawa, S.; Kitamura, S.; Tatsuta, K. *Tetrahedron Lett.* **32** (1991) 5363.
56. Ovchinnikov, M.V.; Bairamova, N.E.; Bakinovskii, L.V.; Kochetkov, N.K. *Bioorg. Khim.* **9** (1983) 401.
57. Bairamova, N.E.; Bakinovskii, L.V.; Kochetkov, N.K. *Izv. Akad. Nauk SSSR, Ser. Khim.* (1985) 1140.
58. Schwartz, D.A.; Lee, H.-H.; Carver, J.P.; Krepinsky, J.J. *Can. J. Chem.* **63** (1985) 1073.
59. Bundle, D.R.; Josephson, S. *Can. J. Chem.* **58** (1980) 2679.
60. Kitahara, T.; Suzuki, N.; Koseki, K.; Mori, K. *Biosci. Biotech. Biochem.* **57** (1993) 1906.
61. Griffith, D.A.; Danishefsky, S.J. *J. Am. Chem. Soc.* **113** (1991) 5863.
62. Shrader, W.D.; Imperiali, B. *Tetrahedron Lett.* **37** (1996) 599.
63. Griffith, D.A.; Danishefsky, S.J. *J. Am. Chem. Soc.* **118** (1996) 9526.
64. Deardroff, D.R.; Matthews, A.J.; McMeekin, D.S.; Craney, C.L. *Tetrahedron Lett.* **27** (1986) 1255.
65. Madhavan, G.V.; Martin, J.C. *J. Org. Chem.* **51** (1986) 1287.
66. Ferrier, R.J.; Prasadi, N. *J. Chem. Soc. C* (1969) 570.
67. Vaverde, S.; Garcia-Ochoa, S.; Martin-Lomas, M. *J. Chem. Soc., Chem. Commun.* (1987) 383.
68. Rollin, P.; Sinaÿ, P. *J. Chem. Soc., Perkin Trans. 1* (1977) 2513.
69. El-Sokkary, R.I.; Silwanis, B.A.; Nashed, M.A.; Paulsen, H. *Carbohydr. Res.* **203** (1990) 319.
70. Maloisel, J.-L.; Vasella, A.; Trost, B.M.; Van Vranken, D.L. *J. Chem. Soc., Chem Commun.* (1991) 1099.
71. Maloisel, J.-L.; Vasella, A. *Helv. Chim. Acta* **75** (1992) 1491.
72. Vasella, A.; Witzig, C.; Husi, R. *Helv. Chim. Acta* **74** (1991) 1362.
73. Liptak, A.; Jodal, I.; Nanasi, P. *Carbohydr. Res.* **44** (1975) 1.
74. Mikami, T.; Asano, H.; Mitsunobu, O. *Chem. Lett.* (1987) 2033.
75. Ek, M.; Garegg, P.J.; Hultberg, H.; Oscarson, S. *J. Carbohydr. Chem.* **2** (1983) 305.
76. Garegg, P.J.; Hultberg, H.; Wallin, S. *Carbohydr. Res.* **108** (1982) 97.
77. Schmidt, R.R.; Michel, J.; Roos, M. *Liebigs Ann. Chem.* (1984) 1343.
78. Takahashi, S.; Terayama, H.; Kuzuhara, H. *Tetrahedron Lett.* **33** (1992) 7565.

5.3.4 *(+)-Biotin*

D-(+)-Biotin (**1**), a biocatalyst of reversible metabolic reactions of carbon dioxide transport in organisms, is one of the water-soluble B-complex group of vitamins[1] and has immense commercial importance in poultry feeds and animal nutrition. Compound **1** was isolated from egg yolk,[2] liver and milk concentrates.[3,4] It is an important vitamin for human nutrition and animal health.[5–8] Its structure was determined[9–11] and confirmed by the first total synthesis.[12] Its absolute configuration by X-ray crystallographic analysis[13] was established. Syntheses of biotin from noncarbohydrate and its analogues from carbohydrate and noncarbohydrate have been reported.[14–44] Syntheses from carbohydrate precursors are discussed in this part.

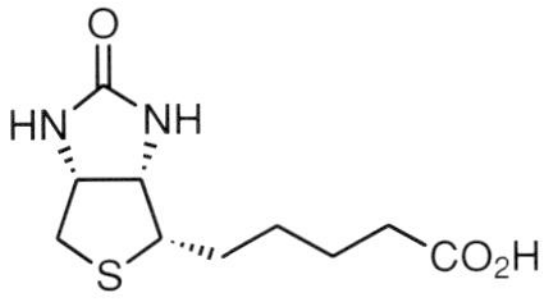

D-(+)-Biotin
1

5.3.4.1 *Synthesis from D-glucose* D-(+)-Biotin (**1**) has been synthesized by conversion of D-glucose into epoxide **2**, followed by treatment with NaN_3 to give **3** (Scheme 1).[45] Mesylation of **3** gave **4**, which was converted into **5** using Ac_2O and $BF_3{\cdot}Et_2O$. Deacetylation of **5** followed by sodium borohydride reduction afforded **6**, which was acetonated and then treated with NaN_3 to produce the diazide **7**. Hydrogenation of **7** followed by imidazolidinone ring formation using $COCl_2$ afforded **8**, which underwent acetylation and removal of the isopropylidene group to afford **9**. Compound **9** was converted in four steps into ester **10**, which was mesylated to give **11**. Treatment of **11** with Na_2S afforded **1**.

5.3.4.2 *Synthesis from D-glucosamine* A shorter sequence than that mentioned above to synthesize the intermediate **10** has been reported from D-glucosamine (Scheme 2).[46] D-Glucosamine (**12**) was converted into **13**, which underwent mesylation followed by treatment with NaN_3 to give the azide **14**. Hydrogenation of **14** gave the amine **15**, which was treated with sodium hydride followed by removal of the isopropylidene group to produce the imidazolidinone **16**. Periodate oxidation of **16** followed by olefination, hydrogenation and then reduction gave **10**.

5.3.4.3 *Synthesis from D-mannose* A total synthesis of D-(+)-biotin (**1**) from D-mannose has also been reported (Scheme 3).[47] Treatment of 2,3:5,6-di-*O*-isopropylidene-α-D-mannofuranose (**17**) with benzoyl chloride followed by selective hydrolysis of the terminal acetal group afforded 1-*O*-benzoyl-2,3-*O*-isopropylidene-α-D-mannofuranose (**18**). Subsequent periodate oxidation and chain extension with the proper phosphorane followed by hydrogenation gave the uronate derivative **19**. Treatment of **19** with $NaOCH_3$ followed by reduction of the resulting aldehyde with sodium borohydride afforded compound **20**.

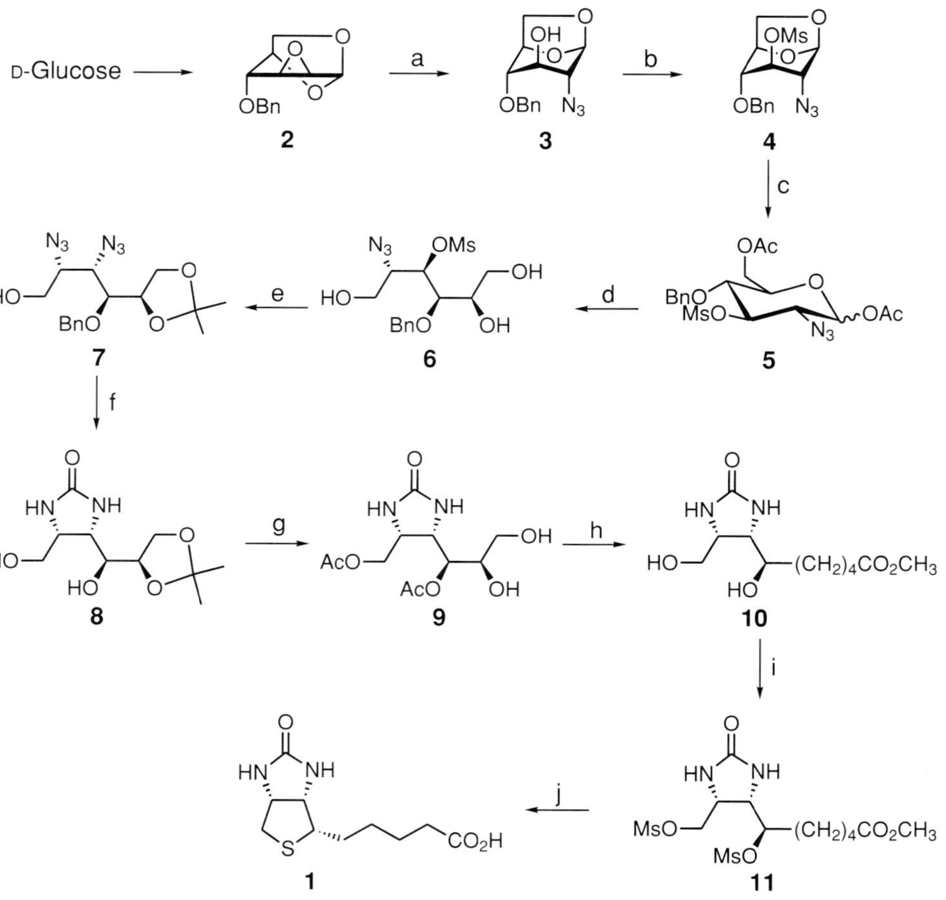

Scheme 1 (*a*) NaN_3, NH_4Cl, $CH_3OCH_2CH_2OH$, H_2O, 120°C, 85%. (*b*) MsCl, Py. (*c*) $BF_3{\cdot}Et_2O$, Ac_2O. (*d*) 1. HCl, CH_3OH; 2. $NaBH_4$, $B(OH)_3$, EtOH. (*e*) 1. $(CH_3)_2C(OCH_3)_2$, DMF, *p*-TsOH; 2. NaN_3, DMF, 80°C. (*f*) 1. H_2, Lindlar, EtOH; 2. $COCl_2$, 45% for four steps. (*g*) Ac_2O, Py; 2. AcOH, H_2O, 70°C. (*h*) 1. $NaIO_4$, EtOH, H_2O; 2. $Ph_3P{=}CHCH{=}CHCO_2CH_3$, CH_2Cl_2; 3. H_2, Pd *on* C, CH_3OH; 4. CH_3ONa, CH_3OH, 40% for six steps. (*i*) MsCl, Py, −10°C. (*j*) 1. Na_2S, DMF, 100°C; 2. NaOH.

Mesylation of **20** followed by treatment with sodium sulfide in HMPA afforded the tetrahydrothiophene derivative **21**. Treatment of **21** with 90% formic acid followed by mesylation and subsequent treatment of the resulting dimesyl **22** with sodium azide afforded the diazido **23**. Hydrogenolysis of **23** in a mixture of methanol and acetic anhydride afforded **24**, which was treated with $Ba(OH)_2$ followed by phosgene to produce **1**.

5.3.4.4 *Synthesis from* D*-arabinose* Two researcher groups[48,49] have synthesized the intermediate **20** from D-arabinose (Scheme 4). The D-arabinose derivative **25** was benzoylated and hydrogenated to give the corresponding hemiacetal, which was then subjected to Wittig reaction to give **26** and **27**. Reduction of **26** followed by debenzoylation led to the diol

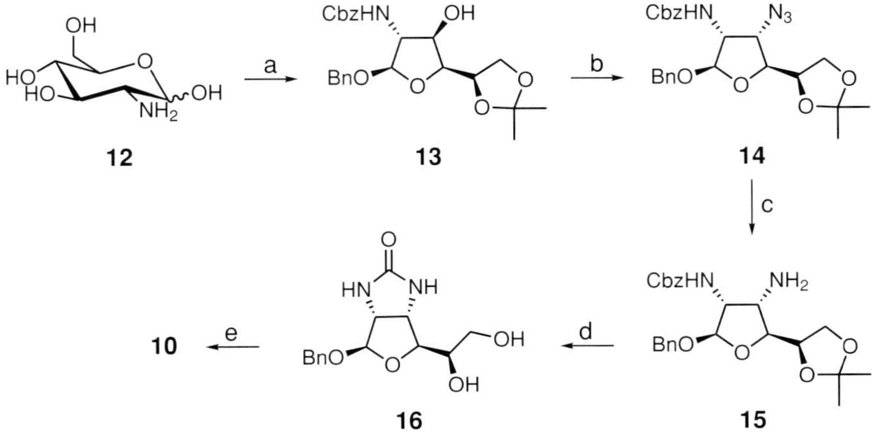

Scheme 2 (*a*) 1. CbzCl, $NaHCO_3$; 2. $(CH_3)_2C(OBn)_2$, *p*-TsOH, DMF, 120°C, 55%. (*b*) 1. *p*-TsCl, Py; 2. NaN_3, DMF. (*c*) H_2, Raney nickel, 61% for three steps. (*d*) 1. NaH, DMF; 2. HOAc, H_2O. (*e*) 1. $NaIO_4$, EtOH, H_2O; 2. $Ph_3P{=}CHCH{=}CHCO_2CH_3$, CH_2Cl_2; 3. H_2, Pd *on* C, CH_3OH; 4. $NaBH_4$, CH_3OH.

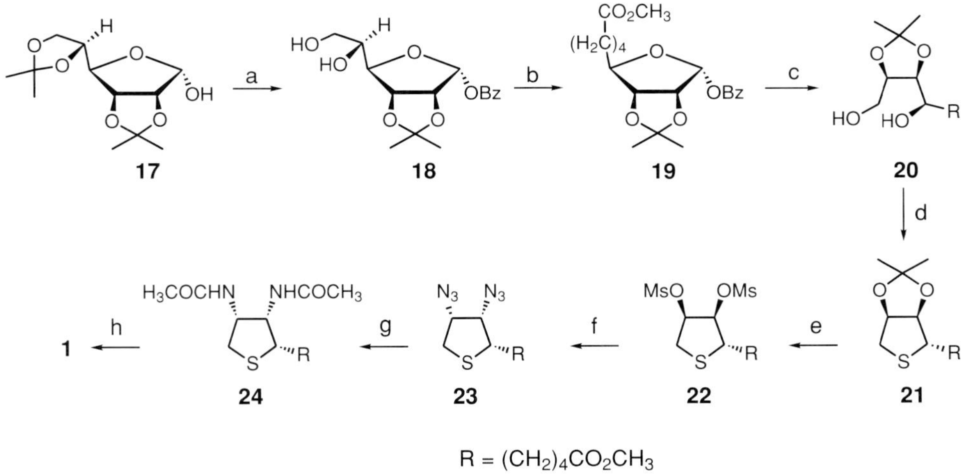

$R = (CH_2)_4CO_2CH_3$

Scheme 3 (*a*) 1. BzCl, Py, 97%; 2. 70% AcOH, 48 h, 20°C, 93%. (*b*) 1. $NaIO_4$, acetone, H_2O; 2. $Ph_3P{=}CHCH{=}CHCO_2CH_3$, CH_2Cl_2, 85%; 3. 10% Pd *on* C, H_2, CH_3OH, 97%. (*c*) 1. $NaOCH_3$, CH_3OH; 2. $NaBH_4$, 85%. (*d*) 1. MsCl, Py, 96%; 2. Na_2S, HMPA, 100°C, 2 h, 75%. (*e*) 1. 90% HCO_2H, 20°C for 15 min, 92%; 2. MsCl, 95%. (*f*) NaN_3, HMPA, 80°C, 7 h, 78%. (*g*) PtO_2, H_2, 3 h, CH_3OH, Ac_2O, 20°C, 60%. (*h*) 1. $Ba(OH)_2$, H_2O, 140°C, 14 h; 2. $COCl_2$, 87%.

intermediate **20**. This sequence suffered, however, from a low yield in the Wittig reaction because of the formation of **27** through intramolecular Michael addition of the primary hydroxyl group to the diene system of **26**. This problem was solved by using the 3,4-*O*-isopropylidene-D-arabinose derivative **28**, which was reacted with the Wittig reagent to give after subsequent hydrogenation the diol **20**.

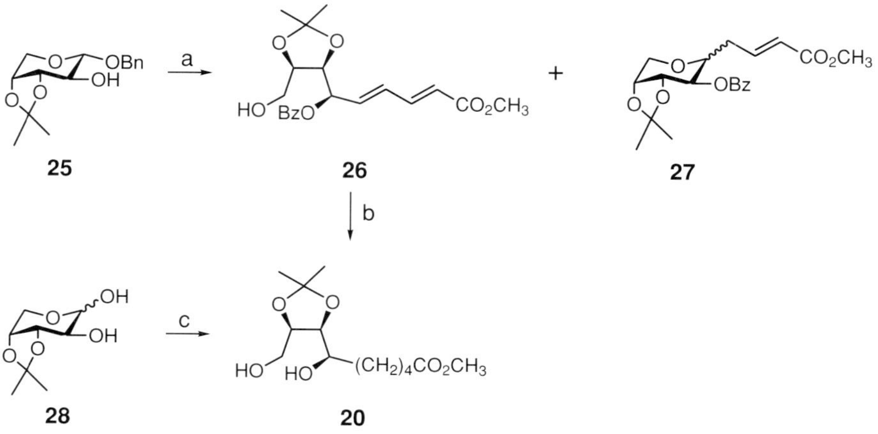

Scheme 4 (*a*) 1. BzCl, Py; 2. H_2, Pd *on* C, dioxane; 3. $Ph_3P{=}CHCH{=}CHCO_2CH_3$, CH_2Cl_2. (*b*) 1. H_2, Pd *on* C, CH_3OH; 2. $NaOCH_3$, CH_3OH, 65% for two steps. (*c*) 1. $Ph_3P{=}CHCH{=}CHCO_2CH_3$, CH_2Cl_2, BzOH; 2. H_2, Pd.

5.3.4.5 *Synthesis from* D*-glucuronolactone* Synthesis of **1** from D-glucurono-6,3-lactone (**29**) has been reported (Scheme 5).[50] Selective reduction of **29** gave L-gulono-1,4-lactone

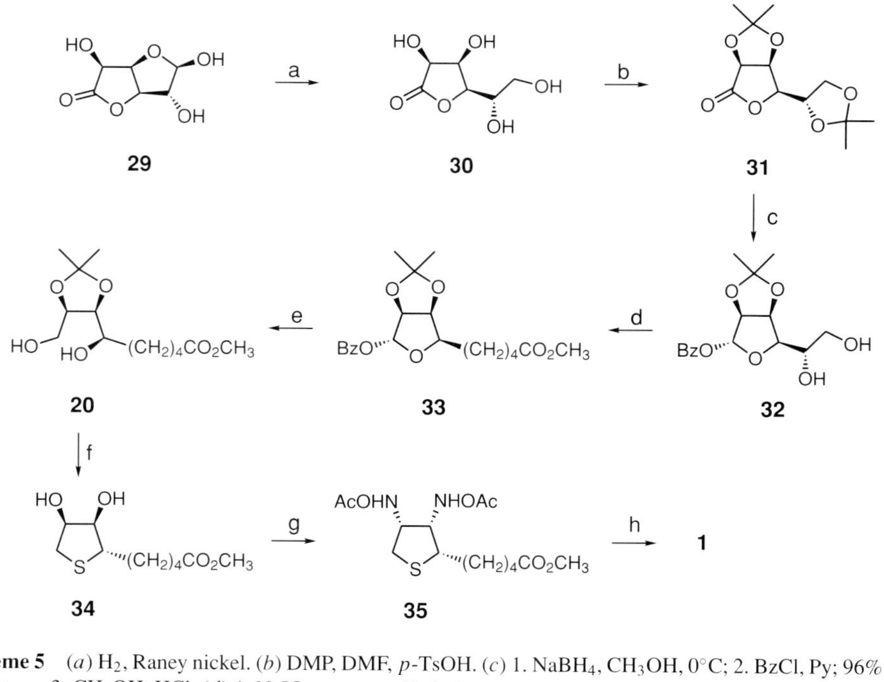

Scheme 5 (*a*) H_2, Raney nickel. (*b*) DMP, DMF, *p*-TsOH. (*c*) 1. $NaBH_4$, CH_3OH, 0°C; 2. BzCl, Py; 96% for two steps; 3. CH_3OH, HCl. (*d*) 1. $NaIO_4$, acetone, H_2O, 0°C; 2. $Ph_3P{=}CHCH{=}CHCO_2CH_3$, CH_2Cl_2; 3. H_2, Pd($NaBH_4$), CH_3OH, 79% for four steps. (*e*) 1. $NaOCH_3$, CH_3OH; 2. $NaBH_4$, CH_3OH, 85% for two steps. (*f*) 1. MsCl, Py; 2. Na_2S, HMPA, 100°C; 3. 90% HCO_2H, 20°C, 66% for three steps. (*g*) 1. MsCl, Py; 2. NaN_3, HMPA, 80°C; 3. PtO_2, CH_3OH, Ac_2O, 45% for three steps. (*h*) 1. $Ba(OH)_2$, H_2O, 140°C; 2. $COCl_2$, 87% for two steps.

(**30**), which was treated with DMP to give **31**. Partial reduction of lactone **31** followed by benzoylation and selective removal of the terminal isopropylidene group afforded the diol **32**. This was subjected to periodate oxidation, followed by Wittig reaction and then hydrogenation to give **33**. Debenzoylation followed by reduction of the resulting hemiacetal gave the diol intermediate **20**, which was converted to **1** via intermediates **34** and **35**.

References

1. Uskokoic, M.R. In *Biotin in Encyclopedia of Chemical Technology*, Vol. 24; Kirk, R.E., Othmer, D.E., Mistry, P.S., Eds.; Wiley: New York, 1984; p. 41.
2. Kogl, F.; Tonnis, B.; Hoppe-Seyl, Z. *Physiol. Chem.* **242** (1936) 43.
3. du Vigneaud, V.; Hofmann, K.; Melville, D.B.; Gyorgy, P. *J. Biol. Chem.* **140** (1941) 643.
4. Melville, D.B.; Hofmann, K.; Hague, E.; du Vigneaud, V. *J. Biol. Chem.* **142** (1942) 615.
5. Dakshinamurti, K. *Vitam. Horm.* **22** (1964) 1.
6. Coggeshall, C.J.; Heggers, P.J.; Robson, C.M.; Baker, H. *Ann. N. Y. Acad. Sci.* **447** (1985) 389.
7. Maebashi, M.; Makino, Y.; Furukawa, Y.; Ohinata, K.; Kimura, S.; Sato, T. *J. Clin. Biochem. Nutr.* **14** (1993) 211.
8. Wang, L.X.; Melean, L.G.; Seeberger, P.H.; Walker, G.C. *Carbohydr. Res.* **333** (2001) 73.
9. du Vigneaud, V.; Hofmann, K.; Melville, D.B.; Rachele, J.R. *J. Biol. Chem.* **140** (1941) 763.
10. du Vigneaud, V.; Melville, D.B.; Fokkers, K.; Wolf, D.E.; Mozingo, R.; Keresztesy, J.C.; Harris, S.A. *J. Biol. Chem.* **146** (1942) 475.
11. Melville, D.B.; Moyer, A.W.; Hofmann, K.; du Vigneaud, V. *J. Biol. Chem.* **146** (1942) 487.
12. Harris, J.A.; Wolf, D.E.; Mozingo, R.; Folkers, K. *Science* **97** (1943) 447.
13. Trotter, J.; Hamilton, Y.A. *Biochemistry* **5** (1966) 713.
14. Shimizu, T.; Seki, M. *Tetrahedron Lett.* **41** (2000) 5099.
15. Chen, F.; Peng, Z.; Shao, L.; Cheng, Y. *Yaoxue Xuebao* **34** (1999) 822.
16. Brieden, W.; Schroer, Eur. Pat. Appl. 827957A1, Mar. 11, 1998; *Chem. Abstr.* **128** (1998) 230190.
17. De Clercq, P.J. *Chem. Rev.* **97** (1997) 1765.
18. Msolenaer, M.J.; Speckamp, W.N.; Hiemstra, H.; Poetsch, E.; Casutt, M. *Angew. Chem., Int. Ed. Engl.* **34** (1995) 2391.
19. Casutt, M.; Poetsch, E.; Speckarnp, W.N. Ger. Offen. BE 3,926,690, 14 Feb. 1991; *Chem. Abstr.* **115** (1991) 8429k.
20. Deroose, F.D.; De Clercq, P.J. *J. Org. Chem.* **60** (1995) 321.
21. Deroose, F.D.; De Clercq, P.J. *Tetrahedron Lett.* **35** (1994) 265.
22. Fusijawa, T.; Nagai, M.; Kaike, Y.; Shimizu, M. *J. Org. Chem.* **59** (1994) 5865.
23. Goldberg, M.W.; Sternbach, L.H. U.S. Patent 2,489,235 and 2,489,238, 1985; *Chem. Abstr.* **46** (1951) l86a, 186g.
24. Aoki, S.; Suzuki, H.; Akiyama, H.; Akano, S. U.S. Patent 3,876,656, 1975; *Chem. Abstr.* **80** (1974) 95951z.
25. De Clercq, P.J.; Deroose, F.D. *Tetrahedron Lett.* **34** (1993) 4365.
26. Alcazar, V.; Tapia, I.; Morau, J.R. *Tetrahedron* **46** (1990) 1057.
27. Bihovaky, R.; Bodepudi, V. *Tetrahedron* **46** (1990) 7667.
28. Corey, E.J.; Mehrotra. M.M. *Tetrahedron Lett.* **29** (1988) 57.
29. Poetsch, E.; Casutt, M. *Chimia* **41** (1987) 148.
30. Lee, H.L.; Baggiolini, E.G.; Uskokivic, M.R. *Tetrahedron* **43** (1987) 4887.
31. Bates, H.A.; Rosenblum, S.R. *J. Org. Chem.* **51** (1986) 3447.
32. Bates, H.A.; Smilowitz, L; Lin, J. *J. Org. Chem.* **50** (1985) 899.
33. Bates, H.A.; Smilowitz, L; Rosenblum, S.B. *J. Chem. Soc., Chem. Commun.* (1985) 353.
34. Kinoshita, H.; Futagami, M.; Inomata, K; Kotabe, H. *Chem. Lett.* (1983) 1275.
35. Chavan, S.P.; Tejwani, R.B.; Ravindranathan, T. *J. Org. Chem.* **66** (2001) 6197.
36. Ohrui, H.; Kuzuhara, H.; Emoto, S. *Agric. Biol. Chem.* **35** (1971) 752.
37. Popsavin, V.; Benedekovic, G.; Popsavin, M.; Miljkovic, D. *Carbohydr. Res.* **337** (2002) 459.
38. Angus, D.I.; Kiefel, J.M.; von Itzstein, M. *Bioorg. Med. Chem.* **8** (2000) 2709.
39. Popsavin, V.; Benedekovic, G.; Popsavin, M. *Tetrahedron Lett.* **43** (2002) 2281.
40. Seki, M.; Hatsuda, M.; Mori, Y.; Yamada, S.-I. *Tetrahedron Lett.* **43** (2002) 3269.
41. Cosper, M.M.; Jameson, G.N.L.; Davydov, R.; Eidsness, M.K.; Hoffman, B.M.; Huynh, B.H.; Johnson, M.K. *J. Am. Chem. Soc.* **124** (2002) 14006.
42. Ugulava, N.B.; Surerus, K.K.; Jarrett, J.T. *J. Am. Chem. Soc.* **124** (2002) 9050.
43. Lo, K.K.-W.; Hui, W.-K.; Ng, D.C.-M. *J. Am. Chem. Soc.* **124** (2002) 9344.

44. Sun, X.-L.; Faucher, K.M.; Houston, M.; Grande, D.; Chailkof, E.L. *J. Am. Chem. Soc.* **124** (2002) 7258.
45. Ogawa, T.; Kawano, T.; Matsui, M. *Carbohydr. Res.* **57** (1977) C31.
46. Ohrui, H.; Sueda, N.; Emoto, S. *Agric. Biol. Chem.* **42** (1978) 865.
47. Ohrui, H.; Emoto, S. *Tetrahedron Lett.* (1975) 2765.
48. Vogel, F.G.M.; Paust, J.; Nürrenbach, A. *Liebigs Ann. Chem.* (1980) 1972.
49. Schmidt, R.R.; Maier, M. *Synthesis* (1982) 747.
50. Ravindranathan, T.; Hiremath, S.V.; Reddy, D.R.; Ramo Rao, A.V. *Carbohydr. Res.* **134** (1984) 332.

5.4 5:6-Fused heterocycles

5.4.1 *Hydroxylated indolizidines*

5.4.1.1 *Castanospermines* (+)-Castanospermine [(1*S*,6*S*,7*R*,8*R*,8a*R*)-1,6,7,8-tetrahydroxyindolizidine, **1**] was isolated from the seeds of the Australian legume *Castanospermum australe*[1] and the dried pod of *Alexa leiopetala*.[2] The analogues (+)-6-*epi*-castanospermine (**2**),[3] 6,7-di-*epi*-castanospermine (**3**)[4] and (+)-7-deoxy-6-*epi*-castanospermine (**4**)[5] were also isolated from *C. australe*. Castanospermine exhibits a potent competitive and reversible inhibition of several glucosidases.[3–17] Also, it has the potential for treating diabetes,[18,19] obesity,[20] cancer[11,19,21–25] and viral infections[26,27] including HIV-1,[28–36] as well as for the processing of oligosaccharide portions of influenza viral hemagglutinin.[37] The structures of **1** and **2** were studied by X-ray crystallography.[1,38]

Syntheses of castanospermine from noncarbohydrates and its unnatural analogues from both carbohydrates and noncarbohydrates have been reported.[39–66]

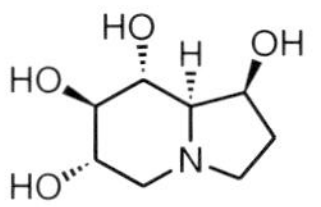

(+)-Castanospermine **1**

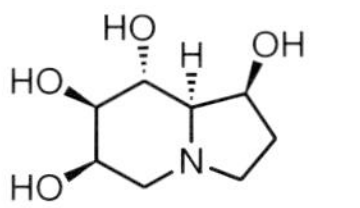

(+)-6-*epi*-Castanospermine **2**

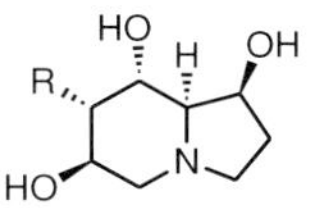

3) 6,7-Di-*epi*-castanospermine, R = OH

4) (+)-7-Deoxy-6-*epi*-castanospermine, R = H

5.4.1.1.1 Synthesis from D-*glucose* The first total synthesis of castanospermine (**1**) has established its absolute stereochemistry (Scheme 1).[67–69] Condensation of 2,3,4-tri-*O*-benzyl-D-glucopyranose (**5**)[70] with benzylamine afforded the respective glucosylamine as an anomeric mixture (77%), which was reduced with $LiAlH_4$ to afford **6**. The amine **6** was triflated and then subjected to epoxide formation to give **7** in 75% yield. Intramolecular cyclization of **7** afforded a mixture of piperidine **9** (45%) and azepane **8** (55%). Hydrogenation of **9** afforded (+)-deoxynojirimycin (**10**). Swern oxidation of **9** furnished the aldehyde **12** in 90% yield, which was condensed with lithio *tert*-butylacetate to give **11** as a 1:1 mixture of diastereomers. The less polar one was hydrogenolyzed, followed by treatment with acid to give **15**, which underwent reduction with DIBAL-H to give (+)-castanospermine (**1**). On the other hand, the aldehyde **12** underwent chelation-controlled Sakurai allylation using allyltrimethylsilane to afford **13**, with excellent stereocontrol. Ozonolysis of **13** followed by sodium borohydride reduction afforded the diol **14**, which was mesylated and subsequently hydrogenated to give **1**[69] in 55% yield from **13**.

A good stereocontrol was also exhibited in the Sakurai allylation of the manno analogue of **12**, whereby (+)-6-*epi*-castanospermine (**2**)[69] was synthesized in 42% yield.

Methyl α-D-glucopyranoside (**16**) has been used as a precursor for the synthesis of (+)-castanospermine (**1**) (Scheme 2).[71] The aldehyde **17**,[72] prepared from **16**, was allylated[73,74] to give a 9:1 ratio of epimers, which upon chromatographic separation and benzylation of the major product afforded **18**, which underwent ring opening to give **19**. Swern oxidation and

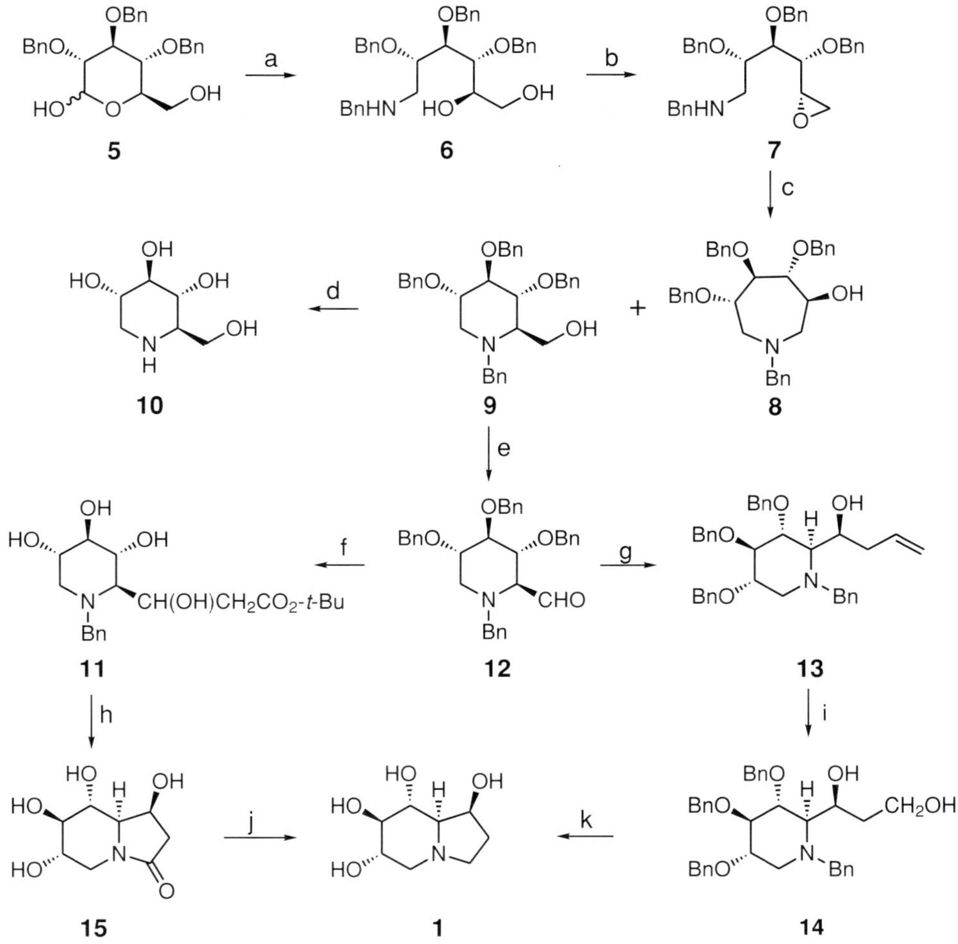

Scheme 1 (*a*) 1. $BnNH_2$, $CHCl_3$, 77%; 2. $LiAlH_4$, THF, reflux, 5 h. (*b*) 1. TFAA, 78%; 2. TBSCl, imidazole; 3. mesylation; 4. TBAF, THF; 5. CH_3ONa, CH_3OH, 75%. (*c*) $NaBH_4$, EtOH, 40°C; **8** (55%), **9** (45%). (*d*) Hydrogenolysis. (*e*) DMSO, $(COCl)_2$, CH_2Cl_2, −78°C; then NEt_3, 90%. (*f*) Lithio *tert*-butylacetate. (*g*) Allyltrimethylsilane, $TiCl_4$, CH_2Cl_2, −85°C, 15 h. (*h*) TFA, H_2O, 60°C, 3 h. (*i*) 1. O_3, CH_2Cl_2, −78°C; 2. $NaBH_4$, EtOH. (*j*) DIBAL-H. (*k*) 1. MsCl, NEt_3, CH_2Cl_2; 2. H_2, 10% Pd *on* C.

ozonolysis of **19** gave **20**, which upon hydrolysis of the acetal group afforded **21**. Treatment of **21** with ammonium formate and sodium cyanoborohydride led to the formation of the tetrabenzyl castanospermine **22** (53%), which underwent debenzylation to afford **1** in 22% overall yield from **17**.

D-Glucose serve as a precursor for the synthesis of **1** by conversion to the tetrabenzyl-gluconolactam **23**,[75–77] whose N-allylation under phase transfer catalysis gave **24** in 93% yield (Scheme 3).[78] Selective removal of the benzyl group of the primary position using ferric chloride and acetic anhydride and subsequent deacetylation followed by oxidation

16) R = H, R′ = CH_2OH
17) R = Bn, R′ = CHO

22) R = Bn
1) R = H

Scheme 2 (*a*) Ref. 72. (*b*) 1. Allyl bromide, Sn, CH_3CN, H_2O, ultrasound, 83%; 2. BnBr, NaH, *n*-Bu_4NI, DMF. (*c*) 1. IDCP, CH_2Cl_2, CH_3OH; 2. Zn, 95% EtOH, reflux, 74% for three steps. (*d*) 1. DMSO, $(COCl)_2$, CH_2Cl_2, −78°C; then NEt_3; 2. O_3, CH_2Cl_2, −78°C; then Ph_3P; 3. THF, 9 M HCl, 90% for three steps. (*e*) HCO_2NH_4, $NaCNBH_3$, CH_3OH, 53%. (*f*) 10% Pd *on* C, CH_3OH, HCO_2H, ~75%.

Scheme 3 (*a*) Refs. 75–77. (*b*) Allyl bromide, 50% aqueous KOH, CH_2Cl_2, TBAI, 93%. (*c*) 1. Ac_2O, $FeCl_3$; then NH_3, CH_3OH; 2. Dess–Martin periodane, 80%. (*d*) $Ph_3P{=}CHCO_2CH_3$, 85%. (*e*) $Cl_2Ru(PCy_3)_2CHCHCPh_2$, toluene, 110°C, 48 h, 70%. (*f*) 1. OsO_4, NMO, $SOCl_2$, TEA, 55%; 2. $NaIO_4$, $RuCl_3$, DCM, water, acetonitrile, 98%. (*g*) $NaBH_4$, DMAC; then 20% aqueous H_2SO_4, Et_2O, 98%. (*h*) 1. $BH_3{\cdot}DMS$; 2. H_2, Pd *on* C.

with periodane[79] afforded the aldehyde **25**, which was treated with $Ph_3P{=}CHCO_2CH_3$ to give **26**. Metathesis catalyzed cyclization of **26** afforded the bicyclic lactam **27**. Oxidation of **27** using OsO_4 with NMO in the presence of $SOCl_2$ followed by oxidation of the respective sulfites gave the sulfates **28** and **29** in a ratio of 1:5. The sulfate **29** was treated with sodium borohydride in dimethylacetamide, where the attack of the hydride ion takes place from the sterically less hindered side to yield a monosulfate whose acid hydrolysis afforded the lactam **30**. Reduction of **30** followed by removal of the protecting groups afforded (+)-castanospermine (**1**).

The olefin **31**,[80] obtained from D-glucose, was converted into the azido diene **32** in 56% overall yield. Intramolecular cycloaddition of **32** produced the indolizine **33**, which could serve as an intermediate for the synthesis of **1** (Scheme 4).[81]

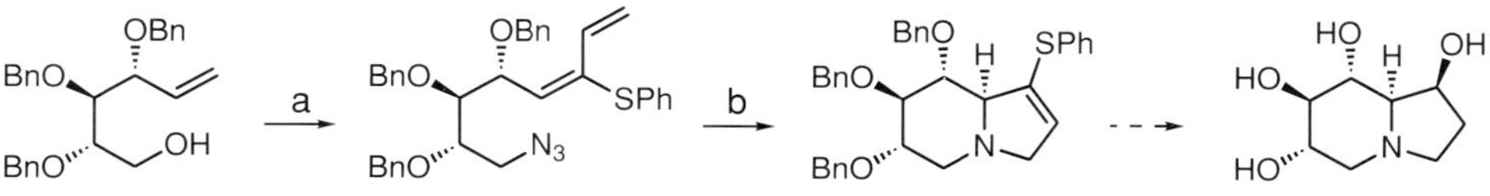

Scheme 4 (*a*) 1. $(PhO)_2PON_3$, $(NCO_2Et)_2$, Ph_3P, 84.8%; 2. O_3, DMS, CH_3OH, −78°C, 2 h, 0°C, 2 h, 90%; 3. BBN, 35°C, 2 h, THF, $TMSC(SPh)CHCH_2BBN$, 74%. (*b*) DMSO, 75°C, 108 h, 55%.

5.4.1.1.2 Synthesis from D-mannose A total synthesis of (+)-castanospermine (**1**) has been achieved starting with D-mannose (Scheme 5).[82] The diacetone **34**,[83] derived from D-mannose, was transformed to the aldehyde **35**, which was epimerized by treatment with K_2CO_3 in methanol to produce **36**. The aldehyde **36** was then converted into the corresponding oxime, followed by hydrogenolysis and protection of the resulting amine to give the carbobenzyloxy derivative **37**. Removal of the terminal isopropylidene and TBS groups followed by selective mesylation of the primary hydroxyl group afforded the monomesylate **38**. Treatment of **38** with sodium methoxide provided **39** (60%), whose oxidation with Collin's reagent gave the aldehyde **40**, which without isolation was reacted with *tert*-butyl lithioacetate to give **41** as a 1:1 mixture of epimers. Hydroxyl protection of **41**, and subsequent hydrogenolysis to the respective amine, followed by a double-cyclization reaction by boiling of the amine in methoxyethanol gave the epimeric indolizidones **42** and **43**. Reduction of **43** with $BH_3{\cdot}THF$ complex followed by treatment with 6 M HCl afforded **1**. Similarly, the other epimer **42** was converted into 1-*epi*-castanospermine (**44**).

5.4.1.1.3 Synthesis from xylose The xylose derivative **47**, obtained from 5,5-bis-benzyloxy-7-oxa-bicyclo[2.2.1]hept-2-ene (**45**), has been used in the synthesis of (+)-castanospermine (**1**) (Scheme 6).[84] Bromination of **45** occurred exclusively on the less hindered convex face of **45**, followed by stereoselective migration of the endo OBn group of the acetal to give **46**, which subsequently converted to **47**. Mesylation of **47** followed by cyclization with ammonia gave **48**, whose protection, hydrolysis, acetylation and cyclization by an intramolecular Wittig–Horner condensation gave **49**. Conversion of **49** into epoxide **50**

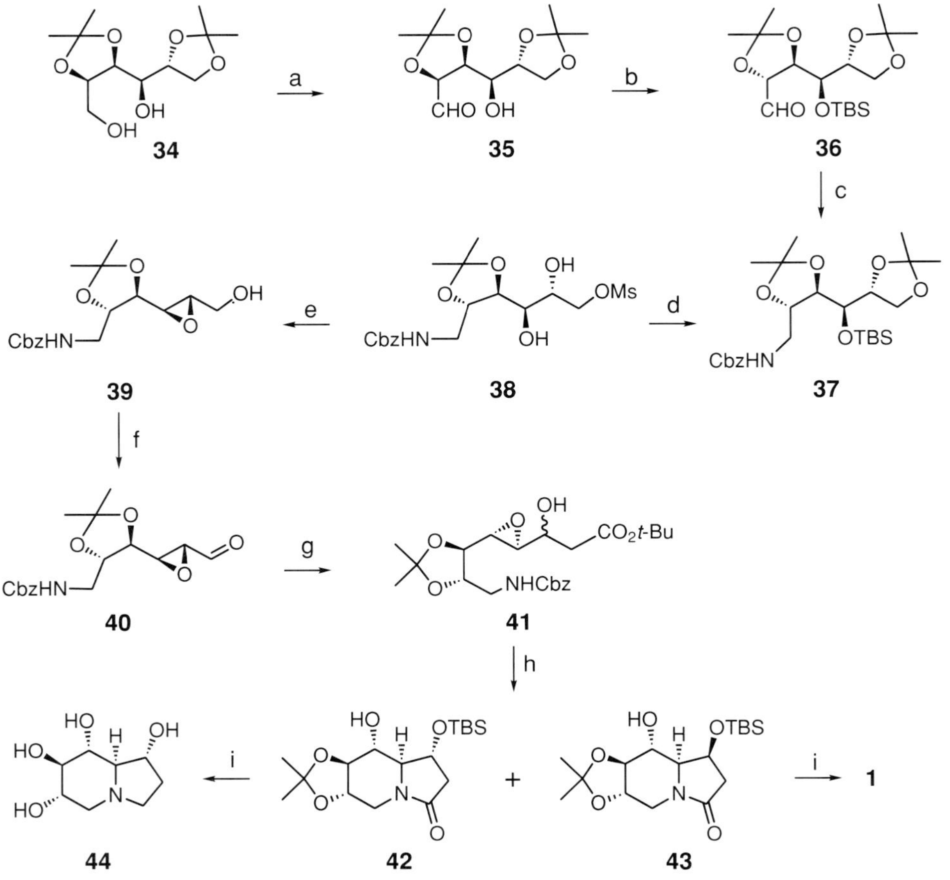

Scheme 5 (*a*) 1. BzCl, Py, rt; 2. TBSCl, imidazole, DMF, 80°C; 3. 1 N NaOH, CH_3OH, rt; 4. DMSO, DCC, TFA, Py, benzene, rt. (*b*) K_2CO_3, CH_3OH, rt. (*c*) 1. $HONH_2{\cdot}HCl$, $NaHCO_3$, EtOH, H_2O, 60°C; 2. $LiAlH_4$, THF, rt; 3. CbzCl, THF, H_2O, 0°C. (*d*) 1. *p*-TsOH, CH_3OH, H_2O, 15°C; 2. TBAF, THF, 0°C; 3. MsCl, Py, 5°C. (*e*) CH_3ONa, CH_3OH, 20°C. (*f*) $2CrO_3{\cdot}Py$, CH_2Cl_2, 5°C. (*g*) *tert*-Butyllithioacetate, THF. (*h*) 1. TBSCl, imidazole, DMF, 80°C; 2. H_2, 10% Pd *on* C, EtOH; 3. $CH_3OCH_2CH_2OH$, reflux. (*i*) 1. $BH_3{\cdot}THF$, THF, reflux; 2. 6 M HCl, THF, reflux.

followed by regioselective opening with H_2O and then acetylation gave the triacetate **51**. Reduction of **51** followed by deprotection gave **1**.

5.4.1.1.4 Synthesis from L-threose A total synthesis of (+)-castanospermine (**1**) has been achieved utilizing the chiral allylic alcohol **52**, obtained from tartaric acid via the respective threose derivative (Scheme 7).[85] Epoxidation[86] of **52** gave **53**, whose epoxide ring was regiospecifically cleaved with Et_2AlNBn_2, followed by protection of the two hydroxyl groups to afford **54**. After deacetylation of **54** by treatment with $LiAlH_4$, the resulting

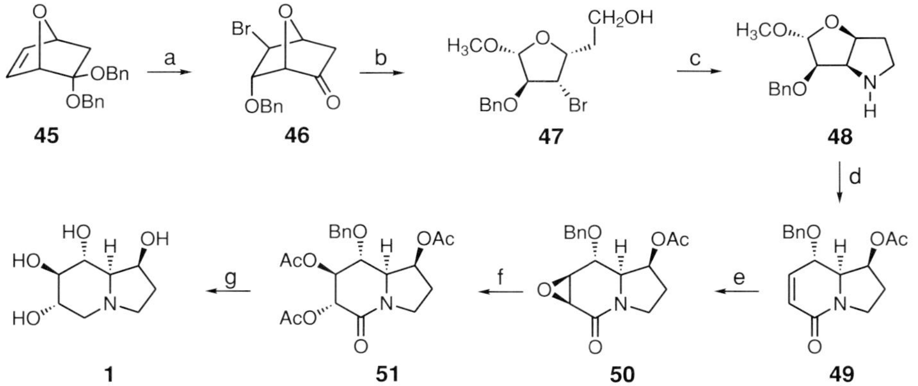

Scheme 6 (*a*) Br_2, CH_2Cl_2, −80°C. (*b*) 1. *m*-CPBA, $NaHCO_3$, CH_2Cl_2, 5–20°C; 2. CH_3OH, $SOC1_2$, 20°C, 24 h; 3. DIBAL-H, THF, −50 to −20°C. (*c*) 1. MsCl, NEt_3, CH_2Cl_2, 0°C; 2. 12% NH_3, EtOH–H_2O (1:1), 70°C, 5 h. (*d*) 1. $C1CH_2COCl$, Py, CH_2Cl_2, −5 to 8°C; 2. Ac_2O, conc. H_2SO_4, 5°C, 2 h; 3. $(EtO)_3P$, 130°C, 7 h; then K_2CO_3, EtOH, 20°C, 12 h; then Ac_2O, Py, DMAP, 20°C, 48 h. (*e*) 1. Br_2, AcOH–Ac_2O (1:2), AgOAc, 9°C; 2. CH_3OH, $SOCl_2$, 20°C, 17 h; then 2-*tert*-butylimino-2-diethylamino-1,3-dimethylperhydro-1,3,2-diazaphosphorine on polystyrene, CH_3CN, 20°C, 35 mm. (*f*) H_2O, 100°C, 4.5 h; then Ac_2O, Py, DMAP, 20°C, 48 h. (*g*) 1. BMS, THF, 20°C, 15 h; 2. H_2, 10 % Pd *on* C, THF–H_2O (5:1), 20°C, 24 h.

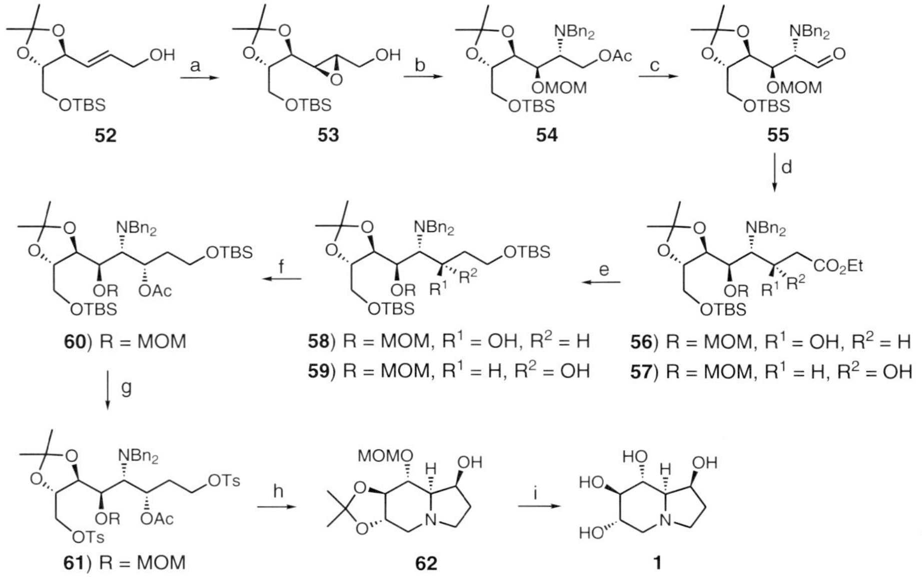

Scheme 7 (*a*) Ref. 86. (*b*) 1. Et_2AlNBn_2, CH_2Cl_2, rt; 2. AcCl, NEt_3, CH_2Cl_2, 0°C; 3. CH_3OCH_2Cl, (*i*-Pr)$_2$NEt, $CHCl_3$, reflux. (*c*) 1. $LiAlH_4$, Et_2O, rt; 2. DMSO, $(COCl)_2$, CH_2Cl_2, −78°C; then NEt_3. (*d*) EtOAc, $LiN(TMS)_2$, THF, −78°C. (*e*) 1. $LiAlH_4$, Et_2O, rt; 2. TBSCl, imidazole, DMF, rt. (*f*) AcOH, Ph_3P, DEAD, benzene, reflux. (*g*) 1. TBAF, THF, rt; 2. *p*-TsCl, Py, rt. (*h*) H_2, $Pd(OH)_2$, CH_3OH; then NEt_3, CH_3OH, reflux. (*i*) HCl, CH_3OH, reflux.

alcohol was oxidized to the aldehyde **55**. The aldehyde **55** was allowed to react with the lithium enolate of ethyl acetate to provide an 89:11 mixture of **56** and **57**. The mixture was subjected to $LiAlH_4$ reduction, followed by protection with *tert*-butyldimethylsilyl chloride to give **58** (68%) and **59** (5%). The major isomer was converted to the desired minor one via Mitsunobu reaction to give **60**. Desilylation of **60** followed by tosylation gave **61**. Hydrogenation of **61** over palladium hydroxide led to the protected castanospermine **62**, which underwent complete deprotection to furnish **1**.

5.4.1.1.5 Synthesis from D-glucono-1,5-lactone An approach to the synthesis of **1** was based on the use of D-glucono-1,5-lactone (Scheme 8).[87] Treatment of 6-*O*-acetyl-2,3,4-tri-*O*-benzyl-D-glucono-1,5-lactone[88] with 2-(3-aminopropylidene)-1,3-dithiane[89] in the presence of K_2CO_3 led to a simultaneous formation of the respective amide and deacetylation process. Subsequent oxidation with lead tetraacetate furnished the lactam **64**, which without purification was cyclized to the indolizidine epimers **65** and **66** by using methanesulfonyl chloride. Oxidation of **66** with singlet oxygen produced an unstable ketone, which was reduced selectively by L-Selectride to give **67**. Reduction of **67** with $LiAlH_4$ followed

Scheme 8 (*a*) 1. 2-(3-Aminopropylidene)-1,3-dithiane, CH_3OH, K_2CO_3; 2. $Pb(OAc)_4$, CH_3CN; then AcOH, 66%. (*b*) NEt_3, MsCl, CH_2Cl_2, 84%. (*c*) O_2, CCl_4, CH_3OH, L-Selectride, THF. (*d*) 1. $LiAlH_4$, THF, 70%; 2. H_2, 10% Pd *on* C, CH_3OH, HCl, 82%.

by hydrogenolysis gave **1** in 5.4% overall yield from **63**. The other epimer **65** was similarly transformed into **68** and then to 1,8a-di-*epi*-castanospermine (**69**) in 6% overall yield **63**.

Another methodology for preparing **1** and **2** has also been developed from D-glucono-1,5-lactone by conversion to the mannoazide **70**,[90,91] whose hydrogenation and subsequent protection of the amine as the 9-phenyl-fluoren-9-yl (Pf) derivative gave **71** (Scheme 9).[92] DIBAL-H reduction of the ester group in **71** to the corresponding alcohol followed by oxidation afforded the aldehyde **72**, which was reacted with vinylmagnesium bromide to give a 1:1 mixture of the diastereomeric alcohols **73**. Oxidation of the hydroxyl group followed by treatment with HBr and subsequent intramolecular cyclization afforded **74**

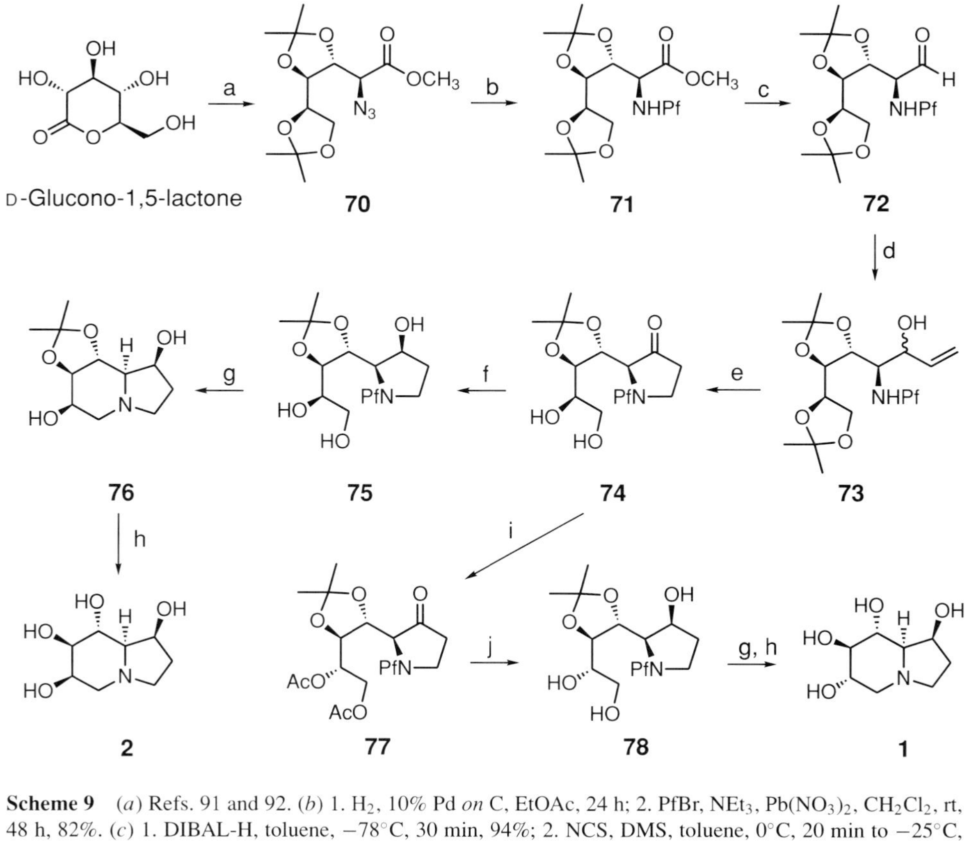

Scheme 9 (*a*) Refs. 91 and 92. (*b*) 1. H_2, 10% Pd *on* C, EtOAc, 24 h; 2. PfBr, NEt_3, $Pb(NO_3)_2$, CH_2Cl_2, rt, 48 h, 82%. (*c*) 1. DIBAL-H, toluene, −78°C, 30 min, 94%; 2. NCS, DMS, toluene, 0°C, 20 min to −25°C, 5.5 h; then NEt_3, 92%. (*d*) $CH_2CHMgBr$, THF, −40°C to rt, 1 h, 91%. (*e*) 1. NCS, DMS, toluene, 0°C, 20 min to −25°C; 2. HBr, Et_2O, H_2O, 0°C, 30 min; then $NaHCO_3$, rt, 2.5 h, 70%. (*f*) $NaBH_4$, EtOH, 0°C, 90 min, 94%. (*g*) 1. *p*-TsCl, DMAP, CH_2Cl_2, 0°C, 1 h to rt, 14 h, 55%; *or* tosylimidazolide, methyl triflate, THF, 0°C, rt, 6 h, 66%; 2. H_2, 10% Pd *on* C, NaOAc, CH_3OH, 20 h; then reflux in CH_3OH, 10 min, 1 N NaOH, 92%. (*h*) TFA, H_2O, dioxane, rt, 24 h; then Dowex 50WX8 resin, 84%. (*i*) 1. Ac_2O, Py, 0°C, 20 h, 80%; 2. Tf_2O, CH_2Cl_2, −15°C, Py; 3. Bu_4NOAc, CH_3CN, 40°C, 50 min; 4. Ac_2O, Py, DMAP, rt, 3 h, 87%. (*j*) $NaBH_4$, CH_3OH, 0°C, K_2CO_3, 0°C, 1 h, 93%.

in 45–58% overall yield from **70**. Reduction of the ketone in **74** with sodium borohydride afforded only the isomer **75**. Selective tosylation of **75** with *N*-methyltosylimidazolium triflate followed by removal of the phenylfluorenyl group and subsequent nucleophilic ring closure gave the protected (+)-6-*epi*-castanospermine **76** whose deprotection afforded **2**. On the other hand, selective acetylation of **74** followed by triflation and then treatment with tetra-*n*-butylammonium acetate in acetonitrile, to invert the configuration of the secondary hydroxyl group, gave the acetate **77**. Reduction of the keto group in **77** with sodium borohydride in the presence of K_2CO_3 gave **78**. Similar sequence of the above reactions led to **1**.

5.4.1.1.6 Synthesis from D*-gulonolactone* The lactam **79**,[93,94] available from D-gulonolactone in 30% overall yield, has been used in the synthesis of (+)-6-*epi*-castanospermine (**2**) (Scheme 10).[95,96] It was benzylated and then treated with $LiAlH_4$ and subsequently desilylated to afford **80**. Swern oxidation of the primary hydroxyl group in **80** followed by Grignard reaction using vinylmagnesium bromide afforded 80% of a 1:1 mixture of the diastereomers **81**. This mixture was reacted with *tert*-butyldimethylsilyl chloride, followed by hydroboration and then oxidation of the double bond with alkaline hydrogen peroxide to give the epimers **82** (25%) and **83** (19%). Mesylation of **82** resulted in spontaneous cyclization to the quaternary ammonium salt, which was subjected to complete deprotection to give **2**. Similarly, the other diastereomer **83** was converted into 1,6-di-*epi*-castanospermine (**84**).

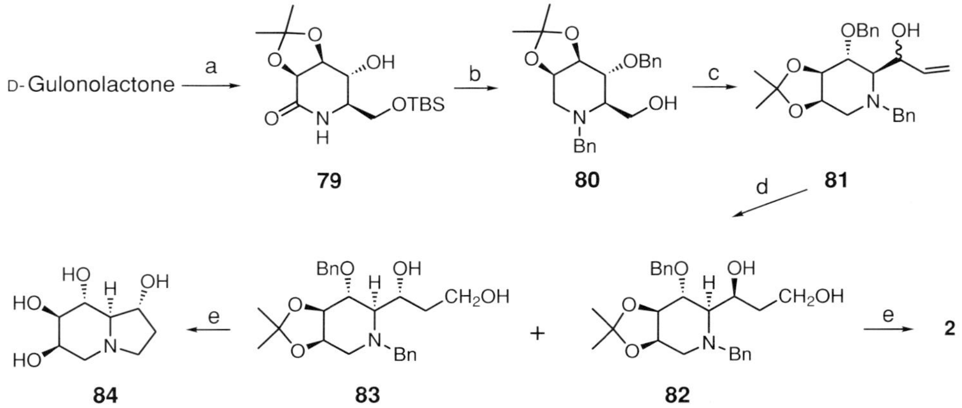

Scheme 10 (*a*) Refs. 93 and 94. (*b*) 1. BnBr, NaH, THF, *n*-Bu_4NI, 92%; 2. $LiAlH_4$, $AlCl_3$, THF, 82%. (*c*) 1. DMSO, $(COCl)_2$, CH_2Cl_2, −40°C; then NEt_3; 2. Vinylmagnesium bromide, THF, 25°C, 80%. (*d*) 1. TBSCl, imidazole, DMF, 73%; 2. BH_3, THF; then alkaline H_2O_2, **82** (25%), **83** (19%). (*e*) 1. MsCl, NEt_3, CH_2Cl_2; 2. H_2, Pd black, CH_3OH, THF, H_2O, 82% for two steps; 3. aqueous TFA–H_2O, 2 h, 85%.

5.4.1.1.7 Synthesis from D*-glucofuranurono-6,3-lactone* D-Glucofuranurono-6,3-lactone has also been used for the synthesis of **1** by conversion to amino lactone **85**[97] that subsequently converted to the hemiketal **86** (97%) (Scheme 11).[98] Catalytic hydrogenation of **86** over PtO_2 gave a 2:7 epimeric mixture of **87** and **88**; **87** was found to be the predominating epimer upon using other reducing reagents. Deprotection of **88** with formic acid followed by $LiAlH_4$ reduction of the resulting lactam afforded the pyrrolidine derivative **89**, which upon treatment with 90% TFA followed by catalytic hydrogenation gave **1**. The epimer **87** was similarly transformed into 1-*epi*-castanospermine (**91**) via **90**.

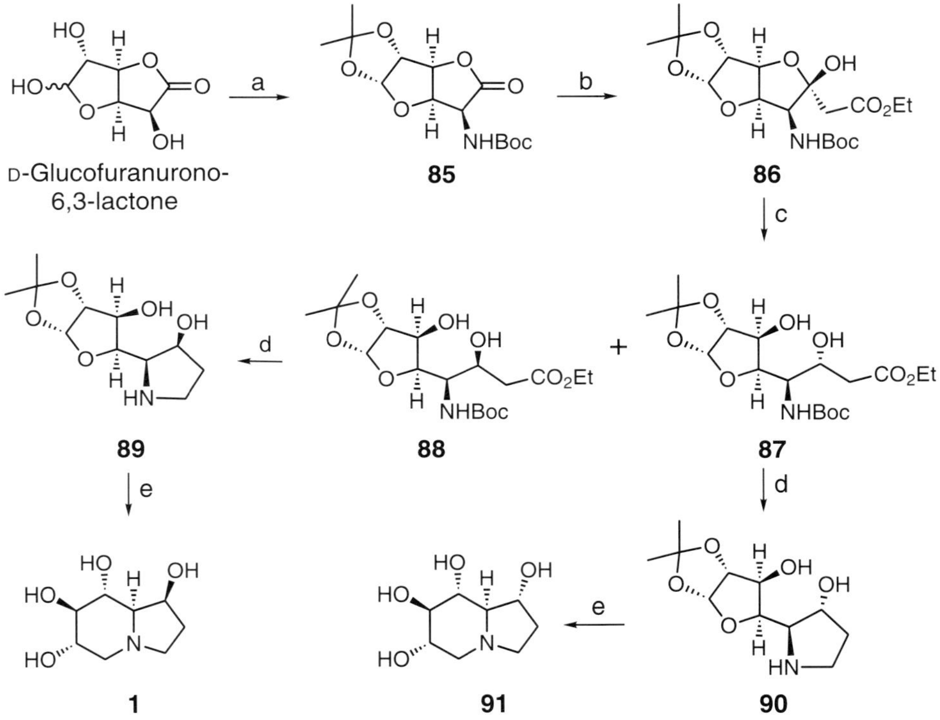

Scheme 11 (*a*) Ref. 98. (*b*) EtOAc, LDA, THF, −78°C, 2.5 h, 97%. (*c*) H_2, PtO_2, EtOAc, 20 h, 100%; *or* $NaBH_4$, EtOH, 0°C, 1 h, 95%. (*d*) 1. HCO_2H, CH_2Cl_2, 0–5°C, 2 h; then 25°C, 6 h; 2. Dowex 1X2 (OH^-) resin, H_2O, 73% for two steps; 3. $LiAlH_4$, THF, reflux, 20 h, 75%. (*e*) 1. TFA, 25°C, 20 h; 2. H_2, 5% Pt *on* C, H_2O, 20 h, 61%.

Alternative approaches for the syntheses of **1** and **91** have also been achieved from D-glucofuranurono-6,3-lactone (Scheme 12).[99,100] 5-*O*-(*tert*-Butyldimethylsilyl)-1,2-*O*-isopropylidene-α-D-glucofuranurono-6,3-lactone (**92**) was converted into **93** and **94** by chain extension using Reformatsky reaction followed by reduction with calcium borohydride. The mixture was converted into compounds **95**, **96** and **97**. Reduction of **95** followed by intramolecular cyclization and protection of the secondary amine afforded **98**, which was deprotected to give **100** and then subjected to an intramolecular cyclization to give 1-*epi*-castanospermine (**91**). On the other hand, deprotection of **97** afforded **101**, which was subjected to hydrogenation followed by intramolecular cyclization to give **91**. Similarly, compound **96** was converted to **1** via **99**.

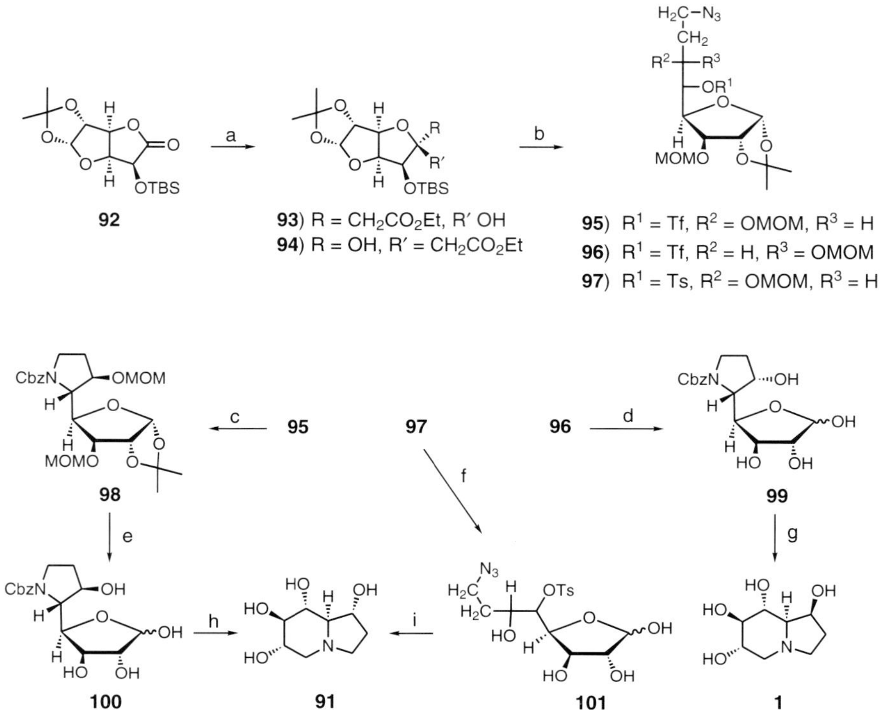

Scheme 12 (*a*) 1. THF, $BrMgCH_2CO_2Et$, Zn, 65°C; 2. $Ca(BH_4)_2$. (*b*) 1. Reduction; 2. *p*-TsCl, Py or Tf_2O, Py; 3. NaN_3, DMF. (*c*) 1. H_2, Pd *on* C, EtOAc; 2. CbzCl, $NaHCO_3$. (*d*) TFA, CH_3CN, H_2O. (*e*) TFA, H_2O. (*f*) TFA, CH_3CN, H_2O. (*g*) H_2, Pd *on* C, TFA. (*h*) Reductive amination, 83%. (*i*) H_2, Pd *on* C; intramolecular cyclization.

References

1. Hohenschutz, L.D.; Bell, E.A.; Jewess, P.J.; Leworthy, D.P.; Pryce, R.J.; Arnold, E.; Clardy, J. *Phytochemistry* **20** (1981) 811.
2. Nash, R.J.; Fellows, L.E.; Dring, J.V.; Stirton, C.H.; Carter, D.; Hegarty, M.P.; Bell, E.A. *Phytochemistry* **27** (1988) 1403.
3. Molyneux, R.J.; Roitman, J.N.; Dunnheim, G.; Szumilo, T.; Elbein, A.D. *Arch. Biochem. Biophys.* **251** (1986) 450.
4. Fleet, G.W.J.; Ramsden, N.G.; Molyneux, R.J.; Jacob, G.S. *Tetrahedron Lett.* **29** (1988) 3603.
5. Molyneux, R.J.; Pan, Y.T.; Tropea, J.E.; Benson, M.; Kaushal, G.P.; Elbein, A.D. *Biochemistry* **30** (1991) 9981, and references cited therein.
6. Saul, R.; Chambers, J.P.; Molyneux, R.J.; Elbein, A.D. *Arch. Biochem. Biophys.* **221** (1983) 593.
7. Pan, Y.T.; Hori. H.; Saul, R.; Sanford, B.A.; Molyneux, R.J.; Elbein, A.D. *Biochemistry* **22** (1983) 3975.
8. Hori, H.; Pan, Y.T.; Molyneux, R.J.; Elbein, A.D. *Arch. Biochem. Biophys.* **228** (1984) 525.
9. Elbein, A.D. *CRC Crit. Rev. Biochem.* **16** (1984) 21.
10. Saul, R.; Ghidoni, J.J.; Molyneux, R.J.; Elbein, A.D. *Proc. Natl. Acad. Sci. USA* **82** (1985) 93.
11. Sasak, V.W.; Ordovas, J.M.; Elbein, A.D.; Berninger, R.W. *Biochem. J.* **232** (1985) 759.
12. Szumilo, T.; Kaushal, G.P.; Elbein, A.D. *Arch. Biochem. Biophys.* **247** (1986) 261.
13. Campbell, B.C.; Molyneux, R.J.; Jones, K.C. *J. Chem. Ecol.* **13** (1987) 1759.
14. Cenci di Bello, I; Mann, D.; Nash, R.J.; Winchester, B. In *Lipid Storage Disorders*, Vol. 150; Salvayre, R., Douste-Blazy, L., Gatt, S., Eds.; Plenum: New York, 1988; p. 635.

15. Fleet, G.W.J.; Fellows, L.E. In *Natural Product Isolation*; Wagman, G.H., Cooper, R., Eds.; Elsevier: Amsterdam, 1988; p. 540.
16. Winchester, B.G.; Cenci di Bello, I.; Richardson, A.C.; Nash, R.J.; Fellows, L.E.; Ramsden, N.G.; Fleet, G.W.J. *Biochem. J.* **269** (1990) 227.
17. Fleet, G.W.J.; Ramsden, N.G.; Nash, R.J.; Fellows, L.E.; Jacob, G.S.; Molyneux, R.J.; Cenci di Bello, I.; Winchester, B. *Carbohydr. Res.* **205** (1990) 269.
18. Rhinehart, B.L.; Robinson, K.M.; Payne, A.J.; Wheatley, M.E.; Fisher, J.L.; Liu, P.S.; Cheng, W. *Life Sci.* **41** (1987) 2325.
19. Trugnan, G.; Rousset, M.; Zweibaum, A. *FEBS Lett.* **195** (1986) 28.
20. Truscheit, E.; Frommer, W.; Junge, B.; Muller, L.; Schmidt, D.D.; Wingender, W. *Angew. Chem., Int. Ed. Engl.* **20** (1981) 744. .
21. Humphries, M.J.; Matsumoto, K.; White, S.L.; Olden, K. *Cancer Res.* **46** (1986) 5215.
22. Dennis, J.W. *Cancer Res.* **46** (1986) 5131.
23. Dennis, J.W.; Laferte, S.; Waghorne, C.; Breitman, M.L.; Kerbel, R.S. *Science* **236** (1987) 582.
24. Ahrens, P.B.; Ankel, H. *J. Biol. Chem.* **262** (1987) 7575.
25. Ostrander, G.K.; Scnibner, N.K.; Rohrschneider, L.R. *Cancer Res.* **48** (1988) 1091.
26. Sunkara, P.S.; Bowlin, T.L.; Liu, P.S.; Sjoerdsma, A. *Biochem. Biophys. Res. Commun.* **148** (1987) 206.
27. Nichols, E.J.; Manger, R.; Hakomori, S.; Herscovics, A.; Rohrschneider, L.R. *Molec. Cell. Biol.* **5** (1985) 3467.
28. Walker, B.D.; Kowalski, M.; Goh, W.C.; Kozarsky, K.; Krieger, M.; Rosen, C.; Rohrschneider, L.; Haseltine, W.A.; Sodroski, J. *Proc. Natl. Acad. Sci. USA* **84** (1987) 8120.
29. Dagani, R. *Chem. Eng. News* June 29 (1987) 25.
30. Gruters, R.A.; Neefjes, J.J.; Tersmette, M.; De Goede, R.E.Y.; Tulp, A.; Huisman, H.G.; Miedema, F.; Ploegh, H.L. *Nature* **330** (1987) 74.
31. Tyms, A.S.; Berrie, E.M.; Ryder, T.A.; Nash, R.J.; Hegarty, M.P.; Taylor, D.L.; Mobberley, M.A.; Davis, J.M.; Bell, E.A.; Jeffries, D.J.; Taylor-Robinson, D.; Fellows, L.E. *Lancet* (1987) 1025.
32. Fleet, G.W.J.; Karpas, A.; Dwek, R.A.; Fellows, L.E.; Tyms, A.S.; Petursson, S.; Namgoong, S.K.; Ramsden, N.G.; Smith, P.W.; Son, J.C.; Wilson, F.; Witty, D.R.; Jacob, G.S.; Rademacher, T.W. *FEBS Lett.* **237** (1988) 128.
33. Karpas, A.; Fleet, G.W.J.; Dwek, R.A.; Petursson, S.; Namgoong, S.K.; Ramsden, N.G.; Jacob, G.S.; Rademacher, T.W. *Proc. Natl. Acad. Sci. USA* **85** (1988) 9229.
34. Montefiori, D.C.; Robinson, W.E.; Mitchell, W.M. *Proc. Natl. Acad. Sci.* **85** (1988) 9248.
35. Ruprecht, R.M.; Mullaney, S.; Andersen, J.; Bronson, R.J. *Acquir. Immune Defic. Syndr.* **2** (1989)149.
36. Sunkara, P.S.; Taylor, D.L.; Kang, M.S.; Bowlin, T.L.; Liu, P.S.; Tyms, A.S.; Sjoerdsma, A. *Lancet* (1989) 1206.
37. Pan, Y.T.; Hori, H.; Saul, R.; Sanford, B.A.; Molyneux, R.J.; Elbein, A.D. *Biochemistry* **22** (1983) 3973.
38. Nash, R.J.; Fellows, L.E.; Girdhar, A.; Fleet, G.W.J.; Peach, J.M.; Watkin, D.J.; Hegarty, M.P. *Phytochemistry* **29** (1990) 1356.
39. Hendry, D.; Hough, L.; Richardson, A.C. *Tetrahedron* **44** (1988) 6153.
40. Bide, R.; Mortezaei, R.; Scilimati, A.; Sih, C.J. *Tetrahedron Lett.* **31** (1990) 4827.
41. Gradnig, G.; Berger, A.; Grassberger, V.; Stütz, A.E.; Legler, G. *Tetrahedron Lett.* **32** (1991) 4889.
42. Reymond, J.-L.; Pinkerton, A.A.; Vogel, P. *J. Org. Chem.* **56** (1991) 2128.
43. Ina, H.; Kibayashi, C. *Tetrahedron Lett.* **32** (1991) 4147.
44. Burgess, K.; Henderson, I. *Tetrahedron* **48** (1992) 4045.
45. Burgess, K.; Chaplin, D.A. *Tetrahedron Lett.* **33** (1992) 6077.
46. Mulzer, J.; Dehmlow, H. *J. Org. Chem.* **57** (1992) 3194.
47. Zhou, P.; Salleh, H.M.; Honek, J.F. *J. Org. Chem.* **58** (1993) 264.
48. Ina, H.; Kibayashi, C. *J. Org. Chem.* **58** (1993) 52.
49. Kim, N.-S.; Kang, C.H.; Cha, J.K. *Tetrahedron Lett.* **35** (1994) 3489.
50. Ahman, J.; Somfai, P. *Tetrahedron* **51** (1995) 9747.
51. Kang, S.H.; Kim, J.S. *Chem. Commun.* (1998) 1353.
52. Izquierdo, I.; Plaza, M.T.; Robles, R.; Mota, A.J. *Tetrahedron: Asymmetry* **9** (1998) 1015.
53. Carretero, J.C.; Arrayás, R.G.; Buezo, N.D.; Garrido, J.L.; Alonso, I.; Adrio, J. *Phosphous Sulfur Silicon* **153–154** (1999) 259.
54. Kim, N.-S.; Kang, C.H.; Cha, J.K. *Tetrahedron Lett.* **35** (1994) 3489.
55. Klitzke, C.F.; Pilli, R.A. *Tetrahedron Lett.* **42** (2001) 5605.
56. Lombardo, M.; Trombini, C. *Tetrahedron* **56** (2000) 323.
57. Batey, R.A.; MacKay, D.B. *Tetrahedron Lett.* **41** (2000) 9935.
58. Socha, D.; Jurczak, M.; Chmielewski, M. *Carbohydr. Res.* **336** (2001) 315.
59. Paolucci, C.; Mattioli, L. *J. Org. Chem.* **66** (2001) 4787.

60. Tayle, P.C.; Winchester, B.G. In *Iminosugars as Glycosidase Inhibitors*; Stütz, A.E., Ed.; Wiley-VCH: Weinheim, 1999; p. 125.
61. Asano, N.; Nash, R.J.; Molyneux, R.J.; Fleet, G.W.J. *Tetrahedron: Asymmetry* **11** (2000) 1645.
62. Svansson, L.; Johnston, B.D.; Gu, J.-H.; Partrick, B.; Pinto, B.M. *J. Am. Chem. Soc.* **122** (2000) 10769.
63. Patil, N.T.; Tilekar, J.N.; Dhavale, D.D. *Tetrahedron Lett.* **42** (2001) 747.
64. Collum, D.B.; McDonald, J.H., III; Still, W.C. *J. Am. Chem. Soc.* **102** (1980) 2118.
65. Rabiczko, J.; Urbańczyk-Lipkowska, Z.; Chmielewski, M. *Tetrahedron* **58** (2002) 1433.
66. de Faria, A.R.; Salvador, E.L.; Correia, C.R.D. *J. Org. Chem.* **67** (2002) 3651.
67. Bernotas, R.C.; Ganem, B. *Tetrahedron lett.* **25** (1984) 165.
68. Bernotas, R.C.; Ganem, B. *Tetrahedron Lett.* **26** (1985) 1123.
69. Hamana, H.; Ikoto, N.; Ganem, B. *J. Org. Chem.* **52** (1987) 5492.
70. Zemplen, G.; Csuros, Z.; Angyal, S. *Chem. Ber.* **70** (1937) 1848.
71. Zhao, H.; Mootoo, D.R. *J. Org. Chem.* **61** (1996) 6762.
72. Hashimoto, H.; Asano, K.; Yoshimura, J. *J. Carbohydr. Res.* **104** (1982) 87.
73. Kim, E.; Gordon, D.M.; Schmid, W.; Whitesides, G.M. *J. Org. Chem.* **58** (1993) 5500.
74. Czernecki, S.; Horns, S.; Valery, J.-M. *J. Org. Chem.* **60** (1995) 650.
75. Overkleeft, H.S.; van Wiltenburg, J.; Pandit, U.K. *Tetrahedron Lett.* **34** (1993) 2527.
76. Overkleeft, H.S.; van Wiltenburg, J.; Pandit, U.K. *Tetrahedron* **50** (1994) 4215.
77. Hoos, R.; Naughton, A.B.; Vasella, A. *Helv. Chim. Acta* **75** (1992) 1802.
78. Overkleeft, H.S.; Pandit, U.K. *Tetrahedron Lett.* **37** (1996) 547.
79. Ireland, R.E.; Liu, L. *J. Org. Chem.* **58** (1993) 2899.
80. Vasella, A.; Bernet, A. *Helv. Chim. Acta* **62** (1979) 1990.
81. Pearson, W.H.; Bergmeier, S.C.; Degan, S.; Lin, K.-C.; Poon, Y.-F.; Schkeryantz, J.M.; Williams, J.P. *J. Org. Chem.* **55** (1990) 5719.
82. Setoi, H.; Takeno, H.; Hashimoto, M. *Tetrahedron Lett.* **26** (1985) 4617.
83. Doane, W.M.; Shasha, B.S.; Russell, C.R.; Rist, C.E. *J. Org. Chem.* **32** (1967) 1080.
84. Reymond, J.-L.; Vogel, P. *Tetrahedron Lett.* **30** (1989) 705.
85. Ina, H.; Kibayashi, C. *Tetrahedron Lett.* **32** (1991) 4147.
86. Iida, H.; Yamazaki, N.; Kibayashi, C. *J. Org. Chem.* **52** (1987) 3337.
87. Miller, S.A.; Chamberlin, A.R. *J. Am. Chem. Soc.* **112** (1990) 8100.
88. Horito, S.; Asano, K.; Umemura, K.; Hashimoto, H.; Yoshimura, J. *Carbohydr. Res.* **121** (1983) 175.
89. Chamberlin, A.R.; Chung, J.Y.L. *J. Org. Chem.* **50** (1985) 4425.
90. Regeling, H.; de Rouville, E.; Chittenden, G.J.F. *Recl. Trav. Chim. Pays-Bas* **106** (1987) 461.
91. Csuk, R.; Hugener, M.; Vasella, A. *Helv. Chim. Acta* **71** (1988) 609.
92. Gerspacher, M.; Rapoport, H. *J. Org. Chem.* **56** (1991) 3700.
93. Fleet, G.W.J.; Ramsden, N.G.; Witty, D.R. *Tetrahedron Lett.* **29** (1988) 2871.
94. Fleet, G.W.J.; Ramsden, N.G.; Witty, D.R. *Tetrahedron* **45** (1989) 319.
95. Fleet, G.W.J.; Ramsden, N.G.; Molyneux, R.J.; Jacob, G.S. *Tetrahedron Lett.* **29** (1988) 3603.
96. Fleet, G.W.J.; Ramsden, N.G.; Nash, R.J.; Fellows, L.E.; Jacob, G.S.; Molyneux, R.J.; Cenci di Bello, I.; Winchester, B. *Carbohydr. Res.* **205** (1990) 269.
97. Anzeveno, P.B.; Creemer, L.J. *Tetrahedron Lett.* **31** (1990) 2085.
98. Anzeveno, P.B.; Angell, P.T.; Creemer, L.J.; Whalon, M.R. *Tetrahedron Lett.* **31** (1990) 4321.
99. GraBberger, V.; Berger, A.; Dax, K.; Fechter, M.; Grading, G.; Slütz, A.E. *Liebigs Ann. Chem.* (1993) 379.
100. Dax, K.; Fechter, M.; Gradnig, G.; GraBberger, V.; Illaszewicz, C., Ungerank, M.; Stütz, A.E. *Carbohydr. Res.* **217** (1991) 59.

5.4.1.2 *(–)-Swainsonine* (–)-Swainsonine [(1*S*,2*R*,8*R*,8a*R*)-1,2,8-trihydroxyindolizidine, **1**] was first isolated from the fungus *Rhizoctonia leguminicola*[1–8] and later found in the Australian *Swainsona canescens*[9–13] plant, cultures of normal and transformed roots of *Swainsona galegifolia*,[9,10] north American plants spotted locoweed *Astragalus lentiginosus*[14–16] and the fungus *Metarhizium anisopline* F-3622.[17–26] (–)-Swainsonine and its analogues are currently the subject of many biological investigations. It is an effective inhibitor of both lysosomal α-mannosidase,[27–50] involved in the cellular degradation of polysaccharides, and mannosidase II,[51–57] a key enzyme in the processing of asparagine-linked glycoproteins.[58] (+)-Swainsonine (**2**) is the most potent inhibitor yet described[59–61] of L-rhamnosidase from *Penicillium decumbers*. (–)-Swainsonine (**1**) has antimetastic,[62–73] antitumor-proliferative[74–78] and anticancer[79–90] activities. It is the first glycoprotein-processing inhibitor to be selected for clinical testing as anticancer drug[91–95] but its high cost has hindered clinical trials and immunoregulating activities.[96–101] Moreover, **1** has other biological effects[102–171] such as murine survival and bone marrow proliferation,[102] modification of glycan structure,[103] activity of intestinal sucrase,[104] rats appetite,[105] aspartate transaminase activity,[106] insulin and lectin binding,[107] inhibition of tyrosinase activity,[108] rat epididymal glycosidases,[109] inhibition of the formation of normal oligosaccharide chain of the G-protein of vesicular stomatitis virus,[110] modulation of ricin toxicity,[111–113] biochemistry and pathology in big,[114] toxicity and lesions production,[115] neuronal lysosomal mannoside storage disease,[116–119] inhibition of mammalian digestive disaccharidases,[120] increasing the high-mannose glycoproteins in cultured mammalian

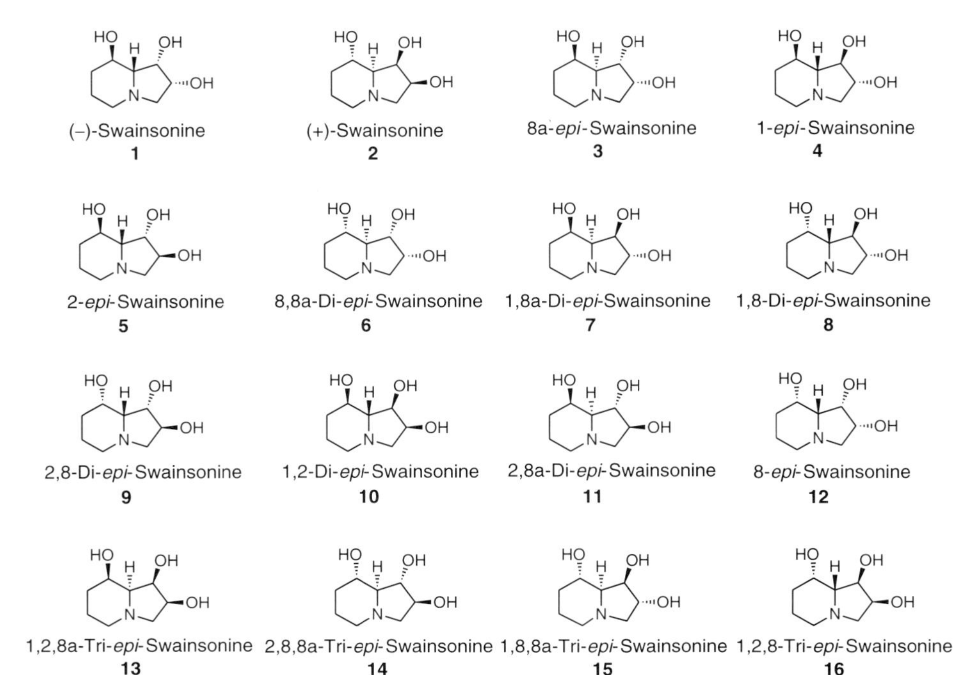

cells,[121] induction of high mountain disease in calves,[122] fucose incorporation in soybean cells,[123] normal human fibroblasts in culture,[124] recycling of the transferring receptor,[125] inhibition of root length elongation[126] and the principal toxin responsible for the induction of locoism.[127]

The absolute configuration of (–)-swainsonine (**1**) was deduced on the basis of its biosynthesis[172] and unambiguous nuclear magnetic resonance assignments.[173] The relative stereochemistry of swainsonine was determined by X-ray crystallography.[174] Noncarbohydrates have been used for the total synthesis of swainsonine and its isomers.[175–205] The first total synthesis of **1** has established its absolute stereochemistry as (1*S*,2*R*,8*R*,8a*R*)-1,2,8-trihydroxyindolizidine.[206,207] Various carbohydrate derivatives have been used for the synthesis of (–)-swainsonine and its analogues.[8] These synthetic approaches will be arranged according to the used carbohydrate derivative.

5.4.1.2.1 Synthesis from D*-glucose* The readily available methyl α-D-glucopyranoside (**17**) has been converted to the amine hydrochloride **18**[208] in 20–25% yield and then to the corresponding 3,6-imino derivative, which upon protection and acid hydrolysis afforded the carbamate **19** in 52% yield. Treatment of **19** with ethanethiol in the presence of conc. HCl furnished the dithioacetal **20** (74%), which upon acetylation, cleavage of the dithioacetal group and subsequent condensation of the resulting aldehyde with [(ethoxycarbonyl)methylene]triphenylphosphorane in acetonitrile gave a 1:1 mixture of the *E*- and *Z*-isomers **21** in 60% yield. Hydrogenation of the isomeric mixture gave a 1:1 mixture of the corresponding hydrogenated derivative **22** and its cyclized product **23**. The lactam **23** was converted upon reduction with BMS, followed by deacetylation, into (–)-swainsonine (**1**) (Scheme 1).[206,207]

A similar methodology has utilized methyl 3-acetamido-4,6-*O*-benzylidene-3-deoxy-α-D-glucopyranoside (**24**),[209] which upon inversion of configuration at C-2 produced the mannoside **25**, whose benzylidene group was cleaved followed by acetylation to afford the

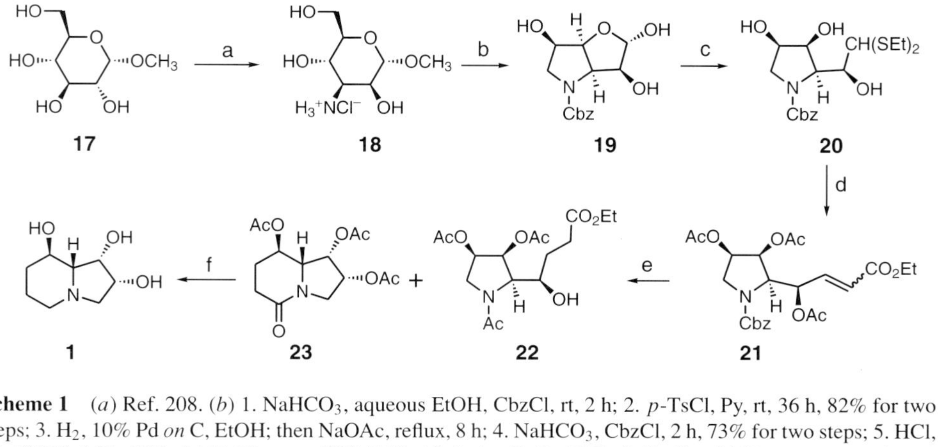

Scheme 1 (*a*) Ref. 208. (*b*) 1. $NaHCO_3$, aqueous EtOH, CbzCl, rt, 2 h; 2. *p*-TsCl, Py, rt, 36 h, 82% for two steps; 3. H_2, 10% Pd *on* C, EtOH; then NaOAc, reflux, 8 h; 4. $NaHCO_3$, CbzCl, 2 h, 73% for two steps; 5. HCl, 95–100°C, 16 h, 52%. (*c*) EtSH, conc. HCl, 74%. (*d*) 1. Ac_2O, Py, 73%; 2. $HgCl_2$, $CdCO_3$, acetone, reflux, 8 h, 30 min, 96%; 3. $Ph_3P{=}CHCO_2Et$, CH_3CN, reflux, 15 min, 86%. (*e*) 10% Pd *on* C, H_2, 2 h, **22** (25%), **23** (25%). (*f*) 1. BMS, THF, under N_2, 71–94%; 2. $NaOCH_3$, CH_3OH, 3 h, 100%.

pentaacetate **26** (Scheme 2).[210–212] Zemplen deacetylation of **26** followed by treatment with ethanethiol in the presence of conc. HCl and subsequent selective protection of the primary hydroxyl group as trityl ether afforded **27** in 32% overall yield from **24**. Benzylation of **27** followed by removal of the trityl group using *p*-toluenesulfonic acid in methanol, tosylation and subsequent intramolecular cyclization afforded the pyrrolidine **28**. Treatment of **28** with mercury(II) chloride and calcium carbonate followed by Horner–Emmons reaction[213] on the resulting aldehyde with diethyl ethoxycarbonylmethylphosphonate afforded a 40:1 mixture of *E*- and *Z*-**29**. Hydrogenation of **29** over Raney nickel afforded **30**. Prolonged heating of **30** with aqueous ethanolic 15 M KOH in a sealed tube afforded the lactam **31** (54%), which was treated with $LiAlH_4$ followed by de-O-benzylation to furnish **1**.

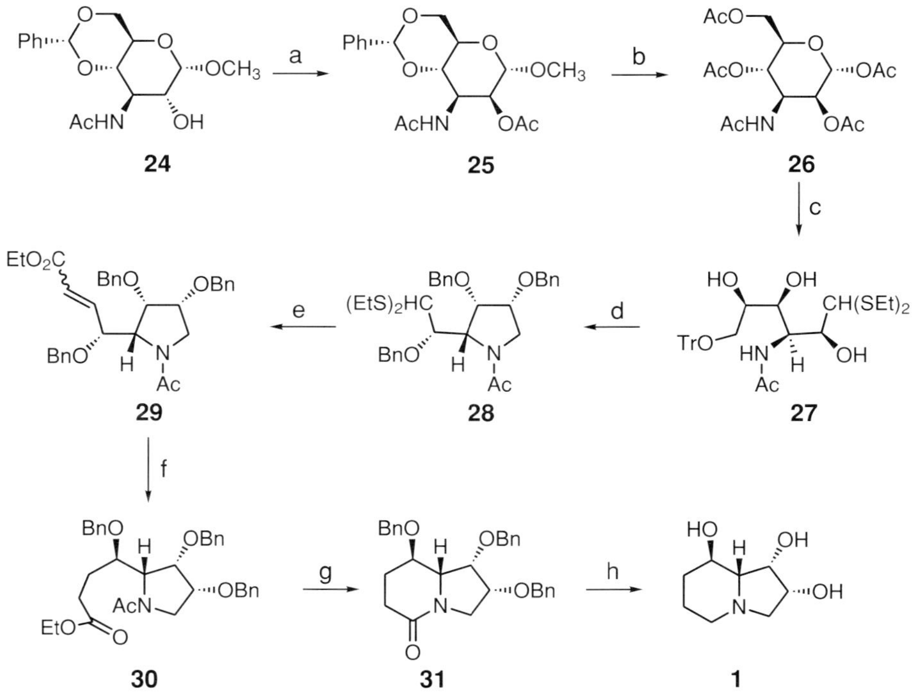

Scheme 2 (*a*) 1. MsCl, Py, 100%; 2. 0.5% HCl, reflux, 1 h, 97%; 3. NaOAc, $CH_3O(CH_2)_2OH$, reflux, 25 h, 61%. (*b*) 1. 2 M HCl, reflux, 13 h, 98%; 2. Ac_2O, Py. (*c*) 1. $NaOCH_3$, CH_3OH; 2. EtSH, conc. HCl; 3. TrCl, Py, DMAP, 55%. (*d*) 1. BnBr, NaH, DMF; 2. *p*-TsOH·H_2O, CH_3OH, 35% two steps; 3. TsCl, Py, 77%; 4. 1,4-dioxane, 1 M NaOH, reflux, 30 min, 93%. (*e*) 1. $HgCl_2$, Ca_2CO_3, CH_3CN; 2. $Et_2P(O)CH_2CO_2Et$, NaH; THF, 4 h, 75% for two steps. (*f*) Raney nickel, H_2, 2 h, *E* (94%) and *Z* (70%). (*g*) 15 M KOH, EtOH, sealed tube, 90°C, 6 days, 54%. (*h*) 1. THF, $LiAlH_4$, reflux, 5 h, 74%; 2. 20% $Pd(OH)_2$ *on* C, cyclohexene, reflux, 44 h, 72%.

Synthesis of (–)-swainsonine (**1**) has been achieved from D-glucal by conversion to **32** and then to **33** (Scheme 3).[214,215] Sharpless asymmetric dihydroxylation followed by kinetic resolution of the α-furfuryl amide **33**[216,217] afforded the optically active dihydropyridone

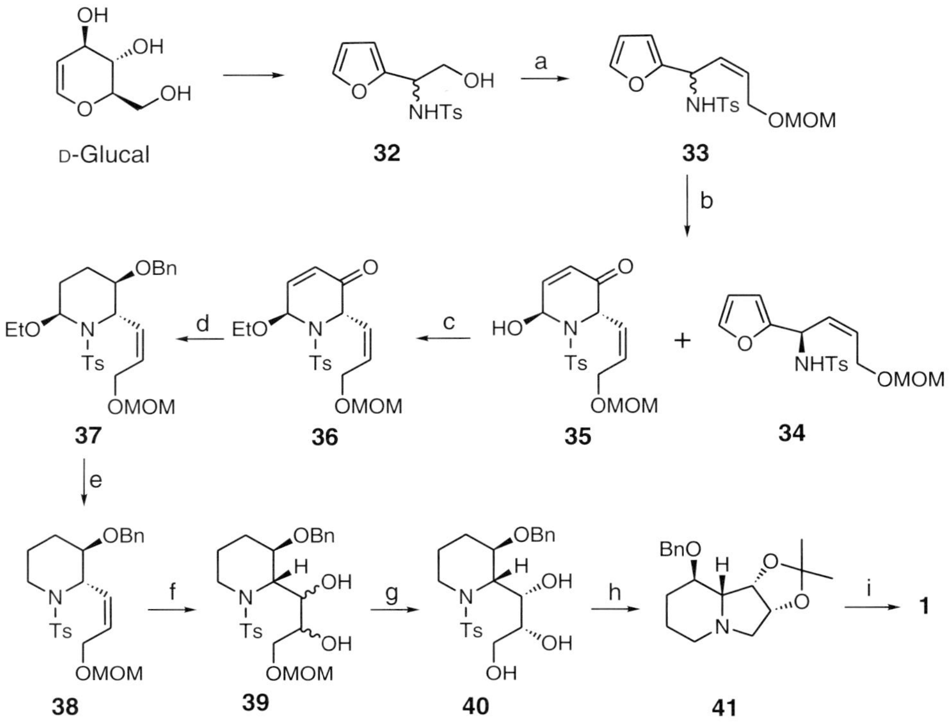

Scheme 3 (*a*) Refs. 216 and 217. (*b*) Ti(O*i*-Pr)$_4$, D-(−)-DIPT, TBHP, silica gel, CaH$_2$, CH$_2$Cl$_2$, 25°C, 2 days, **34** (46%), **35** (42%); separation. (*c*) HC(OEt)$_3$, BF$_3$·OEt$_2$, MS 4 Å, ether, rt, 97%. (*d*) 1. NaBH$_4$, CH$_3$OH, −40 to 30°C, 88%; 2. BnBr, NaH, Bu$_4$NI, THF, 96%. (*e*) NaBH$_4$, HCO$_2$H, −5 to 0°C, 90%. (*f*) OsO$_4$, NMO, DHQ·CLB, trace CH$_3$SO$_2$NH$_2$, acetone–H$_2$O, ultrasonication, 73% and 7%. (*g*) *p*-TsOH, *t*-BuOH, reflux, 90%. (*h*) 1. Na, naphthalene, DMF, −60°C; 2. Ph$_3$P, CCl$_4$, NEt$_3$, DMF, 50%; 3. DMP, *p*-TsOH, CH$_2$Cl$_2$, 94%. (*i*) Deprotection, 57%.

35 (42%) in addition to the unreacted α-furfuryl amide isomer **34**. Treatment of **35** with triethyl orthoformate in the presence of a catalytic amount of BF$_3$·OEt$_2$ gave **36** in 97% yield. Reduction of **36** with sodium borohydride in methanol gave the corresponding alcohol, which upon benzylation afforded **37**. Subsequent reaction with sodium borohydride in formic acid furnished **38** in 90% yield. Dihydroxylation of **38** proceeded smoothly to form a 10:1 mixture of **39**. Removal of the MOM group gave **40**, which upon deprotection of the tosyl group followed by intramolecular cyclization afforded the benzyl derivative of swainsonine, whose debenzylation was not successful; however, debenzylation of its acetonide derivative **41** can be achieved, which was followed by acid hydrolysis to afford **1**.

5.4.1.2.2 Synthesis from D-mannose Synthesis of (−)-swainsonine (**1**) has been also accomplished by utilizing D-mannose as a starting material (Scheme 4).[218–220] D-Mannose was converted into **42** in 81% overall yield. Double inversion of configuration at C-4 of **42**

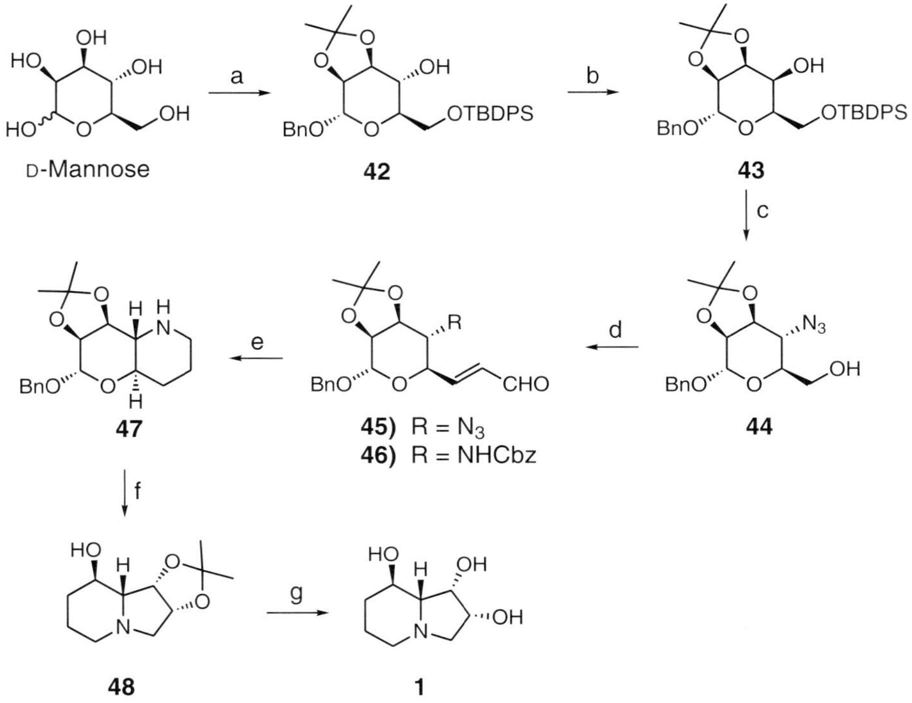

Scheme 4 (*a*) 1. BnOH, HCl, 83%; 2. TBDPSCl, imidazole, DMF, rt, 6 h, 97%; 3. acetone, DMP, CSA, 100%. (*b*) PCC, powdered MS 3 Å, CH_2Cl_2, rt, 2 h; 2. $NaBH_4$, EtOH, 81% for two steps. (*c*) 1. Tf_2O, Py, CH_2Cl_2, −50 to −20°C; 2. NaN_3, DMF, rt, 68%, for two steps; 3. TBAF, THF, rt, 4 h, 97%. (*d*) 1. Pd black, CH_3OH, H_2, rt, 1 h, 100%; then $NaHCO_3$, CbzCl, ether, 1.5 h, 80%; or PCC, powdered MS 3 Å, CH_2Cl_2, rt, 45 min; 2. $Ph_3P{=}CHCHO$, 45 min. (*e*) 1. 10% Pd *on* C, CH_3OH, H_2, rt, 6 h; 2. Pd black, CH_3OH, 48 h, 61%. (*f*) Pd black, AcOH, H_2, rt, 3 days, 87% from azide, 60% from **46**. (*g*) TFA (80%), D_2O, rt, 50 h, 74%; then ion-exchange chromatography (CG-120 H^+), elution with aqueous NH_3.

was achieved by oxidation–reduction reaction to give the taloside **43**, which upon triflation and subsequent displacement with azide ion and then removal of the silyl group with fluoride ion furnished the mannoazide **44** in 53% yield from **42**. Hydrogenation of **44** followed by protection of the resulting amine with benzyl chloroformate and subsequent oxidation of the primary hydroxyl group and then condensation with $Ph_3P{=}CHCHO$ afforded **46**. Oxidation of **44** followed by treatment with $Ph_3P{=}CHCHO$ furnished **45**. Prolonged hydrogenation of either **45** or **46** over palladium black in methanol afforded the protected swainsonine **48** via **47**. Removal of the isopropylidene group from **48** with TFA followed by ion-exchange chromatography afforded **1**.

Different routes for the synthesis of **47** from D-mannose were also reported (Scheme 5).[221] Thus, the protected derivative **49** was obtained from benzyl-α-D-mannopyranoside. The sulfonate group in **49** was subjected to nucleophilic displacement with allylmagnesium chloride followed by desilylation and Swern oxidation to give **50**. Lemieux–Johnson degradation[222] of **50** followed by treatment with diazomethane afforded **53** in 78% overall

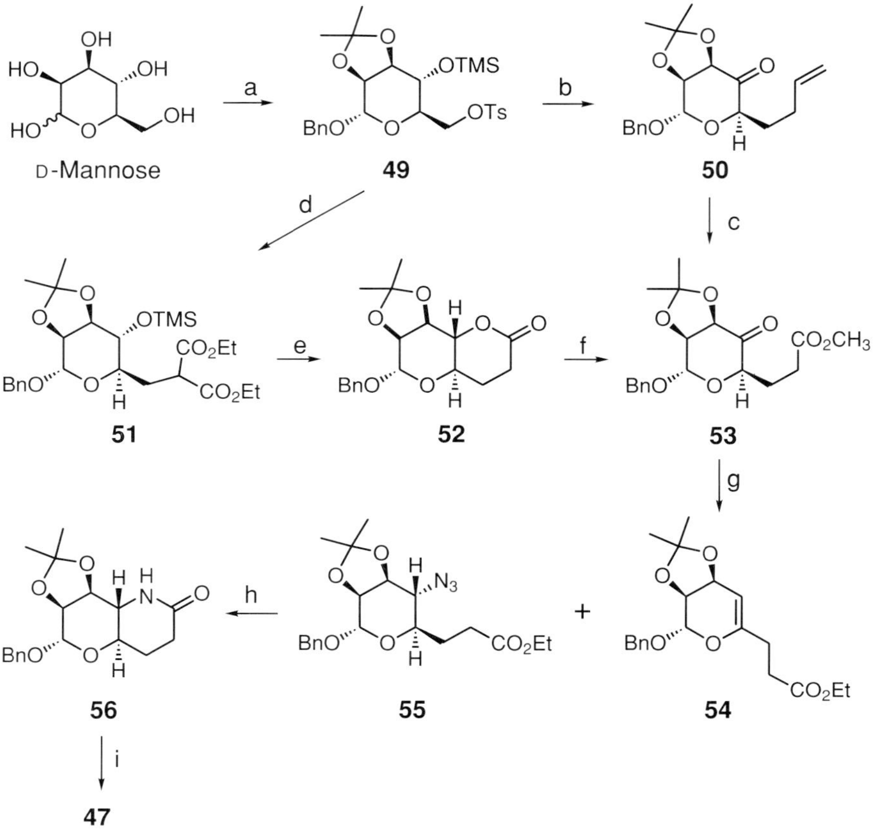

Scheme 5 (*a*) 1. BnOH, HCl, 83%; 2. *p*-TsCl, Py, rt, 75%; 3. DMP, NSA, acetone, rt, 94%; 4. TMSCl, NEt_3, THF, rt, 94%. (*b*) 1. AllMgCl, ether, 88%; 2. $Bu_4NF{\cdot}3H_2O$, THF, rt, 98%; 3. $(COCl)_2$, DMSO, NEt_3, CH_2Cl_2, −60°C, 95%. (*c*) 1. $NaIO_4$, $RuO_2{\cdot}xH_2O$, CCl_4, CH_3CN, H_2O, rt, 15 h, 96%; 2. CH_2N_2, Et_2O, 100%. (*d*) $NaCH(CO_2Et)_2$, toluene, reflux, 87%. (*e*) DMSO, NaCl, H_2O, 145°C, 15 h, 70%. (*f*) 1. 1 M KOH, CH_3OH, rt, 96%; 2. PCC, CH_2Cl_2, 15 h, rt, 95%. (*g*) 1. $NaBH_4$, EtOH, rt, 97%; 2. Tf_2O, Py, −20°C, 88%; 3. NaN_3, DMF, 15 h, rt, 97%. (*h*) 1. H_2, Pd black, rt, 6 h; 2. toluene, reflux, 1 h, 97% for two steps. (*i*) $LiAlH_4$, THF, rt, 15 h, 89%.

yield from **49**. Alternatively, compound **53** was obtained from **49** in 56% yield upon displacement of the sulfonate group with sodium diethyl malonate to give **51**, followed by decarboxylation to give **52**, whose saponification, esterification and then oxidation gave **53**. Reduction of **53** followed by triflation and then displacement of the triflate group with sodium azide afforded **54** and **55**. Reductive cyclization of **55** followed by $LiAlH_4$ reduction of the resulting lactam **56** gave **47** in 48% overall yield from **53**.

Alternatively, methyl 6-*O*-benzoyl-2,3-*O*-isopropylidene-α-D-talopyranoside (**57**), derived from D-mannose,[223] was utilized for the synthesis of swainsonine but in low yield (Scheme 6).[224] Mesylation of **57** followed by removal of the isopropylidene group with TFA, displacement of the mesylate group with azide ion, acetonation with DMP in acetone and

Scheme 6 (*a*) Ref. 223. (*b*) 1. MsCl, Py, 88%; 2. TFA, CH_3OH, rt, 30 min, 98%; 3. NaN_3, DMF, 110–115°C, 3 h, 77%; 4. DMP, *p*-TsOH, acetone; 5. KOH, CH_3OH, 98%. (*c*) 1. SO_3·Py, DMSO, NEt_3, 10 min; 2. $Ph_3P{=}CHCO_2CH_3$, THF, rt, 4 days, 56% for two steps. (*d*) 1. H_2, Pd black, CH_3OH; then CH_3OH, reflux, 12 h, 34%; 2. BH_3, THF, ice cooling, 30 min, 78%. (*e*) 1. BCl_3, CH_3Cl, −78°C, 1.5 h to rt, 16 h; 2. $NaCNBH_3$, H_2O–CH_3OH (1:1), 0.1 M HCl, rt, 24 h, 1.8%.

subsequent debenzoylation afforded **58** in 65% overall yield from **57**. Oxidation of the primary hydroxyl group with pyridine–SO_3 followed by condensation with $Ph_3P{=}CHCO_2CH_3$ afforded the olefin **59**. Hydrogenation of **59** followed by refluxing in methanol and subsequent reduction of the resulting lactam with BH_3 in THF afforded **60** in 27% yield. Reaction of **60** with boron trichloride followed by reduction with sodium cyanoborohydride gave **1**.

Ring transformation of the mannopyranoside derivative **61**, obtained from noncarbohydrate, to **1** has been developed (Scheme 7).[225,226] Radical cyclization of the thiocarbonylimidazolo derivative of **61** gave **62**, which upon oxidation and reduction afforded **63** (30%) and **64** (55%). The latter was benzylated to give **65**, which was converted into (*E*)-oxime **67** via **66**. Beckmann rearrangement of **67** followed by desilylation furnished **68**. Cyclization of **68** to indolizidine skeleton followed by debenzylation, reduction of the lactam with BMS and hydrolysis afforded (–)-swainsonine (**1**).

A synthesis of the intermediate **71**, as a precursor to (–)-swainsonine (**1**), has been reported (Scheme 8).[218] Prolonged hydrogenation of the azide **69**, obtained from D-mannose in eight steps, in methanol and then in acetic acid afforded the pyrrolidine **70** in 90% yield. Protection of the secondary amine in **70** with benzyl chloroformate followed by sodium periodate oxidation and subsequent sodium borohydride reduction gave **71**.

The aldehyde **73**[227] was prepared from diacetone D-mannose **72** in 80% overall yield (Scheme 9).[228] Treatment of **73** with allyltrimethylsilane followed by benzylation gave the benzyl ether **74**. Opening of the furanoside ring gave the hydroxyalkene **75** (78%), which underwent hydroboration followed by oxidation of the resulting diol to afford the ketoaldehyde **76**. Removal of the *p*-methoxybenzyl group with DDQ provided **77**, which

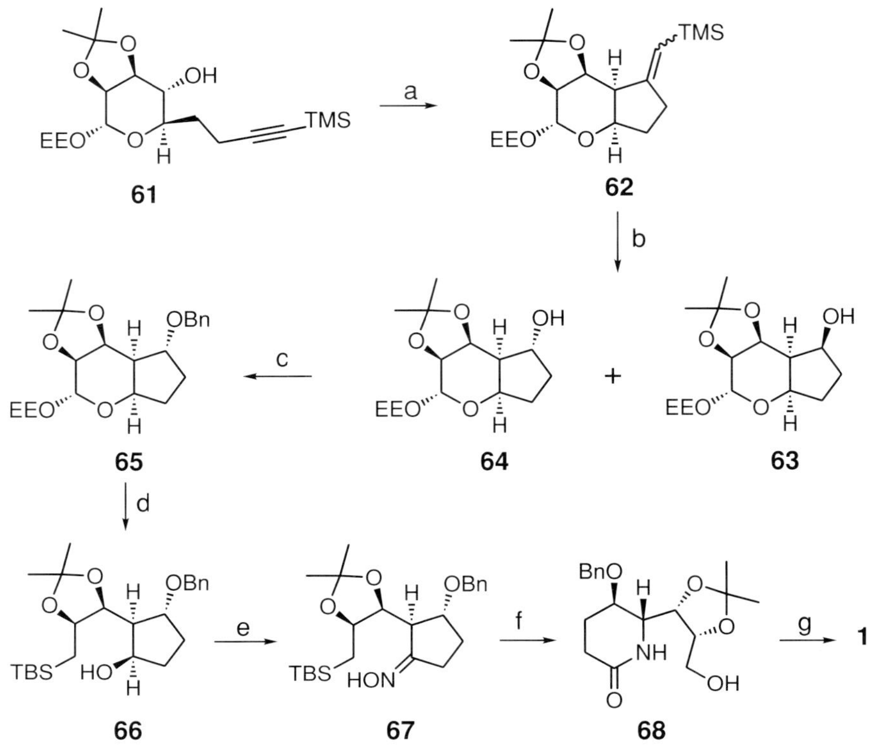

Scheme 7 (*a*) 1. Im_2CS, $ClCH_2CH_2Cl$, DMAP, rt, 5 h, 68%; 2. Bu_3SnH, AIBN, benzene, reflux, 30 min, 92%. (*b*) 1. OsO_4, *t*-BuOH, Py, rt, 4.5 h; then $NaIO_4$; 2. $NaBH_4$, CH_3OH, **63** (96%); *or* Na, liquid NH_3, THF–EtOH (1:1), −78°C, 20 min, rt, 1 h, **63** (30%), **64** (55%). (*c*) BnBr, Bu_4NI, NaH, THF, 0°C, 98%. (*d*) 1. 2 N HCl, THF, rt, 2 h, 98%; 2. $NaBH_4$, CH_3OH, CH_2Cl_2, 100%; 3. TBSCl, NEt_3, DMAP, CH_2Cl_2, rt, 10 h, 100%. (*e*) 1. NMO, MS 4 Å, $(n\text{-}Pr)_4NRuO_4$, rt, 30 min; then $NH_2OH{\cdot}HCl$, Py, rt, 30 min, 86%. (*f*) 1. $SOCl_2$, 1 h, 83%; 2. Bu_4NF, THF, rt, 1 h, 100%. (*g*) 1. MsCl, NEt_3, DMAP, CH_2Cl_2, 0°C, 1 h; then K_2CO_3, 1,4-dioxane, 90°C, 1 h, 96%; 2. H_2, 20% $Pd(OH)_2$ *on* C, EtOH, rt, 1 h, 99%; 3. BMS, THF, rt, 1 h; then K_2CO_3, 65°C, 2 h, 99%; then hydrolysis.

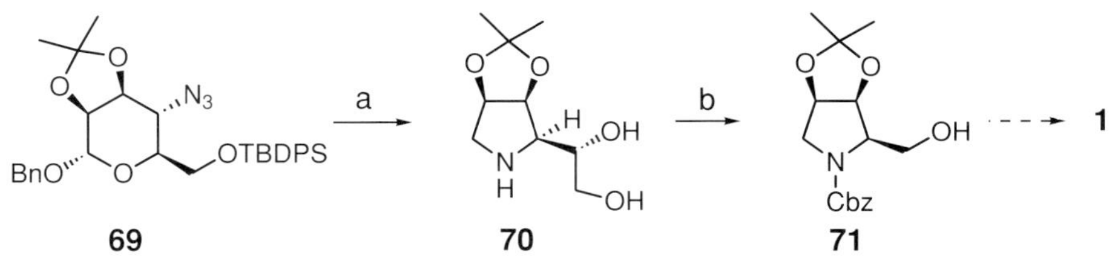

Scheme 8 (*a*) 1. Pd black, CH_3OH, H_2, rt, 1 h, 100%; 2. Pd black, AcOH, H_2. (*b*) 1. CbzCl, $NaHCO_3$; 2. $NaIO_4$; 3. $NaBH_4$.

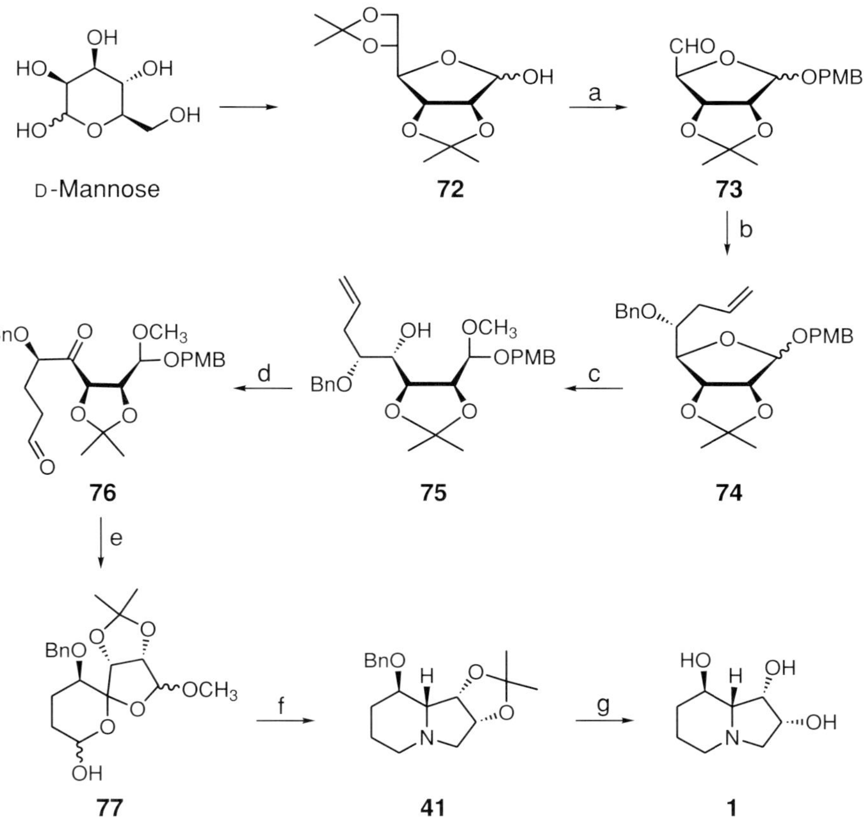

Scheme 9 (*a*) 1. PMBCl, NaH, *n*-Bu_4NI, DMF; 2. AcOH; 3. $NaIO_4$. (*b*) 1. Allyltrimethylsilane, $BF_3 \cdot OEt_2$, CH_2Cl_2, $-78°C$, 2 h, 77%; 2. BnBr, NaH, *n*-Bu_4NI, DMF, 97%. (*c*) 1. IDCP, CH_2Cl_2, CH_3OH; 2. Zn, 95% EtOH, reflux, 78%. (*d*) 1. BH_3, THF, 0°C, 1 h to rt, 18 h; then Na_2O_2, 86%; 2. $(COCl)_2$, DMSO, NEt_3, CH_2Cl_2, 84%. (*e*) DDQ, NEt_3, CH_2Cl_2, H_2O, 79%. (*f*) NH_4HCO_3, $NaCNBH_3$, CH_3OH, rt, 24 h, 69%. (*g*) 1. 10% Pd *on* C, CH_3OH, HCO_2H; 2. HCl, THF, H_2O, 80%.

was converted into the indolizidine **41**. Removal of the protecting groups from **41** led to **1** in 80% yield.

An open chain derivative of D-mannose has also been used for the synthesis of (–)-swainsonine (**1**) (Scheme 10).[229] The oxime **78**[230] was reduced with $LiAlH_4$, followed by protection of the resulting amine and treatment with MsCl to produce **79**. Partial hydrolysis of the terminal isopropylidene group in **79** with *p*-toluenesulfonic acid at room temperature, followed by epoxide ring formation via displacement of the mesylate group, and then oxidation of the primary hydroxyl group with Collins reagent afforded the corresponding aldehyde, which was subjected to the Wittig reaction with $Ph_3P{=}CHCO_2Et$ to give the *trans*-α,β-unsaturated ester **80**. The latter was reduced smoothly with sodium borohydride to afford **81** in 58% yield. Hydrogenation of **81** followed by heating in ethanol, to effect spontaneous double cyclization, afforded the lactam **82**. Reduction of the lactam carbonyl group in **82** using sodium borohydride gave **48**, which underwent acid hydrolysis to furnish **1**.

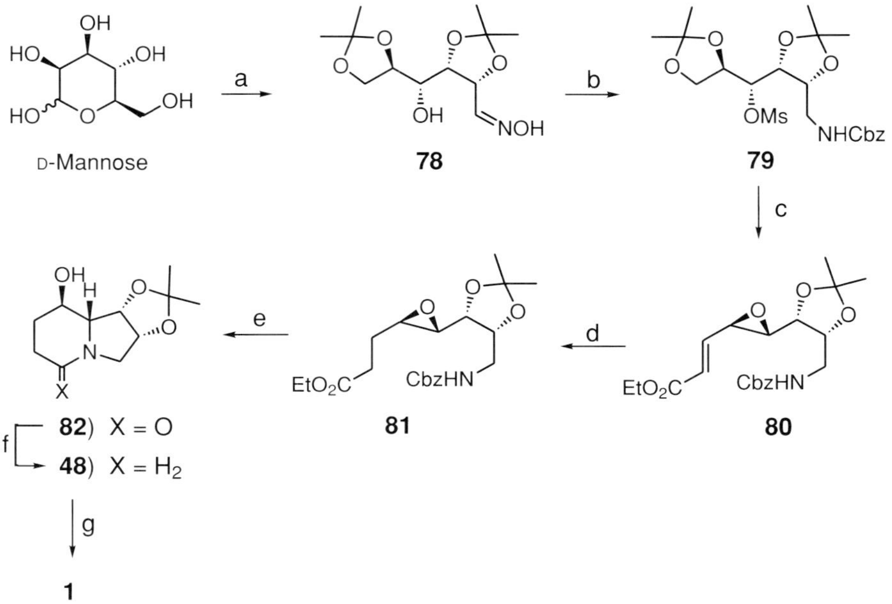

Scheme 10 (*a*) Refs. 230. (*b*) 1. $LiAlH_4$, THF, rt; 2. CbzCl, aqueous THF, 0°C; 3. MsCl, Py, 0°C, 95% for three steps. (*c*) 1. *p*-TsOH, CH_3OH–H_2O, rt, 3 days; 2. Amberlite IRA-400 (OH^-) resin, 43%, recovery 33%; 3. Collins reagent, CH_2Cl_2, 5°C; 4. $Ph_3P{=}CHCO_2Et$, THF, 0°C, 43% for two steps. (*d*) $NaBH_4$ (10 equiv.), EtOH–THF (10:1), reflux, 1 h, 58%. (*e*) H_2, 10% Pd *on* C, EtOH; then EtOH, reflux, 4 h, 60%. (*f*) $NaBH_4$ (10 equiv.), EtOH–THF (10:1), reflux, 1 h, 60%. (*g*) 6 N HCl, THF, rt, 75%.

Compound **79** was also converted to the epoxide **83** in two steps, which upon hydrogenation followed by treatment with di-*tert*-butyl dicarbonate afforded the pyrrolidine derivative **84**. Periodate oxidation of the diol **84** afforded the aldehyde **85**, which could serve for the synthesis of **1** (Scheme 11).[231,232]

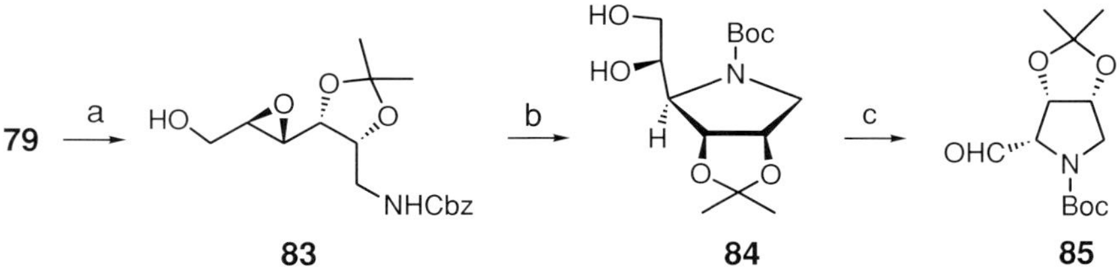

Scheme 11 (*a*) 1. *p*-TsOH–H_2O, CH_3OH, H_2O, rt; 2. epoxide formation. (*b*) 10% Pd *on* C, H_2, EtOH, rt, 5 h; then $(Boc)_2O$, NEt_3, THF, 81% from **83**. (*c*) $NaIO_4$, THF, rt, 1.5 h.

The acyclic dimesylate derivative **86** was also used for the synthesis of **1** (Scheme 12).[233–235] Compound **86** was converted to the azide derivative **87** and then into the 4,5-anhydro-1-azido-2,3-*O*-isopropylidene-D-talitol **88** in three steps. Triflation of **88** followed by two-carbon elongation with lithium *tert*-butyl acetate afforded **89**. Intramolecular

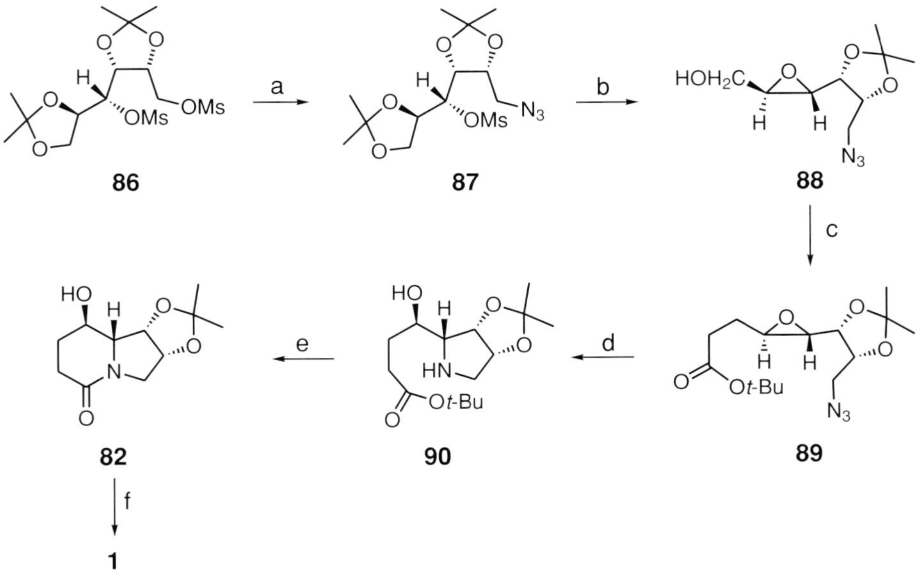

Scheme 12 (*a*) NaN_3, DMF, H_2O, 62%. (*b*) 1. Aqueous CH_3OH, CSA, 56%; 2. $Ba(OCH_3)_2$, CH_3OH, 95%. (*c*) 1. Tf_2O, Py; 2. $LiCH_2CO_2t$-Bu, THF, 60% for two steps. (*d*) H_2, Pd *on* C, EtOH, 80%. (*e*) $NaOCH_3$, CH_3OH, reflux, 92%. (*f*) 1. BMS, 70%; 2. TFA, H_2O, 86%.

double cyclization of **89** furnished the lactam **82** via intermediate **90**. Reduction of **82** with BMS followed by acid hydrolysis gave **1**.

5.4.1.2.3 Synthesis from D-lyxose A methodology using an intramolecular cyclization to an enantiomerically pure cyclic acyliminum ion intermediate has been developed for the synthesis of (–)-swainsonine (**1**) (Scheme 13).[236] Treatment of D-lyxose with 1-methoxycyclohexene followed by prolonged heating with Ag_2CO_3–Celite[237] in benzene afforded the lactone **91**, which was converted into the hydroxy lactam **92** in 32% overall yield from D-lyxose. The formation of the indolizidine ring system was achieved, in 60% yield, by mesylation of **92** in the presence of triethylamine, followed by stirring overnight in CH_3CN to produce **93**. Introduction of a C-8=C-8a double bond, followed by removal of the lactam carbonyl group from **94** with Meerwein's reagent and then reduction of the resultant iminium ion **95** from the less hindered convex face with $NaBH_3CN$, led to the required stereochemistry at the ring junction in lactam **96**. Conversion of **96** into the unstable ketone **97** followed by reduction with $NaBH_4$ or $LiAlH_4$ under a variety of conditions gave a mixture of epimers at C-8, favoring the formation of 8-*epi*-swainsonine. However, treatment of the ketone **97** with Na/NH_3 followed by removal of the cyclohexylidene ketal afforded **1**.

5.4.1.2.4 Synthesis from D-ribose D-Ribose has been used for a facile synthesis of the pyrrolidine derivative **103**, an intermediate for the preparation of **1** and some of its analogues (Scheme 14).[238] Thus, D-ribonolactone could be converted to the benzylidene derivative **98**,[239,240] whose reduction with $LiAlH_4$ followed by treatment with MOMCl gave compounds **99** (53%), **100** (10.5%) and **101** (5%). The major product **99** was mesylated and the

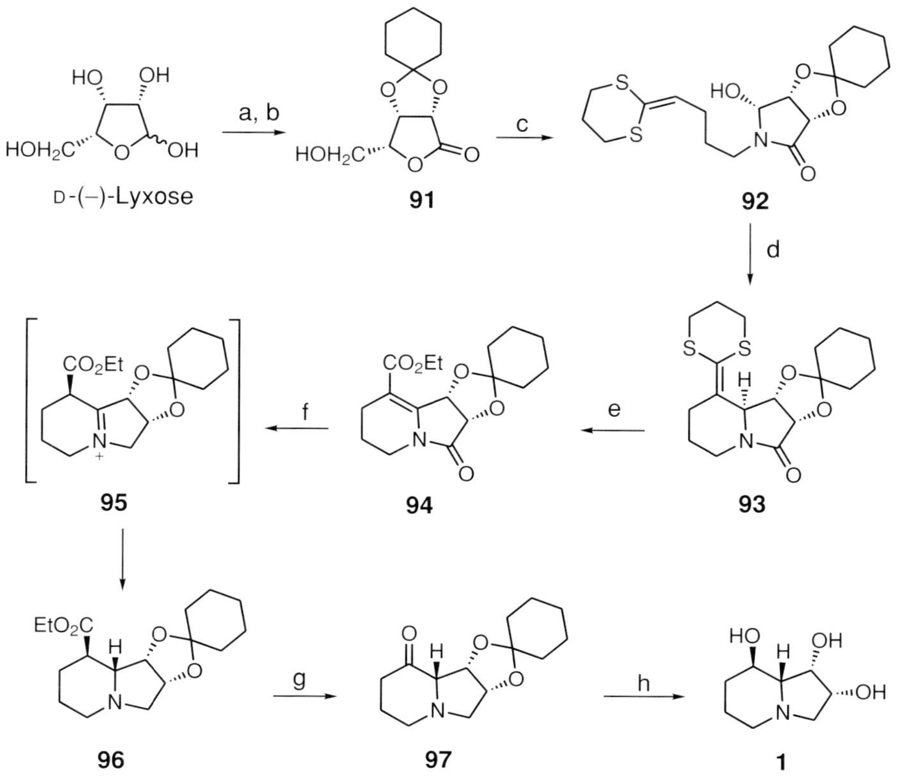

Scheme 13 (*a*) 1-Methoxycyclohexene, $BF_3{\cdot}OEt_2$, THF, 78%. (*b*) Ag_2CO_3, Celite, PhH, 65%. (*c*) 1. $H_2NCH_2CH_2CH_2CH{=}C$(1,3-dithiopropane), CH_3OH; 2. $Pb(OAc)_4$, CH_3CN, 63%. (*d*) MsCl, NEt_3, CH_2Cl_2; then CH_3CN, overnight, rt, 60%. (*e*) 1. NBS, EtOH, CH_3CN; 2. DBU, THF, 71% for two steps. (*f*) 1. Et_3OBF_4, CH_2Cl_2; 2. $NaCNBH_3$, CH_3OH, 86% for two steps. (*g*) 1. LDA, THF, O_2, 76%; 2. $LiAlH_4$, THF; 3. $NaIO_4$, H_2O. (*h*) 1. Na/NH_3, H_2O, THF, 45% for four steps; 2. 6 M HCl, 95%.

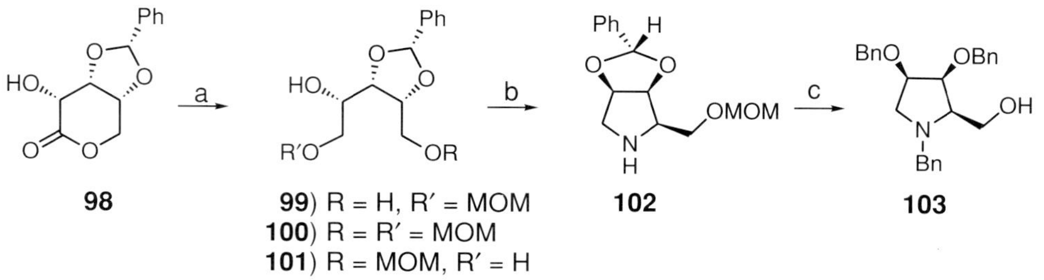

Scheme 14 (*a*) 1. $LiAlH_4$, THF, rt, 5 h, 92%; 2. ClMOM, CH_2Cl_2, −10 to −20°C, 32 h, **99** (53%), **100** (10.5%), **101** (5%). (*b*) 1. MsCl, Py, 0°C, 15 h, 100%; 2. NaN_3, DMF, 110–120°C, 2.5 h, 69%; 3. H_2, Pd black, EtOH, 73%. (*c*) 1. BnBr, K_2CO_3, acetone, rt, 2 h, 95%; 2. 10% aqueous HCl, CH_3OH, 40°C, 2 h, 64%; 3. BnBr, NaH, DMF, THF, rt, 4 h, 92%; 4. 10% aqueous HCl, CH_3OH, 70°C, 2 h, 96%.

terminal mesyloxy group was displaced with azide ion, followed by hydrogenation to produce the pyrrolidine **102** (50%), which was N-benzylated, debenzylidenated, O-benzylated followed by removal of MOM to give **103** in 13% overall yield from **98**.

5.4.1.2.5 Synthesis from D-erythrose The synthesis of **1** has been achieved from D-erythrose via its 2,3-*O*-isopropylidene derivative **104**[241–243] (Scheme 15).[244] Reaction of **104** with $EtO_2C(CH_2)_3P^+Ph_3Br^-$ and $KN(TMS)_2$ followed by tosylation of the generated primary hydroxyl group afforded the olefinic ester **105**. Displacement of the tosyloxy group with NaN_3 and subsequent intramolecular 1,3-dipolar cycloaddition afforded **107** via the triazoline intermediate **106**. Mild hydrolysis of **107** followed by cyclization in boiling toluene provided the lactam **109**, via an acyl group migration and subsequent dehydration of the possible intermediate **108**. This was then treated with borane and alkaline hydrogen peroxide to produce the swainsonine acetonide as a single diastereomer, whose deisopropylidenation gave (–)-swainsonine (**1**). Two patents[245,246] have also described a similar methodology for the synthesis of **1**.

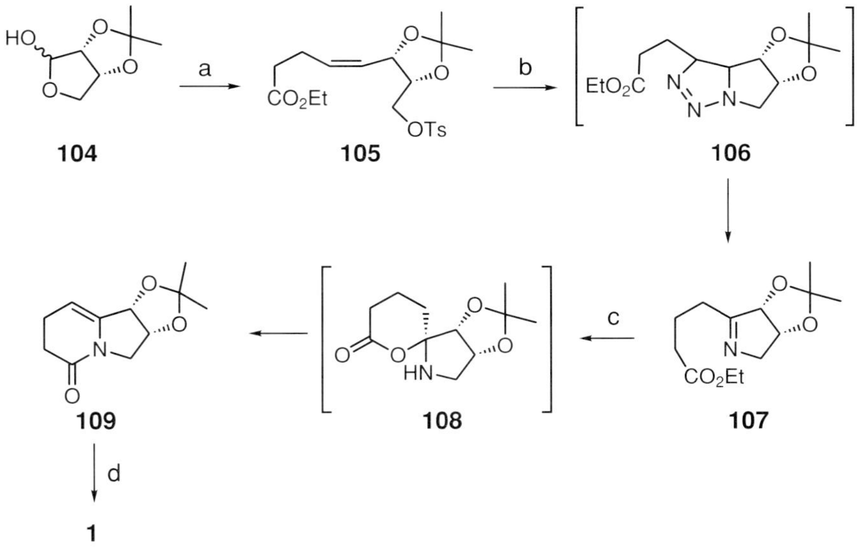

Scheme 15 (*a*) 1. $EtO_2C(CH_2)_3P^+Ph_3Br^-$, $KN(TMS)_2$, THF, −78 to 0°C; 2. *p*-TsCl, NEt_3, CH_2Cl_2. (*b*) NaN_3, DMF, 70–100°C, 81% for three steps. (*c*) 1. K_2CO_3, aqueous CH_3OH, rt, 12 h, 74%; 2. toluene, reflux in Dean–Stark trap, 30 h, 87%. (*d*) 1. BH_3, THF, 0°C to rt, overnight; then H_2O_2, NaOH, EtOH, reflux, 2 h, 79%; 2. 6 N HCl, THF, rt, overnight, 85%.

An analogous strategy utilizing D-erythrose led to an efficient synthesis of **1** (Scheme 16).[247] 2,3-*O*-Isopropylidene-D-erythrose (**104**) was treated with Wittig reagent to give the olefin **110**. This was subjected to a Mitsunobu reaction to afford the azide intermediate **111**, whose intramolecular cycloaddition in refluxing benzene produced the bicyclic iminium ion **112**. Treatment of **112** with *tert*-butylamine gave **113**, which upon hydroboration using the modification of Schultz method[248] afforded the acetonides **48** as a major product in addition to **114** (7%). Aqueous acid hydrolysis of **48** afforded **1** in 39% overall yield from **104**.

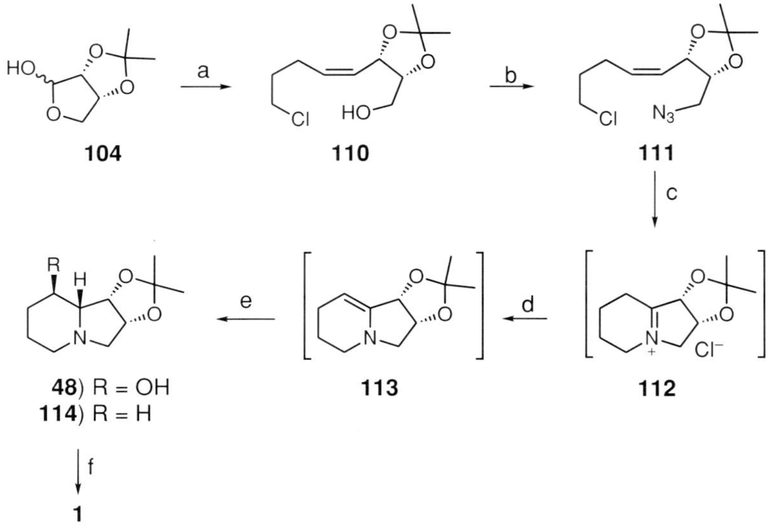

HO O O O a Cl HO O O b Cl N_3 O O

104 110 111

c

R H O O N e N O O d N + O O Cl^-

48) R = OH
114) R = H

113 112

f

1

Scheme 16 (*a*) $Cl(CH_2)_4P^+Ph_3Br^-$, $KN(TMS)_2$, THF, −78 to 23°C, 2 h, 86%. (*b*) $(PhO)_2P(O)N_3$, PPh_3, DEAD, THF, 23°C, 1 h, 76%. (*c*) PhH, reflux, 26 h. (*d*) *t*-$BuNH_2$, $KN(TMS)_2$. (*e*) BH_3–THF, 23°C, 10 h; then NaOAc, CH_3OH, H_2O_2, 23°C, 12 h, **48** (70%), **114** (7%). (*f*) 1. 6 N HCl, THF, 23°C, 12 h; 2. IRA-400 ion-exchange chromatography, 85%.

Aldehyde **85**, as an intermediate for the synthesis of **1**, was prepared from **104** (Scheme 17).[249] Olefination[250] of **104** generated the ketene *S*,*S*-acetal, which was converted into the desired azide **115** in 65% overall yield. Thermolysis of **115** followed by reduction and then N-protection gave the pyrrolidine **117** via the intermediate imine **116**. Cleavage[251] of the dithioacetal in **117** with $Tl(O_2CCF_3)_3$ afforded **85** in 94% yield.

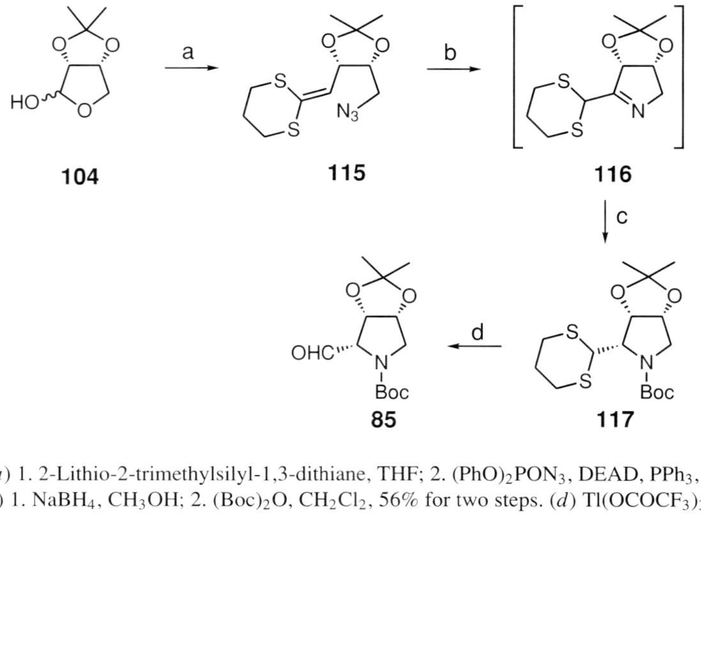

O O HO O a S S O O N_3 b S S O O N

104 115 116

c

O O OHC N Boc d S S O O N Boc

85 117

Scheme 17 (*a*) 1. 2-Lithio-2-trimethylsilyl-1,3-dithiane, THF; 2. $(PhO)_2PON_3$, DEAD, PPh_3, 65%. (*b*) Octane, 126°C. (*c*) 1. $NaBH_4$, CH_3OH; 2. $(Boc)_2O$, CH_2Cl_2, 56% for two steps. (*d*) $Tl(OCOCF_3)_3$, Et_2O, H_2O, 94%.

5.4.1.2.6 Synthesis from D*-erythronolactone* The D-erythronolactone **118**[243,252] was readily prepared from D-isoascorbic acid and is commercially available. It is an attractive precursor for the synthesis of natural products (Scheme 18).[253] Thus, aminolysis of **118** followed by cyclization[254] using hydrazine gave the pyrrolidinone **119**. Treatment of **119** with $TBSOCH_2CH_2CH_2Br$ afforded **120**, which was treated with Lawesson's reagent to furnish **121**. This was condensed with CH_3NO_2 and desilylated with HF to furnish **122**. Tosylation of the primary hydroxyl group followed by cyclization with NaI afforded **123** in 28% yield, a precursor for **1**.

D-Isoascorbic acid →(a) **118** →(b) **119** →(c) **120** →(d) **121** →(e) **122** →(f) **123**

Scheme 18 (*a*) Peroxide degradation, Refs. 251 and 252. (*b*) 1. K-phthalimide; 2. NH_2NH_2; 3. 150°C. (*c*) $TBSOCH_2CH_2CH_2Br$, NaH, DMF, 70%. (*d*) Lawesson's reagent, 61%. (*e*) 1. CH_3I, THF; 2. CH_3NO_2, K_2CO_3; 3. 40% HF, CH_3OH, 46%. (*f*) 1. *p*-TsCl, NEt_3; 2. NaI, reflux, CH_3CN, 28%.

118 →(a) **124** →(b) **125** →(c) **126** + **127** →(d) **128** →(e) **82** →(f) **1**

Scheme 19 (*a*) 1. DIBAL-H, CH_2Cl_2, −78°C, 2 h; 2. $CH_2{=}CHMgBr$, THF, −78 to 0°C, 6 h; 3. TBSCl, imidazole, THF–DMF (1:1), 0°C, 45 min, 73% (*anti*/*syn* 97:3). (*b*) $CH_3C(OCH_3)_3$, $EtCO_2H$, toluene, reflux, 24 h, 99%. (*c*) AD-Mix b, *t*-BuOH, $CH_3SO_2NH_2$, H_2O, 0–25°C, 18 h, then separate. (*d*) 1. TBAF, THF, 0°C, 1.5 h, 84%; 2. MsCl, Py, DMAP, 2°C, 16 h, 90%; 3. NaN_3, DMSO, 80°C, 36 h, 75%. (*e*) 1. H_2, $Pd(OH)_2$, CH_3OH, 6 h, 75%; 2. $NaOCH_3$, CH_3OH, reflux, 60 h. (*f*) 1. BMS, THF, 0°C, 30 min, rt, 2 h, 94%; 2. 6 N HCl, THF, rt, 12 h, Dowex 1X8-200 (OH^-) resin, 96%.

Multigram quantities of (–)-swainsonine (**1**) have also been prepared from 2,3-*O*-isopropylidene-D-erythronolactone (**118**) (Scheme 19).[255–257] Reaction of **118** with Grignard reagent gave the allylic alcohol **124** in 73% overall yield (*anti*/*syn* 97:3). Under Johnson orthoester Claisen rearrangement condition,[258] **124** was converted into the *E*-isomer **125**, which was submitted to the Sharpless dihydroxylation[259–261] to afford the lactones **126** and **127** in 9 and 70% yield, respectively. Removal of the silyl group from **127** followed by mesylation and subsequent selective displacement of the primary mesylate group with sodium azide produced **128**. Hydrogenation of **128** followed by treatment with sodium methoxide effected a reductive double cyclization to give the bicyclic lactam **82** in 75% yield. Reduction of **82** followed by acid hydrolysis of the isopropylidene group and subsequent purification over ion-exchange column gave **1** in 20% overall yield from lactone **118**.

A stereoselective iodoamination of an unsaturated trichloroacetimidate derivative has been used as a key step for the synthesis of **1** starting with lactone **118** (Scheme 20).[262] Reduction of **118** by DIBAL-H followed by olefination gave a 15:1 mixture of the *cis*- and

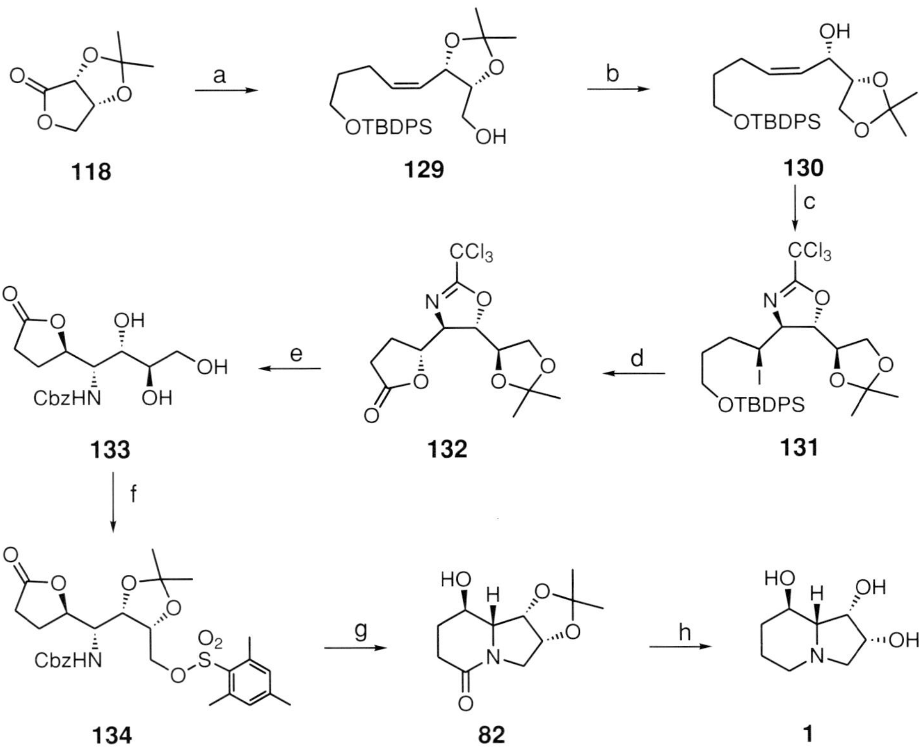

Scheme 20 (*a*) 1. DIBAL-H, CH_2Cl_2, −78°C; 2. $TBDPSO(CH_2)_4P^+Ph_3I^-$, *n*-BuLi, HMPA, THF, 0°C, 77% for two steps. (*b*) *p*-TsOH, acetone, rt, 93%. (*c*) 1. Cl_3CCN, DBU, CH_3CN, CH_2Cl_2, 0°C; 2. DBU (1 equiv.), IBr, CH_3CN, −60 to −50°C, 85–90%. (*d*) 1. NH_4F, CH_3OH, 45°C, 90%; 2. $(COCl)_2$, DMSO, CH_2Cl_2, −78°C; then NEt_3; 3. $NaClO_2$, 2-methyl-2-butene, NaH_2PO_4, aqueous *t*-BuOH, rt; 4. Ag_2CO_3, benzene, 65–70°C, 62% for three steps. (*e*) 1. TFA, H_2O, rt; 2. CbzCl, K_2CO_3, CH_3OH, 0°C, 90% for two steps. (*f*) 1. 2-Mesitylenesulfonyl chloride, NEt_3, CH_2Cl_2, 0°C, 84%; 2. DMP, *p*-TsOH, acetone, rt, 97%. (*g*) H_2, 10% Pd *on* C, K_2CO_3, rt to reflux, 97%. (*h*) 1. BMS, THF, rt; 2. H_2O_2, NaOH, reflux, 97%; 6 N HCl, rt, 92%.

trans-olefins **129** in 77% yield. Rearrangement of the location of the acetonide group in **129** to the terminal position was achieved in acetone in the presence of *p*-TsOH to give **130**. The trichloroacetimidate of **130** was subjected to a stereoselective iodoamination using iodine monobromide to afford *trans*-oxazoline **131**. Removal of the silyl group in **131** followed by oxidation of the primary hydroxyl group to the corresponding carboxylic acid and heating with silver carbonate provided the lactone **132** in 57% overall yield. Complete deprotection of **132** followed by protection of the generated amino group gave **133**. Selective sulfonylation of the primary hydroxyl group in **133** followed by reaction of the hydroxy groups with DMP afforded the acetonide **134**, which was hydrogenated in the presence of potassium carbonate to produce **82**. Reduction of **82** followed by deprotection afforded **1** in 23% overall yield from **118**.

5.4.1.2.7 Synthesis from hydroxymethyl butyrolactones (+)-Swainsonine (**2**) has been prepared from the hydroxymethyl butyrolactone **135** (Scheme 21).[263] The lactone **135**,[264]

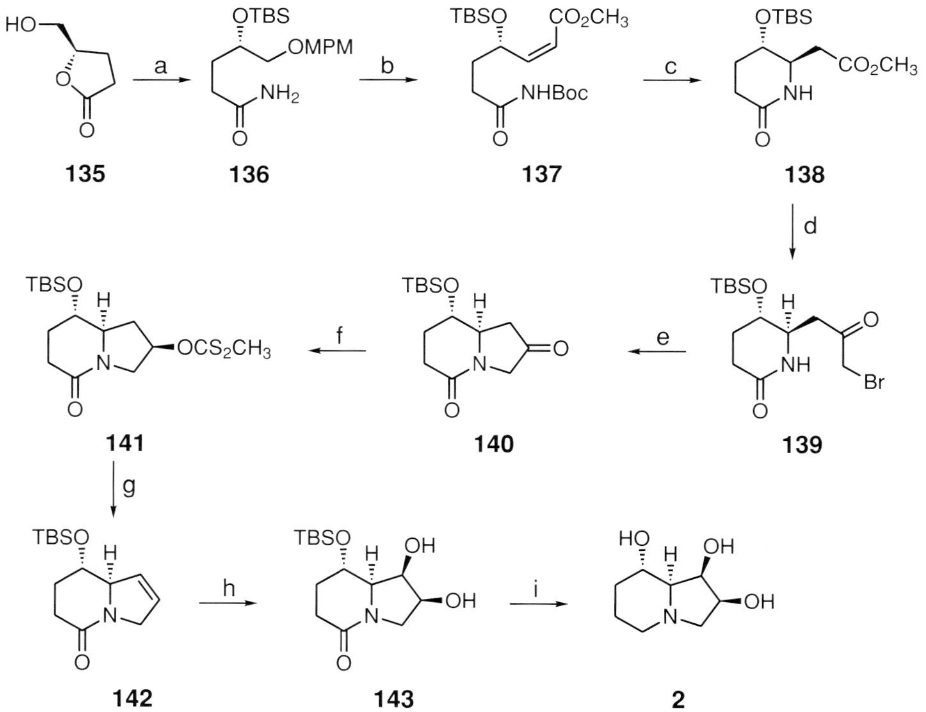

Scheme 21 (*a*) 1. NaH, MPMCl, THF, DMF, 76%; 2. NH_4OH, Et_2O, 0°C, 79%; 3. TBSCl, imidazole, DMF, 91%. (*b*) 1. KH, Boc-S, THF, −30 to 5°C, 81%; 2. DDQ, CH_2Cl_2, H_2O, 94%; 3. DMSO, $(COCl)_2$, CH_2Cl_2; then NEt_3, 81%; 4. $(CF_3CH_2O)_2P(O)CH_2CO_2CH_3$, 18-crown-6, KHMDS, toluene, −78°C, 85%, *Z*/*E* 4.3:1; separation. (*c*) 1. TMSI, $CHCl_3$, 65%; 2. *t*-BuOK, THF, −55°C, 80%. (*d*) $LiCHBr_2$, THF, −90°C; then BuLi, −90°C, 59%. (*e*) 1. K_2CO_3, CH_3OH, 92%; 2. MsCl, NEt_3, CH_2Cl_2, 94%; 3. KH, THF, 87%; 4. *p*-TsOH, acetone, 77%. (*f*) 1. $NaBH_4$, CH_3OH, 0°C, 98%; 2. NaH, THF, CS_2; then CH_3I, 98%, β/α 6.7:1. (*g*) 180°C, 68%. (*h*) OsO_4, NMO, acetone, H_2O, rt, 82%. (*i*) 1. TFA, THF, H_2O; then Ac_2O, Py, CH_2Cl_2, 84%, α/β 1:6.9; separation; 2. BH_3–THF, reflux; K_2CO_3, CH_3OH; then 2 M HCl, reflux, 85%.

prepared from L-glutamic acid, was converted to the amide **136** in 55% overall yield. Protection of the amino group in **136** followed by removal of the MPM group, Swern oxidation and subsequent Wadsworth–Emmons-type reaction afforded the olefin **137**. Subsequent intramolecular conjugate addition afforded exclusively the diastereomer **138**. One carbon elongation of **138** furnished the bromo ketone **139**, which failed to undergo base-catalyzed intramolecular cyclization. Bromo ketone **139** was converted to ketone **140**, which was reduced by sodium borohydride, followed by treatment with CS_2, NaH and CH_3I to give the xanthate **141**. Pyrolysis of **141** gave the olefin **142**, whose dihydroxylation occurred from the opposite face to give **143** as the major isomer. Acetylation and separation of the mixture followed by treatment with BH_3, alkaline hydrolysis of the acetates and acid treatment afforded **2**.

The lactone **145**, obtained from lactone **144** in two steps, gave upon reduction, followed by mesylation and selective displacement of the primary *O*-mesyl group with azide ion, the azide **146**. Reductive cyclization and benzylation of **146** furnished **147**, whose primary hydroxyl group was oxidized to produce the aldehyde **148**. On the other hand, protection of

144 **145** **146** **147**

152) R = R′ = Bn
153) R + R = $C(CH_3)_3$, R′ = Bn

150) R = R′ = Bn
151) R + R = $C(CH_3)_3$, R′ = Bn

148) R = Bn
149) R + R = $C(CH_3)_3$

154) R = Bn
155) R + R = $C(CH_3)_3$

12

156) R = Bn
157) R + R = $C(CH_3)_3$

Scheme 22 (*a*) 1. OsO_4, NMO, aqueous acetone; 2. DMP, acetone, *p*-TsOH. (*b*) 1. $LiAlH_4$, THF; 2. MsCl, Py; 3. NaN_3, DMF, 130°C. (*c*) 1. H_2, Pd black, EtOH; 2. BnBr, K_2CO_3, acetone. (*d*) Oxidation; *or* 1. MOMCl, *N*,*N*-diethylaniline; 2. 10% HCl, CH_3OH, 40°C; 3. NaH, BnBr, DMF–THF; then 10% HCl, CH_3OH, 70°C; 4. oxidation. (*e*) 1. AllMgCl, THF, −78°C, 1 h; *or* allylMgCl, CuI, THF–DMS (5:1), −78°C to rt, 1 h; *or* allyltrimethylsilane, $TiCl_4$, CH_2Cl_2, −78°C, 2 h; 2. NaH, THF–DMF (1:1), BnBr, rt, 2 h. (*f*) BH_3, THF, 45°C, 1 h; then 3 N NaOH, H_2O_2, 60°C, 1 h; then 10% aqueous HCl, 60°C, 5 min; then 10% aqueous NaOH. (*g*) 1. MsCl, NEt_3, CH_2Cl_2, rt, 6 h, 2. 10% aqueous HCl–CH_3OH, 70°C, 1 h; *and/or* H_2, 10% Pd *on* C, EtOH, HCl, CH_3OH, rt, 6 h.

the primary hydroxyl group in **147** with MOM followed by removal of the isopropylidene group and subsequent benzylation, removal of MOM group and oxidation afforded **149**. Reaction of **149** with Grignard reagent gave **151** and **153** in a 1:1.3 ratio in 80% yield, while the reaction with the organocopper reagent afforded a 1:3.2 ratio of **151** and **153** in 71% yield. However, addition of allylmagnesium chloride to **148** gave a 3:1 ratio of **150** and **152** in 85% yield whose opposite diastereoselectivity was observed when the organocopper reagent was used (**150**:**152** 1:3.8, 56% yield). On the other hand, condensation of allyltrimethylsilane with **148** in the presence of $TiCl_4$ produced only **152** (48%) and with **149** gave a 1:1.8 ratio of **151** and **153**, respectively. The high diastereoselectivity could be explained by cyclic chelate formation between $TiCl_4$ and the α-aminocarbonyl group of **148** and **149**, in which the nucleophile approaches from the less hindered side to yield **152** and **153**. Compounds **150** and **151** were converted to the alcohols **156** and **157**, respectively, and then to **1**. Similarly, **154** and **155** were converted to **12** (Scheme 22).[265,266]

5.4.1.2.8 Synthesis from D-glucoheptonolactone Syntheses of (+)-swainsonine (**2**) and dehydro-(+)-swainsonine (**165**) have been achieved (Scheme 23)[59–61] by reduction of D-glucoheptonolactone, followed by one-carbon elongation and acetonation to afford the

Glucoheptonolactone

158

159

160

161

162

163

164

165

2

Scheme 23 (*a*) 1. $NaBH_4$, H_2O; 2. NaCN, H_2O, rt, 68 h, reflux, 23 h, 19%; 3. H_2SO_4, acetone, rt, 16 h, 73%. (*b*) 1. $LiBH_4$, THF; 2. MsCl, Py, DMAP, 91% for two steps. (*c*) $BnNH_2$, 110°C, 2 days, 93%. (*d*) 1. *p*-TsOH, CH_3OH, 68%; 2. MsCl, Py, DMAP, 91%; 3. H_2, Pd black, EtOH, NaOAc, 62%. (*e*) 80% AcOH–H_2O, 85%. (*f*) Im_2CS, toluene; then TBSOTf, Py, CH_2Cl_2, 72%. (*g*) $(EtO)_3P$, heat, 76%. (*h*) 1. H_2, Pd black, EtOAc, 89%; 2. TFA–D_2O (1:1), 74%. (*i*) TFA–D_2O (1:1), 80%.

triacetonide lactone **158**. Reduction of **158** followed by mesylation gave **159**, which upon boiling with benzylamine gave the pyrrolidine **160**. Removal of the terminal isopropylidene group in **160** followed by regioselective mesylation and intramolecular cyclization provided the bicyclic diacetonide **161** in 5% overall yield from glucoheptonolactone. Regioselective hydrolysis of **161** gave **162** whose reaction with 1,1′-thiocarbonylimidazole and then *tert*-butyldimethylsilyl triflate furnished the thionocarbonate **163**. Corey–Winter fragmentation of **163** gave olefin **164**, which upon hydrogenation followed by complete deprotection afforded **2** in 31% overall yield from **161**. Complete deprotection of **164** afforded **165**.

References

1. Guengerich, F.P.; DiMari, S.J.; Broquist, H.P. *J. Am. Chem. Soc.* **95** (1973) 2055.
2. Schneider, M.J.; Ungemach, F.S.; Broquist, H.P.; Harris, T.M. *Tetrahedron* **39** (1983) 29.
3. Daniel, L.R.; Hagler, W.M., Jr.; Croom, W.J.J. *Biodeterior. Res.* (1994) 85; *Chem. Abstr.* **122** (1995) 310336w.
4. Harris, C.M.; Campbell, B.C.; Molyneux, R.J.; Harris, T.M. *Tetrahedron Lett.* **29** (1988) 4815.
5. Bartlett, H.S.; Wilson, M.E.; Croom, J.J.; Broquist, H.P.; Hagler, W.M., Jr.; *Biodeterior. Res. 1* (1987) 135; *Chem. Abstr.* **110** (1989) 36551j.
6. Broquist, H.P.; Mason, P.S.; Wickwire, B.; Homann, R.; Schneider, M.J.; Harris, T.M. *Plant Toxicol. Proc.* (**1985**) 301.
7. Broquist, H.P. *J. Toxicol. Toxin Rev.* **5** (1986) 241.
8. El Nemr, A. *Tetrahedron* **56** (2000) 8579.
9. Colegate, S.M.; Dorling, P.R.; Huxtable, C.R. *Aust. J. Chem.* **32** (1979) 2257.
10. Broquist, H.P. *Annu. Rev. Nutr.* **5** (1985) 391.
11. Colegate, S.M.; Dorling, P.R.; Huxtable, C.R. *Plant Toxicol.* (1985) 249.
12. Ermayanti, T.M.; McComb, J.A.; O'Brien, P.A. *J. Exp. Bot.* **45** (1994) 633.
13. Ermayanti, T.M.; McComb, J.A.; O'Brien, P.A. *Phytochemistry* **36** (1994) 313.
14. Molyneux, R.J.; James, L.F. *Science* **216** (1982) 190.
15. Molyneux, R.J.; James, L.F.; Panter, K.E.; Ralphs, M.H. *Phytochem. Anal.* **2** (1991) 125.
16. Davis, D.; Schwarz, P.; Hernandez, T.; Mitchell, M.; Warnock, B.; Elbein, A.D. *Plant Physiol.* **76** (1984) 972.
17. Hino, M.; Nakayama, O.; Tsurumi, Y.; Adachi, K.; Shibata, T.; Terano, H.; Kohsaka, M.; Aoki, H.; Imanaka, H. *J. Antibiot.* **38** (1985) 926.
18. Donaldson, M.J.; Bucke, C.; Adlard, M.W. *Microb. Util. Renewable Resour.* **7** (1990) 228.
19. Patrick, M.; Adlard, M.W.; Keshavarz, T. *Biotechnol. Lett.* **15** (1993) 997.
20. Sim, K.L.; Perry, D. *Mycol. Res.* **99** (1995) 1078.
21. Patrick, M.S.; Adlard, M.W.; Keshavarz, T. *Biotechnol. Lett.* **17** (1995) 433.
22. Patrick, M.S.; Adlard, M.W.; Keshavarz, T. *Enzyme Microb. Technol.* **18** (1996) 428.
23. Sim, K.L.; Perry, D. *Glycoconjugate J.* **14** (1997) 661.
24. Tamerler-Yildir, C.; Adlard, M.W.; Keshavarz, T. *Biotechnol. Lett.* **19** (1997) 919.
25. Tamerler, C.; Ullah, M.; Adlard, M.W.; Keshavarz, T. *FEMS Microbiol. Lett.* **168** (1998) 17.
26. Tamerler, C.; Keshavarz, T. *Biotechnol. Lett.* **21** (1999) 501.
27. El Ashry, E.S.H.; Rashed, N.; Shobier, A.H.S. *Pharmazie* **55** (2000) 331.
28. Molyneux, R.J.; James, L.F. *Mycotoxins Phytoalexins* (1991) 637.
29. Elbein, A.D. *FASEB J.* **5** (1991) 3055.
30. Winchester, B.; Al Daher, S.; Carpenter, N.C.; Cenci di Bello, I.; Choi, S.S.; Fairbanks, A.J.; Feelt, G.W.J. *Biochem. J.* **290** (1993) 743.
31. Liao, Y.F.; Lal, A.; Moremen, K.W. *J. Biol. Chem.* **271** (1996) 28348.
32. Cenci di Bello, I.; Fleet, G.W.J.; Namgoong, S.K.; Tadano, K.; Winchester, B. *Biochem. J.* **259** (1989) 855.
33. Nagahashi, G.; Tu., S.I.; Fleet, G.; Namgoong, S.K. *Plant Physiol.* **92** (1990) 413.
34. McGee, C.M.; Murray, D.R. *J. Plant Physiol.* **116** (1984) 467.
35. Tulsiani, D.R.P.; Broquist, H.P.; Touster, O. *Arch. Biochem. Biophys.* **236** (1985) 427.
36. Dorling, P.R.; Huxtable, C.R.; Colegate, S.M.; Winchester, B.G. *Plant Toxicol.* (1985) 255; Dorling, P.R.; Huxtable, C.R.; Colegate, S.M. *Biochem. J.* **191** (1980) 649.
37. Elbein, A.D.; Szumilo, T.; Sanford, B.A.; Sharpless, K.B.; Adams, C. *Biochemistry* **26** (1987) 2502.

38. Winkler, J.R.; Segal, H.L. *J. Biol. Chem.* **259** (1984) 15369.
39. Cenci di Bello, I; Dorling, P.R.; Winchester, B. *Biochem. J.* **215** (1983) 693.
40. Segal, H.L.; Winkler, J.R. *Prog. Clin. Biol. Res.* **180** (1985) 491.
41. Haeuw, J.F.; Strecker, G.; Wieruszeski, J.M.; Montreuil, J.; Michalski, J.C. *Eur. J. Biochem.* **202** (1991) 1257.
42. Jauhiainen, A.; Vanha-Perttula, T. *Int. J. Biochem.* **19** (1987) 267.
43. Tulsiani, D.R.P.; Touster, O. *J. Biol. Chem.* **262** (1987) 6506.
44. Vlasova, A.L.; Ushakova, N.A.; Preobrazhenskaya, M.E. *Biokhimiya* **56** (1991) 1479; *Chem. Abstr.* **115** (1991) 229146t.
45. Greenaway, V.A.; Naish, S.; Jessup, W.; Dean, R.T. *Biochem. Soc. Trans.* **10** (1982) 533.
46. Winchester, B. *Biochem. Soc. Trans.* **12** (1984) 522.
47. Glick, M.C.; De Santis, R.; Santer, U.V. *Prog. Clin. Biol. Res.* **175** (1985) 229.
48. Greenaway, V.A.; Jessup, W.; Dean, R.T.; Dorling, P.R. *Biochim. Biophys. Acta* **762** (1983) 569.
49. De Gasperi, R.; Al Daher, S.; Winchester, B.G.; Warren, C.D. *Biochem. J.* **286** (1992) 55.
50. Stannard, B.S.; Gesundheit, N.; Thotakura, N.R.; Gyves, P.W.; Ronin, C.; Weintraub, B.D. *Biochem. Biophys. Res. Commun.* **165** (1989) 788.
51. Elbein, A.D.; Solf, R.; Dorling, P.R.; Vosbeck, K. *Proc. Natl. Acad. Sci. USA* **78** (1981) 7393.
52. Kaushal, G.P.; Szumilo, T.; Pastuszak, I.; Elbein, A.D. *Biochemistry* **29** (1990) 2168.
53. Elbein, A.D.; Dorling, P.R.; Vosbeck, K.; Horisberger, M. *J. Biol. Chem.* **257** (1982) 1573.
54. Tulsiani, D.R.P.; Harris, T.M.; Touster, O. *J. Biol. Chem.* **257** (1982) 7936.
55. Tulsiani, D.R.P.; Touster, O. *Arch. Biochem. Biophys.* **224** (1983) 594.
56. Dorling, P.R.; Colegate, S.M.; Huxtable, C.R. *Toxicon* **3** (1983) 93.
57. Pastuszak, I.; Kaushal, G.P.; Wall, K.A.; Pan, Y.T.; Sturm, A.; Elbein, A.D. *Glycobiology 1* (1990) 71.
58. Hubbard, S.C.; Ivatt, R.J. *Annu. Rev. Biochem.* **50** (1981) 555.
59. Bell, A.A.; Pickering, L.; Watson, A.A.; Nash, R.J.; Griffiths, R.C.; Jones, M.G.; Fleet, G.W.J. *Tetrahedron Lett.* **37** (1996) 8561.
60. Davis, B.; Bell, A.A.; Nash, R.J.; Watson, A.A.; Griffiths, R.C.; Jones, M.G.; Smith, C.; Fleet, G.W.J. *Tetrahedron Lett.* **37** (1996) 8565.
61. Bell, A.A.; Nash, R.J.; Fleet, G.W.J. *Tetrahedron: Asymmetry* **7** (1996) 595.
62. Humphries, M.J.; Matsumoto, K.; White, S.L.; Molyneux, R.J.; Olden, K. *Cancer Res.* **48** (1988) 1410.
63. Dennis, J.W.; White, S.L.; Freer, A.M.; Dime, D. *Biochem. Pharmacol.* **46** (1993) 1459.
64. Bowen, D.; Southerland, W.M.; Bowen, C.D.; Hughes, D.E. *Anticancer Res.* **17** (1997) 4345.
65. Chen, X.; Liu, B.; Ji, Y.; Li, J.; Zhu, Z.; Yin, H.; Lin, J. *Shanghai Yixue* **21** (1998) 256; *Chem. Abstr.* **129** (1998) 254444k.
66. Olden, K.; Mohla, S.; Newton, S.A.; White, S.L.; Humphries, M.J. *Ann. N. Y. Acad. Sci.* **551** (1988) 421.
67. Myc, A.; Kunicka, J.E.; Melamed, M.R.; Darzynkiewicz, Z. *Cancer Res.* **49** (1989) 2879.
68. Spearman, M.A.; Damen, J.E.; Kolodka, T.; Greenberg, A.H.; Jamieson, J.C.; Wright, J.A. *Cancer Lett.* **57** (1991) 7.
69. Welch, D.R.; Lobl, T.J.; Seftor, E.A.; Wack, P.J.; Aeed, P.A.; Yohem, K.H.; Seftor, R.E.B.; Hendrix, M.J.C. *Int. J. Cancer* **43** (1989) 449.
70. Yagel, S.; Feinmesser, R.; Waghorne, C.; Lala, P.K.; Breitman, M.L.; Dennis, J.W. *Int. J. Cancer* **44** (1989) 685.
71. Bowen, D.; Adir, J.; White, S.L.; Bowen, C.D.; Matsumoto, K.; Olden, K. *Anticancer Res.* **13** (1993) 841.
72. Newton, S.A.; White, S.L.; Humphries, M.J.; Olden, K. *J. Natl. Cancer Inst.* **81** (1989) 1024.
73. Liu, B.; Lin, Y.; Yin, H.; Chen, X. *Zhonghua Zhongliu Zazhi* **20** (1998) 168; *Chem. Abstr.* **130** (1999) 20275z.
74. Dennis, J.W. *Cancer Res.* **46** (1986) 5131; Taylor, J.B.; Strickland, J.R. *J. Anim. Sci.* **76** (1998) 2857.
75. White, S.L.; Nagai, T.; Akiyama, S.K.; Reeves, E.J.; Grzegorzewski, K.; Olden, K. *Cancer Commun.* **3** (1991) 83.
76. Myc, A.; DeAngelis, P.; Lassota, P.; Melamed, M.R.; Darzynkiewicz, Z. *Clin. Exp. Immunol.* **84** (1991) 406.
77. Dennis, J.W.; Koch, K.; Beckner, D. *J. Natl. Cancer Inst.* **81** (1989) 1028.
78. Seftor, R.E.B.; Seftor, E.A.; Grimes, W.J.; Liotta, L.A.; Stetler-Stevenson, W.G.; Welch, D.R.; Hendrix, M.J.C. *Melanoma Res. 1* (1991) 43.
79. Galustian, C.; Foulds, S.; Dye, J.F.; Guillou, P.J. *Immunopharmacology* **27** (1994) 165.
80. Roberts, J.D.; Klein, J.L.D.; Palmantier, R.; Dhume, S.T.; George, M.D.; Olden, K. *Cancer Detect. Prev.* **22** (1998) 455.
81. Olden, K.; Breton, P.; Grzegorzewski, K.; Yasuda, Y.; Gause, B.L.; Oredipe, O.A.; Newton, S.A.; White, S.L. *Pharmacol. Ther.* **50** (1991) 285.
82. Olden, K.; Newton, S.A.; Nagai, T.; Yasuda, Y; Grzegorzewski, K.; Breton, P.; Oredipe, O.; White, S.L. *Pigm. Cell Res. Suppl.* **2** (1992) 219.

83. Dennis, J.W. *Semin. Cancer Biol.* **2** (1991) 411.
84. Yagita, M.; Noda, I.; Maehara, M.; Fujieda, S.; Inoue, Y.; Hoshino, T.; Saksela, E. *Int. J. Cancer* **52** (1992) 664.
85. Korczak, B.; Dennis, J.W. *Int. J. Cancer* **53** (1993) 634.
86. Bowlin, T.L.; McKown, B.J.; Kang, M.S.; Sunkara, P.S. *Cancer Res.* **49** (1989) 4109.
87. Dennis, J.W.; Koch, K.; Yousefi, S.; VanderElst, I. *Cancer Res.* **50** (1990) 1867.
88. Breton, P.; Asseffa, A; Grzegorzewski, K.; Akiyama, S.K.; White, S.L.; Cha, J.K.; Olden, K. *Cancer Commun.* **2** (1990) 333.
89. Pulverer, G.; Beuth, J.; Ko, H.L.; Yassin, A.; Ohshima, Y.; Roszkowski, K.; Uhlenbruck, G. *J. Cancer Res. Clin. Oncol.* **114** (1988) 217.
90. Humphries, M.J.; Matsumoto, K.; White, S.L.; Olden, K. *Cancer Res.* **46** (1986) 5215.
91. Grzegorzewski, K.; Newton, S.A.; Akiyama, S.K.; Sharrow, S.; Olden, K.; White, S.L. *Cancer Commun. 1* (1989) 373.
92. Klein, J.L.D.; Roberts, J.D.; George, M.D.; Kurtzberg, J.; Breton, P.; Chermann, J.C.; Olden, K. *Br. J. Cancer* **80** (1999) 87.
93. Das, P.C.; Roberts, J.D.; White, S.L.; Olden, K. *Oncol. Res.* **7**, (1995), 425.
94. Goss, P.E.; Baker, M.A.; Carver, J.P.; Dennis, J.W. *Clin. Cancer Res.* **1**, (1995), 935.
95. Goss, P.E.; Reid, C.L.; Bailey, D.; Dennis, J.W. *Clin. Cancer Res.* **3**, (1997), 1077.
96. Motohiro, H.; Kunio, N.; Hiroshi, T.; Junji, H.; Masanobu, K.; Hatsuo, A.; Hiroshi, I. Eur. Pat. EP 104826, 1982; *Chem. Abstr.* **101** (1984) 28283x.
97. White, S.L.; Schweitzer, K.; Humphries, M.J.; Olden, K. *Biochem. Biophys. Res. Commun.* **150** (1988) 615.
98. Yagita, M.; Saksela, E. *Scand. J. Immunol.* **31** (1990) 275.
99. Kino, T.; Inamura, N.; Nakahara, K.; Kiyoto, S.; Goto, T.; Terano, H.; Kohsaka, M.; Aoki, H.; Imanaka, H. *J. Antibiot.* **38** (1985) 936.
100. Stegelmeier, B.L.; Snyder, P.W.; James, L.F.; Panter, K.E.; Molyneux, R.J.; Gardner, D.R.; Ralphs, M.H.; Pfister, J.A. *Toxic Plants Other Nat. Toxicants* (1998) 285.
101. Dennis, J.W.; Shah, R.N.; Ziser, L. WO Pat. 9846602; *Chem. Abstr.* **129** (1998) 306525j.
102. Oredipe, O.A.; White, S.L.; Grzegorzewski, K.; Gause, B.L.; Cha, J.K.; Miles, V.A.; Olden, K. *J. Natl. Cancer Inst.* **83** (1991) 1149.
103. Peyrieras, N.; Bause, E.; Legler, G.; Vasilov, R.; Claesson, L.; Peterson, P.; Ploegh, H. *EMBO J.* **2** (1983) 823.
104. Pan, Y.T.; Ghidoni, J.; Elbein, A.D. *Arch. Biochem. Biophys.* **303** (1993) 134.
105. Pritchard, D.H.; Huxtable, C.R.R.; Dorling, P.R. *Res. Vet. Sci.* **48** (1990) 228.
106. Baker, D.C.; James, L.F.; Hartley, W.J.; Panter, K.E.; Maynard, H.F.; Pfister, J. *Am. J. Vet. Res.* **50** (1989) 1396.
107. Kovacs, P.; Csaba, G. *Acta Microbiol. Hung.* **40** (1993) 351.
108. Takahashi, H.; Parsons, P.G. *J. Invest. Dermatol.* **98** (1992) 481.
109. Skudlarek, M.D.; Orgebin-Crist, M.C. *J. Reprod. Fertil.* **84** (1988) 611.
110. Kang, M.S.; Elbein, A.D. *J. Virol.* **46** (1983) 60.
111. Muldoon, D.F.; Stohs, S.J. *J. Appl. Toxicol.* **14** (1994) 81.
112. Foddy, L.; Feeney, J.; Hughes, R.C. *Biochem. J.* **233** (1986) 697.
113. Foddy, L.; Hughes, R.C. *Carbohydr. Res.* **151** (1986) 293.
114. Tulsiani, D.R.P.; Broquist, H.P.; James, L.F.; Touster, O. *Plant Toxicol.* (1985) 279.
115. Stegelmeier, B.L.; Molyneux, R.J.; Elbein, A.D.; James, L.F. *Vet. Pathol.* **32** (1995) 289.
116. Stegelmeier, B.L.; James, L.F.; Panter, K.E.; Gardner, D.R.; Ralphs, M.H.; Pfister, J.A. *J. Anim. Sci.* **76** (1998) 1140.
117. Huxtable, C.R.; Dorling, P.R. *Acta Neuropathol.* **68** (1985) 65.
118. Huxtable, C.R.; Dorling, P.R.; Walkley, S.U. *Acta Neuropathol.* **58** (1982) 27.
119. De Balogh, K.K.I.M.; Dimande, A.P.; Van Der Lugt, J.J.; Molyneux, R.J.; Naude, T.W.; Welman, W.G. *Toxic Plants Other Nat. Toxicants* (1998) 428.
120. Scofield, A.M.; Fellows, L.E.; Nash, R.J.; Fleet, G.W.J. *Life Sci.* **39** (1986) 645.
121. Elbein, A.D.; Pan, Y.T.; Solf, R.; Vosbeck, K. *J. Cell. Physiol.* **115** (1983) 265.
122. James, L.F.; Panter, K.E.; Broquist, H.P.; Hartly, W.J. *Vet. Hum. Toxicol.* **33** (1991) 217.
123. Khan, F.A.; Basu, D. *Indian J. Biochem. Biophys.* **21** (1984) 203.
124. Houri, J.J.; Ogier-Denis, E.; Bauvy, C.; Aubery, M.; Sapin, C.; Trugnan, G.; Codogno, P. *Eur. J. Biochem.* **205** (1992) 1169.
125. Kalathur S.; Seetharam, S.; Dahms, N.M.; Seetharam, B. *Arch. Biochem. Biophys.* **315** (1994) 8.
126. Sem, A.; Sambhara, S.; Chadwick, B.S.; Miller, R.G. *J. Cell. Physiol.* **148** (1991) 485.
127. Caroll, M.; Bird, M.M. *Int. J. Biochem.* **23** (1991) 1285.
128. Hori, H.; Kaushal, G.P.; Elbein, A.D. *Plant Physiol.* **77** (1985) 687.

129. Chotai, K.; Jennings, C.; Winchester, B.; Dorling, P. *J. Cell. Biochem.* **21** (1983) 107.
130. Chen, W.; Li, X.; Shao, D.; Gu, J. *Shanghai Yike Daxue Xuebao* **20** (1993) 415; *Chem. Abstr.* **121** (1994) 176613h.
131. Stevens, K.L.; Molyneux, R.J. *J. Chem. Ecol.* **14** (1988) 1467.
132. Tulsiani, D.R.P.; Broquist, H.P.; James, L.F.; Touster, O. *Arch. Biochem. Biophys.* **232** (1984) 76.
133. Croom, W.J.J.; Hagler, W.M., Jr.; Froetschel, M.A.; Johnson, A.D. *J. Anim. Sci.* **73** (1995) 1499.
134. Warren, C.D.; Daniel, P.F.; Bugge, B.; Evans, J.E.; James, L.F.; Jeanloz, R.W. *J. Biol. Chem.* **263** (1988) 15041.
135. Kang, M.S.; Bowlin, T.L.; Vijay, I.K.; Sunkara, S.P. *Carbohydr. Res.* **248** (1993) 327.
136. Chae, B.S.; Ahn, Y.K.; Kim, J.H. *Yakhak Hoechi* **42** (1998) 75; *Chem. Abstr.* **128** (1998) 278760r.
137. Chae, B.S.; Ahn, Y.K.; Kim, J.H. *Arch. Pharmacol Res.* **20** (1997) 545.
138. Chrispeels, M.J.; Vitale, A. *Plant Physiol.* **78** (1985) 704.
139. Daniel, P.F.; Warren, C.D.; James, L.F. *Biochem. J.* **221** (1984) 601.
140. Molyneux, R.J.; McKenzie, R.A.; O'Sullivan, B.M.; Elbein, A.D. *J. Nat. Prod.* **58** (1995) 878.
141. McNally, A.K.; DeFife, K.M.; Anderson, J.M. *Am. J. Pathol.* **149** (1996) 975.
142. Opara, K.N.; Okenu, D.M.N. *J. Parasit. Dis.* **20** (1996) 145.
143. Shao, M.; Chin, C.C.Q.; Caprioli, R.M.; Wold, F. *J. Biol. Chem.* **262** (1987) 2973.
144. Schwarz, P.M.; Elbein, A.D. *J. Biol. Chem.* **260** (1985) 14452.
145. Abraham, D.J.; Sidebothom, R.; Winchester, B.G.; Dorling, P.R.; Dell, A. *FEBS Lett.* **163** (1983) 110.
146. Tropea, J.E.; Swank, R.T.; Segal, H.L. *J. Biol. Chem.* **263** (1988) 4309.
147. Bosch, J.V.; Tlusty, A.; McDowell, W.; Legler, G.; Schwarz, R.T. *Virology* **143** (1985) 342.
148. Dreyer, D.L.; Jones, K.C.; Molyneux, R.J. *J. Chem. Ecol.* **11** (1985) 1045.
149. Goussault, Y.; Warren, C.D.; Bugge, B.; Jeanloz, R.W. *Glycoconjugate J.* **3** (1986) 239.
150. Arumugham, R.G.; Tanzer, M.L. *Biochem. Biophys. Res. Commun.* **116** (1983) 922.
151. Huxtable, C.R.; Dorling, P.R. *Vet. Pathol.* **20** (1983) 727.
152. Broquist, H.P.; Mason, P.S.; Hagler, W.M., Jr.; Harris, T.M. *Appl. Environ. Microbiol.* **48** (1984) 386.
153. Ziegler, C.; Mersmann, G. *Biochim. Biophys. Acta* **799** (1984) 203.
154. McLawhon, R.W.; Berry-Kravis, E.; Dawson, G. *Biochem. Biophys. Res. Commun.* **134** (1986) 1387.
155. Koettgen, E.; Beiswenger, M.; James, L.F.; Bauer, C. *Gastroenterology* **95** (1988) 100.
156. Tulsiani, D.R.P.; Broquist, H.P.; James, L.F.; Touster, O. *Arch. Biochem. Biophys.* **264** (1988) 607.
157. Tulsiani, D.R.P.; Skudlarek, M.D.; Orgebin-Crist, M.C. *Biol. Reprod.* **43** (1990) 130.
158. Tulsiani, D.R.P.; Touster, O. *J. Biol. Chem.* **258** (1983) 7578.
159. Winkler, J.R.; Segal, H.L. *J. Biol. Chem.* **259** (1984) 1958.
160. Dorling, P.R.; Huxtable, C.R.; Cenci di Bello, I.; Winchester, B. *Biochem. Soc. Trans.* **11** (1983) 717.
161. Tulsiani, D.R.P.; Touster, O. *Arch. Biochem. Biophys.* **296** (1992) 556.
162. Costa, J.; Ricardo, C.P.P. *Plant Sci.* **101** (1994) 137.
163. Pulsipher, G.D.; Galyean, M.L.; Hallford, D.M.; Smith, G.S.; Kiehl, D.E. *J. Anim. Sci.* **72** (1994) 1561.
164. Stegelmeier, B.L.; James, L.F.; Panter, K.E.; Molyneux, R.J. *Vet. Hum. Toxicol.* **37** (1995) 336.
165. Daniel, P.F.; Newburg, D.S.; O'Neil, N.E.; Smith, P.W.; Fleet, G.W.J. *Glycoconjugate J.* **6** (1989) 229.
166. Larsson, O.; Engstroem, W. *Biochem. J.* **260** (1989) 597.
167. Van Kemenade, F.J.; Rotteveel, F.T.M.; van den Broek, L.A.G.M.; Baars, P.A.; van Lier, R.A.W.; Miedema, F. *J. Leukocyte Biol.* **56** (1994) 159.
168. Wang, S.; Panter, K.E.; Holyoak, G.R.; Molyneux, R.J.; Liu, G.; Evans, R.C.; Bunch, T.D. *Anim. Reprod. Sci.* **56** (1999) 19.
169. Lehrman, M.A.; Zeng, Y. *J. Biol. Chem.* **264** (1989) 1584.
170. Donaldson, M.J.; Broby, H.; Adlard, M.W.; Bucke, C. *Phytochem. Anal.* *1* (1990) 18.
171. Novikoff, P.M.; Touster, O.; Novikoff, A.B.; Tulsiani, D.P. *J. Cell Biol.* **101** (1985) 339.
172. Schneider, M.J; Ungemach, F.S.; Broquist, H.P.; Harris, T.M. *J. Am. Chem. Soc.* **104** (1982) 6863.
173. Kardono, L.B.S.; Kinghorn, A.D.; Molyneux, R.J. *Phytochem. Anal.* **2** (1991) 120.
174. Skelton, B.W.; White, A.H. *Aust. J. Chem.* **33** (1980) 435.
175. Adams, C.E.; Walker, F.J.; Sharpless, K.B. *J. Org. Chem.* **50** (1985) 420.
176. Setoi, H.; Takeno, H.; Hashimoto, M. *Heterocycles* **24** (1986) 1261.
177. Ikota, N.; Hanaki, A. *Heterocycles* **26** (1987) 2369.
178. Dener, J.M.; Hart, D.J.; Ramesh, S. *J. Org. Chem.* **53** (1988) 6022.
179. Ikota, N. *Chem. Pharm. Bull.* **41** (1993) 1717.
180. Naruse, M.; Aoyagi, S.; Kibayashi, C. *J. Org. Chem.* **59** (1994) 1358.
181. Hunt, J.A.; Roush, W.R. *Tetrahedron Lett.* **36** (1995) 501.
182. Angermann, J.; Homann, K.; Reissig, H.U.; Zimmer, R. *Synlett* (1995) 1014.
183. Pearson, W.H.; Hembre, E.J. *J. Org. Chem.* **61** (1996) 5546.
184. Hembre, E.J.; Pearson, W.H. *Tetrahedron* **53** (1997) 11021.
185. Hunt, J.A.; Roush, W.R. *J. Org. Chem.* **62** (1997) 1112.

186. Ferreira, F.; Greck, C.; Genet, J.P. *Bull. Soc. Chim. Fr.* **134** (1997) 615.
187. Holmes, A.B.; Bourdin, B.; Collins, I.; Davison, E.C.; Rudge, A.J.; Stork, T.C.; Warner, J.A. *Pure Appl. Chem.* **69** (1997) 531.
188. Mukai, C.; Sugimoto, Y.; Miyazawa, K.; Yamaguchi, S.; Hanaoka, M. *J. Org. Chem.* **63** (1998) 6281.
189. Martin-López, M.J.; Rodriguez, R.; Bermejo, F. *Tetrahedron* **54** (1998) 11623.
190. Tayle, P.C.; Winchester, B.G. In *Iminosugars as Glycosidase Inhibitors*; Stütz, A.E., Ed.; Wiley-VCH: Weinheim, 1999; p. 125.
191. Asano, N.; Nash, R.J.; Molyneux, R.J.; Fleet, G.W.J. *Tetrahedron: Asymmetry* **11** (2000) 1645.
192. Svansson, L.; Johnston, B.D.; Gu, J.-H.; Partrick, B.; Pinto, B.M. *J. Am. Chem. Soc.* **122** (2000) 10769.
193. Razavi, H.; Polt, R. *J. Org. Chem.* **65** (2000) 5693.
194. Lombardo, M.; Trombini, C. *Tetrahedron* **56** (2000) 323.
195. Batey, R.A.; MacKay, D.B. *Tetrahedron Lett.* **41** (2000) 9935.
196. Klitzke, C.F.; Pilli, R.A. *Tetrahedron Lett.* **42** (2001) 5605.
197. Socha, D.; Jurczak, M.; Chmielewski, M. *Carbohydr. Res.* **336** (2001) 315.
198. Paolucci, C.; Mattioli, L. *J. Org. Chem.* **66** (2001) 4787.
199. Pearson, W.H.; Guo, L. *Tetrahedron Lett.* **42** (2001) 8267.
200. Buschmann, N.; Rückert, A.; Blechert, S. *J. Org. Chem.* **67** (2002) 4325.
201. Patil, N.T.; Tilekar, J.N.; Dhavale, D.D. *Tetrahedron Lett.* **42** (2001) 747.
202. Rabiczko, J.; Urbanczyk-Lipkowska, Z.; Chmielewski, M. *Tetrahedron* **58** (2002) 1433.
203. Pearson, W.H.; Guo, L.; Jewell, T.M. *Tetrahedron Lett.* **43** (2002) 2175.
204. De Faria, A.R.; Salvador, E.L.; Correia, C.R.D. *J. Org. Chem.* **67** (2002) 3651.
205. Lindsay, K.B.; Pyne, S.G. *J. Org. Chem.* **67** (2002) 7774.
206. Ali, M.H.; Hough, L.; Richardson, A.C. *J. Chem. Soc., Chem. Commun.* (1984) 447.
207. Ali, M.H.; Hough, L.; Richardson, A.C. *Carbohydr. Res.* **136** (1985) 225.
208. Richardson, A.C. *J. Chem. Soc.* (1962) 373.
209. Guthrie, R.D.; Johnson, L.F. *J. Chem. Soc.* (1961) 4166.
210. Suami, T.; Tadano, K.; Iimura, Y. *Chem. Lett.* (1984) 513.
211. Suami, T.; Tadano, K.; Iimura, Y. *Carbohydr. Res.* **136** (1985) 67.
212. Suami, T.; Tadano, K. JP Pat. 60,218,389, 1985; *Chem. Abstr.* **105** (1986) 78823b.
213. Wardworth, W.S., Jr.; Emmons, W.D. *J. Am. Chem. Soc.* **83** (1961) 1733.
214. Zhou, W.S.; Xie, W.G.; Lu, Z.H.; Pan, X.F. *Tetrahedron Lett.* **36** (1995) 1291.
215. Zhou, W.S.; Xie, W.G.; Lu, Z.H.; Pan, X.F. *J. Chem. Soc., Perkin Trans. 1* (1995) 2599.
216. Zhou, W.S.; Lu, Z.H.; Wang, Z.M. *Tetrahedron Lett.* **32** (1991) 1467.
217. Zhou, W.S.; Lu, Z.H.; Wang, Z.M. *Tetrahedron* **49** (1993) 2641.
218. Fleet, G.W.J.; Gough, M.J.; Smith, P.W. *Tetrahedron Lett.* **25** (1984) 1853.
219. Bashyal, B.P.; Fleet, G.W.J.; Gough, M.J.; Smith, P.W. *Tetrahedron.* **43** (1987) 3083.
220. Takatani, T.; Tsutsumi, H.; Yasuda, N. JP Pat. 60,166,680, 1985; *Chem. Abstr.* **105** (1986) 134289w.
221. Bermej Gonzalez, F.; Lopez Barba, A. Ruano Espina, M. *Bull. Chem. Soc. Jpn.* **65** (1992) 567.
222. Pappo, R.; Allen, D.S., Jr.; Lemieux, R.U.; Johanson, W.S. *J. Org. Chem.* **21** (1956) 478.
223. Evans, M.E. *Carbohydr. Res.* **54** (1977) 105.
224. Yasuda, N.; Tsutsumi, H., Takaya, T. *Chem. Lett.* (1984) 1201.
225. Honda, T.; Hoshi, M.; Tsubuki, M. *Heterocycles* **34** (1992) 1515.
226. Honda, T.; Hoshi, M.; Kanai, K.; Tsubuki, M. *J. Chem. Soc., Perkin. Trans. 1* (1994) 2091.
227. Schmit, O. In *Methods in Carbohydrate Chemistry*, Vol. II; Academic Press: London, 1963; p. 319.
228. Zhao, H.; Hans, S.; Cheng, X.; Mootoo, D.R. *J. Org. Chem.* **66** (2001) 1761.
229. Setoi, H.; Takeno, H.; Hashimoto, M. *J. Org. Chem.* **50** (1985) 3948.
230. Vasella, A. *Helv. Chim. Acta* **60** (1977) 1273.
231. Setoi, H.; Kayakiri, H.; Takeno, H.; Hashimoto. M. *Chem. Pharm. Bull.* **35** (1987) 3995.
232. Fleet, G.W.J.; Nicholas, S.J.; Smith, P.W.; Evans, S.V.; Fellows, L.E.; Nash, R.J. *Tetrahedron Lett.* **26** (1985) 3127.
233. Carpenter, N.M.; Fleet, G.W.J., Cenci di Bello, I.; Winchester, B. *Tetrahedron Lett.* **30** (1989) 7261.
234. Fleet, G.W.J. U.S. Patent 5023340, 1990; *Chem. Abstr.* **115** (1991) 136472f.
235. Fleet, G.W.J., Son, J.C.; Green, D.St.C.; Cenci di Bello, I.; Winchester, B. *Tetrahedron* **44** (1988) 2649.
236. Miller, S.A.; Chamberlin, A.R. *J. Am. Chem. Soc.* **112** (1990) 8100.
237. Fetizon, M.; Golfier, M. *Angew. Chem.* **81** (1969) 423.
238. Ikota, N.; Hanaki, A. *Chem. Pharm. Bull.* **36** (1988) 1143.
239. Bagget, N.; Buchanan, J.G.; Fatah, M.Y.; Mucullough, K.J.; Webber, J.M. *Chem. Commun.* (1985) 1836.
240. Dho, J.C.; Fleet, G.W.J.; Peach, J.M.; Prout, K.; Smith, P.W. *Tetrahedron Lett.* **27** (1986) 3203.
241. Cohen, N.; Banner, B.L.; Loprest, R.J.; Wong, F.; Rosenberger, M.; Liu, Y.-Y.; Thom, E.; Liebman, A.A. *J. Am. Chem. Soc.* **150** (1983) 3661.
242. Kiso, M.; Hasegawa, A.; *Carbohydr. Res.* **52** (1976) 95.

243. Cohen, N.; Banner, B.L.; Laurenzano, A.J.; Carozza, L. *Org. Synth.* **63** (1985) 127.
244. Bennett, R.B., III; Choi, J.R.; Montgomery, W.D.; Cha, J.K. *J. Am. Chem. Soc.* **111** (1989) 2580.
245. Cha, J.K.; Bennett, R.B., III. WO Pat. 9,006,311, 1988; *Chem. Abstr.* **113** (1990) 212415r.
246. Tropper, F.; Shah, R.N.; Sharma, P. WO Pat. 9,921,858, 1999; *Chem. Abstr.* **130** (1999) 338281p.
247. Pearson, W.H.; Lin, K.C. *Tetrahedron Lett.* **31** (1990) 7571.
248. Lee, Y.K.; Schultz, A.G. *J. Org. Chem.* **44** (1979) 719.
249. Moss, W.O.; Bradbury, R.H.; Hales, N.J.; Gallagher, T. *J. Chem. Soc., Perkin Trans. 1* (1992) 1901.
250. Seebach, D.; Kolb, M.; Gröbel, B.-T. *Chem. Ber.* **106** (1973) 2277.
251. Ho, T.-L.; Wong, C.M. *Can. J. Chem.* **50** (1972) 3740.
252. Cohen, N.; Banner, B.L.; Laurenzano, A.J.; Carrozza, L. *Org. Synth.* **7** (1990) 297.
253. Howard, A.S; Michael, J.P. In *The Alkaloids*, Vol. 28; Brossi, A., Ed.; Academic Press: New York, 1986; p. 183.
254. Hanessian, S. *J. Org. Chem.* **34** (1969) 675.
255. Pearson, W.H.; Hember, E.J.J. Org. Chem. *J. Org. Chem.* **61** (1996) 7217.
256. Cha, J.K.; Bennett, R.B., U.S. Patent 5,187,279, 1990; *Chem. Abstr.* **118** (1993) 255177k.
257. Pearson, W.H.; Hember, E.J. U.S. Patent 5,919,952, 1999; *Chem. Abstr.* **131** (1999) 73836K.
258. Johnson, W.S.; Wethermann, L.; Bartlett, W.R.; Brocksom, T.J.; Li, T.; Faulkner, D.J.; Petersen, M.R. *J. Am. Chem. Soc.* **92** (1970) 741.
259. Sharpless, K.B.; Amberg, W.; Bennani, Y.L.; Grispino, G.A.; Hartung, J.; Jeong, K.S.; Kwong, H.L.; Morikawa, K.; Wang, Z.M.; Xu, D.; Zhang, X.L. *J. Org. Chem.* **57** (1992) 2768.
260. Wang, Z.M.; Zhang, X.L.; Sharpless, K.B. *Tetrahedron Lett.* **33** (1992) 6407.
261. Keinan, E.; Sinha, S.C.; Sinha-Bagchi, A.; Wang, Z.M.; Zhang, X.L.; Sharpless, K.B. *Tetrahedron Lett.* **33** (1992) 6411.
262. Kang, S.H.; Kim, G.T. *Tetrahedron Lett.* **36** (1995) 5049.
263. Oishi, T.; Iwakuma, T.; Hirama, M.; Ito, S. *Synlett* (1995) 404.
264. Taniguchi, M.; Koga, K.; Yamada, S. *Tetrahedron* **30** (1974) 3547.
265. Ikota, N.; Hanaki, A. *Chem. Pharm. Bull.* **38** (1990) 2712.
266. Ikota, N.; Hanaki, A. *Chem. Pharm. Bull.* **35** (1987) 2140.

5.4.1.3 *Lentiginosine* The alkaloids (+)-lentiginosine (8-deoxy-2,8a-di-*epi*-swainsonine, **1**) and 2-*epi*-lentiginosine (8-deoxy-8a-*epi*-swainsonine, **2**) were isolated from the leaves of spotted locoweed *Astragalus lentiginosus* var. *diphysus*[1] and *Rhizoctonia lenguminicola*,[2] respectively. 2-*epi*-Lentiginosine has been demonstrated to be a biosynthetic precursor to swainsonine.[3] It has a selective and powerful inhibition of amyloglucosidases,[1,4–6] a twice powerful inhibitor than castanospermine. Syntheses of lentiginosine and its isomers from noncarbohydrates have been reported.[7–21]

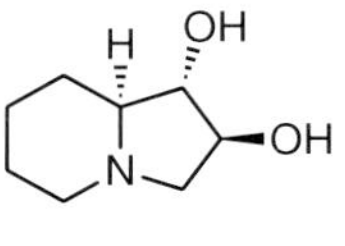

(+)-Lentiginosine
1

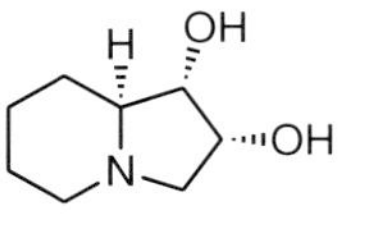

2-*epi*-Lentiginosine
2

5.4.1.3.1 *Synthesis from* D*-xylose* A synthetic pathway to (+)-lentiginosine (**1**) has been carried out by starting with 1,2-*O*-isopropylidene-D-xylofuranose (**3**) (Scheme 1).[22] Thus, its benzylation, followed by heating in methanol containing HCl, and further benzylation of the free hydroxyl group furnished the tribenzyl derivative **4**. Acid hydrolysis of **4** followed by amination afforded **5**, which upon reduction with $LiAlH_4$ followed by oxidative degradation with PCC gave the optically pure lactam **6**. Removal of the MPM and benzyl groups followed by protection of the NH by $(Boc)_2O$ and then silylation gave **7**. Nucleophilic addition of $BnO(CH_2)_4MgBr$ to **7** followed by reductive deoxygenation with Et_3SiH in the presence of $BF_3{\cdot}OEt_2$ afforded **8** with a high stereoselectivity (98:2). The major component

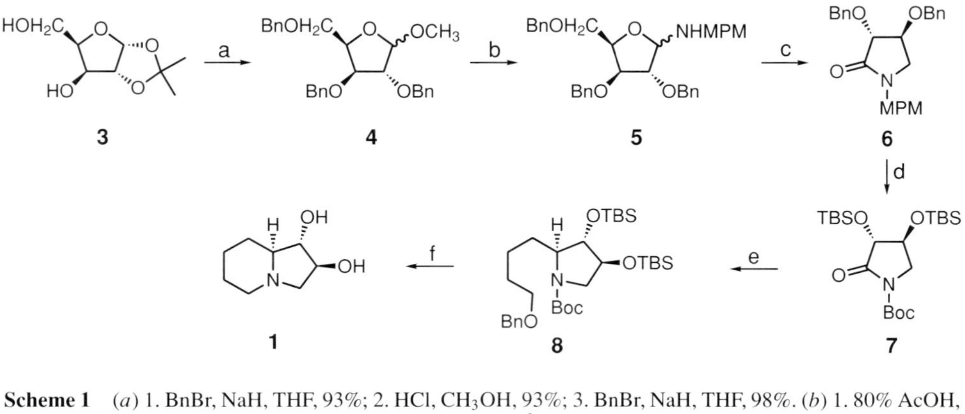

Scheme 1 (*a*) 1. BnBr, NaH, THF, 93%; 2. HCl, CH_3OH, 93%; 3. BnBr, NaH, THF, 98%. (*b*) 1. 80% AcOH, 100°C, 91%; 2. $MPMNH_2$, benzene, $CHCl_3$, 70°C, MS 4 Å, 100%. (*c*) 1. $LiAlH_4$, THF, 83%; 2. PCC, MS 4 Å, CH_2Cl_2, 58%. (*d*) 1. CAN, CH_3CN, H_2O, 81%; 2. $(Boc)_2O$, NEt_3, CH_2Cl_2, 96%; 3. Pd black, 4.4% HCO_2H, CH_3OH, 40°C, 96%; 4. TBSCl, imidazole, DMF, 94%. (*e*) 1. $BnO(CH_2)_4MgBr$, THF, −78°C; 2. Et_3SiH, $BF_3{\cdot}OEt_2$, CH_2Cl_2, −78°C, 55% for two steps. (*f*) 1. Pd black, 4.4% HCO_2H, CH_3OH, 40°C, 94%; 2. *p*-TsCl, Py, 70%; 3. $BF_3{\cdot}OEt_2$, CH_2Cl_2, −20 to 0°C; then KOH, CH_3OH, 74%.

was hydrogenated, followed by tosylation of the resulting primary hydroxyl group and subsequent deprotection with $BF_3{\cdot}OEt_2$ and cyclization under basic conditions to give **1** in 7% overall yield from **3**.

5.4.1.3.2 Synthesis from D-erythrose The D-erythrose derivative **9** has been used for the synthesis of (1*S*,2*R*,8a*R*)-1,2-dihydroxyindolizidine (**13**) by transformation into the azide **10** (Scheme 2).[23] Intramolecular cycloaddition of **10** in boiling benzene produced the bicyclic iminium ion **11**, which underwent sodium borohydride reduction to give **12**, followed by acid hydrolysis of the isopropylidene group to provide **13** in 49% overall yield from **10**.

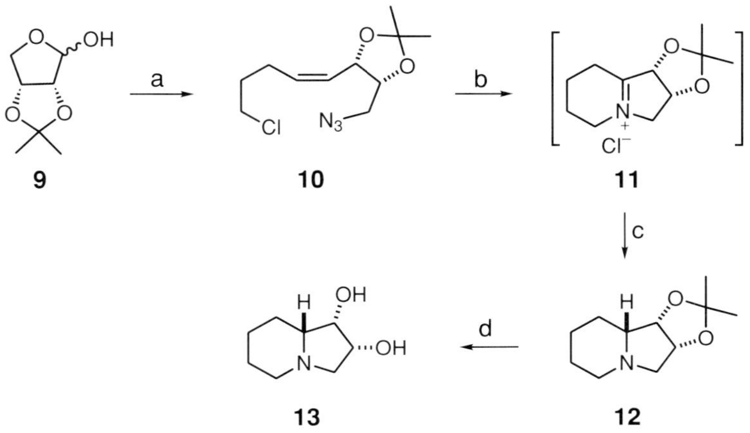

Scheme 2 (*a*) 1. $ClCH_2CH_2CH_2CH_2P^+Ph_3Br^-$, $KN(TMS)_2$, THF, −78 to 23°C, 2 h; 2. $(PhO)_2P(O)N_3$, PPh_3, DEAD, THF, 23°C, 1 h. (*b*) Benzene, reflux, 26 h. (*c*) $NaBH_4$, CH_3OH, 0°C, 1 h, 90%. (*d*) 6 N HCl, THF, 23°C, 12 h, 54%.

5.4.1.3.3 Synthesis from D-mannitol Total synthesis of (−)-lentiginosine [(−)-**1**] was achieved from D-mannitol via **14**[24] whose diol was cleaved with lead tetraacetate to give the respective aldehyde, which was reduced to an alcohol and then converted into azide **15** (Scheme 3).[25] The acetonide group in **15** was cleaved by using TFA to give **16**. The addition of allyltributylstannane to the crude aldehyde obtained from an oxidative cleavage of the diol **16** gave a homoallylic alcohol **17** in a highly diastereoselective manner. Mesylation of **17** afforded **18**, which underwent azide reduction to give the cyclized amine **19**. The secondary amine in **19** was converted into an acrylamide **20**, which underwent ring formation to give the lactam **21**. Hydrogenation of **21** followed by reduction of the crude amide with $LiAlH_4$ gave (−)-**1**.

D-Mannitol has also been used for the synthesis of (−)-8a-*epi*-lentiginosine (**28**) via the dihydrofuran epoxide **22**[26] (Scheme 4).[27] Epoxide ring opening in **22** with sodium azide followed by reduction and protection with di-*tert*-butylcarbonate provided the Boc-aminoalcohol **23**, which was treated with mercury(II) trifluoroacetate to afford **24** via a highly stereoselective cyclization. Dodecane-1-thiol affected the deoxymercuration of **24**, which upon benzylation gave the α-vinylpyrrolidine derivative **25**. This was treated with TFA, followed by acylation with $CH_2{=}CH{-}CH_2COCl$ to give the amide **26**. A ruthenium

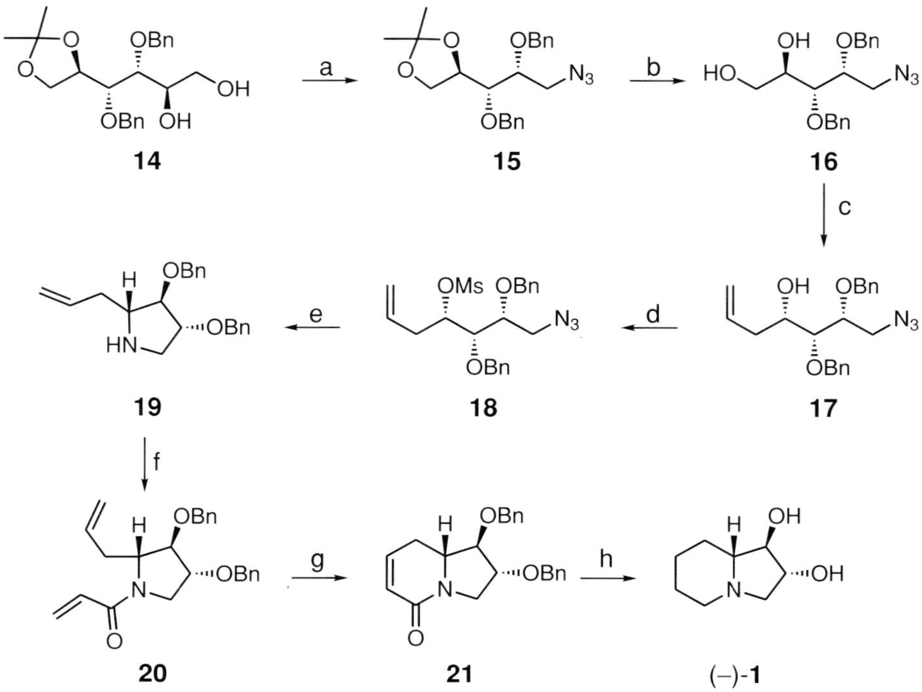

Scheme 3 (*a*) 1. $Pb(OAc)_4$, CH_2Cl_2, 3 h; 2. $NaBH_4$, EtOH, 3 h; 3. *p*-TsCl, NEt_3, CH_2Cl_2, 12 h; 4. NaN_3, DMF, 80°C, 8 h, 80%. (*b*) TFA, THF–H_2O (4:1), 65°C, 8 h. (*c*) 1. $Pb(OAc)_4$, CH_2Cl_2, 3 h; 2. $SnCl_4$, allyltributyltin, CH_2Cl_2, −78°C, 1 h. (*d*) MsCl, NEt_3, CH_2Cl_2, 6 h. (*e*) $LiAlH_4$, THF, reflux, 65°C, 12 h. (*f*) Acryloyl chloride, NEt_3, CH_2Cl_2, 12 h. (*g*) Bis(tricyclohexylphosphane)benzylideneruthenium dichloride (10 mol%), toluene, reflux, 24 h. (*h*) 1. 10% Pd *on* C, H_2, 24 h; 2. $LiAlH_4$, THF, reflux, 6 h.

carbene complex effected intramolecular cyclization of **26** to **27**. Reduction of the carbonyl group followed by hydrogenation afforded **28** in 13.4% overall yield from **22**.

5.4.1.3.4 Synthesis from D-isoascorbic acid Syntheses of **2** and its antipode **37** from the commercially available D-isoascorbic acid have been reported by conversion to the enantiomerically pure acetonide lactone **29**,[28] on a large scale in 75% yield (Scheme 5).[29] Lactone **29** was treated with amino vinylsilane **30** to give the amide **31** in 82% yield, which was used in two different ways. Mitsunobu conditions were not successful to convert **31** to **32**. However, the conversion has been achieved in 88% yield by mesylation and then subjection to intramolecular cyclization. Treatment of **32** with Lawesson's reagent afforded the respective thioamide, which was treated with $BF_3{\cdot}OEt_3$ followed by direct reduction with $LiBEt_3H$ to provide the 2-(ethylthio)pyrrolidine **35**. Cyclization of **35** afforded the single stereoisomeric tetrahydroindolizine **36**. Catalytic hydrogenation of **36** followed by deprotection provided 2-*epi*-lentiginosine (**2**).

Oxidation of **31** followed by treatment with Ac_2O afforded the acetoxy lactam **33**, which was treated with $BF_3{\cdot}OEt_2$ to provide the tetrahydroindolizinone **34**. Hydrogenation of **34** followed by reduction and then heating with 2 M HCl afforded **37**.

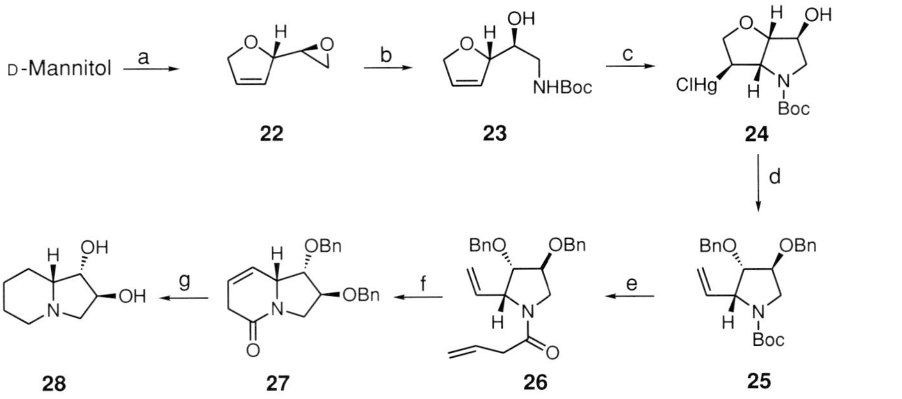

Scheme 4 (*a*) Ref. 26. (*b*) 1. NaN_3, CH_3OH, H_2O, rt, 60 h, sodium dihydrogen phosphate nonhydrate; 2. $LiAlH_4$, THF, N_2, rt, 1 h; then reflux, 2 h; 3. $(Boc)_2O$, THF, rt, 24 h, 89%. (*c*) 1. $Hg(TFA)_2$, THF, 0°C for 30 min, rt; 2. NaCl–H_2O (1:1), 77%. (*d*) 1. *n*-$C_{12}H_{23}SH$, CH_3OH, rt, N_2, 4 h, 67%; 2. NaH, BnBr, DMF, THF, N_2, −10°C, 5 h, 96%. (*e*) 1. TFA, CH_2Cl_2, rt, 30 min, THF, CH_3OH; 2. $ClCOCH_2CH{=}CH_2$, NEt_3, 0°C to rt, 4 h, 90%. (*f*) Bis(tricyclohexylphosphane)benzylideneruthenium dichloride (4%), benzene, reflux, 2 h, 80%. (*g*) 1. $LiAlH_4$, THF, reflux, 5 h, 60%; 2. H_2, 10% Pd *on* C, CH_3OH, HCl, rt, 8 h; then NaOH, 91%.

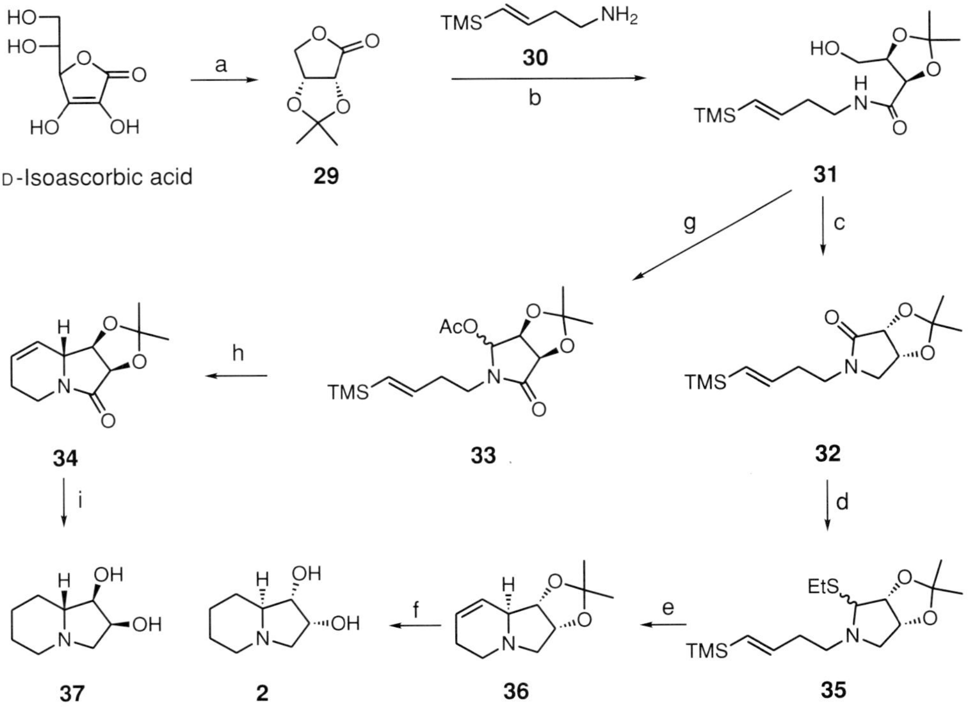

Scheme 5 (a) Ref. 28. (*b*) $(CH_3)_3Al$, CH_2Cl_2, hexane, rt, 82%. (*c*) 1. MsCl, NEt_3, CH_2Cl_2, 0°C to rt; 2. excess NaH, THF, rt, 88%. (*d*) 1. $(ArPS_2)_2$, HMPA, 100°C, 80%; 2. $BF_3{\cdot}OEt_2$, 2,6-di-*tert*-butylpyridine, CH_2Cl_2, rt; 3. $LiBEt_3H$, THF, −78°C, 84%. (*e*) $Cu(OSO_2CF_3)_2$, THF, reflux, 73%. (*f*) 1. H_2, Pd *on* C, EtOAc, 24 h, 72%; 2. 2 M HCl, 16 h, 80°C, 77%. (*g*) 1. $SO_3{\cdot}Py$, DMSO, rt, 74%; 2. Ac_2O, DMAP, CH_2Cl_2, −20°C, 95%. (*h*) $BF_3{\cdot}OEt_2$, rt, 72%. (*i*) 1. H_2, Pd *on* C, EtOAc, rt, 86%; 2. $LiAlH_4$, Et_2O, reflux. 78%; 3. 2 M HCl, 80°C, 72%.

References

1. Pastuszak, I.; Molyneux Russell, J.; James, L.F.; Elbein, A.D. *Biochemistry* **29** (1990) 1886.
2. Harris, T.M.; Harris, C.M.; Hill, J.E.; Ungemach, F.S.; Broquist, H.P.; Wickwire, B.M. *J. Org. Chem.* **52** (1987) 3094.
3. Harris, C.M.; Campbell, B.C.; Molyneux, R.J.; Harris, T.M. *Tetrahedron Lett.* **29** (1988) 4815.
4. Brandi, A.; Cicchi, S.; Corero, F.M.; Frignoli, R.; Goti, A.; Picasso, S.; Vogel, P. *J. Org. Chem.* **60** (1995) 6806.
5. Cardona, F.; Goti, A.; Brandi, A.; Scarselli, M.; Niccolai, N.; Mangani, S. *J. Mol. Model.* **3** (1997) 249.
6. El Ashry, E.S.H.; Rashed, N.; Shobier, A.H.S. *Pharmazie* **55** (2000) 331.
7. Mccaig, A.E.; Meldrum, K.P.; Wightman, R.H. *Tetrahedron* **54** (1998) 9429.
8. Yoda, H.; Kitayama, H.; Katagiri, T.; Takabe, K. *Tetrahedron: Asymmetry* **4** (1993) 1455.
9. Cordero, F.M.; Cicchi, S.; Goti, A.; Brandi, A. *Tetrahedron Lett.* **35** (1994) 949.
10. Giovannini, R.; Marcantoni, E.; Petrini, M. *J. Org. Chem.* **60** (1995) 5706.
11. Nukui, S.; Sodeoka, M.; Sasai, H.; Shibasaki, M. *J. Org. Chem.* **60** (1995) 398.
12. Harris, C.M.; Schneider, M.J.; Ungemach, F.S.; Hill, J.E.; Harris, T.M. *J. Am. Chem. Soc.* **110** (1988) 940.
13. Gurjar, M.K.; Ghosh, L.; Syamala, M.; Jayasree, V. *Tetrahedron Lett.* **35** (1994) 8871.
14. Mccaig, A.E.; Wightman, R.H. *Tetrahedron Lett.* **34** (1993) 3939.
15. Goti, A.; Cardona, F.; Brandi, A. *Synlett* (1996) 761.
16. El Nemr, A. *Tetrahedron* **56** (2000) 8579.
17. Takahata, H.; Banba, Y.; Momose, T. *Tetrahedron: Asymmetry* **3** (1992) 999.
18. Heitz, M.P.; Overman, L.E. *J. Org. Chem.* **54** (1989) 2591.
19. Colegate, S.M.; Dorling, P.R.; Huxtable, C.R. *Aust. J. Chem.* **37** (1984) 1503.
20. Yoda, H.; Katoh, H.; Ujihara, Y.; Takabe, K. *Tetrahedron Lett.* **42** (2001) 2509.
21. Rabiczko, J.; Urbańczyk-Lipkowska, Z.; Chmielewski, M. *Tetrahedron* **58** (2002) 1433.
22. Yoda, H.; Kawauchi, M.; Takabe, K. *Synlett* (1998) 137.
23. Pearson, W.H.; Lin, K.C. *Tetrahedron Lett.* **31** (1990) 7571.
24. Chandrasekhar, M.; Chandra, K.L.; Singh, V.K. *Tetrahedron Lett.* **43** (2000) 2773.
25. Chandra, K.L.; Chandrasekhar, M.; Singh, V.K. *J. Org. Chem.* **67** (2002) 4630.
26. Ceré, V.; Mazzini, C.; Paolucci, C.; Pollicino, S.; Fava, A. *J. Org. Chem.* **58** (1993) 4567.
27. Paolucci, C.; Musiani, L.; Venturelli, F.; Fava, V. *Synthesis* (1997) 1415.
28. Cohen, N.; Banner, B.L.; Laurenzano, A.J.; Carozza, L. *Org. Synth.* **63** (1985) 127.
29. Heitz, M.-P.; Overman, L.E. *J. Org. Chem.* **54** (1989) 2591.

5.4.1.4 *Slaframine* The indolizidine alkaloid (–)-slaframine [(1*S*,6*S*,8a*S*)-1-acetoxy-6-amino-octahydroindolizine, **1**] is a fungal metabolite produced by the mold *Rhizoctonia leguminicola*.[1–5] Slaframine has been shown to be responsible for excess salivation in cattle when they graze on fungus infested feeds.[5] It is known to undergo oxidative activation *in vivo* to a potent and neurotoxic muscarinic agent[6] and oxidized in the liver to an active metabolite, which is a muscarinic agonist.[7] Beyond its potential in the treatment of diseases involving cholinergic dysfunction, slaframine has been under active investigation for its potential beneficial effects on ruminant digestive function.[8,9] Slaframine can also stimulate pancreatic secretion and it has been proposed as a possible drug candidate for the alleviation of the symptoms of cystic fibrosis sufferers.[10] Unfortunately, slaframine is an air-sensitive compound which is not easily obtained in significant quantities by fermentation. Both the biosynthetic origins and metabolism of slaframine have been investigated.[11]

Syntheses of slaframine (**1**) and its analogues from noncarbohydrates have been reported.[12–31]

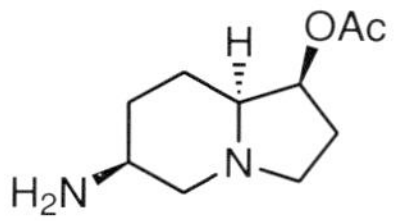

(–)-Slaframine
1

5.4.1.4.1 *Synthesis from* D-*threose* A synthesis of (–)-slaframine (**1**) has been reported from the 3-deoxy-D-threose derivative **2** by reaction with the chiral ylide **4**, obtained from **3**, in the presence of $KN(TMS)_2$ to produce **5** in 89% yield (Scheme 1).[32] Subsequent tosylation followed by opening of the oxazoline ring by reduction[33] afforded the alcohol **6**. Selective N-tosylation of **6** followed by treatment with sodium azide furnished **7** in 93% yield. Mesylation and subsequent reduction produced the protected slaframine derivative **8** in 45–55% overall yield. Complete deprotection of **8** followed by O-acetylation afforded **1**, whose further acetylation gave *N*-acetyl-slaframine (**9**).

5.4.1.4.2 *Synthesis from* D-*mannitol* D-Mannitol has been used as a source for D-(*R*)-glyceraldehyde acetonide (**11**),[34] which is a building block for various natural products. Thus, it was used for the synthesis of (–)-slaframine (**1**) and its enantiomer **24** (Schemes 2 and 3).[35] Reductive amination of **11** with 4-aminobutyraldehyde diethyl acetal (**10**) afforded the amino ketal **12** in 94% yield. Protection of the secondary amine in **12** followed by selective acid hydrolysis of the diethyl ketal afforded the respective aldehyde, which was directly treated with phenylsulfonyl-*p*-tolylsulfinylmethane to give a 1:1 mixture of the γ-hydroxy-α,β-unsaturated sulfone **13** in 89% yield from **12**. Removal of the protecting groups with TFA gave the ammonium salts **14**, which underwent intramolecular cyclization with NEt_3 to give a mixture of hydroxylated pyrrolidine stereoisomers **15**. Treatment of **15** with DMP followed by protection of the remaining secondary hydroxyl group with TIPSOTf afforded **16** as a 1:1 mixture of *cis*-pyrrolidine (59% yield from **13**)

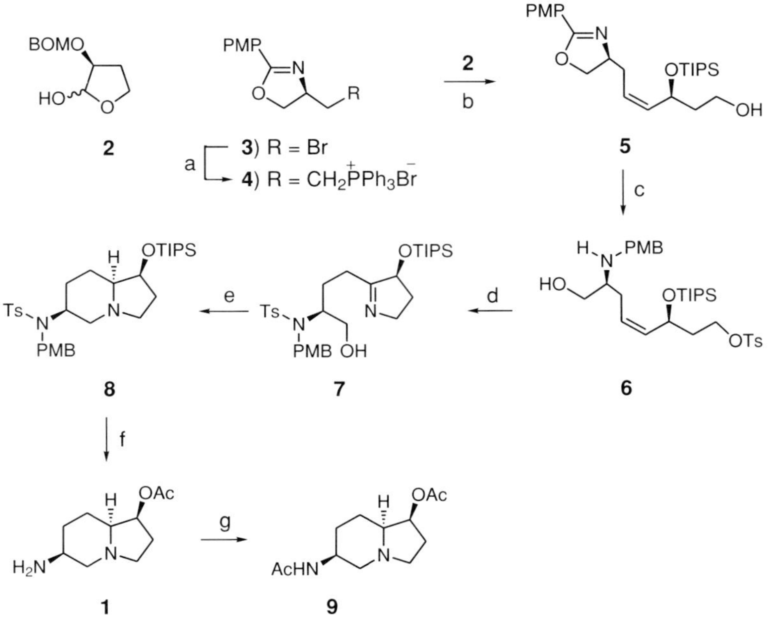

Scheme 1 (*a*) PPh_3, CH_3CN, reflux. (*b*) $KN(TMS)_2$, $-78°C$, 1 h, 89% for two steps. (*c*) 1. *p*-TsCl, NEt_3, cat. DMAP, CH_2Cl_2, 96%; 2. DIBAL-H (5 equiv.), THF, 0°C, 1 h, 97%. (*d*) 1. *N*-Tosyl-*N*-methylpyrrolidine, perchlorate, CH_2Cl_2, 0°C, 1 h, 72%; 2. NaN_3 (5 equiv.), DMF, 60°C, toluene, reflux, 93%. (*e*) 1. MsCl, NEt_3, CH_2Cl_2, 0°C, 63%; 2. $NaBH_4$, EtOH, 0°C, K_2CO_3, reflux, 73%. (*f*) 1. CAN, H_2O–CH_3CN (1:15), 68%; 2. TBAF, THF, 93%; 3. Na, NH_3, THF, 100%; 4. HCl, AcOH, 68%. (*g*) Ac_2O, Py, 87%.

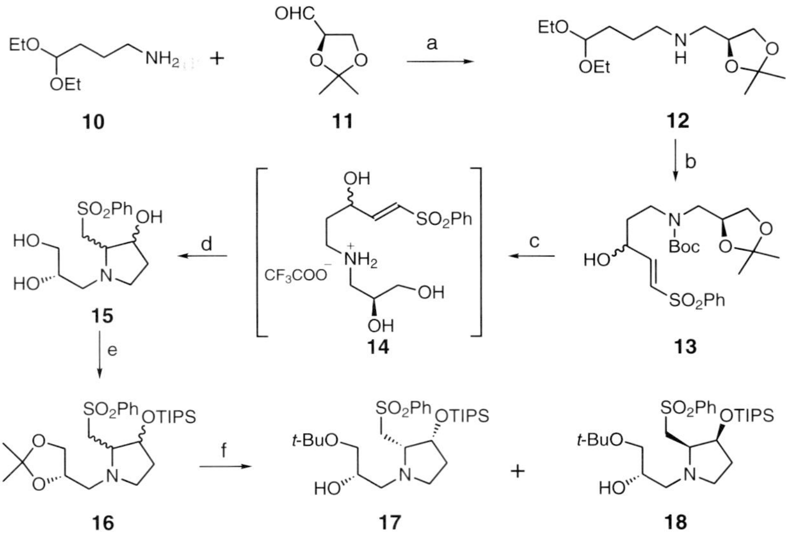

Scheme 2 (*a*) 1. CH_3OH, MS 3 Å, rt; 2. $NaBH_4$, rt. (*b*) 1. $(Boc)_2O$, CH_2Cl_2, rt; 2. AcOH–H_2O (2:1), rt; 3. $PhO_2SCH_2SOC_6H_4$–CH_3-*p*, piperidine, CH_2Cl_2, 0°C, 89%. (*c*) TFA (10 equiv.), CH_2Cl_2, rt. (*d*) NEt_3, THF, $-78°C$. (*e*) 1. DMP, CH_2Cl_2, *p*-TsOH, rt; 2. TIPSOTf, 2,6-lutidine, CH_2Cl_2, rt. (*f*) $Al(CH_3)_3$ (10 equiv.), CH_2Cl_2, rt.

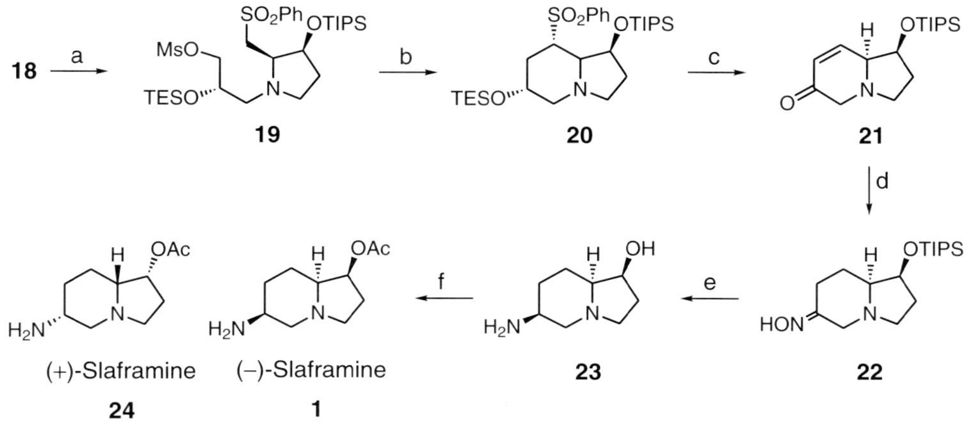

Scheme 3 (*a*) 1. TFA, H_2O, rt, 99%; 2. MsCl, NEt_3, CH_2Cl_2, 0°C; 3. TESCl, imidazole, CH_2Cl_2, rt, 72%. (*b*) LHMDS, THF, 0°C. (*c*) 1. TFA, H_2O, rt; 2. $(COCl)_2$, DMSO, NEt_3, CH_2Cl_2, −78°C, 75%. (*d*) 1. H_2, PtO_2, EtOAc, rt; 2. $NH_2OH{\cdot}HCl$, Py, CH_3OH, rt, 87%. (*e*) H_2, PtO_2, EtOH, conc. HCl; then Dowex (OH^-) resin, 98%. (*f*) AcOH, HCl, 69%.

along with the *trans*-isomer in a minor amount (15%). Treatment of **16** with $(CH_3)_3Al$ led to *tert*-butyl ether **17** (46%) and **18** (44%).

Acid hydrolysis of **18** followed by selective mesylation of the resulting primary hydroxyl group and subsequent silylation of the secondary hydroxyl group with TESCl in the presence of imidazole afforded the pyrrolidine **19** in 69% yield from **18**. Treatment of **19** with LHMDS led to a ring closure to give **20**. Selective deprotection of the TES group under acidic conditions followed by Swern oxidation of the resulting alcohol and *in situ* basic elimination of the sulfone, promoted by the presence of NEt_3, afforded **21**. Catalytic hydrogenation of **21** followed by condensation with hydroxyl amine hydrochloride afforded the oxime **22** (87%) as a mixture of *E*- and *Z*-isomers, which underwent catalytic hydrogenation and subsequent removal of the silyl ether to furnish the deacetyl slaframine **23** (98%). Treatment of **23** with acetic acid containing HCl afforded (−)-slaframine (**1**). Similar reaction sequence converted **17** into the unnatural isomer (+)-slaframine (**24**).

References

1. Rainey, D.P.; Smalley, E.B.; Crump, M.H.; Strong, F.M. *Nature* (*London*) **205** (1965) 203.
2. Aust, S.D.; Broquist, H.P. *Nature (London)* **205** (1965) 204.
3. Aust, S.D.; Broquist, H.P.; Rinehart, K.L. *J. Am. Chem. Soc.* **88** (1966) 2879.
4. Whitlock, B.J.; Rainey, D.P.; Riggs, N.V.; Strong, F.M. *Tetrahedron Lett.* **7** (1966) 3819.
5. Gardiner, R.A.; Rinehart, K.L.; Snyder, J.J.; Broquist, H.P. *J. Am. Chem. Soc.* **90** (1968) 5639.
6. Guengerich, F.P.; Broquist, H.P. In *Bioorganic Chemistry*, Vol. 2; Van Tamelen, E.E., Ed.; Academic Press: New York, 1979; p. 97.
7. Broquist, H.P. *Ann. Rev. Nutr.* **5** (1985) 391.
8. Froetschel, M.A.; Amos, H.E.; Evans, J.J.; Croom, W.J., Jr.; Hagler, W.M., Jr. *J. Am. Chem. Soc.* **111** (1989) 827.
9. Jacques, K.; Harmon, D.L.; Croom, W.J., Jr.; Hagler, W.M., Jr. *J. Dairy Sci.* **72** (1989) 443.
10. Aust, S.D. *Biochem. Pharmacol.* **18** (1969) 929.
11. Harris, C.M.; Schneider, M.J.; Ungemach, F.S.; Hill, J.E.; Harris, T.M. *J. Am. Chem. Soc.* **110** (1988) 940.

12. Pearson, W.H.; Bergmeier, S.C.; Williams, J.P. *J. Org. Chem.* **57** (1992) 3977.
13. Pearson, W.H.; Bergmeier, S.C. *J. Org. Chem.* **56** (1991) 1976.
14. Heidt, P.C.; Bergmeier, S.C.; Pearson, W.H. *Tetrahedron Lett.* **31** (1990) 5441.
15. Knight, D.W.; Harris, T.M. *J. Org. Chem.* **49** (1984) 3681.
16. Knight, D.W.; Sibley, A.W. *J. Chem. Soc., Perkin Trans. 1* (1997) 2179.
17. Szeto, P.; Lathburg, D.C.; Gallagher, T. *Tetrahedron Lett.* **36** (1995) 6957.
18. Sibi, M.P.; Christensen, J.W.; Li, B.; Renhowe, P.A. *J. Org. Chem.* **57** (1992) 4329.
19. Gobao, R.A.; Bremmer, M.L.; Weinreb, S.M. *J. Am. Chem. Soc.* **104** (1982) 7065.
20. Cartwright, D.; Gardiner, R.A.; Rinehart, K.L., Jr. *J. Am. Chem. Soc.* **92** (1970) 7615.
21. Gensler, W.J.; Hu, M.W. *J. Org. Chem.* **38** (1973) 3848.
22. Kang, S.H.; Kim, J.S.; Youn, J.-H. *Tetrahedron Lett.* **39** (1998) 9047.
23. Knapp, S.; Gibson, F.S. *J. Org. Chem.* **57** (1992) 4802.
24. Gmeiner, P.; Junge, D.; Kärtner, A. *J. Org. Chem.* **59** (1994) 6766.
25. Gmeiner, P.; Kärtner, A.; Junge, D. *Tetrahedron Lett.* **34** (1993) 4325.
26. Hua, D.H.; Park, J.-G.; Katsuhira, T.; Bharathi, S.N. *J. Org. Chem.* **58** (1993) 2144.
27. Sibi, M.P.; Christensen, J.W. *J. Org. Chem.* **64** (1999) 6434.
28. Wasserman, H.H.; Vu, C.B. *Tetrahedron Lett.* **35** (1994) 9779.
29. Gmeiner, P.; Junge, D. *J. Org. Chem.* **60** (1995) 3910.
30. Dartmann, M.; Flitsch, W.; Krebs, B.; Pandl, K.; Westfechtel, A. *Liebigs Ann. Chem.* (1988) 695.
31. Knight, D.W.; Sibley, A.W. *Tetrahedron Lett.* **34** (1993) 6607.
32. Choi, J.-R.; Han, S.; Cha, J.K. *Tetrahedron Lett.* **32** (1991) 6469.
33. Merers, A.I.; Himmelsbach, R.J.; Reuman, M. *J. Org. Chem.* **48** (1983) 4053.
34. Niu, C.; Pettersson, T.; Miller, M.J. *J. Org. Chem.* **61** (1996) 1014.
35. Carretero, J.C.; Arrayás, R.G. *Synlett* (1999) 49.

5.4.2 *Miscellaneous*

5.4.2.1 *Kifunensine* Kifunensine (**1**) (FR900494) is a unique cyclic oxamide derivative of 1-amino-substituted mannojirimycin. It was isolated from actinomycete *Kitasatospora kifunense* No. 9482.[1] Kifunensine has a promising immunomodulatory activity, α-mannosidase (Jack bean) inhibition[1] activity, specific inhibition of mannosidase I, and the processing of viral glucoproteins of the influenza in Madin–Darby canine kidney cells.[2] Its structure has been determined[3,4] on the basis of chemical and physical evidences as well as X-ray crystal analysis. The unnatural 8-*epi*-kifunensine (**2**) and 8a-*epi*-kifunensine (**3**) have been synthesized[5–8] from glucose derivatives, but the natural kifunensine (**1**) has been synthesized from D-mannosamine.

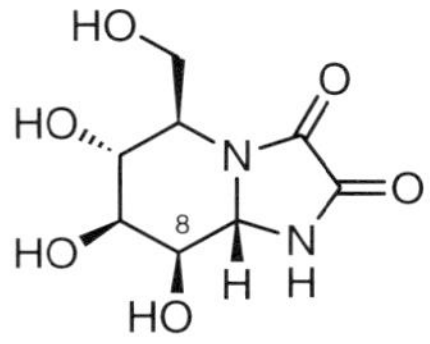

Kifunensine **1** 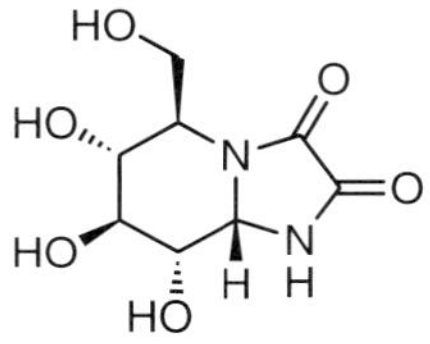 8-*epi*-Kifunensine **2** 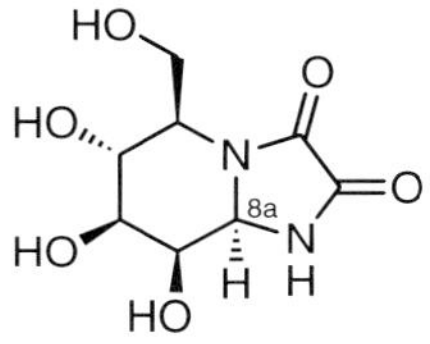 8a-*epi*-Kifunensine **3**

The first syntheses of kifunensine (**1**) and 8a-*epi*-kifunensine (**3**) were achieved from D-mannosamine (**4**) (Scheme 1).[7,8] Selective N-acylation of **4** with oxamic acid in the presence of DCC and HOBT in DMF followed by silylation of the primary hydroxyl group afforded **5**, which was reduced with sodium borohydride, followed by acetonation with acetone in the presence of $BF_3{\cdot}OEt_2$ to give the diacetonide **6**. Desilylation of **6** with TBAF gave **7**, which upon subjection to Collins oxidation of the primary hydroxyl group afforded the oxamide-aldehyde **10**, which was directly treated with NH_3/CH_3OH at room temperature to afford the kifunensine diacetonide **11** in 76% yield along with its 8a-epimer **12** in 4% yield from **7**. Removal of the protecting groups from **11** and **12** afforded **1** and **3**, respectively. Treatment of **10** with 30% CH_3NH_2/CH_3OH instead of ammonia afforded the respective *N*-methylkifunensine diacetonide whose deisopropylidenation gave **13** in 81% yield from **7** and the (8a*R*)-epimer was not obtained.

The synthesis of kifunensine (**1**) was also achieved[9] from the 5-deoxy-5-azidomannolactone derivative **8**, obtained from *cis*-cyclohexadienediols by microbial oxidation using *Pseudomonas putida* 39D,[10–12] by isopropylidenation followed by reduction to furnish **9** in 30% yield. Treatment of **9** with dimethyl oxalate followed by methanolic ammonia furnished intermediate **10** (55%), which underwent oxidation of the primary alcohol to the aldehyde followed by cyclization in methanolic ammonia to afford **11**. Removal of the diacetonide groups from **11** with 75% trifluoroacetic acid gave (+)-**1**.

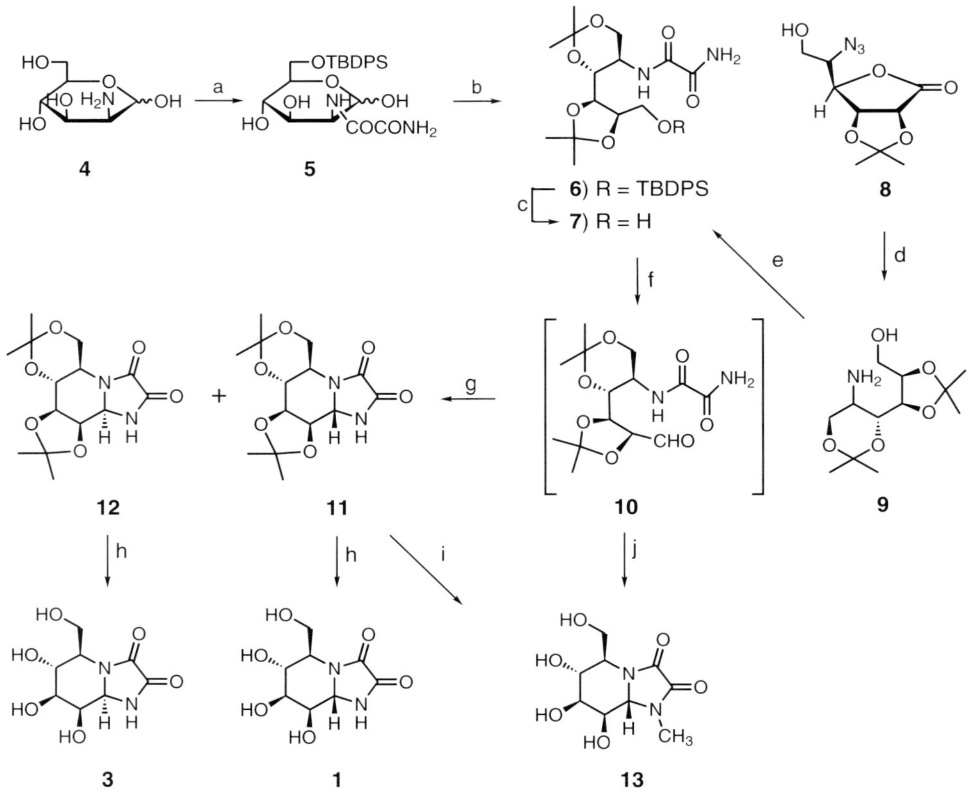

Scheme 1 (*a*) 1. H_2NCOCO_2H, DCC, NEt_3, HOBT, DMF, rt, 15 h; 2. *t*-$BuPh_2SiCl$, imidazole, DMF, 0°C, 3 h, 66% for two steps. (*b*) 1. $NaBH_4$, CH_3OH, rt, 30 min, 92%; 2. acetone, $BF_3{\cdot}OEt_2$, –20°C, 86%. (*c*) TBAF, THF, –20°C, 20 min, 17°C, 1.5 h, 100%. (*d*) 1. DMP, CSA, $(CH_2Cl)_2$, reflux; 2. $LiAlH_4$, Et_2O, rt, 2 h, 30% for two steps. (*e*) $(CO_2CH_3)_2$, CH_3OH, reflux; then NH_3–CH_3OH, 10 min, 55%. (*f*) $CrO_3{\cdot}2Py$, CH_2Cl_2, 30 min. (*g*) 6 N NH_3, CH_3OH, rt, 20 h, **11** (76%), **12** (4%) for two steps. (*h*) 75% aqueous TFA, rt, 3 h, 94%. (*i*) 1. CH_3I, K_2CO_3, acetone, reflux, 1.5 h, 62%; 2. 75% aqueous TFA, rt, 5 h, 82%. (*j*) 1. 30% CH_3NH_2, CH_3OH, 80% from **9**; 2. 75% aqueous TFA, 84%.

References

1. Iwami, M.; Nakayama, O.; Terano, H.; Kohsaka, M.; Aoki, H.; Imanaka, H. *J. Antibiot.* **40** (1987) 612.
2. Elbein, A.D.; Tropea, J.E.; Mitchell, M.; Kaushal, G.P. *J. Biol. Chem.* **265** (1990) 15599.
3. Kayakiri, H.; Takase, S.; Shibata, T.; Okamoto, M.; Terano, H.; Hashimoto, M.; Tada, T.; Koda, S. *J. Org. Chem.* **54** (1989) 4015.
4. Kayakiri, H.; Takase, S.; Shibata, T.; Hashimoto, M.; Tada, T.; Koda, S. *Chem. Pharm. Bull.* **39** (1991) 1378.
5. Kayakiri, H.; Oku, T.; Hashimoto, M. *Chem. Pharm. Bull.* **38** (1990) 293.
6. Kayakiri, H.; Oku, T.; Hashimoto, M. *Chem. Pharm. Bull.* **39** (1991) 1397.
7. Kayakiri, H.; Kasahara, C.; Oku, T.; Hashimoto, M. *Tetrahedron Lett.* **31** (1990) 225.
8. Kayakiri, H.; Kasahara, C.; Nakamura, K.; Oku, T.; Hashimoto, M. *Chem. Pharm. Bull.* **39** (1991) 1392.
9. Rouden, J.; Hudlicky, T. *J. Chem. Soc., Perkin Trans. 1* (1993) 1095.
10. Gibson, D.T.; Koch, G.R.; Kallio, R.E. *Biochemistry* **7** (1968) 2653.
11. Gibson, D.T.; Hensley, M.; Yoshioka, H.; Mabry, J.J. *Biochemistry* **9** (1970) 1626.
12. Hudlicky, T.; Olivo, H.L. *Tetrahedron Lett.* **32** (1991) 6077.

5.4.2.2 *Nagstatin* Nagstatin (**1**) was isolated[1,2] from the fermentation broth of *Streptomyces*. It is a strong inhibitor of *N*-acetyl-β-D-glucosaminidase (IC_{50} 1.2 ηg/mL). Its absolute configuration was studied.[3] A number of unnatural analogues of nagstatin **2–7** have been synthesized.[4,5]

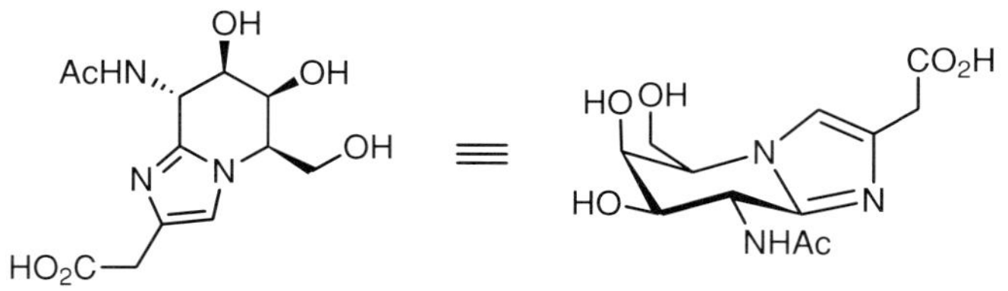

Nagstatin

1

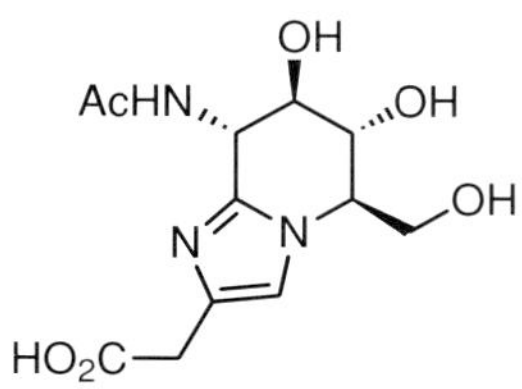

2

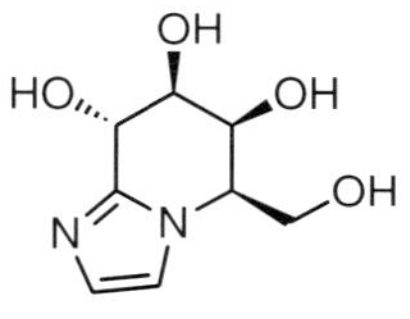

3 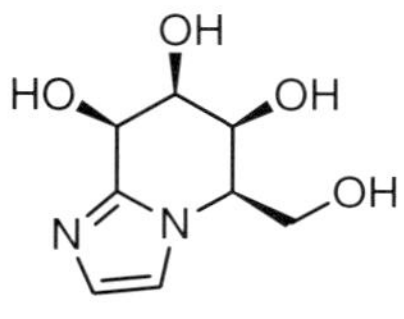 **4** 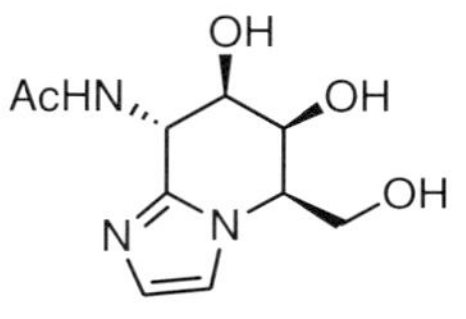 **5**

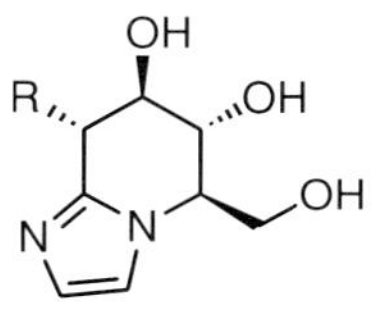

6) R = OH
7) R = NHAc

The first total synthesis of nagstatin (**1**) was reported from L-ribofuranose derivative **8** by conversion into **9** whose cyclization gave the key intermediates **10** and **11**[5] (Scheme 1).[6] Compound **10** was silylated with TBSOTf in the presence of 2,6-lutidine to produce **13**, which was fully brominated with 2,4,4,6-tetrabromo-2,5-cyclohexadien-1-one to produce **14**. Selective debromination with *tert*-butyllithium afforded the monobromo derivative **15**, which was treated with allyl bromide in the presence of *n*-butyllithium and copper iodide to afford **16**. Ozonolysis of **16** caused a concomitant oxidation at C-9 position. However, dihydroxylation using OsO_4 and NMO followed by oxidation and esterification afforded **17**. De-O-silylation of **17** with fluoride ion and subsequent conversion to the azide **18** under Mitsunobu's conditions took place with retention of configuration. Alternatively, in a similar manner, but with inversion of configuration at C-2, the other isomer **11** was used to give the azide **18**. Hydrogenolysis of **18** followed by N-acetylation afforded the acetylamine derivative **19**, which was treated with HCl to give **1** in 6% overall yield from **8**.

Scheme 1 (*a*) Ref. 5. (*b*) TBSOTf, 2,6-lutidine, CH_2Cl_2, –10°C, 30 min, 100%. (*c*) 2,4,4,6-Tetrabromo-2,5-cyclohexadien-1-one, $NaHCO_3$, CH_2Cl_2, 92%. (*d*) *t*-BuLi, THF, –78°C; then H_2O, 89%. (*e*) *n*-BuLi, THF, –78°C, 30 min, CuI, allyl bromide, 88%. (*f*) 1. OsO_4, NMO, THF, H_2O, 4 h, 97%; 2. Ag_2CO_3, PhH, reflux, 12 h, 45%, recovery 40% of **16**; 3. $NaIO_4$, CH_3OH–H_2O, 1 h; 4. $TMSCHN_2$, THF, CH_3OH, 10 min, 63%. (*g*) 1. TBAF, THF, 1 h; 2. HN_3, *n*-Bu_3P, DEAD, THF, toluene, 30 min, 63%. (*h*) 1. H_2, 10% Pd *on* C, AcOH, 15 h; 2. Ac_2O, CH_3OH, 2 h, 55%. (*i*) HCl, 84%.

References

1. Aoyagi, T.; Suda, H.; Uotani, K.; Kojima, F.; Aoyama, T.; Horiuchi, K.; Hamada, M.; Takeuchi, T. *J. Antibiot.* **45** (1992) 1404.
2. Aoyama, T.; Nagannawa, H.; Suda, H.; Uotani, K.; Aoyagi, T.; Takeuchi, T. *J. Antibiot.* **45** (1992) 1557.
3. Tatsuta, K.; Miura, S.; Ohta, S.; Gunji, H. *J. Antibiot.* **48** (1995) 286.
4. Tatsuta, K.; Ikeda, Y.; Miura, S. *J. Antibiot.* **49** (1996) 836.
5. Tatsuta, K.; Miura, S.; Ohta, S.; Gunji, H. *Tetrahedron Lett.* **36** (1995) 1085.
6. Tatsuta, K.; Miura, S. *Tetrahedron Lett.* **36** (1995) 6721.

5.4.2.3 *Calystegines* Calystegines (**1–7**) are alkaloids of the polyhydroxylated nortropane family, and have been isolated from the root secretions of *Calystegia sepium*, a member of the *Convolvulacae sepium*.[1] In addition, calystegines have been found in a variety of fruits and vegetables.[2–4] They are subdivided into three groups based on the number of hydroxyl groups present: calystegine A (three hydroxyl groups), calystegine B (four hydroxyl groups) and calystegine C (five hydroxyl groups). Several derivatives of the calystegines have also been isolated, containing a glycosyl moiety, an *N*-methyl group or an amino group instead of the tertiary hydroxyl group.[3] The most abundant calystegines in plants are calystegine A_3 (**1**), A_5 (**2**), B_2 (**4**), B_3 (**5**), B_4 (**6**) and C_1 (**7**).[5–7] Calystegines **1**, **4**, **5** and **6** are inhibitors of trehalases from various origins.[8] They might act as nutritional mediators of specific plant–bacterium and have been found to stimulate the growth of a nitrogen-fixing bacterium, *Rhizobium meliloti*, by serving as a source of carbon and nitrogen. These compounds have displayed an inhibitory activity toward β-glucosidase and α-galactosidase.[9,10] All of them are polyhydroxylated nortropane skeleton having an aminoketal function at the bridgehead position.[11,12] They exist only as bicyclic compounds in the chair conformation. Syntheses of calystegine analogues from noncarbohydrates have been reported.[2,13–18] The only carbohydrate derivatives used for the synthesis of calystegines are those of D-glucose.

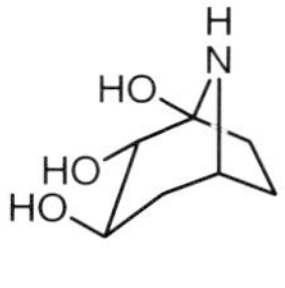

Calystegine A_3 **1** | Calystegine A_5 **2** | Calystegine B_1 **3** | (+)-Calystegine B_2 **4**

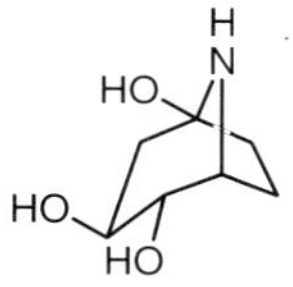

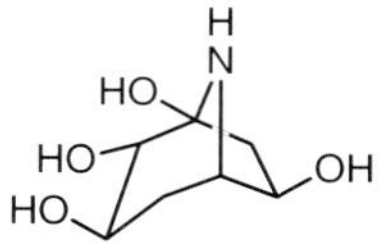

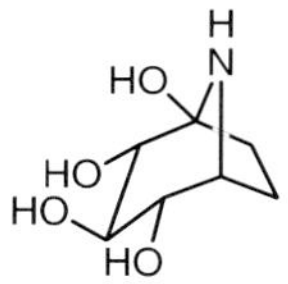

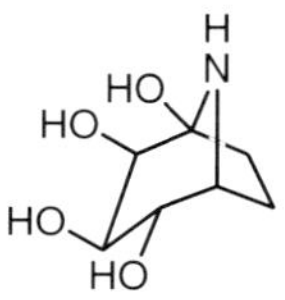

Calystegine B_3 **5** | Calystegine B_4 **6** | Calystegine C_1 **7**

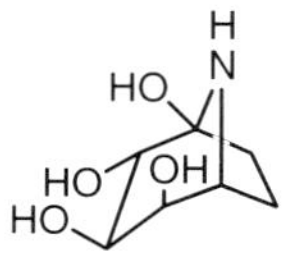

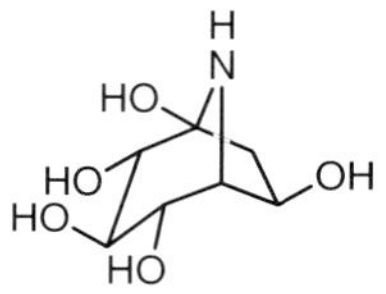

A methodology for the conversion of D-glucose to (+)- and (−)-calystegine B_2 has been achieved via a ring enlargement of a polysubstituted cyclohexanone **10** (Schemes 1–3).[19,20] The latter could be obtained from methyl α-D-glucopyranoside that can be readily transformed to **8**. Dehydroiodination of **8** by NaH in THF followed by benzylation afforded the olefin **9** in 70% yield. Ferrier rearrangement of **9** afforded the polysubstituted cyclohexanone **10** in 90% yield.[21–23] Protection of the hydroxyl group in **10** with TBS afforded **11**, which was treated with LDA followed by TMSCl to produce **13** as the major product, along with its regioisomer **12** (Scheme 1).

Scheme 1 (*a*) 1. PhCHO, $ZnCl_2$; 2. NaH, BnBr, DMF; 3. HCl, CH_3OH, 50%; 4. I_2, PPh_3, imidazole, toluene, 85%. (*b*) NaH, THF; then BnBr, *n*-Bu_4NI, 70%; *or* NaH, BnBr, DMF, 0°C, 60%. (*c*) $Hg(OAc)_2$, acetone, H_2O, 1% AcOH, 90%. (*d*) TBSOTf, 2,6-lutidine, CH_2Cl_2, 94%. (*e*) LDA, TMSCl, THF, –70°C.

Cyclopropanation of the major product **13** with Et_2Zn and CH_2I_2 afforded **14** (Scheme 2). Ring enlargement of **14** was achieved by treatment with $FeCl_3$[24] to give the β-chloroketone **15**, which, without purification, was treated with sodium acetate in boiling methanol to afford **16** (49%). Catalytic hydrogenation of **16** followed by reduction afforded a separable 6:4 mixture of the diastereoisomeric alcohols **17** and **18**. Treatment of **18** with methanesulfonyl chloride in pyridine afforded the mesylate **19**, which underwent S_N2 displacement reaction with azide ion to furnish **20**. This was desilylated with fluoride ion followed by oxidation of the resulting hydroxyl group with PCC to afford the ketone **21** in 68% from **18**. Azide reduction and full deprotection were accomplished by hydrogenolysis of **21** to produce (–)-calystegine B_2 [(–)-**4**] via the aminoketone salt **22**.

On the other hand, deprotection of the silyl group in **16** with fluoride ion furnished **23**, which was mesylated to give **24** (Scheme 3).[19,20] Subsequent displacement of the mesylate group with sodium azide or hydrogenation of the olefin in **24** led to its degradation. However, reduction of the ketone group in **24** with DIBAL-H in diethyl ether afforded **25**, which was subjected to S_N2 displacement with sodium azide to give the azides **27** (55%) and **26** (10%). Oxidation of **27** with Dess–Martin triacetoxyperiodane reagent[25] afforded the ketoazide **28** in 86% yield, which under the same conditions used in the former scheme gave (+)-calystegine B_2 (**4**). These results confirmed the assignment of the stereochemistry in the natural calystegine B_2 as **4**. Biological tests showed that **4** is catabolized by *R. meliloti*, whereas (–)-**4** is not.

The functionalized oxazoline fused with the seven-membered carbocycle **36** has been found to be a key precursor for the synthesis of calystegines (Scheme 4).[26] It has been obtained from methyl α-D-glucopyranoside (**29**) by conversion to **30** in 65% overall yield.[27] Compound **30** was oxidized, and reacted with $Ph_3P{=}CHCO_2Et$ to give the ethylenic ester

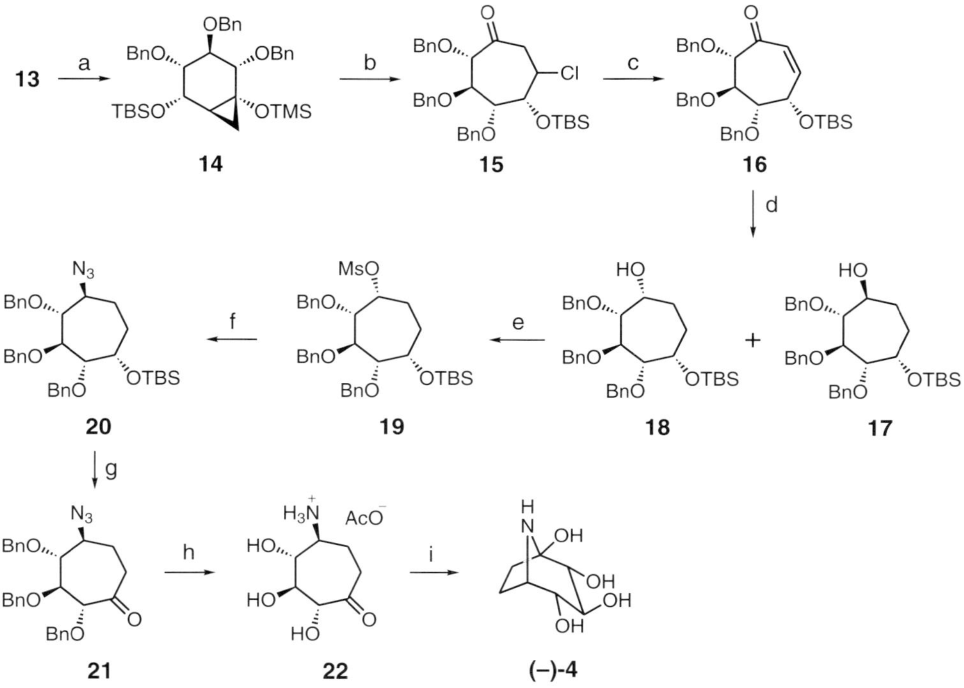

Scheme 2 (*a*) Et_2Zn, CH_2I_2, toluene, 0°C. (*b*) $FeCl_3$, DMF, 70°C. (*c*) NaOAc, CH_3OH, reflux. (*d*) 1. H_2, 10% Pd *on* C, EtOH, 90%; 2. $NaBH_4$, dioxane, 20°C, 83%. (*e*) MsCl, DMAP, Py. (*f*) NaN_3, DMF, 80°C, 80% from **18**. (*g*) 1. TBAF, THF, 90%; 2. CH_2Cl_2, PCC, 94%. (*h*) H_2, 10% Pd *on* C, AcOH, H_2O. (*i*) NaOH, H_2O, pH 11; *or* Permutite 50, aqueous NH_3.

31 in 81% yield. Catalytic hydrogenation of **31** and subsequent $LiAlH_4$ reduction produced **32** in 95% yield. Iodination of **32** and subsequent elimination of hydrogen iodide with potassium *tert*-butoxide afforded **33** in 77% yield, which upon acetolysis and deacetylation furnished the 6-deoxy-6-vinyl-D-glucopyranose **34** in 48% overall yield from **30**. Oximation of **34** afforded the oxime **35**, which underwent intramolecular olefinic nitrile oxide cycloaddition to afford the isoxazolines **36** (50%) and **37** (3%). Substitution of the hydroxyl group at C-5 in **36** with azido group via zinc azide mediated Mitsunobu substitution[28] led to the azido-isoxazoline **38** in 79% yield. Hydrogenolysis of **38** furnished the enantiomerically pure hydroxymethyl calystegine B_2 (**39**) in 45% yield.

(+)- and (–)-Calystegine B_2 have also been synthesized from **36** (Scheme 5).[29] Protection of the hydroxyl group in **36** followed by hydrogenolysis of the isoxazoline ring afforded **40** and **41** in 65 and 56% overall yield, respectively. Swern oxidation of **40** and **41** with an excess of reagents afforded a diastereoisomeric mixture of the α-chloroketones **42** (84%) and **43** (88%), respectively, via the corresponding α-formyl-α-chloroketone intermediate. Reductive dechlorination of the α-chloroketones **42** and **43** with zinc in ethanol gave the ketones **44** and **45**, respectively. Diisobutylaluminum hydride reduction of **44** and **45** afforded the corresponding alcohol **46** and **47**; the reduction occurred quantitatively with

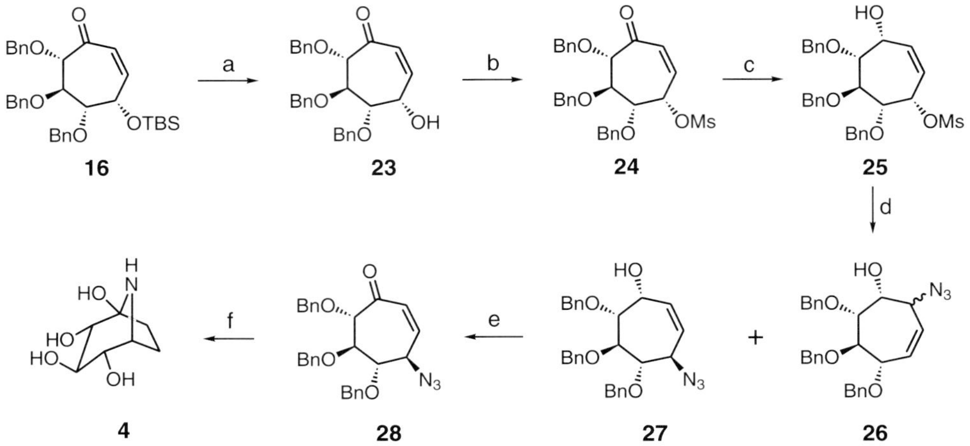

Scheme 3 (*a*) TBAF, THF, 73%. (*b*) MsCl, Py, 72%. (*c*) DIBAL-H, Et_2O, –60°C. (*d*) NaN_3, DMF, rt, **26** (10%), **27** (55%). (*e*) Dess–Martin reagent, Py, CH_2Cl_2, rt, 86%. (*f*) 1. H_2, 10% Pd *on* C, AcOH, H_2O; 2. Permutite 50, aqueous NH_3.

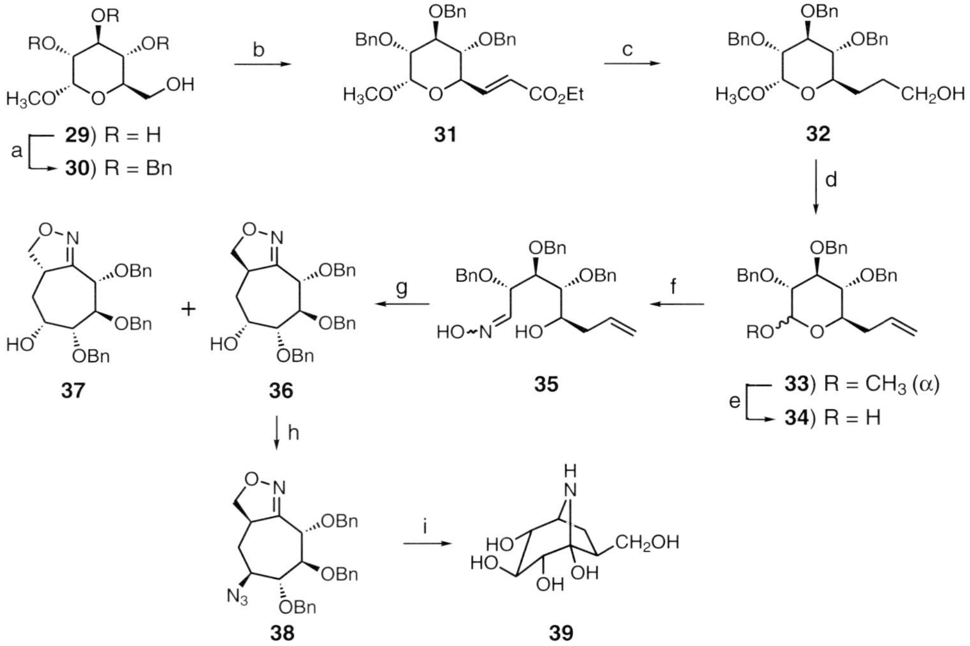

Scheme 4 (*a*) Ref. 27, 65%. (*b*) DMSO, $(COCl)_2$, –78°C, NEt_3, –78 to –40°C; then $Ph_3P{=}CHCO_2Et$, –40°C to rt, 81%. (*c*) 1. H_2, Raney nickel, CH_3OH, rt; 2. $LiAlH_4$, Et_2O, 0°C to rt, 95%. (*d*) 1. I_2, PPh_3, CH_2Cl_2, Py, 0°C; 2. *t*-BuOK, THF, rt, 77%. (*e*) 1. Ac_2O, H_2SO_4, $CHCl_3$, rt; 2. CH_3ONa, CH_3OH, rt, 81%. (*f*) $HONH_2{\cdot}HCl$, $NaOCH_3$, CH_3OH, reflux, 3 h, 94. (*g*) 1.75 M aqueous NaOCl, CH_2Cl_2, 20°C, 20 h, recovered oximes (16%), **36** (50%), **37** (3%). (*h*) $ZnN_6{\cdot}2Py$, PPh_3, diisopropyl azodicarboxylate, 20°C, 1 h, 79%. (*i*) Pd black, 80% aqueous AcOH, H_2, 20°C, 3 days; SiO_2 chromatography, 45%.

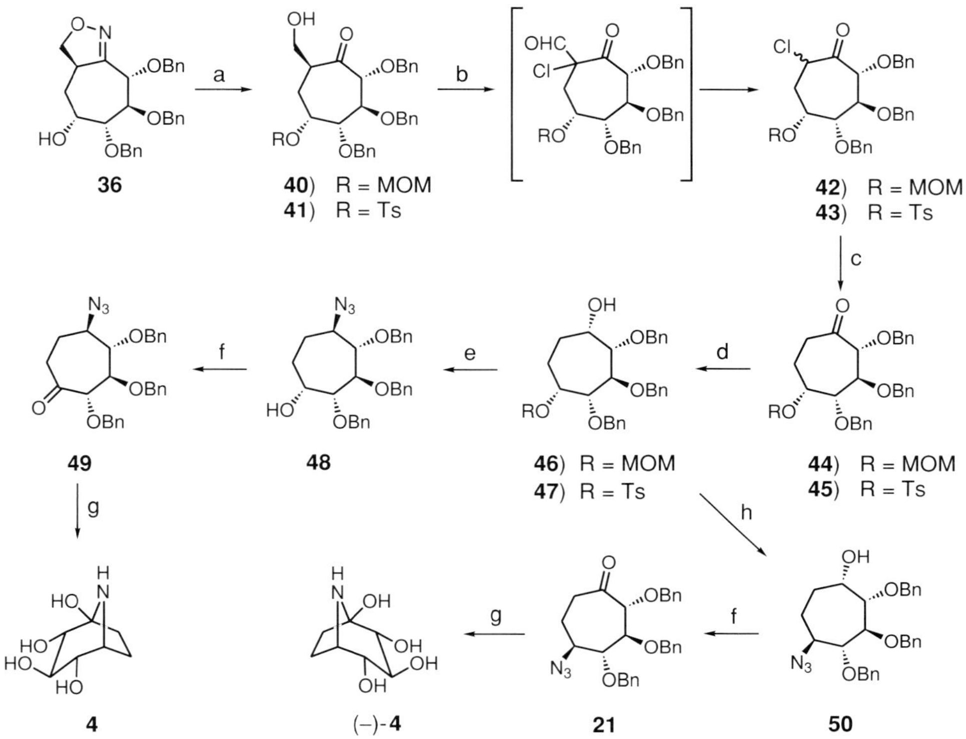

Scheme 5 (*a*) 1. CH_3OCH_2Cl, $(i\text{-Pr})_2NEt$, CH_2Cl_2, 20°C, 81%; *or p*-TsCl, Py, 20°C, 75%; 2. H_2, Raney nickel, CH_3OH, H_2O, $B(OH)_3$, **40** (80%), **41** (75%). (*b*) Excess DMSO, $(COCl)_2$, NEt_3, –60°C, **42** (84%), **43** (88%). (*c*) Zn, TMEDA, AcOH, EtOH, 2 h, **44** (89%) **45** (75%). (*d*) DIBAL-H, Et_2O, –50°C, 95%. (*e*) 1. $ZnN_6{\cdot}2Py$, PPh_3, DIAD; 2. CH_3OH, H^+, 81% for two steps. (*f*) DMSO, $(COCl)_2$, CH_2Cl_2, NEt_3, 95%. (g) Pd black, aqueous AcOH, H_2, 60%. (*h*) NaN_3, DMF, 80°C, 78%; DMSO, $(COCl)_2$, CH_2Cl_2, NEt_3, 95%.

>95% stereoselectivity. The cycloheptitol **46** was treated with ZnN_6 followed by acid hydrolysis of the MOM group to afford the azidoalcohol **48** in 81% yield, which underwent Swern oxidation to produce **49**. This was hydrogenated to furnish (+)-calystegine B_2 (**4**) in 46% overall yield from **46**. On the other hand, **47** was converted into the azide **50**, which was then subjected to Swern oxidation to give **21**, followed by hydrogenation to produce (–)-calystegine B_2 [(–)-**4**] in 44.5% from **47**.

Alternatively, methyl-α-D-glucopyranoside was converted to the 6-iodoglucopyranoside **51**,[26,29] and then subjected to zinc dust in the presence of benzylamine followed by the addition of allyl bromide to give **52** in 85:15 mixture of diastereomers (Scheme 6).[30] The major isomer was protected as the benzyl carbamate **53**, which underwent ring closure using Grubbs catalyst to afford **55** in 97% yield. The intermediate **55**[31] has also been prepared from 2,3,4-tri-*O*-benzyl-D-glucopyranose[32] by reaction with benzylamine, followed by treatment of the resulting amine with allylmagnesium bromide and CbzCl to give **54**, which was treated with I_2 and Ph_3P to produce **55** in low yield. Hydroboration of **55** with BMS followed by oxidative treatment gave a mixture of regioisomeric alcohols **57** and **56** (2.6:1)

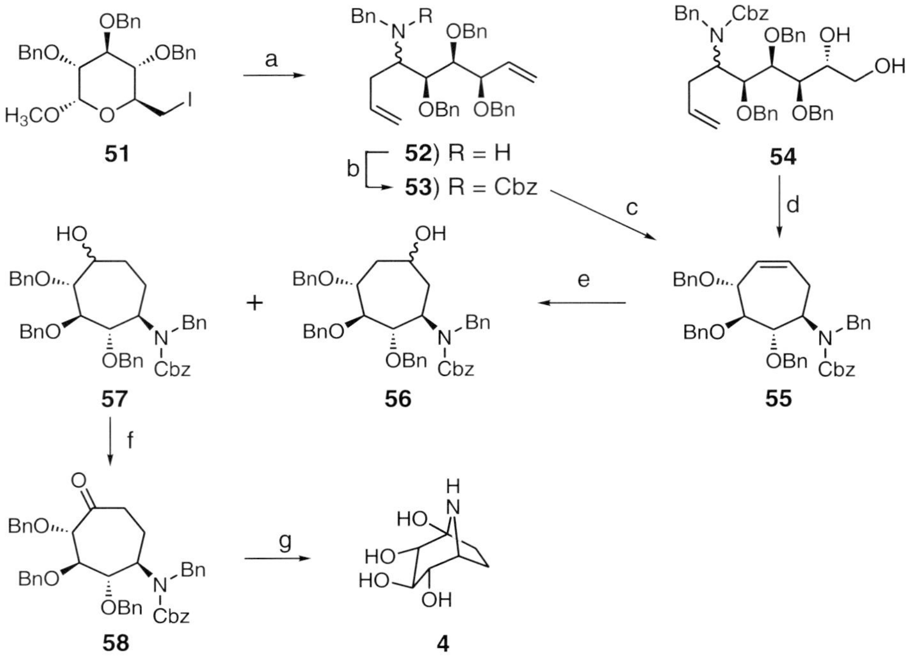

Scheme 6 (*a*) Zn, THF, $BnNH_2$; then $BrCH_2CH{=}CH_2$, 73%. (*b*) CbzCl, $NaHCO_3$, AcOEt, 94%. (*c*) Grubbs catalyst, CH_2Cl_2, rt, 97%. (*d*) I_2, Ph_3P, imidazole, 15%. (*e*) 1. DMS·BH_3, Et_2O, –30 to 0°C; 2. 30% H_2O_2, 2 N NaOH, 84%, **57** (60.5%), **56** (23.5%). (*f*) PCC, CH_2Cl_2, 93%. (*g*) 1. H_2, Pd *on* C, AcOH, H_2O; 2. NH_4OH, 79%.

in 84% combined yield. Oxidation of the major isomer **57** with pyridinium chlorochromate provided the ketone **58** in 93% yield. Finally, hydrogenolysis of **58** afforded (+)-calystegine B_2 (**4**) in 79% yield.

Similar methodology has been used for the synthesis of B_3 (**5**) and B_4 (**6**) from methyl D-galactopyranoside and methyl D-mannopyranoside, respectively.[33]

References

1. Tepfer, D.; Goldmann, A.; Pamboukdian, N.; Maille, N.; Lepingle, A.; Chevalier, D.; Denarié, J.; Rosenberg, C. *J. Bacteriol.* **170** (1988) 1153.
2. Asano, N.; Nash, R.J.; Molyneux, R.J.; Fleet, G.W.J. *Tetrahedron: Asymmetry* **11** (2000) 1645.
3. Watson, A.A.; Davies, D.R.; Asano, N.; Winchester, B.; Kato, A.; Molyneux, R.J.; Stegelmeier, B.L.; Nash, R.J. In *ACS Symposium Series 745: Natural and Selected Synthetic Toxins-Biological Implications*; Tu, A.T., Gaffield, W., Eds.; American Chemical Society: Washington, DC, 2000; p. 129.
4. Stütz, A.E., Ed. *Iminosugars as Glycosidase Inhibitors-Nojirimycin and Beyond*; Wiley-VCH: Weinheim, 1999.
5. Rothe, G.; Garske, U.; Dräger, B. *Plant Sci.* **160** (2001) 1043.
6. Bekkouche, K.; Daali, Y.; Cherkaoui, S.; Veuthey, J.-L.; Christen, P. *Phytochemistry* **58** (2001) 455.
7. Keiner, R.; Dräger, B. *Plant Sci.* **150** (2000) 171.
8. Asano, N.; Kato, A.; Kizu, H.; Matsui, K.; Watson, A.A.; Nash, R. *J. Carbohydr. Res.* **293** (1996) 195.

9. Molyneux, R.J.; Pan, Y.T.; Goldmann, A.; Tepfer, D.A.; Elbein, A.D. *Arch. Biochem. Biophysics* **304** (1993) 81.
10. Asano, N.; Kato, A.; Kizu, H.; Matsui, K. *Eur. J. Biochem.* **229** (1995) 369.
11. Goldmann, A.; Milat, M.L.; Ducrot, P.H.; Lallemand, J.Y.; Maille, M.; Lepingle, A.; Charpin, I.; Tepfer, D. *Phytochemistry* **29** (1990) 2125.
12. Ducrot, P.H.; Lallemand, J.Y. *Tetrahedron Lett.* **31** (1990) 3879.
13. Molyneux, R.J.; Nash, R.J.; Asano, N. In *Alkaloids: Chemical and Biological Perspectives*, Vol. II; Pelletier, S.W., Ed.; Elsevier Science: Oxford, 1996; p. 303.
14. Boyer, F.-D.; Ducrol, P.-H.; Henryon, V.; Soulie, J.; Lallemand, J.-Y. *Synlett* (1992) 357.
15. Johnson, C.R.; Bis, S.J. *J. Org. Chem.* **60** (1995) 615.
16. Faitg, T.; Soulié, J.; Lallemand, J.-Y.; Ricard, L. *Tetrahedron: Asymmetry* **10** (1999) 2165.
17. García-Moreno, M.I.; Benito, J.M.; Ortiz Mellet, C.; García Fernández, J.M. *J. Org. Chem.* **66** (2001) 7604.
18. van Hooft, P.A.V.; Litjens, R.E.J.N.; van der Marel, G.A.; van Boeckel, C.A.A.; van Boom, J.H. *Org. Lett.* **3** (2001) 731.
19. Boyer, F.-D.; Lallemand, J.-Y. *Synlett* (1992) 969.
20. Boyer, F.-D.; Lallemand, J.-Y. *Tetrahedron* **50** (1994) 10443.
21. Köhn, A.; Schmidt, R.R. *Liebigs Ann. Chem.* (1987) 1045.
22. Chrétien, F. *Synth. Commun.* **19** (1989) 1015.
23. Blattner, R.; Ferrier, R.J.; Haines, R.S. *J. Chem. Soc., Perkin Trans. 1* (1985) 2413.
24. Ito, Y.; Fujii, S.; Nakatsuka, M.; Kawamoto, F.; Seagusa, T. *Org. Synth.* **59** (1981) 113.
25. Dess, D.B.; Martin, J.C. *J. Am. Chem. Soc.* **113** (1991) 7277.
26. Duclos, O.; Duréault, A.; Depezay, J.C. *Tetrahedron Lett.* **33** (1992) 1059.
27. Bernet, B.; Vasella, A. *Helv. Chim. Acta* **62** (1979) 1990.
28. Viaud, M.C.; Rollin, P. *Synthesis* (1990) 130.
29. Duclos, O.; Mondange, M.; Duréault, A.; Depezay, J.C. *Tetrahedron Lett.* **33** (1992) 8061.
30. Boyer, F.-D.; Hanna, I. *Tetrahedron Lett.* **42** (2001) 1275.
31. Marco-Contelles, J.; Opazo, E.D. *J. Org. Chem.* **67** (2002) 3705.
32. Tatsuta, K.; Niwata, Y.; Umezawa, K.; Toshima, K.; Nakata, M. *J. Antibiot.* **44** (1991) 456.
33. Skaanderup, P.R.; Madsen, R. *J. Org. Chem.* **68** (2003) 2115.

5.4.2.4 *(–)-Mesembrine* (–)-Mesembrine (**12**) is an octahydroindole alkaloid isolated from certain plants of the *Sceletium* genus, namely *S. namaquense*, *S. strictum* and *S. tortuosum*.[1,2] The mesembrine structure has been established as **12**.[3–5] Mesembrine was synthesized from noncarbohydrate,[6–33] and its synthesis from D-mannitol has been presented herein.

The synthesis of (–)-mesembrine (**12**) has also been achieved from D-mannitol, which could be readily converted to (*S*)-(–)-benzyl 2,3-epoxypropyl ether (**1**)[34] (Scheme 1).[35]

D-Mannitol
a
b
c
d
e
f
g
h
i
j
k
l
OBn
BnO
OH
CN
OCH3
1
2
3
4) X = CH2OBn, Y = H
5) X = CH2OH, Y = H
6) X, Y = H, OH
7) X = Y = H
8
9
10
11
HO
HN
CH3
N
H3C
(–)-Mesembrine
12

Scheme 1 (*a*) Ref. 34. (*b*) 3,4-Dimethoxybenzyl cyanide, LDA, THF, –78°C to rt. (*c*) 1. 10% KOH, EtOH, reflux overnight; 2. 10% HCl, EtOH, rt, 64%. (*d*) LDA, crotyl bromide, THF, –78° to rt. (*e*) Conc. HCl, EtOH, reflux, 3 h. (*f*) 20% KOH, CH_3OH, CO_2 gas; then $NaIO_4$. (*g*) $NaBH_4$; then acid work-up, 75% for four steps. (*h*) $PdCl_2$, CuCl, wet DMF, O_2, 1 week, 73%. (*i*) *t*-BuOK, THF, reflux, overnight; then acid work-up, 66%. (*j*) 40% aqueous CH_3NH_2, sealed tube, 180°C, 1 h, **10** (41%), **11** (7%). (*k*) $(NCO_2Et)_2$, Ph_3P, THF, 10 min, 85%. (*l*) Li, liquid NH_3, 77%.

Condensation of **1** with 3,4-dimethoxybenzyl cyanide afforded the epimeric cyano alcohols **2**, which on alkaline hydrolysis gave the epimeric γ-lactones **3** in 64% yield. Treatment of **3** with crotyl bromide afforded the α,α-disubstituted lactone **4**. Acid-catalyzed debenzylation of **4** afforded the alcohol **5**, which on sequential saponification, periodate cleavage and reduction gave the lactone **7** via **6**. Palladium-catalyzed oxidation of **7** afforded **8**, which underwent intramolecular cyclization to produce **9** in 66% yield. Treatment of **9** with methylamine gave **10** (41%) and **11** (7%); however, the former can be cyclized into the latter in 85% yield. Reduction of **11** furnished **12** in 77% yield.

References

1. Bodendorf, K.; Krieger, W. *Arch. Pharm.* **290** (1957) 441.
2. Ishi, O.; Kugita, H. *Chem. Pharm. Bull.* **18** (1970) 299.
3. Popelak, A.; Haack, E.; Lettenbauer, G.; Spingler, H. *Die Naturwiss.* **47** (1960) 156.
4. Popelak, A.; Haack, E.; Lettenbauer, G.; Spingler, H. *Die Naturwiss.* **47** (1960) 231.
5. Jeffs, P.W.; Hawks, R.L.; Farrier, D.S. *J. Am. Chem. Soc.* **91** (1969) 3831.
6. Shamma, M.; Rodriguez, H.R. *Tetrahedron Lett.* **6** (1965) 4847.
7. Shamma, M.; Rodriguez, H.R. *Tetrahedron* **24** (1968) 6583.
8. Curphey, T.J.; Kim, H.L. *Tetrahedron Lett.* **9** (1968) 1441.
9. Stevens, R.V.; Wentland, M.P. *J. Am. Chem. Soc.* **90** (1968) 5580.
10. Keely, S.L., Jr.; Tahk, F.C. *J. Am. Chem. Soc.* **90** (1968) 5584.
11. Oh-ishi, T.; Kugita, H. *Tetrahedron Lett.* **9** (1968) 5445.
12. Oh-ishi, T.; Kugita, H. *Chem. Pharm. Bull.* **18** (1970) 299.
13. Yamada, S.; Otani, G. *Tetrahedron Lett.* **12** (1971) 1133.
14. Wijnberg, J.B.P.A.; Speckamp, W.N. *Tetrahedron* **34** (1978) 2579.
15. Starauss, H.F.; Wiechers, A. *Tetrahedron Lett.* **20** (1979) 4495.
16. Martin, S.F.; Puckette, T.A.; Colapret, J.A. *J. Org. Chem.* **44** (1979) 3391.
17. Sanchez, I.H.; Tallabs, F.R. *Chem. Lett.* (1981) 891.
18. Takano, S.; Imamura, Y.; Ogasawara, K. *Chem. Lett.* (1981) 1385.
19. Keck, G.E.; Webb, R.R., II. *J. Org. Chem.* **47** (1982) 1302.
20. Kochhar, K.S.; Pinnick, H.W. *Tetrahedron Lett.* **24** (1983) 4785.
21. Winkler, J.D.; Mullar, C.L.; Scott, R.D. *J. Am. Chem. Soc.* **110** (1988) 4831.
22. Meyers, A.I.; Hanreich, R.; Wanner, K.T. *J. Am. Chem. Soc.* **107** (1985) 7776.
23. Takano, S.; Samizu, K.; Ogasawara, K. *Chem. Lett.* (1990) 1239.
24. Kosugi, H.; Miura, Y.; Kanna, H.; Uda, H. *Tetrahedron: Asymmetry* **4** (1993) 1409.
25. Honda, T.; Kimura, N.; Tsubuki, M. *Tetrahedron: Asymmetry* **4** (1993) 21.
26. Yoshimitsu, T.; Ogasawara, K. *Heterocycles* **42** (1996) 135.
27. Denmark, S.E.; Marcin, L.R. *J. Org. Chem.* **62** (1997) 1675.
28. Mori, M.; Kuroda, S.; Zhang, C.-S.; Sato, Y. *J. Org. Chem.* **62** (1997) 3263.
29. Rajagopalan, P. *Tetrahedron Lett.* **38** (1997) 1893.
30. Langlois, Y.; Dalko, P.I.; Brun, V. *Tetrahedron Lett.* **39** (1998) 8979.
31. Ogasawara, K.; Yamada, O. *Tetrahedron Lett.* **39** (1998) 7747.
32. Rigby, J.H.; Dong, W. *Org. Lett.* **2** (2000) 1673.
33. Kulkarni, M.G.; Rasne, R.M.; Davawala, S.I.; Doke, A.K. *Tetrahedron Lett.* **43** (2002) 2297.
34. Anisuzzaman, A.K.M.; Owen, L.N. *J. Chem. Soc. (C)* (1967) 1021.
35. Takano, S.; Imamura, Y.; Ogasawara, K. *Tetrahedron Lett.* **22** (1981) 4479.

5.4.2.5 *Streptolidine* Streptolidine (**1**) is an amino acid containing a guanidine moiety and it is a constituent of a number of antibiotics produced by *Streptomyces*. It was first isolated[1] from the hydrolyzate of streptothricin antibiotics (**3**). The amino acids roseonine and geamine that are isolated from roseothricin[2] and geomycin,[3] respectively, were identified as the same substance streptolidine (**1**). The chemical structure of this amino acid **1** was studied by degradation,[4] and its absolute configuration was established by X-ray crystallography.[5] A retro-synthetic analysis of **3** led to the lactam **2** that could be prepared from **1**.

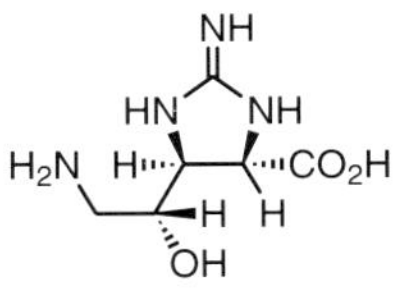

Streptolidine **1** — Streptolidine lactam **2** — Streptothricin antibiotics (n = 1 or 2) **3**

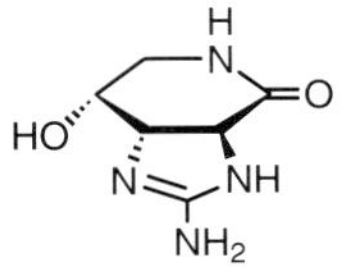

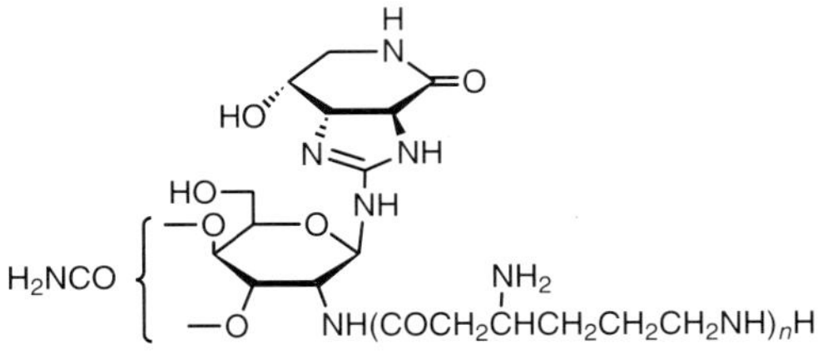

5.4.2.5.1 Synthesis from D-ribose D-Ribose has been used for the synthesis of streptolidine (**1**) (Scheme 1).[6–8] The diazide **4**,[9] obtained from D-ribose, was reduced with $LiAlH_4$, followed by acetylation to produce the aziridine **5**, which was treated with NaN_3 to afford a mixture of the 3-azidoarabinoside derivative **6** (42%) and its xylo isomer (8%). Catalytic hydrogenation of **6** followed by acetylation afforded **7**, whose hydrolysis and subsequent oxidation gave the lactone **8**. On the other hand, hydrolysis of **7** with aqueous TFA followed by acetylation furnished **10**, which was de-O-acetylated followed by oxidation to give the γ-lactone **8**. Deacylation of **8** gave **9** whose treatment with base and excess BrCN and then hydrochloric acid afforded **1**.

5.4.2.5.2 Synthesis from D-xylose D-Xylose has also been used as a starting material for the synthesis of **1**, by conversion into a mixture of tri-*O*-mesyl-α- and -β-D-xylofuranosides (**11**) in a ratio 91:9 (Scheme 2).[10] The mixture was treated with sodium azide to afford a mixture of triazide **13**(α/β) and the diazide **12**(α/β) in addition to the respective monoazide derivative. Compound **13** was also obtained by further reaction of **12** with sodium azide. Hydrogenolysis of the triazide **13** followed by protection of the resulting triamine with benzyloxycarbonyl chloride afforded **14** in 83% yield. Acid hydrolysis of the glycosidic linkage in **14** gave the tricarbamate **15** whose conversion into streptolidine (**1**) had been performed through oxidation, deprotection and guanidination, as shown in the former scheme.

5.4.2.5.3 Synthesis from D-mannitol The synthesis of streptolidine (**1**) has been reported from D-mannitol by conversion to 3,4-anhydro-1,2:5,6-di-*O*-isopropylidene-D-iditol (**16**)[11,12] (Scheme 3).[13] Azidolysis of **16** followed by mesylation of the resulting hydroxyl group afforded **17**. This resulting compound was treated with sodium azide in DMSO to afford **18**, which was hydrogenolyzed and then treated with benzylchloroformate to afford **19**. De-O-isopropylidenation of **19** followed by periodate

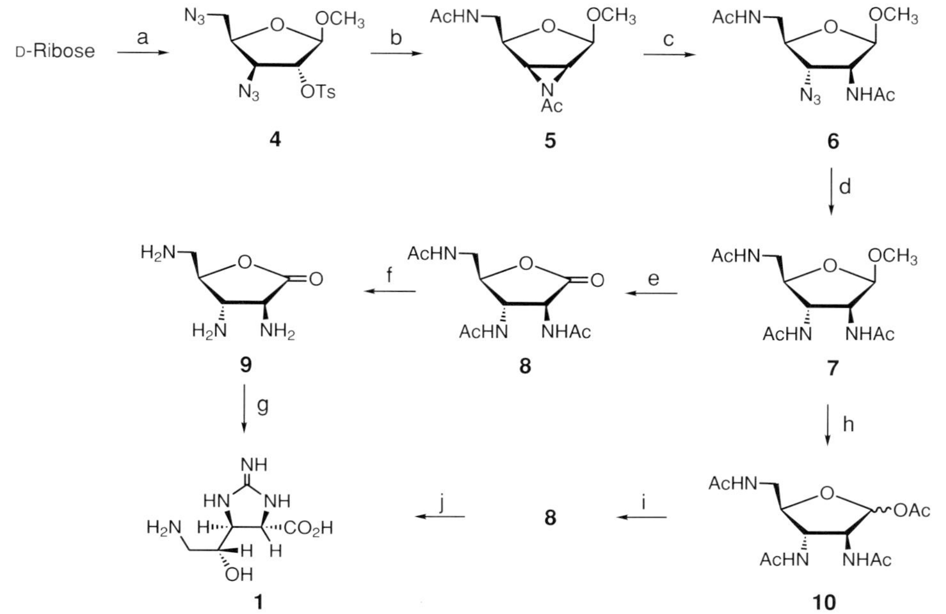

Scheme 1 (*a*) Ref. 9. (*b*) $LiAlH_4$, THF; Ac_2O, CH_3OH, rt, 45% for two steps. (*c*) NaN_3, DMF, 140°C, 45 min, **6** (42%). (*d*) 1. H_2, Pd *on* C, CH_3OH; 2. Ac_2O, CH_3OH, 95% for two steps. (*e*) 1. 2 N *p*-TsOH, dioxane, 35 and 40% recovered starting material; 2. CrO_3, AcOH, 52%. (*f*) HBr, AcOH, 100%. (*g*) 1. Aqueous 1 N NaOH; then BrCN, rt; 2. 6 N HCl, reflux; 3. Dowex 50X8 (NH_4^+) resin. elution with 0.2 N NH_4OH, 7.5%. (*h*) 1. Aqueous TFA, 100°C, 60 min; 2. Ac_2O, Py, 81% for two steps. (*i*) 1. $NaOCH_3$, CH_3OH, rt; 2. CrO_3, AcOH, conc. H_2SO_4, 24%; then Amberlite IRC-50 (H^+) resin. IR-4B (free) resin. (*j*) 1. 6 N HCl, 100°C, 1 h, quantitative; 2. Poly-Hünig-base (diisopropylaminomethylpolystyrene), BrCN, CH_3OH, rt; 5. 6 N HCl, 100°C, 20 min, 67%.

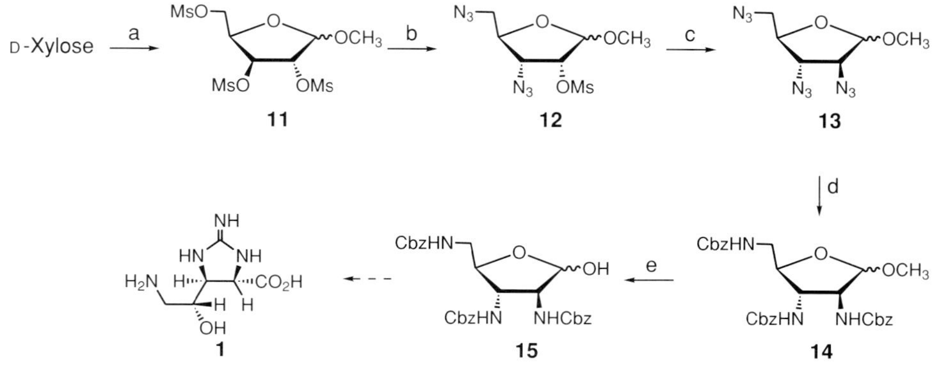

Scheme 2 (*a*) 1. CH_3OH, HCl, rt, overnight; 2. MsCl, Py, 5 h, 68%. (*b*) NaN_3, DMF, 100–110°C, 2 h, 135–140°C, 2 h. (*c*) NaN_3, DMF, 140–150°C. (*d*) EtOH, Pd black, H_2; then NEt_3, CbzCl, 4 h, 83%. (*e*) Dioxane, 2 M aqueous *p*-TsOH, reflux, 1.5 h, 30%.

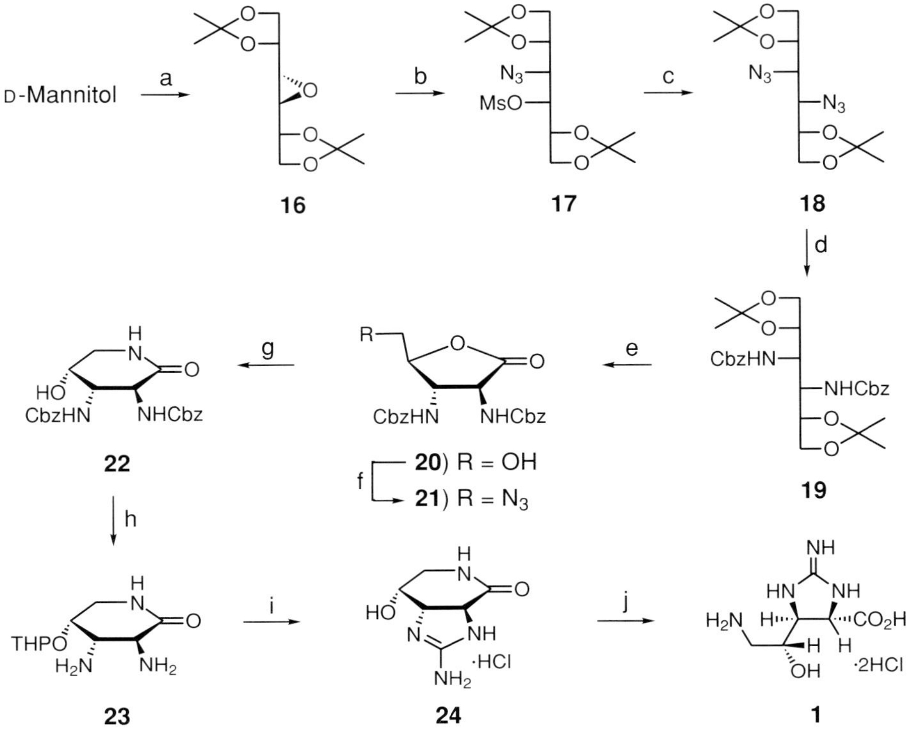

Scheme 3 (*a*) Refs. 11 and 12. (*b*) 1. 80% aqueous methyl cellosolve, NaN_3, NH_4Cl, 120°C, 7 h; 2. MsCl, Py, rt, 1 h, 60%. (*c*) NaN_3, DMSO, 120°C, 3.5 h, 47%. (*d*) H_2, Pd black, Py, –20°C, CbzCl, 1.5 h, 77%. (*e*) AcOH, H_2O, 60°C, 1 h; then $NaIO_4$ (1.2 equiv.), acetone, H_2O; then Br_2, dioxane, H_2O, 1 h, 59%. (*f*) 1. MsCl, Py, rt, 1 h, 89%; 2. NaN_3, DMSO, 100°C, 0.5 h, 68%. (*g*) H_2, CH_3OH, Raney nickel, 4 h, 63%; *or* CH_3OH, H_2, Pd black, 5 h; *N*-(benzyloxycarbonyloxy)succinimide, DMF, H_2O, rt, 1 h, 60%. (*h*) 1. DHP, *p*-TsOH, DMF, 37–40°C, 2 h, NEt_3, 97%; 2. H_2, CH_3OH, 10% Pd *on* C, 4 h, 77%. (*i*) 1. BrCN, H_2O, rt, 5 h, 51%; 2. Ag_2CO_3, H_2O, 10 min, 2 M HCl, rt, 0.5 h, 87%. (*j*) 3 M HCl, rt, 22 h, 83%.

oxidation with 1.2 equiv. of sodium periodate and subsequent oxidation with bromine afforded the lactone **20**. Mesylation of the primary hydroxyl group in **20** followed by S_N2 displacement with sodium azide afforded **21**. Hydrogenolysis of the azide group in **21** afforded the lactam **22** (63%), which was treated with dihydropyran in the presence of *p*-toluenesulfonic acid followed by hydrogenation to give the lactam **23**. Treatment of **23** with cyanogen bromide followed by silver carbonate and hydrochloric acid afforded the hydrochloride salt of the streptolidine lactam **24**. Hydrolysis of **24** furnished the dihydrochloride of **1** in 1.3% overall yield from **16**.

References

1. Carter, H.E.; Clark, R.K.; Kohn, P.; Rothrock, J.M.; Taylor, W.R.; West, C.A.; Whitfield, G.B.; Jackson, H.G. *J. Am. Chem. Soc.* **76** (1954) 566.
2. Nakanishi, K.; Ito, T.; Hirata, Y. *J. Am. Chem. Soc.* **76** (1954) 2845.

3. Brockmann, H.; Musso, H. *Chem. Ber.* **88** (1955) 648.
4. Carter, H.E.; Sweeley, C.C.; Daniels, E.E.; McNary, J.E.; Schaffiner, C.P.; West, C.A.; Van Tamelen, E.E.; Dyer, J.R.; Whaley, H.A. *J. Am. Chem. Soc.* **83** (1961) 4296.
5. Bycroft, B.W.; King, T. *J. Chem. Commun.* (1972) 652.
6. Kusumoto, S.; Tsuji, S.; Shiba, T. *Bull. Chem. Soc. Jpn.* **47** (1974) 2690.
7. Kusumoto, S.; Tsuji, S.; Shiba, T. *Tetrahedron Lett.* **15** (1974) 1417.
8. Goto, T.; Ohgi, T. *Tetrahedron Lett.* **15** (1974) 1413.
9. Hildesheim, J.; Cleophax, J.; Sepulchre, A.M.; Gero, S.D. *Carbohydr. Res.* **9** (1969) 315.
10. Kusumoto, S.; Tsuji, S.; Shima, K. *Bull. Chem. Soc. Jpn.* **49** (1976) 3611.
11. Aspinall, G.O.; Cheetham, N.W.H.; Frdova, J.; Tam, S.G. *Carbohydr. Res.* **36** (1974) 257.
12. Tipson, R.S.; Cohen, A. *Carbohydr. Res. 1* (1965) 338.
13. Kinoshita, M.; Suzuki, Y. *Bull. Chem. Soc. Jpn.* **50** (1977) 2375.

5.5 6:6-Fused heterocycles

5.5.1 *Hydroxylated quinuclidines*

There are a large number of alkaloids that contain the quinuclidine nucleus such as the sarpagine, ajmaline[1] and cinchona[2] families. The quinuclidines **1–3** have been used for the synthesis of such alkaloids.[3,4] They have pharmacological activities.[5] Thus, several reports have highlighted the potential of chiral hydroxylated quinuclidines in propping the active site of muscarinic receptors.[6–8] Substituted quinuclidines may provide selective Vaughan Williams class III antiarrhythmic effects.[6–8]

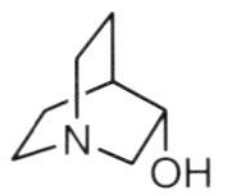

S-Quinuclidinol — **1**

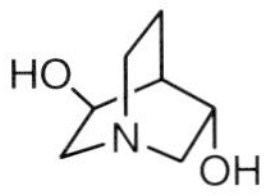

(3*S*,5*S*)-Quinuclidine-3,5-diol — **2**

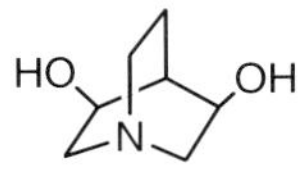

(3*S*,5*R*)-Quinuclidine-3,5-diol (*meso*-quinuclidinediol) — **3**

5.5.1.1 *Synthesis from* D*-glucose* Methyl α-D-glucopyranoside has been used as a chiral precursor for the synthesis of *S*-quinuclidinol (**1**), via its conversion to the 3,4-unsaturated derivative **4**[9] in 38% yield, which subsequently transformed to the branched derivative **5** in 75% yield[10] (Scheme 1).[11] Mesylation of the primary hydroxyl group gave the mesylate **6** (86%), which on treatment with sodium azide followed by hydrogenation and then cyclization with lithium diisopropylamide gave the lactam **7** in 51% yield from **5**. Reduction of **7** followed by protection of the secondary amine with benzyl chloroformate and the resulting carbamate was treated with $TiCl_4$, which followed by treatment with DBU gave **8**. Ozonolysis of **8** and subsequent borohydride reduction afforded the diol **9** (87%), which upon selective mesylation gave **10** (84%). The secondary hydroxyl group in **10** was protected as a silyl ether, followed by deprotection of the Cbz group, cyclization and then desilylation to give **1**.[12,13]

Another approach to the synthesis of **1** has started with the introduction of two-carbon chains at C-3 of D-glucose (Scheme 2).[14] Oxidation of diacetone D-glucose (**11**) with pyridinium chlorochromate followed by treatment with [(carbomethoxy)methylene]triphenylphosphorane, hydrogenation and subsequent reduction with $LiAlH_4$ gave **12**[15] in 79% overall yield. Mesylation of **12** followed by nucleophilic displacement of the mesylate group with azide ion gave **13** (87%). Removal of the terminal isopropylidene group in **13** gave **14**, which upon periodate oxidation and subsequent hydrogenation led to the respective cyclized product whose imino function was protected to furnish the carbamate **15** in 48% overall yield from **12**. Methanolysis of **15** and subsequent conversion to the xanthate gave **16**, whose Barton deoxygenation[16,17] and subsequent hydrolysis

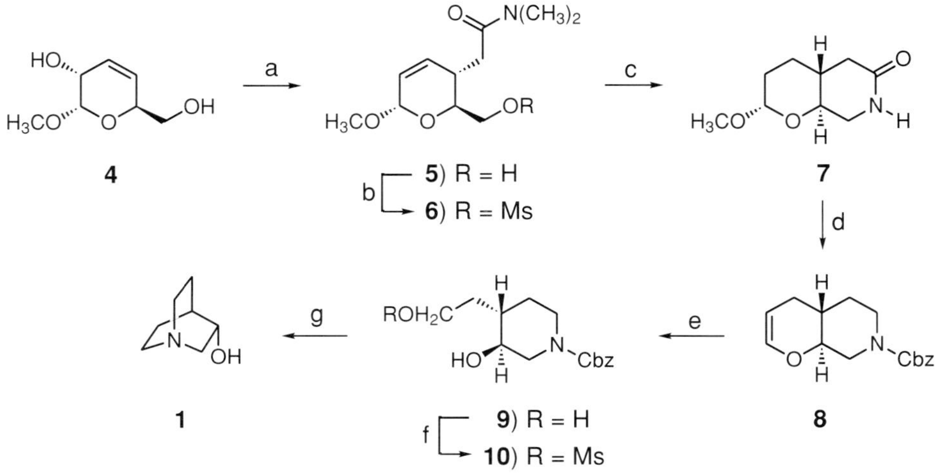

Scheme 1 (*a*) $CH_3C(OCH_3)_2N(CH_3)_2$, diglyme, 160°C; then K_2CO_3, CH_3OH, 1 h, 75%. (*b*) MsCl, Py, 0°C, 2 h, 86%. (*c*) 1. NaN_3, DMF, 65°C, 8 h, 89%; 2. H_2, Pd black, EtOH; then LDA (1.1 equiv.), –40°C, THF, 68%. (*d*) 1. $LiAlH_4$, THF; then CbzCl, aqueous $NaHCO_3$, ether, 75%; 2. $TiCl_4$ in $CDCl_3$, –20°C, 10 min; then DBU, 44%. (*e*) O_3, CH_3OH, CH_2Cl_2, –65°C, 20 min; then $NaBH_4$, 87%. (*f*) MsCl, Py, –10°C, 84%. (*g*) 1. $CF_3SO_3Si(CH_3)_2t$-Bu, 2,6-lutidine, CH_2Cl_2, –20°C, 84%; 2. H_2, Pd black, EtOH, 2 h, 20°C, 67%; 3. TFA, EtOH, 6 h, 50°C.

and reduction of the resulting lactol afforded **9** which was cyclized to *S*-quinuclidinol (**1**). On the other hand, treatment of **15** with ethyl mercaptan in the presence of aqueous TFA followed by benzylation, treatment with mercuric chloride, sodium borohydride reduction and then mesylation gave **17**. The intramolecular cyclization after removal of the protecting groups afforded *meso*-quinuclidinediol (**3**).

Alternatively, compound **14** was oxidized with sodium periodate, followed by reduction with sodium borohydride in ethanol to afford **18** (Scheme 3).[18] Hydrolysis of the isopropylidene group in **18** with 50% aqueous trifluoroacetic acid followed by hydrogenation of the azide group and then intramolecular reductive amination and protection of the imine group afforded **19**. Selective mesylation of the primary hydroxyl group in **19** followed by silylation and subsequent removal of the benzyloxycarbonyl group and then cyclization with sodium acetate afforded **20**. Deprotection of **20** led to *meso*-quinuclidine-3,5-diol (**3**).

5.5.1.2 *Synthesis from* D-*arabinose* (3*S*,5*S*)-Quinuclidine-3,5-diol (**2**) has been synthesized from D-arabinose by conversion to the furanoside **21** in 59% overall yield (Schemes 4).[19] Oxidation of the C-3 hydroxyl group with pyridinium chlorochromate followed by treatment with [(methoxycarbonyl)methylene]triphenylphosphorane and subsequent hydrogenation gave the corresponding ester, which upon reduction, mesylation and nucleophilic displacement of the mesylate group by azide ion afforded the branched azidoethyl lyxofuranoside **22**. Two pathways were used to convert **22** into **2**. Acid hydrolysis of

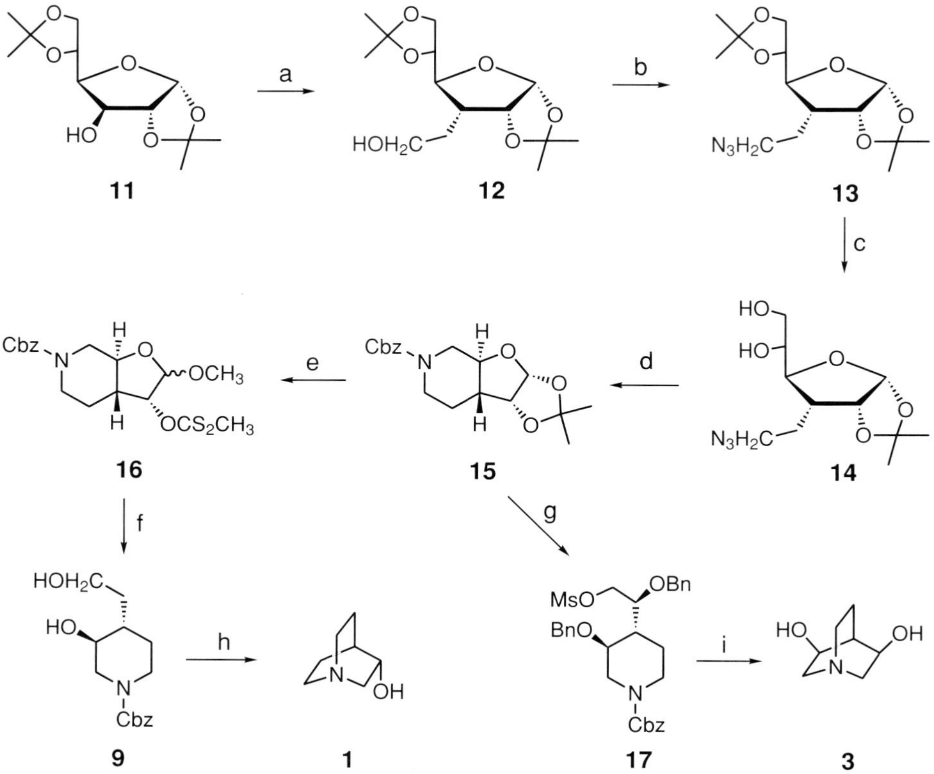

Scheme 2 (*a*) PCC, CH_2Cl_2, 20°C; then $Ph_3PCHCO_2CH_3$, benzene, reflux; then H_2, Pd *on* C, CH_3OH; then $LiAlH_4$, THF, 79%. (*b*) 1. MsCl, Py, 0°C; 2. NaN_3, DMF, 40°C, 87%. (*c*) AcOH, CH_3OH, H_2O, 40°C, 90%. (*d*) 1. $NaIO_4$, CH_3OH, H_2O; 2. H_2, Pd balck, AcOH; then CbzCl, Et_2O; H_2O, $NaHCO_3$, 20°C, 66% for two steps. (*e*) 1. Dowex (H^+) resin, CH_3OH, 67%; 2. NaH, CS_2, CH_3I, THF, 20°C. (*f*) 1. Bu_3SnH, xylene, AIBN, 110°C, 87%; 2. TFA, 20°C, 87%; then $NaBH_4$, EtOH, H_2O, 50%. (*g*) 1. EtSH, aqueous TFA; 92%; 2. dibenzylation; 3. HgCl; 4. $NaBH_4$, MsCl, 52% for four steps. (h) 1. MsCl, Py, 0°C; 2. Pd black, H_2, EtOH; 3. NaOAc. (*i*) 1. Pd black, H_2, EtOH; 2. NaOAc, 64% for two steps; 3. Pd black, H_2, AcOH, 81%.

22 followed by azide reduction and then intramolecular reductive amination and secondary amine protection gave the carbamate **23**. Mesylation of the primary hydroxyl group in **23** afforded **24**. Removal of Cbz group followed by treatment with sodium acetate smoothly cyclized to produce **2**.

On the other hand, removal of the silyl group in **22** with fluoride ion and subsequent mesylation afforded **25**. Hydrogenation of **25** followed by cyclization in the presence of sodium acetate and subsequent protection of the amino group with Cbz, treatment with ethanethiol in aqueous TFA, followed by dibenzylation, mercuric chloride catalyzed hydrolysis, sodium borohydride reduction and mesylation afforded **26**. Selective hydrogenolysis of the carbamate group in **26** followed by intramolecular cyclization afforded the 3,5-di-*O*-benzyl ether of quinuclidinediol. Removal of the benzyl groups afforded **2**.

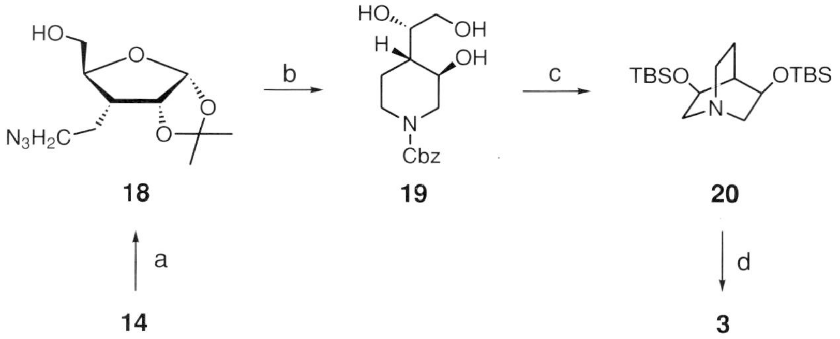

Scheme 3 (*a*) 1. $NaIO_4$; 2. $NaBH_4$, EtOH, 85% for two steps. (*b*) 1. 50% aqueous TFA, 86%; 2. H_2, 10% Pd *on* C, 50°C; then CbzCl, 62%. (*c*) 1. MsCl, Py, –30°C, 61%; 2. TBSOTf, 2,6-lutidine, 86%; 3. H_2, Pd black; 4. NaOAc, 88% for two steps. (*d*) Aqueous TFA.

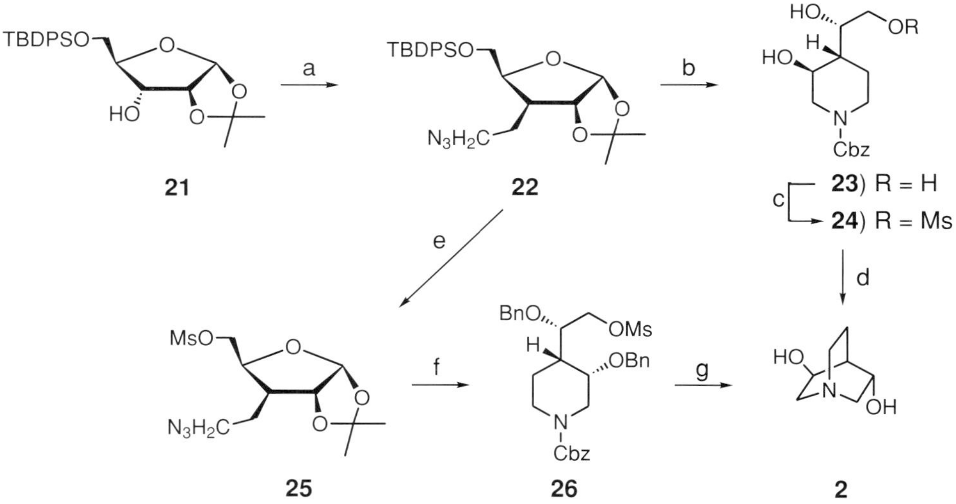

Scheme 4 (*a*) 1. Py·CrO_3; 2. $Ph_3P{=}CHCO_2CH_3$; 3. H_2, Pd *on* C, 81% for three steps; 4. $LiAlH_4$; 5. MsCl, Py; then NaN_3, 41% from D-arabinose. (*b*) 1. Acid hydrolysis, 80%; 2. H_2, Pd; then CbzCl, $NaHCO_3$, 77%. (*c*) MsCl, Py, CH_2Cl_2, 87%. (*d*) 1. H_2, Pd *on* C; 2. NaOAc, 42% from **22** and 17% from D-arabinose. (*e*) 1. TBAF; 2. MsCl, 94%. (*f*) 1. H_2, Pd; 2. NaOAc; then CbzCl, 86% from **22**; 3. EtSH, 89% TFA; 4. BnBr, NaH; 5. $HgCl_2$; 6. $NaBH_4$; 7. MsCl, Py. (*g*) 1. Selective hydrogenolysis, 2. NaOAc; 3. hydrolysis, 98%; 36% from **22** and 15% from D-arabinose.

References

1. Koskinen, A.; Lounasmaa, M. *Prog. Chem. Nat. Prod.* **43** (1983) 267.
2. Woodward, R.B.; Wendler, N.L.; Brutschy, F.J. *J. Am. Chem. Soc.* **67** (1945) 860.
3. Stotter, P.L.; Friedman, M.D.; Doresy, G.O.; Shiely, R.W.; Williams, R.F.; Winter, D.E. *Heterocycles* **25** (1987) 251.
4. Stotter, P.L.; Hill, K.A.; Friedman, M.D. *Heterocycles* **25** (1987) 259.
5. Mashkovsky, M.D.; Yakhontov, L.N. *Prog. Drug Res.* **13** (1969) 294.

6. Rzeszotarski, W.J.; McPherson, D.W.; Ferkany, J.W.; Kinnier, W.J.; Noronha-Blob, L.; Kirkiern-Rzeszotarski, A. *J. Med. Chem.* **31** (1988) 1463.
7. Carroll, F.I.; Abraham, P.; Parham, K.; Griffiths, R.C.; Ahmad, A.; Richard, M.M.; Padilla, F.N.; Witkin, J.M.; Chiang, P.K. *J. Med. Chem.* **30** (1987) 805.
8. Saunders, J.; MacLeod, A.M.; Merchant, K.; Showell, A.; Snow, R.J.; Street, L.J.; Baker, R. *J. Chem. Soc., Chem. Commun.* (1989) 1618.
9. Holder, N.L.; Fraser-Reid, B. *Can. J. Chem.* **51** (1973) 3357.
10. Corey, E.J.; Shibasaki, M.; Knolle, J. *Tetrahedron Lett.* **18** (1977) 1625.
11. Fleet, G.W.J.; Jamess, K.; Lunn, R.J. *Tetrahedron Lett.* **27** (1986) 3053.
12. Baker, R.W.; Pauling, P.J. *J. Chem. Soc., Perkin Trans. 1* (1972) 2340.
13. Lambercht, G. *Arch. Pharm.* **309** (1976) 235.
14. Fleet, G.W.J.; Jamess, K.; Lunn, R.J.; Mathews, C.J. *Tetrahedron Lett.* **27** (1986) 3057.
15. Rosenthal, A.; Nguyen, L.B. *J. Org. Chem.* **34** (1969) 1029.
16. Iacono, S.; Rasmussen, J.R. *Org. Synth.* **64** (1985) 57.
17. Barton, D.H.R.; Motherwell, W.B. *Pure Appl. Chem.* **53** (1981) 15.
18. Fleet, G.W.J.; Mathews, C.J.; Seijas, J.A.; Vazquez Tato, M.P.; Brown, D.J. *J. Chem. Soc., Perkin Trans. 1* (1989) 1067.
19. Fleet, G.W.J.; Mathews, C.J.; Seijas, J.A.; Vazquez Tato, M.P.; Brown, D.J. *J. Chem. Soc., Perkin Trans. 1* (1989) 1065.

5.5.2 *Biopterins*

(–)-Biopterin (**1**) is one of the potent natural pteridines isolated from human urine as the growth factor of *Crithidia fasciculata*.[1] It has attracted much attention as a precursor of (6*R*)-tetrahydrobiopterin, which was known as a coenzyme of aromatic amino acid monooxygenase.[2]

L-*erythro*-Biopterin (**2**) is a widespread naturally occurring enzyme cofactor identified in the phenylalanine-to-tyrosine conversion.[3] It is widely distributed in microorganisms, insects, algae, amphibian and mammals.[3,4] It is the most abundant naturally occurring pterin found in human urine.[5] Its 5,6,7,8-tetrahydro analogue functions as an essential enzyme cofactor in a number of hydroxylation and oxygenase reactions: conversion of tyrosine to dopa,[6,7] melanin synthesis[8,9] and hydroxylation of both tryptophan[10–13] and dihydroorotic acid.[14] Biological oxidation or dehydrogenation reactions including the 17-α-hydroxylation of progesterone, the biosynthesis of the prostaglandins,[15] the conversion of long-chain alkyl ethers of glycerol to fatty acids, the introduction of unsaturation into the carotenes and fatty acids, sterol biosynthesis, as well as oxidation of long-chain saturated fatty acids involve tetrahydropteridine cofactors. The tetrahydrobiopterin and the related tetrahydropterins have been postulated to play a critical role in cellular electron transport, including photosynthesis.[3,4]

The fluorescent compounds that were isolated from *Euglena gracilis* have been given the structures **3–6**, one of which stimulated ferredoxin-dependent oxygen reduction by isolating *Euglena* chloroplasts in the dark.[16,17]

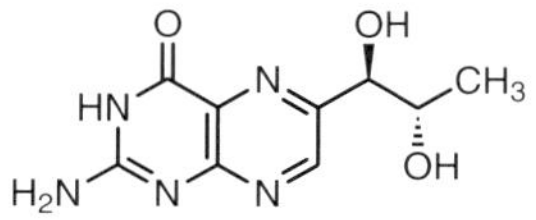

(–)-Biopterin
1

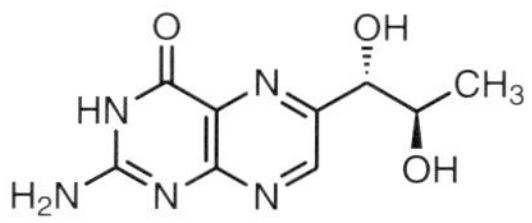

L-*erythro*-Biopterin
2

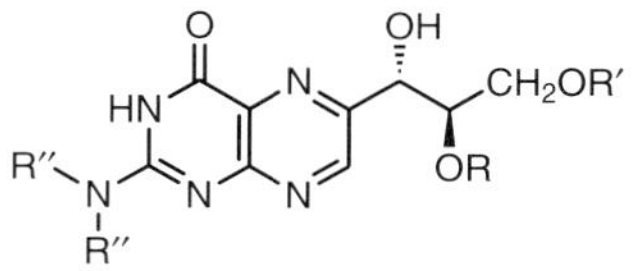

3) Euglenapterin, R, R′ = H, R″ = CH_3
4) R = H, R′ = PO_3H_2, R″ = CH_3
5) R, R′ = Cyclophosphate, R″ = CH_3

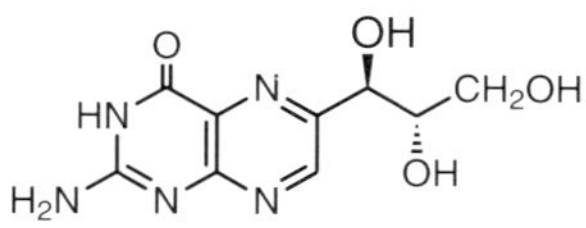

Neopterin
6

5.5.2.1 *Synthesis from L-rhamnose* A synthesis of L-*erythro*-biopterin (**2**) has been started with 5-deoxy-L-arabinose (**9**),[18] which was obtained from the naturally occurring L-rhamnose (Scheme 1).[19,20] L-Rhamnose could be readily converted to L-rhamnose diethylmercaptal (**7**), whose oxidation gave **8**. Degradation of the latter by the action of dil.

NH_4OH gave **9**. Oxidation of **9** with cupric acetate afforded 5-deoxy-L-arabinosone **10**, followed by reaction with acetone oxime and 5% NH_4OH to produce the α-keto aldoxime **11**. Reaction of **11** with $EtOCOCH(NH_2)CN$ in ethanol afforded **12**, which was treated with sodium methoxide and guanidine hydrochloride to afford **13**. Reduction with sodium dithionite in aqueous solution buffered to pH 7 gave L-*erythro*-biopterin (**2**).

Scheme 1 (*a*) EtSH, conc. HCl, 10 min, 81%. (*b*) Dioxane, 10–20°C, *m*-CPBA, 3 h, 97%. (*c*) NH_4OH, rt, 16 h; Amberlite IR-120 and IR-4B resins, 96%, 75.5% overall yield from L-rhamnose. (*d*) H_2O, $(CH_3CO_2)_2Cu{\cdot}xH_2O$, 1 h; Dowex 50WX4 resin, 40%. (*e*) 5% NH_4OH, acetone oxime, 50°C, 6 h, 47%. (*f*) EtOH, $EtOCOCH(NH_2)CN$, rt, 36 h. (*g*) CH_3ONa, guanidine hydrochloride 76%. (*h*) $Na_2S_2O_4$, 1-propanol–water (1:1), pH 7, 83%.

5.5.2.2 *Synthesis from D-ribose* D-Ribose has been used as a precursor for the synthesis of (–)-biopterin (**1**) by transformation to the 2,3-*O*-cyclohexylidene acetal **14** whose reaction with methylmagnesium iodide afforded 3,4-*O*-cyclohexylidene-6-deoxy-L-allitol (**15**); the formation of only one isomer was due to the chelation control and steric hindrance caused by the cyclohexylidene group (Scheme 2).[21] Compound **15** was treated with sodium periodate in ether–water mixture to provide 2,3-*O*-cyclohexylidene-5-deoxy-L-ribose (**16**). Deprotection of **16** followed by treatment with phenyl hydrazine in methanol afforded **17**. Transformation of the hydrazone **17** to **1** was done by employing Viscontini's procedure,[22] whereby **17** was acetylated and treated with 2,5,6-triamino-4-pyrimidinol in the presence of sodium dithionite and sodium acetate, followed by oxidation with iodine and then acetylation to afford the triacetylbiopterin **18**. Deacetylation of **18** afforded **1** in 17% overall yield from D-ribose.

5.5.2.3 *Synthesis from L-xylose* L-Xylose has been used as a starting material for the synthesis of euglenapterin (**3**) (Scheme 3).[23] Its oxidation to L-xylosone (**19**)[24] followed by reaction with acetone oxime in aqueous solution produced **20**. Condensation of **20** with ethyl

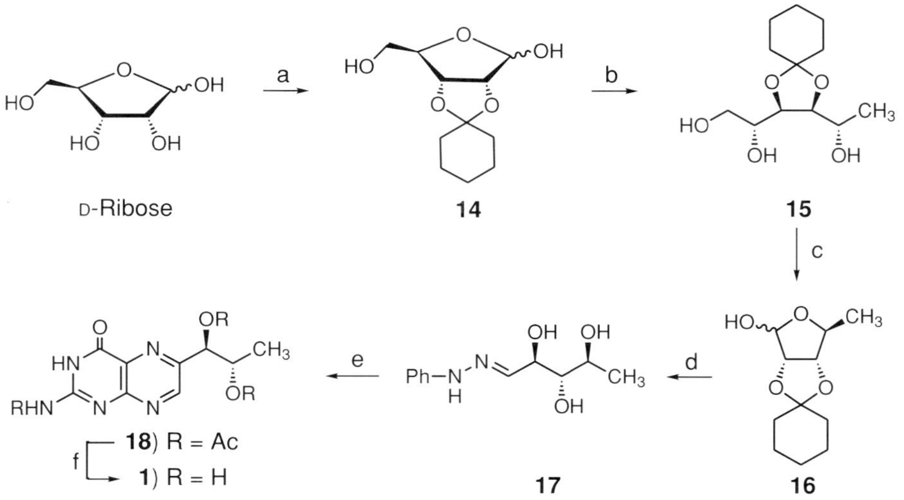

Scheme 2 (*a*) 1,1-Dimethoxycyclohexane, *p*-TsOH, rt, 12 h, 95%. (*b*) CH_3MgI, THF, 5°C, 1 h to rt, 13 h, 92%. (*c*) $NaIO_4$, ether–H_2O, rt, 1 h, 76%. (*d*) 1. 1% aqueous H_2SO_4, 70–80°C, 10 h; 2. $PhNHNH_2$, CH_3OH, AcOH, rt, 2 h, 66% for two steps. (*e*) Py, Ac_2O, rt, 2 h; then CH_3OH–Py (10:9), $Na_2S_2O_4$, $NaOAc{\cdot}3H_2O$; then sulfate of 2,5,6-triamino-4-pyrimidinol, 40–50°C, 1 day, I_2, CH_3OH; then Ac_2O, Py, 100°C, 4 h, 56%. (*f*) 3 N HCl, 100°C, 30 min, 70%.

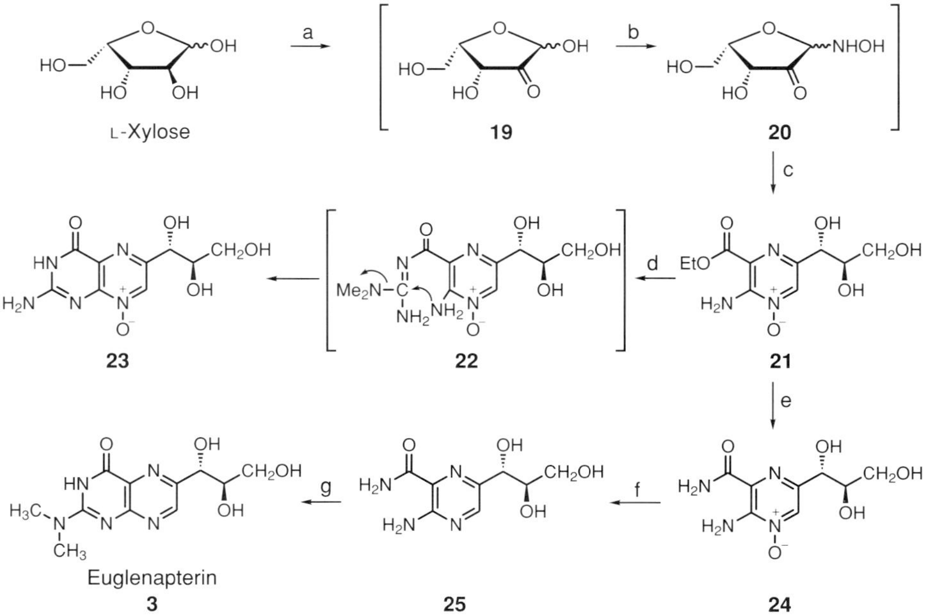

Scheme 3 (*a*) 1. $Cu(OAc)_2$, CH_3OH, reflux, 18 min; 2. Dowex 50WX4 resin, 43%. (*b*) H_2O, NH_4OH, pH 7, $(CH_3)_2C{=}NOH$, 40°C, 2 h. (*c*) $EtO_2CCH(NH_2)CN$, rt, EtOH, H^+, 42% for two steps. (*d*) $NH_2C({=}NH)N(CH_3)_2$. (*e*) Liquid NH_3, 3 h. (*f*) 22% aqueous EtOH, Raney nickel (W-2), 16 h, H_2. (*g*) $(EtO)_2C[N(CH_3)_2]_2$, DMF, tetramethylurea diethyl acetal, 4 h, 77%.

α-aminocyanoacetate gave 2-amino-3-(carboethoxy)-5-(L-*threo*-trihydroxypropyl)pyrazine 1-oxide (**21**). Treatment of **21** with $NH_2C(=NH)N(CH_3)_2$ afforded **23** via **22**. Difficulties have been encountered with the conversion of **23** into **3**. However, treatment of **21** with liquid ammonia afforded **24** (95%), which was smoothly reduced to 2-amino-3-carbamoyl-5-(L-*threo*-trihydroxypropyl)pyrazine (**25**). Finally, treatment of **25** with tetramethylurea diethyl acetal in DMF at room temperature furnished **3** in 11% overall yield from L-xylose.

References

1. Patterson, E.L.; Broquist, H.P.; Albrecht, A.M.; Von Salz, M.H.; Stockstad, E.L.R. *J. Am. Chem. Soc.* **77** (1955) 3167.
2. Kaufman, S. *J. Biol. Chem.* **226** (1957) 511.
3. Rembold, H.; Gyure, W.L. *Angew. Chem., Int. Ed. Engl.* **11** (1972) 1061.
4. Pfleiderer, W. *Angew. Chem., Int. Ed. Engl.* **3** (1964) 114.
5. Fukushima, T.; Shiota, T. *J. Biol. Chem.* **247** (1972) 4549.
6. Nagatsu, T.; Levitt, M.; Udenfriend, S. *J. Biol. Chem.* **239** (1964) 2910.
7. Weiner, N.; Lioyd, T. *Mol. Pharmacol.* **7** (1971) 569.
8. Kokolis, N.; Ziegler, I. *Z. Naturforsch. Teil B* **23** (1968) 860.
9. Ziegler, I. *Z. Naturforsch. Teil B* **18** (1963) 551.
10. Gal, E.M.; Armstrong, J.C.; Ginsberg, B. *Neurochem. J.* **13** (1966) 643.
11. Hosoda, S.; Glick, D. *J. Biol. Chem.* **241** (1966) 192.
12. Lovenberg, W.; Jequier, E.; Sjoerdsma, A. *Science* **155** (1967) 217.
13. Noguchi, T.; Nishino, M.; Kido, R. *Biochem. J.* **131** (1973) 375.
14. Kidder, G.W.; Nolan, L.L. *Biochem. Biophys. Res. Commun.* **53** (1973) 929.
15. Samuelsson, B. *Fed. Proc., Fed. Am. Soc. Exp. Biol.* **31** (1972) 1442.
16. Elstner, E.; Heupel, A. *Arch. Biochem. Biophys.* **173** (1976) 614.
17. Böhme, M.; Pfleiderer, W.; Elstner, E.; Richter, W. *Angew. Chem., Int. Ed. Engl.* **19** (1980) 473.
18. Hough, L.; Taylor, T.J. *J. Chem. Soc.* (1955) 3544.
19. Taylor, E.C.; Jacobi, P.A. *J. Am. Chem. Soc.* **96** (1974) 6781.
20. Taylor, E.C.; Jacobi, P.A. *J. Am. Chem. Soc.* **98** (1976) 2301.
21. Mori, K.; Kikuchi, H. *Liebigs Ann. Chem.* (1989) 1267.
22. Viscontini, M.; Frei, W.F. *Helv. Chim. Acta* **55** (1972) 574.
23. Jacobi, P.A.; Martinelli, M.; Taylor, E.C. *J. Org. Chem.* **46** (1981) 5416.
24. Weidenhagen, R.Z. *Wirtshaftsgruppe Zuckerind.* **87** (1937) 711.

5.5.3 *Isoquinolines*

5.5.3.1 *Calycotomine* Calycotomine (**1**) is a naturally occurring compound[1] and it has been synthesized from D-ribonolactone (Scheme 1).[2] Condensation of D-ribonolactone (**2**) with 2-(3,4-dimethoxyphenyl)ethylamine (**3**) gave **4**, whose acetylation furnished the corresponding per-*O*-acetyl compound. The latter was cyclized[3] with PCl_5 and the resulting imine was oxidized with *m*-CPBA to afford the nitrone **5**. Hydrogenation of **5** over Adams catalyst in strongly acidic solution afforded **6**, which was N-acetylated *in situ*, and then followed by mild methanolysis of the resulting per-*O*-acetyl groups to give 87% yield of the corresponding polyol derivative. Subsequent sodium metaperiodate oxidation afforded aldehyde **7** in 81% yield. Sodium borohydride reduction of **7** gave the *N*-acetylcalycotomine, which underwent deacetylation to give (*R*)-(–)-calycotomine (**1**) in 97% yield.

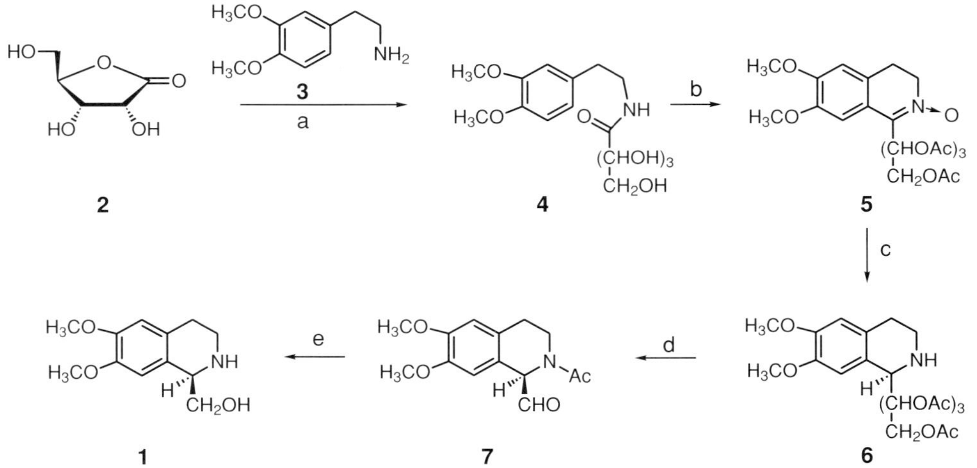

Scheme 1 (*a*) 1,4-Dioxane, reflux. (*b*) 1. Ac_2O, Py, 100%; 2. PCl_5, CH_2Cl_2, 0°C; 3. *m*-CPBA, 30°C. (*c*) AcOH, HCl. (*d*) 1. Ac_2O, NaOAc, 61% from **2**; 2. $NaOCH_3$, CH_3OH, 87%; 3. $NaIO_4$, H_2O, 81%. (*e*) 1. $NaBH_4$; 2. NaOH, EtOH, 97%.

References

1. Kametani, T. *The Chemistry of Isoquinoline Alkaloids*; Elsevier: Amsterdam, 1969.
2. Czarnocki, Z. *J. Chem. Res. (S)* (1992) 334.
3. Dornyei, G.; Szantay, C. *Acta Chim. Acad. Sci. Hung.* **89** (1976) 161.

5.5.3.2 *Decumbensines* The (α-hydroxybenzyl)isoquinoline alkaloids decumbensine (**1**) and *epi*-α-decumbensine (**2**) were isolated from natural sources.[1] Compound **2** has been synthesized from D-ribonolactone (Scheme 1).[2] Condensation[3,4] of D-ribonolactone (**4**) with 2-(3,4-methylenedioxyphenyl)ethylamine (**3**) afforded the amide **5**, which was cyclized to give **6**. Compound **6** was N-methylated by using formaldehyde–sodium cyanoborohydride, whereby the formation of the oxaziridine derivative of **6** was avoided and the title compound **2** was obtained as the only product in 78% yield.

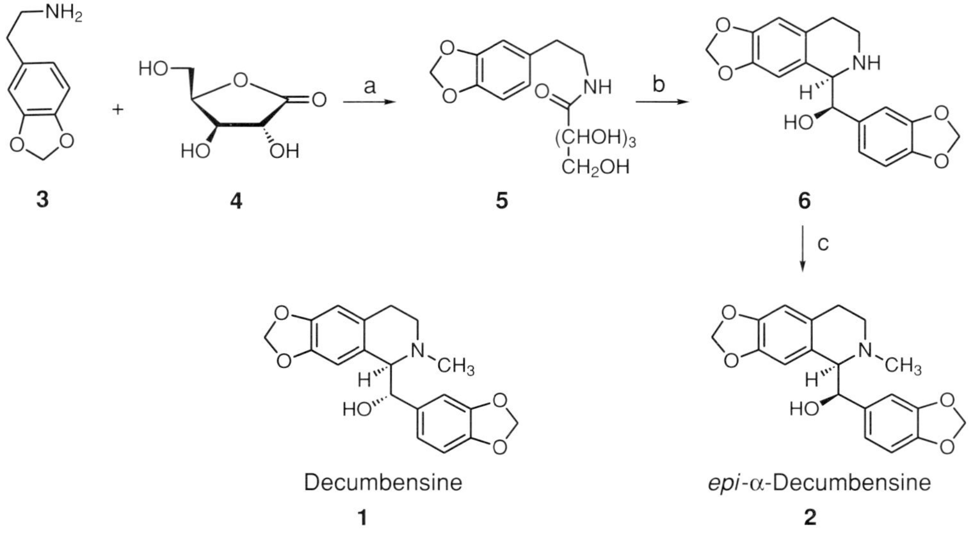

Scheme 1 (*a*) Refs. 3 and 4, 92%. (*b*) 1. Ac_2O, Py, 87%; 2. PCl_5, CH_2Cl_2; 3. *m*-CPBA, 69% for two steps; 4. H_2, Adams catalyst, AcOH–HCl (15:1 v:v); Ac_2O, NaOAc (52%); 5. $NaOCH_3$, CH_3OH, 80%; $NaIO_4$; 6. *n*-BuLi, 91.4%. (*c*) HCHO, $NaBH_3CN$, 78%.

References

1. Zhang, J.-S.; Xu, R.-S.; Quirion, J.C. *J. Nat. Prod.* **51** (1988) 1241.
2. Czarnocki, Z. *J. Chem. Res. (S)* (1992) 402.
3. Bhat, K.L.; Chen, S.-Y.; Jouillie, M.M. *Heterocycles* **23** (1985) 691.
4. Bhat, K.L.; Chen, S.-Y.; Jouillie, M.M. *Aldrichim. Acta* **22** (1989) 49.

5.5.3.3 *Laudanosine and glaucine* The isoquinoline alkaloids (–)-laudanosine (**1**) and (–)-glaucine (**2**) are naturally occurring products[1] that have been synthesized from L-ascorbic acid (Scheme 1).[2] L-Ascorbic acid was converted into L-(+)-gulono-1,4-lactone (**3**).[3] Reaction of **3** with 2-(3,4-dimethoxyphenyl)ethylamine (**4**) afforded the amide **5** (91%), whose cyclization gave **6**, which was converted, via **7**, to the diastereomers **8** and **9** in a ratio of 13:87. The predominant epimer **9** was treated with sodium methoxide to give **10**,

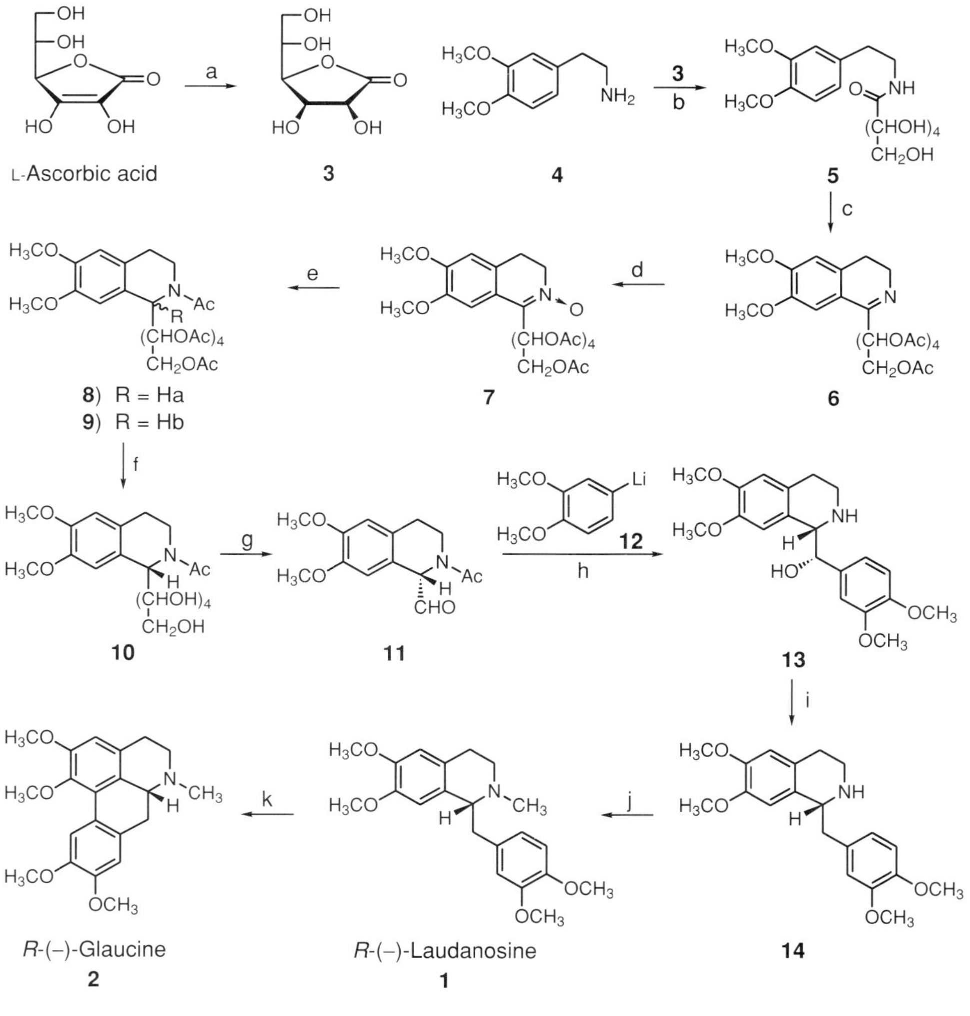

Scheme 1 (*a*) $PdCl_2$, H_2, 1 N HCl, 50°C, 72 h, 85%. (*b*) Dioxane, reflux, 3 h, 91%. (*c*) 1. Ac_2O, Py, 0°C, 2 h; then rt, 3 days, 83%; 2. PCl_5, 0°C, CH_2Cl_2, 3 h. (*d*) *m*-CPBA, CH_2Cl_2, 15 min, 54% for two steps. (*e*) AcOH, HCl, H_2, platinum(II) oxide, 10°C, 1 h, quantitative yield, **8**:**9** 13:87. (*f*) CH_3OH, $NaOCH_3$, rt, 83%. (*g*) $NaIO_4$, ethylene glycol, H_2O, 89%. (*h*) THF, –78°C, 20 min, –20°C, 30 min, 77%. (*i*) 1. $SOCl_2$, Py, THF, –78 to –10°C; 2. THF, $LiAlH_4$. (*j*) CH_2O, CH_3OH, overnight, $NaBH_4$, 0–5°C, 51% from **13**. (*k*) Cr_2O_3, CH_2Cl_2, TFAA, TFA, $BF_3{\cdot}OEt_2$, rt, 48 h; then $NaBH_4$, 83%.

which was then subjected to periodate oxidation to give **11** (89%). Reaction of **11** with the aryl lithium **12** gave **13**, which underwent deoxygenation of the benzylic alcohol to afford the norlaudanosine **14**. This was directly converted to (*R*)-(–)-laudanosine (**1**) by treatment with formaldehyde and subsequent borohydride reduction. Chromium(III) oxide converted **1** into (*R*)-(–)-glaucine (**2**) in 83% yield.

References

1. Kametani, T. *The Chemistry of Isoquinoline Alkaloids*; Elsevier: Amsterdam, 1969.
2. Czarnocki, Z.; Mieczkowski, J.B.; Ziółkowski, M. *Tetrahedron: Asymmetry* **7** (1996) 2711.
3. Andrews, G.C.; Crawfordt, T.C.; Bacon, B.E. *J. Org. Chem.* **46** (1981) 2976.

6 Multi-fused heterocycles

This last chapter of the book discusses some heteroyohimbine alkaloids isolated from the bark and leaves of pharmacological plants. They are indoloquinolizidines: xylopinine, antirhine, allo-yohimbane and ajmalicine. Also discussed are the indolocarbazole alkaloids isolated from some organisms and algae, such as staurosporine. These compounds are important because of their strong protein kinase C inhibitory activity and their use as antiproliferative agents. Included in here are also phenanthridone alkaloids isolated from the roots of Amaryllidaceae such as pancratistatin, narciclasine and lycoricidine, which possess a wide spectrum of biological activities, and the Amaryllidaceae alkaloids, whose unique biological activities have made them attractive synthetic targets, from both carbohydrate and noncarbohydrate starting materials. The last part of this chapter deals with the synthesis of ecteinascidins, isolated from marine tunicate and is undergoing clinical trials.

6.1 Indoloquinolizidines

6.1.1 *Xylopinine*

Xylopinine (**1**) is a naturally occurring compound[1] and has been synthesized from the aldehyde **4**, which was obtained by condensation of D-ribonolactone (**2**) and 3,4-dimethoxyphenethylamine (**3**) as mentioned before for the synthesis of calycotomine (Scheme 1).[2] Treatment of **4** with a 5 M excess of 3,4-dimethoxyphenyllithium afforded **5** (71%), apparently as a result of the nucleophilic attack of the organolithium reagent on the aldehyde and the amide carbonyls. The synthesis of (*S*)-(–)-xylopinine (**1**) was

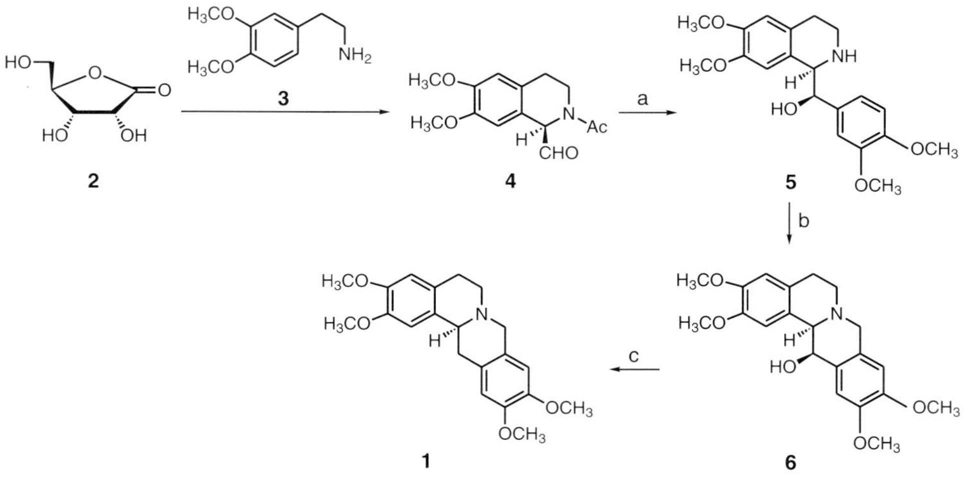

Scheme 1 (*a*) Excess 3,4-dimethoxyphenyllithium, –70°C, 71%. (*b*) Mannich condensation with formaldehyde. (*c*) Deoxygenation, 65% from **5**.

accomplished by condensation[3] of **5** with formaldehyde, followed by deoxygenation of the resulting product **6** to give **1** in 65% yield from **5**.

References

1. Kametani, T. *The Chemistry of Isoquinoline Alkaloids*; Elsevier: Amsterdam, 1969.
2. Czarnocki, Z. *J. Chem. Res. (S)* (1992) 334.
3. Czarnocki, Z.; MacLean, D.B.; Szarek, W.A. *Bull. Soc. Chim. Belg.* **95** (1986) 749.

6.1.2 *Antirhine*

(–)-Antirhine (**1**) is the major alkaloid of *Antirhea jutaminosa*, having a *trans*-3α-H,15β-H structure with a *cis* C/D ring junction.[1,2] A synthesis of **1** has been achieved starting with 2-deoxy-D-ribose (Scheme 1).[3,4] Treatment of **2** with tryptamine **3** in boiling benzene afforded the 2-oxa-8-aza-bicyclo[3.3.1]nonane derivative **4** in quantitative yield. Compound **4** was oxidized into the ketone whose subjection to the Wadsworth–Emmons reaction furnished the unsaturated ester **5** in 70% yield from **4**. Acidic treatment of **5** afforded indoloquinolizidinones **6** and **7**. Catalytic hydrogenation of the major isomer **6** afforded the two saturated lactones **8** and **9** in a 1:1 mixture, which were separated by HPLC, followed by reduction into the corresponding indoloquinolizidines **10** and **11**. The isomer **10** was then regioselectively converted[5] into **12**, followed by treatment with *m*-CPBA to give **1**.

Scheme 1 (*a*) Benzene, reflux, 3 h, 95%. (*b*) 1. SO_3·Py, 70%; 2. Wadsworth–Emmons reaction. (*c*) Toluene, AcOH, reflux, Dean-Stark, 48 h, 90%. (*d*) 10% Pd *on* C, CH_3OH, H_2, 90%. (*e*) $LiAlH_4$, THF, reflux, 4 h, 80%. (*f*) *o*-NO_2–C_6H_4–SeCN, Bu_3P, THF, reflux, 4 h, 80%. (*g*) *m*-CPBA, CH_2Cl_2, 45%.

References

1. Johns, S.R.; Lamberton, J.A.; Occoclowitz, J.L. *Aust. J. Chem.* **20** (1968) 1463.
2. Johns, S.R.; Lamberton, J.A.; Occoclowitz, J.L. *J. Chem. Soc., Chem. Commun.* (1967) 229.
3. Pancrazi, A.; Kervagoret, J.; Kuong-Huu, Q. *Tetrahedron Lett.* **32** (1991) 4303.
4. Pancrazi, A.; Kervagoret, J.; Kuong-Huu, Q. *Tetrahedron Lett.* **32** (1991) 4483.
5. Takano, S.; Takahashi, M.; Ogasawara, K. *J. Am. Chem. Soc.* **102** (1980) 4282.

6.1.3 *Allo-yohimbane*

The Yohimbe alkaloids such as alloyohimbane (**1**)[1] and the antihypertensive drug reserpine[2] constitute attractive goals for synthetic chemists. Syntheses of Yohimbe alkaloids from non-carbohydrates have been reported.[3,4] Synthesis of **1** from cellulose has also been achieved (Scheme 1).[5] Pyrolysis of cellulose gave the levoglucosenone[6] (1,6-anhydro-3,4-dideoxy-β-D-glycero-hex-3-enopyrano-2-ulose, **2**), which underwent Diels–Alder cycloaddition with 1,3-butadiene to afford **3**. This was treated with hydrazine hydrate followed by treatment of the resulting hydrazone with sodium hydride in dimethylsulfoxide to produce the vinyl ether **4**. Acetylation of **4** followed by hydrolysis in THF in the presence of 1 N HCl and the oxidation of the resulting hemiacetal with Jones reagent gave **5** in 63% overall yield from **2**. Deacetylation of **5** followed by condensation with tryptamine in a mixture of THF and diisopropylamine afforded **6**, which was cleaved with HIO_4. Subsequent reduction of the resulting aldehyde with sodium borohydride gave **7**. This was treated with Ph_3P and CCl_4

Cellulose a b c d e f g h i 2 3 4 5 6 7 8 9 (–)-Allo-yohimbane 1

Scheme 1 (*a*) Pyrolysis, Ref. 6. (*b*) 1,3-Butadiene, 140°C, 10 h, 98%. (*c*) $NH_2NH_2{\cdot}H_2O$, NEt_3, EtOH, 60°C, 20 min; then $NaCH_2SOCH_3$, DMSO, 93%. (*d*) 1. Ac_2O, Py; 2. 1 N HCl–THF (1:5), 50°C, 2 h; 3. Jones reagent, 0°C, 5 min, 90%. (*e*) $NaOCH_3$, CH_3OH, tryptamine, $(i\text{-Pr})_2NEt$–THF (6:1). (*f*) 1. HIO_4, THF–H_2O (3:1), –20°C for 1 min; 2. $NaBH_4$, CH_3OH, rt, 5 min, 93% for two steps. (*g*) 1. Benzene, Ph_3P, CCl_4, 55°C, 52%; 2. $LiN(TMS)_2$, THF, –78°C, 91%. (*h*) 1. $POCl_3$, 100°C, 1.5 h; 2. $NaBH_4$, CH_3OH, H_2O, 73% for two steps. (*i*) H_2, Pd *on* C, CH_2Cl_2–CH_3OH (1:1), 92%.

to give the corresponding chloride, which was treated with lithium bis(trimethylsilyl)amine in THF to afford the lactam **8**. Heating of **8** in $POCl_3$ followed by reduction of the resulting product with $NaBH_4$ gave **9**. Hydrogenation of **9** afforded (–)-**1**.

References

1. Kuehne, M.E.; Muth, R.S. *J. Org. Chem.* **56** (1991) 2701.
2. Stork, G. *Pure Appl. Chem.* **61** (1989) 251.
3. Gomez-Pardo, D.; d'Angelo, J. *Tetrahedron Lett.* **33** (1992) 6637.
4. Saxton, J.E. *Nat. Prod. Rep.* **8** (1991) 251.
5. Isobe, M.; Fukami, N.; Goto, T. *Chem. Lett.* (1985) 71.
6. Shafizadeh, F.; Chin, P.P.S. *Carbohydr. Res.* **58** (1977) 79.

6.1.4 *Ajmalicine*

Ajmalicine (raubasine, **1**)[1,2] and 19-*epi*-ajmalicine (mayumbine, **2**) are the most known members of heteroyohimbine (indoloquinolizidine) alkaloids – also including tetrahydroalstonine (**3**) and rauniticine (**4**). They were isolated from the bark and leaves of *Pseudocinchona mayumbensis* (*Corynanthe mayumbensis*) and named mayumbine.[3] Later, the structure of mayumbine was revised and shown to be that of **2**.[4] Its biogenetic pathway was later established in detail using cell-free extracts from *Catharanthus roseus*.[4,5] Compound **1** is a potent peripheral and central vasodilating agent[6,7] with a clinically demonstrated effect in reducing platelet aggregation.[8] The syntheses of these naturally occurring alkaloids have been achieved in racemic and optically active forms from noncarbohydrates.[9–16]

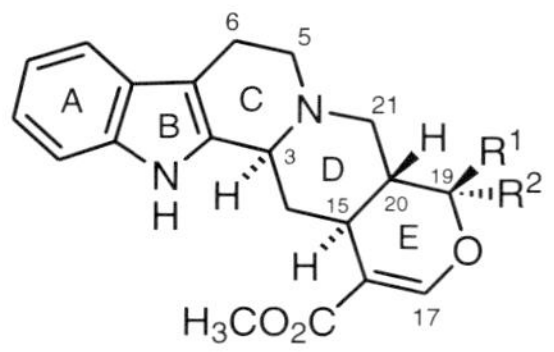

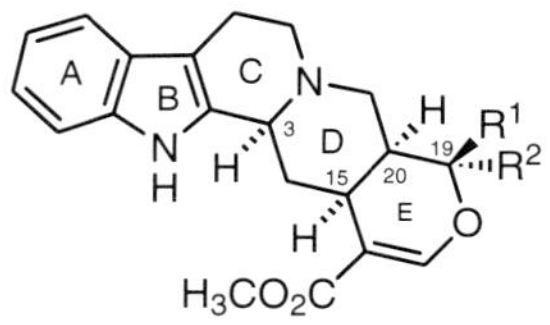

1) (–)-Ajmalicine (Raubasine)
$R^1 = H, R^2 = CH_3$

2) (+)-19-*epi*-Ajmalicine (Mayumbine)
$R^1 = CH_3, R^2 = H$

3) (–)-Tetrahydroalstonine
$R^1 = H, R^2 = CH_3$

4) Rauniticine
$R^1 = CH_3, R^2 = H$

6.1.4.1 *Synthesis from* D-*glucose* Syntheses of ajmalicine (**1**) and 19-*epi*-ajmalicine (**2**) have been achieved from D-glucose (Scheme 1).[17] D-Glucose pentaacetate was converted into **5**[18] in 30% overall yield. Reductive deoxygenation of the keto group with tosylhydrazine and $NaBH_3CN$ followed by deacetylation afforded **6**, which underwent chlorination of the primary hydroxyl group followed by radical-mediated dehalogenation and subsequent ozonolysis of the vinyl group to afford **7**. DBU effected epimerization of **7** to provide **8**, which was coupled with tryptamine under reductive amination condition and protected with $(Boc)_2O$ to give **9**. Acid hydrolysis of the glycosidic linkage followed by oxidation with PCC afforded **10**. The last required stereogenic center at C-3 was obtained by applying the Bischler–Napieralski reaction followed by catalytic reduction to furnish **11**, which underwent a methoxycarbonylation with Mander's reagent ($CNCO_2CH_3$) to give **12** as a single isomer. Lactone **12** was treated with DIBAL-H followed by acid-catalyzed dehydration of the resulting lactol to afford **2** in 8% overall yield from **5**.

Compound **10** has also been used for the synthesis of (–)-ajmalicine (**1**) by epimerization of the methyl group at C-19 to afford **13**, which underwent intramolecular cyclization to give **14**. The stereogenic center at C-3 and the methoxycarbonylation at C-16 were created as mentioned above to give **15**, which underwent reduction, dehydration and N-deprotection to produce **1**.

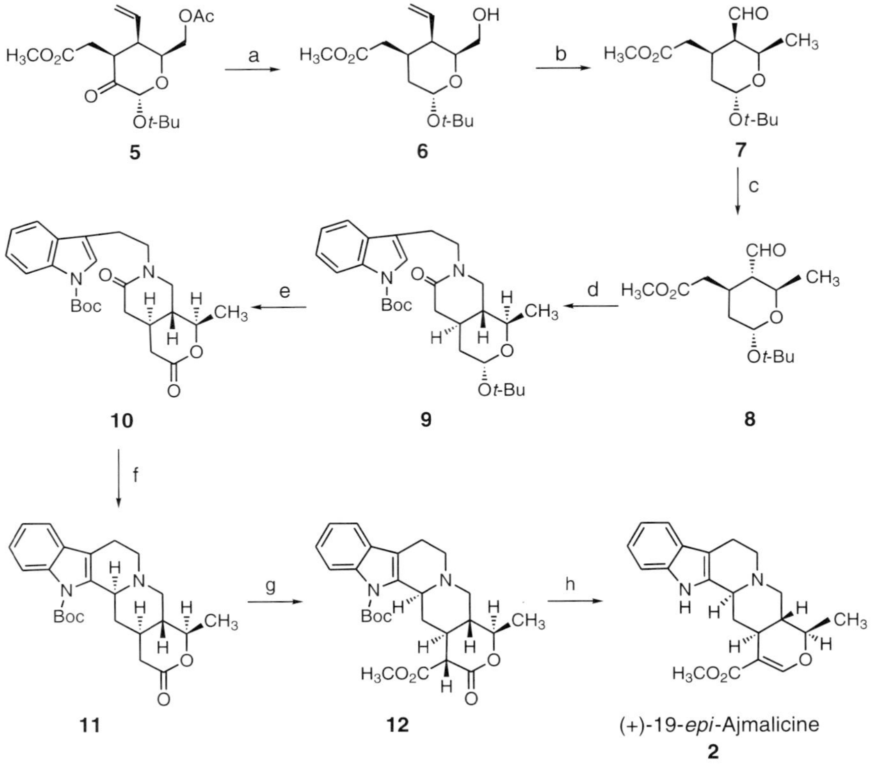

Scheme 1 (*a*) 1. $TsNHNH_2$, EtOH; 2. $NaBH_3CN$; 3. $NaOAc \cdot 3H_2O$, EtOH; 4. NaOAc, CH_3OH, 68% for four steps. (*b*) 1. Ph_3P, CCl_4, 90%; 2. *n*-BuSnH, AIBN, toluene, 94%; 3. O_3, CH_2Cl_2; then NEt_3, 86%. (*c*) DBU, DMF, –12°C, 69%. (*d*) 1. Tryptamine, CH_2Cl_2; then CH_3OH, $NaBH_4$, 93%; 2. $(Boc)_2O$, DMAP, CH_2Cl_2, 99%. (*e*) 1. *p*-TsOH, THF, H_2O, 97%; 2. PCC, MS 3 Å, CH_2Cl_2, 75%. (*f*) 1. $POCl_3$, C_6H_6; 2. H_2, PtO_2, CH_3OH; 3. DMAP, CH_2Cl_2; 4. $(Boc)_2O$, DMAP, CH_2Cl_2, 77% for four steps. (*g*) LDA, THF, 0°C; then HMPA, $CNCO_2CH_3$, –78°C, 72%. (*h*) 1. DIBAL-H, THF, 87%; 2. TFA, 76%.

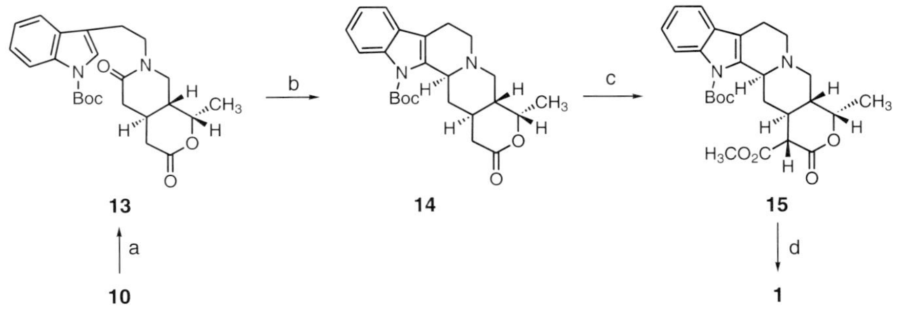

Scheme 2 (*a*) 1. $Ba(OH)_2$, H_2O (1 equiv.), THF–H_2O (10:1), 0°C, 1 h; 2. PPh_3, DEAD, THF, 0°C, 10 min, 92% for two steps. (*b*) 1. $POCl_3$, C_6H_6; 2. H_2, PtO_2, CH_3OH; 3. DMAP, CH_2Cl_2; 4. $(Boc)_2O$, DMAP, CH_2Cl_2, 77% for four steps. (*c*) LDA, THF, 0°C; then HMPA, $CNCO_2CH_3$, –78°C, 72%. (*d*) DIBAL-H, THF; then TFA, 84%.

6.1.4.2 *Synthesis from D-mannose* An intermediate for the synthesis of (–)-ajmalicine (**1**) from D-mannose as a chiral starting material has been reported (Scheme 3).[19] The unsaturated ester **16**,[20–22] obtained from methyl α-D-mannopyranoside, was reduced to the branched-chain methyl glycoside **17**. Subsequent hydrolysis of the benzylidene group in acid medium afforded the furanoside **18** in 75% yield. Oxidation by $NaIO_4$ yielded the respective aldehyde, which was in turn reduced to the primary alcohol **19** in 82% yield. Hydrolysis of **19** gave **20**, which was condensed with *N*-benzyltryptamine (**21**), via the Pictet–Spengler reaction, to afford the chiral 3-epimeric-substituted tetrahydro-β-carbolines **22** in 56% yield. The two epimers **22** were then treated with MsCl or *p*-TsCl in pyridine to afford the ammonium salts **23** or **24**, which were hydrogenolyzed to give the indolo[2,3-*a*]quinolizidines **25** and its epimer.

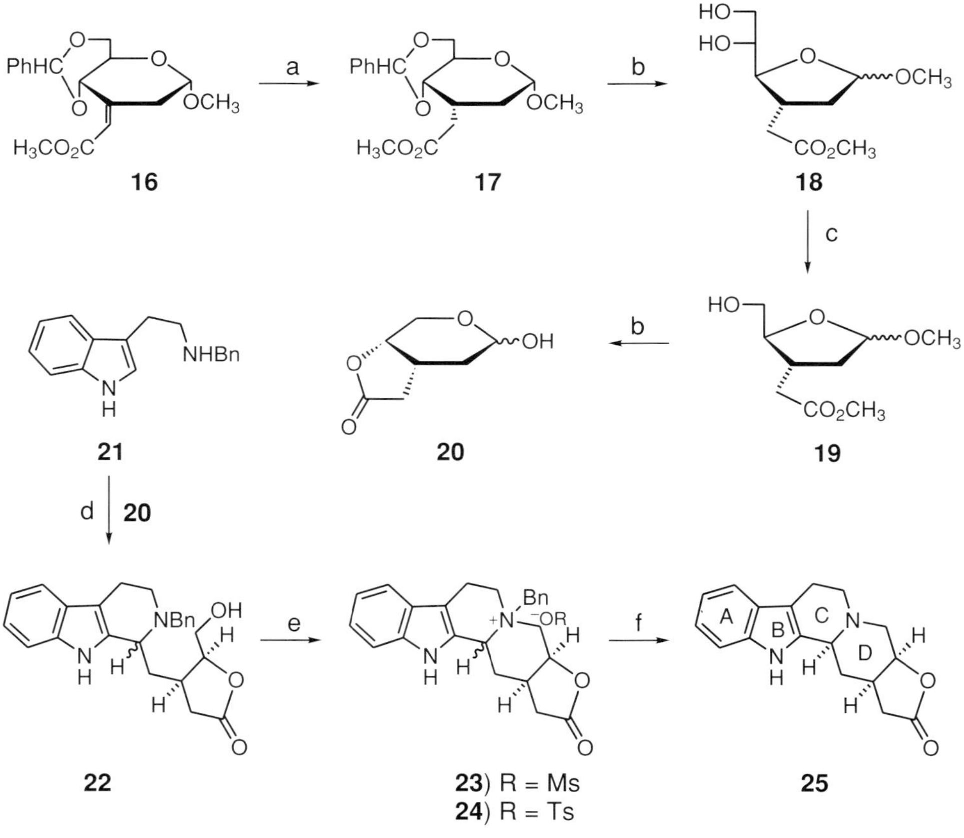

23) R = Ms
24) R = Ts

Scheme 3 (*a*) Raney nickel, H_2, CH_3OH. (*b*) Acid medium, 75%. (*c*) 1. $NaIO_4$; 2. reduction, 82%. (*d*) Pictet–Spengler reaction, 56%. (*e*) MsCl, Py; *or p*-TsCl, Py. (*f*) H_2, Pd, 60%.

6.1.4.3 *Synthesis from D-erythritol* A total synthesis of (–)-ajmalicine (**1**) and (–)-tetrahydroalstonine (**3**) has been achieved from erythritol (Schemes 4 and 5).[23,24] The D-erythritol derivative **26**, obtained from L-tartaric acid, was reduced to provide the

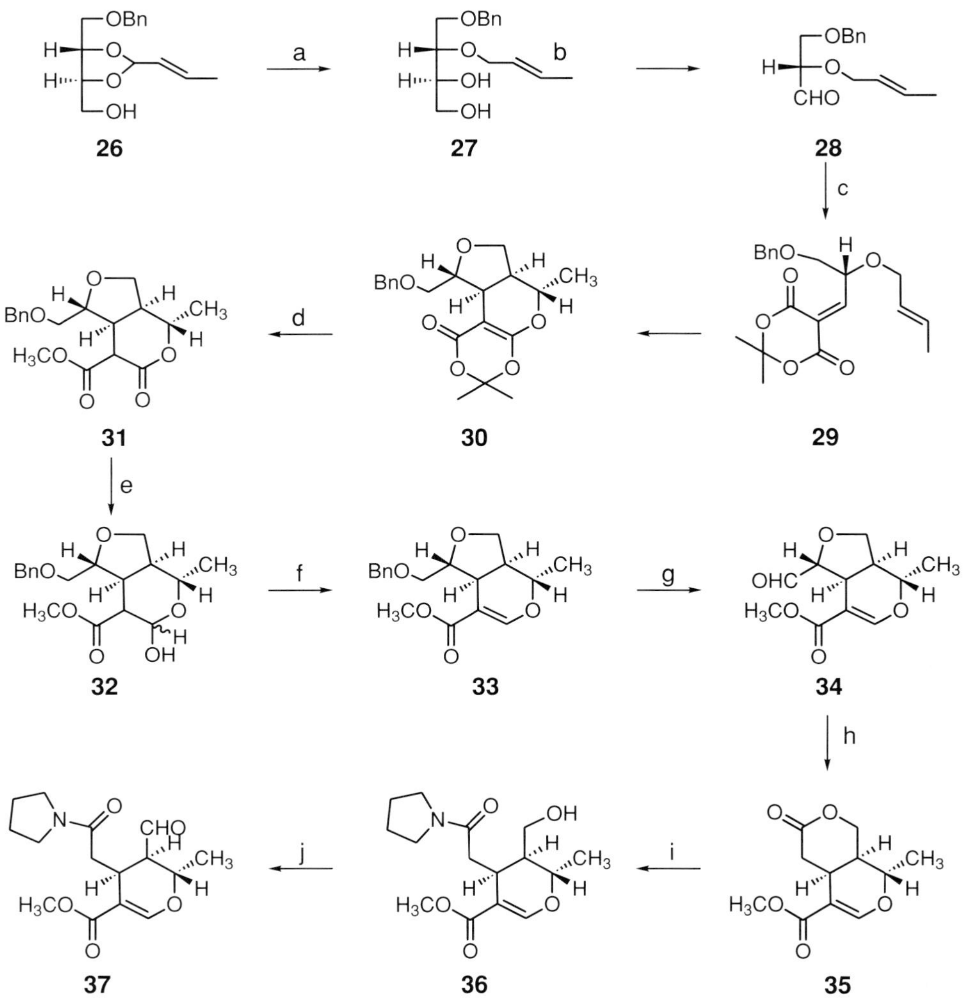

Scheme 4 (*a*) DIBAL-H, toluene, 0°C. (*b*) $NaIO_4$, aqueous CH_3OH, 0°C to rt. (*c*) Meldrum's acid, $(CH_2NH_3)_2$, $(AcO^-)_2$, CH_3OH, 0°C to rt. (*d*) CH_3OH, reflux. (*e*) $LiEt_3BH$. (*f*) *p*-TsOH. (*g*) 1. Debenzylation; 2. $(COCl)_2$, DMSO, CH_2Cl_2, –78°C; then NEt_3. (*h*) 1. Zn, THF, HCl; 2. Ag_2CO_3 *on* Celite, 82% from **33**. (*i*) Dimethylaluminum pyrrolidinide, benzene, 0°C to rt, 98%. (*j*) $(COCl)_2$, DMSO, CH_2Cl_2, –78°C; then NEt_3.

diol **27**, which underwent periodate oxidation followed by condensation of the resulting aldehyde **28** with Meldrum's acid to produce **29** which was converted[25] to **30**. This was heated under reflux with methanol to give the methyl ester **31** in 50% yield from **27**, followed by lithium triethylborohydride and subsequent dehydration with *p*-TsOH to furnish the acrylate **33** (80%) via the lactol **32**. Debenzylation of **33** followed by Swern oxidation afforded the aldehyde **34**, which was treated with zinc in THF and HCl followed by oxidation with silver carbonate on Celite (Fetizon reagent) to furnish the δ-lactone **35** in 82% overall yield. The lactone **35** was treated with dimethylaluminum pyrrolidinide in benzene to give the tertiary amide **36** (98%), which underwent a Swern oxidation of the primary hydroxyl group to furnish the aldehyde **37**.

Compound **37** was condensed with tryptamine perchlorate in the presence of sodium cyanoborohydride to provide the secondary amine **38**. On the other hand, the aldehyde **37** was stirred with silica gel in methylene chloride to afford the epimer, which on reductive condensation with tryptamine perchlorate gave **39**. Lactamization of **38** and **39** gave the lactams **40** and **41**, respectively. Reaction of **40** and **41** with Lawesson's reagent[26] gave the corresponding thio-lactams **42** and **43**, which were treated with *p*-nitrobenzyl bromide to afford the crude salts **44** and **45**. Reduction of **44** and **45** with sodium borohydride afforded (–)-tetrahydroalstonine (**3**) and (–)-ajmalicine (**1**), respectively.

38) R = Hα
39) R = Hβ

40) R = Hα
41) R = Hβ

3) (–)-Tetrahydroalstonine, R = Hα
1) (–)-Ajmalicine, R = Hβ

44) R = Hα
45) R = Hβ

42) R = Hα
43) R = Hβ

Scheme 5 (*a*) Tryptamine perchlorate, $NaCNBH_3$ *or* silica gel, CH_2Cl_2, rt, 12 h. (*b*) Tryptamine perchlorate, $NaCNBH_3$, **39**. (*c*) Toluene, reflux, $(i\text{-}Pr)_2NEt$, Py, **40** (84%), **41** (30%). (*d*) Lawesson's reagent, benzene (**42**, **43**). (*e*) $p\text{-}NO_2C_6H_4Br$, CH_3CN, 60°C. (*f*) $NaBH_4$, CH_3OH, –50°C.

References

1. Siddiqui, S.; Siddiqui, R.H. *J. Ind. Chem. Soc.* **16** (1939) 421.
2. Aynilian, G.H.; Farnsworth, N.R. *Lioydia* **37** (1974) 299.
3. Raymond-Harnet, M.; Acad, C.R. *Sci. Paris* **232** (1951) 2354.
4. Melchio, J.; Bouquet, A.; Pais, M.; Goutarel, R. *Tetrahedron Lett.* **18** (1977) 315.
5. Stockigt, J. In *Indole and Biogenetic Related Alkaloids*; Phillipson, I.D., Zenk, M.H., Eds.; Academic Press: London, 1980; p. 113.
6. Bories, J.; Merland, J.-J.; Thiebot, J. *Neurobiology* **14** (1977) 33.
7. Anand, N. In *Comprehensive Medicinal Chemistry*, Vol. 1; Hansch, C., Ed.; Pergamon Press: Oxford; 1990; p. 127.
8. Neuman, J.; Engel, A.M. F.; Neuman, M.P. *Arzneim. Fortsch.* **36** (1986) 1394.

9. Stockigt, J.; Hofle, G.; Pfitzner, A. *Tetrahedron Lett.* **22** (1981) 1925.
10. Winterfeldt, E.; Radunz, H.E.; Korth, T. *Chem. Ber.* **101** (1968) 3172.
11. Winterfeldt, E.; Gaskell, A.; Korth, T.; Radunz, H.E. Walkowiak, M. *Chem. Ber.* **102** (1969) 3558.
12. Gutzwiller, J.; Pizzolato, G.; Uskokovic, M. *J. Am. Chem. Soc.* **93** (1971) 5907.
13. Wenkert, E.; Chang, G.T.; Chawla, H.P.S.; Cochran, D.W.; Hagamar, E.W.; King, J.C.; Orito, K. *J. Am. Chem. Soc.* **98** (1976) 3645.
14. Djakoure, L.A.; Jarreau, F.X.; Goutarel, R. *Tetrahedron* **31** (1975) 2695.
15. Palmisano, G.; Danieli, B.; Lesma, G.; Riva, R. *J. Chem. Soc., Perkin Trans. 1* (1985) 925.
16. Hatakeyama, S.; Ochi, N.; Numata, H.; Takano, S. *J. Chem. Soc., Chem Commun.* (1988) 1202.
17. Hanessian, S.; Faucher, A.-M. *J. Org. Chem.* **56** (1991) 2947.
18. Hanessian, S.; Faucher, A.-M.; Léger, S. *Tetrahedron* **46** (1990) 231.
19. Kervagoret, J.; Nemlin, J.; Khuong-Huu, Q.; Pancrazi, A. *J. Chem. Soc., Chem. Commun.* (1983) 1120.
20. Rosenthal, A.; Catsoulacos, P. *Can. J. Chem.* **46** (1968) 2868.
21. Klemer, A.; Rodemeyer, G. *Chem. Ber.* **107** (1974) 2612.
22. Horton, D.H.; Weckerle, W. *Carbohydr. Res.* **44** (1975) 222.
23. Takano, S.; Satoh, S.; Ogasawara, K. *J. Chem. Soc., Chem. Commun.* (1988) 59.
24. Takano, S.; Satoh, S.; Ogasawara, K. *Heterocycles* **30** (1990) 583.
25. Takano, S. *Pure Appl. Chem.* **59** (1987) 353.
26. Scheibye, S.; Kristensen, J.; Lawesson, S.O. *Tetrahedron* **35** (1979) 1339.

6.2 Indolocarbazole alkaloids

In 1977 an unusual natural product was isolated from *Streptomyces staurosporeus* during a search for new alkaloids present in actinomycetes. It was given the name AM-2282,[1] and then it was renamed staurosporine (**1**).[2,3] The absolute configuration of the alkaloid was assigned from circular dichroism measurements as structure **3**.[4] The structure of AM-2282 was established by single-crystal X-ray analysis of its methanol solvate and shown to possess an indolocarbazole subunit wherein the two indole nitrogens are bridged by glycosyl linkages.[5] Later, the absolute configuration of staurosporine was revised to

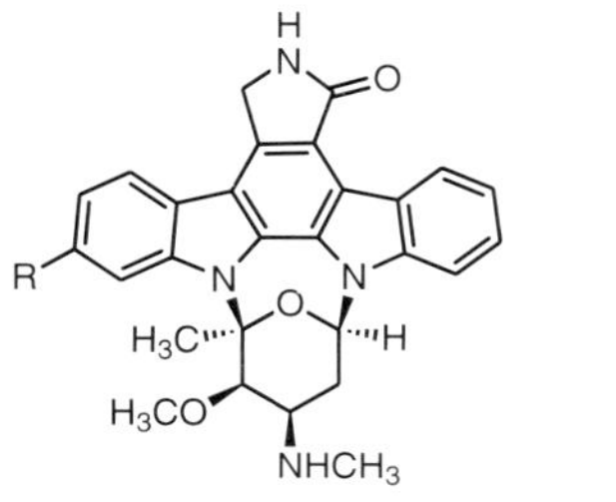

1) (+)-Staurosporine (AM-2282), R = H
2) TAN-999, R = OCH_3

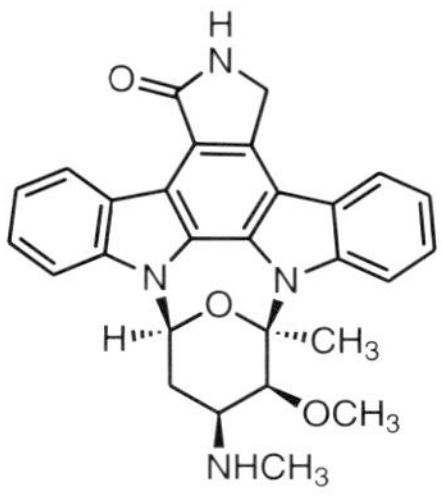

3
(−)-*ent*-Staurosporine

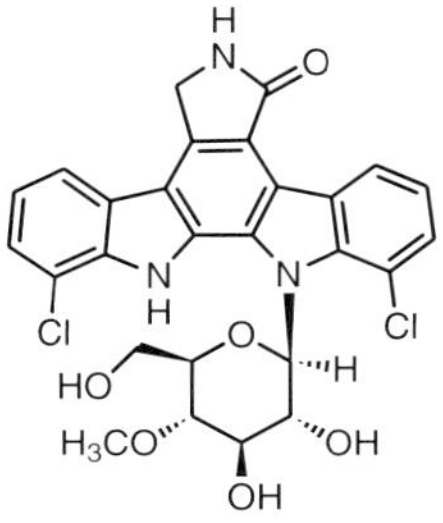

4
Rebeccamycin

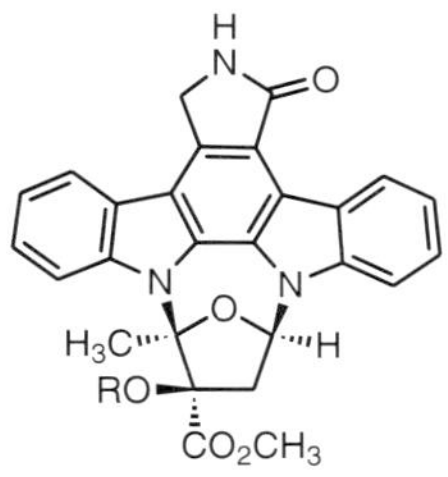

5) (+)-K252a, R = H
6) K252b, R = CH_3

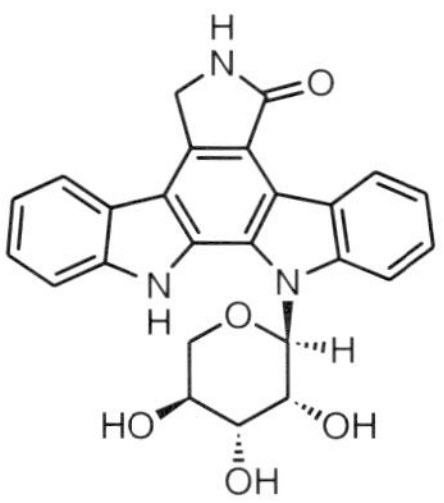

7
(+)-K252d

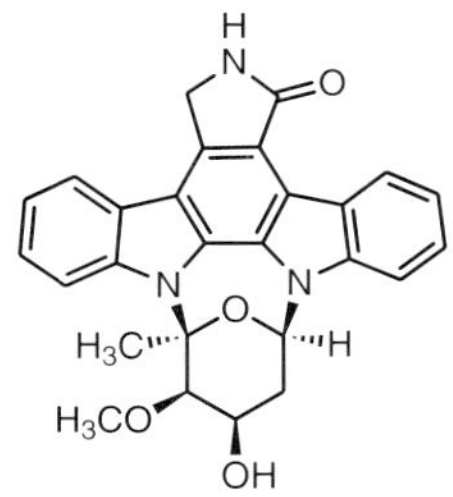

8
(+)-RK286c

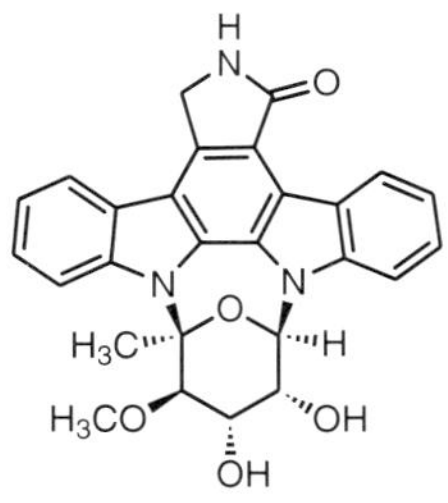

9
(+)-MLR-52

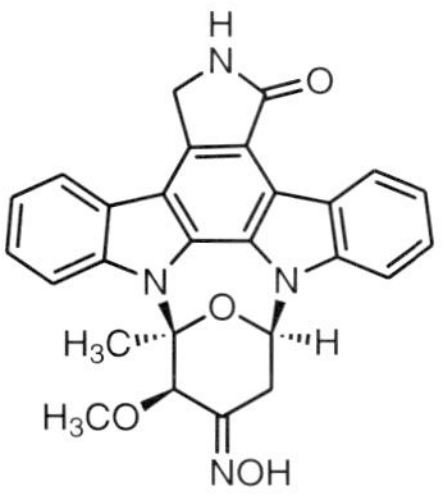

10
TAN-1030a

structure **1** by means of X-ray crystallographic analysis of 4′-*N*-methylstaurosporine methiodide.[6] Since the isolation of staurosporine, approximately 60 members of this class of compounds have been isolated from various soil organisms, blue-green algae and slime molds.

The indolocarbazoles became the focus of intensive investigations that have revealed their potential as chemotherapeutic agents against cancer,[7] blood platelet aggregation,[8] antiproliferative agents,[9] Alzheimer's disease[10,11] and other neurodegenerative disorders.[12]

Rebeccamycin (**4**) has reached a late stage of clinical evaluation as an anticancer agent. It also induces topoisomerase I mediated DNA cleavage.[13] (+)-K252a (**5**) was isolated independently by two Japanese groups[14,15] and its structure as well as the structures of K252b (**6**) and (+)-K252d (**7**) have been elucidated.[16,17]

Staurosporine (**1**) and K252a (**5**) are potent inhibitors of protein kinase C[7]; **1** is one of the most known potent inhibitors with an IC_{50} value of 1.3 ± 0.2 nM.[2,18] PKC is a family of cytosolic serine/threonine phosphorylating isoenzymes that plays a key role in several crucial cellular processes such as signal transduction, cell differentiation and cell growth.[19–25] Consequently, inhibitors of PKC might serve as anticancer agents and could be of value in studying the mechanism of action of the kinases.

(+)-RK286c (**8**) was isolated from *Streptomyces* sp. AM-2282 and it was found to be a weak inhibitor of PKC compared to staurosporine (**1**), but it has a comparable platelet aggregation inhibitory activity.[5] Each of staurosporine (**1**) and (+)-MLR-52 (**9**) possesses immunosuppressive activity[18] and reverses mutidrug resistance.[26,27]

TAN-999 (**2**) and TAN-1030a (**10**) are produced by *Nocardiopsis dassonvillei* C-71425 and *Streptomyces* sp. C-71799, respectively. They exhibited macrophage-activating properties[28] and their structures have been studied using NMR analysis.[29]

Syntheses of staurosporine (**1**) and its analogues from non-carbohydrates have received much attention during the last two decades.[30–55] It is apparent that carbohydrates are incorporated in such a ring system.

6.2.1 *Synthesis from* L*-glucal*

The key step for the synthesis of staurosporine (**1**) has utilized a reaction of a carbohydrate derivative with an indole derivative (Schemes 1 and 2).[56–58] Triisopropylsilyl-L-glucal **11** was converted to its trichloroacetimidate and thence to oxazoline **12**.[59] The oxazoline ring was opened to give **13** that cyclized and protected as its BOM derivative **14**. The TIPS protecting group was replaced by a PMB one and then treated with 2,2-dimethyldioxirane to produce the epoxide **15** and its β-isomer. The mixture of epoxides was treated with the sodium salt of **16** to furnish the indole glycoside **17** in 47% yield; the minor β-epoxide was less effective donor than **15**. Barton deoxygenation[60,61] of the $C_2{}'$ hydroxyl function afforded **18**, which underwent removal of the PMB and SEM protecting groups to provide **19**. Photolytic oxidative cyclization followed by iodination afforded the iodo derivative **20**, which was subjected to β-elimination using DBU in THF to furnish the olefin **21**.

Treatment of **21** with potassium *tert*-butoxide and iodine produced **22**. Radical deiodination of **22** followed by hydrogenation and subsequent treatment with sodium methoxide in methanol provided **23**. Selective protection of the oxazolidinone ring in **23** with Boc followed by protection of imide with BOMCl afforded **24**, which was treated with cesium

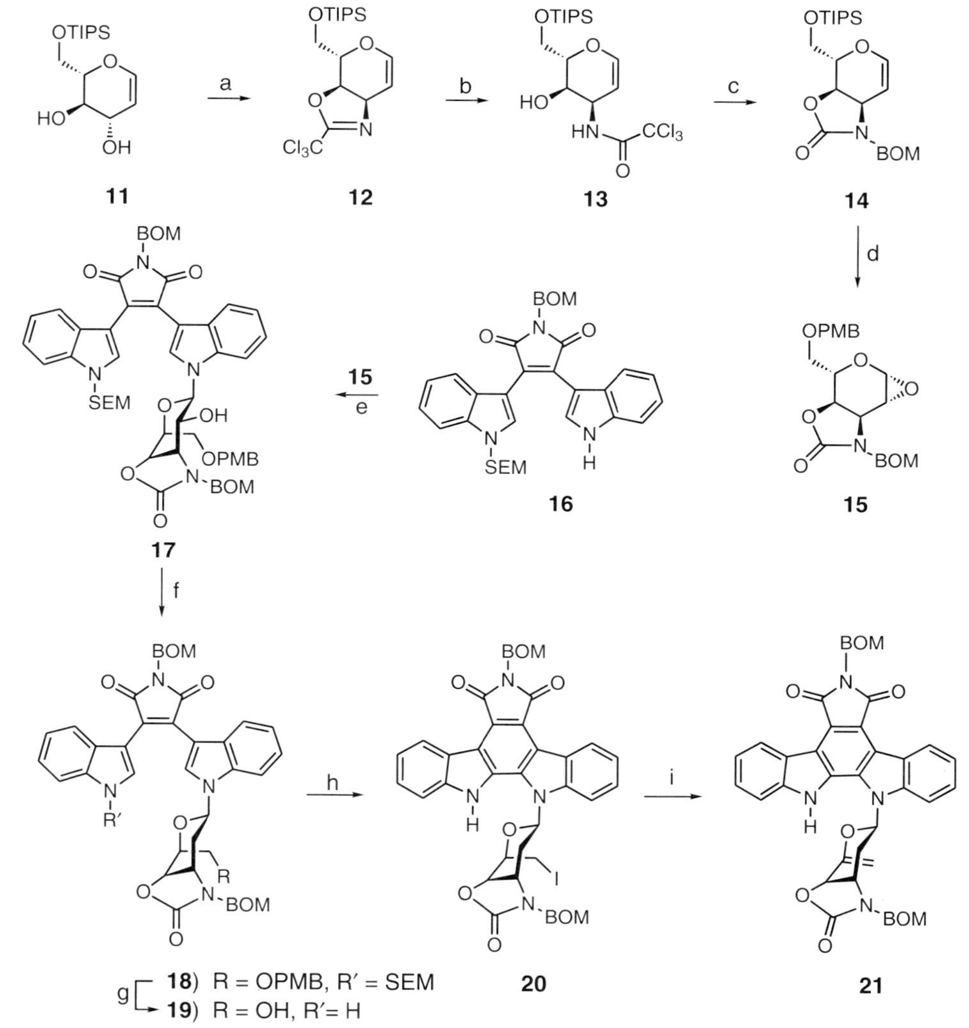

Scheme 1 (*a*) NaH, CH_2Cl_2, 0°C; then Cl_3CCN, 0°C to rt; then $BF_3{\cdot}OEt_2$, –78°C 78%. (*b*) Cat. *p*-TsOH, H_2O, Py, 80°C, 80%. (*c*) 1. NaH, CH_2Cl_2, 0°C to rt, 92%; 2. NaH, DMF; then BOMCl, 40°C, 65%. (*d*) 1. TBAF, THF, 0°C, 95%; 2. NaH, DMF, 0°C to rt; then PMBCl, 0°C to rt; 92%; 3. dimethyldioxirane, CH_2Cl_2, 0°C, 100%. (*e*) **16**, NaH, THF, rt; then **15**, rt to reflux, 47% of **17**. (*f*) 1. ClCSCl, DMAP, Py, CH_2Cl_2, reflux; then C_6F_5OH, reflux, 79%; 2. *n*-Bu_3SnH, AIBN, benzene, reflux, 74%. (*g*) 1. DDQ, CH_2Cl_2, H_2O, 0°C to rt, 97%; 2. TBAF, THF, reflux, 91%. (*h*) 1. *hν*, cat. I_2, air, benzene, rt, 73%; 2. I_2, $P(Ph)_3$, imidazole, CH_2Cl_2, 0°C to rt, 84%. (*i*) THF, DBU, rt, 89%.

carbonate in methanol to give **25** in 93% yield. Methylation of **25** followed by removal of the BOM and Boc protecting groups afforded 7-oxostaurosporine (**26**), which was treated with sodium borohydride followed by PhSeH to afford a 1:1 mixture of staurosporine (**1**) and its iso analogue.

23) R = R′ = H
24) R = Boc, R′ = BOM

26) R = CH_3, X = O
1) R = CH_3, X = H_2

Scheme 2 (*a*) *t*-BuOK, I_2, THF, CH_3OH, rt, 65%. (*b*) 1. *n*-Bu_3SnH, AIBN, PhH, reflux, 99%; 2. H_2, $Pd(OH)_2$, EtOAc, CH_3OH, rt; then $NaOCH_3$, CH_3OH, 92%. (*c*) 1. $(Boc)_2O$, THF, cat. DMAP, rt, 81%; 2. NaH, DMF, rt; then BOMCl, 82%. (*d*) Cs_2CO_3, CH_3OH, rt, 93%. (*e*) 1. NaH, $(CH_3)_2SO_4$, THF, DMF, rt, 86%; 2. H_2, $Pd(OH)_2$, EtOAc, CH_3OH, rt; then $NaOCH_3$, CH_3OH, 84%; 3. TFA, CH_2Cl_2, rt, 97%. (*f*) $NaBH_4$, EtOH, rt; then PhSeH, cat. *p*-TsOH, CH_2Cl_2, rt, 39% of **1**, 39% of epimer and 15% of recovered **26**.

6.2.2 *Synthesis from 2-deoxy-D-ribose*

Stereoselective synthesis of (+)-K252a (**5**) has been reported from a pentose derivative (Schemes 3 and 4).[62] Esterification of the indole 3-acetic acid (**27**) with allyl bromide afforded the corresponding ester, which underwent regiospecific bromination with NBS to give the corresponding 2-bromoindole **28** whose glycosylation with 1-chloro-2-deoxy-3,5-di-*O*-*p*-toluoyl-α-D-*erythro*-pentofuranose (**29**)[63] gave β-*N*-glycoside **30** as the sole product. After deprotection of the allyl ester group in **30**, the resulting acid was condensed with tryptamine under conventional conditions to give **31**. Regioselective oxidation of **31** with 2 equiv. of DDQ gave the ketone **32**, which upon acetylation afforded **33**. This underwent smooth cyclization with DBU to furnish **34**, whose exposure to sunlight in the presence of diisopropylethylamine led to a nonoxidative photocyclization to provide the

Scheme 3 (*a*) 1. Allyl bromide, K_2CO_3, DMF, 23°C, 100 min, 99%; 2. NBS, CCl_4, 23°C, 90 min, 80%. (*b*) NaH, CH_3CN, 23°C, 10 min; then added **29**, 23°C, 30 min, 97%. (*c*) 1. $Pd(PPh_3)_4$, Ph_3P, Py, CH_2Cl_2, 23°C, 1 h; 2. WSCD, tryptamine, CH_2Cl_2, 23°C, 15 min, 72% for two steps. (*d*) DDQ, THF, H_2O, 0°C, 30 min, 93%. (*e*) 2,6-Lutidine, DMAP, Ac_2O, 60°C, 8 h, 78%. (*f*) DBU, MS 4 Å, THF, 60°C, 2.5 h, 92%. (*g*) (i-Pr)$_2$NEt, $h\nu$, CH_2Cl_2, 23°C, 5 h, 96%. (*h*) KOH, H_2O, CH_3OH, THF, 23°C, 45 min, 97%. (*i*) I_2, Ph_3P, imidazole, THF, 23°C, 1 h, 82%. (*j*) PhSeSePh, $NaBH_4$, EtOH, THF, 23°C, 93%. (*k*) Ac_2O, Py, DMAP, 23°C, 10 min, 98%. (*l*) m-CPBA, THF, 23°C, 10 min. (*m*) NEt_3, DHP, 80°C, 30 min, 91% for two steps. (*n*) KI, I_2, DBU, THF, 23°C, 40 min, 93%.

desired indolocarbazole **35** in 96% yield. Deacetylation of **35** with KOH afforded the diol **36**, which underwent selective iodination[64] to furnish the corresponding iodide **37** which was converted into **40** via the intermediates **38** and **39**. Treatment of **40** with NEt_3 and DHP afforded the olefin **41**, which on treatment with iodine, potassium iodide and DBU gave **42** in 93% yield.

44) R^1 = OAc, R^2 = CN
45) R^1 = CN, R^2 = OAc
46) R^1 = OAc, R^2 = $CONH_2$

Scheme 4 (*a*) 1. *n*-Bu_3SnH, AIBN, CH_3CN, reflux, 50 min, 98%; 2. K_2CO_3, CH_3OH, 23°C, 10 min, 90%; DCC, Cl_2CHCO_2H, DMSO, 23°C, 15 min, 99%. (*b*) 1. HCN, Py, CH_3CN, 0°C, 15 min; 2. Ac_2O, DMAP, 23°C, 30 min, 99% of **44**. (*c*) HCl, HCO_2H, 23°C, 19 h, 88%. (*d*) KOH, H_2O, CH_3OH, THF, 100°C, 10 h; CH_2N_2, THF, 65% for two steps.

Radical-mediated deiodination, methanolysis of the acetate and subsequent oxidation of the resulting alcohol furnished the ketone **43** in 87% total yield. Transformation of **43** into the corresponding cyanohydrin acetate under ordinary conditions resulted in the formation of a diastereomeric mixture **44** and **45**. However, only the kinetically favored cyanohydrin acetate **44** was obtained when treated with hydrogen cyanide and pyridine. Treatment of the nitrile **44** with HCl gas afforded the amide **46**, which was subjected to alkaline hydrolysis followed by esterification of the resulting carboxylic acid with diazomethane to furnish **5** in 11% overall yield from **27**.

References

1. Omura, S.; Iwai, Y.; Hirano, A.; Nakagawa, A.; Awaya, J.; Tsuchiya, H.; Takahashi, Y.; Masuma, R. *J. Antibiot.* **30** (1977) 275.
2. Furusaki, A.; Hashiba, N.; Matsumoto, T.; Hirano, A.; Iwai, Y.; Omura, S. *J. Chem. Soc., Chem. Commun.* (1978) 800.
3. Furusaki, A.; Hashiba, N.; Matsumoto, T.; Hirano, A.; Iwai, Y.; Omura, S. *Bull. Chem. Soc. Jpn.* **5** (1982) 3681.
4. Takahashi, H.; Osada, H.; Uramoto, M.; Isono, K. *J. Antibiot.* **43** (1990) 168.
5. Tamaoki, T.; Nomoto, H.; Takahashi, I.; Kato, Y.; Morimoto, M.; Tomita, F. *Biochem. Biophys. Res. Commun.* **135** (1986) 397.
6. Funato, N.; Takayanagi, H.; Konda, Y.; Toda, Y.; Hariyage, Y.; Iwai, Y.; Omura, S. *Tetrahedron Lett.* **35** (1994) 1251.
7. Omura, S.; Sasaki, Y.; Iwai, Y.; Takeshima, H. *J. Antibiot.* **48** (1995) 535.
8. Oka, S.; Kodama, M.; Takeda, H.; Tomizuka, N.; Suzuki, H. *Agric. Biol. Chem.* **50** (1986) 2723.
9. Bradshaw, D.; Hill, C.H.; Nixon, J.S.; Wilkinson, S.E. *Agents Actions* **38** (1993) 135.
10. Masliah, E.; Cole, G.M.; Hansen, L.A.; Mallory, M.; Albright, T.; Terry, R.D.; Saitoh, T.J. *Neurosci. J.* **11** (1991) 2759.
11. Gandy, S.; Czernik, A.J.; Greengard, P. *Proc. Natl. Acad. Sci. USA* **85** (1988) 6218.
12. Knüsel, B.; Hefti, F. *J. Neurochem.* **59** (1992) 1987.
13. Yamashita, Y.; Fujii, N.; Murkata, C.; Ashizawa, T.; Okabe, M.; Nakano, H. *Biochemistry* **31** (1992) 12069.
14. Sezaki, M.; Sasaki, T.; Nakazawa, T.; Takeda, U.; Iwata, M.; Watanabe, T.; Koyama, M.; Kai, F.; Shomura, T.; Kojima, M. *J. Antibiot.* **38** (1985) 1439.

15. Kase, H.; Iwahashi, K.; Matsuda, Y. *J. Antibiot.* **39** (1986) 1059.
16. Nakanishi, S.; Matsuda, Y.; Iwahashi, K.; Kase, H. *J. Antibiot.* **39** (1986) 1066.
17. Yasuzawa, T.; Iida, T.; Yoshida, M.; Hirayama, M.; Takahashi, M.; Shirahata, K.; Sano, H. *J. Antibiot.* **39** (1986) 1072.
18. McAlpine, J.B.; Karwowski, J.P.; Jackson, M.; Mullaly, M.M.; Hochlowski, J.E.; Premachandran, U. *J. Antibiot.* **47** (1994) 281.
19. Nishizuka, Y. *Nature* **308** (1984) 693.
20. Nishizuka, Y. *Science* **223** (1986) 305.
21. Berridge, M.J. *Annu. Rev. Biochem.* **56** (1987) 159.
22. Houslay, M.D. *Eur. J. Biochem.* **195** (1991) 9.
23. Stabel, S.; Parker, P. *J. Pharmacol. Ther.* **51** (1992) 71.
24. Hug, H.; Sarre, T.F. *Biochem. J.* **291** (1993) 329.
25. Castagna, M.; Takai, Y.; Kaibuchi, K.; Sano, K.; Kikkawa, U.; Nishizuka, Y. *J. Biol. Chem.* **257** (1982) 7847.
26. Sato, W.; Yusa, K.; Naito, M.; Tsuruo, T. *Biochem. Biophys. Res. Commun.* **173** (1990) 1252.
27. Wakusawa, S.; Inoko, K.; Miyamoto, K.; Kajita, S.; Hasegawa, T.; Hariyama, K.; Koyama, M. *J. Antibiot.* **46** (1993) 353.
28. Tanida, S.; Takizawa, M.; Takahashi, T.; Tsubotani, S.; Harada, S. *J. Antibiot.* **42** (1989) 1619.
29. Tsubotani, S.; Tanida, S.; Harada, S. *Tetrahedron* **47** (1991) 3565.
30. Magnus, P.D.; Exon, C.; Sear, N.L. *Tetrahedron* **39** (1983) 3725.
31. Hughes, I.; Raphael, R.A. *Tetrahedron Lett.* **24** (1983) 1441.
32. Sarstedt, B.; Winterfeldt, E. *Heterocycles* **20** (1983) 469.
33. Joyce, R.P.; Gainor, J.A.; Weinreb, S.M. *J. Org. Chem.* **52** (1987) 1177.
34. Santos, M.M.M.; Lobo, A.M.; Prabhakar, S.; Marques, M.B. *Tetrahedron Lett.* **45** (2004) 2347.
35. Bergman, J.; Pelcman, B. *J. Org. Chem.* **54** (1989) 824.
36. Hughes, I.; Nolan, W.P.; Raphael, R.A. *J. Chem. Soc., Perkin Trans. 1* (1990) 2475.
37. Davis, P.D.; Bit, R.A.; Hurst, S.A. *Tetrahedron Lett.* **31** (1990) 2353.
38. Davis, P.D.; Bit, R.A. *Tetrahedron Lett.* **31** (1990) 5201.
39. Davis, P.D.; Hill, C.H.; Thomas, W.A.; Whitcombe, I.W.A. *J. Chem. Soc., Chem. Commun.* (1991) 182.
40. Moody, C.J.; Rahimtooda, K.F.; Porter, B.; Ross, B.C. *J. Org. Chem.* **57** (1992) 2105.
41. Harris, W.; Hill, C.-H.; Malsher, P. *Tetrahedron Lett.* **34** (1993) 8361.
42. Xie, G.; Lown, J.W. *Tetrahedron Lett.* **35** (1994) 5555.
43. Brüning, J.; Hache, T.; Winterfeldt, E. *Synthesis* (1994) 25.
44. Wood, J.L.; Stoltz, B.M.; Dietrich, H.-J. *J. Am. Chem. Soc.* **117** (1995) 10413.
45. Lowinger, T.B.; Chu, J.; Spence, P.L. *Tetrahedron Lett.* **36** (1995) 8383.
46. Wood, J.L.; Stoltz, B.M.; Goodman, S.N. *J. Am. Chem. Soc.* **118** (1996) 10656.
47. Stoltz, B.M.; Wood, J.L. *Tetrahedron Lett.* **37** (1996) 3929.
48. Ohkubo, M.; Nishimura, T.; Jona, H.; Honma, T.; Ito, S.; Morishima, H. *Tetrahedron* **53** (1997) 5937.
49. Wood, J.L.; Stoltz, B.M.; Dietrich, H.-J.; Pflum, D.A.; Petsch, D.T. *J. Am. Chem. Soc.* **119** (1997) 9641.
50. Wood, J.L.; Stoltz, B.M.; Goodman, S.N.; Onwueme, K. *J. Am. Chem. Soc.* **119** (1997) 9652.
51. Weinreb, S.M. *Heterocycles* **21** (1984) 309.
52. Shankar, B.B.; McCombie, S.W. *Tetrahedron Lett.* **35** (1994) 3005.
53. Chen, S.-Y.; Uang, B.-J.; Liao, F.-L.; Wang, S.-L. *J. Org. Chem.* **66** (2001) 5627.
54. Faul, M.M.; Sullivan, K.A. *Tetrahedron Lett.* **42** (2001) 3271.
55. Braña, M.F.; Añorbe, L.; Tarrason, G.; Mitjans, F.; Piulats, J. *Bioorg. Med. Chem. Lett.* **11** (2001) 2701.
56. Link, J.T.; Gallant, M.; Danishefsky, S.J. *J. Am. Chem. Soc.* **115** (1993) 3782.
57. Link, J.T.; Raghavan, S.; Danishefsky, S.J. *J. Am. Chem. Soc.* **117** (1995) 552.
58. Link, J.T.; Raghavan, S.; Gallant, M.; Danishefsky, S.J.; Chou, T.C.; Ballas, L.M. *J. Am. Chem. Soc.* **118** (1996) 2825.
59. Schmidt, R.R. *Angew. Chem., Int. Ed. Engl.* **25** (1986) 212.
60. Barton, D.H.R.; Jasberenyi, J.C. *Tetrahedron Lett.* **30** (1989) 2619.
61. Gervay, J.; Danishefsky, S.J. *J. Org. Chem.* **56** (1991) 5448.
62. Kobayashi, Y.; Fujimoto, T.; Fukuyama, T. *J. Am. Chem. Soc.* **121** (1999) 6501.
63. Moffer, M. *Chem. Ber.* **93** (1960) 2777.
64. Classon, B.; Liu, Z.; Samuelsson, B. *J. Org. Chem.* **53** (1988) 6126.

6.3 Phenanthridone alkaloids

Highly oxygenated phenanthridone alkaloids, also named Amaryllidaceae alkaloids (**1–12**), were isolated from the roots of promising pharmacological plants, but with low natural abundance. (+)-Trianthine (**1**) was isolated from *Pancratium triathum*[1] and found to be antipodal to (–)-zephyranthine (**2**), isolated from *Zephyranthes candida*.[2] Pancratistatin (**3**), 7-deoxypancratistatin (**4**) and 2-*O*-β-D-glucosyl-pancratistatin (**6**) were isolated from the roots of the Hawaiian *Pancratium littorale* Jacq.[3,4] They have the potential as clinically useful antitumor agents.[5–10] Compound **3** has been used in herbal folk medicine since ancient Greek time.[11] Compound **4** has been shown in *in vitro* antiviral assays to have a better therapeutic index than does **3** because of decreased toxicity.[12]

Lycoricidine (**8**), narciclasine (**9**) and 4-*O*-glucosyl-narciclasine (**10**) were found in Amaryllidaceae plants, *Lycoris radiate*,[13] *Pancratium litorale*,[3] *Pancratium maritimum*,[14] and in several *Narcissus* species.[3,13–19] They showed strong growth-inhibiting action in the rice seedling test, and they exhibit antitumor activity against *Ehrlich carcinoma* (38–106% life extension, 10.75–12.5 mg/kg dose against murine P388 lymphocytic leukemia, P.S. System; 53–84% life extension, 0.38–3 mg/kg against murine M5076 ovary sarcoma).[20–22] They have attracted considerable interest because of their range and potency of biological effects, including inhibition of protein synthesis and potent, *in vitro*, antitumor activity.[23–29] They have unique structural features, which contain four to six contiguous stereogenic centers in the C ring of the phenanthridone skeleton. The absolute structure of **9** has been studied.[30] The promising biological activity and limited availability of these alkaloids have stimulated considerable synthetic work. Many syntheses of phenanthridone alkaloids and

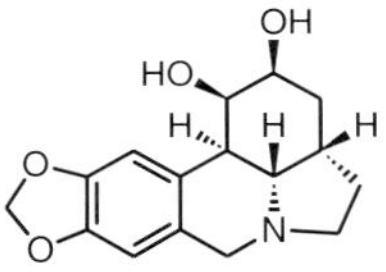

1) (+)-Trianthine

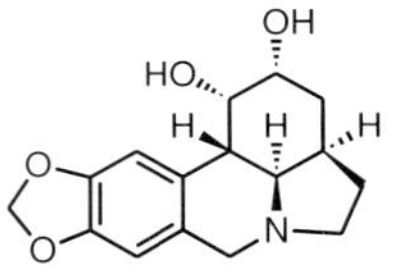

2) (–)-Zephyranthine

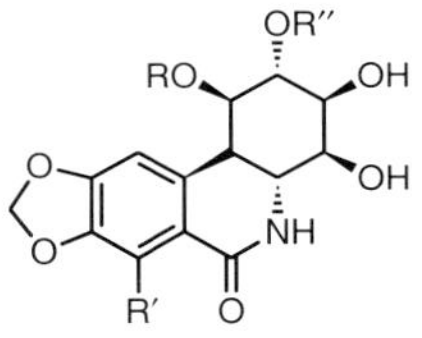

3) Pancratistatin, R = R″ = H, R′ = OH
4) (+)-7-Deoxypancratistatin, R = R″ = R″ = H
5) Telastaside,
R′ = OH, R″ = H, R = β-D-glucopyranosylamine
6) 2-*O*-β-D-Glucosyl-pancratistatin,
R = H, R′ = OH, R″ = β-D-glucopyranosyl
7) Pancratiside,
R′ = OH, R″ = H, R = β-D-glucopyranosyl

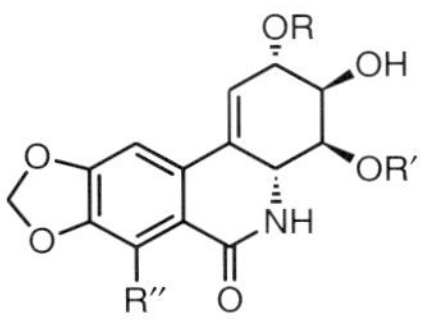

8) (+)-Lycoricidine, R = R′ = R″ = H
9) (+)-Narciclasine, R = R′ = H, R″ = OH
10) Glucosyl narciclasine,
R = H, R′ = β-D-glucopyranosyl, R″ = OH
11) Glucosyl kalbreclasine,
R = β-D-glucopyranosyl, R′ = H, R″ = OH

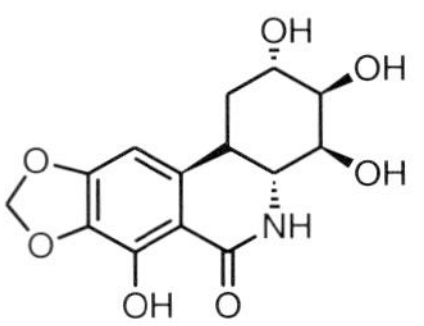

12) Dihydronarciclasine

their analogues from noncarbohydrates have been reported.[31–67] In addition, the stereogenic centers of phenanthridone alkaloids and their analogues have attracted the attention to use carbohydrates as chiral precursors for their syntheses, which are represented here.

6.3.1 *Synthesis from D-galactose*

Methyl-α-D-galactopyranoside (**13**) was used for the synthesis of suitable intermediates for the synthesis of (+)-pancratistatin (**3**) and (+)-narciclasine (**9**) (Scheme 1).[68] Protection of the primary hydroxyl group in **13** as the TIPS ether followed by per-benzylation afforded **14**, and then removal of the TIPS group followed by Swern oxidation afforded **15**. Reaction of **15** with **16** gave a mixture of diastereomers **17**, which underwent bromination with Ph_3PBr_2 to produce **18**. Heating of **18** in dry pyridine afforded **19** as an *E* and *Z* mixture in 75%

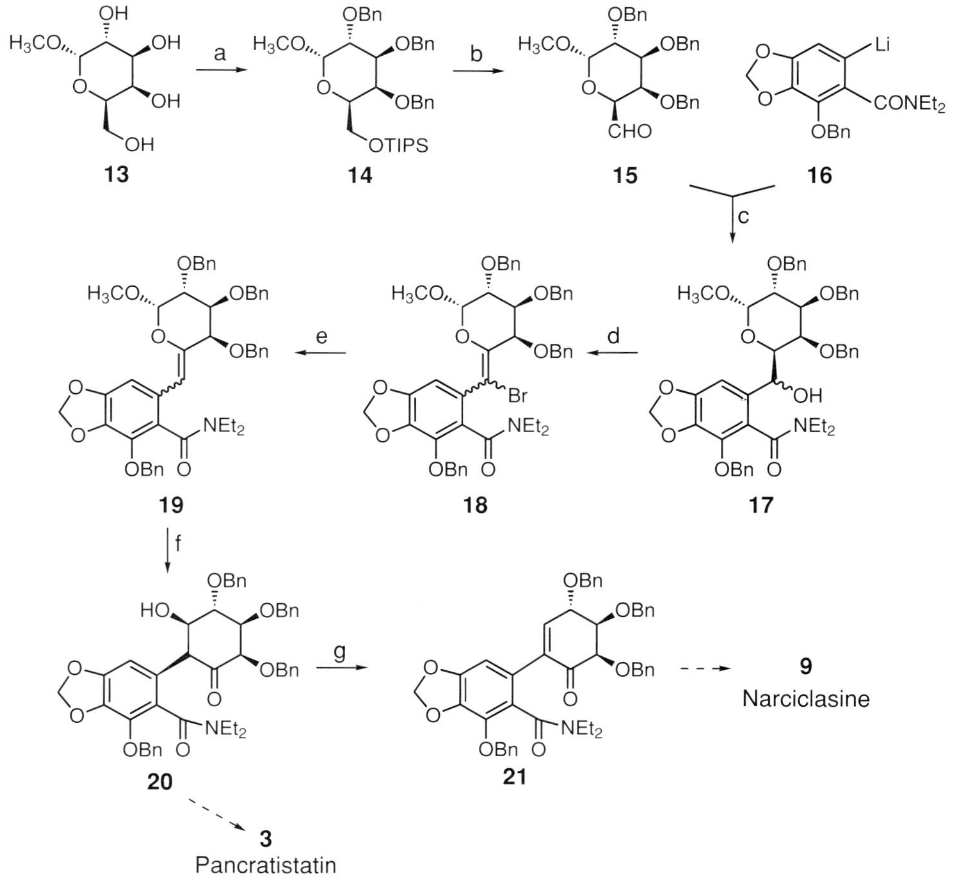

Scheme 1 (*a*) 1. TIPSCl, imidazole, DMF, 85%; 2. NaH, BnBr, DMF, 84%. (*b*) 1. TBAF, THF, 95%; 2. $(COCl)_2$, DMSO, CH_2Cl_2, –78°C; then NEt_3, 96%. (*c*) THF, TMEDA, –78°C, 50–70%. (*d*) Ph_3PBr_2, K_2CO_3, CH_2Cl_2, 75%. (*e*) Py, reflux, 2 h, 75%. (*f*) $HgCl_2$, $H_2O–CH_3CN$ (1:2), reflux, 70–80%. (*g*) Ac_2O, DMAP, Py, 70%.

yield. Subjection of **19** to classical Ferrier conditions[69] afforded **20** and a small amount of **21**. The conversion of **20** to **21** was achieved with acetic anhydride in pyridine. Both **20** and **21** are precursors for **3** and **9**, respectively.

6.3.2 *Synthesis from D-glucose*

Various schemes have been developed for the synthesis of phenanthridene ring systems from D-glucose. Thus, a synthesis of (+)-lycoricidine (**8**) has been reported (Scheme 2),[70,71] starting from D-glucose by conversion into **22**,[72] which upon protection with MOMCl

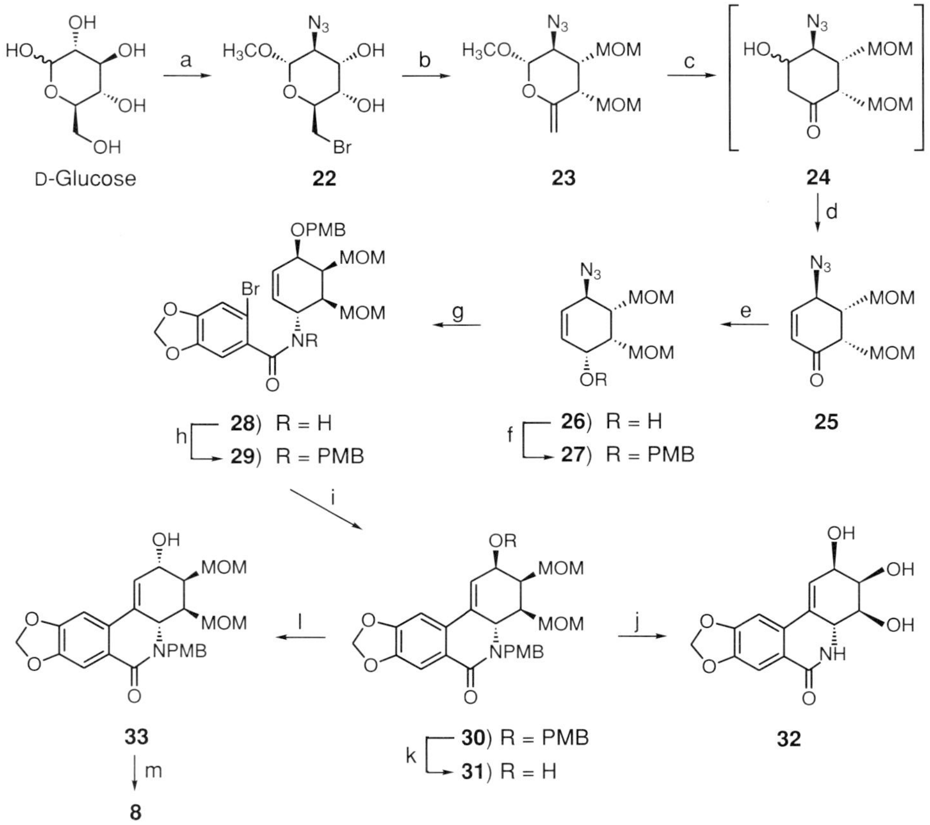

Scheme 2 (*a*) Ref. 72. (*b*) 1. MOMCl, (i-Pr)$_2$NEt, CH_2Cl_2, reflux, 15 h, 87%; 2. DBU, toluene, reflux, 15 h, 73%. (*c*) $Hg(OCOCF_3)_2$ (1 mol%), acetone–H_2O (2:1), rt, 20 h. (*d*) MsCl, NEt_3, CH_2Cl_2, 0°C, 1.5 h, 69%. (*e*) $NaBH_4$, $CeCl_3{\cdot}7H_2O$, CH_3OH, 0°C, 30 min, 86%. (*f*) PMBCl, NaH, DMF, rt, 18 h, 69%. (*g*) $LiAlH_4$, ether, 0°C min; then 6-bromopiperonylic acid, $(EtO)_2P(O)CN$, NEt_3, DMF, 0°C, 15 min, 89%. (*h*) PMBCl, NaH, DMF, rt, 5 h, 100%. (*i*) $Pd(OAc)_2$ (10 mol%), 1,2-bis(diphenylphosphino)ethane (40 mol%), TlOAc (2 equiv.), DMF, 140°C, 7 h, 68%. (*j*) 1. DDQ, CH_2Cl_2, H_2O, rt, 0°C, 3 h, 53%; 2. aqueous HCl, THF, 50°C, 20 h; then Ac_2O, Py, rt, 3 h, 92%; 3. TFA, $CHCl_3$, rt, 1.5 h, 79%; 4. $NaOCH_3$, CH_3OH, rt, 2 h; then Amberlite IR-120B (H^+) resin, 86%. (*k*) DDQ, CH_2Cl_2, H_2O, rt, 0°C, 3 h, 53%. (*l*) 1. BzOH, Ph_3P, DEAD, THF, rt, 15 min, 78%; 2. CH_3ONa, CH_3OH, rt, 1.5 h; then Amberlite IR-120B (H^+) resin, 99%. (*m*) 1. Aqueous HCl, THF, 23 h at 50°C; then Ac_2O, Py, rt, 3 h, 51%; 2. TFA, $CHCl_3$, rt, 1.5 h, 53%; 3. CH_3OH–THF (5:1), $NaOCH_3$, 0°C, 1 h; then Amberlite IR-120B (H^+) resin, 100%.

afforded a mixture of two halogenated compounds; a substantial halide exchange occurred during the reaction. This mixture was heated in toluene in the presence of DBU to afford the 5-enopyranoside **23** via a dehydrohalogenation reaction. Ferrier rearrangement[73,74] of **23** with mercuric(II) trifluoroacetate followed by dehydration with methanesulfonyl chloride and NEt_3 afforded the enone **25**, via the intermediate **24**, which has the three contiguous chiral centers of **8**. The $NaBH_4$–$CeCl_3$ reduction of the carbonyl group in **25** afforded **26**, followed by protection of the generated hydroxyl group as a *p*-methoxybenzyl ether to give **27**. The azido function in **27** was reduced to provide the corresponding amine, which was condensed with 6-bromopiperonylic acid[75] to give the bromo enamide **28**. The N-protected compound **29** underwent an intramolecular palladium-catalyzed cyclization using thallium(I) acetate to afford the diastereoisomer **30**. Removal of the PMB group in **30** afforded **31**. The generated hydroxyl group in **31** underwent a Mitsunobu reaction[76] using benzoic acid as a nucleophile to provide the corresponding inverted benzoate, which upon subsequent debenzoylation afforded **33**. Complete deprotection of **33** afforded **8** in 1.8% overall yield from **22**. On the other hand, complete deprotection of **31** afforded (+)-2-*epi*-lycoricidine (**32**).

An intermediate to the synthesis of 7-deoxypancratistatin (**4**) has been prepared from methyl-α-D-glucopyranoside (**34**) (Scheme 3).[77] Thus, 2,3-di-*O*-benzyl-6-iodo-6-deoxy-α-D-glucopyranoside (**35**) was obtained from **34**, in 55% overall yield,[78] which was treated with excess NaH and 6-iodo-3,4-methylenedioxybenzylchloride (**36**) to afford **37** (92%) which underwent Ferrier rearrangement[79] to furnish a diastereomeric mixture of β-hydroxyketones **38** (59%). Silyl protection of **38** followed by reduction of the ketone under aprotic conditions, mesylation and fluorodesilylation furnished **39** and **40** in 70%

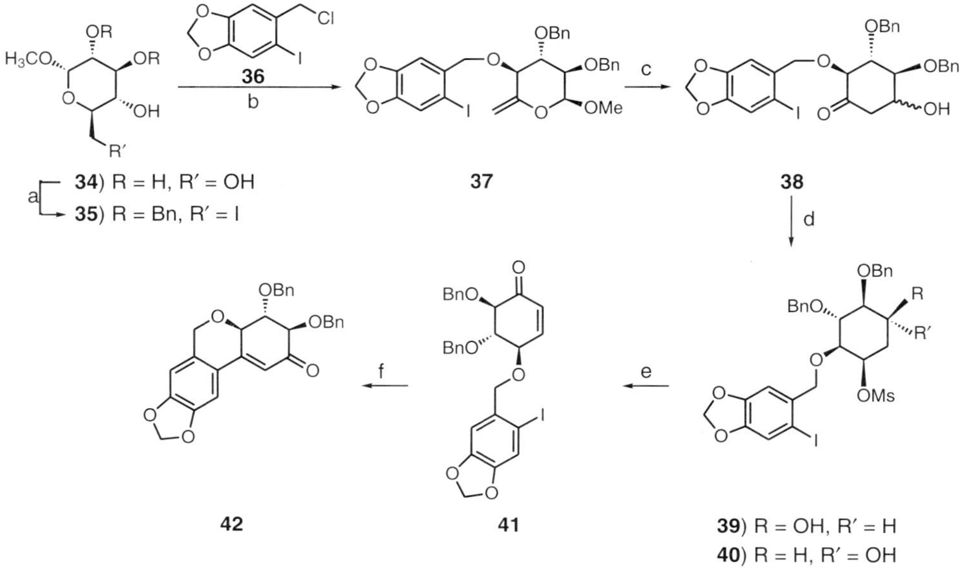

Scheme 3 (*a*) Ref. 78, 55% for four steps. (*b*) Excess NaH, DMF, rt, 92%. (*c*) Cat. $Hg(OCOCF_3)_2$, aqueous acetone, 59%. (*d*) 1. TBSOTf, lutidine; 2. $LiAlH(Ot\text{-}Bu)_3$; 3. MsCl, NEt_3; 4. TBAF, THF, 70% for four steps. (*e*) $(COCl)_2$, DMSO, NEt_3, CH_2Cl_2, –78°C, 92%. (*f*) $Pd(OAc)_2$, PPh_3, NEt_3, $AgNO_3$, CH_3CN, reflux, 70%.

overall yield from **38**. Swern oxidation of the mixture of **39** and **40** gave **41**. Intramolecular palladium catalyzed conjugate addition reactions of **41** produced **42** in 70% yield as an intermediate for the synthesis of **4**.

A precursor to (+)-pancratistatin (**3**) has been obtained from the aldehyde **45**, obtained from diacetone D-glucose (Scheme 4).[80] Treatment of the aryl bromide **43** with *tert*-butyllithium afforded the aryl lithium **44**, which was treated with $ZnCl_2$ to give the corresponding aryl zinc which was condensed with the aldehyde **45**[81] to produce **46** in 74% yield as a single diastereomer. Deoxygenation of **46** afforded **47** in 98% yield, which underwent acid hydrolysis of the isopropylidene group followed by reduction with $NaBH_3CN$ to afford **48** in 78% overall yield from **46**. Selective silylation of the primary hydroxyl group

43) Y = Br
44) Y = Li
45
46) R = OH
47) R = H
48) R = R′ = H
49) R = BOM, R′ = TBDPS
50) R = Ac, R′ = CH_2OH
51) R = H, R′ = CO_2CH_3
52
53

Scheme 4 (*a*) 1. *t*-BuLi, –78°C; 2. $ZnCl_2$. (*b*) MsCl, NEt_3; $LiAlH_4$, 98%. (*c*) 1. 20% aqueous HNO_3–DMF (1:1), 50°C, 3.5 h, 86%; 2. $NaBH_3CN$, TFA, 93%, EtOH, THF, 25°C. (*d*) 1. TBDPSCl, imidazole, 87%; 2. $BnOCH_2Cl$, $(i\text{-Pr})_2NEt$, 85%. (*e*) 1. TBAF; 2. AcCl, NEt_3, DMAP; 3. TBAF, 81% for three steps. (*f*) 1. PDC, DMF, 25°C, 19 h; 2. NaOH, CH_3OH–H_2O (3:1), 0°C, 30 min; 3. CH_2N_2, 50%. (*g*) 1. TBSCl, imidazole, CH_2Cl_2, 25°C, 24 h, 93%; 2. $CH_2C(OLi)Ot\text{-Bu}$, THF, –78°C, 4 h, 85%; 3. TBAF, THF, 25°C, 30 min, 83%. (*h*) Ag_2O (30 equiv.), CH_2Cl_2, 25°C, 24 h, 89%.

in **48** with TBDPSCl followed by protection of the secondary hydroxyl groups with $BnOCH_2Cl$ in the presence of Hünig's base afforded **49** in 74% yield and 43% overall yield from **45**. Acetylation of **49** and then selective removal of the TBDPS group afforded **50**. Oxidation of the hydroxyl group of **50** with PDC followed by deacetylation with sodium hydroxide in methanol and esterification of the resulting carboxylate with CH_2N_2 afforded the ester **51** in 50% yield. Protection of the phenolic group in **51** as the TBS ether followed by homologation with CH_3CO_2t-Bu and LDA and then removal of the silyl group with fluoride ion afforded the ester **52** in 66% yield. Oxidative cyclization of **52** with excess Ag_2O afforded **53** in 89% yield. Compound **53** possesses three of the stereogenic centers of **3**, and is therefore a suitable precursor for **3**.

Compound **47** from the former scheme was also used for the synthesis of the narciclasine alkaloids (Scheme 5).[82] Its treatment with excess ethanethiol and magnesium bromide afforded the dithioacetal **54** in 86% yield. Protection of the hydroxyl groups in **54** followed by hydrolysis of the dithioacetal afforded the corresponding aldehyde, which was treated with nitromethane to give a mixture of diastereomers **55** (1.8:1 ratio) in 80% yield. Treatment of the mixture with excess TBSOTf resulted in the silylation of the hydroxyl group. Subsequent selective deprotection of the phenolic TBS group afforded **56**. Oxidation of the mixture of diastereomers **56** with silver(I) oxide afforded **57**, whose treatment with DMAP afforded **60** and **61** in 29 and 57% yield, respectively. The minor product **60** possesses five of the six stereogenic centers of pancratistatin (**3**).

Alternatively, compound **54** was converted to the corresponding benzylidene acetal, upon reaction with benzaldehyde dimethylacetal, whose dithioacetal was hydrolyzed to give **58**, which was reacted with nitromethane to afford **59** in 83% yield and >99:1 diastereoselectivity. Treatment of **59** with ethanethiol and stannous chloride effected removal of the benzylidene acetal, without dehydration of the β-hydroxynitro functionality, to afford the corresponding triol in 82% yield. Protection of the triol with excess TBSOTf afforded a 91% yield of the corresponding TBS ether. Selective removal of the phenolic TBS group was effected by treatment with CSA in methanol to afford **56** in 88% yield, which upon similar sequence of reactions as shown above afforded **60** as the sole cyclization product in 90% yield.

Another method for using the aldehyde **45** for the synthesis of (+)-lycoricidine (**8**) has also been developed (Scheme 6).[83,84] The aldehyde **45** was condensed with CH_3NO_2 to give **62**, which was converted into olefin **63**.[85] Coupling of **63** with **64** in THF in the presence of CO_2 afforded a mixture of **65**. Removal of the isopropylidene group followed by intramolecular cyclization with K_2CO_3 afforded **67** via intermediate **66**. Hydrogenation of **67** over palladium led to removal of the benzyl group and reduction of the nitro group to afford the amine **68**, which underwent rearrangement to give the lactam 7-deoxypancratistatin (**4**). This was dehydrated to give **8**.

6.3.3 *Synthesis from* L*-arabinose*

L-Arabinose has also been used for the synthesis of (+)-lycoricidine (**8**) (Scheme 7).[86] Condensation of **71** with the aldehyde **70**, prepared[87] from 2,3,4-tri-*O*-benzyl-L-arabinose **69**, afforded the adduct **72**, as a complex mixture of diastereomers. Desilylation of **72** with TBAF followed by treatment with CSA in benzene furnished the lactone **73**, whose immediate esterification and oxidation afforded the ketone **74** in 87% yield. Oxidative

Scheme 5 (*a*) EtSH (10 equiv.), $MgBr_2 \cdot OEt_2$ (10 equiv.), ether, 0°C to rt, 21 h, 86%. (*b*) 1. TBSOTf (2.9 equiv.), 2,6-lutidine (3 equiv.), CH_2Cl_2, 0°C, 15 min; then rt, 2 h, 79%; 2. $HgCl_2$ (4 equiv.), HgO (5 equiv.), CH_3CN, H_2O (5 equiv.), CH_3CN–H_2O (10:1), rt, 1 h, 80%; 3. CH_3NO_2 (10 equiv.), *t*-BuOK (1 equiv.), 0°C, 45 min, 80%. (*c*) 1. $(CH_3O)_2CHPh$ (5 equiv.), CSA (0.2 equiv.), benzene, rt, 20 min (100%); 2. $HgCl_2$ (4 equiv.), HgO (5 equiv.), CH_3CN–H_2O (10:1), rt, 20 min, 87%. (*d*) 1. TBSOTf (4 equiv.), 2,6-lutidine (10 equiv.), CH_2Cl_2, 0°C, 30 min; then rt, 67 h; flash chromatography, 95%; 2. CSA (0.4 equiv.), CH_3OH, rt, 4 h, 79%. (*e*) Ag_2O (5 equiv.), ultrasound, $CDCl_3$, 22–55°C, 14 h, 98%. (*f*) 1. CH_3NO_2 (10 equiv.), *t*-BuOK (1 equiv.), THF, 0°C, 35 min, 83%. (*g*) DMAP (5 equiv.), CH_2Cl_2, rt, 5 h, **61** (57%), **60** (29%). (*h*) 1. EtSH (10 equiv.), $SnCl_2$ (1.0 equiv.), CH_2Cl_2, rt, 20 h, 82%; 2. TBSOTf (7 equiv.), 2,6-lutidine (8 equiv.), CH_2Cl_2, 0°C to rt, 44 h, 91%; 3. CSA (0.4 equiv.), CH_3OH, rt, 4 h, 88%; 4. same as (*e*), 96%; 5. same as (*g*), 90%.

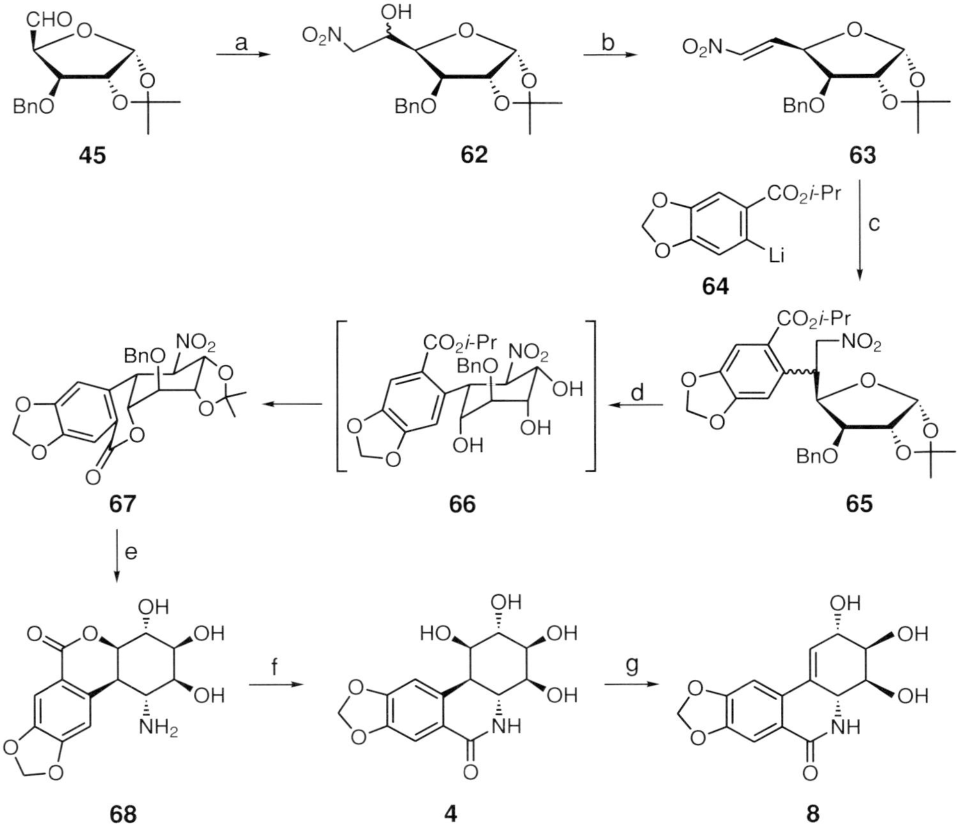

Scheme 6 (*a*) CH_3NO_2. (*b*) Ref. 85. (*c*) THF, CO_2, EtOH. (*d*) 1. AcOH; 2. K_2CO_3. (*e*) Pd, H_2. (*f*) K_2CO_3. (*g*) $SOCl_2$, Py.

cleavage of the olefinic moiety in **74** and direct treatment of the resulting keto aldehyde with DBU resulted in a smooth intramolecular aldol reaction, which upon addition of benzylamine and subsequent treatment with cyanoborohydride gave the phenanthridone **75**, as a single stereoisomer. Finally, **75** was dehydrated via the intermediacy of the corresponding iodide to afford (+)-tetrabenzyllycoricidine (**76**).

Alternatively, the dithioacetal **77** was treated with TBSCl in the presence of DMAP to afford **78** (Scheme 8).[88] Deblocking of the dithioacetal afforded the aldehyde **79**, which underwent Corey–Fuchs aldehyde-to-acetylene conversion[89] to give the dibromoolefin **80**, in high yield, which was then converted[90] to silylacetylene **81**. Reduction of **81** with catalytic hydrogenation afforded the *Z*- and *E*-isomers of vinylsilane **82**. Desilylated of **82** under mild condition followed by Swern oxidation gave the key intermediate **83**. Cyclization of **83** afforded the aminocyclitol **84** as a single stereoisomer. Coupling of 6-iodopiperonyl chloride (**85**) with aminocyclitol **84** afforded the *N*-acylsulfonamide **86**, which was cyclized to give **87**, a protected derivative of **8**, in 7% overall yield from **77**.

69 70 71 72 73 74 75 (+)-Tetrabenzyllycoricidine 76

Scheme 7 (*a*) 1. $Ph_3P{=}CH_2$, THF; 2. $(COCl)_2$, DMSO, CH_2Cl_2, –78°C; then NEt_3, 65%. (*b*) *sec*-BuLi, THF, –78°C, 95%. (*c*) 1. TBAF, THF, 0–25°C; 2. CSA, benzene, 90°C, 77%. (*d*) 1. LiOH, THF, CH_3OH; then CH_2N_2, Et_2O; 2. $(COCl)_2$, DMSO, NEt_3, CH_2Cl_2, 87%. (*e*) 1. O_3, CH_3OH, –78°C; then DMS; 2. DBU, THF, 25°C; 3. $BnNH_2$, PPTs, 52%. (*f*) 1. $CH_3P(OPh)_3I$, HMPA, 100°C; 2. DBU, THF, 85%.

6.3.4 *Synthesis from* D*-lyxose*

D-Lyxose has been used for the synthesis of (–)-lycoricidine (Scheme 9).[91] Reaction of D-lyxose with benzyl alcohol and *p*-toluenesulfonic acid followed by acetonation afforded benzyl 2,3-isopropylidene-D-lyxopyranoside,[92] which was silylated with TBSCl to afford **88** in 69% overall yield from D-lyxose. Removal of the benzyl group with lithium in liquid ammonia followed by condensation of the crude lactol with *O*-benzylhydroxylamine afforded the *O*-benzyloxime **89** as a 2.5:1 mixture of *E*- and *Z*-oximes in 93% yield. Oxidation of the primary hydroxyl group in **89** followed by introduction of a terminal alkyne group afforded **90** in 50% yield. Coupling of **90** with 6-bromopiperonal gave the alkyne aldehyde **91** (91%), which underwent removal of the TBS group, oxidation of the aldehyde function, esterfication of the formed acid to produce **92** in 21% overall yield from D-lyxose. Treatment of **92** with thiophenol afforded **93** in 91% yield as a single diastereomer. Reductive cleavage of the N–OBn bond in **93**, cyclization of the resulting amino ester and removal

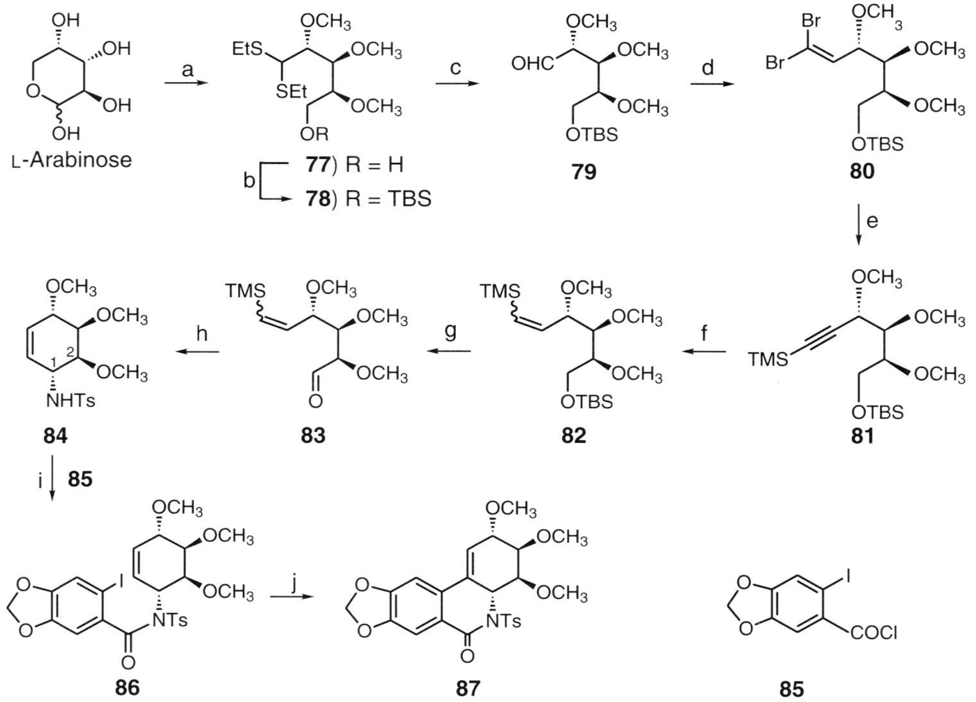

Scheme 8 (*a*) EtSH, H^+. (*b*) TBSCl, DMAP, NEt_3, imidazole, 86%. (*c*) HgO, $HgCl_2$, acetone, H_2O, 50°C, 100%. (*d*) *t*-BuOK, PPh_3, $CHBr_3$, toluene, –20°C to rt, 72%; *or* PPh_3, CBr_4, CH_2Cl_2, NEt_3, –78°C, 75%. (*e*) 1. *n*-BuLi, THF, TMEDA, –78°C; 2. TMSCl, 81%; *or* 1. *n*-BuLi, ether, 0°C; 2. NH_4Cl, H_2O, 88%; 3. BuLi, THF, –78°C; 4. TMSCl, 85%. (*f*) 5% Pd, $BaSO_4$, Py, H_2, rt, 12–16 h, 96%. (*g*) 1. HOAc, H_2O, rt, 12 h, 100%; 2. $(COCl)_2$, DMSO, CH_2Cl_2, –78°C; then NEt_3, 99%. (*h*) 1. TSNSO, $ClCH_2CH_2Cl$, 80°C, 24 h; 2. $BF_3 \cdot OEt_2$, 0°C to rt, 36%. (*i*) NEt_3, DMAP, 77%. (*j*) $Pd(DIPHOS)_2$, TlOAc, DMF, 68°C, 36 h, 50%.

of the thiophenyl group were effected in one step using SmI_2 to furnish the lactam **95** (76%) and the intermediate **94** (15%). The latter could be resubjected to the SmI_2 reduction to give **95** in 73% yield. Removal of the isopropylidene group from **95** with TFA furnished (–)-lycoricidine (**96**) in 11.1% overall yield from D-lyxose.

6.3.5 *Synthesis from D-gulonolactone*

D-Gulonolactone has been used for the syntheses of (+)-lycoricidine (**8**) and (+)-narciclasine (**9**) (Scheme 10).[93] 2,3-*O*-Isopropylidene-D-gulonolactone (**97**)[94] underwent oxidative cleavage of the diol, followed by Corey–Fuchs reaction on the resulting aldehyde to afford the dibromoalkene **98** (80%). Subsequent reduction of the lactone to the corresponding lactol followed by condensation with $BnONH_2$ gave the oxime **99**, which underwent debromination with *n*-BuLi to furnish alkyne **100**. Palladium-mediated coupling of **100** with iodo ester **101**[95] gave **103**, which underwent radical cyclization with PhSH to give **105**, followed by treatment with SmI_2 which led to the reductive cleavage of the

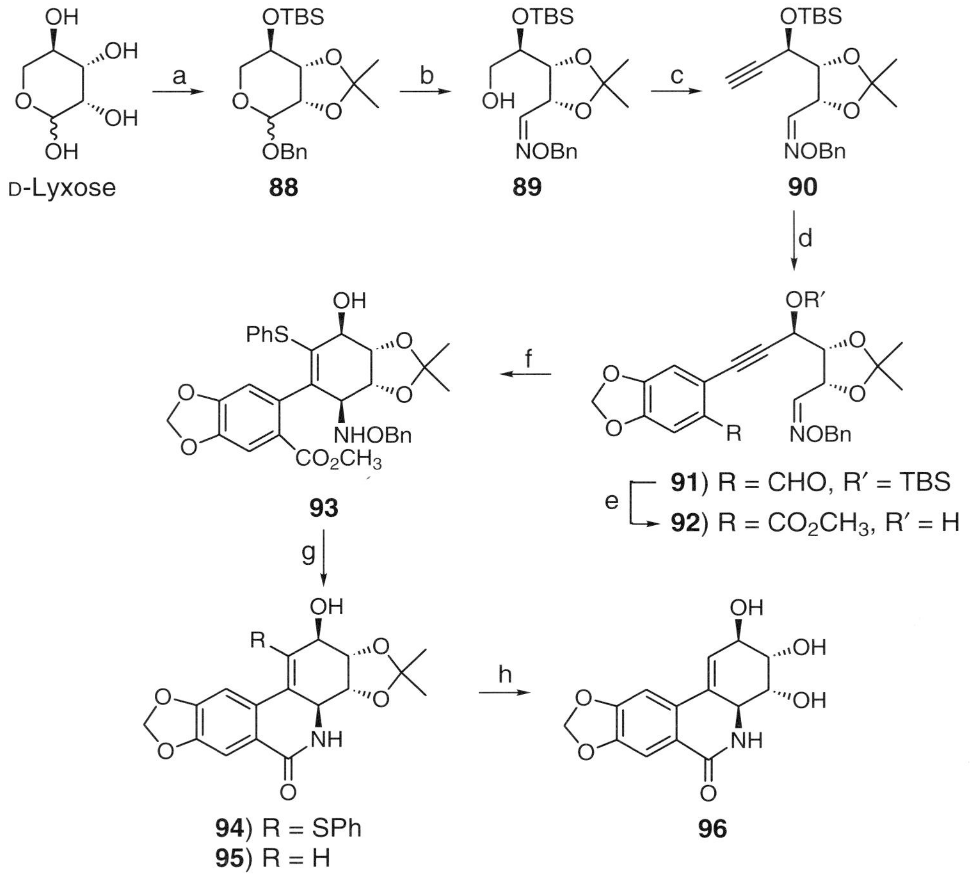

Scheme 9 (*a*) 1. BnOH, *p*-TsOH, 81%; 2. DMP, acetone, *p*-TsOH, 90%; 3. TBSCl, imidazole, 95%. (*b*) 1. Li, liquid NH_3; 2. $BnONH_2HCl$, Py, 93% for two steps. (*c*) 1. TPAP, NMO, MS 4 Å; 2. CBr_4, PPh_3, NEt_3, 55% for two steps; 3. *n*-BuLi, 91%. (*d*) $Pd(OAc)_2$, NEt_3, PPh_3, CuI, bromopiperonal, 91%. (*e*) 1. HF·Py, 88%; 2. MnO_2, NaCN, HOAc, CH_3OH, 81%. (*f*) PhSH, toluene, 27°C, *hv*, 91%. (*g*) SmI_2, THF, **95** (76%), **94** (15%). (*h*) TFA, 77%.

N–OBn bond, cyclization to the lactam and removal of the sulfide group. Finally, removal of the isopropylidene group from **105** afforded **8** in 44% overall yield.

The tosylate derivative **102** was condensed with alkyne **100** to produce compound **104** (89%), which can be under similar sequence of reactions used above, but with removal of tosyl group with SmI_2 afforded **107**, which was cyclized to the lactam **108**, whose de-O-methylation gave (+)-narciclasine (**9**).

D-Gulonolactone was also used for the synthesis of 7-deoxypancratistatin (**4**) via an approach based upon a radical cyclization process (Scheme 11).[96] Reaction of the trichloroacetimidate of the iodopiperonol **110** with **109**[97] in the presence of trifluoromethanesulfonic acid in THF afforded **111** in 75% yield. Reduction of **111** with L-Selectride followed by reaction with *O*-benzylhydroxylamine afforded **112** in 96% yield, which was successively

97 98 99

100

103) R = H
104) R = OTs

101) R = H
102) R = OTs

105) R = H
106) R = OTs
107) R = OH
108) R = OCH_3

8) (+)-Lycoricidine, R = H
9) (+)-Narciclasine, R = OH

Scheme 10 (*a*) 1. $NaIO_4$, CH_2Cl_2; 2. CBr_4, PPh_3, NEt_3, 80%. (*b*) 1. L-Selectride, Et_2O, –78°C; 2. $BnONH_2{\cdot}HCl$, Py, 90%. (*c*) *n*-BuLi, Et_2O, –90°C, 93%. (*d*) $Pd(OAc)_2$, PPh_3, CuI, NEt_3, THF, 95%. (*e*) PhSH, *hv*, toluene, 27°C, 90%. (*f*) 1. SmI_2, THF, H_2O, 0°C, 86%; 2. TFA, 90%. (*g*) Same as (*f*), 94%. (*h*) 1. CH_3I, K_2CO_3, DMF, 96%; 2. $(CH_3)_3Al$, THF, –15 to 65°C; 72%.

silylated with TBSOTf followed by selective desilylation with HF–pyridine to furnish **113** (84%). Oxidation of **113** with TPAP and NMO followed by treatment of the resulting product with 1-amino-2-phenylaziridine in ethanol afforded the aziridinylimine **114** (83%). Radical cyclization of **114** with Ph_3SnH and AIBN in benzene afforded the cyclized compound **115** in 78% yield. Cleavage of the N–O bond using SmI_2 in THF and direct quenching with TFAA furnished the trifluoroacetamide **116** (88%), which was oxidized with PCC to produce the lactone **117** in 83% yield. Removal of the protecting groups with $BF_3{\cdot}OEt_2$ followed by rearrangement of the lactone gave 7-deoxypancratistatin (**4**) in 25% overall yield from **110**.

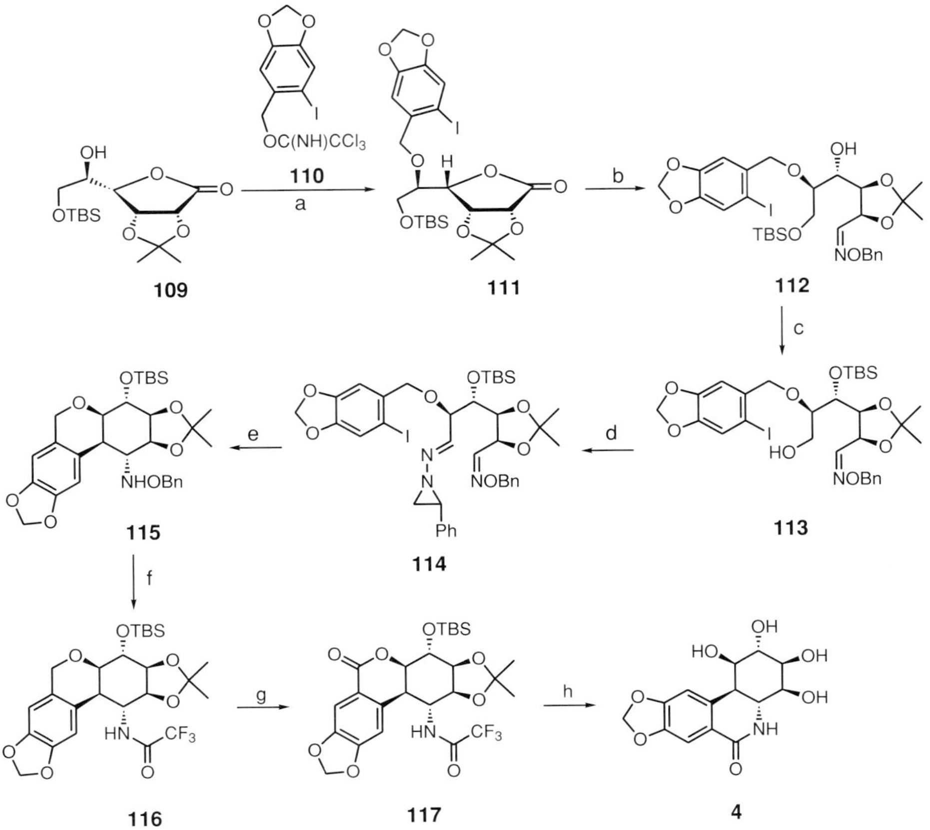

Scheme 11 (*a*) TFA, THF, 0°C. (*b*) 1. L-Selectride, CH_2Cl_2, –78°C; 2. $BnOH_2N{\cdot}HCl$, Py, 96% for two steps. (*c*) 1. TBSOTf, 2,6-lutidine, CH_2Cl_2, 0°C; 2. HF·Py, THF, 84% for two steps. (*d*) 1. TPAP, NMO, MS 4 Å; 2. 1-amino-2-phenylaziridine, EtOH, 0°C, 83% for two steps. (*e*) Ph_3SnH, AIBN, benzene, 78%. (*f*) SmI_2, THF, then TFAA, 88%. (*g*) PCC, CH_2Cl_2, 50°C, 83%. (*h*) 1. $BF_3{\cdot}OEt_2$, CH_2Cl_2; 2. K_2CO_3, CH_3OH, 88% for two steps.

Alternatively, the di-*O*-TBS derivative of D-gulonolactone **118** has also been used for the synthesis of 7-deoxypancratistatin (**4**) (Schemes 12 and 13).[98,99] Reduction of **118** with DIBAL-H followed by treatment with *O*-benzylhydroxylamine produced the oxime **119** in 89% yield. This was protected with MOMCl in the presence of *N,N*-diisopropylethylamine and subsequent selective desilylation of the primary position with HF·Py to afford **120** in 62% yield. Oxidation of **120** gave the corresponding aldehyde and then the carboxylic acid **121**, which underwent Mitsunobu esterification with **122** to afford **123** in 80% yield. Treatment of **123** with *n*-butyllithium followed by oxidation of the resulting rearranged benzylic alcohol afforded **124** in 72% yield, which was converted into **125**. Treatment of **125** with 1,1′-thiocarbonyldiimidazole afforded the cyclic radical precursor **126**, whose cyclization afforded **127** (25%). Treatment of **127** with TFAA followed by treatment with SmI_2 afforded the key intermediate **128**. Removal of the acetonide and MOM ether gave the corresponding triol, which was treated with K_2CO_3 in dry methanol to afford **4**.

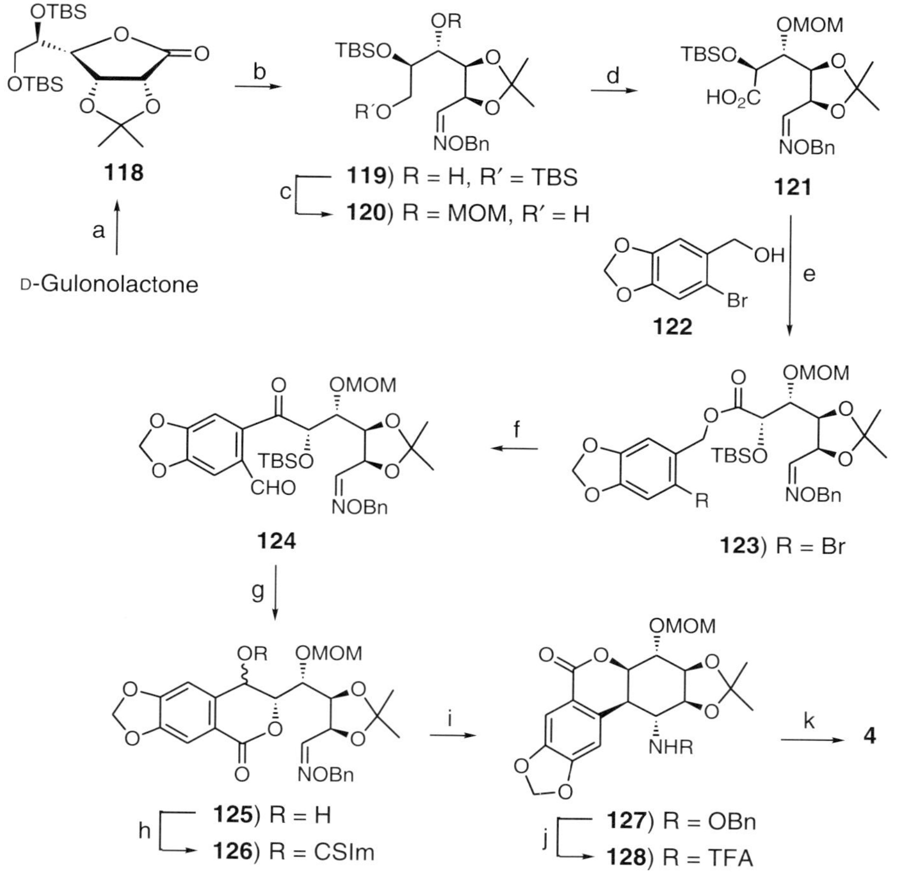

Scheme 12 (*a*) 1. Acetone, DMP, *p*-TsOH, rt, 24 h; 2. AcOH–H_2O (7:1), 12 h; 3. DMF, imidazole, TBSCl, 20 h, 90% for three steps. (*b*) 1. DIBAL-H, CH_2Cl_2, –78°C, 1 h; 2. $BnONH_2$·HCl, Py, rt, 23 h, 89%. (*c*) 1. MOMCl, DIPEA, $(i\text{-Pr})_2NEt$, 55–60°C, 15 h; 2. HF·Py, THF, rt, 3 h, 62%. (*d*) 1. TPAP, NMO, 4-methylmorpholine *N*-oxide, CH_2Cl_2, rt, MS 4 Å, 1 h, 95%; 2. $NaClO_2$, KH_2PO_4, *t*-BuOH, 2-methyl-2-butene, –5°C, 84%. (*e*) DEAD, rt, Ph_3P, THF, 80%. (*f*) 1. *n*-BuLi, –98 to –78°C, THF, 1.5 h, 2. TPAP, NMO, 45 min, 72%. (*g*) 1. HF, Py, THF, rt, 24 h; 2. Dess–Martin reagent, 3 h, EtOAc; 3. $NaBH_4$, CH_3OH, –5°C, rt, 57%. (*h*) TCDI, DMA, DCE, rt, $ClCH_2CH_2Cl$, 60%. (*i*) Bu_3SnH, AIBN, 65°C, 3.5 h, toluene, 25%. (*j*) 1. TFAA, Py; 2. SmI_2, THF, –23°C, 67%. (*k*) 1. Dowex 50WX8 (H^+) resin, 70°C, 5.5 h; 2. K_2CO_3, CH_3OH, 70°C, 5.5 h; then Dowex (H^+) resin.

On the other hand, compound **124** was converted to the TBS-protected lactol **129** via TBC deprotection, followed by reprotection of the resulting lactol with TBSCl. Sodium borohydride reduction of **129** followed by treatment with 1,1′-thiocarbonyldiimidazole afforded **130**, which upon radical cyclization yielded **131** (72%) as a single stereoisomer (Scheme 13). Acylation of **131** with TFAA followed by removal of the silyl group and subsequent oxidation afforded **132** in 81% yield. Cleavage of the N–OBn bond in **132** with SmI_2 in THF afforded the amide **128** in 86% yield.[100] Cleavage of the N–OBn in **132** under hydrogenolysis conditions and aluminum or sodium amalgam was not successful.

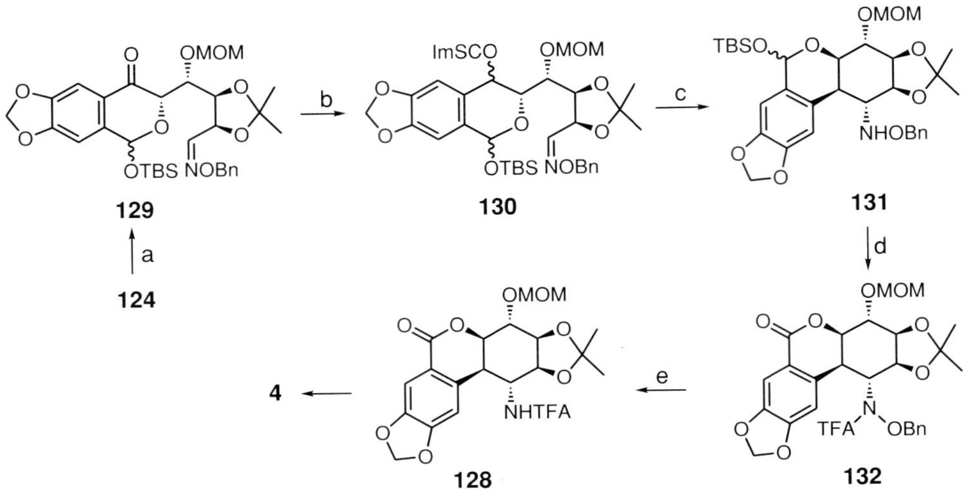

Scheme 13 (*a*) 1. HF, Py, THF, rt; 2. TBSCl, rt, 19 h, CH_2Cl_2, 75%. (*b*) 1. $NaBH_4$, CH_3OH, rt, 2 h; 2. TCDI, DCE, $ClCH_2CH_2Cl$, 77%. (*c*) Bu_3SnH, AIBN, 4 h, toluene, 72%. (*d*) 1. TFAA, Py, CH_2Cl_2; 2. TBAF, THF, CH_2Cl_2; TPAP, NMO, MS 4 Å, 45 min, 81%. (*e*) SmI_2, THF, 1 h, –23°C, 86%.

References

1. Frederik, D.M.; Murav'eva, D.A. *Khim. Prir. Soedin.* **4** (1982) 534.
2. Ozeki, S. *Chem. Pharm. Bull.* **12** (1964) 253.
3. Pettit, G.R.; Gaddamidi, V.; Cragg, G.M.; Herald, D.L.; Sagawa, Y. *J. Chem. Soc., Chem. Commun.* (1984) 1693.
4. Ghosal, S.; Singh, S.; Kumar, Y.; Srivastava, R.S. *Phytochemistry* **28** (1989) 611.
5. Pettit, G.R.; Gaddamidi, V.; Herald, D.L.; Singh, S.B.; Cragg, G.M.; Schmidt, J.M.; Boettner, F.E.; Williams, M.; Sagawa, Y. *J. Nat. Prod.* **49** (1986) 995.
6. Gabrielson, B.; Monath, T.P.; Huggins, J.W.; Kirsi, J.J.; Hollingshead, M.; Shannon, W.M.; Pettit, G.R. In *Natural Products as Antiviral Angents*; Chu, C.K., Cutler, H.G., Eds.; Plenum: New York, 1992; p. 121.
7. Torres-Labanderia, J.J.; Davignon, P.; Pitha, J. *J. Pharm. Sci.* **80** (1990) 384.
8. Jimenez, A.; Sanchez, L.; Vazquez, D. *FEBS Lett.* **60** (1975) 66.
9. Jimenez, A.; Santos, A.; Alonso, G.; Vazquez, D. *Biochim. Biophys. Acta* **518** (1978) 95.
10. Rivera, G.; Gosalbez, M.; Ballesta, J.P.G. *Biochem. Biophys. Res. Commun.* **94** (1980) 800.
11. Hartwell, J.L. *Lloydia* **30** (1967) 379.
12. Gabrielsen, B.; Monath, T.P.; Huggins, J.W.; Kefauver, D.F.; Pettit, G.R.; Groszek, G.; Hollingshead, M.; Kirisi, J.J.; Shannon, W.M.; Schubert, E.M.; Dare, J.; Ugarkar, B.; Ussery, M.A.; Phelan, M.J. *J. Nat. Prod.* **55** (1992) 1569.
13. Okamoto, T.; Torii, Y.; Isogai, Y. *Chem. Pharm. Bull.* **16** (1968) 1860.
14. Abou-Donia, A.H.; De Giulio, A.; Evidente, A.; Gaber, M.; Habib, A.; Lanzeta, R.; Seif El Din., A. *Phytochemistry* **30** (1991) 3435.
15. Piozzi, F.; Fugante, C.; Mondelli, R.; Ceriotti, G. *Tetrahedron* **24** (1968) 1119.
16. Ceriotti, G. *Nature* **213** (1967) 595.
17. Martin, S.F.; Tso, H.H. *Heterocycles* **35** (1993) 85.
18. Carrasco, L.; Fresno, M.; Vazquez, D. *Fed. Eur. Biochem. Soc. Lett.* **52** (1975) 236.
19. Abou-Donia, A.H.; De Giulio, A.; Evidente, A.; Gaber, M.; Habib, A.-A.; Lanzetta, R.; Seif El Din, A.A. *Phytochemistry* **30** (1991) 3445.
20. Evidente, A. *Planta Med.* **57** (1991) 293.
21. Mondon, A.; Krohn, K. *Chem. Ber.* **108** (1975) 445.
22. Pettit, G.R.; Cragg, G.M.; Sing, S.B.; Duke, J.A.; Doubek, D.L. *J. Nat. Prod.* **53** (1990) 176.
23. Pettit, G.R.; Pettit, G.R., III; Backhaus, R.A.; Boyd, M.R.; Meerow, A.W. *J. Nat. Prod.* **56** (1993) 1682.

24. Tsuda, Y.; Sano, T.; Taga, J.; Isobe, K.; Toda, J.; Takagi, S.; Yamaki, M.; Murata, M.; Irie, H.; Tanaka, H. *J. Chem. Soc., Perkin Trans. 1* (1979) 1358.
25. Hudlicky, T.; Tian, X.; Königsberger, K.; Maurya, R.; Rouden, J.; Fan, B. *J. Am. Chem. Soc.* **118** (1996) 10752.
26. Pettit, G.R.; Gaddamidi, V.; Cragg, G.M. *J. Nat. Prod.* **47** (1984) 1018.
27. Fitzgerald, R.; Hartwell, J.L.; Leiter, J. *J. Natl. Cancer Inst.* **20** (1958) 763.
28. Immirzi, A. *J. Chem. Soc., Chem. Commun.* (1972) 240.
29. Jimenez, A.; Sanchez, L.; Vazquez, D. *FEBS Lett.* **55** (1975) 53.
30. Oppolzer, W.; Spivey, A.C.; Bochet, C.G. *J. Am. Chem. Soc.* **116** (1994) 3139.
31. Weller, T.; Seebach, D. *Tetrahedron Lett.* **23** (1982) 935.
32. Magnus, P.; Sebhat, I.K. *Tetrahedron* **54** (1998) 15509.
33. Keck, G.E.; Boden, E.; Sonnewald, U. *Tetrahedron Lett.* **22** (1981) 2615.
34. Ohta, S.; Kimoto, S. *Tetrahedron Lett.* **16** (1975) 2279.
35. Quelet, R.; Dran, R. *Compt. Rend.* **258** (1964) 1826.
36. Aceña, J.L.; Arjona, O.; Iradier, F.; Plument, J. *Tetrahedron Lett.* **37** (1996) 105.
37. Gauthier, D.R., Jr.; Bender, S.L. *Tetrahedron Lett.* **37** (1996) 13.
38. Lopes, R.S.C.; Lopes, C.C.; Heathcock, C.H. *Tetrahedron Lett.* **33** (1992) 6775.
39. Banwell, M.G.; Cowden, C.J.; Gable, R.W. *J. Chem. Soc., Perkin Trans. 1* (1994) 3515.
40. Banwell, M.G.; Cowden, C.J.; Mackay, M.F. *J. Chem. Soc., Chem. Commun.* (1994) 61.
41. Ugarkar, B.G.; Dare, J.; Schubert, E.M. *Synthesis* (1987) 715.
42. Ohta, S.; Kimoto, S. *Chem. Pharm. Bull.* **24** (1976) 2969.
43. Ohta, S.; Kimoto, S. *Chem. Pharm. Bull.* **24** (1976) 2977.
44. Gonzalez, D.; Martinot, T.; Hudlicky, T. *Tetrahedron Lett.* **40** (1999) 3077.
45. Tian, X.; Hudlicky, T.; Königsberger, K. *J. Am. Chem. Soc.* **117** (1995) 3643.
46. Danishefsky, S.; Lee, J.Y. *J. Chem. Soc.* **111** (1989) 4829.
47. Doyle, T.J.; Hendrix, M.; Van Derveer, D.; Javanmard, S.; Haseltine, J. *Tetrahedron* **53** (1997) 11153.
48. Doyle, T.J.; Hendrix, M.; Haseltine, J. *Tetrahedron Lett.* **35** (1994) 8295.
49. Doyle, T.J.; Van Derveer, D.; Haseltine, J. *Tetrahedron Lett.* **36** (1995) 6197.
50. Hudlicky, T.; Tian, X.; Königsberger, K.; Maurya, R.; Rouden, J.; Fan, B. *J. Am. Chem. Soc.* **118** (1996) 10752.
51. Tian, X.; Maurya, R.; Königsberger, K.; Hudlicky, T. *Synlett* (1995) 1125.
52. Hudlicky, T.; Olivo, H.F.; Mckibben, B. *J. Am. Chem. Soc.* **116** (1994) 5108.
53. Hudlicky, T.; Olivo, H.F. *Tetrahedron Lett.* **32** (1991) 6077.
54. Hudlicky, T.; Olivo, H.F. *J. Am. Chem. Soc.* **114** (1992) 9694.
55. Trost, B.M.; Pulley, S.R. *J. Am. Chem. Soc.* **117** (1995) 10143.
56. Magnus, P.; Sebhat, I.K. *J. Am. Chem. Soc.* **120** (1998) 5341.
57. Rinner, U.; Hillebrenner, H.L.; Adams, D.S.R.; Hudlicky, T.; Pettit, G.R. *Bioorg. Med. Chem. Lett.* **14** (2004) 2911.
58. Rigby, J.H.; Mateo, M.E. *J. Am. Chem. Soc.* **119** (1997) 12655.
59. Martin, S.F. In *The Alkaloids*, Vol. 30; Brossi, A.A., Ed.; Academic Press: New York, 1987; p. 251.
60. Fuganti, C.; Mazza, M. *J. Chem. Soc., Chem. Commun.* (1972) 239.
61. Kim, S.; Ko, H.; Kim, E.; Kim, D. *Org. Lett.* **4** (2002) 1343.
62. Pandey, G.; Murugan, A.; Balakrishnan, M. *Chem. Commun.* (2002) 624.
63. Rinner, U.; Siengalewicz, P.; Hudlicky, T. *Org. Lett.* **4** (2002) 115.
64. Rigby, J.H.; Maharoof, U.S.M.; Mateo, M.E. *J. Am. Chem. Soc.* **122** (2000) 6624.
65. Pettit, G.R.; Melody, N.; Herald, D.L.J. *Org. Chem.* **66** (2001) 2583.
66. Elango, S.; Yan, T.-H. *Tetrahedron* **58** (2002) 7335.
67. Hudlicky, T.; Rinner, U.; Gonzalez, D.; Akgun, H.; Schilling, S.; Siengalewicz, P.; Martinot, T.A.; Pettit, G.R. *J. Org. Chem.* **67** (2002) 8726.
68. Park, T.K.; Danishefsky, S.J. *Tetrahedron Lett.* **36** (1995) 195.
69. Ferrier, R.J. *J. Chem. Soc., Perkin Trans. 1* (1979) 1455.
70. Chida, N.; Ohtsuka, M.; Ogawa, S. *Tetrahedron Lett.* **32** (1991) 4525.
71. Chida, N.; Ohtsuka, M.; Ogawa, S. *J. Org. Chem.* **58** (1993) 4441.
72. Hanessian, S.; Masse, R. *Carbohydr. Res.* **35** (1974) 175.
73. Chida, N.; Ohtsuka, M.; Nakazawa, K.; Ogawa, S. *J. Org. Chem.* **56** (1991) 2976.
74. Gemal, A.L.; Luche, J.-L. *J. Am. Chem. Soc.* **103** (1981) 5454.
75. Yamada, S.; Kasai, Y.; Shioiri, T. *Tetrahedron Lett.* **14** (1973) 1595.
76. Mitsunobu, O. *Synthesis* (1981) 1.
77. Friestad, G.K.; Branchaud, B.P. *Tetrahedron Lett.* **38** (1997) 5933.
78. Kohn, A.; Schmidt, R.R. *Liebigs Ann. Chem.* (1987) 1045.
79. Chida, N.; Ohtsuka, M.; Ogura, K.; Ogawa, S. *Bull. Chem. Soc. Jpn.* **64** (1991) 2118.

80. Angle, S.R.; Louie, M.S. *Tetrahedron Lett.* **34** (1993) 4751.
81. Wolfrom, M.L.; Hanessian, S. *J. Org. Chem.* **27** (1962) 1800.
82. Angle, S.R.; Wada, T. *Tetrahedron Lett.* **36** (1997) 7955.
83. Paulsen, H.; Stubbe, M. *Liebigs Ann. Chem.* (1983) 535.
84. Paulsen, H.; Stubbe, M. *Tetrahedron Lett.* **23** (1982) 3171.
85. Funabashi, M.; Yoshimura, J. *J. Chem. Soc., Perkin Trans. 1* (1979) 1425.
86. Thompson, R.C.; Kallmerten, J. *J. Org. Chem.* **55** (1990) 6076.
87. Tejima, S.; Fletcher, H.G. *J. Org. Chem.* **28** (1963) 2999.
88. McIntosh, M.C.; Weinreb, S.M. *J. Org. Chem.* **58** (1993) 4823.
89. Corey, E.J.; Erickson, B.W. *J. Org. Chem.* **36** (1971) 3553.
90. Corey, E.J.; Fuchs, P.L. *Tetrahedron Lett.* **36** (1972) 3769.
91. Keck, G.E.; Wager, T.T. *J. Org. Chem.* **61** (1996) 8366.
92. Keck, G.E.; Kachensky, D.F.; Enholm, E.J. *J. Org. Chem.* **50** (1985) 4317.
93. Keck, G.E.; Wager, T.T.; Rodriquez, J.F.D. *J. Am. Chem. Soc.* **121** (1999) 5176.
94. Fleet, G.W.J.; Ramsden, N.G.; Witty, D.R. *Tetrahedron* **45** (1989) 319.
95. Bogucki, D.E.; Charlton, J.L. *J. Org. Chem.* **60** (1995) 588.
96. Keck, G.E.; Wager, T.T.; McHardy, S.F. *J. Org. Chem.* **63** (1998) 9164.
97. Fleet, G.W.J.; Ramsden, N.G.; Witty, O.R. *Tetrahedron* **45** (1989) 319.
98. Keck, G.E.; McHardy, S.F.; Murry, J.A. *J. Org. Chem.* **64** (1999) 4465.
99. Keck, G.E.; McHardy, S.F.; Murry, J.A. *J. Am. Chem. Soc.* **117** (1995) 7289.
100. Keck, G.E.; McHardy, S.F.; Wager, T.T. *Tetrahedron Lett.* **36** (1995) 7419.

6.4 Ecteinascidins

The ecteinascidins were isolated from the marine tunicate *Ecteinascidia turbinasa*.[1,2] Their structures and extremely potent antitumor activities have promoted the attention; the first of these compounds was ecteinascidin 743 (**1**),[3–5] which has been advanced to clinical trials.

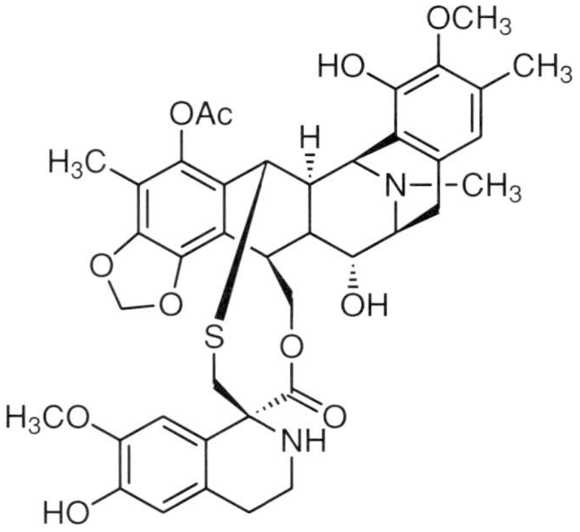

Ecteinascidin 743

1

Synthesis of the key intermediate polycycle **14**, bearing four chiral centers and the two aromatic rings of ecteinascidin 743 (**1**), has been reported from D-glucose (Schemes 1 and 2).[6] The epoxide **2**,[7] prepared from D-glucose, was treated with *p*-toluenesulfonamide and Cs_2CO_3 followed by mesylation to afford **3**, which underwent acidic hydrolysis followed by treatment with $SnCl_4$ to furnish the α-isomer **4**. Base-induced ring closure of mesylate **4** and subsequent protection as a TBS provided aziridine **5**. This was treated with **6** to form **7**. Protection of the NH of sulfonamide **7** with the Boc group and removal of the TBS afforded **8**. The second nitrogen atom was incorporated in **8** by conversion of the free secondary hydroxyl group to a triflate whose treatment with lithium azide produced **9**.

Removal of the MOM and Boc groups followed by regioselective bromination with Py–HBr in dichloromethane followed by cyclization through an iminium ion intermediate furnished the bicycle **10** as a single isomer. Deprotection of the benzyl ether and reduction of the azide group afforded the amino diol **11**, which was treated with $BrCH_2CO_2Ph$ and propylene oxide and subsequent $Pb(OAc)_4$ oxidation of the amine to provide **12** (80%). Intermolecular addition of the phenolic compound **13** to **12** in the presence of TFA occurred at the sterically less hindered convex face to provide the key intermediate **14** (Scheme 2).[6]

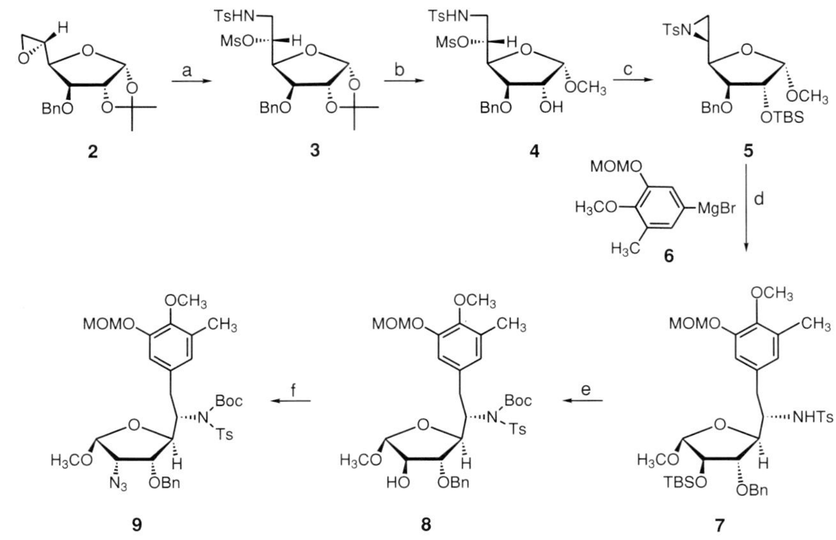

Scheme 1 (*a*) 1. *p*-$TsNH_2$, Cs_2CO_3, DMF, 80°C; 2. MsCl, NEt_3, CH_2Cl_2, 0°C, 96% for two steps. (*b*) 1. HCl, CH_3OH, reflux, 99%; 2. $SnCl_4$, CH_2Cl_2, rt, 88%. (*c*) 1. NaOH, CH_3OH, rt, 91%; 2. TBSCl, imidazole, DMF, rt, 98%. (*d*) CuI, THF, 0°C to rt, 91%. (*e*) 1. $(Boc)_2O$, DMAP, CH_3CN, rt, 96%; 2. TBAF, THF, rt, 98%. (*f*) 1. Tf_2O, Py, CH_2Cl_2, 0°C; 2. LiN_3, DMF, 80°C, 90% for two steps.

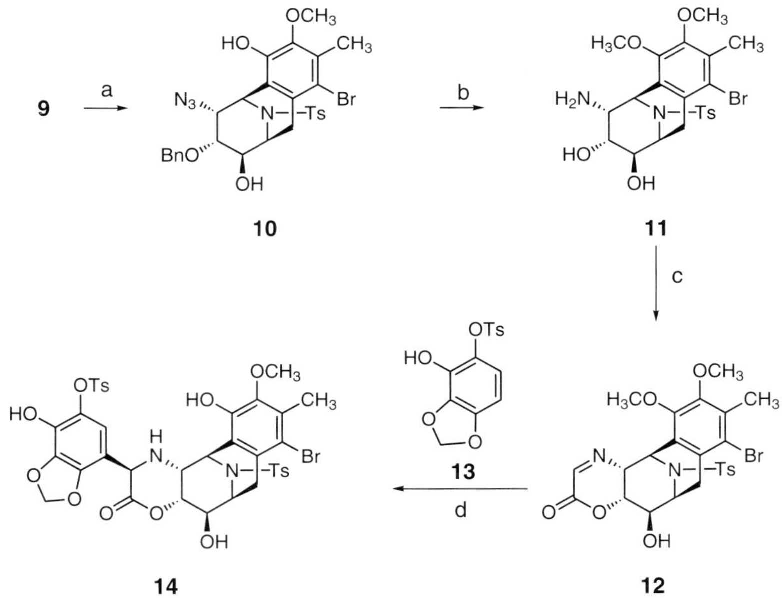

Scheme 2 (*a*) 1. TMSBr, CH_2Cl_2, rt; 2. TFA, CH_2Cl_2, rt, 97% for two steps; 3. Py–HBr, CH_2Cl_2, rt, 89%; 4. TFA, H_2O, 70°C, 88%. (*b*) 1. CH_3I, K_2CO_3, acetone, reflux, 89%; 2. BCl_3, CH_2Cl_2, –78 to 0°C, 87%; 3. Rh *on* C, H_2, EtOAc, rt, 82%. (*c*) 1. $BrCH_2CO_2Ph$, MS 4 Å, propylene oxide, CH_3CN, 80°C, 92%; 2. $Pb(OAc)_4$, benzene, 80°C, 80%. (*d*) TFA, CH_2Cl_2, 0°C, 89%.

References

1. Rinehart, K.L.; Holt, T.G.; Fregeau, N.L.; Stroh, J.G.; Kreifer, P.A.; Sun, F.; Li, L.H.; Martin, D.G. *J. Org. Chem.* **55** (1990) 4512.
2. Sakai, R.; Jares-Erijman, E.A.; Manzanares, I.; Elipe, M.V.S.; Rinehart, K.L. *J. Am. Chem. Soc.* **118** (1996) 9017.
3. Sakai, R.; Rinehart, K.L.; Guan, Y.; Wang, A.H.-J. *Proc. Natl. Acad. Sci. USA* **89** (1992) 11456.
4. Guan, Y.; Sakai, R.; Rinehart, K.L.; Wang, A.H.-J. *J. Biomol. Struct. Dyn.* **10** (1993) 793.
5. Corey, E.J.; Gin, D.Y.; Kania, R.S. *J. Am. Chem. Soc.* **118** (1996) 9202.
6. Endo, A.; Kann, T.; Fukuyama, T. *Synlett* (1999) 1103.
7. Schmit, O. *Methods in Carbohydrate Chemistry*, Vol. II; Academic Press: New York, 1963; p. 190.

Natural Source (Natural Product) Index

A

M

N

O

P

T

U

X

Z

Biological Activities Index

M

N

O

P

R

S

Index

A

B

C

D

E

F

H

I

J

K

L

M

T